The Linearized Theory of Elasticity

William S. Slaughter

The Linearized Theory
of Elasticity

With 153 Figures

Birkhäuser
Boston • Basel • Berlin

William S. Slaughter
Department of Mechanical Engineering
University of Pittsburgh
Pittsburgh, PA 15261
USA

Library of Congress Cataloging-in-Publication Data
Slaughter, William S.
    The linearized theory of elasticity / William S. Slaughter.
      p. cm.
    Includes bibliographical references and index.
    ISBN 0-8176-4117-3 (alk. paper)
      1. Elasticity. I. Title.
    QA931 .S584 2001
    532'.0535—dc21   2001037385

Printed on acid-free paper.
© 2002 Birkhäuser Boston

ISBN 0-8176-4117-3
ISBN 3-7643-4117-3   SPIN 10716247

Production managed by Louise Farkas; manufacturing supervised by Jacqui Ashri.
Typeset by the author in TeX.
Printed and bound by Maple-Vail Book Manufacturing Group, York, PA.
Printed in the United States of America.

9 8 7 6 5 4 3 2 1

Birkhäuser   Boston Basel Berlin
A member of BertelsmannSpringer Science+Business Media GmbH

To my parents

Contents

Preface

This book is derived from notes used in teaching a first-year graduate-level course in elasticity in the Department of Mechanical Engineering at the University of Pittsburgh. This is a modern treatment of the linearized theory of elasticity, which is presented as a specialization of the general theory of continuum mechanics. It includes a comprehensive introduction to tensor analysis, a rigorous development of the governing field equations with an emphasis on recognizing the assumptions and approximations inherent in the linearized theory, specification of boundary conditions, and a survey of solution methods for important classes of problems. Two- and three-dimensional problems, torsion of noncircular cylinders, variational methods, and complex variable methods are covered.

This book is intended as the text for a first-year graduate course in mechanical or civil engineering. Sufficient depth is provided such that the text can be used without a prerequisite course in continuum mechanics, and the material is presented in such a way as to prepare students for subsequent courses in nonlinear elasticity, inelasticity, and fracture mechanics. Alternatively, for a course that is preceded by a course in continuum mechanics, there is enough additional content for a full semester of linearized elasticity.

It is anticipated that students will mostly have undergraduate mechanical or civil engineering backgrounds, with the mathematical training that entails. Such students have usually not been exposed to modern real analysis or to abstract vector spaces, for instance. This has necessarily had an impact on the manner in which the material in this book is presented. An attempt has been made not to introduce a surfeit of unfamiliar mathematical notation. For example, the reader will not find any mathematical expressions like

$$(\mathbb{R} \times \mathbb{R}) \ni (x, y) \mapsto x^2 + y^2 \in \mathbb{R} \,.$$

Additionally, it is deemed worthwhile to spend a little extra time on indicial notation and tensors—students who do not master these concepts will increasingly find it impossible to follow the rest of the material.

When is the best time to introduce the linearizing assumptions? This is an important question when teaching linear elasticity. Traditionally, the linearization has been introduced as soon as possible [e.g., Sokolnikoff (1956) and Timoshenko and Goodier (1970)]. This approach has the virtue of allowing one to move on to solution methods very quickly. An alternative is to develop completely the nonlinear theory of elasticity prior to linearizing [e.g., Atkin and Fox (1980) and Spencer (1980)]. This gives students a broad framework that will serve them well when they take other courses that address related topics such as fluid dynamics and inelasticity, but scarcely leaves time to learn how to solve the important linear elasticity problems that arise in engineering. Perhaps the best of all worlds is one in which students first take an introductory course in continuum mechanics, followed by specialized classes in elasticity, fluid dynamics, inelasticity, and so forth. Unfortunately, the realities of manpower and teaching loads mean that adding an additional introductory course in continuum mechanics is often not a practical option. Consequently, an attempt has been made here to strike a happy middle ground. The introduction of linearizing assumptions in this book is delayed long enough to provide students with a context from which they can see the relationships that exist between linear elasticity and other related subjects and still have time in a one-semester course to explore some of the important classes of problems and solution methods.

In the analysis of kinematics and measures of stress, referential (Lagrangian) and spatial (Eulerian) formulations have been presented separately. The viewpoint taken is that linear elasticity is most naturally seen as a linearization of the referential formulation, with fields in the linearized theory viewed as being over the reference configuration of the body. If desired, the sections in which the spatial formulations are presented can be omitted with minimal disruption.

The so-called "Gibbs notation" for tensor analysis has been used instead of the "Ricci notation" favored by many authors in continuum mechanics [e.g., Truesdell and Noll (1992)]. For example, the bilinear form of a second-order tensor $\mathbf{T}$ with respect to the vectors $\mathbf{u}$ and $\mathbf{v}$ (in that order) is given as $\mathbf{u} \cdot \mathbf{T} \cdot \mathbf{v}$ rather than $\mathbf{u} \cdot (\mathbf{Tv})$. It is the author's opinion that the Gibbs notation makes it easier for students who are new to the subject to grasp the concepts that are most important at this level, even though it may obscure some of the more subtle issues involving the composition of linear operators, Cartesian products, abstract vector spaces, and the like. Similarly, the dyad (or tensor product) formed by two vectors $\mathbf{u}$ and $\mathbf{v}$ is given as $\mathbf{uv}$ rather that $\mathbf{u} \otimes \mathbf{v}$, so that the dyadic representation of the second-order tensor $\mathbf{T}$ in an orthonormal vector basis is $\mathbf{T} = T_{ij}\hat{\mathbf{e}}_i\hat{\mathbf{e}}_j$ rather than $\mathbf{T} = T_{ij}\hat{\mathbf{e}}_i \otimes \hat{\mathbf{e}}_j$.

As much as is practical, results are presented in both a basis-independent tensorial form and a basis-dependent scalar component form. For instance, the traction-stress relation derived in Chapter 4 is given as

$$\mathbf{t} = \hat{\mathbf{n}} \cdot \boldsymbol{\sigma} \,, \quad t_i = \hat{n}_j \sigma_{ji} \,.$$

In doing this, an orthonormal vector basis and, when necessary, a Cartesian coordinate system are presumed. It is felt that students are overwhelmed by a too early introduction to general curvilinear coordinates and, since they are not required for the applications covered in this book, they have been relegated to an appendix. Cylindrical and spherical coordinate systems are treated explicitly, rather than as special cases of general curvilinear coordinates. The tensor notation reinforces the fact that the underlying physical principles are valid in any coordinate system.

The mechanics of materials, as presented to sophomore engineering majors in a typical undergraduate program in the United States, is briefly reviewed in Chapter 1. This material sets the stage, in some sense, for what follows, but may be omitted. Chapter 2 acquaints the student with the notation and conventions that are to be used, introduces the concept of indicial notation, and develops the tensor analysis. The foundations for the linearized theory of elasticity are developed in Chapters 3 to 6. The remaining chapters cover solution methods for a variety of classes of problems ranging from two-dimensional antiplane strain problems to three-dimensional problems involving dissimilar inclusions. The order in which they are covered is somewhat arbitrary, except that Chapter 11 on complex variable methods assumes that Chapter 7 on two-dimensional problems has been covered.

Pittsburgh, Pennsylvania *William S. Slaughter*

List of Figures

List of Tables

Chapter 1

Review of Mechanics of Materials

The purpose of this chapter is to reacquaint the reader with some of the fundamental concepts of mechanics of materials, as it is taught to undergraduate engineering students. Mechanics of materials (sometimes referred to as "strength of materials") is distinguished from the theory of elasticity by its relative disregard for mathematical rigor. In the usual undergraduate treatment of the subject, so-called "engineering" approximations and assumptions are casually proposed and accepted—their validity is based on experimental verification. For a great many engineering applications, this is perfectly acceptable. As long as one does not venture beyond the boundaries of common experience in utilizing the results from mechanics of materials, their accuracy can be relied upon.

Difficulties arise, however, as soon as one wishes to extend the scope of the problems commonly analyzed using a mechanics of materials approach. In order to do this, one needs a clear understanding of all assumptions and approximations so that rational judgments about whether or not they are appropriate can be made. It is also important to define exactly what are the parameters of a problem, that is, what distinguishes one problem from another. The proliferation of commercial finite element analysis codes, coupled with their robustness and *apparent* ease of use, adds to the importance of these issues. They will be addressed during the development of the theory of elasticity in subsequent chapters. During the course of this book, results from elasticity theory will on occasion be compared and contrasted with those from mechanics of materials.

The material in this chapter is abridged, abrupt, and unelaborative. For more information on the mechanics of materials, there are many excellent textbooks, including those by Beer and Johnston (1992), Popov (1999), Hibbeler (1997), and Boresi et al. (1993).

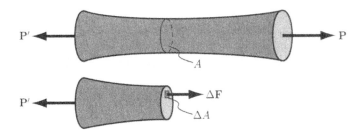

FIGURE 1.1. Normal force acting on a cross section of a cylinder in axial tension.

1.1   Forces and Stress

Solid bodies subject to external forces can be thought of as transmitting these forces from one point to another. This transmission of forces leads to internal forces, the intensity of which are characterized by stresses.

1.1.1   Normal Stress

Consider the solid cylinder with variable (noncircular) cross section, subject to equal and opposite axial tensile loads given by the force vectors $\mathbf{P}$ and $\mathbf{P}'$, shown in Figure 1.1. Make an imaginary cut through the cylinder along a cross section, perpendicular to the line of action of $\mathbf{P}$ and $\mathbf{P}'$ and with area A, and look at the free-body diagram of the portion of the cylinder on one side of the cut. Acting on each incremental element ΔA of the surface A, there is a resultant force vector $\Delta \mathbf{F}$ that is assumed to be normal to the surface. This is called a *normal force*.

The *normal stress* at a point on the surface of the cross section is the normal force per unit area defined by the limit $\sigma = \lim_{\Delta A \to 0} \Delta F/\Delta A$. The requirement that the free body be in equilibrium means that $\int_A \sigma \, dA = P$, where P is the magnitude of $\mathbf{P}$. It follows that the average normal stress on the cross section is

$$\sigma_{\text{avg}} = \frac{1}{A} \int_A \sigma \, dA = \frac{P}{A} \, . \tag{1.1.1}$$

In practice, if the common line of action of the force vectors $\mathbf{P}$ and $\mathbf{P}'$ is coincident with the centroidal axis of the cylinder (centric loading), then the normal stress distribution on A is assumed to be uniform and thus equal to the average stress.

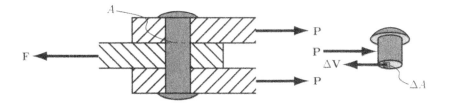

FIGURE 1.2. *Shear force acting on section of a rivet.*

1.1.2 Shear Stress

Consider a rivet connecting three plates as shown in Figure 1.2. Make an imaginary cut through the rivet along a cross section, with area A, that lies in the plane between the top two plates and consider the free-body diagram of the portion of the rivet above the cut. The top plate exerts a force $\mathbf{P}$ on the rivet that is parallel to the surface A. On each incremental element ΔA of the surface, there is a resultant force vector $\Delta \mathbf{V}$ that is assumed to be tangential to the surface and to act in the opposite direction as $\mathbf{P}$. This is called a *shear force*.

The *shear stress* at a point on the surface is the shear force per unit area defined by the limit $\tau = \lim_{\Delta A \to 0} \Delta V / \Delta A$. In order for the free body to be in equilibrium, $\int_A \tau \, dA = P$ and the average shear stress is

$$\tau_{\text{avg}} = \frac{1}{A} \int_A \tau \, dA = \frac{P}{A} \, . \tag{1.1.2}$$

It is not appropriate to assume that the shear stress distribution on A is uniform, for reasons that will become clear later.

1.1.3 Components of Stress

Consider a body subjected to several loads $\mathbf{P}_1$, $\mathbf{P}_2$, ... , as shown in Figure 1.3. The internal forces at a point Q inside the body are characterized in the following way. Pass a section through Q using a plane that is perpendicular to the x-direction, that is, parallel to the yz-plane. The free body formed to the left of the section is subject to some of the original loads and to normal and shear forces distributed over the section. On an area ΔA surrounding the point Q, the resultant normal force is $\Delta \mathbf{F}^x$, the resultant shear force in the y-direction is $\Delta \mathbf{V}_y^x$, and the resultant shear force in the z-direction is $\Delta \mathbf{V}_z^x$. Three components of stress are defined at the point Q inside the body by

$$\sigma_x = \lim_{\Delta A \to 0} \frac{\Delta F^x}{\Delta A} \, , \quad \tau_{xy} = \lim_{\Delta A \to 0} \frac{\Delta V_y^x}{\Delta A} \, , \quad \tau_{xz} = \lim_{\Delta A \to 0} \frac{\Delta V_z^x}{\Delta A} \, . \tag{1.1.3}$$

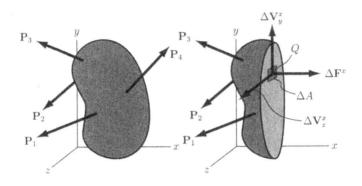

FIGURE 1.3.   Internal force acting on a surface normal to the x-axis.

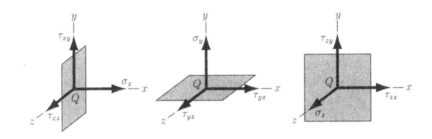

FIGURE 1.4.   Components of stress.

By taking sections through Q that are perpendicular to the y-direction and the z-direction, respectively, the components of stress σ_y, τ_{yx}, and τ_{yz} and σ_z, τ_{zx}, and τ_{zy} are similarly defined (see Figure 1.4). The notation is such that the first subscript of the shear components of stress indicates the surface on which the component is defined and the second subscript indicates the direction in which it acts. Thus, τ_{zx} acts in the x-direction on the plane perpendicular to the z-direction. By considering a differential cubic volume, with components of stress acting on the cubic faces as defined above, it is easily shown from balance of moments that $\tau_{xy} = \tau_{yx}$, $\tau_{yz} = \tau_{zy}$, and $\tau_{zx} = \tau_{xz}$. Thus, only six of the stress components are independent. It will be shown later that the normal and shear stresses on *any* surface that passes though the point Q can be determined from these six components of stress. Therefore, the six independent components of stress completely describe the internal forces at a given point Q in the body.

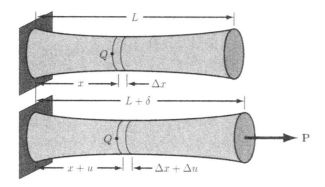

FIGURE 1.5. *Variable cross section cylinder under axial loading.*

1.2 Stress and Strain

Central to the theory of solid mechanics, as opposed to rigid-body dynamics say, is the acceptance that bodies are deformable. The deformation of a body is related to internal forces, that is, stresses, and is characterized locally by the strains.

1.2.1 Longitudinal Strain Under Axial Loading

Consider the solid cylinder with a variable (noncircular) cross section under axial loading shown in Figure 1.5. At a point Q in the member, define an element of the cylinder of length Δx. Denoting by Δu the change in length of the element under the given loading, the change in length per unit length, or *longitudinal strain*, at point Q is defined as

$$\varepsilon = \lim_{\Delta x \to 0} \frac{\Delta u}{\Delta x} = \frac{du}{dx}. \tag{1.2.1}$$

If the cylinder has a *uniform cross section*, then the longitudinal strain is uniform along the length of the cylinder and is given by $\varepsilon = \delta/L$, where L is the original length of the rod and δ is the total change in length.

1.2.2 Hooke's Law and Young's Modulus

It is seen experimentally that the relation between normal stress and longitudinal strain is a material property—it is independent of test specimen geometry. The stress-strain diagram shown in Figure 1.6 is representative of a ductile material. If, when an axial load is applied to a specimen and

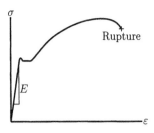

FIGURE 1.6. Stress-strain diagram for a ductile material.

then removed, the normal stress and longitudinal strain in the body return to zero, the intervening deformation is said to be *elastic*. For most
engineering materials, if the longitudinal strain is sufficiently small, the deformation is elastic and the relation between normal stress and longitudinal
strain is approximately linear—one may write

$$\sigma = E\varepsilon \ . \tag{1.2.2}$$

This relation is known as *Hooke's law* and the coefficient E is a material
property called *Young's modulus*.

1.2.3   Poisson's Ratio

Axial extension is accompanied by contraction in the transverse directions.
Thus, during elastic deformation, there is *lateral strain* $\varepsilon_y = \varepsilon_z$ in addition
to the the axial strain $\varepsilon_x = \sigma_x/E$ given by Hooke's law. The lateral and
axial strains are related by *Poisson's ratio*;

$$\nu = \left| \frac{\text{lateral strain}}{\text{axial strain}} \right| = -\frac{\varepsilon_y}{\varepsilon_x} = -\frac{\varepsilon_z}{\varepsilon_x} \ . \tag{1.2.3}$$

Poisson's ratio ν is another material property.

1.2.4   Shear Strain

Consider the rectangular member subject to the uniform shear stress τ
shown in Figure 1.7. The *(orthogonal) shear strain* is defined as the reduction in the initially right angle $\angle BAC = \pi/2$, so that

$$\gamma = \frac{\pi}{2} - \angle B'A'C' \ . \tag{1.2.4}$$

For most engineering materials, the relation between shear stress and strain
during elastic deformation is approximately linear and one may write

$$\tau = G\gamma \ , \tag{1.2.5}$$

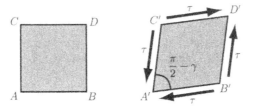

FIGURE 1.7. *Orthogonal shear strain.*

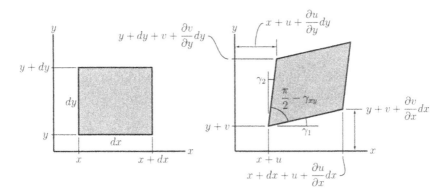

FIGURE 1.8. *Engineering components of strain.*

where G is a material property called the *shear modulus*. It can be shown by considering deformation in different coordinate systems that Young's modulus E, Poisson's ratio ν, and the shear modulus G are not independent. Instead, they are related by

$$G = \frac{E}{2(1+\nu)} \ . \tag{1.2.6}$$

1.2.5 Components of Strain

In general, during deformation one will have different longitudinal strains along different directions and different shear strains between different orthogonal angles. Consider the two-dimensional differential element shown in Figure 1.8. The results found here easily generalize to three dimensions. The element is initially rectangular with sides of length dx and dy, where dx and dy are as small as necessary. During deformation, a material point initially at (x, y) is displaced to $(x + u, y + v)$; u and v are the *displace-*

ments of the material point in the x- and y-directions. The sides initially of length dx and dy change length during deformation and are no longer orthogonal, forming angles γ_1 and γ_2 with the x- and y-directions. The displacements of the material point initially at $(x+dx,y)$ are given, by the first two terms of the Taylor expansion, as $u+(\partial u/\partial x)\,dx$ in the x-direction and $v+(\partial v/\partial x)\,dx$ in the y-direction. Displacements of other points of the element are given similarly, as shown in Figure 1.8.

The longitudinal strain in the x-direction is the change in length per unit length $\varepsilon_x = (L - L_0)/L_0$, where $L_0 = dx$ and, assuming $\gamma_1 \ll 1$, $L \approx dx\,(1 + \partial u/\partial x)$. Therefore,

$$\varepsilon_x = \frac{dx\left(1 + \frac{\partial u}{\partial x}\right) - dx}{dx} = \frac{\partial u}{\partial x} \ . \tag{1.2.7}$$

Generalizing to three dimensions, where w is the displacement in the z-direction, the longitudinal strains in the x-, y-, and z-directions are

$$\varepsilon_x = \frac{\partial u}{\partial x}, \quad \varepsilon_y = \frac{\partial v}{\partial y}, \quad \varepsilon_z = \frac{\partial w}{\partial z} \ . \tag{1.2.8}$$

The shear strain between the x- and y-directions is $\gamma_{xy} = \gamma_1 + \gamma_2$. If $\partial u/\partial x \ll 1$ and $\partial v/\partial x \ll 1$, then one can see from Figure 1.8 that

$$\gamma_1 = \tan^{-1}\left(\frac{\frac{\partial v}{\partial x}}{1 + \frac{\partial u}{\partial x}}\right) \approx \tan^{-1}\left(\frac{\partial v}{\partial x}\right) \approx \frac{\partial v}{\partial x} \ . \tag{1.2.9}$$

Similarly, if $\partial u/\partial y \ll 1$ and $\partial v/\partial y \ll 1$, then

$$\gamma_2 = \tan^{-1}\left(\frac{\frac{\partial u}{\partial y}}{1 + \frac{\partial v}{\partial y}}\right) \approx \tan^{-1}\left(\frac{\partial u}{\partial y}\right) \approx \frac{\partial u}{\partial y} \ . \tag{1.2.10}$$

Generalizing to three dimensions, this gives

$$\gamma_{xy} = \frac{\partial u}{\partial y} + \frac{\partial v}{\partial x}, \quad \gamma_{yz} = \frac{\partial v}{\partial z} + \frac{\partial w}{\partial y}, \quad \gamma_{zx} = \frac{\partial w}{\partial x} + \frac{\partial u}{\partial z} \ . \tag{1.2.11}$$

Note that each of the relations in (1.2.8) and (1.2.11) is an approximation, valid only for small (infinitesimal) values of the partial differentiations of the displacements.

1.2.6   Generalized Hooke's Law

In order to express the components of strain as functions of the components of stress during elastic deformation, consider separately the effect of each stress component and sum the results;

$$\varepsilon_x = \frac{1}{E}[\sigma_x - \nu(\sigma_y + \sigma_z)], \quad \gamma_{xy} = \frac{\tau_{xy}}{G},$$

$$\varepsilon_y = \frac{1}{E}[\sigma_y - \nu(\sigma_z + \sigma_x)], \quad \gamma_{yz} = \frac{\tau_{yz}}{G}, \tag{1.2.12}$$

$$\varepsilon_z = \frac{1}{E}[\sigma_z - \nu(\sigma_x + \sigma_y)], \quad \gamma_{zx} = \frac{\tau_{zx}}{G}.$$

This *superposition principle* applies when the strains are linearly dependent on the stresses and it is assumed that the deformation resulting from any given load is unaffected by the presence of other loads.

1.2.7 Bulk Modulus

Consider the elastic deformation of a cube initially with edges of unit length aligned with the coordinate axes. This cube will initially have unit volume, $V_0 = 1$. After deformation, the edges of the cube will have changed length and, neglecting the effect of shear strains, the volume will have changed to

$$V = (1 + \varepsilon_x)(1 + \varepsilon_y)(1 + \varepsilon_z). \tag{1.2.13}$$

Expanding this expression and neglecting higher-order terms, since strains are small,

$$V \approx 1 + \varepsilon_x + \varepsilon_y + \varepsilon_z. \tag{1.2.14}$$

The change in volume per unit volume, or *dilatation*, is thus given by

$$e = \frac{V - V_0}{V_0} = \varepsilon_x + \varepsilon_y + \varepsilon_z. \tag{1.2.15}$$

Using (1.2.12), the dilatation is related to the normal components of stress by

$$e = \frac{1 - 2\nu}{E}(\sigma_x + \sigma_y + \sigma_z). \tag{1.2.16}$$

In the special case of *hydrostatic pressure*, the normal components of stress are all equal to $-p$ (by convention, the pressure p is taken to be positive in compression) and the shear stresses are zero. Under these conditions, it follows from (1.2.16) that the dilatation is

$$e = -\frac{p}{K}, \tag{1.2.17}$$

where the *bulk modulus* K is a material property related to Young's modulus and Poisson's ratio by

$$K = \frac{E}{3(1 - 2\nu)}. \tag{1.2.18}$$

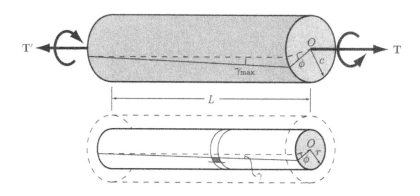

FIGURE 1.9. Torsion of a circular cylinder.

1.3 Torsion of Circular Cylinders

Consider deformation of the circular cylinder shown in Figure 1.9. The centroidal axis of the cylinder is the axis running through the centroids O of each cross section. The cylinder of length L and outer radius c has equal and opposite torques $\mathbf{T}$ and $\mathbf{T}'$ applied to either end. The total angle of twist between the two ends of the cylinder caused by the applied torque is ϕ. During deformation, it as assumed that each cross section remains planar and undistorted; each cross section rotates through some angle but neither warps nor changes shape. It is further assumed that the amount each cross section rotates is proportional to its distance from the end of the cylinder.

To determine the distribution of shear strains in the cylinder, consider a small square element formed by two adjacent circles of radius $r \le c$ centered on the centroidal axis of the cylinder and two straight lines parallel to the centroidal axis and intersecting the two circles (Figure 1.9). As the cylinder is twisted, the lines parallel to the centroidal axis are skewed so that they form an angle γ with their initial direction. This then is the shear strain at that point in the cylinder. For small angles ϕ and γ, the arc lengths $L\gamma$ and $r\phi$ are approximately equal so that

$$\gamma = \frac{r\phi}{L} \, . \tag{1.3.1}$$

Thus, the shear strain is a linear function of the distance r from the centroidal axis and can alternatively be expressed as

$$\gamma = \frac{r}{c}\gamma_{\max} \, , \tag{1.3.2}$$

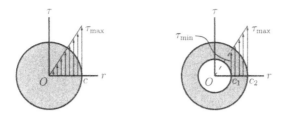

FIGURE 1.10. Shear stress distribution during torsion.

where $\gamma_{\max} = c\phi/L$ is the maximum shear strain in the cylinder, which occurs at the outer surface of the cylinder. All other components of strain are assumed to be zero.

For elastic deformation, the shear stress in the cylinder is related to the shear strain by $\tau = G\gamma$ so that, from (1.3.1),

$$\tau = \frac{r\phi G}{L} .$$
(1.3.3)

Thus, the shear stress is also a linear function of the distance r from the centroidal axis and is alternatively expressed as

$$\tau = \frac{r}{c}\tau_{\max} ,$$
(1.3.4)

where $\tau_{\max}$ is the maximum shear stress in the cylinder. This relation is shown in Figure 1.10. The resultant torque on each cross section about the centroidal axis is found by integrating the shear stress τ times the moment arm r over the area A;

$$T = \int_A \tau r\, dA = \frac{G\phi}{L} \int_A r^2\, dA .$$
(1.3.5)

However, $J \equiv \int_A r^2\, dA$ is the polar moment of inertia of the cross section with respect to its centroid O. Therefore, solving for $\tau = r\phi G/L$,

$$\tau = \frac{Tr}{J} , \qquad \tau_{\max} = \frac{Tc}{J} .$$
(1.3.6)

The polar moment of inertia of a circle of radius c is $J = \frac{1}{2}\pi c^4$. For a hollow circular cylinder of inner radius c_1 and outer radius c_2, the polar moment of inertia is $J = \frac{1}{2}\pi(c_2^4 - c_1^4)$. From (1.3.3) and (1.3.6), the total twist of the cylinder is

$$\phi = \frac{TL}{JG} .$$
(1.3.7)

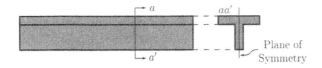

FIGURE 1.11. Beam with a plane of mirror symmetry.

Example 1.1
Consider a carbon steel $(G = 80$ GPa) solid circular cylinder of radius $c = 2$ cm and length $L = 0.2$ m subjected to a torque $T = 1$ kN $\cdot$ m. Find the maximum shear stress in the cylinder and the total angle of twist.

   Solution. The polar moment of inertia $J = \frac{1}{2}\pi(0.02 \text{ m})^4 = 2.51 \times 10^{-7}$ m^4. The maximum shear stress in the cylinder is given by (1.3.6),

$$\tau_{\max} = \frac{(1000 \text{ N} \cdot \text{m})(0.02 \text{ m})}{2.51 \times 10^{-7} \text{ m}^4} = 79.7 \text{ MPa} .$$

From (1.3.7), the total angle of twist of the cylinder is

$$\phi = \frac{(1000 \text{ N} \cdot \text{m})(0.2 \text{ m})}{(2.51 \times 10^{-7} \text{ m}^4)(80 \times 10^9 \text{ Pa})} = 9.96 \times 10^{-3} \text{ rad} = 0.571° .$$

1.4   Bending of Prismatic Beams

Beams are a common structural element in many engineering designs. Consequently, a large proportion of the study of mechanics of materials centers around their analysis. The synopsis given here is limited to prismatic beams with straight centroidal axes. A beam is said to be prismatic if its cross sections are each the same and are not rotated relative to one another. Assume also that each cross section has a line of mirror symmetry, which collectively define a plane of symmetry for the beam (Figure 1.11), and that the loads applied to the beam are such that this plane of symmetry is maintained when the beam is deformed.

1.4.1   Shear Force and Bending Moment

The combination of the loads applied to a beam and the reactions exerted on the beam by its supports cause the beam to deform and stresses to arise at points within the beam. Suppose that the loads applied to the beam

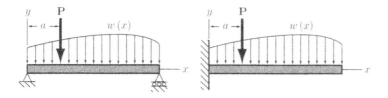

FIGURE 1.12. *A simply supported beam and a cantilevered beam.*

FIGURE 1.13. *Positive shear force and bending moment.*

are in the plane of symmetry of the beam and perpendicular to the beam's centroidal axis. Two examples of such a beam, one simply supported and the other cantilevered, are shown in Figure 1.12. The loads exerted on each beam are composed of a concentrated transverse force of magnitude P acting a distance a from the end of the beam and a distributed force per unit length $w(x)$, where x is the distance from the end of the beam. The simply supported beam is constrained so that its deflection is zero at its ends, whereas the cantilevered beam is constrained so that its deflection and slope are zero at the supported end.

The normal and shear stresses that act on a cross section of a beam are statically equivalent to a bending moment $\mathbf{M}$ and a shear force $\mathbf{V}$ that can be determined by considering equilibrium of a free-body diagram of the portion of the beam to one side of the cross section. The bending moment and shear force acting on either side of an arbitrary cross section are shown in Figure 1.13, where the magnitudes of $\mathbf{M}$ and $\mathbf{V}$ are equal to the those of $\mathbf{M}'$ and $\mathbf{V}'$. A sign convention is introduced in which the magnitude M of the bending moment and the magnitude V of the shear force are positive as shown in Figure 1.13.

The shear force V and the bending moment M are, in general, functions of the distance x along the beam. It can readily be shown that the rate of change of the shear force at a point along the beam is equal to minus the distributed force at that point,

$$\frac{dV}{dx} = -w(x) , \qquad (1.4.1)$$

and that the rate of change of the bending moment at a point along the

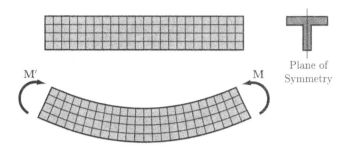

FIGURE 1.14. Pure bending of a prismatic beam.

beam is equal to the shear force at that point,

$$\frac{dM}{dx} = V(x) . \tag{1.4.2}$$

As an alternative to considering free-body diagrams, these relations can sometimes be used to determine the shear force and bending moment distributions in a beam.

1.4.2　Pure Bending

A beam (or a portion of a beam) is said to be in a state of pure bending if $dM/dx = V = 0$. Consider pure bending of a prismatic beam possessing a plane of symmetry, as shown in Figure 1.14. The bending moment M in the beam is constant. During deformation, it is assumed that plane cross sections initially perpendicular to the centroidal axis of the beam remain planar and that deformation is uniformly distributed along the beam. As a result of pure bending, the top surface is shortened, the bottom surface is lengthened, and there is some intermediate surface, called the *neutral surface*, that does not change length. In the plane of symmetry, lines that were initially parallel to the neutral surface are deformed into circular arcs with a common center of curvature C and angle of arc θ (Figure 1.15).

Let the x-axis lie along the intersection of the plane of symmetry and the neutral surface in the undeformed beam, and let the y-axis lie in the plane of symmetry (Figure 1.15). After deformation, the intersection of the plane of symmetry and the neutral surface forms a circular arc of radius ρ which, since the neutral surface does not change length, satisfies the relation

$$L_0 = \rho\theta , \tag{1.4.3}$$

where L_0 is the initial length of the beam. At a distance y above the neutral

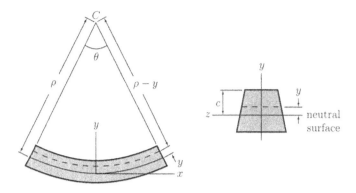

FIGURE 1.15. Deformations in a symmetric beam in pure bending.

surface, the length is reduced to

$$L = (\rho - y)\theta .$$ (1.4.4)

The change in length along the x-direction is $L - L_0 = -y\theta$ so that the longitudinal strain $\varepsilon_x = (L - L_0)/L_0$ and its maximum absolute value are

$$\varepsilon_x = -\frac{y}{\rho} , \quad |\varepsilon|_{\max} = \frac{c}{\rho} ,$$ (1.4.5)

where c denotes the largest perpendicular distance from the neutral surface to a point in the cross section (which happens to be the distance from the neutral surface to the top surface of the beam in Figure 1.15).

For elastic deformation, Hooke's law and (1.4.5) combine to give the axial normal stress

$$\sigma_x = -\frac{yE}{\rho} .$$ (1.4.6)

The normal stress is a linear function of y as shown in Figure 1.16. In order for the beam to be in equilibrium, the resultant normal force on each cross section must be zero, $\int_A \sigma_x \, dA = 0$. Using (1.4.6) and noting that E and ρ are constant, this condition requires that

$$\int_A y \, dA = 0 .$$ (1.4.7)

Thus, the neutral surface passes through the centroid of each cross section. The other nontrivial equilibrium condition is that the resultant moment on each cross section about the z-axis must equal the applied moment;

$$\int_A (-y\sigma_x) \, dA = M .$$ (1.4.8)

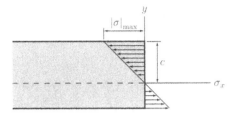

FIGURE 1.16. Normal stress distribution.

Using (1.4.6), this leads to $EI/\rho = M$, where $I = \int_A y^2 \, dA$ is the moment of inertia of the cross section with respect to the z-axis. Thus, substituting back into (1.4.6),

$$\sigma_x = -\frac{My}{I}, \quad |\sigma|_{\max} = \frac{|M|c}{I}. \tag{1.4.9}$$

It is assumed that all other components of stress are negligible.

1.4.3   Transverse Loading

In most applications, beams are not in a state of pure bending. Rather, they are subject to transverse loads as shown in Figure 1.12 such that the shear force V and the bending moment M are, in general, nonzero functions of x. It is assumed that the axial normal stress σ_x at a point along the beam is still given by (1.4.9). However, there is also a distribution of shear stress τ_{xy} on each cross section associated with the shear force V.

As shown in Figure 1.17, at any given distance above or below the neutral surface, the shear stress will typically vary through the thickness of the beam. The mechanics of materials does not tell one how the shear stress varies with z. However, the average shear stress acting on a cross section, at a particular distance y above the neutral surface, can be determined by considering equilibrium of the portion of the beam above that particular value of y, that is, with cross section A' as shown in Figure 1.17. One finds that

$$\tau_{\text{avg}} = \frac{1}{t} \int_{-t/2}^{t/2} \tau_{xy} \, dz = \frac{VQ}{It}, \tag{1.4.10}$$

where t is the thickness of the beam at that distance above the neutral surface and

$$Q = \int_{A'} y \, dA \tag{1.4.11}$$

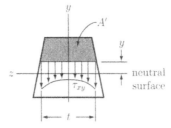

FIGURE 1.17. Shear stress distribution.

is the first moment of the area A'. It is readily shown that $Q = A'\bar{y}$, where $\bar{y}$ is the distance from the centroid of the area A' to the neutral surface.

1.4.4 Deflection

From geometry, the radius of curvature ρ at a point along the beam is related to the first and second derivatives of the deflection $v(x)$ of the beam at that point. Assuming that $|dv/dx| \ll 1$,

$$\frac{1}{\rho} \approx \frac{d^2v}{dx^2} \ . \tag{1.4.12}$$

Thus, from (1.4.6) and (1.4.9), the deflection is related to the bending moment by

$$\frac{d^2v}{dx^2} = \frac{M(x)}{EI} \ . \tag{1.4.13}$$

If the moment distribution $M(x)$ is known, this equation can be integrated, with appropriate boundary conditions as determined by the beam's supports, to obtain the deflection curve for the beam. The contribution of shear forces to the curvature is neglected.

Example 1.2
Consider the cantilevered beam shown in Figure 1.18. The beam has length L, thickness b, and height h and a transverse load of magnitude P applied to the end. Determine the maximum absolute value of the normal stress, the maximum average shear stress, and the equation for the deflection curve of the beam.

Solution. The bending moment distribution in this beam is

$$M = -P(L - x) ,$$

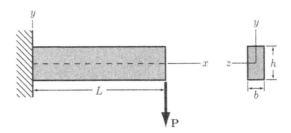

FIGURE 1.18. Cantilevered beam.

so that, from (1.4.9), the normal stress distribution on the cross section a distance x from the end of the beam is given by

$$\sigma_x = \frac{P(L-x)y}{I} \ .$$

The maximum absolute value of the bending moment $|M|_{max} = PL$ occurs at $x = 0$ and the maximum distance from the neutral surface is $c = h/2$. Therefore, the maximum absolute value of the normal stress in the beam is

$$|\sigma|_{max} = \frac{PLh}{2I} \ .$$

For the rectangular cross section shown, the moment of inertia is $I = bh^3/12$.

The shear force in the beam, $V = P$, is constant and the first moment of the area of that portion of the cross section greater than a distance y above the neutral surface is $Q = b(h^2 - 4y^2)/8$. Thus, the through-the-thickness average of the shear stress is

$$\tau_{avg} = \frac{P}{8I}(h^2 - 4y^2) \ .$$

The maximum average shear stress,

$$|\tau_{avg}|_{max} = \frac{Ph^2}{8I} \ ,$$

occurs at the neutral surface where $y = 0$.

The deflection curve is found by integrating (1.4.13):

$$\frac{d^2v}{dx^2} = -\frac{P(L-x)}{EI},$$

$$\frac{dv}{dx} = -\frac{P}{EI}\left(Lx - \frac{1}{2}x^2\right) + C_1,$$

$$v = -\frac{P}{EI}\left(\frac{1}{2}Lx^2 - \frac{1}{6}x^3\right) + C_1x + C_2,$$

where C_1 and C_2 are constants of integration. The boundary conditions for this cantilevered beam are $v = 0$ and $dv/dx = 0$ at $x = 0$, so that $C_1 = C_2 = 0$. Thus, the deflection of the beam is given by

$$v = -\frac{Px^2}{6EI}(3L - x).$$

The maximum deflection,

$$|v|_{max} = \frac{PL^3}{3EI},$$

occurs at the end $x = L$ of the beam.

Problems

1.1 In a standard tensile test, a 10-mm-diameter rod made of carbon steel is subjected to a tensile force of magnitude 15 kN. An elongation of 2.86×10^{-2} mm and a decrease in diameter of 3.06×10^{-3} mm are observed in a 30-mm gauge length of the rod. Determine the Young's modulus, Poisson's ratio, shear modulus, and bulk modulus for the material.

1.2 Consider uniaxial tension of a uniform rod of length L and cross-sectional area A and made of a material with Young's modulus E. Show that, for elastic deformation, the increase in length δ of the rod is given in terms of the axial load P by $\delta = PL/AE$.

1.3 Two wooden members of 60×100-mm uniform rectangular cross section are joined by a simple glued joint as shown. The ultimate stresses for the joint are $\sigma_U = 1.26$ MPa in tension and $\tau_U = 1.50$ MPa in shear. If the magnitude of the tensile axial load applied to the members is $P = 6$ kN, determine the factor of safety for the joint when (a) $\alpha = 20°$, (b) $\alpha = 35°$, and (c) $\alpha = 45°$. For each of these values of α, also determine whether the joint will fail in tension or in shear if P is increased until rupture occurs.

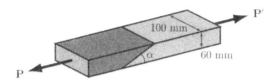

1.4 Two solid circular cylindrical rods are joined at B and loaded as shown below. Rod AB is made of brass ($E = 105$ GPa) and rod BC is made of steel ($E = 200$ GPa). Determine (a) the change in length of the composite shaft ABC and (b) the deflection of point B.

1.5 The rigid bar AD is supported as shown by two steel ($E = 200$ GPa) wires of 2 mm diameter and a pin and bracket at D. Knowing that the wires were initially taut, determine (a) the additional tension in each wire when a vertical load of magnitude $P = 200$ N is applied at B and (b) the corresponding deflection of point B.

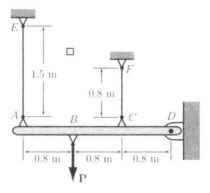

1.6 The solid spindle AB has a diameter of 100 mm and is made of steel with shear modulus $G = 77$ GPa and allowable shear stress $\tau_{all} = 80$ MPa. The sleeve CD has an outer diameter of 250 mm and a wall thickness of 25 mm and is made of aluminum with $G = 26$ GPa and $\tau_{all} = 70$ MPa. Determine the largest angle through which A may be rotated.

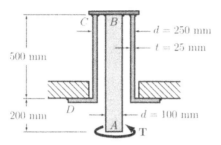

1.7 Determine the magnitude and location of the maximum deflection of the beam AB in terms of b, L, E, I, and P. You may assume that $a > b$.

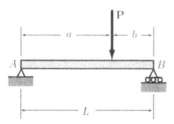

1.8 For the beam and loading shown below, where w is the force per unit length, determine (a) the equation of the elastic deflection curve of the beam, (b) the slope of the beam at A, and (c) the deflection of the beam at its midpoint.

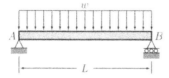

Chapter 2

Mathematical Preliminaries

The principal objective of this chapter is to introduce the theory of *tensor analysis*. Other mathematical topics, such as the calculus of variations and the theory of functions of a complex variable, will be reviewed in subsequent chapters as they are needed. Three-dimensional Euclidean vector and point spaces are implicitly assumed throughout.

A *reference frame* is required in order to observe an event in space and time. Physical phenomena are composed of a collection of such events. In general, different reference frames are in relative motion with respect to one another and events may appear differently in different reference frames. For example, a particle may be accelerating in one reference frame and stationary in another. It is assumed that physical laws for characterizing material behavior should be the same in all reference frames (Malvern 1969; Truesdell and Noll 1992; Ogden 1984). The consequences of this *principle of material frame-indifference* are examined in Chapter 5. Here, consideration is confined to a single fixed frame of reference. Within this reference frame, different coordinate systems can be defined which may have different orientations and/or origins, but are *not* in relative motion. A physical phenomenon is described in terms of a given coordinate system, but the phenomenon itself is independent of whatever coordinate system might be used to describe it. This independence will be implicit in the formulation of physical laws as *tensor equations*. Tensors are physical quantities that will be defined in such a way that if a tensor equation is true in one coordinate system, then it is true in all coordinate systems.

2.1 Scalars and Vectors

2.1.1 Scalar Quantities

Physical quantities, such as temperature, density, and time, that are characterized by a single real number are called *scalars*. Scalar quantities will

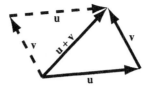

FIGURE 2.1. Graphical representation of the sum of two vectors.

be represented by italicized Latin or Greek characters (a or α, for example). The value of a scalar may in general be either positive or negative; the absolute value of a scalar a will be denoted by $|a|$.

2.1.2 Vector Quantities

Vectors are physical quantities that are completely described by a magnitude and a direction. For example, displacement, velocity, acceleration, and force are vector quantities. Such a quantity can be represented graphically as an arrow pointing in the direction of the vector with length proportional to the vector's magnitude. A vector will be denoted by a boldface character, **v** say. Other common notations for vectors include $\vec{v}$, $\bar{v}$, and $\underline{v}$. The magnitude of a vector **v** will be represented by $|\mathbf{v}|$. Two vectors are said to be equal if they have the same direction and magnitude. Thus, a vector is unchanged if it is moved parallel to itself. However, in some applications, the point of application or the line of action of a vector is important. For example, the point of application of a force on a body will affect the deformation. Thus, two vectors may be equal without being physically equivalent in the context of any given application.

Two vectors **u** and **v** of the same kind (two velocities, for example) may be represented graphically as arrows with the tail of one placed at the head of the other (Figure 2.1). Then, the sum of **u** and **v** is defined as the vector **u** + **v** extending from the tail of **u** to the head of **v**. The commutative law for vector addition,

$$\mathbf{u} + \mathbf{v} = \mathbf{v} + \mathbf{u} \,, \qquad (2.1.1)$$

is demonstrated in Figure 2.1. The associative law for vector addition,

$$(\mathbf{u} + \mathbf{v}) + \mathbf{w} = \mathbf{u} + (\mathbf{v} + \mathbf{w}) \,, \qquad (2.1.2)$$

can similarly be established. The *rule of vector addition* states that the physical effect of the quantity represented by the sum **u** + **v** must equal the sum of the physical effects of **u** and **v** taken separately. Note that not all physical entities representable as directed lines are vectors—in order to be vectors, they must also obey this rule of vector addition. For instance, a finite rotation about an axis can be thought of as a directed line parallel to the axis of rotation and of magnitude equal to the angle of rotation, but this quantity would not be a vector because it does not obey the rule of vector addition.

The *negative of a vector* is a vector of equal magnitude in the opposite direction and subtraction of one vector from another is defined as the addition of the negative of the vector,

$$\mathbf{u} - \mathbf{v} = \mathbf{u} + (-\mathbf{v}) , \qquad (2.1.3)$$

where $-\mathbf{v}$ is the negative of $\mathbf{v}$. A vector subtracted from itself is the *zero vector*,

$$\mathbf{v} - \mathbf{v} = \mathbf{0} , \qquad (2.1.4)$$

a vector of zero magnitude and undefined direction.

A vector $\mathbf{v}$ may be multiplied by a scalar c to yield a new vector $c\mathbf{v}$ of magnitude $|c||\mathbf{v}|$ that is in the same direction as $\mathbf{v}$ if c is positive and in the opposite direction if c is negative. If c is zero, then the product is the zero vector $\mathbf{0}$. The associative law,

$$a(b\mathbf{v}) = (ab)\mathbf{v} = b(a\mathbf{v}) , \qquad (2.1.5)$$

and the distributive laws,

$$(a + b)\mathbf{v} = a\mathbf{v} + b\mathbf{v} , \qquad (2.1.6)$$
$$c(\mathbf{u} + \mathbf{v}) = c\mathbf{u} + c\mathbf{v} , \qquad (2.1.7)$$

for multiplication of vectors by scalars are easily shown.

The division of a vector by a scalar is defined as multiplication of the vector by the reciprocal of the scalar,

$$\frac{\mathbf{v}}{c} = \left(\frac{1}{c}\right)\mathbf{v} . \qquad (2.1.8)$$

The result is undefined if $c = 0$.

Two vectors are *orthogonal* if, when represented graphically as arrows and placed tail to tail, they form a right (i.e., 90°) angle. Three vectors are said to be *mutually orthogonal* if each is orthogonal to both of the other two.

A dimensionless vector of unit magnitude is called a *unit vector*. A caret (^) will be used to distinguish unit vectors from other vectors. Any nonzero vector may be uniquely expressed as the product of its magnitude and a unit vector. For instance, if $\hat{\boldsymbol{\mu}}$ is the unit vector in the direction of $\mathbf{v}$, then $\mathbf{v} = |\mathbf{v}|\hat{\boldsymbol{\mu}}$. Conversely, the unit vector in the same direction as $\mathbf{v}$ is $\hat{\boldsymbol{\mu}} = \mathbf{v}/|\mathbf{v}|$.

Scalar components

In the vector equation $\mathbf{v} = \mathbf{a} + \mathbf{b} + \mathbf{c} + \cdots$, the vectors $\mathbf{a}$, $\mathbf{b}$, $\mathbf{c}$, ... whose sum is $\mathbf{v}$ are called *vector components* of $\mathbf{v}$. Given three noncoplanar *base vectors*, a vector $\mathbf{v}$ can be uniquely resolved into the sum of three vector components that are parallel to the given base vectors. The set of base vectors is called a *vector basis*. If the base vectors are mutually orthogonal unit vectors, then the vector basis is said to be *orthonormal*. An orthonormal basis can be defined, for instance, with unit base vectors in the positive directions of a set of *rectangular Cartesian coordinate axes* (see Section 2.4). An orthonormal basis can also be defined with unit base vectors in the positive directions of a set of *orthogonal curvilinear coordinate axes*, such as those in cylindrical or spherical coordinates (see Section 2.5), the difference being that in a curvilinear coordinate system, the directions of the base vectors will vary with the values of the coordinate variables. Further discussion of coordinate systems will come later. For now, it is sufficient to assume that a vector basis can be defined. Unless stated otherwise, whenever base vectors or vector bases are referred to, it is implied that the vector basis is orthonormal.

Let $\hat{\mathbf{e}}_1$, $\hat{\mathbf{e}}_2$, and $\hat{\mathbf{e}}_3$ be the base vectors of an orthonormal basis (Figure 2.2).[1] The orthonormal basis formed by these base vectors is denoted by $\{\hat{\mathbf{e}}_i\} \equiv \{\hat{\mathbf{e}}_1, \hat{\mathbf{e}}_2, \hat{\mathbf{e}}_3\}$, where it will sometimes be convenient to think of $\{\hat{\mathbf{e}}_i\}$ as the column matrix of base vectors

$$\{\hat{\mathbf{e}}_i\} = \begin{bmatrix} \hat{\mathbf{e}}_1 \\ \hat{\mathbf{e}}_2 \\ \hat{\mathbf{e}}_3 \end{bmatrix} . \tag{2.1.9}$$

Orthonormal bases will usually be defined so that they form *right-handed systems*.[2]

If $\mathbf{v}$ is a vector, then the vector components of $\mathbf{v}$ *in the orthonormal basis* $\{\hat{\mathbf{e}}_i\}$ are defined as the three vectors, parallel to the base vectors, whose sum is $\mathbf{v}$. Each vector component can be expressed as the product of a *scalar component* and a unit base vector. Thus, the vector $\mathbf{v}$ can be expressed as

$$\mathbf{v} = v_1\hat{\mathbf{e}}_1 + v_2\hat{\mathbf{e}}_2 + v_3\hat{\mathbf{e}}_3 = \sum_{i=1}^{3} v_i\hat{\mathbf{e}}_i , \tag{2.1.10}$$

where $v_1\hat{\mathbf{e}}_1$, $v_2\hat{\mathbf{e}}_2$, and $v_3\hat{\mathbf{e}}_3$ are the vector components of $\mathbf{v}$ and v_1, v_2, and v_3 are the scalar components of $\mathbf{v}$ in $\{\hat{\mathbf{e}}_i\}$ (Figure 2.2). The scalar components are the orthogonal projections of the vector onto the directions

---

[1] The reader may be more familiar with the notation $\mathbf{i}$, $\mathbf{j}$, and $\mathbf{k}$ for base vectors in the directions of the Cartesian coordinate axes x, y, and z. However, until Section 2.4, the discussion is applicable to *any* orthonormal basis. The advantages of the particular

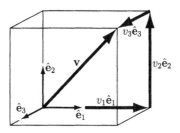

FIGURE 2.2. An orthonormal basis and the corresponding vector components of a vector.

of the base vectors [e.g., $v_1 = |\mathbf{v}| \cos(\mathbf{v}, \hat{\mathbf{e}}_1)$] and may be negative.[3] Note that the scalar components of $\mathbf{v}$ will be different in a different orthonormal basis. For two vectors $\mathbf{u}$ and $\mathbf{v}$ and a scalar c, it follows that

$$\mathbf{u} \pm \mathbf{v} = \sum_{i=1}^{3} (u_i \pm v_i)\hat{\mathbf{e}}_i \ , \qquad (2.1.11)$$

$$c\mathbf{v} = \sum_{i=1}^{3} cv_i\hat{\mathbf{e}}_i \ ; \qquad (2.1.12)$$

in the orthonormal basis $\{\hat{\mathbf{e}}_i\}$, the scalar components of $\mathbf{u} \pm \mathbf{v}$ are $u_1 \pm v_1$, $u_2 \pm v_2$, and $u_3 \pm v_3$ and the scalar components of $c\mathbf{v}$ are cv_1, cv_2, and cv_3.

Scalar and vector products

The *scalar product* (or *dot product*) of two vectors $\mathbf{u}$ and $\mathbf{v}$ is a scalar whose value is given by

$$\mathbf{u} \cdot \mathbf{v} = |\mathbf{u}||\mathbf{v}| \cos(\mathbf{u}, \mathbf{v}) \ . \qquad (2.1.13)$$

The scalar product $\mathbf{u} \cdot \mathbf{v}$ can be interpreted as either the orthogonal projection of $\mathbf{u}$ onto the direction of $\mathbf{v}$ times $|\mathbf{v}|$ or the orthogonal projection of $\mathbf{v}$ onto the direction of $\mathbf{u}$ times $|\mathbf{u}|$ (Figure 2.3). The commutative law

$$\mathbf{u} \cdot \mathbf{v} = \mathbf{v} \cdot \mathbf{u} \ , \qquad (2.1.14)$$

and the distributive laws

$$(a\mathbf{u}) \cdot (b\mathbf{v}) = ab(\mathbf{u} \cdot \mathbf{v}) \ , \qquad (2.1.15)$$

$$\mathbf{w} \cdot (\mathbf{u} + \mathbf{v}) = (\mathbf{w} \cdot \mathbf{u}) + (\mathbf{w} \cdot \mathbf{v}) \ , \qquad (2.1.16)$$

notation chosen here will become apparent shortly.

[2]If you imagine the index finger of your right hand sweeping through the 90° angle formed by $\hat{\mathbf{e}}_1$ and $\hat{\mathbf{e}}_2$, then your thumb points in the direction of $\hat{\mathbf{e}}_3$.

[3]The notation $\cos(\mathbf{u}, \mathbf{v})$ represents the cosine of the angle between zero and 180° formed by the vectors $\mathbf{u}$ and $\mathbf{v}$ when placed tail to tail; similarly for $\sin(\mathbf{u}, \mathbf{v})$.

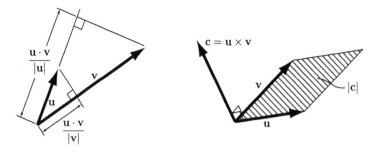

FIGURE 2.3.   Graphical interpretation of scalar and vector products.

for scalar products of vectors follow immediately from the scalar product definition (2.1.13). Note that $\mathbf{u} \cdot \mathbf{u} = |\mathbf{u}|^2$. If $\mathbf{u}$ and $\mathbf{v}$ are nonzero, then $\mathbf{u} \cdot \mathbf{v} = 0$ if and only if $\mathbf{u}$ and $\mathbf{v}$ are orthogonal, which provides an alternate definition of orthogonality for vectors.

For orthonormal base vectors, $\hat{\mathbf{e}}_1 \cdot \hat{\mathbf{e}}_1 = 1$, $\hat{\mathbf{e}}_1 \cdot \hat{\mathbf{e}}_2 = 0$, etc. Thus, in terms of the corresponding scalar components,

$$
\begin{aligned}
\mathbf{u} \cdot \mathbf{v} &= (u_1\hat{\mathbf{e}}_1 + u_2\hat{\mathbf{e}}_2 + u_3\hat{\mathbf{e}}_3) \cdot (v_1\hat{\mathbf{e}}_1 + v_2\hat{\mathbf{e}}_2 + v_3\hat{\mathbf{e}}_3) \\
&= u_1v_1(\hat{\mathbf{e}}_1 \cdot \hat{\mathbf{e}}_1) + u_1v_2(\hat{\mathbf{e}}_1 \cdot \hat{\mathbf{e}}_2) + \cdots \\
&= u_1v_1 + u_2v_2 + u_3v_3 \\
&= \sum_{i=1}^{3} u_iv_i \;.
\end{aligned}
\tag{2.1.17}
$$

Note also from the definition (2.1.13), that the scalar components of a vector $\mathbf{v}$ in the orthonormal basis $\{\hat{\mathbf{e}}_i\}$ are $v_1 = \mathbf{v} \cdot \hat{\mathbf{e}}_1$, $v_2 = \mathbf{v} \cdot \hat{\mathbf{e}}_2$, and $v_3 = \mathbf{v} \cdot \hat{\mathbf{e}}_3$ and that the orthogonal projection of $\mathbf{v}$ onto the direction of an arbitrary unit vector $\hat{\boldsymbol{\mu}}$ is $\mathbf{v} \cdot \hat{\boldsymbol{\mu}}$.

The *vector product* (or *cross product*) $\mathbf{c} = \mathbf{u} \times \mathbf{v}$ of two vectors $\mathbf{u}$ and $\mathbf{v}$ is a vector orthogonal to both $\mathbf{u}$ and $\mathbf{v}$ with magnitude

$$
|\mathbf{c}| = |\mathbf{u}||\mathbf{v}| \sin(\mathbf{u}, \mathbf{v}) \;.
\tag{2.1.18}
$$

The magnitude $|\mathbf{c}|$ of the vector product can be interpreted as the area of the parallelogram with edges formed by the vectors $\mathbf{u}$ and $\mathbf{v}$. The sense of $\mathbf{c}$ is given by the right-hand rule—if you imagine letting the index finger of your right hand sweep through the angle $< 180°$ from $\mathbf{u}$ to $\mathbf{v}$, then your thumb points in the direction of $\mathbf{c}$ (Figure 2.3). The distributive laws

$$
(a\mathbf{u}) \times (b\mathbf{v}) = ab(\mathbf{u} \times \mathbf{v}) \;,
\tag{2.1.19}
$$

$$
\mathbf{w} \times (\mathbf{u} + \mathbf{v}) = (\mathbf{w} \times \mathbf{u}) + (\mathbf{w} \times \mathbf{v}) \;,
\tag{2.1.20}
$$

follow. Note, however, that the vector product obeys neither an associative law,

$$\mathbf{u} \times (\mathbf{v} \times \mathbf{w}) \neq (\mathbf{u} \times \mathbf{v}) \times \mathbf{w} , \qquad (2.1.21)$$

nor a commutative law, since

$$\mathbf{u} \times \mathbf{v} = -(\mathbf{v} \times \mathbf{u}) . \qquad (2.1.22)$$

The vector product of a vector with itself is the zero vector, $\mathbf{u} \times \mathbf{u} = \mathbf{0}$. If $\mathbf{u}$ and $\mathbf{v}$ are nonzero, then $\mathbf{u} \times \mathbf{v} = \mathbf{0}$ if and only if $\mathbf{u}$ and $\mathbf{v}$ are parallel.

For a right-handed orthonormal basis $\{\hat{\mathbf{e}}_i\}$, it follows that $\hat{\mathbf{e}}_1 \times \hat{\mathbf{e}}_1 = \mathbf{0}$, $\hat{\mathbf{e}}_1 \times \hat{\mathbf{e}}_2 = \hat{\mathbf{e}}_3$, $\hat{\mathbf{e}}_1 \times \hat{\mathbf{e}}_3 = -\hat{\mathbf{e}}_2$, etc. Thus, in terms of the corresponding scalar components,

$$\begin{aligned}
\mathbf{u} \times \mathbf{v} &= (u_1\hat{\mathbf{e}}_1 + u_2\hat{\mathbf{e}}_2 + u_3\hat{\mathbf{e}}_3) \times (v_1\hat{\mathbf{e}}_1 + v_2\hat{\mathbf{e}}_2 + v_3\hat{\mathbf{e}}_3) \\
&= u_1v_1(\hat{\mathbf{e}}_1 \times \hat{\mathbf{e}}_1) + u_1v_2(\hat{\mathbf{e}}_1 \times \hat{\mathbf{e}}_2) + \cdots \\
&= (u_2v_3 - u_3v_2)\hat{\mathbf{e}}_1 + (u_3v_1 - u_1v_3)\hat{\mathbf{e}}_2 + (u_1v_2 - u_2v_1)\hat{\mathbf{e}}_3 \\
&= \det \begin{bmatrix} \hat{\mathbf{e}}_1 & \hat{\mathbf{e}}_2 & \hat{\mathbf{e}}_3 \\ u_1 & u_2 & u_3 \\ v_1 & v_2 & v_3 \end{bmatrix} , \qquad (2.1.23)
\end{aligned}$$

the determinant of the 3×3 matrix whose rows are composed of the unit base vectors and the scalar components of $\mathbf{u}$ and $\mathbf{v}$.

Vector transformation rule

The *vector transformation rule* gives the relation between the scalar components of a vector for two different orthonormal bases, $\{\hat{\mathbf{e}}_i\}$ and $\{\hat{\mathbf{e}}_i'\}$ say (Figure 2.4). The vector $\mathbf{v}$ is understood to represent some physical quantity that is independent of the vector basis, but the scalar components $v_i = \mathbf{v} \cdot \hat{\mathbf{e}}_i$ and $v_i' = \mathbf{v} \cdot \hat{\mathbf{e}}_i'$, with i equal to 1, 2, and 3, depend on the orthonormal bases. To examine the relation between v_i and v_i', consider the vector representation (2.1.10), $\mathbf{v} = \sum_{i=1}^{3} v_i\hat{\mathbf{e}}_i$. Taking the dot product of both sides with $\hat{\mathbf{e}}_1'$, one gets $\mathbf{v} \cdot \hat{\mathbf{e}}_1' = \sum_{i=1}^{3} v_i\hat{\mathbf{e}}_i \cdot \hat{\mathbf{e}}_1'$, which, since $\mathbf{v} \cdot \hat{\mathbf{e}}_1' = v_1'$, can be rewritten as

$$v_1' = v_1(\hat{\mathbf{e}}_1 \cdot \hat{\mathbf{e}}_1') + v_2(\hat{\mathbf{e}}_2 \cdot \hat{\mathbf{e}}_1') + v_3(\hat{\mathbf{e}}_3 \cdot \hat{\mathbf{e}}_1') . \qquad (2.1.24)$$

Similarly, taking the dot product of both sides of (2.1.10) with $\hat{\mathbf{e}}_2'$ and $\hat{\mathbf{e}}_3'$, and defining the *direction cosines*,[4]

$$\ell_{ij} \equiv \hat{\mathbf{e}}_i' \cdot \hat{\mathbf{e}}_j = \cos(\hat{\mathbf{e}}_i', \hat{\mathbf{e}}_j) , \qquad (2.1.25)$$

---

[4]Note that, in general, $\ell_{ij} \neq \ell_{ji}$. It is important to be consistent in following the convention that $\ell_{ij} = \hat{\mathbf{e}}_i' \cdot \hat{\mathbf{e}}_j$, and not $\hat{\mathbf{e}}_i \cdot \hat{\mathbf{e}}_j'$, as is the convention in some texts. The first index will *always* refer to the "primed" orthonormal basis. Another notation used by some authors gives the direction cosines as $a_i^j = \ell_{ij}$ (Malvern 1969).

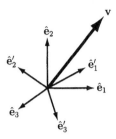

FIGURE 2.4.   Change of orthonormal basis.

where i and j represent 1, 2, and 3, leads to the transformation rule for the scalar components v_i' in terms of the components v_i:

$$v_1' = \ell_{11}v_1 + \ell_{12}v_2 + \ell_{13}v_3 \ ,$$
$$v_2' = \ell_{21}v_1 + \ell_{22}v_2 + \ell_{23}v_3 \ , \qquad (2.1.26a)$$
$$v_3' = \ell_{31}v_1 + \ell_{32}v_2 + \ell_{33}v_3 \ .$$

By representing the vector $\mathbf{v}$ in terms of the scalar components v_i', that is, $\mathbf{v} = \sum_{i=1}^{3} v_i'\hat{\mathbf{e}}_i'$, the transformation rule for expressing v_i in terms of v_i' can similarly be derived:

$$v_1 = \ell_{11}v_1' + \ell_{21}v_2' + \ell_{31}v_3' \ ,$$
$$v_2 = \ell_{12}v_1' + \ell_{22}v_2' + \ell_{32}v_3' \ , \qquad (2.1.26b)$$
$$v_3 = \ell_{13}v_1' + \ell_{23}v_2' + \ell_{33}v_3' \ .$$

In matrix form, (2.1.26) can be written as

$$[\mathbf{v}]' = [L][\mathbf{v}] \ , \quad [\mathbf{v}] = [L]^{\mathsf{T}}[\mathbf{v}]' \ , \qquad (2.1.26c)$$

where, when matrix equations of this sort are used, it is understood that

$$[\mathbf{v}] = \begin{bmatrix} v_1 \\ v_2 \\ v_3 \end{bmatrix} \ , \quad [\mathbf{v}]' = \begin{bmatrix} v_1' \\ v_2' \\ v_3' \end{bmatrix} \ , \quad [L] - \begin{bmatrix} \ell_{11} & \ell_{12} & \ell_{13} \\ \ell_{21} & \ell_{22} & \ell_{23} \\ \ell_{31} & \ell_{32} & \ell_{33} \end{bmatrix} \ . \qquad (2.1.27)$$

$[\mathbf{v}]$ is the column matrix whose elements are the scalar components of $\mathbf{v}$ in the orthonormal basis $\{\hat{\mathbf{e}}_i\}$, $[\mathbf{v}]'$ is the column matrix whose elements are the scalar components of $\mathbf{v}$ in the orthonormal basis $\{\hat{\mathbf{e}}_i'\}$, and $[L]$ is the 3×3 matrix whose elements are the direction cosines. $[L]^{\mathsf{T}}$ is the transpose of $[L]$.

Since the vector bases $\{\hat{\mathbf{e}}_i\}$ and $\{\hat{\mathbf{e}}_i'\}$ are both orthonormal, the transformation is said to be *orthogonal*. It is clear from (2.1.26c) that for an orthogonal transformation,

$$[L]^\mathsf{T} = [L]^{-1} \,, \tag{2.1.28}$$

where $[L]^{-1}$ is the inverse of $[L]$. Taking the determinant of both sides of (2.1.28) and recalling the results from linear algebra that $\det[A]^\mathsf{T} = \det[A]$ and $\det[A]^{-1} = 1/\det[A]$ yields

$$\det[L] = \frac{1}{\det[L]} \,, \tag{2.1.29}$$

which implies that

$$\det[L] = \pm 1 \,. \tag{2.1.30}$$

If $\{\hat{\mathbf{e}}_i\}$ and $\{\hat{\mathbf{e}}_i'\}$ are both right-handed orthonormal bases, the transformation is said to be *proper orthogonal*, in which case it can be shown that

$$\det[L] = +1 \,. \tag{2.1.31}$$

From now on, transformations will be assumed to be proper orthogonal unless stated otherwise.

The base vectors are vectors in their own right, of course, and have different scalar components in different vector bases (e.g., the "i^{th} scalar component" of $\hat{\mathbf{e}}_1'$ in the orthonormal basis $\{\hat{\mathbf{e}}_i\}$ is $\hat{\mathbf{e}}_1' \cdot \hat{\mathbf{e}}_i = \ell_{1i}$). Thus, $\hat{\mathbf{e}}_1' = \sum_{i=1}^{3} \ell_{1i}\hat{\mathbf{e}}_i$, and

$$\{\hat{\mathbf{e}}_i'\} = [L]\{\hat{\mathbf{e}}_i\} \,, \quad \{\hat{\mathbf{e}}_i\} = [L]^\mathsf{T}\{\hat{\mathbf{e}}_i'\} \,. \tag{2.1.32}$$

Note that $\{\hat{\mathbf{e}}_i\}$ and $\{\hat{\mathbf{e}}_i'\}$ are sets of base vectors, whereas $[\mathbf{v}]$ and $[\mathbf{v}]'$ are column matrices formed by the scalar components of $\mathbf{v}$ in $\{\hat{\mathbf{e}}_i\}$ and $\{\hat{\mathbf{e}}_i'\}$, respectively. In this regard, despite their similar appearance, (2.1.26c) and (2.1.32) are fundamentally different.

2.2 Indicial Notation

Indicial notation is a collection of time saving conventions. By carefully developing these conventions and then *slavishly following the convention rules*, time will be saved *and* the proper results will be obtained. Sloppiness in applying the rules is guaranteed to lead to mistakes.

2.2.1   Range Convention

When an equation is given involving quantities with Latin character sub-
scripts (indices), with each character appearing once in each term of the
equation, it is understood that the equation holds over the range of values
$\{1, 2, 3\}$ for each index. In the following examples, an indicial notation
equation is given on the left and the equivalent *expanded* equations are
given on the right:

$$x_i = 0.32 \qquad \Longleftrightarrow \qquad \begin{cases} x_1 = 0.32 \\ x_2 = 0.32 \\ x_3 = 0.32 \end{cases}, \qquad (2.2.1)$$

$$p_r = \Psi_r + d_r \qquad \Longleftrightarrow \qquad \begin{cases} p_1 = \Psi_1 + d_1 \\ p_2 = \Psi_2 + d_2 \\ p_3 = \Psi_3 + d_3 \end{cases}, \qquad (2.2.2)$$

$$a_{pq} = B_q c_p \qquad \Longleftrightarrow \qquad \begin{cases} a_{11} = B_1 c_1 \\ a_{12} = B_2 c_1 \\ a_{13} = B_3 c_1 \\ a_{21} = B_1 c_2 \\ a_{22} = B_2 c_2 \\ \quad \vdots \end{cases} . \qquad (2.2.3)$$

These indices, for example p and q in (2.2.3), are called *free indices*. The
same free indices *must* appear once and only once in every term of an
indicial notation equation. The only exception is when a term is a numeric
value, as in (2.2.1), in which case the value is understood to carry over to
each of the expanded equations. If there are n free indices in an indicial
notation equation, then there are 3^n corresponding expanded equations.
For an indicial notation equation to be true, *all* of the expanded equations
must be true and vice versa.

2.2.2   Summation Convention

If an index appears twice in a term of an indicial notation equation, summation over the range of the index is implied. Such an index is called
a *dummy index*. Examples with equivalent expanded equations are given
below:

$$g_{iji} = k_j \qquad \Longleftrightarrow \qquad \sum_{i=1}^{3} g_{iji} = k_j$$

$$\Longleftrightarrow \qquad \begin{cases} g_{111} + g_{212} + g_{313} = k_1 \\ g_{121} + g_{222} + g_{323} = k_2 \\ g_{131} + g_{232} + g_{333} = k_3 \end{cases}, \qquad (2.2.4)$$

$$b_{ij}c_ix_j = 0 \quad \Longleftrightarrow \quad \sum_{i=1}^{3}\sum_{j=1}^{3} b_{ij}c_ix_j = 0$$

$$\Longleftrightarrow \quad \begin{cases} b_{11}c_1x_1 + b_{12}c_1x_2 + b_{13}c_1x_3 \\ + b_{21}c_2x_1 + b_{22}c_2x_2 + b_{23}c_2x_3 \\ + b_{31}c_3x_1 + b_{32}c_3x_2 + b_{33}c_3x_3 = 0 \end{cases} . \quad (2.2.5)$$

A term in an equation with n dummy index pairs expands to 3^n terms.

2.2.3 Guidelines

Two indicial notation equations are equivalent if and only if they correspond to the same expanded equations. Thus, the relative position of indices in an indicial notation equation is important. For example, $a_{pq} = B_p c_q$ is not equivalent to $a_{pq} = B_q c_p$ (2.2.3) and $b_{ij}c_jx_i = 0$ is not equivalent to $b_{ij}c_ix_j = 0$ (2.2.5), since in each case the expanded forms of the equations are different. The actual Latin character used does not matter, so long as the relative position is maintained; $a_{ij} = B_j c_i$ is equivalent to $a_{pq} = B_q c_p$ (2.2.3). In the case of dummy indices, this means, for example, that $g_{iji} = k_j$ (2.2.4) could equivalently be written as $g_{pjp} = k_j$ (this is why they are called dummy indices). Some guidelines for the use of indicial notation equations are as follows:

1. The same free indices *must* appear in every term of an equation. For example, $y_{ij} = d_{ji} - e_j$ would *not* be a valid indicial notation equation because the index i does not appear in the last term.

2. A character already in use as a free index should never be used to represent a dummy index pair. For example, it would *not* be valid to express $g_{iji} = k_j$ (2.2.4) as $g_{jjj} = k_j$, since in the later expression it is unclear which indices of the first term are to be summed over.

3. The same character should never be used to represent more than one dummy index pair in any given term of an equation.

4. An index character should appear no more than twice in any given term of an equation (as a result of 2 and 3).

In short, there is no ambiguity in how a valid indicial notation equation would be expanded. The first thing one should do when faced with an indicial notation equation is to identify the free and dummy indices and compute how many equations are represented and how many terms each equation will have.

Example 2.1
Rewrite the vector transformation rule (2.1.26) using indicial notation.

Solution. For a vector $\mathbf{v}$, the scalar components in the orthonormal bases $\{\hat{\mathbf{e}}_i\}$ and $\{\hat{\mathbf{e}}_i'\}$ are related by

$$v_i' = \ell_{ij} v_j , \quad v_i = \ell_{ji} v_j' , \tag{2.1.26d}$$

where it is understood that the range and summation conventions apply.

2.2.4 Contraction

If a given indicial notation equation with at least two free indices is true, then a new indicial notation equation can be formed by changing any two of the free indices into a pair of dummy indices. For example, given the equation

$$a_{pq} = B_q c_p \quad \Longleftrightarrow \quad \begin{cases} a_{11} = B_1 c_1 \\ a_{12} = B_2 c_1 \\ a_{13} = B_3 c_1 \\ a_{21} = B_1 c_2 \\ a_{22} = B_2 c_2 \\ \quad \vdots \end{cases} , \tag{2.2.6}$$

a new equation can be formed by replacing q with p, which can be symbolized as $q \to p$:

$$a_{pp} = B_p c_p \quad \Longleftrightarrow \quad a_{11} + a_{22} + a_{33} = B_1 c_1 + B_2 c_2 + B_3 c_3 . \tag{2.2.7}$$

It is evident from the expanded equations on the right that if (2.2.6) is true, then (2.2.7) is also true. Replacing p with q ($p \to q$) yields precisely the same result. One says that *contraction* over the indices p and q (either $q \to p$ or $p \to q$) in (2.2.6) yields (2.2.7). Of course, (2.2.7) does *not* imply (2.2.6)—contraction does not work in reverse.

Some other examples of the application of contraction are

$$\sigma_{ij} = C_{ijkl} \varepsilon_{kl} \quad \Rightarrow \quad \sigma_{ii} = C_{iikl} \varepsilon_{kl} , \tag{2.2.8}$$

in which $j \to i$ on the left-hand side implies the result on the right-hand side, and

$$D_{ijk} = b_{ij} c_k \quad \Rightarrow \quad \begin{cases} D_{ijj} = b_{ij} c_j \\ D_{iik} = b_{ii} c_k \\ D_{iji} = b_{ij} c_i \end{cases} , \tag{2.2.9}$$

in which each of three possible distinct contractions $k \to j$, $j \to i$, and $k \to i$ have been applied.

2.2.5 Kronecker Delta

The *Kronecker delta* δ_{ij} is an indicial notation symbol representing the components of the *identity matrix* $[I]$:

$$[I] \equiv \begin{bmatrix} 1 & 0 & 0 \\ 0 & 1 & 0 \\ 0 & 0 & 1 \end{bmatrix} = \begin{bmatrix} \delta_{11} & \delta_{12} & \delta_{13} \\ \delta_{21} & \delta_{22} & \delta_{23} \\ \delta_{31} & \delta_{32} & \delta_{33} \end{bmatrix} , \qquad \delta_{ij} \equiv \begin{cases} 1 & \text{if } i = j \\ 0 & \text{if } i \neq j \end{cases} . \qquad (2.2.10)$$

In other words, $\delta_{11} = \delta_{22} = \delta_{33} = 1$ and $\delta_{12} = \delta_{13} = \delta_{21} = \delta_{23} = \delta_{31} = \delta_{32} = 0$. Note that, from the summation convention,

$$\delta_{ii} = \delta_{11} + \delta_{22} + \delta_{33} = 3 . \qquad (2.2.11)$$

One can relate the Kronecker delta to the base vectors in an orthonormal basis $\{\hat{\mathbf{e}}_i\}$ by noting that

$$\hat{\mathbf{e}}_i \cdot \hat{\mathbf{e}}_j = \delta_{ij} . \qquad (2.2.12)$$

This result can be used to derive the expression for the dot product of two vectors $\mathbf{u}$ and $\mathbf{v}$ in terms of their scalar components (taking care not to have the same character represent two dummy index pairs in one term):

$$\begin{aligned} \mathbf{u} \cdot \mathbf{v} &= (u_i \hat{\mathbf{e}}_i) \cdot (v_j \hat{\mathbf{e}}_j) \\ &= u_i v_j (\hat{\mathbf{e}}_i \cdot \hat{\mathbf{e}}_j) \\ &= u_i v_j \delta_{ij} \\ &= \delta_{11} u_1 v_1 + \delta_{12} u_1 v_2 + \cdots + \delta_{33} u_3 v_3 \\ &= u_1 v_1 + u_2 v_2 + u_3 v_3 \\ &= u_i v_i . \end{aligned} \qquad (2.2.13)$$

This illustrates an important property of δ_{ij} called the *substitution property*. If δ_{ij} appears in a term of an indicial notation equation and i (or j) is one of a pair of dummy indices, then the other index of the pair can be changed to j (or i) and δ_{ij} dropped from the term. For example,

$$\begin{aligned} A_{jkm} \delta_{lk} &= A_{jlm} , \\ A_{jkm} \delta_{km} &= A_{jkk} = A_{jmm} , \\ c_{ijk} \delta_{jr} \delta_{sk} &= c_{irs} , \\ d_{ij} \delta_{ik} \delta_{rs} &= d_{kj} \delta_{rs} . \end{aligned} \qquad (2.2.14)$$

To convince oneself that this is true, just imagine expanding these equations and remember that δ_{ij} equals zero unless $i = j$, in which case it equals one.

2.2.6   Orthogonality Condition

Recall that if $\{\hat{\mathbf{e}}_i\}$ and $\{\hat{\mathbf{e}}'_i\}$ are both orthonormal bases, then $\hat{\mathbf{e}}_i \cdot \hat{\mathbf{e}}_j = \delta_{ij}$, $\hat{\mathbf{e}}'_i \cdot \hat{\mathbf{e}}'_j = \delta_{ij}$, and the direction cosines $\ell_{ij} \equiv \hat{\mathbf{e}}'_i \cdot \hat{\mathbf{e}}_j$ form an orthogonal transformation matrix $[L]$ that satisfies the orthogonality condition (2.1.28). This orthogonality condition can equivalently be stated as

$$[L][L]^\mathsf{T} = [L]^\mathsf{T}[L] = [I] , \qquad (2.2.15)$$

where $[I]$ is the identity matrix. Alternatively, in terms of the direction cosines and the Kronecker delta and using indicial notation, the orthogonality condition is

$$\ell_{ik}\ell_{jk} = \ell_{ki}\ell_{kj} = \delta_{ij} . \qquad (2.2.16)$$

It is instructive to show how (2.2.16) can be derived via a more direct route; since $\ell_{ik} = \hat{\mathbf{e}}'_i \cdot \hat{\mathbf{e}}_k$,

$$\ell_{ik}\ell_{jk} = (\hat{\mathbf{e}}'_i \cdot \hat{\mathbf{e}}_k)\ell_{jk} = \hat{\mathbf{e}}'_i \cdot (\ell_{jk}\hat{\mathbf{e}}_k) . \qquad (2.2.17)$$

Therefore, since $\ell_{jk}\hat{\mathbf{e}}_k = \hat{\mathbf{e}}'_j$,

$$\ell_{ik}\ell_{jk} = \hat{\mathbf{e}}'_i \cdot \hat{\mathbf{e}}'_j = \delta_{ij} . \qquad (2.2.18)$$

The condition $\ell_{ki}\ell_{kj} = \delta_{ij}$ can similarly be shown.

2.2.7   Permutation Symbol

The *permutation symbol* e_{ijk} is an indicial notation symbol representing 27 numbers whose value is 1 if i, j, and k are an even permutation of 1, 2, and 3 and -1 if i, j, and k are an odd permutation of 1, 2, and 3. The value is zero otherwise, that is, if the value of any two indices are equal. Explicitly, $e_{123} = e_{231} = e_{312} = 1$, $e_{321} = e_{213} = e_{132} = -1$, and $e_{111} = e_{112} = e_{113} = e_{211} = \cdots = 0$. It follows directly from the definition of the permutation symbol that, for arbitrary values of i, j, and k,

$$e_{ijk} = e_{jki} = e_{kij} = -e_{jik} = -e_{ikj} = -e_{kji} . \qquad (2.2.19)$$

Furthermore, it can easily be seen that the permutation symbol is related to the base vectors in a right-handed orthonormal basis $\{\hat{\mathbf{e}}_i\}$ by

$$\hat{\mathbf{e}}_i \times \hat{\mathbf{e}}_j = e_{ijk}\hat{\mathbf{e}}_k . \qquad (2.2.20)$$

This relation can be used to derive the following result for the vector product of two vectors $\mathbf{u}$ and $\mathbf{v}$,

$$\mathbf{u} \times \mathbf{v} = (u_i\hat{\mathbf{e}}_i) \times (v_j\hat{\mathbf{e}}_j) = u_iv_j(\hat{\mathbf{e}}_i \times \hat{\mathbf{e}}_j) = e_{ijk}u_iv_j\hat{\mathbf{e}}_k . \qquad (2.2.21)$$

It is shown below that this is consistent with the result (2.1.23) derived earlier for the vector product of two vectors.

The permutation symbol is also related to the determinant of 3×3 matrices; by expanding the indicial notation expression $e_{ijk} a_i b_j c_k$ and keeping only the six nonzero terms, it is seen explicitly that

$$e_{ijk} a_i b_j c_k = \det \begin{bmatrix} a_1 & a_2 & a_3 \\ b_1 & b_2 & b_3 \\ c_1 & c_2 & c_3 \end{bmatrix} = \det \begin{bmatrix} a_1 & b_1 & c_1 \\ a_2 & b_2 & c_2 \\ a_3 & b_3 & c_3 \end{bmatrix} . \qquad (2.2.22)$$

This shows that the results (2.1.23) and (2.2.21) for the vector product of two vectors are equivalent. The relation (2.2.22) can alternatively be given as

$$\det[A] = e_{ijk} A_{1i} A_{2j} A_{3k} = e_{ijk} A_{i1} A_{j2} A_{k3} , \qquad (2.2.23)$$

where A_{ij} are the components of the 3×3 matrix $[A]$.

It can be shown using (2.2.22) that the permutation symbol is related to the Kronecker delta by

$$e_{ijk} = \det \begin{bmatrix} \delta_{i1} & \delta_{i2} & \delta_{i3} \\ \delta_{j1} & \delta_{j2} & \delta_{j3} \\ \delta_{k1} & \delta_{k2} & \delta_{k3} \end{bmatrix} = \det \begin{bmatrix} \delta_{i1} & \delta_{j1} & \delta_{k1} \\ \delta_{i2} & \delta_{j2} & \delta_{k2} \\ \delta_{i3} & \delta_{j3} & \delta_{k3} \end{bmatrix} , \qquad (2.2.24)$$

and, consequently, that

$$e_{ijk} e_{pqr} = \det \begin{bmatrix} \delta_{ip} & \delta_{iq} & \delta_{ir} \\ \delta_{jp} & \delta_{jq} & \delta_{jr} \\ \delta_{kp} & \delta_{kq} & \delta_{kr} \end{bmatrix} . \qquad (2.2.25)$$

Some other useful identities relating the Kronecker delta and the permutation symbol, which follow by successive contractions of (2.2.25), the substitution property of the Kronecker delta, and the identity $\delta_{ii} = 3$ are

$$e_{ijk} e_{pqk} = \delta_{ip} \delta_{jq} - \delta_{iq} \delta_{jp} , \qquad (2.2.26)$$
$$e_{ijk} e_{pjk} = 2\delta_{ip} , \qquad (2.2.27)$$
$$e_{ijk} e_{ijk} = 6 . \qquad (2.2.28)$$

It can also be shown that

$$e_{lmn} \det[A] = e_{ijk} A_{il} A_{jm} A_{kn} , \qquad (2.2.29)$$

$$\det[A] = \frac{1}{6} e_{ijk} e_{lmn} A_{il} A_{jm} A_{kn} , \qquad (2.2.30)$$

where A_{ij} are the components of the 3×3 matrix $[A]$.

Example 2.2

Show that, if $A_{ij} = A_{ji}$, then $e_{ijk}A_{ij} = 0$.

Solution. Trivially,

$$e_{ijk}A_{ij} = \frac{1}{2}(e_{ijk}A_{ij} + e_{ijk}A_{ij}) \,.$$

Since $A_{ij} = A_{ji}$, this can be rewritten as

$$e_{ijk}A_{ij} = \frac{1}{2}(e_{ijk}A_{ij} + e_{ijk}A_{ji}) \,,$$

but from (2.2.19), one can substitute $e_{ijk} = -e_{jik}$ in the second term on the right-hand side so that

$$e_{ijk}A_{ij} = \frac{1}{2}(e_{ijk}A_{ij} - e_{jik}A_{ji}) \,.$$

Finally, since i and j are dummy indices, the change in indices $i \to j$ and $j \to i$ can be made in the second term on the right-hand side:

$$e_{ijk}A_{ij} = \frac{1}{2}(e_{ijk}A_{ij} - e_{ijk}A_{ij}) \,,$$

thus showing that

$$e_{ijk}A_{ij} = 0 \,.$$

This result could also be shown explicitly by expanding $e_{ijk}A_{ij}$ into three expressions (one free index k), each with nine terms (two dummy index pairs i and j), and substituting the values of the permutation symbol and recalling that $A_{12} = A_{21}$, $A_{23} = A_{32}$, and $A_{31} = A_{13}$. However, this option will become less and less attractive as the complexity of the expanded equations increases (imagine trying to do this if there were $3^3 = 27$ equations, each with $3^4 = 91$ terms!). Thus, one should become comfortable manipulating indicial notation equations without trying to expand them. This will be important throughout this text.

2.3   Tensors

A *vector operator* F is a rule for assigning exactly one member of a set of vectors $\mathcal{B}$ to each member of a set of vectors $\mathcal{A}$. This is denoted as $F: \mathcal{A} \to \mathcal{B}$, and one says that F is a *mapping* from $\mathcal{A}$ into $\mathcal{B}$. For every vector $\mathbf{u}$ in $\mathcal{A}$, the vector operator F assigns a unique vector $\mathbf{v}$ that belongs to $\mathcal{B}$, which can be expressed as $\mathbf{v} = F(\mathbf{u})$. In this expression, $\mathbf{u}$ is the argument of the vector operator F and $\mathbf{v}$ is its value, sometimes referred to as the *image* of $\mathbf{u}$ under F.

The vector operator $F: \mathcal{A} \to \mathcal{B}$ is said to be a *linear vector operator* if and only if, for all scalars a and b and all vectors $\mathbf{u}$ and $\mathbf{w}$ in $\mathcal{A}$ such that $a\mathbf{u} + b\mathbf{w}$ is also in $\mathcal{A}$,

$$F(a\mathbf{u} + b\mathbf{w}) = aF(\mathbf{u}) + bF(\mathbf{w}) . \tag{2.3.1}$$

Note, by setting $a = b = 0$, that the linear vector operator maps the zero vector into itself, $F(\mathbf{0}) = \mathbf{0}$. The linear vector operator is defined without reference to a particular vector basis. Since the argument and the value of the vector operator are vectors, which are independent of vector basis, the mapping given by the vector operator must also be independent of vector basis. However, the mapping will always be characterized in terms of scalar components defined with respect to some convenient vector basis.

In any given orthonormal basis $\{\hat{\mathbf{e}}_i\}$, where $\mathbf{u} = u_i\hat{\mathbf{e}}_i$ and $\mathbf{v} = v_i\hat{\mathbf{e}}_i$, the most general linear mapping of the scalar components of $\mathbf{u}$ into the scalar components of $\mathbf{v}$, that also maps the zero vector into itself, can be given as $v_i = T_{ij}u_j$. Thus, the linear vector operator $F: \mathcal{A} \to \mathcal{B}$ implies the existence of nine scalar quantities T_{ij} such that

$$\mathbf{v} = F(\mathbf{u}) \quad \Longleftrightarrow \quad v_i = T_{ij}u_j . \tag{2.3.2}$$

The nine scalar quantities T_{ij} and the indicial notation equation in (2.3.2) mapping the scalar components of $\mathbf{u}$ into scalar components of $\mathbf{v}$ completely describe the linear vector operation $\mathbf{v} = F(\mathbf{u})$ in the orthonormal basis $\{\hat{\mathbf{e}}_i\}$. However, in another arbitrary orthonormal basis $\{\hat{\mathbf{e}}'_i\}$, where $\mathbf{u} = u'_i\hat{\mathbf{e}}'_i$ and $\mathbf{v} = v'_i\hat{\mathbf{e}}'_i$, the linear vector operator $F: \mathcal{A} \to \mathcal{B}$ implies the existence of nine *different* scalar quantities T'_{ij} such that

$$\mathbf{v} = F(\mathbf{u}) \quad \Longleftrightarrow \quad v'_i = T'_{ij}u'_j , \tag{2.3.3}$$

where u'_i and v'_i are related to u_i and v_i by the vector transformation rule (2.1.26). In order to completely characterize the linear vector operator, for all orthonormal bases, one needs, in addition to the indicial notation equation in (2.3.2) and the nine scalar quantities T_{ij} in some orthonormal basis $\{\hat{\mathbf{e}}_i\}$, a transformation rule for finding the corresponding T'_{ij} in *any* other orthonormal basis.

To derive the necessary transformation rule for the nine scalar quantities T_{ij}, recall from the vector transformation rule (2.1.26) that $v'_i = \ell_{ip}v_p$. Substituting into (2.3.3),

$$\ell_{ip}v_p = T'_{ij}u'_j , \tag{2.3.4}$$

but from (2.3.2), $v_p = T_{pq}u_q$;

$$\ell_{ip}T_{pq}u_q = T'_{ij}u'_j . \tag{2.3.5}$$

Now applying the transformation rule to **u**, it follows that $u_q = \ell_{jq}u'_j$ and

$$\ell_{ip}T_{pq}\ell_{jq}u'_j = T'_{ij}u'_j , \tag{2.3.6}$$

which can be rearranged to give

$$(T'_{ij} - \ell_{ip}\ell_{jq}T_{pq})u'_j = 0 . \tag{2.3.7}$$

Condition (2.3.7) can only hold for arbitrary vectors **u** (that belong to the set $\mathcal{A}$ in $T: \mathcal{A} \to \mathcal{B}$) if[5]

$$\boxed{T'_{ij} = \ell_{ip}\ell_{jq}T_{pq} .} \tag{2.3.8a}$$

It can similarly be shown that the inverse transformation is

$$\boxed{T_{ij} = \ell_{pi}\ell_{qj}T'_{pq} .} \tag{2.3.8b}$$

The nine scalar quantities T_{ij} that transform under changes of orthonormal basis according to (2.3.8) together make up a physical entity known as a *tensor of order two*, or *second-order tensor*, that is denoted by **T**. One says that T_{ij} are the scalar components of the second-order tensor **T** in the orthonormal basis $\{\hat{\mathbf{e}}_i\}$. A second-order tensor is any physical entity that can be completely described, with respect to an orthonormal basis, by nine scalar components that transform according to (2.3.8).[6] Notice the similarity of the *transformation rule for second-order tensors* (2.3.8) and the vector transformation rule (2.1.26). The second-order tensor **T** together with the relation (2.3.2) completely characterize the linear vector operator $F: \mathcal{A} \to \mathcal{B}$.

The linear vector operator $F: \mathcal{A} \to \mathcal{B}$ also implies the existence of nine scalar quantities S_{ij} such that

$$\mathbf{v} = F(\mathbf{u}) \quad\Longleftrightarrow\quad v_i = u_j S_{ji} . \tag{2.3.9}$$

By comparing (2.3.2) and (2.3.9), it is seen that $S_{ji} = T_{ij}$. It follows that the nine scalar quantities S_{ij} satisfy the transformation rule (2.3.8) and are thus the scalar components of a second-order tensor **S** that is different than **T** (it will be seen later that **S** is the transpose of **T**). The second-order tensor **S** together with the relation (2.3.9) also completely characterize the

---

[5] Equations of particular importance and relevance to the material that follows are enclosed in boxes for the reader's convenience.

[6] Note that the direction cosines ℓ_{ij} are *not* the components of a second-order tensor.

linear vector operator $F: \mathcal{A} \to \mathcal{B}$. To distinguish between these different characterizations, one can introduce the following *dot product* notation, which will be further discussed in Section 2.3.3:

$$\mathbf{v} = F(\mathbf{u}) = \mathbf{T} \cdot \mathbf{u} = \mathbf{u} \cdot \mathbf{S} \quad \Longleftrightarrow \quad v_i = T_{ij}u_j = u_j S_{ji} \ . \qquad (2.3.10)$$

Note that $\mathbf{w} = \mathbf{u} \cdot \mathbf{T}$ is a different linear vector operation and that, in general, $\mathbf{T} \cdot \mathbf{u} \neq \mathbf{u} \cdot \mathbf{T}$.

It can easily be shown (see Problem 2.10) that $(\mathbf{u} \cdot \mathbf{T}) \cdot \mathbf{v} = \mathbf{u} \cdot (\mathbf{T} \cdot \mathbf{v})$, where $\mathbf{u}$ and $\mathbf{v}$ are arbitrary members of the set of vectors $\mathcal{A}$, which leads to the following definition:

$$\mathbf{u} \cdot \mathbf{T} \cdot \mathbf{v} \equiv (\mathbf{u} \cdot \mathbf{T}) \cdot \mathbf{v} = \mathbf{u} \cdot (\mathbf{T} \cdot \mathbf{v}) \ , \qquad (2.3.11)$$

which is known as a *bilinear vector operation* on the ordered pair of vectors $(\mathbf{u}, \mathbf{v})$. The pair of vectors is *ordered* because, in general, $\mathbf{u} \cdot \mathbf{T} \cdot \mathbf{v} \neq \mathbf{v} \cdot \mathbf{T} \cdot \mathbf{u}$. The result is a scalar which, in an orthonormal basis $\{\hat{\mathbf{e}}_i\}$, is given by

$$\mathbf{u} \cdot \mathbf{T} \cdot \mathbf{v} = T_{ij}u_i v_j \ , \qquad (2.3.12)$$

where T_{ij}, u_i, and v_j are the scalar components in $\{\hat{\mathbf{e}}_i\}$ of $\mathbf{T}$, $\mathbf{u}$, and $\mathbf{v}$, respectively.

Let $G: (\mathcal{A} \times \mathcal{A}) \to \mathcal{S}$ be a rule for assigning exactly one member of a set of scalars $\mathcal{S}$ to each member of a set $\mathcal{A} \times \mathcal{A}$ of ordered pairs of vectors, where $\phi = G(\mathbf{u}, \mathbf{v})$ is the image under G of $(\mathbf{u}, \mathbf{v})$. The *Cartesian product* $\mathcal{A} \times \mathcal{A}$ means that both the first and second element of the ordered pair of vectors is a member of $\mathcal{A}$. Then, G is a *bilinear vector operator* if and only if

$$\begin{aligned} G(a\mathbf{u}, b\mathbf{v}) &= (ab)G(\mathbf{u}, \mathbf{v}) \ , \\ G(\mathbf{u} + \mathbf{w}, \mathbf{v}) &= G(\mathbf{u}, \mathbf{v}) + G(\mathbf{w}, \mathbf{v}) \ , \qquad (2.3.13) \\ G(\mathbf{u}, \mathbf{v} + \mathbf{w}) &= G(\mathbf{u}, \mathbf{v}) + G(\mathbf{u}, \mathbf{w}) \ , \end{aligned}$$

for all scalars a and b and all vectors $\mathbf{u}$, $\mathbf{v}$, and $\mathbf{w}$ in $\mathcal{A}$ such that $(a\mathbf{u}, b\mathbf{v})$, $(\mathbf{u} + \mathbf{w}, \mathbf{v})$, and $(\mathbf{u}, \mathbf{v} + \mathbf{w})$ are in $\mathcal{A} \times \mathcal{A}$. It follows that the bilinear vector operator is completely characterized by a second-order tensor $\mathbf{T}$, where

$$\phi = G(\mathbf{u}, \mathbf{v}) = \mathbf{u} \cdot \mathbf{T} \cdot \mathbf{v} \quad \Longleftrightarrow \quad \phi = T_{ij}u_i v_j \ . \qquad (2.3.14)$$

This is an alternate interpretation for second-order tensors.[7]

---

[7]Some authors [e.g., Truesdell and Noll (1992)] use a different notation in which (2.3.10) is replaced by

$$\mathbf{v} = F(\mathbf{u}) = \mathbf{T}\mathbf{u} = \mathbf{S}^{\mathsf{T}}\mathbf{u} \quad \Longleftrightarrow \quad v_i = T_{ij}u_j = u_j S_{ji} \ ,$$

and (2.3.14) is replaced by

$$\phi = G(\mathbf{u}, \mathbf{v}) = \mathbf{u} \cdot \mathbf{T}\mathbf{v} \quad \Longleftrightarrow \quad \phi = T_{ij}u_i v_j \ .$$

2.3.1 Tensor Algebra

Two second-order tensors $\mathbf{A}$ and $\mathbf{B}$ are equal if and only if $\mathbf{A} \cdot \mathbf{u} = \mathbf{B} \cdot \mathbf{u}$ and $\mathbf{u} \cdot \mathbf{A} = \mathbf{u} \cdot \mathbf{B}$, for all vectors $\mathbf{u}$. It follows that if A_{ij} and B_{ij} are the scalar components of $\mathbf{A}$ and $\mathbf{B}$, then $\mathbf{A} = \mathbf{B}$ if and only if $A_{ij} = B_{ij}$ in *every* orthonormal basis $\{\hat{\mathbf{e}}_i\}$. Conversely, if $A_{ij} = B_{ij}$ in one orthonormal basis, then it follows from the transformation rule (2.3.8) for second-order tensors that $A'_{ij} = B'_{ij}$ in every other orthonormal basis $\{\hat{\mathbf{e}}'_i\}$ and, consequently, $\mathbf{A} = \mathbf{B}$.

The second-order *zero tensor* $\mathbf{0}^{(2)}$ is defined such that

$$\mathbf{0}^{(2)} \cdot \mathbf{u} = \mathbf{u} \cdot \mathbf{0}^{(2)} = \mathbf{0} \, , \tag{2.3.15}$$

for all vectors $\mathbf{u}$. It follows that $\mathbf{A} = \mathbf{0}^{(2)}$ if and only if

$$A_{ij} = 0 \tag{2.3.16}$$

in every orthonormal basis $\{\hat{\mathbf{e}}_i\}$. Conversely, it follows from the transformation rule (2.3.8) for second-order tensors that if (2.3.16) is true in one orthonormal basis, then it is true in all orthonormal bases and $\mathbf{A} = \mathbf{0}^{(2)}$. From now on, if the meaning is clear from the context in which it appears, the second-order zero tensor will be denoted simply by $\mathbf{0}$.

Tensor addition and multiplication by a scalar are defined such that if

$$\mathbf{C} = \alpha \mathbf{A} + \beta \mathbf{B} = \mathbf{A}\alpha + \mathbf{B}\beta \, , \tag{2.3.17}$$

where α and β are scalars, then $\mathbf{C}$ is a second-order tensor such that

$$\begin{aligned} \mathbf{C} \cdot \mathbf{u} &= \alpha(\mathbf{A} \cdot \mathbf{u}) + \beta(\mathbf{B} \cdot \mathbf{u}) \, , \\ \mathbf{u} \cdot \mathbf{C} &= \alpha(\mathbf{u} \cdot \mathbf{A}) + \beta(\mathbf{u} \cdot \mathbf{B}) \, , \end{aligned} \tag{2.3.18}$$

for all vectors $\mathbf{u}$. It follows that $\mathbf{C} = \alpha \mathbf{A} + \beta \mathbf{B}$ if and only if the scalar components of $\mathbf{C}$ are given by

$$C_{ij} = \alpha A_{ij} + \beta B_{ij} \tag{2.3.19}$$

in every orthonormal basis $\{\hat{\mathbf{e}}_i\}$. Conversely, it follows from the transformation rule (2.3.8) that if (2.3.19) is true in one orthonormal basis, then it is true in all orthonormal bases and $\mathbf{C} = \alpha \mathbf{A} + \beta \mathbf{B}$.

The following properties of tensor addition and multiplication by a scalar are easily established from the above definitions. If $\mathbf{A}$, $\mathbf{B}$, and $\mathbf{C}$ are second-order tensors and α and β are scalars, then

$$\mathbf{A} + \mathbf{B} = \mathbf{B} + \mathbf{A} \,, \tag{2.3.20}$$
$$\mathbf{A} + (\mathbf{B} + \mathbf{C}) = (\mathbf{A} + \mathbf{B}) + \mathbf{C} \,, \tag{2.3.21}$$
$$\mathbf{A} + \mathbf{0} = \mathbf{A} \,, \tag{2.3.22}$$
$$\alpha(\beta\mathbf{A}) = (\alpha\beta)\mathbf{A} \,, \tag{2.3.23}$$
$$1\mathbf{A} = \mathbf{A} \,, \tag{2.3.24}$$
$$(\alpha + \beta)\mathbf{A} = \alpha\mathbf{A} + \beta\mathbf{A} \,, \tag{2.3.25}$$
$$\alpha(\mathbf{A} + \mathbf{B}) = \alpha\mathbf{A} + \alpha\mathbf{B} \,. \tag{2.3.26}$$

In addition, every second-order tensor $\mathbf{A}$ has an additive inverse $-\mathbf{A} = (-1)\mathbf{A}$, also a second-order tensor, such that

$$\mathbf{A} + (-\mathbf{A}) = \mathbf{0} \,. \tag{2.3.27}$$

The properties (2.3.20)–(2.3.27) establish that the set of all second-order tensors is a vector space (Knowles 1998).

2.3.2 Higher-Order Tensors

Consider a linear operator $F \colon \mathcal{A} \to \mathcal{B}$ where $\mathcal{A}$ is a set of vectors and $\mathcal{B}$ is a set of second-order tensors. The second-order tensor $\mathbf{V} = F(\mathbf{u})$ is the image of the vector $\mathbf{u}$ under the linear operator F. The linear operator $F \colon \mathcal{A} \to \mathcal{B}$ implies the existence of $3^3 = 27$ quantities T_{ijk} such that

$$\mathbf{V} = F(\mathbf{u}) \quad \Longleftrightarrow \quad V_{ij} = T_{ijk} u_k \,, \tag{2.3.28}$$

where V_{ij} and u_k are the scalar components of $\mathbf{V}$ and $\mathbf{u}$ in the orthonormal basis $\{\hat{\mathbf{e}}_i\}$. It follows from the invariance of the linear operation with respect to changes in orthonormal basis that the T_{ijk} obey a transformation rule analogous to that for second-order tensors. The T_{ijk} are the scalar components in the orthonormal basis $\{\hat{\mathbf{e}}_i\}$ of a tensor $\mathbf{T}$ of order 3. The rules of tensor algebra defined for second-order tensors are extended to tensors of order 3.

By proceeding in this manner, tensors of ever-increasing order are defined with respect to linear operations on sets of previously defined tensors of lesser order. In general, a linear operator $F \colon \mathcal{A} \to \mathcal{B}$ that maps a set $\mathcal{A}$ of tensors of order r into a set $\mathcal{B}$ of tensors of order s implies the existence of 3^n scalar quantities $T_{i_1 i_2 \cdots i_n}$ with n indices, where $n = r + s$, such that

$$\mathbf{V} = F(\mathbf{U}) \quad \Longleftrightarrow \quad V_{j_1 j_2 \cdots j_s} = T_{j_1 j_2 \cdots j_s k_1 k_2 \cdots k_r} U_{k_1 k_2 \cdots k_r} \,, \tag{2.3.29}$$

where $V_{j_1 j_2 \cdots j_s}$ are the scalar components of the tensor $\mathbf{V}$ of order s and $U_{k_1 k_2 \cdots k_r}$ are the scalar components of the tensor $\mathbf{U}$ of order r. It follows from the invariance of the linear operation that the scalar quantities

$T_{i_1 i_2 \cdots i_n}$ obey the *tensor transformation rule*

$$
\begin{aligned}
T'_{i_1 i_2 \cdots i_n} &= \ell_{i_1 p_1} \ell_{i_2 p_2} \cdots \ell_{i_n p_n} T_{p_1 p_2 \cdots p_n} \, , \\
T_{i_1 i_2 \cdots i_n} &= \ell_{p_1 i_1} \ell_{p_2 i_2} \cdots \ell_{p_n i_n} T'_{p_1 p_2 \cdots p_n} \, ,
\end{aligned}
\tag{2.3.30}
$$

where $T'_{i_1 i_2 \cdots i_n}$ are the corresponding quantities in any other orthonormal basis $\{\hat{\mathbf{e}}'_i\}$. The $T_{i_1 i_2 \cdots i_n}$ are the scalar components in the orthonormal basis $\{\hat{\mathbf{e}}_i\}$ of a tensor $\mathbf{T}$ of order n. A *tensor of order* n is any physical entity that can be completely described, with respect to an orthonormal basis, by 3^n scalar components that transform according to (2.3.30). The rules of tensor algebra are extended to tensors of order n. Multiplication by a scalar does not change the order of a tensor and only tensors of the same order may be summed. The set of all tensors of order n is a vector space.

Tensors (except for tensors of order 0) will be denoted by boldface characters. No special notation will be used to distinguish between tensors of different order—the context in which a tensor appears will be relied upon to make it clear what its order is. It is clear from the tensor transformation rule (2.3.30) that tensors of order 0 are equivalent to scalars, and tensors of order 1 are equivalent to vectors. It will be seen that many important physical quantities, such as the various measures of stress and strain, are tensors of order 2.

2.3.3   Dyadic Notation

The *tensor product* $\mathbf{ab}$ of two vectors $\mathbf{a}$ and $\mathbf{b}$ is a mathematical object called a *dyad* (Budiansky 1983), defined by the *dot product* relations[8]

$$
\begin{aligned}
\mathbf{ab} \cdot \mathbf{u} &= \mathbf{a}(\mathbf{b} \cdot \mathbf{u}) \, , \\
\mathbf{u} \cdot \mathbf{ab} &= (\mathbf{u} \cdot \mathbf{a})\mathbf{b} \, ,
\end{aligned}
\tag{2.3.31}
$$

where $\mathbf{u}$ is an arbitrary vector. Note that $\mathbf{ab} \cdot \mathbf{u} \neq \mathbf{u} \cdot \mathbf{ab}$. Note also that $\mathbf{ab} \neq \mathbf{ba}$, since $\mathbf{ab} \cdot \mathbf{u} \neq \mathbf{ba} \cdot \mathbf{u}$. The dot product of a dyad and a vector is another vector and it follows from the definition (2.3.31) that

$$
\begin{aligned}
\mathbf{ab} \cdot (\alpha \mathbf{u} + \beta \mathbf{v}) &= \alpha(\mathbf{ab} \cdot \mathbf{u}) + \beta(\mathbf{ab} \cdot \mathbf{v}) \, , \\
(\alpha \mathbf{u} + \beta \mathbf{v}) \cdot \mathbf{ab} &= \alpha(\mathbf{u} \cdot \mathbf{ab}) + \beta(\mathbf{v} \cdot \mathbf{ab}) \, ,
\end{aligned}
\tag{2.3.32}
$$

for all vectors $\mathbf{u}$ and $\mathbf{v}$ and all scalars α and β. Thus, the dot product of a dyad and a vector is a linear vector operator and, consequently, a dyad $\mathbf{S} = \mathbf{ab}$ is a second-order tensor with scalar components $S_{ij} = a_i b_j$.

---

[8]Some authors (Truesdell and Noll 1992) prefer the notation $\mathbf{a} \otimes \mathbf{b}$ to represent the tensor product formed by two vectors $\mathbf{a}$ and $\mathbf{b}$ and the notation $(\mathbf{a} \otimes \mathbf{b})\mathbf{u}$ instead of $\mathbf{ab} \cdot \mathbf{u}$.

Alternatively, it can be shown that $\mathbf{S} = \mathbf{ab}$ is a tensor by showing that its scalar components $S_{ij} = a_i b_j$ obey the transformation rule for second-order tensors (2.3.8). Applying the vector transformation rule to the scalar components of $\mathbf{a}$ and $\mathbf{b}$, one has that $a'_i = \ell_{ip} a_p$ and $b'_j = \ell_{jq} b_q$, from which it follows that

$$S'_{ij} = a'_i b'_j = \ell_{ip} \ell_{jq} a_p b_q = \ell_{ip} \ell_{jq} S_{pq} \ . \tag{2.3.33}$$

Note however that, although every dyad is a second-order tensor, not every second-order tensor can be expressed as the tensor product of two vectors since a second-order tensor has nine independent scalar components in general and two vectors have only six independent scalar components. A linear vector operation such as $\mathbf{v} = \mathbf{ab} \cdot \mathbf{u}$ is restricted in the sense that it maps every vector $\mathbf{u}$ into a vector $\mathbf{v}$ that is parallel to $\mathbf{a}$.

A sum of dyads of the form

$$\mathbf{T} = \mathbf{ab} + \mathbf{cd} + \mathbf{ef} + \cdots \tag{2.3.34}$$

is called a *dyadic*. It follows from (2.3.31) and the rules of tensor algebra that

$$\begin{aligned}
\mathbf{T} \cdot \mathbf{v} &= \mathbf{a}(\mathbf{b} \cdot \mathbf{v}) + \mathbf{c}(\mathbf{d} \cdot \mathbf{v}) + \mathbf{e}(\mathbf{f} \cdot \mathbf{v}) + \cdots \ , \\
\mathbf{v} \cdot \mathbf{T} &= (\mathbf{v} \cdot \mathbf{a})\mathbf{b} + (\mathbf{v} \cdot \mathbf{c})\mathbf{d} + (\mathbf{v} \cdot \mathbf{e})\mathbf{f} + \cdots \ .
\end{aligned} \tag{2.3.35}$$

Any dyadic can be expressed in terms of an arbitrary orthonormal basis $\{\hat{\mathbf{e}}_i\}$; with $\mathbf{a} = a_i \hat{\mathbf{e}}_i$, $\mathbf{b} = b_i \hat{\mathbf{e}}_i$, $\mathbf{c} = c_i \hat{\mathbf{e}}_i$, $\ldots$, it follows that

$$\begin{aligned}
\mathbf{T} &= a_i b_j \hat{\mathbf{e}}_i \hat{\mathbf{e}}_j + c_i d_j \hat{\mathbf{e}}_i \hat{\mathbf{e}}_j + e_i f_j \hat{\mathbf{e}}_i \hat{\mathbf{e}}_j + \cdots \\
&= (a_i b_j + c_i d_j + e_i f_j + \cdots) \hat{\mathbf{e}}_i \hat{\mathbf{e}}_j \ .
\end{aligned} \tag{2.3.36}$$

Hence, $\mathbf{T}$ can always be written in the form

$$\mathbf{T} = T_{ij} \hat{\mathbf{e}}_i \hat{\mathbf{e}}_j \tag{2.3.37}$$

in terms of the nine independent quantities T_{ij}.

Dyadic representation of tensors

Every second-order tensor $\mathbf{T}$ can be expressed as a dyadic. In particular, if T_{ij} are the scalar components of $\mathbf{T}$ in the orthonormal basis $\{\hat{\mathbf{e}}_i\}$, then $\mathbf{T}$ can be expressed in the form (2.3.37). This is consistent with the notation introduced in (2.3.10) in which

$$\begin{aligned}
\mathbf{v} = \mathbf{T} \cdot \mathbf{u} &\iff v_i = T_{ij} u_j \ , \\
\mathbf{w} = \mathbf{u} \cdot \mathbf{T} &\iff w_j = u_i T_{ij} \ ,
\end{aligned} \tag{2.3.38}$$

where $\mathbf{u} = u_i\hat{\mathbf{e}}_i$, $\mathbf{v} = v_i\hat{\mathbf{e}}_i$, and $\mathbf{w} = w_i\hat{\mathbf{e}}_i$ are vectors. Applying the dot product relations (2.3.31) that define a dyad and the rules of tensor algebra, it follows that

$$
\begin{aligned}
\mathbf{T}\cdot\mathbf{u} &= (T_{ij}\hat{\mathbf{e}}_i\hat{\mathbf{e}}_j)\cdot(u_k\hat{\mathbf{e}}_k)\\
&= T_{ij}u_k(\hat{\mathbf{e}}_i\hat{\mathbf{e}}_j\cdot\hat{\mathbf{e}}_k)\\
&= T_{ij}u_k[\hat{\mathbf{e}}_i(\hat{\mathbf{e}}_j\cdot\hat{\mathbf{e}}_k)]\\
&= T_{ij}u_k\hat{\mathbf{e}}_i\delta_{jk}\\
&= T_{ij}u_j\hat{\mathbf{e}}_i
\end{aligned}
\tag{2.3.39}
$$

and similarly that

$$
\mathbf{u}\cdot\mathbf{T} = u_iT_{ij}\hat{\mathbf{e}}_j\ .
\tag{2.3.40}
$$

Thus, $\mathbf{v} = \mathbf{T}\cdot\mathbf{u}$ and $\mathbf{w} = \mathbf{u}\cdot\mathbf{T}$ if and only if $v_i = T_{ij}u_j$ and $w_j = u_iT_{ij}$.

The form of the representation (2.3.37) is invariant under changes of orthonormal basis; if $\hat{\mathbf{e}}_i = \ell_{pi}\hat{\mathbf{e}}'_p$, then $T_{ij} = \ell_{mi}\ell_{nj}T'_{mn}$ and

$$
\begin{aligned}
\mathbf{T} &= T_{ij}\hat{\mathbf{e}}_i\hat{\mathbf{e}}_j\\
&= (\ell_{mi}\ell_{nj}T'_{mn})(\ell_{pi}\hat{\mathbf{e}}'_p)(\ell_{qj}\hat{\mathbf{e}}'_q)\\
&= \ell_{mi}\ell_{pi}\ell_{nj}\ell_{qj}T'_{mn}\hat{\mathbf{e}}'_p\hat{\mathbf{e}}'_q\\
&= \delta_{mp}\delta_{nq}T'_{mn}\hat{\mathbf{e}}'_p\hat{\mathbf{e}}'_q\\
&= T'_{mn}\hat{\mathbf{e}}'_m\hat{\mathbf{e}}'_n\ .
\end{aligned}
\tag{2.3.41}
$$

Representing tensors in this manner is referred to as *dyadic notation*. Note that $\hat{\mathbf{e}}_p\cdot\mathbf{T} = T_{pj}\hat{\mathbf{e}}_j$, $\mathbf{T}\cdot\hat{\mathbf{e}}_q = T_{iq}\hat{\mathbf{e}}_i$, and

$$
\hat{\mathbf{e}}_p\cdot\mathbf{T}\cdot\hat{\mathbf{e}}_q = T_{pq}\ ,
\tag{2.3.42}
$$

which is another way in which the scalar components of a second-order tensor could be defined.

Dyadic notation generalizes to higher-order tensors. For example, the mathematical object denoted by $\mathbf{abc}$, where $\mathbf{a}$, $\mathbf{b}$, and $\mathbf{c}$ are vectors, is defined by the dot product relations

$$
\begin{aligned}
\mathbf{abc}\cdot\mathbf{v} &= \mathbf{ab}(\mathbf{c}\cdot\mathbf{v})\ ,\\
\mathbf{v}\cdot\mathbf{abc} &= (\mathbf{v}\cdot\mathbf{a})\mathbf{bc}\ ,
\end{aligned}
\tag{2.3.43}
$$

where $\mathbf{v}$ is an arbitrary vector. Such objects composed of three or more vectors are called *polyads*. It follows, through an argument similar to that for second-order tensors, that the dyadic notation representation of a third-order tensor $\mathbf{D}$ is

$$
\mathbf{D} = D_{ijk}\hat{\mathbf{e}}_i\hat{\mathbf{e}}_j\hat{\mathbf{e}}_k\ ,
\tag{2.3.44}
$$

where D_{ijk} are the scalar components of $\mathbf{D}$ in the orthonormal basis $\{\hat{\mathbf{e}}_i\}$. The dyadic notation representation of a tensor $\mathbf{R}$ of order n is

$$\mathbf{R} = R_{i_1 i_2 \cdots i_n} \hat{\mathbf{e}}_{i_1} \hat{\mathbf{e}}_{i_2} \cdots \hat{\mathbf{e}}_{i_n} , \tag{2.3.45}$$

where $R_{i_1 i_2 \cdots i_n}$ are the scalar components of $\mathbf{R}$ in the orthonormal basis $\{\hat{\mathbf{e}}_i\}$. Dyadic notation is very useful for manipulation of tensors in tensor calculus, as will be seen.

Dot products of tensors

The dot product of two dyads, say $\mathbf{ab}$ and $\mathbf{cd}$, is defined by

$$\mathbf{ab} \cdot \mathbf{cd} = (\mathbf{b} \cdot \mathbf{c})\mathbf{ad} \tag{2.3.46}$$

and the result is another dyad and hence a second-order tensor. It follows that the dot product of two second-order tensors $\mathbf{A}$ and $\mathbf{B}$ is a second-order tensor given in an orthonormal basis $\{\hat{\mathbf{e}}_i\}$ by

$$\begin{aligned}
\mathbf{A} \cdot \mathbf{B} &= (A_{ij}\hat{\mathbf{e}}_i\hat{\mathbf{e}}_j) \cdot (B_{kl}\hat{\mathbf{e}}_k\hat{\mathbf{e}}_l) \\
&= A_{ij}B_{kl}(\hat{\mathbf{e}}_i\hat{\mathbf{e}}_j \cdot \hat{\mathbf{e}}_k\hat{\mathbf{e}}_l) \\
&= A_{ij}B_{kl}(\hat{\mathbf{e}}_j \cdot \hat{\mathbf{e}}_k)\hat{\mathbf{e}}_i\hat{\mathbf{e}}_l \\
&= A_{ij}B_{kl}\delta_{jk}\hat{\mathbf{e}}_i\hat{\mathbf{e}}_l \\
&= A_{ij}B_{jl}\hat{\mathbf{e}}_i\hat{\mathbf{e}}_l .
\end{aligned} \tag{2.3.47}$$

Note that $\mathbf{A} \cdot \mathbf{B} \neq \mathbf{B} \cdot \mathbf{A}$. It can be shown (see Problem 2.10) that

$$\begin{aligned}
(\mathbf{A} \cdot \mathbf{B}) \cdot \mathbf{u} &= \mathbf{A} \cdot (\mathbf{B} \cdot \mathbf{u}) , \\
\mathbf{u} \cdot (\mathbf{A} \cdot \mathbf{B}) &= (\mathbf{u} \cdot \mathbf{A}) \cdot \mathbf{B} ,
\end{aligned} \tag{2.3.48}$$

for all vectors $\mathbf{u}$, which is an alternate way to define the dot product of two second-order tensors.

Dot products between higher-order tensors are similarly defined. If $\mathbf{R}$ is a tensor of order $m \geq 1$ and $\mathbf{S}$ is a tensor of order $n \geq 1$, with scalar components in an orthonormal basis $\{\hat{\mathbf{e}}_i\}$ given by $R_{i_1 i_2 \cdots i_m}$ and $S_{p_1 p_2 \cdots p_n}$, respectively, then

$$\mathbf{R} \cdot \mathbf{S} = R_{i_1 i_2 \cdots i_{m-1} i_m} S_{i_m p_2 p_3 \cdots p_n} \hat{\mathbf{e}}_{i_1} \hat{\mathbf{e}}_{i_2} \cdots \hat{\mathbf{e}}_{i_{m-1}} \hat{\mathbf{e}}_{p_2} \cdots \hat{\mathbf{e}}_{p_n} \tag{2.3.49}$$

is a tensor of order $m + n - 2$. Note that, in general, $\mathbf{R} \cdot \mathbf{S} \neq \mathbf{S} \cdot \mathbf{R}$, unless $\mathbf{R}$ and $\mathbf{S}$ are vectors. The rules of thumb are that the sequence in which the vectors are assembled to compose the polyads in a dot product operation must not be changed and the dot product operation is treated as though it operates on the two vector elements directly adjacent to the dot product symbol.

Double dot products of tensors

The *double dot product* of two dyads, say **ab** and **cd**, is a scalar defined by

$$\mathbf{ab} : \mathbf{cd} = (\mathbf{a} \cdot \mathbf{c})(\mathbf{b} \cdot \mathbf{d}) \,. \tag{2.3.50}$$

It follows that the double dot product of two second-order tensors **A** and **B** is a scalar given in an orthonormal basis $\{\hat{\mathbf{e}}_i\}$ by

$$
\begin{aligned}
\mathbf{A} : \mathbf{B} &= (A_{ij}\hat{\mathbf{e}}_i\hat{\mathbf{e}}_j) : (B_{kl}\hat{\mathbf{e}}_k\hat{\mathbf{e}}_l) \\
&= A_{ij}B_{kl}(\hat{\mathbf{e}}_i\hat{\mathbf{e}}_j : \hat{\mathbf{e}}_k\hat{\mathbf{e}}_l) \\
&= A_{ij}B_{kl}(\hat{\mathbf{e}}_i \cdot \hat{\mathbf{e}}_k)(\hat{\mathbf{e}}_j \cdot \hat{\mathbf{e}}_l) \\
&= A_{ij}B_{kl}\delta_{ik}\delta_{jl} \\
&= A_{ij}B_{ij} \,.
\end{aligned} \tag{2.3.51}
$$

Double dot products between higher-order tensors are similarly defined. For example, the double dot product between a fourth-order tensor **C** and a second-order tensor $\boldsymbol{\varepsilon}$ is a second-order tensor given, in an orthonormal basis $\{\hat{\mathbf{e}}_i\}$, by

$$
\begin{aligned}
\mathbf{C} : \boldsymbol{\varepsilon} &= (C_{ijkl}\hat{\mathbf{e}}_i\hat{\mathbf{e}}_j\hat{\mathbf{e}}_k\hat{\mathbf{e}}_l) : (\varepsilon_{rs}\hat{\mathbf{e}}_r\hat{\mathbf{e}}_s) \\
&= C_{ijkl}\varepsilon_{rs}(\hat{\mathbf{e}}_i\hat{\mathbf{e}}_j\hat{\mathbf{e}}_k\hat{\mathbf{e}}_l : \hat{\mathbf{e}}_r\hat{\mathbf{e}}_s) \\
&= C_{ijkl}\varepsilon_{rs}\hat{\mathbf{e}}_i\hat{\mathbf{e}}_j(\hat{\mathbf{e}}_k \cdot \hat{\mathbf{e}}_r)(\hat{\mathbf{e}}_l \cdot \hat{\mathbf{e}}_s) \\
&= C_{ijkl}\varepsilon_{rs}\hat{\mathbf{e}}_i\hat{\mathbf{e}}_j\delta_{kr}\delta_{ls} \\
&= C_{ijkl}\varepsilon_{kl}\hat{\mathbf{e}}_i\hat{\mathbf{e}}_j \,.
\end{aligned} \tag{2.3.52}
$$

In general, if **R** is a tensor of order $m \geq 2$ and **S** is a tensor of order $n \geq 2$, with scalar components in an orthonormal basis $\{\hat{\mathbf{e}}_i\}$ given by $R_{i_1 i_2 \cdots i_m}$ and $S_{p_1 p_2 \cdots p_n}$, respectively, then

$$\mathbf{R} : \mathbf{S} = R_{i_1 i_2 \cdots i_{m-2} i_{m-1} i_m} S_{i_{m-1} i_m p_3 p_4 \cdots p_n} \hat{\mathbf{e}}_{i_1}\hat{\mathbf{e}}_{i_2} \cdots \hat{\mathbf{e}}_{i_{m-2}}\hat{\mathbf{e}}_{p_3} \cdots \hat{\mathbf{e}}_{p_n} \tag{2.3.53}$$

is a tensor of order $m + n - 4$. Note that, in general, $\mathbf{R} : \mathbf{S} \neq \mathbf{S} : \mathbf{R}$, unless **R** and **S** are second-order tensors.

Contraction

The *contraction* of a dyad **ab**, where **a** and **b** are vectors is a scalar defined as

$$\mathbf{a} \cdot \mathbf{b} \,. \tag{2.3.54}$$

It follows that the contraction of a second-order tensor **T** is a scalar; in an orthonormal basis $\{\hat{\mathbf{e}}_i\}$, the contraction of a second-order tensor $\mathbf{T} = T_{ij}\hat{\mathbf{e}}_i\hat{\mathbf{e}}_j$ is

$$T_{ij}\hat{\mathbf{e}}_i \cdot \hat{\mathbf{e}}_j = T_{ii} \,. \tag{2.3.55}$$

For a polyad of order n, there are $\frac{1}{2}n(n-1)$ possible contractions, each of which result in a polyad of order $n-2$. For example, the polyad **abc**, where **a**, **b**, and **c** are vectors, has three possible contractions:

$$(\mathbf{a} \cdot \mathbf{b})\mathbf{c}\,, \quad (\mathbf{a} \cdot \mathbf{c})\mathbf{b}\,, \quad (\mathbf{b} \cdot \mathbf{c})\mathbf{a}\,. \tag{2.3.56}$$

It follows that, in any given orthonormal basis $\{\hat{\mathbf{e}}_i\}$, a tensor of order 3, $\mathbf{D} = D_{ijk}\hat{\mathbf{e}}_i\hat{\mathbf{e}}_j\hat{\mathbf{e}}_k$, has three possible contractions:

$$D_{iik}\hat{\mathbf{e}}_k\,, \quad D_{ijj}\hat{\mathbf{e}}_i\,, \quad D_{iji}\hat{\mathbf{e}}_j\,, \tag{2.3.57}$$

each of which is a vector.

Outer and inner products of tensors

The *outer product* of a tensor **A** of order r and a tensor **B** of order s, with scalar components in an orthonormal basis $\{\hat{\mathbf{e}}_i\}$ given by $A_{i_1 i_2 \cdots i_r}$ and $B_{i_1 i_2 \cdots i_s}$, respectively, is a tensor $\mathbf{C} = \mathbf{AB}$ of order $n = r + s$ with scalar components

$$C_{j_1 j_2 \cdots j_r k_1 k_2 \cdots k_s} = A_{j_1 j_2 \cdots j_r} B_{k_1 k_2 \cdots k_s}\,. \tag{2.3.58}$$

It follows that since $A_{i_1 i_2 \cdots i_r}$ and $B_{i_1 i_2 \cdots i_s}$ obey the tensor transformation rule (2.3.30), so will the scalar components $C_{i_1 i_2 \cdots i_n}$ of **C**. The tensor product of two vectors is an example of an outer product. As another example, the outer product $\mathbf{U} = \mathbf{Tv}$ of a second-order tensor $\mathbf{T} = T_{ij}\hat{\mathbf{e}}_i\hat{\mathbf{e}}_j$ and a vector $\mathbf{v} = v_i\hat{\mathbf{e}}_i$ is a third-order tensor $\mathbf{U} = U_{ijk}\hat{\mathbf{e}}_i\hat{\mathbf{e}}_j\hat{\mathbf{e}}_k = T_{ij}v_k\hat{\mathbf{e}}_i\hat{\mathbf{e}}_j\hat{\mathbf{e}}_k$. Note that, in general, $\mathbf{AB} \neq \mathbf{BA}$.

An *inner product* of two tensors is formed by one or more contractions between the two tensors in an outer product. Consider, for example, the outer product $\mathbf{AB} = A_{ij}B_{pq}\hat{\mathbf{e}}_i\hat{\mathbf{e}}_j\hat{\mathbf{e}}_p\hat{\mathbf{e}}_q$ of two second-order tensors **A** and **B**. The result **AB** is a tensor of order 4 from which a single contraction between components of **A** and **B** yields four distinct inner products:

$$A_{ij}B_{iq}\hat{\mathbf{e}}_j\hat{\mathbf{e}}_q\,, \quad A_{ij}B_{pj}\hat{\mathbf{e}}_i\hat{\mathbf{e}}_p\,, \quad A_{ij}B_{pi}\hat{\mathbf{e}}_j\hat{\mathbf{e}}_p\,, \quad A_{ij}B_{jq}\hat{\mathbf{e}}_i\hat{\mathbf{e}}_q\,, \tag{2.3.59}$$

each of which is a second-order tensor with scalar components that obey the tensor transformation rule (2.3.30). A second contraction between components of **A** and **B** yields two more distinct inner products, $A_{ij}B_{ij}$ and $A_{ij}B_{ji}$, each of which is a scalar. The dot product and the double dot product are particular types of inner product.

Tensor equations such as $\mathbf{R} = \alpha\mathbf{S} + \beta\mathbf{T}$, $\mathbf{w} = \mathbf{u} \cdot \mathbf{T}$, $\boldsymbol{\sigma} = \mathbf{C} : \boldsymbol{\varepsilon}$, and $\mathbf{S} = \mathbf{ab}$ have the advantage of being defined without having to specify a vector basis. In this sense, the representation of physical relations as tensor equations is preferable to indicial notation and matrix notation representations. On the other hand, it will often be more convenient for a particular application to use indicial and/or matrix notations. With this in mind, note that if it can be established that a tensor equation is true in one vector basis, then it must be true in all vector bases.

2.3.4   Isotropic Tensors

A tensor **T** of order n is *isotropic* if the components $T_{i_1 i_2 \cdots i_n}$ in one orthonormal basis $\{\hat{e}_i\}$ and the components $T'_{i_1 i_2 \cdots i_n}$ in another orthonormal basis $\{\hat{e}'_i\}$ are such that $T_{i_1 i_2 \cdots i_n} = T'_{i_1 i_2 \cdots i_n}$ for *all* $\{\hat{e}_i\}$ and $\{\hat{e}'_i\}$. All scalars (tensors of order 0) are trivially isotropic.

The only isotropic vector (tensor of order 1) is the zero vector **0**. To prove this, first note from the vector transformation rule (2.1.26) that **0** is trivially isotropic. To establish that this is the only possible isotropic vector, suppose that the vector **v** is isotropic and consider the change of orthonormal basis from $\{\hat{e}_i\}$ to $\{\hat{e}'_i\}$ given by the matrix of direction cosines

$$[L] = \begin{bmatrix} 0 & 1 & 0 \\ -1 & 0 & 0 \\ 0 & 0 & 1 \end{bmatrix} , \tag{2.3.60}$$

corresponding to a $90°$ rotation about $\hat{e}_3$. By applying the vector transformation rule, it is seen that in this case, the scalar components of **v** in $\{\hat{e}'_i\}$ are related to those in $\{\hat{e}_i\}$ by

$$v'_1 = v_2 , \quad v'_2 = -v_1 , \quad v'_3 = v_3 . \tag{2.3.61}$$

However, since **v** is isotropic,

$$v'_1 = v_1 , \quad v'_2 = v_2 , \quad v'_3 = v_3 . \tag{2.3.62}$$

Both (2.3.61) and (2.3.62) are true if and only if $v_1 = v_2 = 0$. For an isotropic vector, $v'_i = v_i$ for *every* proper orthogonal transformation $[L]$, yet by just considering this one possibility, it has already be shown that two of the scalar components of an isotropic vector must be zero. Consider now the transformation

$$[L] = \begin{bmatrix} 1 & 0 & 0 \\ 0 & 0 & -1 \\ 0 & 1 & 0 \end{bmatrix} , \tag{2.3.63}$$

corresponding to a $90°$ rotation about $\hat{e}_1$, from which it follows that

$$v'_1 = v_1 , \quad v'_2 = -v_3 , \quad v'_3 = v_2 . \tag{2.3.64}$$

Both (2.3.62) and (2.3.64) are true if and only if $v_2 = v_3 = 0$. Thus, the only way that a vector **v** can be isotropic is if its scalar components are all zero. Thus, the only isotropic vector is the zero vector **0**.

All isotropic tensors of order 2 have scalar components of the form $\alpha \delta_{ij}$, where α is a scalar and δ_{ij} is the Kronecker delta. One can show first that the Kronecker delta represents the scalar components of an isotropic tensor of order 2:

$$\delta'_{ij} = \ell_{im} \ell_{jn} \delta_{mn} = \ell_{in} \ell_{jn} = \delta_{ij} . \tag{2.3.65}$$

This tensor is the *identity tensor* $\mathbf{I} = \delta_{ij}\hat{\mathbf{e}}_i\hat{\mathbf{e}}_j = \hat{\mathbf{e}}_i\hat{\mathbf{e}}_i$. By considering different specific transformations, as was done in proving that the only isotropic vector is the zero vector, one can then establish that *all* isotropic tensors of order 2 must have the form $\alpha\mathbf{I}$.

All isotropic tensors of order 3 have scalar components of the form αe_{ijk}, where α is a scalar and e_{ijk} is the permutation symbol. One can first show that the permutation symbol represents the scalar components of an isotropic tensor of order 3:

$$e'_{ijk} = \ell_{ip}\ell_{jq}\ell_{kr}e_{pqr} = e_{ijk}\det[L] = e_{ijk} . \qquad (2.3.66)$$

(There is no commonly accepted symbol for the tensor whose scalar components are e_{ijk}; $\mathbf{e}$ will be used later for something different.) The procedure for proving *all* isotropic tensors of order 3 have scalar components of the form αe_{ijk} is similar to that used above.

Finally, all fourth-order isotropic tensors have scalar components of the form

$$C_{pqrs} = \lambda\delta_{pq}\delta_{rs} + \mu\delta_{pr}\delta_{qs} + \kappa\delta_{ps}\delta_{qr} , \qquad (2.3.67)$$

where λ, μ, and κ are scalars. That tensors of this form are isotropic follows immediately from the fact that the Kronecker deltas represent scalar components of isotropic second-order tensors. Again, the procedure for proving that *all* isotropic tensors of order 4 have scalar components of this form is as used above.

2.3.5   Tensors of Order 2

Tensors of order 2 appear very frequently in elasticity (and continuum mechanics in general) and deserve special comment. Given here are some properties of second-order tensors which will be used later. When dealing with tensors of order 2 and under, it is sometimes convenient, especially when manipulating actual scalar components in a given orthonormal basis, to use a *matrix notation* in which $[\mathbf{A}]$ is the 3×3 matrix composed of the scalar components of the second-order tensor $\mathbf{A}$ in an orthonormal basis $\{\hat{\mathbf{e}}_i\}$. For example, if $\mathbf{C} = \mathbf{A} \cdot \mathbf{B}$, where $\mathbf{A}$ and $\mathbf{B}$ and therefore $\mathbf{C}$, are second-order tensors, then it is seen that in any given orthonormal basis, $[\mathbf{C}] = [\mathbf{A}][\mathbf{B}]$. The transformation rule for second-order tensors, in indicial notation and matrix notation, is

$$\begin{aligned} A'_{ij} &= \ell_{im}\ell_{jn}A_{mn} , & [\mathbf{A}]' &= [L][\mathbf{A}][L]^{\mathsf{T}} , \\ A_{ij} &= \ell_{mi}\ell_{nj}A'_{mn} , & [\mathbf{A}] &= [L]^{\mathsf{T}}[\mathbf{A}]'[L] . \end{aligned} \qquad (2.3.68)$$

Similarly, if $\mathbf{v} = \mathbf{T} \cdot \mathbf{u}$ and $\mathbf{w} = \mathbf{u} \cdot \mathbf{T}$, where $\mathbf{T}$ is a second-order tensor and $\mathbf{u}$, $\mathbf{v}$, and $\mathbf{w}$ are vectors, then $[\mathbf{v}] = [\mathbf{T}][\mathbf{u}]$ and $[\mathbf{w}] = [\mathbf{u}]^{\mathsf{T}}[\mathbf{T}]$, where $[\mathbf{u}]$,

[v], and [w] are the column matrices composed of the scalar components of **u**, **v**, and **w**, and $[\mathbf{u}]^\mathsf{T}$ is the row matrix that is the transpose of [**u**]. It is emphasized that, without substantial modification, indicial notation and matrix notation relations are not applicable to vector bases that are not orthonormal (see the Appendix). Tensor notation relations do not suffer from this shortcoming.

Transpose

The *transpose* $\mathbf{A}^\mathsf{T}$ of a second-order tensor **A** is defined such that

$$\mathbf{u} \cdot \mathbf{A} \cdot \mathbf{v} = \mathbf{v} \cdot \mathbf{A}^\mathsf{T} \cdot \mathbf{u} , \tag{2.3.69}$$

for all vectors **u** and **v**. Since **u** and **v** are arbitrary, it follows immediately that

$$\mathbf{u} \cdot \mathbf{A} = \mathbf{A}^\mathsf{T} \cdot \mathbf{u} , \quad \mathbf{A} \cdot \mathbf{v} = \mathbf{v} \cdot \mathbf{A}^\mathsf{T} , \tag{2.3.70}$$

for all vectors **u** and **v**. It can be shown (see Problem 2.11) that

$$(\mathbf{A} \cdot \mathbf{B})^\mathsf{T} = \mathbf{B}^\mathsf{T} \cdot \mathbf{A}^\mathsf{T} \tag{2.3.71}$$

and that $(\mathbf{A}^\mathsf{T})^\mathsf{T} = \mathbf{A}$.

In an orthonormal basis $\{\hat{\mathbf{e}}_i\}$ in which $\mathbf{A} = A_{ij}\hat{\mathbf{e}}_i\hat{\mathbf{e}}_j$, $\mathbf{A}^\mathsf{T} = A_{ij}^\mathsf{T}\hat{\mathbf{e}}_i\hat{\mathbf{e}}_j$, $\mathbf{u} = u_i\hat{\mathbf{e}}_i$, and $\mathbf{v} = v_i\hat{\mathbf{e}}_i$, the defining equation (2.3.69) is expressed in indicial notation and matrix notation as

$$A_{ij}u_iv_j = A_{ij}^\mathsf{T}v_iu_j , \quad [\mathbf{u}]^\mathsf{T}[\mathbf{A}][\mathbf{v}] = [\mathbf{v}]^\mathsf{T}[\mathbf{A}^\mathsf{T}][\mathbf{u}] . \tag{2.3.72}$$

If follows that, in an orthonormal basis, the matrix of components of $\mathbf{A}^\mathsf{T}$ is equal to the transpose of the matrix of components of **A**,

$$A_{ij}^\mathsf{T} = A_{ji} , \quad [\mathbf{A}^\mathsf{T}] = [\mathbf{A}]^\mathsf{T} . \tag{2.3.73}$$

In dyadic notation, $\mathbf{A}^\mathsf{T} = A_{ji}\hat{\mathbf{e}}_i\hat{\mathbf{e}}_j$.

A second-order tensor **A** is said to be *symmetric* if

$$\mathbf{A} = \mathbf{A}^\mathsf{T} , \quad A_{ij} = A_{ji} , \quad [\mathbf{A}] = [\mathbf{A}]^\mathsf{T} \tag{2.3.74}$$

and *skew-symmetric* if

$$\mathbf{A} = -\mathbf{A}^\mathsf{T} , \quad A_{ij} = -A_{ji} , \quad [\mathbf{A}] = -[\mathbf{A}]^\mathsf{T} . \tag{2.3.75}$$

It follows that a symmetric tensor has only six independent scalar components and that a skew-symmetric tensor has only three independent scalar components. The diagonal elements of a skew-symmetric tensor's matrix of scalar components in an orthonormal basis are necessarily zero. A tensor that is symmetric in one vector basis is symmetric in all vector bases, so symmetry is a tensor property; likewise for skew-symmetric tensors.

Trace

The *trace* tr $\mathbf{A}$ of a second-order tensor $\mathbf{A}$ is a scalar defined as the contraction of $\mathbf{A}$ (see page 48). Thus, in an orthonormal basis, the trace is the sum of the diagonal elements of the tensor's matrix of scalar components,

$$\text{tr}\,\mathbf{A} = A_{ij}\hat{\mathbf{e}}_i \cdot \hat{\mathbf{e}}_j = A_{ii} . \qquad (2.3.76)$$

It follows from its definition that tr $\mathbf{A}$ has the same value in all vector bases. Nevertheless, it is instructive to show that tr $\mathbf{A}$ is invariant under changes in orthonormal basis:

$$A'_{ii} = \ell_{im}\ell_{in}A_{mn} = \delta_{mn}A_{mn} = A_{mm} . \qquad (2.3.77)$$

Note that tr $\mathbf{A}^\mathsf{T} = \text{tr}\,\mathbf{A}$. If $\mathbf{A}$ is skew-symmetric, then tr $\mathbf{A} = 0$.

Example 2.3
Show that $\mathbf{A} : \mathbf{B} = \text{tr}(\mathbf{A}^\mathsf{T} \cdot \mathbf{B}) = \text{tr}(\mathbf{A} \cdot \mathbf{B}^\mathsf{T})$.

Solution. One can establish the equivalence $\mathbf{A} : \mathbf{B} = \text{tr}(\mathbf{A}^\mathsf{T} \cdot \mathbf{B})$ in an orthonormal basis $\{\hat{\mathbf{e}}_i\}$. Since it is a tensor equation, if it is true in $\{\hat{\mathbf{e}}_i\}$, it must be true in all vector bases. Using dyadic notation, with $\mathbf{A}^\mathsf{T} = A_{ji}\hat{\mathbf{e}}_i\hat{\mathbf{e}}_j$ and $\mathbf{B} = B_{rs}\hat{\mathbf{e}}_r\hat{\mathbf{e}}_s$, it follows that

$$\mathbf{A}^\mathsf{T} \cdot \mathbf{B} = A_{ji}\hat{\mathbf{e}}_i\hat{\mathbf{e}}_j \cdot B_{rs}\hat{\mathbf{e}}_r\hat{\mathbf{e}}_s = A_{ji}B_{rs}\delta_{jr}\hat{\mathbf{e}}_i\hat{\mathbf{e}}_s = A_{ji}B_{js}\hat{\mathbf{e}}_i\hat{\mathbf{e}}_s .$$

Thus,

$$\text{tr}(\mathbf{A}^\mathsf{T} \cdot \mathbf{B}) = A_{ji}B_{js}\hat{\mathbf{e}}_i \cdot \hat{\mathbf{e}}_s = A_{ji}B_{js}\delta_{is} = A_{ji}B_{ji} = \mathbf{A} : \mathbf{B} .$$

Similarly,

$$\text{tr}(\mathbf{A} \cdot \mathbf{B}^\mathsf{T}) = \text{tr}(A_{ij}B_{sj}\hat{\mathbf{e}}_i\hat{\mathbf{e}}_s) = A_{ij}B_{ij} = \mathbf{A} : \mathbf{B} .$$

Determinant

The *determinant* of a second-order tensor $\mathbf{A}$ is a scalar defined as

$$\det\mathbf{A} \equiv \frac{[(\mathbf{A} \cdot \mathbf{u}) \times (\mathbf{A} \cdot \mathbf{v})] \cdot (\mathbf{A} \cdot \mathbf{w})}{(\mathbf{u} \times \mathbf{v}) \cdot \mathbf{w}} , \qquad (2.3.78)$$

where $\mathbf{u}$, $\mathbf{v}$, and $\mathbf{w}$ are three arbitrary, noncoplanar vectors (Antman 1995). The value of (2.3.78) is independent of $\mathbf{u}$, $\mathbf{v}$, and $\mathbf{w}$. If $[\mathbf{A}]$ is the matrix of scalar components of $\mathbf{A}$ in an orthonormal basis, then it can be shown (see Problem 2.13) that the determinant of $\mathbf{A}$ is equal to the determinant of its matrix of scalar components:

$$\det\mathbf{A} = \det[\mathbf{A}] . \qquad (2.3.79)$$

Since the definition (2.3.78) is a tensor equation, the value of $\det \mathbf{A}$ is the same in all vector bases. Nevertheless, it is again instructive to show that $\det[\mathbf{A}]$ is invariant under changes in orthonormal basis. Recall the basic results from linear algebra that $\det([\mathbf{A}][\mathbf{B}]) = \det[\mathbf{A}]\det[\mathbf{B}]$ and $\det[\mathbf{A}]^\mathsf{T} = \det[\mathbf{A}]$. Also, recall from (2.1.31) that for orthogonal transformations, $\det[L] = \pm 1$. Therefore,

$$
\begin{aligned}
\det[\mathbf{A}] &= \det\left([L]^\mathsf{T}[\mathbf{A}]'[L]\right) \\
&= \det[L]^\mathsf{T}\det[\mathbf{A}]'\det[L] \\
&= \det[\mathbf{A}]' \ .
\end{aligned}
\tag{2.3.80}
$$

Since, in an orthonormal basis, $\det(\mathbf{A}\cdot\mathbf{B}) = \det[\mathbf{A}\cdot\mathbf{B}]$, $[\mathbf{A}\cdot\mathbf{B}] = [\mathbf{A}][\mathbf{B}]$ and $\det([\mathbf{A}][\mathbf{B}]) = \det[\mathbf{A}]\det[\mathbf{B}]$, it follows that

$$
\det(\mathbf{A}\cdot\mathbf{B}) = \det\mathbf{A}\det\mathbf{B} \ .
\tag{2.3.81}
$$

Similarly, since $\det\mathbf{A}^\mathsf{T} = \det[\mathbf{A}^\mathsf{T}]$, $[\mathbf{A}^\mathsf{T}] = [\mathbf{A}]^\mathsf{T}$, and $\det[\mathbf{A}]^\mathsf{T} = \det[\mathbf{A}]$, it follows that

$$
\det\mathbf{A}^\mathsf{T} = \det\mathbf{A} \ .
\tag{2.3.82}
$$

The important results (2.3.81) and (2.3.82) can also be established directly from the definition (2.3.78), obviating the need to introduce any particular vector basis. If $\det\mathbf{A} = 0$, then $\mathbf{A}$ is said to be *singular*; otherwise, $\mathbf{A}$ is said to be *nonsingular*. Given a second-order tensor $\mathbf{A}$, the equation $\mathbf{A}\cdot\mathbf{u} = \mathbf{0}$ has a nontrivial solution $\mathbf{u} \neq \mathbf{0}$ if and only if $\mathbf{A}$ is singular.

Inverse

If $\mathbf{A}$ is a nonsingular second-order tensor, then it can be shown that there exists a unique second-order tensor $\mathbf{A}^{-1}$, called the *inverse* of $\mathbf{A}$, such that

$$
\mathbf{A}\cdot\mathbf{A}^{-1} = \mathbf{A}^{-1}\cdot\mathbf{A} = \mathbf{I} \ ,
\tag{2.3.83}
$$

where $\mathbf{I}$ is the identity tensor (Knowles 1998). Conversely, if the inverse of $\mathbf{A}$ exists, then $\mathbf{A}$ must be nonsingular. It follows that if $\mathbf{v} = \mathbf{A}\cdot\mathbf{u}$ and $\mathbf{w} = \mathbf{u}\cdot\mathbf{A}$, then $\mathbf{u} = \mathbf{A}^{-1}\cdot\mathbf{v}$ and $\mathbf{u} = \mathbf{w}\cdot\mathbf{A}^{-1}$. It can be shown (see Problem 2.11) that

$$
(\mathbf{A}\cdot\mathbf{B})^{-1} = \mathbf{B}^{-1}\cdot\mathbf{A}^{-1} \ ,
\tag{2.3.84}
$$

where $\mathbf{A}$ and $\mathbf{B}$ are nonsingular second-order tensors.

In an orthonormal basis $\{\hat{\mathbf{e}}_i\}$ in which $\mathbf{A} = A_{ij}\hat{\mathbf{e}}_i\hat{\mathbf{e}}_j$ and $\mathbf{A}^{-1} = A_{ij}^{-1}\hat{\mathbf{e}}_i\hat{\mathbf{e}}_j$, the indicial and matrix notation representations of (2.3.83) are

$$
A_{ik}A_{kj}^{-1} = A_{ik}^{-1}A_{kj} = \delta_{ij} \ , \quad [\mathbf{A}][\mathbf{A}^{-1}] = [\mathbf{A}^{-1}][\mathbf{A}] = [I] \ .
\tag{2.3.85}
$$

Thus, in an orthonormal basis, the matrix of scalar components of $\mathbf{A}^{-1}$ is equal to the inverse of the matrix of scalar components of $\mathbf{A}$:

$$[\mathbf{A}^{-1}] = [\mathbf{A}]^{-1} . \tag{2.3.86}$$

By taking the determinant of $\mathbf{A} \cdot \mathbf{A}^{-1} = \mathbf{I}$ and noting that $\det(\mathbf{A} \cdot \mathbf{A}^{-1}) = \det \mathbf{A} \det \mathbf{A}^{-1}$ and $\det \mathbf{I} = 1$, it follows that

$$\det \mathbf{A}^{-1} = \frac{1}{\det \mathbf{A}} , \tag{2.3.87}$$

which exists since $\mathbf{A}$ is nonsingular.

Orthogonality

A second-order tensor $\mathbf{A}$ is *orthogonal* if and only if

$$\mathbf{A} \cdot \mathbf{A}^{\mathsf{T}} = \mathbf{A}^{\mathsf{T}} \cdot \mathbf{A} = \mathbf{I} . \tag{2.3.88}$$

In an orthonormal basis $\{\hat{\mathbf{e}}_i\}$ in which $\mathbf{A} = A_{ij}\hat{\mathbf{e}}_i\hat{\mathbf{e}}_j$, the equivalent indicial and matrix notation equations are

$$A_{ik}A_{jk} = A_{ki}A_{kj} = \delta_{ij} , \quad [\mathbf{A}][\mathbf{A}]^{\mathsf{T}} = [\mathbf{A}]^{\mathsf{T}}[\mathbf{A}] = [I] . \tag{2.3.89}$$

Comparing (2.3.83) and (2.3.88), it is seen that $\mathbf{A}$ is orthogonal if and only if $\mathbf{A}^{\mathsf{T}} = \mathbf{A}^{-1}$. A tensor that is orthogonal in one vector basis is orthogonal in all vector bases, so orthogonality is a tensor property. By taking the determinant of $\mathbf{A} \cdot \mathbf{A}^{\mathsf{T}} = \mathbf{I}$ and noting that $\det(\mathbf{A} \cdot \mathbf{A}^{\mathsf{T}}) = (\det \mathbf{A})^2$ and $\det \mathbf{I} = 1$, it follows that if $\mathbf{A}$ is orthogonal, then

$$\det \mathbf{A} = \pm 1 . \tag{2.3.90}$$

An orthogonal second-order tensor $\mathbf{A}$ is said to be *proper orthogonal* if $\det \mathbf{A} = +1$.

Integer powers of a second-order tensor

The *integer powers* of a second-order tensor $\mathbf{A}$ are defined recursively by

$$\mathbf{A}^n \equiv \mathbf{A}^{n-1} \cdot \mathbf{A} , \tag{2.3.91}$$

where n is a positive integer and $\mathbf{A}^0 \equiv \mathbf{I}$. It follows for instance that $\mathbf{A}^2 = \mathbf{A} \cdot \mathbf{A}$ and $\mathbf{A}^3 = \mathbf{A}^2 \cdot \mathbf{A} = \mathbf{A} \cdot \mathbf{A} \cdot \mathbf{A}$. If $\mathbf{A}$ is nonsingular, then the definition can be extended to negative integer powers of $\mathbf{A}$ by including the recursive relation

$$\mathbf{A}^{-n} \equiv \mathbf{A}^{-n+1} \cdot \mathbf{A}^{-1} , \tag{2.3.92}$$

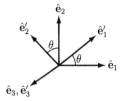

FIGURE 2.5.   Two-dimensional transformation of base vectors.

where n is a positive integer. Thus, for example, $\mathbf{A}^{-2} = \mathbf{A}^{-1} \cdot \mathbf{A}^{-1}$ and $\mathbf{A}^{-3} = \mathbf{A}^{-2} \cdot \mathbf{A}^{-1} = \mathbf{A}^{-1} \cdot \mathbf{A}^{-1} \cdot \mathbf{A}^{-1}$. Equivalently, $\mathbf{A}^{-n} = \left(\mathbf{A}^{-1} \right)^n = \left(\mathbf{A}^n \right)^{-1}$. It follows from (2.3.81) and (2.3.87) that

$$\det \mathbf{A}^n = (\det \mathbf{A})^n , \qquad (2.3.93)$$

for all integer values of n. If $\mathbf{A}$ is symmetric, then $\mathbf{A}^n$ is also symmetric.

Positive definite second-order tensors

A second-order tensor $\mathbf{A}$ is said to be *positive definite* if $\mathbf{u} \cdot \mathbf{A} \cdot \mathbf{u} > 0$ for all vectors $\mathbf{u} \neq \mathbf{0}$. The tensor $\mathbf{A}$ is *positive semidefinite* if $\mathbf{u} \cdot \mathbf{A} \cdot \mathbf{u} \geq 0$ for all vectors $\mathbf{u}$ and there exists at least one vector $\mathbf{a} \neq \mathbf{0}$ for which $\mathbf{a} \cdot \mathbf{A} \cdot \mathbf{a} = 0$. The tensor $\mathbf{A}$ is *negative definite* or *negative semidefinite* as $-\mathbf{A}$ is positive definite or positive semidefinite, respectively. Finally, the tensor $\mathbf{A}$ is said to be *indefinite* if it is none of the above, that is, if there exists at least one vector $\mathbf{a}$ and at least one vector $\mathbf{b}$ for which $\mathbf{a} \cdot \mathbf{A} \cdot \mathbf{a} > 0$ and $\mathbf{b} \cdot \mathbf{A} \cdot \mathbf{b} < 0$. Note that $\mathbf{A}$ cannot be either positive or negative definite unless it is nonsingular; if $\mathbf{A}$ is singular, then there exists at least one vector $\mathbf{a} \neq \mathbf{0}$ such that $\mathbf{A} \cdot \mathbf{a} = \mathbf{0}$ and therefore $\mathbf{a} \cdot \mathbf{A} \cdot \mathbf{a} = 0$.

Special tensors for two-dimensional problems

A special family of second-order tensors, which will be important in the examination of two-dimensional problems, have scalar components in the orthonormal basis $\{ \hat{\mathbf{e}}_i \}$ given by

$$[\mathbf{A}] = \begin{bmatrix} A_{11} & A_{12} & 0 \\ A_{21} & A_{22} & 0 \\ 0 & 0 & A_{33} \end{bmatrix} . \qquad (2.3.94)$$

If transformations are restricted to rotations about the base vector $\hat{\mathbf{e}}_3$ as shown in Figure 2.5, so that the transformation matrix has the form

$$[L] = \begin{bmatrix} \cos\theta & \sin\theta & 0 \\ -\sin\theta & \cos\theta & 0 \\ 0 & 0 & 1 \end{bmatrix} , \qquad (2.3.95)$$

then the matrix of scalar components of $\mathbf{A}$ in $\{\hat{\mathbf{e}}_i'\}$ will have the same zero elements as (2.3.94):

$$[\mathbf{A}]' = \begin{bmatrix} A_{11}' & A_{12}' & 0 \\ A_{21}' & A_{22}' & 0 \\ 0 & 0 & A_{33}' \end{bmatrix} , \qquad (2.3.96)$$

with $A_{33}' = A_{33}$. The remaining components A_{ij}' are given by

$$\begin{aligned} A_{11}' &= A_{11} \cos^2 \theta + A_{22} \sin^2 \theta + (A_{12} + A_{21}) \cos \theta \sin \theta , \\ A_{22}' &= A_{11} \sin^2 \theta + A_{22} \cos^2 \theta - (A_{12} + A_{21}) \cos \theta \sin \theta , \\ A_{12}' &= -(A_{11} - A_{22}) \cos \theta \sin \theta + A_{12} \cos^2 \theta - A_{21} \sin^2 \theta , \\ A_{21}' &= -(A_{11} - A_{22}) \cos \theta \sin \theta - A_{12} \sin^2 \theta + A_{21} \cos^2 \theta . \end{aligned} \qquad (2.3.97)$$

If $\mathbf{A}$ is symmetric, then

$$\begin{aligned} A_{11}' &= A_{11} \cos^2 \theta + A_{22} \sin^2 \theta + 2A_{12} \cos \theta \sin \theta , \\ A_{22}' &= A_{11} \sin^2 \theta + A_{22} \cos^2 \theta - 2A_{12} \cos \theta \sin \theta , \\ A_{12}' &= A_{21}' = -(A_{11} - A_{22}) \cos \theta \sin \theta + A_{12}(\cos^2 \theta - \sin^2 \theta) . \end{aligned} \qquad (2.3.98)$$

2.3.6 Eigenvalue Problems

The eigenvalues and eigenvectors of symmetric second-order tensors, defined below, have important physical significance in the characterization of deformation and internal forces, as will be seen in subsequent chapters. A general treatment is presented here that will be applied to specific cases later. This general treatment includes the definition of two scalar measures of a symmetric second-order tensor, the normal and the shear, which require the specification of either one direction for the normal or two orthogonal directions for the shear. Again, the physical significance of these scalar measures will become apparent during subsequent chapters.

Normal of a symmetric second-order tensor

Given a symmetric second-order tensor $\mathbf{A}$, with scalar components A_{ij} in the orthonormal basis $\{\hat{\mathbf{e}}_i\}$, one can define the *normal of* $\mathbf{A}$ *in the direction* $\hat{\boldsymbol{\mu}}$ as

$$N_{\mathbf{A}}(\hat{\boldsymbol{\mu}}) \equiv \hat{\boldsymbol{\mu}} \cdot \mathbf{A} \cdot \hat{\boldsymbol{\mu}} , \qquad N_{\mathbf{A}}(\hat{\mu}_1, \hat{\mu}_2, \hat{\mu}_3) \equiv A_{ij} \hat{\mu}_i \hat{\mu}_j , \qquad (2.3.99)$$

where $\hat{\boldsymbol{\mu}}$, with scalar components $\hat{\mu}_i$, is an arbitrary unit vector. $N_{\mathbf{A}}$ is a scalar and one may ask which directions $\hat{\boldsymbol{\mu}}$ correspond to extrema of $N_{\mathbf{A}}$ and what are the values of these extrema?

To find the unit vectors $\hat{\mu}$ that correspond to extrema of $N_{\mathbf{A}}$, subject to the constraint that $\hat{\mu} \cdot \hat{\mu} = 1$, consider instead the following function of $\hat{\mu}$ and λ:

$$\tilde{N}_{\mathbf{A}}(\hat{\mu}, \lambda) \equiv \hat{\mu} \cdot \mathbf{A} \cdot \hat{\mu} - \lambda(\hat{\mu} \cdot \hat{\mu} - 1) ,$$

$$\tilde{N}_{\mathbf{A}}(\hat{\mu}_1, \hat{\mu}_2, \hat{\mu}_3, \lambda) \equiv A_{ij}\hat{\mu}_i\hat{\mu}_j - \lambda(\hat{\mu}_i\hat{\mu}_i - 1) . \qquad (2.3.100)$$

The new independent variable λ introduced in (2.3.100) is known as a *Lagrange multiplier*. It can be seen that setting $\partial\tilde{N}_{\mathbf{A}}/\partial\lambda = 0$ enforces the constraint $\hat{\mu} \cdot \hat{\mu} = 1$. Thus, values of $\hat{\mu}$ and λ that correspond to extrema of $\tilde{N}_{\mathbf{A}}$ correspond to extrema of $N_{\mathbf{A}}$ that satisfy this constraint.

By recalling that $\mathbf{A}$ is symmetric and noting that $\partial\hat{\mu}_i/\partial\hat{\mu}_j = \delta_{ij}$, it can be shown (see Problem 2.14) that the conditions $\partial\tilde{N}_{\mathbf{A}}/\partial\hat{\mu}_i = 0$ for extremizing $\tilde{N}_{\mathbf{A}}$ lead to the equation

$$\mathbf{A} \cdot \hat{\mu} = \lambda\hat{\mu} , \quad A_{ij}\hat{\mu}_j = \lambda\hat{\mu}_i , \qquad (2.3.101)$$

to be solved for the unit vector $\hat{\mu}$ corresponding to an extrema of $\tilde{N}_{\mathbf{A}}$. This equation represents what is known as the *eigenvalue problem* for $\mathbf{A}$, which can be rewritten as

$$(\mathbf{A} - \lambda\mathbf{I}) \cdot \hat{\mu} = \mathbf{0} , \quad (A_{ij} - \lambda\delta_{ij})\hat{\mu}_j = 0 . \qquad (2.3.102)$$

The trivial solution $\hat{\mu} = \mathbf{0}$ does not satisfy the constraint that $\hat{\mu} \cdot \hat{\mu} = 1$ and is of no relevance, and (2.3.102) has nontrivial solutions ($\hat{\mu} \neq \mathbf{0}$) if and only if

$$\det(\mathbf{A} - \lambda\mathbf{I}) = 0 , \quad \det \begin{bmatrix} A_{11} - \lambda & A_{12} & A_{13} \\ A_{21} & A_{22} - \lambda & A_{23} \\ A_{31} & A_{32} & A_{33} - \lambda \end{bmatrix} = 0 . \quad (2.3.103)$$

This determinant can be expressed as a cubic equation in λ, called the *characteristic equation* of $\mathbf{A}$;

$$\lambda^3 - I_{\mathbf{A}}\lambda^2 + II_{\mathbf{A}}\lambda - III_{\mathbf{A}} = 0 , \qquad (2.3.104)$$

where the coefficients are given in terms of the scalar components of $\mathbf{A}$ by

$$I_{\mathbf{A}} = A_{ii} = A_{11} + A_{22} + A_{33} ,$$

$$\begin{aligned} II_{\mathbf{A}} &= \frac{1}{2}(A_{ii}A_{jj} - A_{ij}A_{ji}) \\ &= A_{11}A_{22} + A_{22}A_{33} + A_{33}A_{11} \\ &\quad - A_{12}A_{21} - A_{23}A_{32} - A_{31}A_{13} , \end{aligned} \qquad (2.3.105a)$$

$$III_{\mathbf{A}} = \det[\mathbf{A}] = e_{ijk}A_{1i}A_{2j}A_{3k} .$$

The three roots $\lambda^{(1)}$, $\lambda^{(2)}$, and $\lambda^{(3)}$ of (2.3.104) are the *eigenvalues* of $\mathbf{A}$ and have the same units as $\mathbf{A}$. It can be shown that since $\mathbf{A}$ is symmetric, the eigenvalues of $\mathbf{A}$ are real; they should be labeled so that $\lambda^{(1)} \geq \lambda^{(2)} \geq \lambda^{(3)}$. The corresponding solutions $\hat{\boldsymbol{\mu}}^{(1)}$, $\hat{\boldsymbol{\mu}}^{(2)}$, and $\hat{\boldsymbol{\mu}}^{(3)}$ of (2.3.102) are the *eigenvectors* of $\mathbf{A}$. Clearly, solutions of (2.3.102) have arbitrary magnitude; in accordance with the original constraint, the eigenvectors should always be normalized to unit magnitude.

By writing the coefficients (2.3.105) in tensor notation,

$$I_{\mathbf{A}} = \operatorname{tr} \mathbf{A} \,,$$

$$II_{\mathbf{A}} = \frac{1}{2} \left[(\operatorname{tr} \mathbf{A})^2 - \operatorname{tr}(\mathbf{A}^2) \right] \,, \qquad (2.3.105\mathrm{b})$$

$$III_{\mathbf{A}} = \det \mathbf{A} \,,$$

it is clear that they are scalar quantities; the coefficients $I_{\mathbf{A}}$, $II_{\mathbf{A}}$, and $III_{\mathbf{A}}$ are the same in all vector bases and are called the *principal invariants* of $\mathbf{A}$.

Substituting the eigenvectors $\hat{\boldsymbol{\mu}}^{(1)}$, $\hat{\boldsymbol{\mu}}^{(2)}$, and $\hat{\boldsymbol{\mu}}^{(3)}$ back into the expression (2.3.99) yields the extrema of $N_{\mathbf{A}}$. Thus, the extremum of $N_{\mathbf{A}}$ corresponding to the direction given by the eigenvector $\hat{\boldsymbol{\mu}}^{(k)}$ is

$$N_{\mathbf{A}}(\hat{\boldsymbol{\mu}}^{(k)}) = \hat{\boldsymbol{\mu}}^{(k)} \cdot \mathbf{A} \cdot \hat{\boldsymbol{\mu}}^{(k)} \,, \quad N_{\mathbf{A}}(\hat{\mu}_1^{(k)}, \hat{\mu}_2^{(k)}, \hat{\mu}_3^{(k)}) = A_{ij}\hat{\mu}_i^{(k)}\hat{\mu}_j^{(k)} \,,$$
$$(2.3.106)$$

where the summation convention does not apply to the superscript k in parentheses. However, from (2.3.101),

$$\mathbf{A} \cdot \hat{\boldsymbol{\mu}}^{(k)} = \lambda^{(k)}\hat{\boldsymbol{\mu}}^{(k)} \,, \quad A_{ij}\hat{\mu}_j^{(k)} = \lambda^{(k)}\hat{\mu}_i^{(k)} \,. \qquad (2.3.107)$$

Substituting (2.3.107) into (2.3.106) and recalling that $\hat{\boldsymbol{\mu}}^{(k)}$ is a unit vector, one finds that $N_{\mathbf{A}}(\hat{\boldsymbol{\mu}}^{(k)}) = \lambda^{(k)}$. Thus, the maximum normal of $\mathbf{A}$ is the eigenvalue $\lambda^{(1)}$,

$$(N_{\mathbf{A}})_{\max} = \lambda^{(1)} \,, \qquad (2.3.108)$$

and is in the direction given by the eigenvector $\hat{\boldsymbol{\mu}}^{(1)}$. The minimum normal of $\mathbf{A}$ is the eigenvalue $\lambda^{(3)}$,

$$(N_{\mathbf{A}})_{\min} = \lambda^{(3)} \,, \qquad (2.3.109)$$

and is in the direction given by the eigenvector $\hat{\boldsymbol{\mu}}^{(3)}$. The eigenvalue $\lambda^{(2)}$ corresponds to a saddle point of $N_{\mathbf{A}}$.

Shear of a symmetric second-order tensor

In addition to the normal of $\mathbf{A}$ (2.3.99), one can define the *shear of* $\mathbf{A}$ *between the orthogonal directions* $\hat{\mu}'$ *and* $\hat{\mu}''$ as

$$S_{\mathbf{A}}(\hat{\mu}', \hat{\mu}'') \equiv \hat{\mu}' \cdot \mathbf{A} \cdot \hat{\mu}'' \,,$$
$$S_{\mathbf{A}}(\hat{\mu}_1', \hat{\mu}_2', \hat{\mu}_3', \hat{\mu}_1'', \hat{\mu}_2'', \hat{\mu}_3'') \equiv A_{ij}\hat{\mu}_i'\hat{\mu}_j'' \,, \tag{2.3.110}$$

where the unit vectors $\hat{\mu}'$ and $\hat{\mu}''$ are constrained to be orthogonal ($\hat{\mu}' \cdot \hat{\mu}'' = 0$). Note that since $\mathbf{A}$ is symmetric, $S_{\mathbf{A}}$ has the following symmetry:

$$S_{\mathbf{A}}(\hat{\mu}', \hat{\mu}'') = S_{\mathbf{A}}(\hat{\mu}'', \hat{\mu}') \tag{2.3.111}$$

and

$$S_{\mathbf{A}}(\hat{\mu}', \hat{\mu}'')$$
$$= S_{\mathbf{A}}(-\hat{\mu}', -\hat{\mu}'') = -S_{\mathbf{A}}(-\hat{\mu}', \hat{\mu}'') = -S_{\mathbf{A}}(\hat{\mu}', -\hat{\mu}'') \,. \tag{2.3.112}$$

If either $\hat{\mu}' = \hat{\mu}^{(k)}$ or $\hat{\mu}'' = \hat{\mu}^{(k)}$ in (2.3.110), then from (2.3.107) and since $\mathbf{A}$ is symmetric and $\hat{\mu}' \cdot \hat{\mu}'' = 0$, the shear $S_{\mathbf{A}} = 0$. Thus, for directions $\hat{\mu}^{(k)}$ corresponding to extrema of the normal of $\mathbf{A}$, one finds that the shear of $\mathbf{A}$ between $\hat{\mu}^{(k)}$ and every direction orthogonal to $\hat{\mu}^{(k)}$ is zero; $\hat{\mu} \cdot \mathbf{A} \cdot \hat{\mu}^{(k)} = 0$ and $\hat{\mu}^{(k)} \cdot \mathbf{A} \cdot \hat{\mu} = 0$ for all $\hat{\mu}$ orthogonal to $\hat{\mu}^{(k)}$.

Using Lagrange multipliers to impose the constraint that $\hat{\mu}'$ and $\hat{\mu}''$ be mutually orthogonal unit vectors (and recalling the labeling convention $\lambda^{(1)} \geq \lambda^{(2)} \geq \lambda^{(3)}$), one can show that the pair of orthogonal directions $(\hat{\mu}', \hat{\mu}'')$ that corresponds to the maximum absolute value of shear (2.3.110) of $\mathbf{A}$ is given in terms of the eigenvectors $\hat{\mu}^{(k)}$ by

$$(\hat{\mu}', \hat{\mu}'') = \left(\frac{\hat{\mu}^{(1)} + \hat{\mu}^{(3)}}{\sqrt{2}}, \frac{\hat{\mu}^{(1)} - \hat{\mu}^{(3)}}{\sqrt{2}} \right) \,; \tag{2.3.113}$$

the maximum shear of $\mathbf{A}$ is between the directions that bisect the eigenvectors $\hat{\mu}^{(1)}$ and $\hat{\mu}^{(3)}$, the directions corresponding to the maximum and minimum normal of $\mathbf{A}$. The maximum absolute value of the shear is therefore

$$|S_{\mathbf{A}}|_{\max} = \left| S_{\mathbf{A}}\left(\frac{\hat{\mu}^{(1)} + \hat{\mu}^{(3)}}{\sqrt{2}}, \frac{\hat{\mu}^{(1)} - \hat{\mu}^{(3)}}{\sqrt{2}} \right) \right| = \frac{1}{2}\left(\lambda^{(1)} - \lambda^{(3)} \right) \,. \tag{2.3.114}$$

The maximum shear is one-half the difference between the maximum and minimum normals.

Principal basis

It can be proven that, for a symmetric tensor $\mathbf{A}$, if the three eigenvalues are distinct ($\lambda^{(1)} \neq \lambda^{(2)} \neq \lambda^{(3)} \neq \lambda^{(1)}$), then the three corresponding eigenvectors are uniquely determined and mutually orthogonal. If two of the eigenvalues are equal ($\lambda^{(1)} = \lambda^{(2)} \neq \lambda^{(3)}$, say), then the eigenvector corresponding to the distinct eigenvalue ($\hat{\boldsymbol{\mu}}^{(3)}$, say) is uniquely determined and the remaining two eigenvectors are restricted to be in the plane normal to the uniquely determined eigenvector, but are otherwise arbitrary. Finally, if all three eigenvalues are equal ($\lambda^{(1)} = \lambda^{(2)} = \lambda^{(3)}$), then the eigenvectors are completely arbitrary. In any event, three mutually orthogonal eigenvectors can always be chosen, whether some are arbitrary or not. The sense of an eigenvector is changed by multiplying by -1. By convention, this should be done as necessary to form a right-handed system, that is, so that $\hat{\boldsymbol{\mu}}^{(1)} \times \hat{\boldsymbol{\mu}}^{(2)} = +\hat{\boldsymbol{\mu}}^{(3)}$.

Let the *principal basis* $\{\hat{\mathbf{e}}_i^*\}_{\mathbf{A}}$ be defined as the right-handed orthonormal basis whose base vectors are the mutually orthogonal eigenvectors of $\mathbf{A}$:

$$\{\hat{\mathbf{e}}_i^*\}_{\mathbf{A}} \equiv \left\{ \hat{\boldsymbol{\mu}}^{(1)}, \hat{\boldsymbol{\mu}}^{(2)}, \hat{\boldsymbol{\mu}}^{(3)} \right\}. \tag{2.3.115}$$

Then, the components of $\mathbf{A}$ in its principal basis $\{\hat{\mathbf{e}}_i^*\}_{\mathbf{A}}$ are given by

$$A_{ij}^* = \hat{\mathbf{e}}_i^* \cdot \mathbf{A} \cdot \hat{\mathbf{e}}_j^* = \hat{\boldsymbol{\mu}}^{(i)} \cdot \mathbf{A} \cdot \hat{\boldsymbol{\mu}}^{(j)}. \tag{2.3.116}$$

Comparing (2.3.116) with the functions $N_{\mathbf{A}}$ and $S_{\mathbf{A}}$ defined in (2.3.99) and (2.3.110) and with the values of $N_{\mathbf{A}}$ and $S_{\mathbf{A}}$ when the directions are given by the eigenvectors, it is immediately seen that

$$[\mathbf{A}]^* = \begin{bmatrix} \lambda^{(1)} & 0 & 0 \\ 0 & \lambda^{(2)} & 0 \\ 0 & 0 & \lambda^{(3)} \end{bmatrix}. \tag{2.3.117}$$

In its principal basis, the matrix of scalar components of a symmetric second-order tensor is diagonal—the off-diagonal components are zero and the diagonal components are equal to the eigenvalues. Conversely, if an orthonormal basis can be found in which the off-diagonal terms of a symmetric second-order tensor are zero, then the base vectors are eigenvectors and the diagonal terms are the eigenvalues. It follows that every symmetric second-order tensor can be expressed in dyadic notation as

$$\mathbf{A} = \lambda^{(1)} \hat{\boldsymbol{\mu}}^{(1)} \hat{\boldsymbol{\mu}}^{(1)} + \lambda^{(2)} \hat{\boldsymbol{\mu}}^{(2)} \hat{\boldsymbol{\mu}}^{(2)} + \lambda^{(3)} \hat{\boldsymbol{\mu}}^{(3)} \hat{\boldsymbol{\mu}}^{(3)}, \tag{2.3.118}$$

where $\lambda^{(k)}$ and $\hat{\boldsymbol{\mu}}^{(k)}$ are the eigenvalues and eigenvectors of $\mathbf{A}$. This is known as the *spectral representation* of $\mathbf{A}$.

By using the result (2.3.117) for the matrix of scalar components of $\mathbf{A}$ in its principal basis, it follows that the principal invariants (2.3.105) can be expressed in terms of the eigenvalues as

$$I_\mathbf{A} = A_{ii}^* = \lambda^{(1)} + \lambda^{(2)} + \lambda^{(3)} ,$$

$$II_\mathbf{A} = \frac{1}{2}(A_{ii}^* A_{jj}^* - A_{ij}^* A_{ji}^*)$$
$$= \lambda^{(1)}\lambda^{(2)} + \lambda^{(2)}\lambda^{(3)} + \lambda^{(3)}\lambda^{(1)} ,$$

$$III_\mathbf{A} = \det[\mathbf{A}]^* = \lambda^{(1)}\lambda^{(2)}\lambda^{(3)} .$$

(2.3.105c)

Thus, the principal invariants are easily found if the eigenvalues are known.

From the spectral representation (2.3.118), it can be seen that the eigenvalues of a symmetric second-order tensor $\mathbf{A}$ can be determined by

$$\lambda^{(k)} = \hat{\boldsymbol{\mu}}^{(k)} \cdot \mathbf{A} \cdot \hat{\boldsymbol{\mu}}^{(k)} , \qquad (2.3.119)$$

where $\hat{\boldsymbol{\mu}}^{(k)}$ are the (mutually orthogonal) eigenvectors of $\mathbf{A}$. Thus, if $\mathbf{A}$ is positive definite, then the eigenvalues of $\mathbf{A}$ are positive. It then follows from (2.3.105c) that the principal invariants of $\mathbf{A}$ are also positive. Conversely, if the eigenvalues of a symmetric second-order tensor $\mathbf{A}$ are positive, then $\mathbf{A}$ is positive definite (and therefore nonsingular).

Real powers of a second-order tensor

Recall that λ is an eigenvalue and $\hat{\boldsymbol{\mu}}$ is a corresponding eigenvector of a symmetric second-order tensor $\mathbf{A}$ if and only if

$$\mathbf{A} \cdot \hat{\boldsymbol{\mu}} = \lambda\hat{\boldsymbol{\mu}} . \qquad (2.3.120)$$

By premultiplying both sides of (2.3.120) by $\mathbf{A}$, one obtains the equality

$$\mathbf{A} \cdot (\mathbf{A} \cdot \hat{\boldsymbol{\mu}}) = \lambda(\mathbf{A} \cdot \hat{\boldsymbol{\mu}}) . \qquad (2.3.121)$$

However, $\mathbf{A} \cdot (\mathbf{A} \cdot \hat{\boldsymbol{\mu}}) = (\mathbf{A} \cdot \mathbf{A}) \cdot \hat{\boldsymbol{\mu}} = \mathbf{A}^2 \cdot \hat{\boldsymbol{\mu}}$ and $\mathbf{A} \cdot \hat{\boldsymbol{\mu}} = \lambda\hat{\boldsymbol{\mu}}$. Therefore, it follows from (2.3.120) that if λ is an eigenvalue and $\hat{\boldsymbol{\mu}}$ is a corresponding eigenvector of a symmetric second-order tensor $\mathbf{A}$, then

$$\mathbf{A}^2 \cdot \hat{\boldsymbol{\mu}} = \lambda^2\hat{\boldsymbol{\mu}} , \qquad (2.3.122)$$

thus showing that λ^2 is an eigenvalue and $\hat{\boldsymbol{\mu}}$ is a corresponding eigenvector of $\mathbf{A}^2$.

If $\mathbf{A}$ is nonsingular, then (2.3.120) can be premultiplied by $\mathbf{A}^{-1}$ to obtain

$$\mathbf{A}^{-1} \cdot (\mathbf{A} \cdot \hat{\boldsymbol{\mu}}) = \lambda(\mathbf{A}^{-1} \cdot \hat{\boldsymbol{\mu}}) . \qquad (2.3.123)$$

However, $\mathbf{A}^{-1} \cdot (\mathbf{A} \cdot \hat{\mu}) = (\mathbf{A}^{-1} \cdot \mathbf{A}) \cdot \hat{\mu} = \hat{\mu}$. Therefore, it follows from (2.3.120) that if λ is an eigenvalue and $\hat{\mu}$ is a corresponding eigenvector of a symmetric, nonsingular second-order tensor $\mathbf{A}$, then

$$\mathbf{A}^{-1} \cdot \hat{\mu} = \frac{1}{\lambda}\hat{\mu} , \tag{2.3.124}$$

thus showing that λ^{-1} is an eigenvalue and $\hat{\mu}$ is a corresponding eigenvector of $\mathbf{A}^{-1}$.

It follows through induction that if λ is an eigenvalue and $\hat{\mu}$ is a corresponding eigenvector of a symmetric, nonsingular second-order tensor $\mathbf{A}$, then λ^n is an eigenvalue and $\hat{\mu}$ is a corresponding eigenvector of $\mathbf{A}^n$ for all integers n. Therefore, if the spectral representation of $\mathbf{A}$ is

$$\mathbf{A} = \sum_{k=1}^{3} \lambda^{(k)} \hat{\mu}^{(k)} \hat{\mu}^{(k)} , \tag{2.3.125}$$

where $\lambda^{(k)}$ are the eigenvalues of $\mathbf{A}$ and $\hat{\mu}^{(k)}$ are the corresponding (mutually orthogonal) eigenvectors, then the spectral representation of $\mathbf{A}^n$ is

$$\mathbf{A}^n = \sum_{k=1}^{3} \left(\lambda^{(k)}\right)^n \hat{\mu}^{(k)} \hat{\mu}^{(k)} , \tag{2.3.126}$$

for all integers n. If $\mathbf{A}$ is positive definite, then the eigenvalues of $\mathbf{A}$ are positive and arbitrary real powers of $\mathbf{A}$ can accordingly be defined as

$$\mathbf{A}^\alpha = \sum_{k=1}^{3} \left(\lambda^{(k)}\right)^\alpha \hat{\mu}^{(k)} \hat{\mu}^{(k)} , \tag{2.3.127}$$

for all real numbers α. The eigenvalues $\left(\lambda^{(k)}\right)^\alpha$ of $\mathbf{A}^\alpha$ are all positive, so $\mathbf{A}^\alpha$ is symmetric, positive definite.

2.4 Tensor Calculus

A *tensor field* is a tensor quantity that varies in space and time. For example, the velocity of a fluid flowing around a fixed obstacle is a first-order tensor (i.e., a vector) field. At an instant in time, the velocities at different points in the flow will, in general, be in different directions and have different magnitudes. If the flow is unsteady, the velocity at a particular point in space will vary with time. The tensor properties discussed previously apply to the value of a tensor field at a particular point and time. For instance, a tensor field may be singular, symmetric, isotropic, and so forth at some points and times and not at others.

A position in space is given relative to a prescribed origin O by the *position vector* from the origin to the point in space. This position vector will usually be denoted by $\mathbf{x}$. The value $\mathbf{T}$ of a tensor field can then be given as a function ψ of the position vector $\mathbf{x}$ and time t,

$$\mathbf{T} = \psi(\mathbf{x}, t) . \tag{2.4.1}$$

In general, the scalar components of a tensor field will also vary with position and time. If $\mathbf{T}$ is the value of a tensor field of order n, then the values $T_{i_1 i_2 \cdots i_n}$ of its scalar components in the orthonormal basis $\{\hat{\mathbf{e}}_i\}$ are given as functions $\psi_{i_1 i_2 \cdots i_n}$ of $\mathbf{x}$ and t:

$$T_{i_1 i_2 \cdots i_n} = \psi_{i_1 i_2 \cdots i_n}(\mathbf{x}, t) . \tag{2.4.2}$$

Since the position vector $\mathbf{x}$ of a point in space is defined relative to the origin O, the functions ψ and $\psi_{i_1 i_2 \cdots i_n}$ are dependent on the choice of origin.

2.4.1   Coordinate Systems

Once an origin and a coordinate system have been defined, the position of a point in space can be characterized by a set of three scalar coordinate variables, (ξ_1, ξ_2, ξ_3) say. In a (rectangular) Cartesian coordinate system (Figure 2.6), the set of coordinate variables is (x_1, x_2, x_3), representing the distances along three straight, mutually orthogonal coordinate axes. In cylindrical and spherical coordinate systems (see Section 2.5), the sets of coordinate variables are (r, θ, z) and (R, θ, ϕ), respectively. The values of a tensor field and its scalar components in the orthonormal basis $\{\hat{\mathbf{e}}_i\}$ can be given as functions of the three scalar coordinate variables and time:

$$\mathbf{T} = \widetilde{\psi}(\xi_1, \xi_2, \xi_3, t) , \quad T_{i_1 i_2 \cdots i_n} = \widetilde{\psi}_{i_1 i_2 \cdots i_n}(\xi_1, \xi_2, \xi_3, t) . \tag{2.4.3}$$

Partial derivatives with respect to the coordinate variables and time are defined in the usual way. For instance, the value at position (ξ_1, ξ_2, ξ_3) and time t of the partial derivative of a tensor field $\mathbf{T}$ with respect to ξ_1 is defined as

$$\frac{\partial \mathbf{T}}{\partial \xi_1} \equiv \lim_{\Delta \xi_1 \to 0} \frac{\widetilde{\psi}(\xi_1 + \Delta \xi_1, \xi_2, \xi_3, t) - \widetilde{\psi}(\xi_1, \xi_2, \xi_3, t)}{\Delta \xi_1} . \tag{2.4.4}$$

The partial derivative of a tensor field of order n is itself a tensor field of order n. Note that unless the unit base vectors are independent of the coordinate variables,

$$\frac{\partial \mathbf{T}}{\partial \xi_q} \neq \frac{\partial T_{i_1 i_2 \cdots i_n}}{\partial \xi_q} \hat{\mathbf{e}}_{i_1} \hat{\mathbf{e}}_{i_2} \cdots \hat{\mathbf{e}}_{i_n} ; \tag{2.4.5}$$

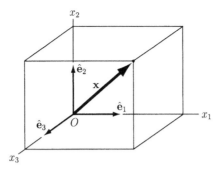

FIGURE 2.6. *Cartesian coordinates and corresponding base vectors.*

the scalar components of the partial derivative of a tensor field are, in general, *not* equal to the partial derivatives of the scalar components of the tensor field.

Cartesian coordinates

In a Cartesian coordinate system, with coordinate variables x_1, x_2, and x_3, the unit base vectors will usually be defined in the positive directions of the coordinate axes (Figure 2.6), so that $\hat{\mathbf{e}}_i$ is a unit vector that points in the direction of increasing x_i. In such an orthonormal basis, the scalar components of the position vector are equal to the coordinate variables,

$$\mathbf{x} = x_i \hat{\mathbf{e}}_i \, . \tag{2.4.6}$$

Unless stated otherwise, it will be assumed from now on that the unit base vectors in a Cartesian coordinate system are aligned with the coordinate axes. These base vectors are constants—they are independent of the co-ordinate variables. Therefore, in a Cartesian coordinate system with the orthonormal basis aligned with the coordinate axes,

$$\frac{\partial \mathbf{T}}{\partial x_q} = \frac{\partial T_{i_1 i_2 \cdots i_n}}{\partial x_q} \hat{\mathbf{e}}_{i_1} \hat{\mathbf{e}}_{i_2} \hat{\mathbf{e}}_{i_n} \, ; \tag{2.4.7}$$

the scalar components of the partial derivative of a tensor field *are* equal to the partial derivatives of the scalar components of the tensor field.

Consider two Cartesian coordinate systems (x_1, x_2, x_3) and (x_1', x_2', x_3'), with corresponding orthonormal bases $\{\hat{\mathbf{e}}_i\}$ and $\{\hat{\mathbf{e}}_i'\}$ defined in the usual way (Figure 2.7). The two coordinate systems will, in general, have different orientations related by the direction cosines $\ell_{ij} = \hat{\mathbf{e}}_i' \cdot \hat{\mathbf{e}}_j$ and different origins O and O'. Let $\mathbf{c} = c_i \hat{\mathbf{e}}_i = c_i' \hat{\mathbf{e}}_i'$ be the vector from the origin O to the origin O'. The position vector of a point in space relative to the origin

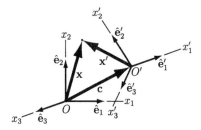

FIGURE 2.7. Two Cartesian coordinate systems.

O is $\mathbf{x} = x_i\hat{\mathbf{e}}_i$ and the position vector of the point in space relative to the origin O' is $\mathbf{x}' = x_i'\hat{\mathbf{e}}_i'$. The two position vectors are related by

$$\mathbf{x}' = \mathbf{x} - \mathbf{c} , \quad \mathbf{x} = \mathbf{x}' + \mathbf{c} , \tag{2.4.8}$$

which can be rewritten in dyadic notation as

$$x_i'\hat{\mathbf{e}}_i' = x_i\hat{\mathbf{e}}_i - c_i'\hat{\mathbf{e}}_i' , \quad x_i\hat{\mathbf{e}}_i = x_i'\hat{\mathbf{e}}_i' + c_i\hat{\mathbf{e}}_i . \tag{2.4.9}$$

Recalling that $\hat{\mathbf{e}}_i = \ell_{ji}\hat{\mathbf{e}}_j'$ and $\hat{\mathbf{e}}_i' = \ell_{ij}\hat{\mathbf{e}}_j$, it follows that the coordinates of a point in space in the two coordinate systems are related by

$$x_i' = \ell_{ij}x_j - c_i' , \quad x_i = \ell_{ji}x_j' + c_i . \tag{2.4.10}$$

Noting the trivial result

$$\frac{\partial x_i}{\partial x_j} = \frac{\partial x_i'}{\partial x_j'} = \delta_{ij} , \tag{2.4.11}$$

it can be seen that, since the direction cosines and the scalar components of $\mathbf{c}$ are constants,

$$\frac{\partial x_i'}{\partial x_j} = \ell_{ij} , \quad \frac{\partial x_i}{\partial x_j'} = \ell_{ji} . \tag{2.4.12}$$

These results will be used when applying the chain rule below.

Partial differentiation of the scalar components of a tensor results in the scalar components of a new tensor. For example, consider the nine numbers given by

$$A_{ij} = \frac{\partial u_i}{\partial x_j} , \tag{2.4.13}$$

where the u_i are the scalar components of a vector field $\mathbf{u}$. Indicial notation equations are understood to be invariant under changes in orthonormal basis, so that (2.4.13) implies that in a new coordinate system (x'_1, x'_2, x'_3) with the corresponding orthonormal basis $\{\hat{e}'_i\}$,

$$A'_{ij} = \frac{\partial u'_i}{\partial x'_j} , \tag{2.4.14}$$

where the u'_i are the scalar components of $\mathbf{u}$ in $\{\hat{e}'_i\}$. Applying the chain rule, this can be rewritten as

$$A'_{ij} = \frac{\partial x_k}{\partial x'_j} \frac{\partial u'_i}{\partial x_k} . \tag{2.4.15}$$

From the vector transformation rule, $u'_i = \ell_{ip} u_p$. Thus, using (2.4.12) and noting that the direction cosines are constants,

$$A'_{ij} = \ell_{jk} \ell_{ip} \frac{\partial u_p}{\partial x_k} = \ell_{ip} \ell_{jk} A_{pk} . \tag{2.4.16}$$

This is the tensor transformation rule for second-order tensors. Thus, $\partial u_i / \partial x_j$ are the scalar components of a second-order tensor.

The *comma convention* is a shorthand notation for partial derivatives with respect to the coordinate variables in a Cartesian coordinate system. This is illustrated by the following examples:

$$\mathbf{T}_{,ij} = \frac{\partial^2 \mathbf{T}}{\partial x_i \partial x_j} , \quad \phi_{,i} = \frac{\partial \phi}{\partial x_i} , \quad A_{ij,kj} = \frac{\partial^2 A_{ij}}{\partial x_k \partial x_j} . \tag{2.4.17}$$

Whenever the comma convention is used in indicial notation equations, it is understood that a Cartesian coordinate system is being used and that the orthonormal basis $\{\hat{e}_i\}$ is aligned with the coordinate axes. The usual range and summation conventions still apply where indices after a comma can be either free indices are part of dummy index pairs; for example,

$$u_i = \phi_{,i} \quad \Longleftrightarrow \quad \begin{cases} u_1 = \dfrac{\partial \phi}{\partial x_1} \\[2mm] u_2 = \dfrac{\partial \phi}{\partial x_2} \\[2mm] u_3 = \dfrac{\partial \phi}{\partial x_3} \end{cases} \tag{2.4.18}$$

and

$$\psi = v_{i,i} \quad \Longleftrightarrow \quad \psi = \frac{\partial v_1}{\partial x_1} + \frac{\partial v_2}{\partial x_2} + \frac{\partial v_3}{\partial x_3} . \tag{2.4.19}$$

2.4.2   Gradient

If $\mathbf{T} = \psi(\mathbf{x}, t)$ is a tensor field of order n, then its *gradient* $\nabla\mathbf{T}$ is a tensor field of order $n + 1$ whose value at position $\mathbf{x}$ and time t is defined by the relation

$$(\nabla\mathbf{T}) \cdot \mathbf{a} = \lim_{\alpha \to 0} \frac{\psi(\mathbf{x} + \alpha\mathbf{a}, t) - \psi(\mathbf{x}, t)}{\alpha} = \frac{d}{d\alpha}\psi(\mathbf{x} + \alpha\mathbf{a}, t)\Big|_{\alpha=0} \, , \quad (2.4.20)$$

where $\mathbf{a}$ is an arbitrary constant vector. If $\mathbf{a}$ is a unit vector, then (2.4.20) is called the *direction derivative* of $\mathbf{T}$ in the direction of $\mathbf{a}$ at $\mathbf{x}$ at t.

Gradient in Cartesian coordinates

Because the unit base vectors in a Cartesian coordinate system are independent of the coordinate variables,

$$\mathbf{x} + \alpha\mathbf{a} = (x_i + \alpha a_i)\hat{\mathbf{e}}_i \, . \quad (2.4.21)$$

Therefore, with the tensor field given by $\mathbf{T} = \psi(\mathbf{x}, t) = \tilde{\psi}(x_1, x_2, x_3, t)$,

$$\begin{aligned}
\frac{d}{d\alpha}\psi(\mathbf{x} + \alpha\mathbf{a}, t)\Big|_{\alpha=0} &= \frac{d}{d\alpha}\tilde{\psi}(x_1 + \alpha a_1, x_2 + \alpha a_2, x_3 + \alpha a_3, t)\Big|_{\alpha=0} \\
&= \frac{\partial\mathbf{T}}{\partial x_i} a_i \\
&= \frac{\partial\mathbf{T}}{\partial x_i}\hat{\mathbf{e}}_i \cdot \mathbf{a} \, .
\end{aligned}$$

$$(2.4.22)$$

It follows from (2.4.20) that, in a Cartesian coordinate system, the gradient of a tensor field $\mathbf{T}$ is[9]

$$\boxed{\nabla\mathbf{T} = \frac{\partial\mathbf{T}}{\partial x_i}\hat{\mathbf{e}}_i = \mathbf{T}_{,i}\,\hat{\mathbf{e}}_i \, .} \quad (2.4.23)$$

Thus, in a Cartesian coordinate system, if ϕ is a scalar, $\mathbf{v}$ is a vector, and $\mathbf{S}$ is a second-order tensor, then

$$\nabla\phi = (\phi)_{,i}\,\hat{\mathbf{e}}_i = \phi_{,i}\,\hat{\mathbf{e}}_i \, , \quad (2.4.24)$$

$$\nabla\mathbf{v} = (v_i\hat{\mathbf{e}}_i)_{,j}\,\hat{\mathbf{e}}_j = v_{i,j}\,\hat{\mathbf{e}}_i\hat{\mathbf{e}}_j \, , \quad (2.4.25)$$

$$\nabla\mathbf{S} = (S_{ij}\hat{\mathbf{e}}_i\hat{\mathbf{e}}_j)_{,k}\,\hat{\mathbf{e}}_k = S_{ij,k}\,\hat{\mathbf{e}}_i\hat{\mathbf{e}}_j\hat{\mathbf{e}}_k \, , \quad (2.4.26)$$

---

[9]Some authors prefer to define the "del operator,"

$$\nabla = \hat{\mathbf{e}}_i\frac{\partial}{\partial x_i} \, ,$$

that is treated like a vector to define the gradient, divergence, and curl of tensors. This is why the operator symbol ∇ is boldface. This leads to slightly different definitions than those used here. In referencing other texts, the reader should be aware of this.

where it has been noted that the invariance of the unit base vectors $\hat{\mathbf{e}}_i$ means that $\hat{\mathbf{e}}_{i,j} = \mathbf{0}$.

Example 2.4
Show that if $R \equiv |\mathbf{x}|$ is the distance from the origin, then $\nabla(R^n) = nR^{n-2}\mathbf{x}$.

Solution. Using a Cartesian coordinate system, the distance from the origin is given in terms of the coordinate variables by $R = (x_i x_i)^{1/2}$. If follows from the expression (2.4.23) for the gradient in a Cartesian coordinate system, and noting $R^n = (x_i x_i)^{n/2}$, that

$$\nabla(R^n) = (R^n)_{,i}\,\hat{\mathbf{e}}_i = \left[(x_i x_i)^{n/2}\right]_{,j}\,\hat{\mathbf{e}}_j \ .$$

Thus, by applying the chain rule (and being careful not to use the same dummy index pair more than once in a term),

$$
\begin{aligned}
\nabla(R^n) &= \frac{n}{2}(x_i x_i)^{\frac{n}{2}-1}(x_k x_k)_{,j}\,\hat{\mathbf{e}}_j \\
&= \frac{n}{2}(x_i x_i)^{\frac{n-2}{2}}(x_{k,j}\,x_k + x_k x_{k,j})\hat{\mathbf{e}}_j \\
&= \frac{n}{2}(x_i x_i)^{\frac{n-2}{2}}(\delta_{kj}x_k + x_k \delta_{kj})\hat{\mathbf{e}}_j \\
&= \frac{n}{2}(x_i x_i)^{\frac{n-2}{2}}(2x_j)\hat{\mathbf{e}}_j \\
&= n(x_i x_i)^{\frac{n-2}{2}}x_j\hat{\mathbf{e}}_j \ .
\end{aligned}
$$

Noting that $(x_i x_i)^{\frac{n-2}{2}} = R^{n-2}$ and $x_j\hat{\mathbf{e}}_j = \mathbf{x}$, it follows that $\nabla(R^n) = nR^{n-2}\mathbf{x}$. Since this tensor equation has been shown to be true in a Cartesian coordinate system, it must be true in all coordinate systems.

2.4.3 Divergence

If $\mathbf{v}$ is a vector field, then its *divergence* $\nabla \cdot \mathbf{v}$ is a scalar field defined as the trace of the gradient of $\mathbf{v}$,

$$\nabla \cdot \mathbf{v} \equiv \mathrm{tr}(\nabla \mathbf{v}) \ . \tag{2.4.27}$$

This trace can be thought of as the contraction of $\nabla \mathbf{v}$. If $\mathbf{T}$ is a tensor field of order $n > 1$, then its divergence $\nabla \cdot \mathbf{T}$ is a tensor field of order $n - 1$ that is a contraction of $\nabla \mathbf{T}$ (a tensor field of order $n + 1$) where the contraction is between ∇ and $\mathbf{T}$. There are n possible such contractions and the choice one makes is a matter of convention. One can fix which of the possible contractions is to be the divergence by defining $\nabla \cdot \mathbf{T}$ such that

$$\mathbf{a} \cdot (\nabla \cdot \mathbf{T}) = \nabla \cdot (\mathbf{a} \cdot \mathbf{T}) \ , \tag{2.4.28}$$

where **a** is an arbitrary constant vector. The divergence has several important physical applications. It will be shown, for instance, that the divergence of the displacement vector is the (approximate) change in volume per unit volume at a point in a body and that the divergence of the stress tensor is related to the static equilibrium of a body.

Divergence in Cartesian coordinates

If $\mathbf{v} = v_i \hat{\mathbf{e}}_i$ is a vector field, then, from (2.4.27) and (2.4.25), its divergence in a Cartesian coordinate system is the scalar field

$$\boldsymbol{\nabla} \cdot \mathbf{v} = v_{i,i} \ . \tag{2.4.29}$$

If $\mathbf{S} = S_{ij}\hat{\mathbf{e}}_i\hat{\mathbf{e}}_j$ is a second-order tensor field, then, in a Cartesian coordinate system,

$$\boldsymbol{\nabla} \cdot (\mathbf{a} \cdot \mathbf{S}) = \boldsymbol{\nabla} \cdot (a_i S_{ij}\hat{\mathbf{e}}_j) = a_i S_{ij,j} = \mathbf{a} \cdot S_{ij,j}\,\hat{\mathbf{e}}_i \ . \tag{2.4.30}$$

It follows from (2.4.28) that

$$\boldsymbol{\nabla} \cdot \mathbf{S} = S_{ij,j}\,\hat{\mathbf{e}}_i \ . \tag{2.4.31}$$

If $\mathbf{R} = R_{ijk}\hat{\mathbf{e}}_i\hat{\mathbf{e}}_j\hat{\mathbf{e}}_k$ is a tensor field of order 3, then, in a Cartesian coordinate system,

$$\boldsymbol{\nabla} \cdot (\mathbf{a} \cdot \mathbf{R}) = \boldsymbol{\nabla} \cdot (a_i R_{ijk}\hat{\mathbf{e}}_j\hat{\mathbf{e}}_k) = a_i R_{ijk,k}\,\hat{\mathbf{e}}_j = \mathbf{a} \cdot R_{ijk,k}\,\hat{\mathbf{e}}_i\hat{\mathbf{e}}_j \ . \tag{2.4.32}$$

It follows from (2.4.28) that

$$\boldsymbol{\nabla} \cdot \mathbf{R} = R_{ijk,k}\,\hat{\mathbf{e}}_i\hat{\mathbf{e}}_j \ . \tag{2.4.33}$$

These results can be summarized by

$$\boxed{\boldsymbol{\nabla} \cdot \mathbf{T} = \mathbf{T}_{,i} \cdot \hat{\mathbf{e}}_i \ .} \tag{2.4.34}$$

2.4.4   Curl

If $\mathbf{v}$ is a vector field, then its *curl* $\boldsymbol{\nabla} \times \mathbf{v}$ is a vector field defined by the relation

$$(\boldsymbol{\nabla} \times \mathbf{v}) \cdot \mathbf{a} = \boldsymbol{\nabla} \cdot (\mathbf{v} \times \mathbf{a}) \ , \tag{2.4.35}$$

where **a** is an arbitrary constant vector. If $\mathbf{T}$ is a tensor field of order $n > 1$, then its curl $\boldsymbol{\nabla} \times \mathbf{T}$ is a tensor field of order n; it will later prove convenient to define this by the relation

$$(\boldsymbol{\nabla} \times \mathbf{T}) \cdot \mathbf{a} = \boldsymbol{\nabla} \times (\mathbf{a} \cdot \mathbf{T}) \ . \tag{2.4.36}$$

The curl of the velocity can be shown to be twice the angular velocity vector and the curl of the second-order strain tensor will be shown to be related to the existence of a corresponding displacement field.

Curl in Cartesian coordinates

If $\mathbf{v} = v_i \hat{\mathbf{e}}_i$ is a vector field, then, in a Cartesian coordinate system,

$$\boldsymbol{\nabla} \cdot (\mathbf{v} \times \mathbf{a}) = \boldsymbol{\nabla} \cdot (e_{ijk} v_j a_k \hat{\mathbf{e}}_i) = e_{ijk} v_{j,i}\, a_k = e_{ijk} v_{j,i}\, \hat{\mathbf{e}}_k \cdot \mathbf{a} . \qquad (2.4.37)$$

It follows from (2.4.35) that

$$\boldsymbol{\nabla} \times \mathbf{v} = e_{ijk} v_{j,i}\, \hat{\mathbf{e}}_k . \qquad (2.4.38)$$

One can introduce the following notation to help remember and apply this result:

$$\boxed{\boldsymbol{\nabla} \times \mathbf{v} = \hat{\mathbf{e}}_i \times \mathbf{v}_{,i} .} \qquad (2.4.39)$$

If $\mathbf{S} = S_{ij} \hat{\mathbf{e}}_i \hat{\mathbf{e}}_j$ is a second-order tensor, then, in a Cartesian coordinate system,

$$\boldsymbol{\nabla} \times (\mathbf{a} \cdot \mathbf{S}) = \boldsymbol{\nabla} \times (a_r S_{rj} \hat{\mathbf{e}}_j) = e_{ijk} a_r S_{rj,i}\, \hat{\mathbf{e}}_k = e_{ijk} S_{rj,i}\, \hat{\mathbf{e}}_k \hat{\mathbf{e}}_r \cdot \mathbf{a} , \qquad (2.4.40)$$

and it follows from (2.4.36) that

$$\boldsymbol{\nabla} \times \mathbf{S} = e_{ijk} S_{rj,i}\, \hat{\mathbf{e}}_k \hat{\mathbf{e}}_r . \qquad (2.4.41)$$

Unfortunately, there is no readily applicable notation such as (2.4.39) to help one remember this result. The result (2.4.41) is *not* given by (2.4.39).

Example 2.5

Show, for an arbitrary scalar field ϕ, that $\boldsymbol{\nabla} \times (\boldsymbol{\nabla} \phi) = \mathbf{0}$.

Solution. If it can be shown that $\boldsymbol{\nabla} \times (\boldsymbol{\nabla} \phi) = \mathbf{0}$ for arbitrary ϕ in a Cartesian coordinate system, then it must be true in all coordinate systems. From (2.4.24), the gradient of a scalar ϕ in a Cartesian coordinate system is the vector $\boldsymbol{\nabla} \phi = \phi_{,j}\, \hat{\mathbf{e}}_j$. Thus, using the notation (2.4.39) for the curl of a vector in a Cartesian coordinate system,

$$\boldsymbol{\nabla} \times (\boldsymbol{\nabla} \phi) = \hat{\mathbf{e}}_i \times (\phi_{,j}\, \hat{\mathbf{e}}_j)_{,i} .$$

In a Cartesian coordinate system, the base vectors $\hat{\mathbf{e}}_j$ are constants, so

$$\boldsymbol{\nabla} \times (\boldsymbol{\nabla} \phi) = (\hat{\mathbf{e}}_i \times \hat{\mathbf{e}}_j) \phi_{,ji} = e_{ijk} \phi_{,ji}\, \hat{\mathbf{e}}_k .$$

However, recall from Example 2.2 that if $A_{ij} = A_{ji}$, then $e_{ijk} A_{ij} = 0$. Therefore, since the order of differentiation in $\phi_{,ij}$ is interchangeable, $e_{ijk} \phi_{,ij} = 0$ for each value of k and $\boldsymbol{\nabla} \times (\boldsymbol{\nabla} \phi) = \mathbf{0}$.

The result from Example 2.5 can be generalized to show that

$$\boldsymbol{\nabla} \times (\boldsymbol{\nabla} \mathbf{T}) = \mathbf{0} , \qquad (2.4.42)$$

for arbitrary tensor fields $\mathbf{T}$ of any order (see Problem 2.25).

2.4.5 Laplacian

If $\mathbf{T}$ is a tensor field of order n, then its *Laplacian* $\nabla^2 \mathbf{T}$ is a tensor field of order n defined as the divergence of the gradient:

$$\nabla^2 \mathbf{T} \equiv \boldsymbol{\nabla} \cdot (\boldsymbol{\nabla} \mathbf{T}) . \tag{2.4.43}$$

The *biharmonic operator* ∇^4 is just notation for the Laplacian applied twice in succession:

$$\nabla^4 \mathbf{T} \equiv \nabla^2 (\nabla^2 \mathbf{T}) . \tag{2.4.44}$$

Laplacian in Cartesian coordinates

If ϕ is a scalar, $\mathbf{v} = v_i \hat{\mathbf{e}}_i$ is a vector, and $\mathbf{S} = S_{ij} \hat{\mathbf{e}}_i \hat{\mathbf{e}}_j$ is a second-order tensor, then, in a Cartesian coordinate system,

$$\nabla^2 \phi = \boldsymbol{\nabla} \cdot (\phi_{,i}\, \hat{\mathbf{e}}_i) = \phi_{,ii} \ , \tag{2.4.45}$$

$$\nabla^2 \mathbf{v} = \boldsymbol{\nabla} \cdot (v_{i,j}\, \hat{\mathbf{e}}_i \hat{\mathbf{e}}_j) = v_{i,jj}\, \hat{\mathbf{e}}_i \ , \tag{2.4.46}$$

$$\nabla^2 \mathbf{S} = \boldsymbol{\nabla} \cdot (S_{ij,k}\, \hat{\mathbf{e}}_i \hat{\mathbf{e}}_j \hat{\mathbf{e}}_k) = S_{ij,kk}\, \hat{\mathbf{e}}_i \hat{\mathbf{e}}_j \ . \tag{2.4.47}$$

If $\mathbf{T}$ is a tensor field of arbitrary order, then, in a Cartesian coordinate system,

$$\boxed{\nabla^2 \mathbf{T} = \mathbf{T}_{,ii} \ , \quad \nabla^4 \mathbf{T} = \mathbf{T}_{,iijj} \ .} \tag{2.4.48}$$

In a Cartesian coordinate system, the scalar components of the Laplacian of a tensor of order n are the Laplacians of the scalar components of the original tensor.

2.4.6 Line, Surface, and Volume Integrals

Consider a continuous curve $\mathcal{C}$ in three-dimensional space that consists of a finite number of smooth arcs joined end to end, along each of which the direction of the tangent line changes continuously. There may be a finite number of "corners," where the direction of the tangent line changes abruptly, corresponding to the points where the smooth arcs are joined. Such a curve is said to be *piecewise smooth* (Courant and Hilbert 1937; Kreyszig 1988; Wylie and Barrett 1995).

Consider two points A and B on the piecewise smooth curve $\mathcal{C}$ and let the portion of the curve with end points A and B be divided into n adjoining segments, each of which is a smooth arc. Label the position vectors, relative to some origin, of the ends of these segments as $\mathbf{x}^{(p)}$, where $p = 0, 1, \ldots, n$, so that the position of A is $\mathbf{x}^A = \mathbf{x}^{(0)}$, the position of B is $\mathbf{x}^B = \mathbf{x}^{(n)}$, and segment p has end points at $\mathbf{x}^{(p-1)}$ and $\mathbf{x}^{(p)}$. Let $\boldsymbol{\xi}^{(p)}$ be the position of an

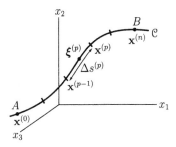

FIGURE 2.8. *Integration along the line* $\mathcal{C}$ *from* A *to* B.

arbitrary point on segment p and let $\Delta s^{(p)}$ be the length of segment p, so that the total length of curve $\mathcal{C}$ from A to B is $\sum_{p=1}^{n} \Delta s^{(p)}$ (Figure 2.8).

If $\mathbf{T} = \boldsymbol{\psi}(\mathbf{x}, t)$ is a tensor field defined throughout some region of space that contains the portion of the curve $\mathcal{C}$ between A and B, then the value at time t of the *line integral of* $\mathbf{T}$ *along* $\mathcal{C}$ *with respect to the arc length* s is defined as

$$\int_A^B \mathbf{T}\, ds \equiv \lim_{n \to \infty} \sum_{p=1}^{n} \boldsymbol{\psi}(\boldsymbol{\xi}^{(p)}, t)\, \Delta s^{(p)}\,, \tag{2.4.49}$$

where it is required that as n becomes infinite, the lengths $\Delta s^{(p)}$ of each segment approach zero. The length $\Delta s^{(p)}$ of segment p is intrinsically positive, so that

$$\int_A^B \mathbf{T}\, ds = \int_B^A \mathbf{T}\, ds\,. \tag{2.4.50}$$

One can also see that if a is a constant,

$$\int_A^B a\mathbf{T}\, ds = a \int_A^B \mathbf{T}\, ds\,, \tag{2.4.51}$$

and if P is a point on the curve $\mathcal{C}$ between A and B,

$$\int_A^P \mathbf{T}\, ds + \int_P^B \mathbf{T}\, ds = \int_A^B \mathbf{T}\, ds\,. \tag{2.4.52}$$

The curve $\mathcal{C}$ is often referred to as the *path of integration*.

A curve can be defined parametrically by giving the position vector of points on the curve as a function of the distance s along the curve from A to the point:

$$\mathbf{x} = \tilde{\mathbf{x}}(s)\,, \quad 0 \leq s \leq L\,, \tag{2.4.53}$$

where L is the length of the curve between A and B. The curve is *closed* if $\tilde{\mathbf{x}}(0) = \tilde{\mathbf{x}}(L)$ so that A and B coincide. It crosses itself if $\tilde{\mathbf{x}}(s)$ has the same value for any other two distinct values of s. A piecewise smooth, closed curve that does not intersect itself will be referred to as a *simple closed curve*. The line integral of $\mathbf{T}$ around a simple closed curve $\mathcal{C}$ with respect to the arc length s,

$$\oint_{\mathcal{C}} \mathbf{T}\, ds , \qquad (2.4.54)$$

has the same definition as (2.4.49), except the end points A and B coincide.

Let $\hat{\boldsymbol{v}}$ be the unit vector that is tangent to the curve $\mathcal{C}$ at $\mathbf{x}$ and points in the direction of integration. The *line integral of* $\mathbf{T}$ *along* $\mathcal{C}$ *with respect to position* $\mathbf{x}$ is defined as

$$\int_A^B \mathbf{T} \cdot d\mathbf{x} \equiv \int_A^B \mathbf{T} \cdot \hat{\boldsymbol{v}}\, ds . \qquad (2.4.55)$$

Since $\hat{\boldsymbol{v}}$ is in the direction of integration,

$$\int_A^B \mathbf{T} \cdot d\mathbf{x} = - \int_B^A \mathbf{T} \cdot d\mathbf{x} . \qquad (2.4.56)$$

If the value of (2.4.55) depends only on the positions of A and B, and not on the path of integration, then the line integral is said to be *path independent*. For instance, consider the case when $\mathbf{T} = \mathbf{I}$, the identity tensor. Then, $\mathbf{I} \cdot d\mathbf{x} = d\mathbf{x}$ and

$$\int_A^B d\mathbf{x} = \mathbf{x}^B - \mathbf{x}^A = \mathbf{r}^{B/A} , \qquad (2.4.57)$$

where $\mathbf{x}^A$ and $\mathbf{x}^B$ are the position vectors of A and B, respectively, and $\mathbf{r}^{B/A}$ is the position vector of B relative to A. Clearly, when the end points of the path of integration coincide (i.e., for integration around a simple closed curve) the value of a path independent integral vanishes.

Recall that if $\hat{\boldsymbol{v}}$ is a unit vector, then $(\boldsymbol{\nabla}\mathbf{T}) \cdot \hat{\boldsymbol{v}}$ is the direction derivative of $\mathbf{T} = \boldsymbol{\psi}(\mathbf{x}, t)$ in the direction of $\hat{\boldsymbol{v}}$ at position $\mathbf{x}$ and time t. Thus, if $\hat{\boldsymbol{v}}$ is the unit tangent vector to the curve $\mathcal{C}$ at $\mathbf{x}$ pointing in the direction of integration, then, with $d\mathbf{x} = \hat{\boldsymbol{v}}\, ds$,

$$\int_A^B (\boldsymbol{\nabla}\mathbf{T}) \cdot d\mathbf{x} = \boldsymbol{\psi}(\mathbf{x}^B, t) - \boldsymbol{\psi}(\mathbf{x}^A, t) , \qquad (2.4.58)$$

where $\mathbf{x}^A$ and $\mathbf{x}^B$ are the position vectors of A and B, respectively. Note that (2.4.58) is also path independent.

A surface is *smooth* if at each point there exists a tangent plane that varies continuously as the point on the surface is varied continuously. A surface composed of a finite number of smooth pieces that are separated by a finite number of piecewise smooth curves is said to be *regular*. Let a regular surface S be subdivided into n smooth pieces. The area of piece p of the surface is $\Delta A^{(p)}$ and $\boldsymbol{\xi}^{(p)}$ is the position of an arbitrary point on piece p. If $\mathbf{T} = \boldsymbol{\psi}(\mathbf{x}, t)$ is a tensor field defined in some region containing S, then the *surface integral of* $\mathbf{T}$ *over the surface* S is

$$\int_S \mathbf{T} \, dA \equiv \lim_{n \to \infty} \sum_{p=1}^{n} \boldsymbol{\psi}(\boldsymbol{\xi}^{(p)}, t) \, \Delta A^{(p)} , \qquad (2.4.59)$$

where it is required that the area and the maximum chord of each piece of the surface approach zero as n becomes infinite.

Volume integrals are similarly defined. Let a region of space $\mathcal{R}$ be subdivided into n pieces. The volume of piece p of the region is $\Delta V^{(p)}$ and $\boldsymbol{\xi}^{(p)}$ is the position of an arbitrary point in piece p. If $\mathbf{T} = \boldsymbol{\psi}(\mathbf{x}, t)$ is a tensor field defined in some region containing $\mathcal{R}$, then the *volume integral of* $\mathbf{T}$ *over the region* $\mathcal{R}$ is

$$\int_{\mathcal{R}} \mathbf{T} \, dV \equiv \lim_{n \to \infty} \sum_{p=1}^{n} \boldsymbol{\psi}(\boldsymbol{\xi}^{(p)}, t) \, \Delta V^{(p)} , \qquad (2.4.60)$$

where it is required that the volume and the maximum chord of each piece of the region approach zero as n becomes infinite.

Divergence theorem

A region in space is *closed* if its boundary is taken to be part of the region and is *bounded* if it can be enclosed within a sphere of sufficiently large radius. Let $\mathcal{R}$ be a closed bounded region in space whose boundary $\partial \mathcal{R}$ is a regular surface (Figure 2.9). The *unit outward normal* $\hat{\mathbf{n}}$ is the unit vector normal to $\partial \mathcal{R}$ at the point $\mathbf{x}$ and pointing out from $\mathcal{R}$.

Let $\mathbf{v}$ be a vector field defined in some region containing $\mathcal{R}$. It is well known from vector analysis (Courant and Hilbert 1937; Kreyszig 1988; Wylie and Barrett 1995) that if $\mathbf{v}$ and $\boldsymbol{\nabla} \cdot \mathbf{v}$ are continuous over $\mathcal{R}$, then

$$\int_{\mathcal{R}} \boldsymbol{\nabla} \cdot \mathbf{v} \, dV = \int_{\partial \mathcal{R}} \mathbf{v} \cdot \hat{\mathbf{n}} \, dA . \qquad (2.4.61)$$

In a Cartesian coordinate system, with $\mathbf{v} = v_i \hat{\mathbf{e}}_i$, this divergence theorem is given in indicial notation by

$$\int_{\mathcal{R}} v_{k,k} \, dV = \int_{\partial \mathcal{R}} v_k \hat{n}_k \, dA . \qquad (2.4.62)$$

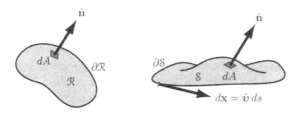

FIGURE 2.9.   Elements of the divergence and Stokes' theorems.

This result is generalized below to tensors of higher order.

Let $\mathbf{T}$ be a tensor field of order $n \geq 1$ defined in some region containing $\mathcal{R}$, with scalar components $T_{i_1 i_2 \cdots i_{n-1} i_n}$ in a Cartesian coordinate system. If $\mathbf{T}$ and $\nabla \cdot \mathbf{T}$ are continuous over $\mathcal{R}$, then the result (2.4.62) can be applied to each of the 3^{n-1} possible combinations of values for the first $n-1$ indices of $T_{i_1 i_2 \cdots i_{n-1} i_n}$, so that, in a Cartesian coordinate system,

$$\int_{\mathcal{R}} T_{i_1 i_2 \cdots i_{n-1} i_n , i_n} \, dV = \int_{\partial \mathcal{R}} T_{i_1 i_2 \cdots i_{n-1} i_n} \hat{n}_{i_n} \, dA \, . \tag{2.4.63}$$

It follows that, since $\nabla \cdot \mathbf{T} = T_{i_1 i_2 \cdots i_{n-1} i_n , i_n} \, \hat{\mathbf{e}}_{i_1} \hat{\mathbf{e}}_{i_2} \cdots \hat{\mathbf{e}}_{i_{n-1}}$ and $\mathbf{T} \cdot \hat{\mathbf{n}} = T_{i_1 i_2 \cdots i_{n-1} i_n} \hat{n}_{i_n} \hat{\mathbf{e}}_{i_1} \hat{\mathbf{e}}_{i_2} \cdots \hat{\mathbf{e}}_{i_{n-1}}$, the divergence theorem for higher-order tensors can be written in tensor notation as

$$\int_{\mathcal{R}} \nabla \cdot \mathbf{T} \, dV = \int_{\partial \mathcal{R}} \mathbf{T} \cdot \hat{\mathbf{n}} \, dA \, , \tag{2.4.64}$$

where this result is applicable to all coordinate systems.

Stokes's theorem

Let $\mathcal{S}$ be a regular surface in space that is bounded by a simple closed curve $\partial \mathcal{S}$. Let $\mathbf{v}$ be a vector field defined in some region containing $\mathcal{S}$. It is well known that if $\mathbf{v}$ and $\nabla \times \mathbf{v}$ are continuous over $\mathcal{S}$, then

$$\int_{\mathcal{S}} \hat{\mathbf{n}} \cdot (\nabla \times \mathbf{v}) \, dA = \oint_{\partial \mathcal{S}} \mathbf{v} \cdot d\mathbf{x} \, , \tag{2.4.65}$$

where $\hat{\mathbf{n}}$ is a unit vector normal to the surface $\mathcal{S}$ at point $\mathbf{x}$. The direction $d\mathbf{x} = \hat{\upsilon} \, ds$ of integration around the simple closed boundary $\partial \mathcal{S}$ is determined by $\hat{\mathbf{n}}$, as shown in Figure 2.9. In a Cartesian coordinate system, with $\mathbf{v} = v_i\hat{\mathbf{e}}_i$, this Stokes's theorem is given in indicial notation by

$$\int_S e_{ijk} v_{j,i} \hat{n}_k \, dA = \oint_{\partial S} v_i \, dx_i \,, \qquad (2.4.66)$$

where $dx_i = \hat{v}_i \, ds$. This result can be generalized to tensors of order 2.

Let $\mathbf{S}$ be a tensor field of order 2 defined in some region containing S, with scalar components S_{rj} in a Cartesian coordinate system. If $\mathbf{S}$ and $\nabla \times \mathbf{S}$ are continuous over S, then the result (2.4.66) can be applied to each of the three possible combinations of values for the index r of S_{rj}, so that, in a Cartesian coordinate system,

$$\boxed{\int_S e_{ijk} S_{rj,i} \hat{n}_k \, dA = \oint_{\partial S} S_{ri} \, dx_i \,.} \qquad (2.4.67)$$

Since $\hat{\mathbf{n}} \cdot (\nabla \times \mathbf{S}) = e_{ijk} S_{rj,i} \hat{n}_k \hat{\mathbf{e}}_r$ and $\mathbf{S} \cdot d\mathbf{x} = S_{ri} \hat{\mathbf{e}}_r \, dx_i$, it follows that

$$\boxed{\int_S \hat{\mathbf{n}} \cdot (\nabla \times \mathbf{S}) \, dA = \oint_{\partial S} \mathbf{S} \cdot d\mathbf{x} \,,} \qquad (2.4.68)$$

where this result is applicable for second-order tensors in all coordinate systems.

Green-Riemann theorem

Let S be a closed bounded region in the $x_1 x_2$-plane of a Cartesian coordinate system, whose boundary ∂S is a simple closed curve. Let $P = f(x_1, x_2, t)$ and $Q = g(x_1, x_2, t)$ be continuous scalar fields defined in some region that contains S, that are continuously differentiable over S. Then,

$$\int_S (Q_{,1} - P_{,2}) \, dA = \oint_{\partial S} (P \, dx_1 + Q \, dx_2) \,, \qquad (2.4.69)$$

the integration being taken around ∂S such that S is on the left as one advances in the direction of integration. This result can be derived from either the divergence theorem or Stokes's theorem.

2.5 Cylindrical and Spherical Coordinates

Cylindrical and spherical coordinate systems are defined by orthogonal curvilinear coordinate axes (lines along which two of the three coordinate variables are constant)—the coordinate axes are not straight lines, but are everywhere mutually orthogonal, nevertheless. One usually forms an orthonormal basis in these coordinate systems by defining the base vectors at a point in space to be unit vectors that are tangent to the coordinate axes. The directions of base vectors defined in this manner will vary with position—they will depend on the coordinate variables. All of the indicial notation relations derived up to the discussion of tensor calculus still apply so long as one uses an orthonormal basis. However, since the base vectors aligned with the coordinate axes in a curvilinear coordinate system are functions of the coordinate variables, the expressions for the gradient, divergence, curl, and Laplacian in terms of the scalar components of a tensor will not be the same as those in Cartesian coordinates. The tensor definitions of these quantities were made without reference to any particular coordinate system and so can be used to derive their scalar component forms in cylindrical and spherical coordinates. See the Appendix for a discussion of general curvilinear coordinate systems, with cylindrical and spherical coordinate systems given as special cases.

2.5.1   Cylindrical Coordinates

A cylindrical coordinate system can be defined with respect to a given Cartesian coordinate system so that the cylindrical coordinate variables (r, θ, z) are related to Cartesian coordinate variables (x_1, x_2, x_3) by the following relations (Figure 2.10):

$$r = \sqrt{x_1^2 + x_2^2}\,, \quad \theta = \tan^{-1}\frac{x_2}{x_1}\,, \quad z = x_3\,, \tag{2.5.1}$$

and

$$x_1 = r\cos\theta\,, \quad x_2 = r\sin\theta\,, \quad x_3 = z\,. \tag{2.5.2}$$

The z-axis is coincident with the x_3-axis and z is the position of the orthogonal projection of a point in space onto the z-axis. The coordinate variable r is the perpendicular distance of the point in space from the z-axis and θ is the angle formed by the plane containing the z-axis and the point in space, with the $x_1 x_3$-plane as shown in Figure 2.10. Points in space with the same value of r define a circular cylindrical surface with radius r and centroidal axis along the z-axis. The coordinate variable θ is undefined for points in space along the z-axis.

An orthonormal basis can be defined in a cylindrical coordinate system with base vectors at a point in space that are tangent to the curvilinear coordinate axes (i.e., lines along which two of the three coordinate variables are constant). Thus, the base vector $\hat{\mathbf{e}}_r$ is a unit vector that points in the direction of increasing r when θ and z are held constant; likewise for $\hat{\mathbf{e}}_\theta$

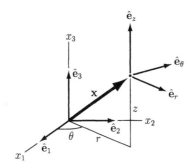

FIGURE 2.10. *Cylindrical coordinates and base vectors.*

and $\hat{\mathbf{e}}_z$ (see Figure 2.10). These base vectors are related to the base vectors aligned with the corresponding Cartesian coordinate system by

$$\hat{\mathbf{e}}_r = \cos\theta\hat{\mathbf{e}}_1 + \sin\theta\hat{\mathbf{e}}_2 \; ,$$
$$\hat{\mathbf{e}}_\theta = -\sin\theta\hat{\mathbf{e}}_1 + \cos\theta\hat{\mathbf{e}}_2 \; , \qquad (2.5.3)$$
$$\hat{\mathbf{e}}_z = \hat{\mathbf{e}}_3 \; .$$

Note that, whereas $\hat{\mathbf{e}}_1$, $\hat{\mathbf{e}}_2$, and $\hat{\mathbf{e}}_3$ point in the same directions regardless of the point in space, $\hat{\mathbf{e}}_r$ and $\hat{\mathbf{e}}_\theta$ are functions of the coordinate variable θ. Also, $\hat{\mathbf{e}}_r$ and $\hat{\mathbf{e}}_\theta$ are undefined at points in space on the z-axis.

Let the orthonormal basis defined by $\hat{\mathbf{e}}_r$, $\hat{\mathbf{e}}_\theta$, and $\hat{\mathbf{e}}_z$ be denoted by $\{\hat{\mathbf{e}}_i'\}$, so that

$$\{\hat{\mathbf{e}}_i'\} = \begin{bmatrix} \hat{\mathbf{e}}_r \\ \hat{\mathbf{e}}_\theta \\ \hat{\mathbf{e}}_z \end{bmatrix} \; . \qquad (2.5.4)$$

The scalar components of a vector $\mathbf{v}$ and a second-order tensor $\mathbf{S}$ in $\{\hat{\mathbf{e}}_i'\}$ are denoted by

$$[\mathbf{v}]' = \begin{bmatrix} v_r \\ v_\theta \\ v_z \end{bmatrix} \; , \quad [\mathbf{S}]' = \begin{bmatrix} S_{rr} & S_{r\theta} & S_{rz} \\ S_{\theta r} & S_{\theta\theta} & S_{\theta z} \\ S_{zr} & S_{z\theta} & S_{zz} \end{bmatrix} \; , \qquad (2.5.5)$$

so that in dyadic notation,

$$\mathbf{v} = v_r\hat{\mathbf{e}}_r + v_\theta\hat{\mathbf{e}}_\theta + v_z\hat{\mathbf{e}}_z \; , \qquad (2.5.6)$$
$$\mathbf{S} = S_{rr}\hat{\mathbf{e}}_r\hat{\mathbf{e}}_r + S_{r\theta}\hat{\mathbf{e}}_r\hat{\mathbf{e}}_\theta + S_{rz}\hat{\mathbf{e}}_r\hat{\mathbf{e}}_z + S_{\theta r}\hat{\mathbf{e}}_\theta\hat{\mathbf{e}}_r + \cdots \; . \qquad (2.5.7)$$

The notation for scalar components of higher-order tensors in $\{\hat{\mathbf{e}}_i'\}$ is similar. The position vector from the origin to a point in space with coordinates (r, θ, z) is

$$\mathbf{x} = r\hat{\mathbf{e}}_r + z\hat{\mathbf{e}}_z \; , \qquad (2.5.8)$$

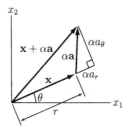

FIGURE 2.11.   Orthogonal projections of the vectors $\mathbf{x}$ and $\mathbf{x} + \alpha\mathbf{a}$ onto the x_1x_2-plane.

where $\hat{\mathbf{e}}_r$ and $\hat{\mathbf{e}}_z$ are the base vectors at the point in space.

The indicial notation and matrix notation conventions and relations up until the discussion of tensor calculus in Section 2.4 still apply so long as it is understood that for scalar components of tensors in the cylindrical coordinate orthonormal basis $\{\hat{\mathbf{e}}_i'\}$, the indicial values 1, 2, and 3 represent r, θ, and z, respectively. For example, the tensor equation $\mathbf{u} = \mathbf{T} \cdot \mathbf{v}$, where $\mathbf{u}$ and $\mathbf{v}$ are vectors and $\mathbf{T}$ is a second-order tensor, implies that in any orthonormal basis $u_i = T_{ij}v_j$ so that

$$u_r = T_{rr}v_r + T_{r\theta}v_\theta + T_{rz}v_z , \qquad (2.5.9)$$

$$u_\theta = T_{\theta r}v_r + T_{\theta\theta}v_\theta + T_{\theta z}v_z , \qquad (2.5.10)$$

$$u_z = T_{zr}v_r + T_{z\theta}v_\theta + T_{zz}v_z . \qquad (2.5.11)$$

For transformation of scalar components from the Cartesian orthonormal basis $\{\hat{\mathbf{e}}_i\}$ to the cylindrical coordinate orthonormal basis $\{\hat{\mathbf{e}}_i'\}$, the matrix of direction cosines $\ell_{ij} \equiv \hat{\mathbf{e}}_i' \cdot \hat{\mathbf{e}}_j$ is

$$[L] = \begin{bmatrix} \cos\theta & \sin\theta & 0 \\ -\sin\theta & \cos\theta & 0 \\ 0 & 0 & 1 \end{bmatrix} \qquad (2.5.12)$$

and the tensor transformation rule is still given by (2.3.30). For a vector $\mathbf{v}$ and a second-order tensor $\mathbf{S}$, the transformations are given in matrix notation by $[\mathbf{v}]' = [L][\mathbf{v}]$ and $[\mathbf{S}]' = [L][\mathbf{S}][L]^{\mathsf{T}}$.

Gradient in cylindrical coordinates

The tensor definition (2.4.20) of the gradient of a tensor is used to derive the specific expression for the gradient in a cylindrical coordinate system. The orthogonal projections of $\mathbf{x}$ and $\mathbf{x} + \alpha\mathbf{a}$ onto the x_1x_2-plane are shown in Figure 2.11. It can be seen that if the point in space with position vector $\mathbf{x}$ has cylindrical coordinates (r, θ, z), then the point in space with position vector $\mathbf{x} + \alpha\mathbf{a}$ has cylindrical coordinates

$$\left(\sqrt{(r+\alpha a_r)^2 + (\alpha a_\theta)^2}\,,\; \theta + \tan^{-1}\frac{\alpha a_\theta}{r+\alpha a_r}\,,\; z + \alpha a_z\right)\,, \qquad (2.5.13)$$

where a_r, a_θ, and a_z are the scalar components of $\mathbf{a}$. If $|\alpha\mathbf{a}| \ll |\mathbf{x}|$, the cylindrical coordinates of the point in space with position vector $\mathbf{x} + \alpha\mathbf{a}$ are approximately[10]

$$\left(r+\alpha a_r,\; \theta + \frac{\alpha a_\theta}{r},\; z + \alpha a_z\right)\,, \qquad (2.5.14)$$

which approach the exact coordinates in the limit as $\alpha \to 0$. It follows that for a tensor field given by $\mathbf{T} = \psi(\mathbf{x},t) = \tilde{\psi}(r,\theta,z,t)$,

$$\frac{d}{d\alpha}\psi(\mathbf{x}+\alpha\mathbf{a},t)\bigg|_{\alpha=0} = \frac{d}{d\alpha}\tilde{\psi}\left(r+\alpha a_r, \theta + \frac{\alpha a_\theta}{r}, z + \alpha a_z, t\right)\bigg|_{\alpha=0}$$

$$= \frac{\partial \mathbf{T}}{\partial r}a_r + \frac{1}{r}\frac{\partial \mathbf{T}}{\partial \theta}a_\theta + \frac{\partial \mathbf{T}}{\partial z}a_z$$

$$= \left(\frac{\partial \mathbf{T}}{\partial r}\hat{\mathbf{e}}_r + \frac{1}{r}\frac{\partial \mathbf{T}}{\partial \theta}\hat{\mathbf{e}}_\theta + \frac{\partial \mathbf{T}}{\partial z}\hat{\mathbf{e}}_z\right)\cdot\mathbf{a}\,, \qquad (2.5.15)$$

so that, from the definition of the gradient (2.4.20),

$$\boxed{\nabla \mathbf{T} = \frac{\partial \mathbf{T}}{\partial r}\hat{\mathbf{e}}_r + \frac{1}{r}\frac{\partial \mathbf{T}}{\partial \theta}\hat{\mathbf{e}}_\theta + \frac{\partial \mathbf{T}}{\partial z}\hat{\mathbf{e}}_z\,.} \qquad (2.5.16)$$

This result is used below to derive expressions for the gradient, divergence, curl, and Laplacian of tensors of various orders (as required in subsequent chapters) in terms of their scalar components.

Gradient of scalars and vectors

For a scalar field f, it follows immediately from (2.5.16) that

$$\boxed{\nabla f = \frac{\partial f}{\partial r}\hat{\mathbf{e}}_r + \frac{1}{r}\frac{\partial f}{\partial \theta}\hat{\mathbf{e}}_\theta + \frac{\partial f}{\partial z}\hat{\mathbf{e}}_z\,.} \qquad (2.5.17)$$

However, for a vector field $\mathbf{u} = u_r\hat{\mathbf{e}}_r + u_\theta\hat{\mathbf{e}}_\theta + u_z\hat{\mathbf{e}}_z$, one must take into account that the cylindrical coordinate base vectors are themselves functions of the cylindrical coordinate variables. From (2.5.3) it is seen that

$$\frac{\partial \hat{\mathbf{e}}_r}{\partial r} = 0\,, \quad \frac{\partial \hat{\mathbf{e}}_r}{\partial \theta} = \hat{\mathbf{e}}_\theta\,, \quad \frac{\partial \hat{\mathbf{e}}_r}{\partial z} = 0\,,$$

$$\frac{\partial \hat{\mathbf{e}}_\theta}{\partial r} = 0\,, \quad \frac{\partial \hat{\mathbf{e}}_\theta}{\partial \theta} = -\hat{\mathbf{e}}_r\,, \quad \frac{\partial \hat{\mathbf{e}}_\theta}{\partial z} = 0\,, \qquad (2.5.18)$$

$$\frac{\partial \hat{\mathbf{e}}_z}{\partial r} = 0\,, \quad \frac{\partial \hat{\mathbf{e}}_z}{\partial \theta} = 0\,, \quad \frac{\partial \hat{\mathbf{e}}_z}{\partial z} = 0\,.$$

---

[10] Actually, this approach to deriving (2.5.16) is dubious on at least a couple of fronts, though the end result is correct. See the Appendix for a proper derivation.

Thus, by substituting $\mathbf{u} = u_r\hat{\mathbf{e}}_r + u_\theta\hat{\mathbf{e}}_\theta + u_z\hat{\mathbf{e}}_z$ into (2.5.16) and using the chain rule, one finds that

$$
\begin{aligned}
\boldsymbol{\nabla}\mathbf{u} = {} & \frac{\partial u_r}{\partial r}\hat{\mathbf{e}}_r\hat{\mathbf{e}}_r + \frac{1}{r}\left(\frac{\partial u_r}{\partial\theta} - u_\theta\right)\hat{\mathbf{e}}_r\hat{\mathbf{e}}_\theta + \frac{\partial u_r}{\partial z}\hat{\mathbf{e}}_r\hat{\mathbf{e}}_z \\
& + \frac{\partial u_\theta}{\partial r}\hat{\mathbf{e}}_\theta\hat{\mathbf{e}}_r + \frac{1}{r}\left(\frac{\partial u_\theta}{\partial\theta} + u_r\right)\hat{\mathbf{e}}_\theta\hat{\mathbf{e}}_\theta + \frac{\partial u_\theta}{\partial z}\hat{\mathbf{e}}_\theta\hat{\mathbf{e}}_z \\
& + \frac{\partial u_z}{\partial r}\hat{\mathbf{e}}_z\hat{\mathbf{e}}_r + \frac{1}{r}\frac{\partial u_z}{\partial\theta}\hat{\mathbf{e}}_z\hat{\mathbf{e}}_\theta + \frac{\partial u_z}{\partial z}\hat{\mathbf{e}}_z\hat{\mathbf{e}}_z \;.
\end{aligned}
\tag{2.5.19}
$$

This result can now be used to derive the results for the divergence.

Divergence of vectors and second-order tensors

Recall that the divergence of a vector is defined to be the contraction of its gradient. Thus, from (2.5.19), the divergence of a vector field $\mathbf{u}$ is

$$
\boldsymbol{\nabla}\cdot\mathbf{u} = \frac{\partial u_r}{\partial r} + \frac{1}{r}\left(\frac{\partial u_\theta}{\partial\theta} + u_r\right) + \frac{\partial u_z}{\partial z}\;.
\tag{2.5.20}
$$

The divergence of higher-order tensors is defined by (2.4.28). For a second-order tensor field $\mathbf{S}$ and an arbitrary constant vector $\mathbf{a}$,

$$
\begin{aligned}
\mathbf{a}\cdot\mathbf{S} = {} & (a_r S_{rr} + a_\theta S_{\theta r} + a_z S_{zr})\hat{\mathbf{e}}_r \\
& + (a_r S_{r\theta} + a_\theta S_{\theta\theta} + a_z S_{z\theta})\hat{\mathbf{e}}_\theta \\
& + (a_r S_{rz} + a_\theta S_{\theta z} + a_z S_{zz})\hat{\mathbf{e}}_z \;.
\end{aligned}
\tag{2.5.21}
$$

Since $\mathbf{a}\cdot\mathbf{S}$ is a vector, the expression $\boldsymbol{\nabla}\cdot(\mathbf{a}\cdot\mathbf{S})$ on the right-hand side of (2.4.28) is given by (2.5.20). However, one must take care with the constant vector $\mathbf{a}$—since the base vectors are not constant, neither are the scalar components of $\mathbf{a}$. Specifically, using (2.5.18), one finds that

$$
\frac{\partial\mathbf{a}}{\partial\theta} = \left(\frac{\partial a_r}{\partial\theta} - a_\theta\right)\hat{\mathbf{e}}_r + \left(\frac{\partial a_\theta}{\partial\theta} + a_r\right)\hat{\mathbf{e}}_\theta + \frac{\partial a_z}{\partial\theta}\hat{\mathbf{e}}_z = 0\;.
\tag{2.5.22}
$$

Therefore,

$$
\begin{aligned}
&\frac{\partial a_r}{\partial r} = 0\;, && \frac{\partial a_r}{\partial\theta} = a_\theta\;, && \frac{\partial a_r}{\partial z} = 0\;, \\
&\frac{\partial a_\theta}{\partial r} = 0\;, && \frac{\partial a_\theta}{\partial\theta} = -a_r\;, && \frac{\partial a_\theta}{\partial z} = 0\;, \\
&\frac{\partial a_z}{\partial r} = 0\;, && \frac{\partial a_z}{\partial\theta} = 0\;, && \frac{\partial a_z}{\partial z} = 0\;.
\end{aligned}
\tag{2.5.23}
$$

With this in mind, it follows from (2.5.20) and (2.5.21) that

$$\nabla \cdot (\mathbf{a} \cdot \mathbf{S}) = a_r \left[\frac{\partial S_{rr}}{\partial r} + \frac{1}{r}\frac{\partial S_{r\theta}}{\partial \theta} + \frac{\partial S_{rz}}{\partial z} + \frac{1}{r}(S_{rr} - S_{\theta\theta}) \right]$$

$$+ a_\theta \left[\frac{\partial S_{\theta r}}{\partial r} + \frac{1}{r}\frac{\partial S_{\theta\theta}}{\partial \theta} + \frac{\partial S_{\theta z}}{\partial z} + \frac{1}{r}(S_{r\theta} + S_{\theta r}) \right]$$

$$+ a_z \left[\frac{\partial S_{zr}}{\partial r} + \frac{1}{r}\frac{\partial S_{z\theta}}{\partial \theta} + \frac{\partial S_{zz}}{\partial z} + \frac{1}{r}S_{zr} \right] , \qquad (2.5.24)$$

and therefore, from the definition (2.4.28), that

$$\nabla \cdot \mathbf{S} = \left[\frac{\partial S_{rr}}{\partial r} + \frac{1}{r}\frac{\partial S_{r\theta}}{\partial \theta} + \frac{\partial S_{rz}}{\partial z} + \frac{1}{r}(S_{rr} - S_{\theta\theta}) \right] \hat{\mathbf{e}}_r$$

$$+ \left[\frac{\partial S_{\theta r}}{\partial r} + \frac{1}{r}\frac{\partial S_{\theta\theta}}{\partial \theta} + \frac{\partial S_{\theta z}}{\partial z} + \frac{1}{r}(S_{r\theta} + S_{\theta r}) \right] \hat{\mathbf{e}}_\theta$$

$$+ \left[\frac{\partial S_{zr}}{\partial r} + \frac{1}{r}\frac{\partial S_{z\theta}}{\partial \theta} + \frac{\partial S_{zz}}{\partial z} + \frac{1}{r}S_{zr} \right] \hat{\mathbf{e}}_z . \qquad (2.5.25)$$

These results can now be used to examine the curl.

Curl of vectors

The curl of a vector field is defined by (2.4.35). If $\mathbf{u}$ is a vector field and $\mathbf{a}$ is an arbitrary constant vector, then

$$\mathbf{u} \times \mathbf{a} = (u_\theta a_z - u_z a_\theta)\hat{\mathbf{e}}_r + (u_z a_r - u_r a_z)\hat{\mathbf{e}}_\theta + (u_r a_\theta - u_\theta a_r)\hat{\mathbf{e}}_z . \quad (2.5.26)$$

Since $\mathbf{u} \times \mathbf{a}$ is a vector, the expression $\nabla \cdot (\mathbf{u} \times \mathbf{a})$ on the right-hand side of the definition (2.4.35) is given by (2.5.20). Using (2.5.23), it follows that

$$\nabla \cdot (\mathbf{u} \times \mathbf{a}) = \left(\frac{1}{r}\frac{\partial u_z}{\partial \theta} - \frac{\partial u_\theta}{\partial z} \right) a_r + \left(\frac{\partial u_r}{\partial z} - \frac{\partial u_z}{\partial r} \right) a_\theta$$

$$+ \left(\frac{\partial u_\theta}{\partial r} + \frac{u_\theta}{r} - \frac{1}{r}\frac{\partial u_r}{\partial \theta} \right) a_z , \qquad (2.5.27)$$

and therefore, from the definition (2.4.35), that

$$\nabla \times \mathbf{u} = \left(\frac{1}{r}\frac{\partial u_z}{\partial \theta} - \frac{\partial u_\theta}{\partial z} \right) \hat{\mathbf{e}}_r + \left(\frac{\partial u_r}{\partial z} - \frac{\partial u_z}{\partial r} \right) \hat{\mathbf{e}}_\theta$$

$$+ \left(\frac{\partial u_\theta}{\partial r} + \frac{u_\theta}{r} - \frac{1}{r}\frac{\partial u_r}{\partial \theta} \right) \hat{\mathbf{e}}_z . \qquad (2.5.28)$$

Laplacian of scalars and vectors

Recall that the Laplacian of a tensor field is defined to be the divergence of its gradient. Using the above results, it follows that for a scalar field f,

$$\nabla^2 f = \frac{\partial^2 f}{\partial r^2} + \frac{1}{r}\frac{\partial f}{\partial r} + \frac{1}{r^2}\frac{\partial^2 f}{\partial \theta^2} + \frac{\partial^2 f}{\partial z^2} , \qquad (2.5.29)$$

and for a vector field $\mathbf{u}$,

$$\nabla^2 \mathbf{u} = \left(\nabla^2 u_r - \frac{2}{r^2}\frac{\partial u_\theta}{\partial \theta} - \frac{1}{r^2} u_r\right)\hat{\mathbf{e}}_r$$

$$+ \left(\nabla^2 u_\theta + \frac{2}{r^2}\frac{\partial u_r}{\partial \theta} - \frac{1}{r^2} u_\theta\right)\hat{\mathbf{e}}_\theta + \left(\nabla^2 u_z\right)\hat{\mathbf{e}}_z . \qquad (2.5.30)$$

2.5.2 Spherical Coordinates

A spherical coordinate system can be defined with respect to a given Cartesian coordinate system so that the spherical coordinate variables (R, θ, ϕ) are related to Cartesian coordinate variables (x_1, x_2, x_3) by the following relations (Figure 2.12):

$$R = \sqrt{x_1^2 + x_2^2 + x_3^2} , \quad \theta = \tan^{-1}\frac{\sqrt{x_1^2 + x_2^2}}{x_3} , \quad \phi = \tan^{-1}\frac{x_2}{x_1} , \quad (2.5.31)$$

and

$$x_1 = R\sin\theta\cos\phi , \quad x_2 = R\sin\theta\sin\phi , \quad x_3 = R\cos\theta . \qquad (2.5.32)$$

The coordinate variable R is the distance of a point in space from the origin, the variable θ is the angle formed by the position vector and the x_3-axis, and ϕ is the angle formed by the plane containing the point in space and the x_3-axis, with the $x_1 x_3$-plane as shown in Figure 2.12. Points in space with the same value of R define a spherical surface with radius R and centroid at the origin. The coordinate variables θ and ϕ are undefined for the point in space at the origin.

An orthonormal basis can be defined in a spherical coordinate system with base vectors at a point in space that are tangent to the curvilinear coordinate axes (i.e., lines along which two of the three coordinate variables are constant). Thus, the base vector $\hat{\mathbf{e}}_R$ is a unit vector that points in the direction of increasing R when θ and ϕ are held constant; likewise for $\hat{\mathbf{e}}_\theta$ and $\hat{\mathbf{e}}_\phi$ (see Figure 2.12). These base vectors are related to the base vectors

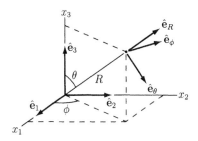

FIGURE 2.12. *Spherical coordinates and base vectors.*

aligned with the corresponding Cartesian coordinate system by

$$\hat{e}_R = \sin\theta\cos\phi\hat{e}_1 + \sin\theta\sin\phi\hat{e}_2 + \cos\theta\hat{e}_3 \,,$$
$$\hat{e}_\theta = \cos\theta\cos\phi\hat{e}_1 + \cos\theta\sin\phi\hat{e}_2 - \sin\theta\hat{e}_3 \,, \qquad (2.5.33)$$
$$\hat{e}_\phi = -\sin\phi\hat{e}_1 + \cos\phi\hat{e}_2 \,.$$

Note that, whereas $\hat{e}_1$, $\hat{e}_2$, and $\hat{e}_3$ point in the same directions regardless of the point in space, $\hat{e}_R$, $\hat{e}_\theta$, and $\hat{e}_\phi$ are functions of the coordinate variables θ and ϕ. Also, $\hat{e}_R$, $\hat{e}_\theta$, and $\hat{e}_\phi$ are undefined at the origin.

Let the orthonormal basis defined by $\hat{e}_R$, $\hat{e}_\theta$, and $\hat{e}_\phi$ be denoted by $\{\hat{e}'_i\}$, so that

$$\{\hat{e}'_i\} = \begin{bmatrix} \hat{e}_R \\ \hat{e}_\theta \\ \hat{e}_\phi \end{bmatrix} . \qquad (2.5.34)$$

The scalar components of a vector $\mathbf{v}$ and a second-order tensor $\mathbf{S}$ in $\{\hat{e}'_i\}$ are denoted by

$$[\mathbf{v}]' = \begin{bmatrix} v_R \\ v_\theta \\ v_\phi \end{bmatrix} \,, \quad [\mathbf{S}]' = \begin{bmatrix} S_{RR} & S_{R\theta} & S_{R\phi} \\ S_{\theta R} & S_{\theta\theta} & S_{\theta\phi} \\ S_{\phi R} & S_{\phi\theta} & S_{\phi\phi} \end{bmatrix} \,, \qquad (2.5.35)$$

so that in dyadic notation,

$$\mathbf{v} = v_R\hat{e}_R + v_\theta\hat{e}_\theta + v_\phi\hat{e}_\phi \,, \qquad (2.5.36)$$
$$\mathbf{S} = S_{RR}\hat{e}_R\hat{e}_R + S_{R\theta}\hat{e}_R\hat{e}_\theta + S_{R\phi}\hat{e}_R\hat{e}_\phi + S_{\theta R}\hat{e}_\theta\hat{e}_R + \cdots . \qquad (2.5.37)$$

The notation for scalar components of higher-order tensors in $\{\hat{e}'_i\}$ is similar. The position vector from the origin to a point in space with coordinates (R,θ,ϕ) is

$$\mathbf{x} = R\hat{e}_R \,, \qquad (2.5.38)$$

where $\hat{e}_R$ is the base vector at the point in space.

The indicial notation and matrix notation conventions and relations up until the discussion of tensor calculus in Section 2.4 still apply so long as it is understood that for scalar components of tensors in the spherical coordinate orthonormal basis $\{\hat{e}'_i\}$, the indicial values 1, 2, and 3 represent R, θ, and ϕ, respectively. For example, the tensor equation $\mathbf{u} = \mathbf{T} \cdot \mathbf{v}$, where $\mathbf{u}$ and $\mathbf{v}$ are vectors and $\mathbf{T}$ is a second-order tensor, implies that in any orthonormal basis, $u_i = T_{ij}v_j$ so that

$$u_R = T_{RR}v_R + T_{R\theta}v_\theta + T_{R\phi}v_\phi \,, \tag{2.5.39}$$

$$u_\theta = T_{\theta R}v_R + T_{\theta\theta}v_\theta + T_{\theta\phi}v_\phi \,, \tag{2.5.40}$$

$$u_\phi = T_{\phi R}v_R + T_{\phi\theta}v_\theta + T_{\phi\phi}v_\phi \,. \tag{2.5.41}$$

For the transformation of scalar components from the Cartesian orthonormal basis $\{\hat{e}_i\}$ to the spherical coordinate orthonormal basis $\{\hat{e}'_i\}$, the matrix of direction cosines $\ell_{ij} \equiv \hat{e}'_i \cdot \hat{e}_j$ is

$$[L] = \begin{bmatrix} \sin\theta\cos\phi & \sin\theta\sin\phi & \cos\theta \\ \cos\theta\cos\phi & \cos\theta\sin\phi & -\sin\theta \\ -\sin\phi & \cos\phi & 0 \end{bmatrix} \tag{2.5.42}$$

and the tensor transformation rule is still given by (2.3.30). For a vector $\mathbf{v}$ and a second-order tensor $\mathbf{S}$, the transformations are given in matrix notation by $[\mathbf{v}]' = [L][\mathbf{v}]$ and $[\mathbf{S}]' = [L][\mathbf{S}][L]^\mathsf{T}$.

Gradient in spherical coordinates

The tensor definition (2.4.20) of the gradient of a tensor is used to derive the specific expression for the gradient in a spherical coordinate system. If the point in space with position vector $\mathbf{x}$ has spherical coordinates (R, θ, ϕ), then the point in space with position vector $\mathbf{x}+\alpha\mathbf{a}$ has spherical coordinates that, if $|\alpha\mathbf{a}| \ll |\mathbf{x}|$, are given approximately by[11]

$$\left(R + \alpha a_R \,,\ \theta + \frac{\alpha a_\theta}{R} \,,\ \phi + \frac{\alpha a_\phi}{R\sin\theta} \right) \,, \tag{2.5.43}$$

where a_r, a_θ, and a_ϕ are the scalar components of $\mathbf{a}$. These approach the exact coordinates in the limit as $\alpha \to 0$. It follows that for a tensor field

---

[11] Again, this approach to deriving (2.5.45) is not rigorous, although the end result is correct. See the Appendix for a proper derivation.

given by $\mathbf{T} = \psi(\mathbf{x}, t) = \tilde{\psi}(R, \theta, \phi, t)$,

$$
\begin{aligned}
\frac{d}{d\alpha}\psi(\mathbf{x}+\alpha\mathbf{a}, t)\Big|_{\alpha=0} &= \frac{d}{d\alpha}\tilde{\psi}\left(R+\alpha a_R\,,\; \theta+\frac{\alpha a_\theta}{R}\,,\; \phi+\frac{\alpha a_\phi}{R\sin\theta}\,,\; t\right)\Big|_{\alpha=0}\\
&= \frac{\partial\mathbf{T}}{\partial R}a_R + \frac{1}{R}\frac{\partial\mathbf{T}}{\partial\theta}a_\theta + \frac{1}{R\sin\theta}\frac{\partial\mathbf{T}}{\partial\phi}a_\phi\\
&= \left(\frac{\partial\mathbf{T}}{\partial R}\hat{\mathbf{e}}_R + \frac{1}{R}\frac{\partial\mathbf{T}}{\partial\theta}\hat{\mathbf{e}}_\theta + \frac{1}{R\sin\theta}\frac{\partial\mathbf{T}}{\partial\phi}\hat{\mathbf{e}}_\phi\right)\cdot\mathbf{a}\,,
\end{aligned}
$$

$$(2.5.44)$$

so that, from the definition of the gradient (2.4.20),

$$
\boxed{\nabla\mathbf{T} = \frac{\partial\mathbf{T}}{\partial R}\hat{\mathbf{e}}_R + \frac{1}{R}\frac{\partial\mathbf{T}}{\partial\theta}\hat{\mathbf{e}}_\theta + \frac{1}{R\sin\theta}\frac{\partial\mathbf{T}}{\partial\phi}\hat{\mathbf{e}}_\phi\,.}
\qquad (2.5.45)
$$

This result is used below to derive expressions for the gradient, divergence, curl, and Laplacian of tensors of various orders (as required in subsequent chapters) in terms of their scalar components.

Gradient of scalars and vectors

For a scalar field f, it follows immediately from (2.5.45) that

$$
\boxed{\nabla f = \frac{\partial f}{\partial R}\hat{\mathbf{e}}_R + \frac{1}{R}\frac{\partial f}{\partial\theta}\hat{\mathbf{e}}_\theta + \frac{1}{R\sin\theta}\frac{\partial f}{\partial\phi}\hat{\mathbf{e}}_\phi\,.}
\qquad (2.5.46)
$$

However, for a vector field $\mathbf{u} = u_R\hat{\mathbf{e}}_R + u_\theta\hat{\mathbf{e}}_\theta + u_\phi\hat{\mathbf{e}}_\phi$, one must take into account that the spherical coordinate base vectors are themselves functions of the spherical coordinate variables. From (2.5.33) it is seen that

$$
\begin{aligned}
\frac{\partial\hat{\mathbf{e}}_R}{\partial R} = 0\,, \qquad & \frac{\partial\hat{\mathbf{e}}_R}{\partial\theta} = \hat{\mathbf{e}}_\theta\,, \qquad & \frac{\partial\hat{\mathbf{e}}_R}{\partial\phi} = \sin\theta\,\hat{\mathbf{e}}_\phi\,,\\
\frac{\partial\hat{\mathbf{e}}_\theta}{\partial R} = 0\,, \qquad & \frac{\partial\hat{\mathbf{e}}_\theta}{\partial\theta} = -\hat{\mathbf{e}}_R\,, \qquad & \frac{\partial\hat{\mathbf{e}}_\theta}{\partial\phi} = \cos\theta\,\hat{\mathbf{e}}_\phi\,,\\
\frac{\partial\hat{\mathbf{e}}_\phi}{\partial R} = 0\,, \qquad & \frac{\partial\hat{\mathbf{e}}_\phi}{\partial\theta} = 0\,, \qquad & \frac{\partial\hat{\mathbf{e}}_\phi}{\partial\phi} = -\sin\theta\,\hat{\mathbf{e}}_R - \cos\theta\,\hat{\mathbf{e}}_\theta\,.
\end{aligned}
$$

$$(2.5.47)$$

Thus, by substituting $\mathbf{u} = u_R\hat{\mathbf{e}}_R + u_\theta\hat{\mathbf{e}}_\theta + u_\phi\hat{\mathbf{e}}_\phi$ into (2.5.45) and using the chain rule, one finds that

$$
\begin{aligned}
\boldsymbol{\nabla}\mathbf{u} = {} & \frac{\partial u_R}{\partial R}\hat{\mathbf{e}}_R\hat{\mathbf{e}}_R + \frac{1}{R}\left(\frac{\partial u_R}{\partial \theta} - u_\theta\right)\hat{\mathbf{e}}_R\hat{\mathbf{e}}_\theta + \frac{1}{R\sin\theta}\left(\frac{\partial u_R}{\partial \phi} - u_\phi\sin\theta\right)\hat{\mathbf{e}}_R\hat{\mathbf{e}}_\phi \\
& + \frac{\partial u_\theta}{\partial R}\hat{\mathbf{e}}_\theta\hat{\mathbf{e}}_R + \frac{1}{R}\left(\frac{\partial u_\theta}{\partial \theta} + u_R\right)\hat{\mathbf{e}}_\theta\hat{\mathbf{e}}_\theta + \frac{1}{R\sin\theta}\left(\frac{\partial u_\theta}{\partial \phi} - u_\phi\cos\theta\right)\hat{\mathbf{e}}_\theta\hat{\mathbf{e}}_\phi \\
& + \frac{\partial u_\phi}{\partial R}\hat{\mathbf{e}}_\phi\hat{\mathbf{e}}_R + \frac{1}{R}\frac{\partial u_\phi}{\partial \theta}\hat{\mathbf{e}}_\phi\hat{\mathbf{e}}_\theta + \frac{1}{R\sin\theta}\left(\frac{\partial u_\phi}{\partial \phi} + u_R\sin\theta + u_\theta\cos\theta\right)\hat{\mathbf{e}}_\phi\hat{\mathbf{e}}_\phi .
\end{aligned}
$$

$$(2.5.48)$$

This result can now be used to derive the results for the divergence.

Divergence of vectors and second-order tensors

Recall that the divergence of a vector is defined to be the contraction of its gradient. Thus, from (2.5.48), the divergence of a vector field $\mathbf{u}$ is

$$
\boldsymbol{\nabla}\cdot\mathbf{u} = \frac{\partial u_R}{\partial R} + \frac{1}{R}\left(\frac{\partial u_\theta}{\partial \theta} + 2u_R + u_\theta\cot\theta\right) + \frac{1}{R\sin\theta}\frac{\partial u_\phi}{\partial \phi} . \qquad (2.5.49)
$$

The divergence of higher-order tensors is defined by (2.4.28). For a second-order tensor field $\mathbf{S}$ and an arbitrary constant vector $\mathbf{a}$,

$$
\begin{aligned}
\mathbf{a}\cdot\mathbf{S} = {} & (a_R S_{RR} + a_\theta S_{\theta R} + a_\phi S_{\phi R})\hat{\mathbf{e}}_R \\
& + (a_R S_{R\theta} + a_\theta S_{\theta\theta} + a_\phi S_{\phi\theta})\hat{\mathbf{e}}_\theta \\
& + (a_R S_{R\phi} + a_\theta S_{\theta\phi} + a_\phi S_{\phi\phi})\hat{\mathbf{e}}_\phi .
\end{aligned}
\qquad (2.5.50)
$$

Since $\mathbf{a}\cdot\mathbf{S}$ is a vector, the expression $\boldsymbol{\nabla}\cdot(\mathbf{a}\cdot\mathbf{S})$ on the right-hand side of (2.4.28) is given by (2.5.49). However, one must take care with the constant vector $\mathbf{a}$—since the base vectors are not constant, neither are the scalar components of $\mathbf{a}$. Specifically, using (2.5.47), one finds that

$$
\frac{\partial\mathbf{a}}{\partial\theta} = \left(\frac{\partial a_R}{\partial \theta} - a_\theta\right)\hat{\mathbf{e}}_R + \left(\frac{\partial a_\theta}{\partial \theta} + a_R\right)\hat{\mathbf{e}}_\theta + \frac{\partial a_\phi}{\partial \theta}\hat{\mathbf{e}}_\phi = 0 , \qquad (2.5.51)
$$

and

$$
\begin{aligned}
\frac{\partial\mathbf{a}}{\partial\phi} = {} & \left(\frac{\partial a_R}{\partial \phi} - a_\phi\sin\theta\right)\hat{\mathbf{e}}_R + \left(\frac{\partial a_\theta}{\partial \phi} - a_\phi\cos\theta\right)\hat{\mathbf{e}}_\theta \\
& + \left(\frac{\partial a_\phi}{\partial \phi} + a_R\sin\theta + a_\theta\cos\theta\right)\hat{\mathbf{e}}_\phi = 0 . \quad (2.5.52)
\end{aligned}
$$

Therefore,

$$\frac{\partial a_R}{\partial R} = 0 , \quad \frac{\partial a_R}{\partial \theta} = a_\theta , \quad \frac{\partial a_R}{\partial \phi} = a_\phi \sin \theta ,$$

$$\frac{\partial a_\theta}{\partial R} = 0 , \quad \frac{\partial a_\theta}{\partial \theta} = -a_R , \quad \frac{\partial a_\theta}{\partial \phi} = a_\phi \cos \theta , \qquad (2.5.53)$$

$$\frac{\partial a_\phi}{\partial R} = 0 , \quad \frac{\partial a_\phi}{\partial \theta} = 0 , \quad \frac{\partial a_\phi}{\partial \phi} = -a_R \sin \theta - a_\theta \cos \theta .$$

With this in mind, it follows from (2.5.49) and (2.5.50) that

$$
\begin{aligned}
\boldsymbol{\nabla} \cdot (\mathbf{a} \cdot \mathbf{S}) = a_R & \left\{ \frac{\partial S_{RR}}{\partial R} + \frac{1}{R} \frac{\partial S_{R\theta}}{\partial \theta} + \frac{1}{R \sin \theta} \frac{\partial S_{R\phi}}{\partial \phi} \right. \\
& \left. + \frac{1}{R} [2S_{RR} - S_{\theta\theta} - S_{\phi\phi} + S_{R\theta} \cot \theta] \right\} \\
+ a_\theta & \left\{ \frac{\partial S_{\theta R}}{\partial R} + \frac{1}{R} \frac{\partial S_{\theta\theta}}{\partial \theta} + \frac{1}{R \sin \theta} \frac{\partial S_{\theta\phi}}{\partial \phi} \right. \\
& \left. + \frac{1}{R} [(S_{\theta\theta} - S_{\phi\phi}) \cot \theta + 2S_{\theta R} + S_{R\theta}] \right\} \\
+ a_\phi & \left\{ \frac{\partial S_{\phi R}}{\partial R} + \frac{1}{R} \frac{\partial S_{\phi\theta}}{\partial \theta} + \frac{1}{R \sin \theta} \frac{\partial S_{\phi\phi}}{\partial \phi} \right. \\
& \left. + \frac{1}{R} [(S_{\theta\phi} + S_{\phi\theta}) \cot \theta + 2S_{\phi R} + S_{R\phi}] \right\} , \qquad (2.5.54)
\end{aligned}
$$

and therefore, from the definition (2.4.28), that

$$
\begin{aligned}
\boldsymbol{\nabla} \cdot \mathbf{S} = & \left\{ \frac{\partial S_{RR}}{\partial R} + \frac{1}{R} \frac{\partial S_{R\theta}}{\partial \theta} + \frac{1}{R \sin \theta} \frac{\partial S_{R\phi}}{\partial \phi} \right. \\
& \left. + \frac{1}{R} [2S_{RR} - S_{\theta\theta} - S_{\phi\phi} + S_{R\theta} \cot \theta] \right\} \hat{\mathbf{e}}_R \\
& + \left\{ \frac{\partial S_{\theta R}}{\partial R} + \frac{1}{R} \frac{\partial S_{\theta\theta}}{\partial \theta} + \frac{1}{R \sin \theta} \frac{\partial S_{\theta\phi}}{\partial \phi} \right. \\
& \left. + \frac{1}{R} [(S_{\theta\theta} - S_{\phi\phi}) \cot \theta + 2S_{\theta R} + S_{R\theta}] \right\} \hat{\mathbf{e}}_\theta \\
& + \left\{ \frac{\partial S_{\phi R}}{\partial R} + \frac{1}{R} \frac{\partial S_{\phi\theta}}{\partial \theta} + \frac{1}{R \sin \theta} \frac{\partial S_{\phi\phi}}{\partial \phi} \right. \\
& \left. + \frac{1}{R} [(S_{\theta\phi} + S_{\phi\theta}) \cot \theta + 2S_{\phi R} + S_{R\phi}] \right\} \hat{\mathbf{e}}_\phi . \qquad (2.5.55)
\end{aligned}
$$

These results can now be used to examine the curl.

Curl of vectors

The curl of a vector field is defined by (2.4.35). If $\mathbf{u}$ is a vector field and $\mathbf{a}$ is an arbitrary constant vector, then

$$\mathbf{u} \times \mathbf{a} = (u_\theta a_\phi - u_\phi a_\theta)\hat{\mathbf{e}}_R + (u_\phi a_R - u_R a_\phi)\hat{\mathbf{e}}_\theta + (u_R a_\theta - u_\theta a_R)\hat{\mathbf{e}}_\phi \ . \tag{2.5.56}$$

Since $\mathbf{u} \times \mathbf{a}$ is a vector, the expression $\nabla \cdot (\mathbf{u} \times \mathbf{a})$ on the right-hand side of the definition (2.4.35) is given by (2.5.49). Using (2.5.53), it follows that

$$
\begin{aligned}
\nabla \times \mathbf{u} = &\frac{1}{R}\left(\frac{\partial u_\phi}{\partial \theta} + u_\phi \cot\theta - \frac{1}{\sin\theta}\frac{\partial u_\theta}{\partial \phi}\right) a_R \\
&+ \left(\frac{1}{R\sin\theta}\frac{\partial u_R}{\partial \phi} - \frac{\partial u_\phi}{\partial R} - \frac{u_\phi}{R}\right) a_\theta \\
&+ \left(\frac{\partial u_\theta}{\partial R} + \frac{u_\theta}{R} - \frac{1}{R}\frac{\partial u_R}{\partial \theta}\right) a_\phi \ ,
\end{aligned}
\tag{2.5.57}
$$

and therefore, from the definition (2.4.35), that

$$
\boxed{
\begin{aligned}
\nabla \times \mathbf{u} = &\frac{1}{R}\left(\frac{\partial u_\phi}{\partial \theta} + u_\phi \cot\theta - \frac{1}{\sin\theta}\frac{\partial u_\theta}{\partial \phi}\right) \hat{\mathbf{e}}_R \\
&+ \left(\frac{1}{R\sin\theta}\frac{\partial u_R}{\partial \phi} - \frac{\partial u_\phi}{\partial R} - \frac{u_\phi}{R}\right) \hat{\mathbf{e}}_\theta \\
&+ \left(\frac{\partial u_\theta}{\partial R} + \frac{u_\theta}{R} - \frac{1}{R}\frac{\partial u_R}{\partial \theta}\right) \hat{\mathbf{e}}_\phi \ .
\end{aligned}
}
\tag{2.5.58}
$$

Laplacian of scalars and vectors

Recall that the Laplacian of a tensor field is defined to be the divergence of its gradient. Using the above results, it follows that for a scalar field f,

$$
\boxed{
\nabla^2 f = \frac{\partial^2 f}{\partial R^2} + \frac{2}{R}\frac{\partial f}{\partial R} + \frac{1}{R^2}\left(\frac{\partial^2 f}{\partial \theta^2} + \cot\theta\frac{\partial f}{\partial \theta} + \frac{1}{\sin^2\theta}\frac{\partial^2 f}{\partial \phi^2}\right) \ ,
}
\tag{2.5.59}
$$

and for a vector field $\mathbf{u}$,

$$
\boxed{
\begin{aligned}
\nabla^2 \mathbf{u} = &\left[\nabla^2 u_R - \frac{2}{R^2}\left(u_R + u_\theta \cot\theta + \frac{\partial u_\theta}{\partial \theta} + \frac{1}{\sin\theta}\frac{\partial u_\phi}{\partial \phi}\right)\right]\hat{\mathbf{e}}_R \\
&+ \left[\nabla^2 u_\theta + \frac{1}{R^2}\left(2\frac{\partial u_R}{\partial \theta} - \frac{1}{\sin^2\theta}u_\theta - 2\frac{\cos\theta}{\sin^2\theta}\frac{\partial u_\phi}{\partial \phi}\right)\right]\hat{\mathbf{e}}_\theta \\
&+ \left[\nabla^2 u_\phi + \frac{1}{R^2\sin\theta}\left(2\frac{\partial u_R}{\partial \phi} + 2\cot\theta\frac{\partial u_\theta}{\partial \phi} - \frac{1}{\sin\theta}u_\phi\right)\right]\hat{\mathbf{e}}_\phi \ .
\end{aligned}
}
\tag{2.5.60}
$$

Problems

2.1 For each of the following, determine whether or not the given expression is a valid indicial notation expression. If it is *valid*, identify the free indices and the dummy index pairs, determine the number of expanded equations represented and how many terms each expanded equation will have, and give the expanded equations (or, if they are too lengthy, at least demonstrate that you know what they are).

(a) $a_{ms} = b_m(c_r - d_r)$

(b) $a_{ms} = b_m(c_s - d_s)$

(c) $t_i = \sigma_{ji}n_j$

(d) $t_i = \sigma_{ji}n_i$

(e) $\sigma_{ij} = 2\mu\varepsilon_{ij} + \lambda\varepsilon_{kk}\delta_{ij}$

(f) $x_i x_i = r^2$

(g) $\varepsilon_{rs} = h_r(d_s - h_s k_{rr})$

(h) $b_{ij}c_j = 3$

2.2 Show each of the following, where δ_{ij} is the Kronecker delta, ℓ_{ij} are the direction cosines, and e_{ijk} is the permutation symbol.

(a) $\delta_{ij}\delta_{jk}\delta_{kp}\delta_{pi} = 3$

(b) $\delta_{ij}e_{ijk} = 0$

(c) $\ell_{ki}\ell_{kj} = \delta_{ij}$

(d) $e_{qrs}d_q d_s = 0$

2.3 Prove the following relations between the permutation symbol e_{ijk} and the Kronecker delta δ_{ij} (**Hint:** recall that $\det[A]^\mathsf{T} = \det[A]$ and $\det[A]\det[B] = \det([A][B])$, and use successive contractions).

(a) $e_{ijk} = \det\begin{bmatrix} \delta_{i1} & \delta_{i2} & \delta_{i3} \\ \delta_{j1} & \delta_{j2} & \delta_{j3} \\ \delta_{k1} & \delta_{k2} & \delta_{k3} \end{bmatrix}$

(b) $e_{ijk}e_{pqr} = \det\begin{bmatrix} \delta_{ip} & \delta_{iq} & \delta_{ir} \\ \delta_{jp} & \delta_{jq} & \delta_{jr} \\ \delta_{kp} & \delta_{kq} & \delta_{kr} \end{bmatrix}$

(c) $e_{ijk}e_{iqr} = \delta_{jq}\delta_{kr} - \delta_{jr}\delta_{kq}$

(d) $e_{ijk}e_{ijr} = 2\delta_{kr}$

(e) $e_{ijk}e_{ijk} = 6$

2.4 Prove the following relations between the permutation symbol e_{ijk} and the determinant of the 3×3 matrix $[A]$ with elements A_{ij}.

(a) $e_{lmn}\det[A] = e_{ijk}A_{il}A_{jm}A_{kn}$

(b) $\det[A] = \frac{1}{6}e_{ijk}e_{lmn}A_{il}A_{jm}A_{kn}$

2.5 Given an orthonormal basis $\{\hat{e}_i\}$ and three arbitrary vectors $\mathbf{u} = u_i\hat{e}_i$, $\mathbf{v} = v_i\hat{e}_i$, and $\mathbf{w} = w_i\hat{e}_i$, show that

(a) $\hat{e}_i \times \hat{e}_j \cdot \hat{e}_k = e_{ijk}$

(b) $\mathbf{u} \times \mathbf{v} \cdot \mathbf{w} = e_{ijk}u_i v_j w_k$

2.6 In the orthonormal basis $\{\hat{e}_i\}$, the dyadic representations of a second-order tensor $\mathbf{A}$ and a vector $\mathbf{u}$ are

$$\mathbf{A} = 5\hat{e}_1\hat{e}_1 - 4\hat{e}_2\hat{e}_1 + 2\hat{e}_3\hat{e}_3 , \quad \mathbf{u} = -2\hat{e}_1 + 3\hat{e}_3 .$$

(a) Find the dyadic representation, in the orthonormal basis $\{\hat{e}_i\}$, of the vector $\mathbf{v} = \mathbf{A} \cdot \mathbf{u}$.

(b) Use the tensor transformation rule to find the matrices $[\mathbf{A}]'$, $[\mathbf{u}]'$, and $[\mathbf{v}]'$ of the scalar components of $\mathbf{A}$, $\mathbf{u}$, and $\mathbf{v}$ in an orthonormal basis $\{\hat{e}_i'\}$ that is obtained by rotating $\{\hat{e}_i\}$ through a 30° clockwise angle about the $\hat{e}_3$-direction, as shown below.

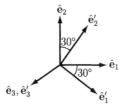

(c) Show that $[\mathbf{v}]' = [\mathbf{A}]'[\mathbf{u}]'$.

(d) Give the dyadic representations of $\mathbf{A}$, $\mathbf{u}$, and $\mathbf{v}$ in the orthonormal basis $\{\hat{e}_i'\}$.

2.7 For three arbitrary vectors $\mathbf{u}$, $\mathbf{v}$, and $\mathbf{w}$, show that

$$\mathbf{u} \times (\mathbf{v} \times \mathbf{w}) = (\mathbf{w} \cdot \mathbf{u})\mathbf{v} - (\mathbf{v} \cdot \mathbf{u})\mathbf{w} .$$

2.8 Given an arbitrary vector $\mathbf{v}$ and an arbitrary unit vector $\hat{\boldsymbol{\mu}}$, show that

$$\mathbf{v} = (\mathbf{v} \cdot \hat{\boldsymbol{\mu}})\hat{\boldsymbol{\mu}} + \hat{\boldsymbol{\mu}} \times (\mathbf{v} \times \hat{\boldsymbol{\mu}})$$

and give physical interpretations of $(\mathbf{v} \cdot \hat{\boldsymbol{\mu}})\hat{\boldsymbol{\mu}}$ and $\hat{\boldsymbol{\mu}} \times (\mathbf{v} \times \hat{\boldsymbol{\mu}})$.

2.9 Given a second-order tensor $\mathbf{A}$, show that $\mathbf{u} \cdot \mathbf{A} \cdot \mathbf{u} = 0$ for all vectors $\mathbf{u}$ if and only if $\mathbf{A}$ is skew-symmetric.

2.10 Given arbitrary vectors $\mathbf{u}$ and $\mathbf{v}$ and arbitrary second-order tensors $\mathbf{A}$ and $\mathbf{B}$, show that

(a) $(\mathbf{u} \cdot \mathbf{A}) \cdot \mathbf{v} = \mathbf{u} \cdot (\mathbf{A} \cdot \mathbf{v})$

(b) $(\mathbf{A} \cdot \mathbf{B}) \cdot \mathbf{u} = \mathbf{A} \cdot (\mathbf{B} \cdot \mathbf{u})$

(c) $\mathbf{u} \cdot (\mathbf{A} \cdot \mathbf{B}) = (\mathbf{u} \cdot \mathbf{A}) \cdot \mathbf{B}$

2.11 Given arbitrary vectors $\mathbf{u}$ and $\mathbf{v}$ and arbitrary second-order tensors $\mathbf{A}$ and $\mathbf{B}$, show that

(a) $(\mathbf{A} \cdot \mathbf{B})^{\mathsf{T}} = \mathbf{B}^{\mathsf{T}} \cdot \mathbf{A}^{\mathsf{T}}$

(b) $(\mathbf{A} \cdot \mathbf{u}) \cdot (\mathbf{B} \cdot \mathbf{v}) = \mathbf{u} \cdot \left(\mathbf{A}^{\mathsf{T}} \cdot \mathbf{B}\right) \cdot \mathbf{v}$

(c) $(\mathbf{A} \cdot \mathbf{B})^{-1} = \mathbf{B}^{-1} \cdot \mathbf{A}^{-1}$

(d) $\left(\mathbf{A}^{\mathsf{T}}\right)^{-1} = \left(\mathbf{A}^{-1}\right)^{\mathsf{T}} \equiv \mathbf{A}^{-\mathsf{T}}$

2.12 If, in an orthonormal basis $\{\hat{\mathbf{e}}_i\}$, A_{ij} are the scalar components of a skew-symmetric second-order tensor $\mathbf{A}$, then the vector $\mathbf{b}$ with scalar components given by

$$b_i = \frac{1}{2} e_{ijk} A_{kj}$$

is called the *axial vector of* $\mathbf{A}$.

(a) Show that the scalar components of $\mathbf{A}$ are given in terms of the scalar components of $\mathbf{b}$ by $A_{pq} = e_{qpi} b_i$.

(b) Give the matrix of scalar components of $\mathbf{A}$ in terms of the scalar components of $\mathbf{b}$.

(c) Show that, for an arbitrary vector $\mathbf{c}$, $\mathbf{A} \cdot \mathbf{c} = \mathbf{b} \times \mathbf{c}$.

2.13 Show that, in an orthonormal basis $\{\hat{\mathbf{e}}_i\}$, the determinant of a second-order tensor $\mathbf{A}$ is equal to the determinant of its matrix of scalar components $[\mathbf{A}]$, that is, that

$$\det \mathbf{A} \equiv \frac{[(\mathbf{A} \cdot \mathbf{u}) \times (\mathbf{A} \cdot \mathbf{v})] \cdot (\mathbf{A} \cdot \mathbf{w})}{(\mathbf{u} \times \mathbf{v}) \cdot \mathbf{w}} = \det[\mathbf{A}] \, ,$$

where $\mathbf{u}$, $\mathbf{v}$, and $\mathbf{w}$ are *arbitrary* vectors.

2.14 Given, in an orthonormal basis, that

$$\widetilde{N}_{\mathbf{A}}(\hat{\mu}_1, \hat{\mu}_2, \hat{\mu}_3, \lambda) = A_{ij}\hat{\mu}_i\hat{\mu}_j - \lambda(\hat{\mu}_i\hat{\mu}_i - 1)$$

and $A_{ij} = A_{ji}$, show that the conditions $\partial \widetilde{N}_{\mathbf{A}}/\partial \hat{\mu}_k = 0$ lead to the equations

$$A_{kj}\hat{\mu}_j = \lambda \hat{\mu}_k \, .$$

(**Hint:** note that $\partial \hat{\mu}_i/\partial \hat{\mu}_k = \delta_{ik}$.)

2.15 Given a nonsingular second-order tensor $\mathbf{F}$, show that $\mathbf{C} = \mathbf{F}^{\mathsf{T}} \cdot \mathbf{F}$ is symmetric and positive definite. Why is it necessary that $\mathbf{F}$ be nonsingular?

2.16 Prove the *Cayley-Hamilton theorem*, which says that a symmetric second-order tensor **A** satisfies its own characteristic equation, that is, that

$$\mathbf{A}^3 - I_\mathbf{A}\mathbf{A}^2 + II_\mathbf{A}\mathbf{A} - III_\mathbf{A}\mathbf{I} = 0 ,$$

where $I_\mathbf{A}$, $II_\mathbf{A}$, and $III_\mathbf{A}$ are the principal invariants of **A**.

2.17 Consider a nonsingular second-order tensor **A** and let $\mathbf{B} = \mathbf{A}^{-1}$. Show that the principal invariants of **B** can be expressed in terms of the principal invariants of **A** by

$$I_\mathbf{B} = \frac{II_\mathbf{A}}{III_\mathbf{A}} , \qquad II_\mathbf{B} = \frac{I_\mathbf{A}}{III_\mathbf{A}} , \qquad III_\mathbf{B} = \frac{1}{III_\mathbf{A}} .$$

[**Hint:** recall that $\det(\mathbf{B} - \lambda\mathbf{I}) = \lambda^3 - I_\mathbf{B}\lambda^2 + II_\mathbf{B}\lambda - III_\mathbf{B}$ and $\det(\alpha\mathbf{C}) = \alpha^3 \det \mathbf{C}$ and note that $(\mathbf{B} - \lambda\mathbf{I}) = \mathbf{A}^{-1}(\mathbf{I} - \lambda\mathbf{A})$.]

2.18 For the function $F = a_{ij}x_ix_j$, where a_{ij} are constants, derive expressions for

(a) $\dfrac{\partial F}{\partial x_r}$ (b) $\dfrac{\partial^2 F}{\partial x_r \partial x_s}$

2.19 Given an arbitrary scalar field ζ, an arbitrary vector field **v**, and the position vector **x**, show

(a) $\boldsymbol{\nabla}(\nabla^2\zeta) = \nabla^2(\boldsymbol{\nabla}\zeta)$ (b) $\boldsymbol{\nabla}(\mathbf{v} \cdot \mathbf{x}) = \mathbf{v} + \mathbf{x} \cdot (\boldsymbol{\nabla}\mathbf{v})$

(c) $\boldsymbol{\nabla} \cdot (\boldsymbol{\nabla} \times \mathbf{v}) = 0$ (d) $\boldsymbol{\nabla} \cdot (\nabla^2\mathbf{v}) = \nabla^2(\boldsymbol{\nabla} \cdot \mathbf{v})$

(e) $\boldsymbol{\nabla} \cdot (\zeta\mathbf{v}) = \zeta\boldsymbol{\nabla} \cdot \mathbf{v} + (\boldsymbol{\nabla}\zeta) \cdot \mathbf{v}$ (f) $\boldsymbol{\nabla} \times (\nabla^2\mathbf{v}) = \nabla^2(\boldsymbol{\nabla} \times \mathbf{v})$

(g) $\nabla^2(\zeta\mathbf{x}) = 2\boldsymbol{\nabla}\zeta + \mathbf{x}\nabla^2\zeta$ (h) $\nabla^2(\mathbf{v} \cdot \mathbf{x}) = 2\boldsymbol{\nabla} \cdot \mathbf{v} + \mathbf{x} \cdot \nabla^2\mathbf{v}$

(i) $\boldsymbol{\nabla} \times (\boldsymbol{\nabla} \times \mathbf{v}) = \boldsymbol{\nabla}(\boldsymbol{\nabla} \cdot \mathbf{v}) - \nabla^2\mathbf{v}$

2.20 Given the dyad **ab** formed by two arbitrary vectors **a** and **b**, show that

$$\nabla^2(\mathbf{ab}) = (\nabla^2\mathbf{a})\mathbf{b} + \mathbf{a}(\nabla^2\mathbf{b}) + 2\boldsymbol{\nabla}\mathbf{a} \cdot (\boldsymbol{\nabla}\mathbf{b})^\mathsf{T} .$$

2.21 Let $R \equiv |\mathbf{x}|$ be the distance from the origin. Show each of the following first by using a Cartesian coordinate system, as in Example 2.4, and then by using a spherical coordinate system.

(a) $\boldsymbol{\nabla}(R^n\mathbf{x}) = R^n\mathbf{I} + nR^{n-2}\mathbf{xx}$ (b) $\boldsymbol{\nabla} \cdot (R^n\mathbf{x}) = (n+3)R^n$

(c) $\boldsymbol{\nabla} \cdot (R^n\mathbf{xx}) = (n+4)R^n\mathbf{x}$ (d) $\nabla^2(R^n\mathbf{x}) = n(n+3)R^{n-2}\mathbf{x}$

2.22 Consider the vector field $\mathbf{h} = \alpha R^{-2} \hat{\mathbf{e}}_R$, where α is a constant, R is the distance from the origin, and $\hat{\mathbf{e}}_R$ is the spherical coordinate base vector pointing out from the origin. Determine the gradient $\nabla \mathbf{h}$, the divergence $\nabla \cdot \mathbf{h}$, and the curl $\nabla \times \mathbf{h}$ of this vector field.

2.23 Recall that, in a Cartesian coordinate system, the curl of a vector field $\mathbf{v} = v_i \hat{\mathbf{e}}_i$ is given in terms of its scalar components by $\nabla \times \mathbf{v} = e_{ijk} v_{j,i} \hat{\mathbf{e}}_k$ and the curl of a second-order tensor $\mathbf{S} = S_{ij} \hat{\mathbf{e}}_i \hat{\mathbf{e}}_j$ is given in terms of its scalar components by $\nabla \times \mathbf{S} = e_{ijk} S_{rj,i} \hat{\mathbf{e}}_k \hat{\mathbf{e}}_r$. Determine the curl of a third-order tensor $\mathbf{T} = T_{ijk} \hat{\mathbf{e}}_i \hat{\mathbf{e}}_j \hat{\mathbf{e}}_k$ in terms of its scalar components.

2.24 Given, in a Cartesian coordinate system, a second-order tensor field $\varepsilon = \varepsilon_{ij} \hat{\mathbf{e}}_i \hat{\mathbf{e}}_j$, derive a dyadic notation expression for $\nabla \times (\nabla \times \varepsilon)$ in terms of the scalar components of ε. Assuming that ε is symmetric, determine explicitly the six independent scalar equations, in a Cartesian coordinate system, that are represented by the tensor notation equation $\nabla \times (\nabla \times \varepsilon) = \mathbf{0}$.

2.25 It was shown in Example 2.5 that $\nabla \times (\nabla \phi) = \mathbf{0}$, where ϕ is an arbitrary scalar field. Prove that $\nabla \times (\nabla \mathbf{T}) = \mathbf{0}$ for an arbitrary tensor field $\mathbf{T}$ of *any* order. [**Hint:** use the definition (2.4.36) with $\mathbf{T} \to \nabla \mathbf{T}$, show that $\mathbf{a} \cdot (\nabla \mathbf{T}) = \nabla (\mathbf{a} \cdot \mathbf{T})$ since $\mathbf{a}$ is a constant, and establish the proof by a process of induction.]

Chapter 3

Kinematics

A body is composed of discrete material quanta in the form of atoms, molecules, grains, polymer chains, and so forth. It is the central premise of continuum mechanics that this discrete nature of material can be neglected in studying the deformation of a body, provided the length scales of interest are large relative to the length scale of the material quanta. A body in this case is modeled as a *continuum* in which physical quantities distributed over the body are represented by fields that are continuous (except possibly at a finite number of surfaces, lines, and/or points).

3.1   Configurations

The continuum hypothesis allows one to define a *material point* as an infinitesimal element of material. A *body* then is composed of an infinite number of material points and the assignment of each of these material points to a unique position in space defines a *configuration* of the body. There are an infinite number of possible configurations of a body. A configuration will be assumed to occupy a contiguous region of space whose boundary consists of a finite number of regular surfaces. In other words, each point in the region occupied by a body's configuration can be connected to any other point in the region by a line in space that lies entirely within the region, and the region has at most a finite number of "voids."

Of course, no two material points can occupy the same position in space at the same time. It is, however, possible for two different configurations of a body to occupy the same region in space. Consider, for example, two cubic configurations of a body that are of equal volume and are related by a 90° rotation about a fourfold symmetry axis. The two configurations occupy the same cubic region in space but have the material points of the body assigned to different positions. Nonetheless, for simplicity, a particular configuration of a body and the region in space which it occupies

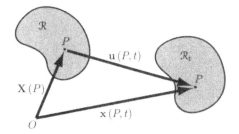

FIGURE 3.1.   The reference and current configurations of a body.

will sometimes be given the same notation, $\mathcal{R}$ for example.

A prescribed *reference configuration* of a body $\mathbb{B}$, occupying the region $\mathcal{R}$ with boundary $\partial\mathcal{R}$, is defined against which other configurations of the body are to be compared. The *current configuration* of the body, occupying the region $\mathcal{R}_t$ with boundary $\partial\mathcal{R}_t$, is the configuration of $\mathbb{B}$ at time t. It is not required that the body ever actually be in the reference configuration. However, an *initial configuration* of $\mathbb{B}$, at some time $t = t_0$ say, is often the natural choice of reference configuration in elasticity problems. In this case, the action of some external agents, the nature of which are not of concern at this juncture, cause the material points of the body to move until, at time t, they are in new positions which define the current configuration of the body.

A body $\mathbb{B}$ can be thought of as a set of material points, so that each material point P in $\mathbb{B}$ is an element of the set, which is indicated by writing $P \in \mathbb{B}$. When $\mathbb{B}$ is in its reference configuration, let the position vector of a material point $P \in \mathbb{B}$ relative to some prescribed origin O be denoted by $\mathbf{X}(P)$. This will be referred to as the *reference position* of P. Then, the region $\mathcal{R}$ occupied by $\mathbb{B}$ when it is in its reference configuration can be thought of as a set of reference positions:

$$\mathcal{R} = \{\, \mathbf{X}(P) \mid P \in \mathbb{B} \,\} \,, \qquad (3.1.1)$$

the set of all $\mathbf{X}(P)$ such that P is in $\mathbb{B}$. The material points of $\mathbb{B}$ occupy new positions in the current configuration. When $\mathbb{B}$ is in its current configuration, that is, at time t, let the position vector of a material point $P \in \mathbb{B}$ relative to the same origin O be denoted by $\mathbf{x}(P,t)$ (see Figure 3.1). This will be referred to as the *current position* of P. Thus, the region $\mathcal{R}_t$ occupied by $\mathbb{B}$ at time t can be thought of as a set of current positions:

$$\mathcal{R}_t = \{\, \mathbf{x}(P,t) \mid P \in \mathbb{B} \,\} \,, \qquad (3.1.2)$$

the set of all $\mathbf{x}(P,t)$ such that P is in $\mathbb{B}$.

In general, it may be convenient to define different coordinate systems over the reference and current configurations of a body in order to characterize reference and current positions of material points. Also, different vector bases may be used to define the scalar components of tensor quantities associated with a material point, depending on whether the reference or current position of the material point is being used. Nevertheless, for the indicial notation relations in the following chapters, a single Cartesian coordinate system, with associated vector basis $\{\hat{\mathbf{e}}_i\}$, will be used for all configurations of a body, unless otherwise noted. The reference Cartesian coordinates of a material point will be denoted by (X_1, X_2, X_3) and the current Cartesian coordinates will be denoted by (x_1, x_2, x_3), so that $\mathbf{X} = X_A \hat{\mathbf{e}}_A$ and $\mathbf{x} = x_i \hat{\mathbf{e}}_i$. Using the same Cartesian coordinate system for all configurations simplifies the discussion to follow, so that the underlying physical significance of the relations defined below will be more readily apparent. Of course, the tensor notation forms of these relations (when properly interpreted) are valid for any choice of coordinate systems and vector bases. For a more general treatment, see, for example, Truesdell and Noll (1992) or Green and Zerna (1968).

Since each material point in a body must occupy a single position in space at a time, at each time t there is a (nonlinear) vector operator $\boldsymbol{\chi}(\cdot, t)\colon \mathcal{R} \to \mathcal{R}_t$ that assigns a unique position vector in $\mathcal{R}_t$ to every position vector in $\mathcal{R}$. Thus, for every reference position $\mathbf{X}$ corresponding to a material point in the reference configuration $\mathcal{R}$,

$$\boxed{\mathbf{x} = \boldsymbol{\chi}(\mathbf{X}, t)\,, \quad x_i = \chi_i(X_1, X_2, X_3, t)} \qquad (3.1.3)$$

is the current position of the same material point in the current configuration $\mathcal{R}_t$. Since no two material points may occupy the same point in space at the same time, (3.1.3) cannot assign the same current position in $\mathcal{R}_t$ to two distinct reference positions in $\mathcal{R}$. Thus, (3.1.3) is a *one-to-one mapping* of $\mathcal{R}$ into $\mathcal{R}_t$.

Since each material point in a body has a unique position in both the reference and current configurations, every current position in $\mathcal{R}_t$ corresponds through (3.1.3) to a reference position in $\mathcal{R}$. Thus, (3.1.3) is a one-to-one mapping of $\mathcal{R}$ *onto* $\mathcal{R}_t$, also known as an *isomorphism*. It follows that the mapping (3.1.3) is invertible so that, at each time t, there is a (nonlinear) vector operator $\overset{-1}{\boldsymbol{\chi}}(\cdot, t)\colon \mathcal{R}_t \to \mathcal{R}$ that assigns a unique position vector in $\mathcal{R}$ to every position vector in $\mathcal{R}_t$. Thus, for every current position $\mathbf{x}$ corresponding to a material point in the current configuration $\mathcal{R}_t$,

$$\boxed{\mathbf{X} = \overset{-1}{\boldsymbol{\chi}}(\mathbf{x}, t)\,, \quad X_A = \overset{-1}{\chi}_A(x_1, x_2, x_3, t)} \qquad (3.1.4)$$

is the reference position of the same material point in the reference configuration $\mathcal{R}$. At each instant in time, the isomorphic mappings (3.1.3) and

(3.1.4) provide a one-to-one correspondence between the reference positions in $\mathcal{R}$ and the current positions in $\mathcal{R}_t$ and ensure that no two material points can occupy the same position in space at the same time. Such a mapping is called a *motion* of the body (relative to the prescribed reference configuration). If distances between material points remain constant during the motion, then the motion is a *rigid-body motion*. Otherwise, the body is said to have experienced *deformation*.

Uppercase and lowercase indices, such as those in X_A and x_i, will serve to emphasize that a physical quantity's derivation lies in a description given with respect to either the reference or current configuration of a body, respectively, as discussed in Section 3.1.2. When the same Cartesian coordinate system is used for all configurations, uppercase and lowercase indices have no other significance.

3.1.1 Material Lines, Surfaces, and Volumes

Consider an infinite subset of the body, $\mathbb{C} \subset \mathbb{B}$, and let $\mathcal{C}$ be the set of reference positions of material points in $\mathbb{C}$,

$$\mathcal{C} = \{\, \mathbf{X}(P) \mid P \in \mathbb{C} \,\} \,. \tag{3.1.5}$$

If $\mathcal{C}$ defines a continuous curve in space, then $\mathbb{C}$ is known as a *material line*. Let $\mathcal{C}_t$ be the set of current positions of material points in $\mathbb{C}$,

$$\mathcal{C}_t = \{\, \mathbf{x}(P,t) \mid P \in \mathbb{C} \,\} \,. \tag{3.1.6}$$

The isomorphic mapping between configurations ensures that, for all times t, $\mathcal{C}_t$ will also define a continuous curve in space. Similarly, $\mathbb{S} \subset \mathbb{B}$ is known as a *material surface* if

$$\mathcal{S} = \{\, \mathbf{X}(P) \mid P \in \mathbb{S} \,\} \tag{3.1.7}$$

defines a connected surface in space, and $\mathbb{B}' \subset \mathbb{B}$ is known as a *material volume* if

$$\mathcal{R}' = \{\, \mathbf{X}(P) \mid P \in \mathbb{B}' \,\} \tag{3.1.8}$$

defines a connected volume in space. At any time t, the sets $\mathcal{S}_t$ and $\mathcal{R}'_t$ composed of the current positions of the material points in $\mathbb{S}$ and $\mathbb{B}'$ will also define a connected surface and a connected volume, respectively.

Clearly, a body $\mathbb{B}$ is itself a material volume and the boundary $\partial \mathbb{B}$ of the body consists of one or more material surfaces (one material surface if the body has no "voids" and more than one material surface if the body does have "voids"). The reference positions of the material points in $\partial \mathbb{B}$ define the boundary $\partial \mathcal{R}$ of the region $\mathcal{R}$ occupied by the body in its reference configuration:

$$\partial \mathcal{R} = \{\, \mathbf{X}(P) \mid P \in \partial \mathbb{B} \,\} \,; \tag{3.1.9}$$

and the current positions of the material points define the boundary $\partial \mathcal{R}_t$ of the region $\mathcal{R}_t$ occupied by the body in its current configuration:

$$\partial \mathcal{R}_t = \{ \mathbf{x}(P,t) \mid P \in \partial \mathbb{B} \} \ . \qquad (3.1.10)$$

It follows that since the same material points comprise the boundary in all configurations, no new boundary surfaces can be created during the motion of a body, which precludes the study of crack growth for instance, and no existing boundary surfaces can disappear. In order to consider such physical phenomena, the previous assumptions must be relaxed.

One can think of reference positions $\mathbf{X}$ as labels of material points— a given reference position always refers to the same material point. In contrast, a given position vector $\mathbf{x}$ refers to a fixed point in space, which is the location of different material points at different times. Thus, for a fixed value of $\mathbf{X}$, the mapping $\chi(\mathbf{X},t)$ gives the position as a function of time of that particular material point whose reference position is $\mathbf{X}$. On the other hand, for a fixed value of $\mathbf{x}$, the mapping $\overset{-1}{\chi}(\mathbf{x},t)$ gives the reference positions of the different material points that occupy the position $\mathbf{x}$ at different times.

3.1.2 Material and Spatial Descriptions

A tensor $\mathbf{T}(P,t)$ represents some physical quantity associated with the material point P of a body at the time t. This representation can be given as a field over the material points in a body either in terms of the material points' positions in the reference configuration $\mathbf{X}(P)$ or in terms of the material points' positions in the current configuration $\mathbf{x}(P,t)$.

- The *material description* of a tensor field $\mathbf{T}$ is as a function of reference position $\mathbf{X}$, or reference Cartesian coordinates X_A, and time t:

$$\mathbf{T} = \mathbf{\Psi}(\mathbf{X},t) = \tilde{\mathbf{\Psi}}(X_1, X_2, X_3, t) \ . \qquad (3.1.11)$$

  For a fixed value of $\mathbf{X}$, $\mathbf{\Psi}(\mathbf{X},t)$ gives the value of $\mathbf{T}$, as a function of time, associated with the fixed material point whose position in the reference configuration is $\mathbf{X}$. This is sometimes referred to as the Lagrangian description.

- The *spatial description* of a tensor field $\mathbf{T}$ is as a function of current position $\mathbf{x}$, or current Cartesian coordinates x_i, and time t:

$$\mathbf{T} = \psi(\mathbf{x},t) = \tilde{\psi}(x_1, x_2, x_3, t) \ . \qquad (3.1.12)$$

  For a fixed value of $\mathbf{x}$, $\psi(\mathbf{x},t)$ gives the value of $\mathbf{T}$, as a function of time, associated with the fixed point in space $\mathbf{x}$, which will be the value of $\mathbf{T}$ associated with different material points at different times. This is sometimes referred to as the Eulerian description.

The material and spatial descriptions of a tensor field $\mathbf{T}$ are related through the mapping (3.1.3) and (3.1.4). Thus, the material description is given in terms of the spatial description and (3.1.3) by

$$\mathbf{\Psi}(\mathbf{X}, t) = \psi[\chi(\mathbf{X}, t), t] \tag{3.1.13a}$$

and

$$\begin{aligned}
&\widetilde{\mathbf{\Psi}}(X_1, X_2, X_3, t) \\
&= \widetilde{\psi}[\chi_1(X_1, X_2, X_3, t), \chi_2(X_1, X_2, X_3, t), \chi_3(X_1, X_2, X_3, t), t] \, . \tag{3.1.13b}
\end{aligned}$$

Similarly, the spatial description is given in terms of the material description and (3.1.4) by

$$\psi(\mathbf{x}, t) = \mathbf{\Psi}\left[\overset{-1}{\chi}(\mathbf{x}, t), t\right] \tag{3.1.14a}$$

and

$$\begin{aligned}
&\widetilde{\psi}(x_1, x_2, x_3, t) \\
&= \widetilde{\mathbf{\Psi}}\left[\overset{-1}{\chi}_1(x_1, x_2, x_3, t), \overset{-1}{\chi}_2(x_1, x_2, x_3, t), \overset{-1}{\chi}_3(x_1, x_2, x_3, t), t\right] \, . \tag{3.1.14b}
\end{aligned}$$

These relations reflect the fact that the value of the tensor quantity $\mathbf{T}$ is the same at any given material point, regardless of whether a material or spatial description is used for the position of that material point.

Strictly speaking, the use of five separate symbols for the tensor quantity $\mathbf{T}$ and the associated functions $\mathbf{\Psi}$ and $\widetilde{\mathbf{\Psi}}$ of the material description and ψ and $\widetilde{\psi}$ of the spatial description is correct. However, to avoid the proliferation of an ever-increasing number of symbols, the same symbol will frequently be used in all five roles. Consequently, to avoid confusion over whether differentiation is with respect to reference or spatial coordinates, use of the comma convention is temporarily suspended, and for a tensor field $\mathbf{T}$, the following notation is defined for differentiation with respect to Cartesian coordinate variables:

$$\frac{\partial \mathbf{T}}{\partial X_A} \equiv \frac{\partial}{\partial X_A} \widetilde{\mathbf{\Psi}}(X_1, X_2, X_3, t) \, , \quad \frac{\partial \mathbf{T}}{\partial x_i} \equiv \frac{\partial}{\partial x_i} \widetilde{\psi}(x_1, x_2, x_3, t) \, . \tag{3.1.15}$$

Define the *material gradient* and *spatial gradient* of $\mathbf{T}$ as

$$\text{Grad}\,\mathbf{T} \equiv \mathbf{\nabla}\mathbf{\Psi}(\mathbf{X}, t) \, , \quad \text{grad}\,\mathbf{T} \equiv \mathbf{\nabla}\psi(\mathbf{x}, t) \, , \tag{3.1.16}$$

respectively, so that in a Cartesian coordinate system,

$$\text{Grad } \mathbf{T} = \frac{\partial \mathbf{T}}{\partial X_A} \hat{\mathbf{e}}_A , \quad \text{grad } \mathbf{T} = \frac{\partial \mathbf{T}}{\partial x_i} \hat{\mathbf{e}}_i . \tag{3.1.17}$$

The material and spatial divergence, curl, etc. are similarly defined.

Partial differentiation with respect to time of the material and spatial descriptions of a tensor are distinguished by the following definitions:

$$\frac{D\mathbf{T}}{Dt} \equiv \frac{\partial}{\partial t} \mathbf{\Psi}(\mathbf{X}, t) , \quad \frac{\partial \mathbf{T}}{\partial t} \equiv \frac{\partial}{\partial t} \psi(\mathbf{x}, t) . \tag{3.1.18}$$

It will be shown shortly how the *material time derivative* D/Dt and the *spatial time derivative* $\partial/\partial t$ are related.

A similar notation will be used for line, surface, and volume integrals in the reference and current configurations. Consider a material line $\mathbb{C}$ that occupies the curve $\mathcal{C}$ in the reference configuration and the curve $\mathcal{C}_t$ in the current configuration. Let dS and ds be the lengths of the differential element $d\mathbb{C}$ of the material line in the reference and current configurations, respectively. Then, the following notations are defined:

$$\int_{\mathcal{C}} \mathbf{T} \, dS \equiv \int_{\mathcal{C}} \mathbf{\Psi}(\mathbf{X}, t) \, dS , \quad \int_{\mathcal{C}_t} \mathbf{T} \, ds \equiv \int_{\mathcal{C}_t} \psi(\mathbf{x}, t) \, ds . \tag{3.1.19}$$

Next, consider a material surface $\mathbb{S}$ that occupies the surfaces $\mathcal{S}$ in the reference configuration and $\mathcal{S}_t$ in the current configuration, and let dA and da be the areas of the differential element $d\mathbb{S}$ of the material surface in the reference and current configurations, respectively. Then,

$$\int_{\mathcal{S}} \mathbf{T} \, dA \equiv \int_{\mathcal{S}} \mathbf{\Psi}(\mathbf{X}, t) \, dA , \quad \int_{\mathcal{S}_t} \mathbf{T} \, da \equiv \int_{\mathcal{S}_t} \psi(\mathbf{x}, t) \, da . \tag{3.1.20}$$

Finally, for a material volume $\mathbb{B}'$ that occupies the regions $\mathcal{R}'$ in the reference configuration and $\mathcal{R}'_t$ in the current configuration,

$$\int_{\mathcal{R}'} \mathbf{T} \, dV \equiv \int_{\mathcal{R}'} \mathbf{\Psi}(\mathbf{X}, t) \, dV , \quad \int_{\mathcal{R}'_t} \mathbf{T} \, dv \equiv \int_{\mathcal{R}'_t} \psi(\mathbf{x}, t) \, dv , \tag{3.1.21}$$

where dV and dv are the volumes of the differential element $d\mathbb{B}'$ of the material volume in the reference and current configurations, respectively.

Often, only one reference configuration and one current configuration (sometimes called the *deformed configuration*) are considered and the explicit dependence on time can be dropped. In this case, the material description of a tensor field $\mathbf{T}$ will be given as $\mathbf{T}(\mathbf{X})$ and its spatial description will be given as $\mathbf{T}(\mathbf{x})$.

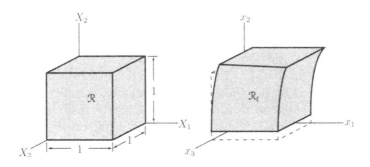

FIGURE 3.2.   Example motion of a unit cube.

3.1.3   Displacement

The *displacement* $\mathbf{u}$ of a material point P at time t is defined as the vector from its position in the reference configuration to its position in the current configuration, so that $\mathbf{u}(P,t) \equiv \mathbf{x}(P,t) - \mathbf{X}(P)$ (see Figure 3.1). Therefore, the material description of the displacement field is given by

$$\boxed{\mathbf{u} = \chi(\mathbf{X},t) - \mathbf{X} , \quad u_A = \chi_A(X_1, X_2, X_3, t) - X_A ,} \qquad (3.1.22\text{a})$$

and the spatial description of the displacement field is given by

$$\boxed{\mathbf{u} = \mathbf{x} - \overset{-1}{\chi}(\mathbf{x},t) , \quad u_i = x_i - \overset{-1}{\chi}_i(x_1, x_2, x_3, t) .} \qquad (3.1.22\text{b})$$

A material point is said to be displaced from its reference position.

Example 3.1

Consider a body which, in the reference configuration $\mathcal{R}$, occupies the unit cube $(0 \leq X_1 \leq 1, 0 \leq X_2 \leq 1, 0 \leq X_3 \leq 1)$ as shown in Figure 3.2. The mapping $x_i = \chi_i(X_1, X_2, X_3)$ for the Cartesian coordinates of a material point in the current configuration in terms of its Cartesian coordinates in the reference configuration is given by

$$x_1 = X_1 + a_1 X_2^2 , \quad x_2 = X_2 + a_2 , \quad x_3 = X_3 + a_3 X_2 X_3 ,$$

where a_1, a_2, and a_3 are positive constants. Derive the inverse expressions for the Cartesian coordinates of a material point in the reference configuration in terms of its Cartesian coordinates in the current configuration.

Solution. By solving the above relation for X_A in terms of x_i, one finds that the inverse mapping $X_A = \overset{-1}{\chi}_A(x_1, x_2, x_3)$ is given by

$$X_1 = x_1 - a_1(x_2 - a_2)^2 , \quad X_2 = x_2 - a_2 , \quad X_3 = \frac{x_3}{1 + a_3(x_2 - a_2)} .$$

Note that, in general, even if an analytic expression for the mapping $x_i = \chi_i(X_1, X_2, X_3)$ is given, it may not be possible to obtain an analytic expression for the inverse mapping $X_A = \overset{-1}{\chi}_A(x_1, x_2, x_3)$, as was possible here. This does not mean that the inverse mapping does not exist [it does exist so long as $x_i = \chi_i(X_1, X_2, X_3)$ is isomorphic], only that an analytic expression for it does not exist.

Example 3.2
The material description of the current temperature distribution in the body described in Example 3.1 is given by

$$\theta = A(1 + X_2 X_3) ,$$

where A is a constant. This gives the current temperature of each material point in terms of each material point's position in the reference configuration. Find the spatial description of the current temperature distribution.

Solution. By substituting the result for $X_A = \overset{-1}{\chi}_A(x_1, x_2, x_3)$ found in Example 3.1, one finds that

$$\theta = A \left\{ 1 + (x_2 - a_2) \left[\frac{x_3}{1 + a_3(x_2 - a_2)} \right] \right\}$$
$$= A \left[1 + \frac{x_3(x_2 - a_2)}{1 + a_3(x_2 - a_2)} \right] .$$

This gives the current temperature in the body as a function of current position.

Example 3.3
Derive the material and spatial descriptions of the displacement vector field **u** for the motion given in Example 3.1.

Solution. Using the mapping $x_i = \chi_i(X_1, X_2, X_3)$ given in Example 3.1, the material description of the displacement field $u_A = \chi_A(X_1, X_2, X_3) - X_A$ is given in terms of scalar components by

$$u_1 = a_1 X_2^2 , \quad u_2 = a_2 , \quad u_3 = a_3 X_2 X_3 .$$

Similarly, from the result $X_A = \overset{-1}{\chi}_A(x_1, x_2, x_3)$ found in Example 3.1, it follows that the spatial description of the displacement field $u_i = x_i - \overset{-1}{\chi}_i(x_1, x_2, x_3)$ is given in terms of scalar components by

$$u_1 = a_1(x_2 - a_2)^2 , \quad u_2 = a_2 , \quad u_3 = \frac{a_3 x_3(x_2 - a_2)}{1 + a_3(x_2 - a_2)} .$$

It is left to the reader to verify that these expressions are consistent with the relations (3.1.13) and (3.1.14) between material and spatial descriptions of tensor fields.

3.1.4   Material Time Derivative

Let the material and spatial descriptions of the tensor quantity $\mathbf{T}$ be given as in (3.1.11) and (3.1.12). Then, the time rate of change of $\mathbf{T}$ associated with the material point P is

$$\frac{d}{dt}\mathbf{\Psi}[\mathbf{X}(P),t] \ . \tag{3.1.23}$$

Therefore, the tensor field representing the time rates of change of $\mathbf{T}$ experienced by all the material points in a body is

$$\frac{D\mathbf{T}}{Dt} \equiv \frac{\partial}{\partial t}\mathbf{\Psi}(\mathbf{X},t) \ . \tag{3.1.24}$$

For this reason, this quantity is known as the *material time derivative* of $\mathbf{T}$. The material time derivative of $\mathbf{T}$ can be expressed in terms of the spatial description of $\mathbf{T}$ by recalling that $\mathbf{\Psi}(\mathbf{X},t) = \psi[\chi(\mathbf{X},t),t]$ and applying the chain rule:

$$\begin{aligned}
\frac{D\mathbf{T}}{Dt} &= \frac{\partial}{\partial t}\psi[\chi(\mathbf{X},t),t] \\
&= \frac{\partial}{\partial t}\tilde{\psi}[\chi_1(X_1,X_2,X_3,t),\chi_2(X_1,X_2,X_3,t),\chi_3(X_1,X_2,X_3,t),t] \\
&= \frac{\partial}{\partial t}\tilde{\psi}(x_1,x_2,x_3,t) + \frac{\partial}{\partial x_i}\tilde{\psi}(x_1,x_2,x_3,t)\frac{\partial}{\partial t}\chi_i(X_1,X_2,X_3,t) \ .
\end{aligned}$$
$$\tag{3.1.25}$$

Using the notation defined in (3.1.15) and (3.1.18), this can be rewritten as

$$\frac{D\mathbf{T}}{Dt} = \frac{\partial\mathbf{T}}{\partial t} + (\text{grad}\,\mathbf{T})\cdot\frac{D\mathbf{x}}{Dt} \ , \qquad \frac{D\mathbf{T}}{Dt} = \frac{\partial\mathbf{T}}{\partial t} + \frac{\partial\mathbf{T}}{\partial x_i}\frac{Dx_i}{Dt} \ . \tag{3.1.26}$$

The material time derivative of the current position vector,

$$\mathbf{v} \equiv \frac{D\mathbf{x}}{Dt} \ , \qquad v_i \equiv \frac{Dx_i}{Dt} \ , \tag{3.1.27}$$

is called the *material velocity field* and represents the current velocities of the material points in a body. The material time derivative in terms of the spatial description of $\mathbf{T}$ can thus be rewritten as

$$\boxed{\frac{D\mathbf{T}}{Dt} = \frac{\partial\mathbf{T}}{\partial t} + (\text{grad}\,\mathbf{T})\cdot\mathbf{v} \ , \qquad \frac{D\mathbf{T}}{Dt} = \frac{\partial\mathbf{T}}{\partial t} + \frac{\partial\mathbf{T}}{\partial x_i}v_i \ .} \tag{3.1.28}$$

The material time derivative of the material velocity field is the *material acceleration field*,

$$\mathbf{a} \equiv \frac{D\mathbf{v}}{Dt} = \frac{\partial\mathbf{v}}{\partial t} + (\text{grad}\,\mathbf{v})\cdot\mathbf{v} \ , \qquad a_i \equiv \frac{Dv_i}{Dt} = \frac{\partial v_i}{\partial t} + \frac{\partial v_i}{\partial x_j}v_j \ , \tag{3.1.29}$$

and represents the current accelerations of the material points in a body.

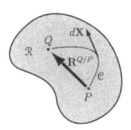 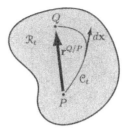

FIGURE 3.3. *Mapping of relative position vectors.*

3.1.5 Deformation Gradient Tensor

Consider two material points P and Q and an arbitrary material line $\mathbb{C}$ connecting P and Q that occupies the piecewise smooth curve $\mathcal{C}$ in the reference configuration (Figure 3.3). The position vector $\mathbf{R}^{Q/P} = \mathbf{X}(Q) - \mathbf{X}(P)$ of Q relative to P in the reference configuration is given in terms of the path integral along the curve $\mathcal{C}$ from $\mathbf{X}(P)$ to $\mathbf{X}(Q)$ by

$$\mathbf{R}^{Q/P} = \int_{\mathcal{C}} d\mathbf{X} , \quad R_A^{Q/P} = \int_{\mathcal{C}} dX_A . \tag{3.1.30}$$

Recall that at a point on the curve, the differential vector $d\mathbf{X} = \hat{\mathbf{\Upsilon}} \, dS$, where $\hat{\mathbf{\Upsilon}}$ is the unit tangent vector to the curve at that point, in the direction of integration, and dS is the length in the reference configuration of the differential element $d\mathbb{C}$ of the material line. The integral in (3.1.30) is path independent; that is, the value of the integral is the same for all material lines $\mathbb{C}$ with end points P and Q.

The material line $\mathbb{C}$ occupies the piecewise smooth curve $\mathcal{C}_t$ in the current configuration (Figure 3.3). The position vector $\mathbf{r}^{Q/P} = \mathbf{x}(Q,t) - \mathbf{x}(P,t)$ of Q relative to P in the current configuration is given in terms of the path integral along $\mathcal{C}_t$ from $\mathbf{x}(P,t)$ to $\mathbf{x}(Q,t)$ by

$$\mathbf{r}^{Q/P} = \int_{\mathcal{C}_t} d\mathbf{x} , \quad r_i^{Q/P} = \int_{\mathcal{C}_t} dx_i . \tag{3.1.31}$$

At a point on the curve, the differential vector $d\mathbf{x} = \hat{\boldsymbol{v}} \, ds$, where $\hat{\boldsymbol{v}}$ is the unit tangent vector to the curve at that point and ds is the length in the current configuration of the differential element $d\mathbb{C}$ of the material line. The integral in (3.1.31) is also path independent.

For an isomorphic mapping between the reference and current configurations of a body, the curve $\mathcal{C}_t$ is uniquely determined by the curve $\mathcal{C}$. Thus, at every material point on the material line $\mathbb{C}$, the differential vector $d\mathbf{x}$

is uniquely determined by the differential vector $d\mathbf{X}$, which, by the chain rule, implies the existence of a nonsingular, second-order tensor,

$$\mathbf{F} \equiv \operatorname{Grad} \mathbf{x} , \quad F_{iA} \equiv \frac{\partial x_i}{\partial X_A} , \tag{3.1.32}$$

such that

$$d\mathbf{x} = \mathbf{F} \cdot d\mathbf{X} , \quad dx_i = F_{iA} \, dX_A \tag{3.1.33}$$

and

$$d\mathbf{X} = \mathbf{F}^{-1} \cdot d\mathbf{x} , \quad dX_A = F_{Ai}^{-1} \, dx_i . \tag{3.1.34}$$

The second-order tensor $\mathbf{F}$ is called the *deformation gradient tensor* and its nonsingularity ensures the existence of its inverse $\mathbf{F}^{-1}$, which is also given by

$$\mathbf{F}^{-1} = \operatorname{grad} \mathbf{X} , \quad F_{Ai}^{-1} = \frac{\partial X_A}{\partial x_i} . \tag{3.1.35}$$

Conversely, if at every material point in a body the deformation gradient tensor exists and is nonsingular, then at every material point on every material line in the body there is a one-to-one mapping between the differential vectors $d\mathbf{x}$ and $d\mathbf{X}$ given by (3.1.33), so that every curve $\mathcal{C}_t$ is uniquely determined by a corresponding curve $\mathcal{C}$. This implies an isomorphic mapping between the reference and current configurations of the body. Therefore, necessary and sufficient conditions for a motion to be isomorphic are that its deformation gradient tensor (3.1.32) exists and is nonsingular at every material point in the body. In terms of the *Jacobian of the motion*,

$$J \equiv \det \mathbf{F} , \tag{3.1.36}$$

necessary and sufficient conditions for an isomorphic motion are that J exists and $J \neq 0$ at every material point in the body.

It follows from (3.1.33) that the position vector of Q relative to P in the *current* configuration (3.1.31) can be rewritten as a path integral along the curve $\mathcal{C}$ occupied by the material line $\mathbb{C}$ in the *reference* configuration:

$$\mathbf{r}^{Q/P} = \int_{\mathcal{C}} \mathbf{F} \cdot d\mathbf{X} , \quad r_i^{Q/P} = \int_{\mathcal{C}} F_{iA} \, dX_A . \tag{3.1.37}$$

Alternatively, the position vector of Q relative to P in the *reference* configuration (3.1.30) can be rewritten as a path integral along the curve $\mathcal{C}_t$ occupied by the material line $\mathbb{C}$ in the *current* configuration:

$$\mathbf{R}^{Q/P} = \int_{\mathcal{C}_t} \mathbf{F}^{-1} \cdot d\mathbf{x} , \quad R_A^{Q/P} = \int_{\mathcal{C}_t} F_{Ai}^{-1} \, dx_i . \tag{3.1.38}$$

The deformation gradient tensor $\mathbf{F} = F_{iA}\hat{\mathbf{e}}_i\hat{\mathbf{e}}_A$ is related to both the reference and current configurations, as indicated by the combination of lowercase and uppercase indices. Tensors of this sort are known as *two-point tensors*. The distinction is not important here since the same coordinate system and vector basis are used for all configurations of a body, but if two different vector bases are used for the reference and current configurations and one of them changes while the other remains the same, then a two-point tensor's scalar components will transform like those of a vector.

Given a reference configuration $\mathcal{R}$ and the material description $\mathbf{F}(\mathbf{X}, t)$ of the deformation gradient tensor, (3.1.37) gives the relative position vector between any two material points in the current configuration (i.e., at time t). Conversely, given a current configuration and the spatial description $\mathbf{F}^{-1}(\mathbf{x}, t)$ of the inverse deformation gradient tensor, (3.1.38) gives the relative position vector between any two material points in the reference configuration. Since two vectors in the same direction with the same magnitude are equal, $\mathbf{F}$ conveys no information about the translation of a body. It does, however, completely describe the rotation and deformation of a body.

Consider a portion of the body (that is, a material volume) that occupies the region $\mathcal{R}'$ in the reference configuration and $\mathcal{R}'_t$ in the current configuration. If $\mathbf{F} = \mathbf{I}$ everywhere in $\mathcal{R}'$, then from (3.1.37) it is seen that the relative position vector between any two material points in $\mathcal{R}'$ is unchanged by the mapping to $\mathcal{R}'_t$—this portion of the body is not rotated and is undeformed, although it may have been translated. By considering the limit as this portion of the body shrinks to a material point, it can be seen that the value of $\mathbf{F}$ at a material point completely describes the local rotation and deformation of the body at that material point.

If the deformation gradient tensor $\mathbf{F}$ has the same value at every material point in a body for some mapping from a reference configuration to a current configuration, then the mapping is said to be a *homogeneous motion* of the body. At any given time t, a mapping $\mathbf{x} = \chi(\mathbf{X}, t)$ is a homogeneous motion if and only if it can be expressed as

$$\mathbf{x} = \mathbf{A} \cdot \mathbf{X} + \mathbf{c}, \quad x_i = A_{iA}X_A + c_i, \tag{3.1.39}$$

where the second-order tensor $\mathbf{A}$ and the vector $\mathbf{c}$ are constants, in which case, $\mathbf{c}$ is a rigid-body translation of the body and the deformation gradient tensor is $\mathbf{F} = \mathbf{A}$. It follows from (3.1.37) and (3.1.38) that, for a homogeneous motion, the relative position vectors $\mathbf{R}^{Q/P}$ and $\mathbf{r}^{Q/P}$ between two material points P and Q in the reference and current configurations are related by

$$\mathbf{r}^{Q/P} = \mathbf{F} \cdot \mathbf{R}^{Q/P}, \quad r_i^{Q/P} = F_{iA}R_A^{Q/P} \tag{3.1.40}$$

and

$$\mathbf{R}^{Q/P} = \mathbf{F}^{-1} \cdot \mathbf{r}^{Q/P} , \quad R_A^{Q/P} = F_{Ai}^{-1} r_i^{Q/P} . \tag{3.1.41}$$

Example 3.4
For the motion given in Example 3.1, find the deformation gradient tensor
$\mathbf{F}$, verify that its Jacobian exists and is nonzero at every material point in
the body, and find the inverse deformation gradient tensor $\mathbf{F}^{-1}$.

Solution. It follows from the definition $F_{iA} \equiv \partial x_i / \partial X_A$ and the mapping $x_i = \chi_i(X_1, X_2, X_3)$ given in Example 3.1 that the scalar components
of $\mathbf{F}$ are given in matrix form by

$$[\mathbf{F}(\mathbf{X})] = \begin{bmatrix} 1 & 2a_1 X_2 & 0 \\ 0 & 1 & 0 \\ 0 & a_3 X_3 & 1 + a_3 X_2 \end{bmatrix} .$$

The Jacobian $J \equiv \det \mathbf{F}$ of the motion is

$$J = 1 + a_3 X_2 .$$

Clearly, J exists and since a_3 is positive and, for every material point in
the body, X_2 is non-negative, it follows that $J \neq 0$ at every point in the
body.

From the relation $F_{Ai}^{-1} \equiv \partial X_A / \partial x_i$ and the result $X_A = \overset{-1}{\chi}_A(x_1, x_2, x_3)$
found in Example 3.1, the scalar components of $\mathbf{F}^{-1}$ are given by

$$[\mathbf{F}^{-1}(\mathbf{x})] = \begin{bmatrix} 1 & -2a_1(x_2 - a_2) & 0 \\ 0 & 1 & 0 \\ 0 & \dfrac{-a_3 x_3}{[1 + a_3(x_2 - a_2)]^2} & \dfrac{1}{1 + a_3(x_2 - a_2)} \end{bmatrix} .$$

Note from the mapping $x_i = \chi_i(X_1, X_2, X_3)$ that the material description
of $\mathbf{F}^{-1}$ is

$$[\mathbf{F}^{-1}(\mathbf{X})] = \begin{bmatrix} 1 & -2a_1 X_2 & 0 \\ 0 & 1 & 0 \\ 0 & \dfrac{-a_3 X_3}{1 + a_3 X_2} & \dfrac{1}{1 + a_3 X_2} \end{bmatrix}$$

and that, as required, $[\mathbf{F}^{-1}] = [\mathbf{F}]^{-1}$.

3.1.6 Compatibility

Recall the identity (2.4.42), which states that, for a tensor field $\mathbf{T}$ of any
order, $\nabla \times (\nabla \mathbf{T}) = \mathbf{0}$. It follows, since $\mathbf{F} \equiv \operatorname{Grad} \mathbf{x}$, that

$$\operatorname{Curl} \mathbf{F} = \mathbf{0} , \quad e_{ABC} \frac{\partial F_{iB}}{\partial X_A} = 0 . \tag{3.1.42}$$

This *compatibility condition* is equivalent to the requirement in a Cartesian coordinate system that

$$\frac{\partial^2 x_i}{\partial X_A \partial X_B} = \frac{\partial^2 x_i}{\partial X_B \partial X_A} \tag{3.1.43}$$

and is a necessary condition that $\mathbf{F}$ must satisfy if it is derived from a (two-times continuously differentiable) mapping $\mathbf{x} = \chi(\mathbf{X}, t)$. It can also be shown that this compatibility condition is sufficient in a simply connected body (see Section 3.4.4) to ensure that there exists a mapping $\mathbf{x} = \chi(\mathbf{X}, t)$ such that $\mathbf{F} = \text{Grad} \, \mathbf{x}$.

3.1.7 Polar Decomposition Theorem

The deformation gradient tensor (actually, any nonsingular second-order tensor) can be uniquely decomposed as either

$$\mathbf{F} = \mathbf{R} \cdot \mathbf{U} , \quad F_{iA} = R_{iB} U_{BA} \tag{3.1.44}$$

or

$$\mathbf{F} = \mathbf{V} \cdot \mathbf{R} , \quad F_{iA} = V_{ij} R_{jA} , \tag{3.1.45}$$

where $\mathbf{R}$ is an orthogonal (two-point) tensor and $\mathbf{U}$ and $\mathbf{V}$ are symmetric, positive definite tensors. The *right Cauchy stretch tensor* $\mathbf{U}$ is given by

$$\boxed{\mathbf{U} \equiv (\mathbf{F}^\mathsf{T} \cdot \mathbf{F})^{1/2} ,} \tag{3.1.46}$$

the *left Cauchy stretch tensor* $\mathbf{V}$ is given by

$$\boxed{\mathbf{V} \equiv (\mathbf{F} \cdot \mathbf{F}^\mathsf{T})^{1/2} ,} \tag{3.1.47}$$

and the *rotation tensor* $\mathbf{R}$ is given by

$$\boxed{\mathbf{R} = \mathbf{F} \cdot \mathbf{U}^{-1} = \mathbf{V}^{-1} \cdot \mathbf{F} .} \tag{3.1.48}$$

As their names indicate, it will be shown that $\mathbf{U}$ and $\mathbf{V}$ each completely characterize the deformation of a body and that $\mathbf{R}$ characterizes a separate rigid-body motion.

To show that the decompositions (3.1.44) and (3.1.45) exist, note first that $\mathbf{F}^\mathsf{T} \cdot \mathbf{F}$ and $\mathbf{F} \cdot \mathbf{F}^\mathsf{T}$ are symmetric, positive definite second-order tensors and therefore have arbitrary real powers defined by (2.3.127), which are also symmetric and positive definite. Thus, $\mathbf{U}$ (3.1.46) and $\mathbf{V}$ (3.1.47) exist and are symmetric, positive definite (and are therefore nonsingular and have inverses). It follows that $\mathbf{R} \cdot \mathbf{U}$ (3.1.48) exists and clearly that $\mathbf{R} \cdot \mathbf{U} = \mathbf{V} \cdot \mathbf{R} = \mathbf{F}$. Finally,

$$\mathbf{R}^\mathsf{T} \cdot \mathbf{R} = (\mathbf{F} \cdot \mathbf{U}^{-1})^\mathsf{T} \cdot (\mathbf{F} \cdot \mathbf{U}^{-1}) = (\mathbf{U}^{-1})^\mathsf{T} \cdot \mathbf{F}^\mathsf{T} \cdot \mathbf{F} \cdot \mathbf{U}^{-1} . \qquad (3.1.49)$$

However, $(\mathbf{U}^{-1})^\mathsf{T} = (\mathbf{U}^\mathsf{T})^{-1} = \mathbf{U}^{-1}$, since $\mathbf{U}$ is symmetric, and $\mathbf{F}^\mathsf{T} \cdot \mathbf{F} = \mathbf{U}^2 = \mathbf{U} \cdot \mathbf{U}$, so that

$$\mathbf{R}^\mathsf{T} \cdot \mathbf{R} = (\mathbf{U}^{-1} \cdot \mathbf{U}) \cdot (\mathbf{U} \cdot \mathbf{U}^{-1}) = \mathbf{I} \cdot \mathbf{I} = \mathbf{I} , \qquad (3.1.50)$$

thus showing that $\mathbf{R}$ is orthogonal.

To show that the decompositions (3.1.44) and (3.1.45) are unique, suppose that two symmetric, positive definite tensors $\widetilde{\mathbf{U}}$ and $\widetilde{\mathbf{V}}$ and an orthogonal tensor $\widetilde{\mathbf{R}}$ exist such that $\mathbf{F} = \widetilde{\mathbf{R}} \cdot \widetilde{\mathbf{U}} = \widetilde{\mathbf{V}} \cdot \widetilde{\mathbf{R}}$. Then, since $\widetilde{\mathbf{U}}$ and $\widetilde{\mathbf{V}}$ are symmetric, it follows that $\mathbf{F}^\mathsf{T} = \widetilde{\mathbf{U}} \cdot \widetilde{\mathbf{R}}^\mathsf{T} = \widetilde{\mathbf{R}}^\mathsf{T} \cdot \widetilde{\mathbf{V}}$ from whence

$$\mathbf{F}^\mathsf{T} \cdot \mathbf{F} = \widetilde{\mathbf{U}} \cdot \widetilde{\mathbf{R}}^\mathsf{T} \cdot \widetilde{\mathbf{R}} \cdot \widetilde{\mathbf{U}} = \widetilde{\mathbf{U}}^2 \qquad (3.1.51)$$

and

$$\mathbf{F} \cdot \mathbf{F}^\mathsf{T} = \widetilde{\mathbf{V}} \cdot \widetilde{\mathbf{R}} \cdot \widetilde{\mathbf{R}}^\mathsf{T} \cdot \widetilde{\mathbf{V}} = \widetilde{\mathbf{V}}^2 . \qquad (3.1.52)$$

Thus, $\widetilde{\mathbf{U}} = (\mathbf{F}^\mathsf{T} \cdot \mathbf{F})^{1/2} = \mathbf{U}$ and $\widetilde{\mathbf{V}} = (\mathbf{F} \cdot \mathbf{F}^\mathsf{T})^{1/2} = \mathbf{V}$, and $\widetilde{\mathbf{R}} = \mathbf{F} \cdot \widetilde{\mathbf{U}}^{-1} = \mathbf{F} \cdot \mathbf{U}^{-1} = \mathbf{R}$.

3.2   Strain Tensors: Referential Formulation

What is desired is a description of the deformation of a body, independent of both translation and rotation. It will be shown that there are several second-order tensors, collectively referred to as *strain tensors*, any one of which completely describes the deformation. In this section, strain tensors are derived that are suitable for use when directions in a body are given with respect to its reference configuration.

In the reference configuration, the length of the previously defined material line $\mathbb{C}$ (Figure 3.3) is

$$S = \int_{\mathfrak{C}} dS , \qquad (3.2.1)$$

whereas its length in the current configuration is

$$s = \int_{\mathfrak{C}_t} ds , \qquad (3.2.2)$$

where $\mathcal{C}$ and $\mathcal{C}_t$ are the curves occupied by $\mathbb{C}$ in the reference and current configurations, respectively. Since $d\mathbf{X} = \hat{\mathbf{\Upsilon}}\,dS$ and $d\mathbf{x} = \hat{v}\,ds$, it follows that

$$
\begin{aligned}
(ds)^2 &= d\mathbf{x} \cdot d\mathbf{x} \\
&= (\mathbf{F} \cdot d\mathbf{X}) \cdot (\mathbf{F} \cdot d\mathbf{X}) \\
&= d\mathbf{X} \cdot (\mathbf{F}^\mathsf{T} \cdot \mathbf{F}) \cdot d\mathbf{X} \\
&= [\hat{\mathbf{\Upsilon}} \cdot (\mathbf{F}^\mathsf{T} \cdot \mathbf{F}) \cdot \hat{\mathbf{\Upsilon}}](dS)^2 \ .
\end{aligned}
\tag{3.2.3}
$$

Recalling that $\mathbf{F}^\mathsf{T} \cdot \mathbf{F} = \mathbf{U}^2$ is symmetric and positive definite, it follows that $\hat{\mathbf{\Upsilon}} \cdot (\mathbf{F}^\mathsf{T} \cdot \mathbf{F}) \cdot \hat{\mathbf{\Upsilon}} > 0$ and one may write

$$
ds = (\hat{\mathbf{\Upsilon}} \cdot \mathbf{C} \cdot \hat{\mathbf{\Upsilon}})^{1/2} dS \ ,
\tag{3.2.4}
$$

where the *right Cauchy-Green deformation tensor* $\mathbf{C}$ is defined as

$$
\boxed{\mathbf{C} \equiv \mathbf{F}^\mathsf{T} \cdot \mathbf{F} = \mathbf{U}^2 \ , \quad C_{AB} \equiv F_{iA} F_{iB} = U_{AC} U_{CB} \ .}
\tag{3.2.5}
$$

Thus, the *current* length of the material line $\mathbb{C}$ can be expressed as an integral over the curve $\mathcal{C}$ occupied by the material line in the *reference* configuration:

$$
s = \int_{\mathcal{C}} (\hat{\mathbf{\Upsilon}} \cdot \mathbf{C} \cdot \hat{\mathbf{\Upsilon}})^{1/2} dS \ , \quad s = \int_{\mathcal{C}} (C_{AB}\,\hat{\Upsilon}_A\,\hat{\Upsilon}_B)^{1/2} dS \ ,
\tag{3.2.6}
$$

where $\hat{\mathbf{\Upsilon}}$ is the unit tangent vector to the curve $\mathcal{C}$ occupied by the material line $\mathbb{C}$ in the reference configuration. Note that the right Cauchy-Green deformation tensor $\mathbf{C}$ is symmetric, positive definite.

Using (3.2.1) and (3.2.6), the difference between the current and reference lengths of a material line $\mathbb{C}$ can be given as an integral over the curve $\mathcal{C}$ occupied by $\mathbb{C}$ in the reference configuration:

$$
s - S = \int_{\mathcal{C}} [(\hat{\mathbf{\Upsilon}} \cdot \mathbf{C} \cdot \hat{\mathbf{\Upsilon}})^{1/2} - 1] dS \ .
\tag{3.2.7}
$$

Given a reference configuration $\mathcal{R}$ and the material description $\mathbf{C}(\mathbf{X}, t)$ of the right Cauchy-Green deformation tensor, (3.2.7) can be used to find the change in length of any material line whose curve $\mathcal{C}$ in the reference configuration is known. The value of the right Cauchy-Green deformation tensor at a material point completely characterizes the local deformation at that material point independent of translation and rotation, and since $\mathbf{C} = \mathbf{U}^2$, so does the value of the right Cauchy stretch tensor.

Consider a material volume that occupies the region $\mathcal{R}'$ in the reference configuration and $\mathcal{R}'_t$ in the current configuration. If $\mathbf{C} = \mathbf{I}$ everywhere in $\mathcal{R}'$, then from (3.2.7) it is seen that the length of any material line in $\mathcal{R}'$ is unchanged by the mapping to $\mathcal{R}'_t$—the motion of this portion of the body is a rigid-body motion. Thus, any one of the following five statements implies the other four: this portion of the body experiences a rigid-body motion, the right Cauchy-Green deformation tensor $\mathbf{C} = \mathbf{I}$ at every point in $\mathcal{R}'$, the right Cauchy stretch tensor $\mathbf{U} = \mathbf{I}$ at every point in $\mathcal{R}'$, the deformation gradient tensor $\mathbf{F}$ is orthogonal at every point in $\mathcal{R}'$, or the deformation gradient tensor $\mathbf{F} = \mathbf{R}$ at every point in $\mathcal{R}'$ where $\mathbf{R}$ is the orthogonal rotation tensor. Furthermore, it can be shown, as a consequence of the compatibility condition (3.1.42) and the properties of orthogonal tensors, that if the deformation gradient tensor $\mathbf{F}$ is orthogonal everywhere in $\mathcal{R}'$, then the motion must be homogeneous. It follows from the result (3.1.39) for homogeneous motions that, at any given time t, the mapping $\mathbf{x} = \chi(\mathbf{X}, t)$ for a rigid-body motion of $\mathcal{R}'$ will be of the form

$$\mathbf{x} = \mathbf{R} \cdot \mathbf{X} + \mathbf{c} , \quad x_i = R_{iA} X_A + c_i , \tag{3.2.8}$$

where the rotation tensor $\mathbf{R}$ is a constant that characterizes the rigid-body rotation and $\mathbf{c}$ is a constant vector that characterizes the rigid-body translation.

Define the *Green-St. Venant (Lagrangian) strain tensor*,

$$\boxed{\mathbf{E} \equiv \frac{1}{2}(\mathbf{C} - \mathbf{I}) , \quad E_{AB} \equiv \frac{1}{2}(C_{AB} - \delta_{AB}) .} \tag{3.2.9}$$

The value of $\mathbf{E}$ at a material point completely describes $\mathbf{C} = 2\mathbf{E} + \mathbf{I}$ and therefore the local deformation of the body at that material point. By construction, the Green-St. Venant strain tensor is a symmetric second-order tensor. This strain tensor is convenient in part because it vanishes for rigid-body motions. However, it is mainly useful for the linearizations that come later. The right Cauchy stretch tensor, the right Cauchy-Green deformation tensor, and the Green-St. Venant strain tensor will be collectively referred to as strain tensors.

Example 3.5
Find the right Cauchy-Green deformation tensor $\mathbf{C}$ and the Green-St. Venant strain tensor $\mathbf{E}$ for the motion of the unit cube given in Example 3.1.

Solution. Using the material description of $\mathbf{F}$ derived in Example 3.4, the material description of the right Cauchy-Green deformation tensor (3.2.5) is

$$[\mathbf{C}(\mathbf{X})] = \begin{bmatrix} 1 & 2a_1 X_2 & 0 \\ 2a_1 X_2 & 1 + 4a_1^2 X_2^2 + a_3^2 X_3^2 & a_3 X_3(1 + a_3 X_2) \\ 0 & a_3 X_3(1 + a_3 X_2) & (1 + a_3 X_2)^2 \end{bmatrix} ,$$

and the material description of the Green-St. Venant strain tensor (3.2.9) is

$$[\mathbf{E}(\mathbf{X})] = \begin{bmatrix} 0 & a_1 X_2 & 0 \\ a_1 X_2 & 2a_1^2 X_2^2 + \frac{1}{2} a_3^2 X_3^2 & \frac{1}{2} a_3 X_3 (1 + a_3 X_2) \\ 0 & \frac{1}{2} a_3 X_3 (1 + a_3 X_2) & a_3 X_2 (1 + \frac{1}{2} a_3 X_2) \end{bmatrix} .$$

3.2.1 Local Physical Interpretation

It is often useful when analyzing problems and solutions to have a local physical interpretation of the measures of strain. By local, it is meant that the values of the strain tensors at a material point will be related to physical quantities such as change in length per unit reference length, or change in volume per unit reference volume, at that point.

Stretch

Consider the equation (3.2.6) for the current length of a material line $\mathbb{C}$ in terms of an integral over the curve $\mathcal{C}$ occupied by the material line in the reference configurations. The value of the integrand,

$$\boxed{\begin{aligned} \Lambda(\hat{\Upsilon}) &= (\hat{\Upsilon} \cdot \mathbf{C} \cdot \hat{\Upsilon})^{1/2} , \\ \Lambda(\hat{\Upsilon}_1, \hat{\Upsilon}_2, \hat{\Upsilon}_3) &= (C_{AB} \hat{\Upsilon}_A \hat{\Upsilon}_B)^{1/2} , \end{aligned}} \tag{3.2.10a}$$

at a material point on the curve $\mathcal{C}$ is the current length per unit reference length, or the *stretch*, of the curve at that point. Since $\mathbf{C} = \mathbf{I} + 2\mathbf{E}$, the stretch is given in terms of the Green-St. Venant strain tensor by

$$\boxed{\begin{aligned} \Lambda(\hat{\Upsilon}) &= (1 + 2\hat{\Upsilon} \cdot \mathbf{E} \cdot \hat{\Upsilon})^{1/2} , \\ \Lambda(\hat{\Upsilon}_1, \hat{\Upsilon}_2, \hat{\Upsilon}_3) &= (1 + 2E_{AB} \hat{\Upsilon}_A \hat{\Upsilon}_B)^{1/2} . \end{aligned}} \tag{3.2.10b}$$

Note that the only information about the material line $\mathbb{C}$ involved in determining the stretch Λ at a material point on $\mathbb{C}$ is the unit tangent vector $\hat{\Upsilon}$ to the reference configuration of the material line at that point. Thus, any two material lines that intersect at a material point and have the same unit tangent vector at that point (i.e., they intersect tangentially) will have the same stretch at that point. Therefore, one need only refer to the *reference material direction* $\hat{\Upsilon}$ in which the stretch is being specified. Given the value of $\mathbf{C}$ or $\mathbf{E}$ at a material point, (3.2.10) gives the stretch at that point in the reference material direction $\hat{\Upsilon}$. Of course, at any given material point, the stretch in different reference material directions will, in general, be different.

Since $\mathbf{C}$ is positive definite, the stretch Λ in any given direction will always be positive. If $0 < \Lambda(\hat{\Upsilon}) < 1$, then the motion results in a reduction in length, or *contraction*, in the reference material direction $\hat{\boldsymbol{\Upsilon}}$. If $\Lambda(\hat{\boldsymbol{\Upsilon}}) >$ 1, then the motion results in an increase in length, or *extension*, in the reference material direction $\hat{\boldsymbol{\Upsilon}}$. If $\Lambda(\hat{\boldsymbol{\Upsilon}}) = 1$, then there is no change in length in the reference material direction $\hat{\boldsymbol{\Upsilon}}$.

Two material lines that intersect tangentially at a material point in the reference configuration will intersect tangentially at the same material point in the current configuration. Thus, at a material point, a reference material direction $\hat{\boldsymbol{\Upsilon}}$ is mapped by the motion into a unique *current material direction* $\hat{\boldsymbol{v}}$. Since $d\mathbf{x} = \hat{\boldsymbol{v}}\,ds$, $d\mathbf{X} = \hat{\boldsymbol{\Upsilon}}\,dS$, and $d\mathbf{x} = \mathbf{F} \cdot d\mathbf{X}$, it follows that

$$\hat{\boldsymbol{v}} = (\mathbf{F} \cdot \hat{\boldsymbol{\Upsilon}})\frac{dS}{ds} \ . \tag{3.2.11}$$

Therefore, the current material direction corresponding to a given reference material direction is given by

$$\boxed{\hat{\boldsymbol{v}} = \frac{\mathbf{F} \cdot \hat{\boldsymbol{\Upsilon}}}{\Lambda(\hat{\boldsymbol{\Upsilon}})} \ , \quad \hat{v}_i = \frac{F_{iA}\hat{\Upsilon}_A}{\Lambda(\hat{\Upsilon}_1, \hat{\Upsilon}_2, \hat{\Upsilon}_3)} \ ,} \tag{3.2.12}$$

where it is noted that the stretch is related to ds and dS by $\Lambda = ds/dS$.

Longitudinal strain

Consider the equation (3.2.7) for the change in length of a material line $\mathbb{C}$ that occupies the curve $\mathcal{C}$ in the reference configuration, as the body goes from the reference to the current configuration. The value of the integrand in this equation,

$$\boxed{\begin{aligned} H(\hat{\boldsymbol{\Upsilon}}) &= (\hat{\boldsymbol{\Upsilon}} \cdot \mathbf{C} \cdot \hat{\boldsymbol{\Upsilon}})^{1/2} - 1 \ , \\ H(\hat{\Upsilon}_1, \hat{\Upsilon}_2, \hat{\Upsilon}_3) &= (C_{AB}\hat{\Upsilon}_A\hat{\Upsilon}_B)^{1/2} - 1 \ , \end{aligned}} \tag{3.2.13a}$$

at a material point on the curve $\mathcal{C}$ is the change in length per unit reference length, or *longitudinal strain*, of the material line at that point. The longitudinal strain is given in terms of the Green-St. Venant strain tensor by

$$\boxed{\begin{aligned} H(\hat{\boldsymbol{\Upsilon}}) &= (1 + 2\hat{\boldsymbol{\Upsilon}} \cdot \mathbf{E} \cdot \hat{\boldsymbol{\Upsilon}})^{1/2} - 1 \ , \\ H(\hat{\Upsilon}_1, \hat{\Upsilon}_2, \hat{\Upsilon}_3) &= (1 + 2E_{AB}\hat{\Upsilon}_A\hat{\Upsilon}_B)^{1/2} - 1 \ . \end{aligned}} \tag{3.2.13b}$$

Note again that the only information about the material line $\mathbb{C}$ involved in determining the longitudinal strain H at a material point on $\mathbb{C}$ is the unit tangent vector $\hat{\boldsymbol{\Upsilon}}$ to the reference configuration of the material line at that point. Given the value of either $\mathbf{C}$ or $\mathbf{E}$ at a material point, (3.2.13) gives the longitudinal strain at that point in the reference material direction $\hat{\boldsymbol{\Upsilon}}$.

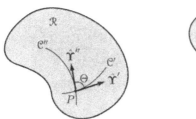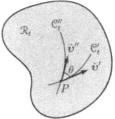

FIGURE 3.4. Shear strain.

Note that the longitudinal strain is simply related to the stretch by $H(\hat{\mathbf{\Upsilon}}) = \Lambda(\hat{\mathbf{\Upsilon}}) - 1$. Thus, in the reference material direction $\hat{\mathbf{\Upsilon}}$, one has that $-1 < H(\hat{\mathbf{\Upsilon}}) < 0$ for a contraction, $H(\hat{\mathbf{\Upsilon}}) > 0$ for an extension, and $H(\hat{\mathbf{\Upsilon}}) = 0$ if there is no change in length.

Shear strain

Consider two material lines $\mathbb{C}'$ and $\mathbb{C}''$ that intersect at the material point P. If the two material lines occupy the curves $\mathcal{C}'$ and $\mathcal{C}''$ in the reference configuration (Figure 3.4), then the angle Θ formed by $\mathbb{C}'$ and $\mathbb{C}''$ where they intersect in the reference configuration is given by

$$\cos\Theta = \hat{\mathbf{\Upsilon}}' \cdot \hat{\mathbf{\Upsilon}}'', \tag{3.2.14}$$

where $\hat{\mathbf{\Upsilon}}'$ and $\hat{\mathbf{\Upsilon}}''$ are the unit tangent vectors to the curves $\mathcal{C}'$ and $\mathcal{C}''$, respectively, at the material point P. If the two material lines occupy the curves $\mathcal{C}'_t$ and $\mathcal{C}''_t$ in the current configuration (Figure 3.4), then the angle θ formed by $\mathbb{C}'$ and $\mathbb{C}''$ in the current configuration is given by

$$\cos\theta = \hat{\boldsymbol{v}}' \cdot \hat{\boldsymbol{v}}'', \tag{3.2.15}$$

where $\hat{\boldsymbol{v}}'$ and $\hat{\boldsymbol{v}}''$ are the unit tangent vectors to the curves $\mathcal{C}'_t$ and $\mathcal{C}''_t$, respectively, at the material point P. A difference in the angles Θ and θ formed by the two material lines is an indication of *shear strain* at that material point.

Recalling the relation (3.2.12) between corresponding reference and current material directions, it follows that

$$\hat{\boldsymbol{v}}' = \frac{\mathbf{F} \cdot \hat{\mathbf{\Upsilon}}'}{\Lambda(\hat{\mathbf{\Upsilon}}')}, \quad \hat{\boldsymbol{v}}'' = \frac{\mathbf{F} \cdot \hat{\mathbf{\Upsilon}}''}{\Lambda(\hat{\mathbf{\Upsilon}}'')}. \tag{3.2.16}$$

Thus, (3.2.15) can be rewritten as

$$\cos\theta = \frac{(\mathbf{F}\cdot\hat{\boldsymbol{\Upsilon}}')\cdot(\mathbf{F}\cdot\hat{\boldsymbol{\Upsilon}}'')}{\Lambda(\hat{\boldsymbol{\Upsilon}}')\Lambda(\hat{\boldsymbol{\Upsilon}}'')} = \frac{\hat{\boldsymbol{\Upsilon}}'\cdot(\mathbf{F}^\mathsf{T}\cdot\mathbf{F})\cdot\hat{\boldsymbol{\Upsilon}}''}{\Lambda(\hat{\boldsymbol{\Upsilon}}')\Lambda(\hat{\boldsymbol{\Upsilon}}'')} \tag{3.2.17}$$

or, using the definition (3.2.5) for the right Cauchy-Green deformation tensor,

$$\cos\theta = \frac{\hat{\boldsymbol{\Upsilon}}'\cdot\mathbf{C}\cdot\hat{\boldsymbol{\Upsilon}}''}{\Lambda(\hat{\boldsymbol{\Upsilon}}')\Lambda(\hat{\boldsymbol{\Upsilon}}'')}. \tag{3.2.18}$$

Given the unit tangent vectors in the reference configuration for any two material lines at the point where they intersect, and the right Cauchy-Green deformation tensor at that point, the angle Θ formed by the two lines in the reference configuration is given by (3.2.14) and the angle θ formed by the two lines in the current configuration is given by (3.2.18). Thus, the change in the angle is a local measure of the shear strain at a material point that depends only on the two reference material directions $\hat{\boldsymbol{\Upsilon}}'$ and $\hat{\boldsymbol{\Upsilon}}''$ and the value of the right Cauchy-Green deformation tensor $\mathbf{C}$ at that point.

If the two reference material directions are orthogonal, so that $\Theta = \pi/2$ and $\hat{\boldsymbol{\Upsilon}}'\cdot\hat{\boldsymbol{\Upsilon}}'' = 0$, then the difference $\gamma = \Theta - \theta = \pi/2 - \theta$ between the angles in the reference and current configurations is the *orthogonal shear strain* between $\hat{\boldsymbol{\Upsilon}}'$ and $\hat{\boldsymbol{\Upsilon}}''$ at that material point. By noting that

$$\cos\theta = \cos(\pi/2 - \gamma) = \sin\gamma, \tag{3.2.19}$$

it follows from (3.2.18) that the orthogonal shear strain between the two orthogonal reference material directions $\hat{\boldsymbol{\Upsilon}}'$ and $\hat{\boldsymbol{\Upsilon}}''$ is given in terms of the right Cauchy-Green deformation tensor by

$$\boxed{\gamma(\hat{\boldsymbol{\Upsilon}}',\hat{\boldsymbol{\Upsilon}}'') = \sin^{-1}\left[\frac{\hat{\boldsymbol{\Upsilon}}'\cdot\mathbf{C}\cdot\hat{\boldsymbol{\Upsilon}}''}{\Lambda(\hat{\boldsymbol{\Upsilon}}')\Lambda(\hat{\boldsymbol{\Upsilon}}'')}\right],} \tag{3.2.20a}$$

$$\gamma(\hat{\Upsilon}_1',\hat{\Upsilon}_2',\hat{\Upsilon}_3',\hat{\Upsilon}_1'',\hat{\Upsilon}_2'',\hat{\Upsilon}_3'')$$
$$= \sin^{-1}\left[\frac{C_{AB}\hat{\Upsilon}_A'\hat{\Upsilon}_B''}{\Lambda(\hat{\Upsilon}_1',\hat{\Upsilon}_2',\hat{\Upsilon}_3')\Lambda(\hat{\Upsilon}_1'',\hat{\Upsilon}_2'',\hat{\Upsilon}_3'')}\right]. \tag{3.2.20b}$$

The orthogonal shear strain is given in terms of the Green-St. Venant strain tensor by

$$\boxed{\gamma(\hat{\boldsymbol{\Upsilon}}',\hat{\boldsymbol{\Upsilon}}'') = \sin^{-1}\left[\frac{2\hat{\boldsymbol{\Upsilon}}'\cdot\mathbf{E}\cdot\hat{\boldsymbol{\Upsilon}}''}{\Lambda(\hat{\boldsymbol{\Upsilon}}')\Lambda(\hat{\boldsymbol{\Upsilon}}'')}\right],} \tag{3.2.20c}$$

$$\gamma(\hat{\Upsilon}_1', \hat{\Upsilon}_2', \hat{\Upsilon}_3', \hat{\Upsilon}_1'', \hat{\Upsilon}_2'', \hat{\Upsilon}_3'')$$

$$= \sin^{-1} \left[\frac{2E_{AB}\hat{\Upsilon}_A'\hat{\Upsilon}_B''}{\Lambda(\hat{\Upsilon}_1', \hat{\Upsilon}_2', \hat{\Upsilon}_3')\Lambda(\hat{\Upsilon}_1'', \hat{\Upsilon}_2'', \hat{\Upsilon}_3'')} \right] . \qquad (3.2.20d)$$

Note that, as with the stretch and the longitudinal strain, the orthogonal shear strain (3.2.20) is seen to be a local measure of deformation that depends only on the value of either $\mathbf{C}$ or $\mathbf{E}$ at a material point and the two orthogonal material directions $\hat{\Upsilon}'$ and $\hat{\Upsilon}''$. The orthogonal shear strain will be the same between any two material lines that intersect at a given material point and have the same (orthogonal) unit tangent vectors $\hat{\Upsilon}'$ and $\hat{\Upsilon}''$ in the reference configuration at that point.

Example 3.6
Consider the motion of the unit cube described in Example 3.1. For the material point whose reference position is given by $(X_1, X_2, X_3) = (1/2, 1/2, 1/2)$, find the longitudinal strain H in the reference material direction $\hat{\Upsilon} = (\hat{e}_1 + \hat{e}_3)/\sqrt{2}$ and the orthogonal shear strain γ between the orthogonal reference material directions $\hat{\Upsilon}' = (\hat{e}_1 + \hat{e}_3)/\sqrt{2}$ and $\hat{\Upsilon}'' = \hat{e}_2$ (note that $\hat{\Upsilon}' \cdot \hat{\Upsilon}'' = 0$ as required).

Solution. Using the expression for $\mathbf{C}$ derived in Example 3.5 with the reference coordinates $(X_1, X_2, X_3) = (1/2, 1/2, 1/2)$, the value of the right Cauchy-Green deformation tensor at this material point is given by

$$[\mathbf{C}] = \begin{bmatrix} 1 & a_1 & 0 \\ a_1 & 1 + a_1^2 + \frac{1}{4}a_3^2 & \frac{1}{2}a_3 \left(1 + \frac{1}{2}a_3\right) \\ 0 & \frac{1}{2}a_3 \left(1 + \frac{1}{2}a_3\right) & \left(1 + \frac{1}{2}a_3\right)^2 \end{bmatrix} .$$

It follows that

$$\hat{\Upsilon} \cdot \mathbf{C} \cdot \hat{\Upsilon} = \frac{1}{2} \begin{bmatrix} 1 & 0 & 1 \end{bmatrix} \begin{bmatrix} 1 & a_1 & 0 \\ a_1 & 1 + a_1^2 + \frac{1}{4}a_3^2 & \frac{1}{2}a_3 \left(1 + \frac{1}{2}a_3\right) \\ 0 & \frac{1}{2}a_3 \left(1 + \frac{1}{2}a_3\right) & \left(1 + \frac{1}{2}a_3\right)^2 \end{bmatrix} \begin{bmatrix} 1 \\ 0 \\ 1 \end{bmatrix}$$

$$= 1 + \frac{1}{2}a_3 + \frac{1}{8}a_3^2$$

and similarly that $\hat{\Upsilon}' \cdot \mathbf{C} \cdot \hat{\Upsilon}' = 1 + \frac{1}{2}a_3 + \frac{1}{8}a_3^2$, $\hat{\Upsilon}'' \cdot \mathbf{C} \cdot \hat{\Upsilon}'' = 1 + a_1^2 + \frac{1}{4}a_3^2$, and $\hat{\Upsilon}' \cdot \mathbf{C} \cdot \hat{\Upsilon}'' = (a_1 + \frac{1}{2}a_3 + \frac{1}{4}a_3^2)/\sqrt{2}$. Substituting these results into the expression for the longitudinal strain (3.2.13) yields

$$H(\hat{\Upsilon}) = \sqrt{1 + \frac{1}{2}a_3 + \frac{1}{8}a_3^2} - 1 ,$$

and into the expression for the orthogonal shear strain (3.2.20) yields

$$\gamma(\hat{\Upsilon}', \hat{\Upsilon}'') = \sin^{-1}\left[\frac{4a_1 + 2a_3 + a_3^2}{\sqrt{(8 + 4a_3 + a_3^2)(4 + 4a_1^2 + a_3^2)}}\right].$$

Principal stretches and principal directions of stretch

In the terminology of Section 2.3.6 on eigenvalue problems, the stretch (3.2.10) in the reference material direction $\hat{\Upsilon}$ is the square root of the normal of the right Cauchy-Green strain tensor $\mathbf{C}$ in that direction (2.3.99),

$$\Lambda(\hat{\Upsilon}) = \sqrt{N_{\mathbf{C}}(\hat{\Upsilon})} \,. \tag{3.2.21}$$

Since the eigenvalues of $\mathbf{C}$ correspond to extrema of $N_{\mathbf{C}}(\hat{\Upsilon})$ and the eigenvectors of $\mathbf{C}$ correspond to the directions in which those extrema occur, it follows that the square roots of the eigenvalues of $\mathbf{C}$ correspond to extrema of Λ and the eigenvectors of $\mathbf{C}$ are the reference material directions in which these extrema occur. Recall also that since $\mathbf{C} = \mathbf{U}^2$, the eigenvalues of $\mathbf{U}$ are the square roots of the eigenvalues of $\mathbf{C}$, and $\mathbf{U}$ and $\mathbf{C}$ have the same eigenvectors. Thus, if $C^{(1)} \geq C^{(2)} \geq C^{(3)}$ are the eigenvalues of $\mathbf{C}$, $U^{(1)} \geq U^{(2)} \geq U^{(3)}$ are the eigenvalues of $\mathbf{U}$, and $\hat{\Upsilon}^{(1)}$, $\hat{\Upsilon}^{(2)}$, and $\hat{\Upsilon}^{(3)} = \hat{\Upsilon}^{(1)} \times \hat{\Upsilon}^{(2)}$ are the corresponding (mutually orthogonal) eigenvectors, then $\Lambda^{(1)} = \sqrt{C^{(1)}} = U^{(1)}$ is the maximum stretch and is in the direction of $\hat{\Upsilon}^{(1)}$, $\Lambda^{(3)} = \sqrt{C^{(3)}} = U^{(3)}$ is the minimum stretch and is in the direction of $\hat{\Upsilon}^{(3)}$, and $\Lambda^{(2)} = \sqrt{C^{(2)}} = U^{(2)}$ corresponds to a saddle point of $\Lambda(\hat{\Upsilon})$ in the direction of $\hat{\Upsilon}^{(2)}$. The extrema $\Lambda^{(1)} \geq \Lambda^{(2)} \geq \Lambda^{(3)}$ are the *principal stretches* for the motion and $\hat{\Upsilon}^{(1)}$, $\hat{\Upsilon}^{(2)}$, and $\hat{\Upsilon}^{(3)}$ are the *principal reference directions of stretch*.

Recall that if $\hat{\Upsilon}^{(k)}$ is an eigenvector of $\mathbf{C}$, then $\hat{\Upsilon}^{(k)} \cdot \mathbf{C} \cdot \hat{\Upsilon} = 0$ for all unit vectors $\hat{\Upsilon}$ that are orthogonal to $\hat{\Upsilon}^{(k)}$. Therefore, the orthogonal shear strain (3.2.20) between a principal direction of stretch and every orthogonal direction is zero.

The *principal strains* $H^{(1)} = \Lambda^{(1)} - 1$, $H^{(2)} = \Lambda^{(2)} - 1$, and $H^{(3)} = \Lambda^{(3)} - 1$ are the extrema of the longitudinal strain $H(\hat{\Upsilon})$ and are in the same directions as the principal stretches. Thus, the *principal reference directions of strain* are the same as the principal reference directions of stretch.

Dilatation

Consider an arbitrary material volume $\mathbb{B}'$ that occupies the region $\mathcal{R}'$ in the reference configuration and the region $\mathcal{R}'_t$ in the current configuration.

The size of the material volume in the reference configuration is

$$V' = \int_{\mathcal{R}'} dV , \qquad (3.2.22)$$

and in the current configuration, it is

$$v' = \int_{\mathcal{R}'_t} dv , \qquad (3.2.23)$$

where dV and dv are differential volume elements, in the reference and current configurations respectively, that correspond to the same differential material volume element $d\mathbb{B}'$. To relate the differential volume elements dV and dv, let dV be a parallelepiped formed by three noncoplanar differential vectors $d\mathbf{X}'$, $d\mathbf{X}''$, and $d\mathbf{X}'''$ in the reference configuration:

$$dV = d\mathbf{X}' \times d\mathbf{X}'' \cdot d\mathbf{X}''' = e_{ABC}\, dX'_A\, dX''_B\, dX'''_C . \qquad (3.2.24)$$

In the current configuration, $d\mathbf{X}'$, $d\mathbf{X}''$, and $d\mathbf{X}'''$ map into $d\mathbf{x}'$, $d\mathbf{x}''$, and $d\mathbf{x}'''$ and the differential volume dV maps into dv, given by

$$dv = d\mathbf{x}' \times d\mathbf{x}'' \cdot d\mathbf{x}''' = e_{ijk}\, dx'_i\, dx''_j\, dx'''_k . \qquad (3.2.25)$$

Recall from (3.1.33) that $dx_i = F_{iA}dX_A$, where F_{iA} are the scalar components of the deformation gradient tensor, so that

$$dv = e_{ijk} F_{iA} F_{jB} F_{kC}\, dX'_A\, dX''_B\, dX'''_C , \qquad (3.2.26)$$

and from the identity (2.2.29) relating the determinant to the permutation symbol,

$$dv = e_{ABC} \det \mathbf{F}\, dX'_A\, dX''_B\, dX'''_C . \qquad (3.2.27)$$

Finally, using the definition $J \equiv \det \mathbf{F}$ for the Jacobian of the motion and the representation (3.2.24) for the differential volume in the reference configuration,

$$dv = J\, dV . \qquad (3.2.28)$$

Thus, if the material description $J(\mathbf{X}, t)$ of the Jacobian is known, the *current* size of the material volume $\mathbb{B}'$ is given in terms of an integral over the region occupied by the material volume in the *reference* configuration by

$$v' = \int_{\mathcal{R}'} J\, dV . \qquad (3.2.29)$$

Recall that the condition for an isomorphic mapping between the reference
and current configurations is that the Jacobian must exist and be nonzero.
It follows from (3.2.29) that if the volumes V' and v' are both to be positive for arbitrary $\mathcal{R}'$, then the Jacobian must also be positive. Thus, the
Jacobian must be positive,

$$J > 0 , \qquad (3.2.30)$$

everywhere in the body.

The change in size of the material volume is

$$v' - V' = \int_{\mathcal{R}'} (J - 1) \, dV . \qquad (3.2.31)$$

The integrand in (3.2.31) is known as the *dilatation*,

$$\boxed{e = J - 1 ,} \qquad (3.2.32)$$

and represents the change in volume per unit reference volume at a material
point. The dilatation is sometimes referred to as the volume strain.

The dilation can also be given in terms of the right Cauchy stretch tensor
(3.1.46), the right Cauchy-Green deformation tensor (3.2.5), and the Green-
St. Venant strain tensor (3.2.9). Note first that

$$J \equiv \det \mathbf{F} = \det(\mathbf{R} \cdot \mathbf{U}) = \det \mathbf{R} \det \mathbf{U} \qquad (3.2.33)$$

and recall that $\det \mathbf{R} = \pm 1$ since $\mathbf{R}$ is orthogonal, and $\det \mathbf{U} > 0$ since
$\mathbf{U}$ is symmetric, positive definite. Thus, $J > 0$ if and only if $\mathbf{R}$ is *proper*
orthogonal ($\det \mathbf{R} = +1$), in which case $J = \det \mathbf{U}$. It follows that

$$\boxed{e = \det \mathbf{U} - 1 = \sqrt{\det \mathbf{C}} - 1 = \sqrt{\det(\mathbf{I} + 2\mathbf{E})} - 1 .} \qquad (3.2.34)$$

Volume is conserved (i.e., the motion is *isochoric*) at a material point if and
only if $e = 0$ at that material point. There is a local reduction in volume
if $-1 < e < 0$ and there is a local increase in volume if $e > 0$.

Example 3.7

For the motion of the unit cube given in Example 3.1, show that the
Jacobian is positive everywhere in the body. Find the dilatation at the
material point whose position in the reference configuration is given by
$(X_1, X_2, X_3) = (1/2, 1/2, 1/2)$ and determine the total change in volume of
the cube.

Solution. From the deformation gradient tensor found in Example 3.4,
the Jacobian $J \equiv \det \mathbf{F}$ is

$$J = 1 + a_3 X_2 .$$

Since $a_3 > 0$ by assumption, and for the unit cube $0 \leq X_2 \leq 1$, the Jacobian is positive at every material point in the body.

Substituting this result into the expression (3.2.32) for the dilatation gives

$$e = a_3 X_2 .$$

It is seen that the dilatation in this case is independent of X_1 and X_3 and the dilatation at $X_2 = 1/2$ is $e = a_3/2$. The total volume change is the integral of the dilatation over the region occupied by the body in the reference configuration:

$$v - V = \int_{\mathcal{R}} e \, dV = a_3 \int_0^1 \int_0^1 \int_0^1 X_2 \, dX_1 dX_2 dX_3 = \frac{1}{2} a_3 .$$

The relation (3.2.28) also allows one to relate volume integrals over a material volume in the reference and current configurations. Consider a material volume that occupies the region $\mathcal{R}'$ in the reference configuration and $\mathcal{R}'_t$ in the current configuration. Then, if $\mathbf{T}$ is a tensor field of order n with scalar components $T_{i_1 i_2 \cdots i_n}$ in a vector basis $\{\hat{\mathbf{e}}_i\}$, it follows that

$$\boxed{\int_{\mathcal{R}'_t} \mathbf{T} \, dv = \int_{\mathcal{R}'} \mathbf{T} J \, dV \; , \quad \int_{\mathcal{R}'_t} T_{i_1 i_2 \cdots i_n} \, dv = \int_{\mathcal{R}'} T_{i_1 i_2 \cdots i_n} J \, dV \; ,} \quad (3.2.35)$$

where dV and dv are the volumes of a differential element of the material volume in the reference and current configurations, respectively.

Nanson's formula

It will also be desirable to have a relation between surface integrals over a material surface in the reference and current configurations. Consider a differential element $d\mathbb{S}$ of a material surface $\mathbb{S}$ which, in the reference configuration, occupies the differential area dA that is a parallelogram whose edges are coincident with two nonparallel differential vectors $d\mathbf{X}'$ and $d\mathbf{X}''$, so that

$$\hat{\mathbf{N}} \, dA = d\mathbf{X}' \times d\mathbf{X}'' \; , \quad \hat{N}_I \, dA = e_{IJK} \, dX'_J \, dX''_K \; , \quad (3.2.36)$$

where $\hat{\mathbf{N}}$ it the unit normal vector to dA. The mapping from the reference to the current configuration maps $d\mathbf{X}'$ and $d\mathbf{X}''$ into $d\mathbf{x}'$ and $d\mathbf{x}''$ so that (3.2.36) is mapped into

$$\hat{\mathbf{n}} \, da = d\mathbf{x}' \times d\mathbf{x}'' \; , \quad \hat{n}_i \, da = e_{ijk} \, dx'_j \, dx''_k \; . \quad (3.2.37)$$

In the current configuration, the differential element $d\mathbb{S}$ of the material surface occupies the differential area da with unit normal vector $\hat{n}$. Since $dx_i = F_{iA}dX_A$, where F_{iA} are the scalar components of the deformation gradient tensor, it follows that

$$\hat{n}_i \, da = e_{ijk} F_{jQ} F_{kR} \, dX'_Q \, dX''_R \, . \tag{3.2.38}$$

Multiplying both sides by F_{iP} and applying the identity (2.2.29) relating the determinant to the permutation tensor,

$$F_{iP}\hat{n}_i \, da = e_{ijk} F_{iP} F_{jQ} F_{kR} \, dX'_Q \, dX''_R = e_{PQR} \det \mathbf{F} \, dX'_Q \, dX''_R \, , \tag{3.2.39}$$

and using the definition $J \equiv \det \mathbf{F}$ for the Jacobian of the motion and the result (3.2.36), one obtains

$$F_{iP}\hat{n}_i \, da = \hat{N}_P J \, dA \, . \tag{3.2.40}$$

Finally, multiplying both sides by F_{Pj}^{-1}, where F_{Pj}^{-1} are the scalar components of the inverse deformation gradient tensor, and recalling that $F_{iP}F_{Pj}^{-1} = \delta_{ij}$,

$$\hat{\mathbf{n}} \, da = \mathbf{F}^{-\mathsf{T}} \cdot \hat{\mathbf{N}} J \, dA \, , \quad \hat{n}_j \, da = F_{Pj}^{-1} \hat{N}_P J \, dA \, . \tag{3.2.41}$$

This result is known as *Nanson's formula*. It is important to note that the unit normal to a material surface is *not* a material direction, which was defined to be the unit tangent vector to a material line. Consider a material line that intersects a material surface at a material point. In general, if the material line and material surface are orthogonal in one configuration, they will not be orthogonal in another configuration due to shear of the material.

The relation (3.2.41) allows one to relate surface integrals over a material surface in the reference and current configurations. Consider a material surface $\mathbb{S}$ that occupies the surface S in the reference configuration and S_t in the current configuration. Then, if $\mathbf{T}$ is a tensor field of order n with scalar components $T_{i_1 i_2 \cdots i_n}$ in a vector basis $\{\hat{\mathbf{e}}_i\}$, and $\hat{\mathbf{N}}$ and $\hat{n}$ are the unit normals to the material surface in the reference and current configurations, respectively, it follows that

$$
\boxed{
\begin{aligned}
\int_{S_t} \mathbf{T} \cdot \hat{\mathbf{n}} \, da &= \int_S \mathbf{T} \cdot \mathbf{F}^{-\mathsf{T}} \cdot \hat{\mathbf{N}} J \, dA \, , \\
\int_{S_t} T_{i_1 i_2 \cdots i_{n-1} i_n} \hat{n}_{i_n} \, da &= \int_S T_{i_1 i_2 \cdots i_{n-1} i_n} F_{R i_n}^{-1} \hat{N}_R J \, dA \, ,
\end{aligned}
}
\tag{3.2.42}
$$

where dA and da are the areas of a differential element $d\mathbb{S}$ of the material surface in the reference and current configurations, respectively.

Conservation of mass

As part of the continuum hypothesis, it is required that every material volume, regardless of how small, must have a positive mass and if a material volume is divided into two components, then the mass of the material volume must equal the sum of the masses of its two components. It follows from these requirements that, at any instant in time, every material point in a body must have a positive mass density (a mass per unit volume) associated with it.

Let the configuration occupied by a body at time $t = t_0$ be the body's reference configuration and let $\rho_0(P)$ and $\rho(P,t)$ be the mass densities associated with a material point P at times t_0 and t, respectively. Consider a portion $\mathbb{B}'$ of the body (i.e., a material volume) that occupies the region $\mathcal{R}'$ in the reference configuration and $\mathcal{R}'_t$ in the current configuration. Assuming that the mass of the material volume is conserved (i.e., mass is neither created nor destroyed during the motion from $\mathcal{R}'$ to $\mathcal{R}'_t$) the total mass of the material volume is

$$\int_{\mathcal{R}'} \rho_0 \, dV = \int_{\mathcal{R}'_t} \rho \, dv . \tag{3.2.43}$$

Using (3.2.35), the second integral can be rewritten as an integral over $\mathcal{R}'$ so that the conservation of mass can be expressed as

$$\int_{\mathcal{R}'} (\rho_0 - \rho J) \, dV = 0 . \tag{3.2.44}$$

This condition for the conservation of mass must hold for arbitrary material volumes $\mathbb{B}'$, so the integrand must be zero and the *continuity condition* for conservation of mass at each material point in a body is

$$\boxed{\rho_0 = J\rho .} \tag{3.2.45}$$

The current mass density at a material point is thus related to the material point's initial mass density at time $t = t_0$.

3.2.2 Strain-Displacement Relations

The Green-St. Venant strain tensor can be expressed in terms of the displacement field in a body as well as in terms of the deformation gradient tensor. Recall that the displacement vector field relates the positions of material points in the reference and current configurations such that $\mathbf{x} = \mathbf{X} + \mathbf{u}$. Therefore, the deformation gradient tensor $\mathbf{F} \equiv \text{Grad}\,\mathbf{x}$ can be given as

$$\mathbf{F} = \mathbf{I} + \text{Grad}\,\mathbf{u} , \qquad F_{iA} = \delta_{iA} + \frac{\partial u_i}{\partial X_A} . \tag{3.2.46}$$

The Green-St. Venant strain tensor (3.2.9) can thus be written as

$$\mathbf{E} = \frac{1}{2}(\mathbf{F}^{\mathsf{T}} \cdot \mathbf{F} - \mathbf{I}) = \frac{1}{2}\left[(\mathbf{I} + \text{Grad } \mathbf{u})^{\mathsf{T}} \cdot (\mathbf{I} + \text{Grad } \mathbf{u}) - \mathbf{I}\right] , \qquad (3.2.47)$$

which simplifies to

$$\boxed{\begin{aligned} \mathbf{E} &= \frac{1}{2}\left[\text{Grad } \mathbf{u} + (\text{Grad } \mathbf{u})^{\mathsf{T}} + (\text{Grad } \mathbf{u})^{\mathsf{T}} \cdot (\text{Grad } \mathbf{u})\right] , \\ E_{AB} &= \frac{1}{2}\left(\frac{\partial u_A}{\partial X_B} + \frac{\partial u_B}{\partial X_A} + \frac{\partial u_C}{\partial X_A}\frac{\partial u_C}{\partial X_B}\right) . \end{aligned}} \qquad (3.2.48)$$

This representation of the Green-St. Venant strain tensor in terms of the material description of the displacement field will be important during the linearization process that follows.

3.3   Strain Tensors: Spatial Formulation

In the referential formulation presented previously, strain tensors were defined that related local measures of strain, such as the longitudinal strain and the shear strain, to material directions given in the reference configuration. However, in some applications, material directions are more conveniently given in the current configuration. Accordingly, a spatial formulation is used to derive alternative strain tensors that relate the local measures of strain to the current material directions. The derivation is similar to that used for the material formulation and is only outlined here.

Considering the differential vector $d\mathbf{X} = \hat{\mathbf{\Upsilon}}\, dS$ in the reference configuration and its image $d\mathbf{x} = \hat{\boldsymbol{v}}\, ds$ in the current configuration, and using the definition (3.1.35) of the inverse deformation gradient tensor $\mathbf{F}^{-1}$, it follows that

$$\begin{aligned} (dS)^2 &= d\mathbf{X} \cdot d\mathbf{X} \\ &= (\mathbf{F}^{-1} \cdot d\mathbf{x}) \cdot (\mathbf{F}^{-1} \cdot d\mathbf{x}) \\ &= d\mathbf{x} \cdot (\mathbf{F}^{-\mathsf{T}} \cdot \mathbf{F}^{-1}) \cdot d\mathbf{x} \\ &= \left[\hat{\boldsymbol{v}} \cdot (\mathbf{F}^{-\mathsf{T}} \cdot \mathbf{F}^{-1}) \cdot \hat{\boldsymbol{v}}\right](ds)^2 . \end{aligned} \qquad (3.3.1)$$

Thus, the length of a material line $\mathbb{C}$ in the *reference* configuration can be expressed as an integral over the curve $\mathcal{C}_t$ occupied by the material line in the *current* configuration:

$$S = \int_{\mathcal{C}_t} (\hat{\boldsymbol{v}} \cdot \mathbf{c} \cdot \hat{\boldsymbol{v}})^{1/2}\, ds , \quad S = \int_{\mathcal{C}_t} (c_{ij}\, \hat{v}_i\, \hat{v}_j)^{1/2}\, ds , \qquad (3.3.2)$$

where the *Cauchy strain tensor* **c** is defined as

$$\boxed{\mathbf{c} \equiv \mathbf{F}^{-\mathsf{T}} \cdot \mathbf{F}^{-1} , \quad c_{ij} \equiv F_{Ai}^{-1} F_{Aj}^{-1} ,} \tag{3.3.3}$$

and $\hat{\boldsymbol{v}}$ is the unit tangent vector to the curve $\mathcal{C}_t$ occupied by the material line $\mathbb{C}$ in the current configuration. Note that the Cauchy strain tensor is symmetric by construction.

The Cauchy strain tensor is related to the left Cauchy stretch tensor **V** (3.1.47). By defining the *left Cauchy-Green deformation tensor* as

$$\boxed{\mathbf{B} \equiv \mathbf{F} \cdot \mathbf{F}^{\mathsf{T}} = \mathbf{V}^2 , \quad B_{ij} = F_{iA} F_{jA} ,} \tag{3.3.4}$$

it follows that $\mathbf{c} = \mathbf{B}^{-1} = \mathbf{V}^{-2}$.

Using (3.3.2), the difference in lengths of a material line $\mathbb{C}$ in the current and reference configurations can be given as an integral over the curve $\mathcal{C}_t$ occupied by $\mathbb{C}$ in the current configuration:

$$s - S = \int_{\mathcal{C}_t} \left[1 - (\hat{\boldsymbol{v}} \cdot \mathbf{c} \cdot \hat{\boldsymbol{v}})^{1/2} \right] ds . \tag{3.3.5}$$

Given a current configuration $\mathcal{R}_t$ and the spatial description $\mathbf{c}(\mathbf{x}, t)$ of the Cauchy strain tensor, (3.3.5) can be used to find the change in length of any material line whose curve $\mathcal{C}_t$ in the current configuration is known. If $\mathbf{c} = \mathbf{I}$ at a material point, then the body is locally undeformed at this material point.

Define the *Almansi-Hamel (Eulerian) strain tensor*,

$$\boxed{\mathbf{e} \equiv \frac{1}{2}(\mathbf{I} - \mathbf{c}) , \quad e_{ij} \equiv \frac{1}{2}(\delta_{ij} - c_{ij}) .} \tag{3.3.6}$$

The Almansi-Hamel strain tensor is symmetric. The value of **e** at a material point completely describes $\mathbf{c} = \mathbf{I} - 2\mathbf{e}$ and therefore the local deformation of the body at that point. The Almansi-Hamel strain tensor vanishes at material points where there is no local deformation. Note that $\mathbf{c} = \mathbf{I}$ and $\mathbf{e} = \mathbf{0}$ at a material point if and only if the right Cauchy-Green deformation tensor $\mathbf{C} = \mathbf{I}$ and the Green-St. Venant strain tensor $\mathbf{E} = \mathbf{0}$ at that material point.

Example 3.8
Find the Cauchy strain tensor **c** and the Almansi-Hamel strain tensor **e** for the motion of the unit cube given in Example 3.1.

Solution. Using the material description of $\mathbf{F}^{-1}$ derived in Example 3.4, the material description of the Cauchy strain tensor (3.3.3) is

$$[\mathbf{c}(\mathbf{X})] = \begin{bmatrix} 1 & -2a_1X_2 & 0 \\ -2a_1X_2 & 1 + 4a_1^2X_2^2 + \dfrac{a_3^2X_3^2}{(1+a_3X_2)^2} & \dfrac{-a_3X_3}{(1+a_3X_2)^2} \\ 0 & \dfrac{-a_3X_3}{(1+a_3X_2)^2} & \dfrac{1}{(1+a_3X_2)^2} \end{bmatrix} ,$$

and the material description of the Almansi-Hamel strain tensor (3.3.6) is

$$[\mathbf{e}(\mathbf{X})] = \begin{bmatrix} 0 & a_1X_2 & 0 \\ a_1X_2 & -2a_1^2X_2^2 - \dfrac{a_3^2X_3^2}{2(1+a_3X_2)^2} & \dfrac{a_3X_3}{2(1+a_3X_2)^2} \\ 0 & \dfrac{a_3X_3}{2(1+a_3X_2)^2} & \dfrac{a_3X_2(1+\frac{1}{2}a_3X_2)}{(1+a_3X_2)^2} \end{bmatrix} .$$

The spatial descriptions of $\mathbf{c}$ and $\mathbf{e}$, which is what one often needs, can be found either by using the spatial description of the inverse deformation gradient tensor $\mathbf{F}^{-1}$ or by applying the mapping $X_A = \overset{-1}{\chi}_A(x_1, x_2, x_3)$ found in Example 3.1 to the material descriptions given above. The result is quite messy and therefore omitted here.

3.3.1 Local Physical Interpretation

The previously defined local measures of strain are given here in terms of current material directions.

Stretch

Consider the equation (3.3.2) for the reference length of a material line $\mathcal{C}$ in terms of an integral over the curve $\mathcal{C}_t$ occupied by the material line in the current configuration. The integrand in this expression is the reference length per unit current length at a material point on the line. Therefore, the stretch, the current length per unit reference length, is the reciprocal of the integrand in this expression:

$$\begin{aligned} \lambda(\hat{\boldsymbol{v}}) &= (\hat{\boldsymbol{v}} \cdot \mathbf{c} \cdot \hat{\boldsymbol{v}})^{-1/2} , \\ \lambda(\hat{v}_1, \hat{v}_2, \hat{v}_3) &= (c_{ij}\hat{v}_i\hat{v}_j)^{-1/2} . \end{aligned} \tag{3.3.7a}$$

Since $\mathbf{c} = \mathbf{I} - 2\mathbf{e}$, the stretch is given in terms of the Almansi-Hamel strain tensor by

$$\boxed{\begin{aligned} \lambda(\hat{\boldsymbol{v}}) &= (1 - 2\hat{\boldsymbol{v}} \cdot \mathbf{e} \cdot \hat{\boldsymbol{v}})^{-1/2} \,, \\ \lambda(\hat{v}_1, \hat{v}_2, \hat{v}_3) &= (1 - 2e_{ij}\hat{v}_i\hat{v}_j)^{-1/2} \,. \end{aligned}} \tag{3.3.7b}$$

The expression (3.3.7) gives the stretch at a material point in the current material direction $\hat{\boldsymbol{v}}$. Note that $\lambda(\hat{\boldsymbol{v}}) = \Lambda(\hat{\boldsymbol{\Upsilon}})$, the stretch (3.2.10) in the reference material direction $\hat{\boldsymbol{\Upsilon}}$, if $\hat{\boldsymbol{v}}$ and $\hat{\boldsymbol{\Upsilon}}$ are corresponding current and reference material directions related by (3.2.12). Also, the relation (3.2.12) between corresponding reference and current material directions can alternatively be given as

$$\boxed{\hat{\boldsymbol{\Upsilon}} = \lambda(\hat{\boldsymbol{v}})\mathbf{F}^{-1} \cdot \hat{\boldsymbol{v}} \,, \quad \hat{\Upsilon}_A = \lambda(\hat{v}_1, \hat{v}_2, \hat{v}_3)F_{Ai}^{-1}\hat{v}_i \,.} \tag{3.3.8}$$

Longitudinal strain

The integrand in the expression (3.3.5) for the change in length of a material line $\mathbb{C}$ is the change in length per unit *current length*. Recall, however, that the longitudinal strain is defined as the change in length per unit *reference length*. By noting from (3.3.1) that

$$ds = \frac{dS}{(\hat{\boldsymbol{v}} \cdot \mathbf{c} \cdot \hat{\boldsymbol{v}})^{1/2}} \,, \tag{3.3.9}$$

the change in length of the material line $\mathbb{C}$ can be rewritten as an integral over the curve $\mathcal{C}$ occupied by $\mathbb{C}$ in the reference configuration:

$$s - S = \int_{\mathcal{C}} \left[(\hat{\boldsymbol{v}} \cdot \mathbf{c} \cdot \hat{\boldsymbol{v}})^{-1/2} - 1 \right] dS \,. \tag{3.3.10}$$

It follows that the longitudinal strain is given in terms of the Cauchy strain tensor $\mathbf{c}$ and the current material direction $\hat{\boldsymbol{v}}$ by

$$\boxed{\begin{aligned} \eta(\hat{\boldsymbol{v}}) &= (\hat{\boldsymbol{v}} \cdot \mathbf{c} \cdot \hat{\boldsymbol{v}})^{-1/2} - 1 \,, \\ \eta(\hat{v}_1, \hat{v}_2, \hat{v}_3) &= (c_{ij}\hat{v}_i\hat{v}_j)^{-1/2} - 1 \end{aligned}} \tag{3.3.11a}$$

and is given in terms of the Almansi-Hamel strain tensor $\mathbf{e}$ by

$$\boxed{\begin{aligned} \eta(\hat{\boldsymbol{v}}) &= (1 - 2\hat{\boldsymbol{v}} \cdot \mathbf{e} \cdot \hat{\boldsymbol{v}})^{-1/2} - 1 \,, \\ \eta(\hat{v}_1, \hat{v}_2, \hat{v}_3) &= (1 - 2e_{ij}\hat{v}_i\hat{v}_j)^{-1/2} - 1 \,. \end{aligned}} \tag{3.3.11b}$$

The expression (3.3.11) gives the longitudinal strain at a material point in the current material direction $\hat{v}$. Note that $\eta(\hat{v}) = H(\hat{\Upsilon})$, the longitudinal strain (3.2.13) in the reference material direction $\hat{\Upsilon}$, if $\hat{v}$ and $\hat{\Upsilon}$ are corresponding current and reference material directions related by (3.2.12).

Example 3.9
Consider the motion of the unit cube described in Example 3.1. For the material point whose reference position is given by $(X_1, X_2, X_3) = (1/2, 1/2, 1/2)$, find the longitudinal strain η in the current material direction $\hat{v} = (\hat{e}_1 + \hat{e}_3)/\sqrt{2}$.

Solution. Using the expression for $\mathbf{c}$ derived in Example 3.8 with the reference coordinates $(X_1, X_2, X_3) = (1/2, 1/2, 1/2)$, the value of the Cauchy strain tensor at this material point is given by

$$[\mathbf{c}] = \begin{bmatrix} 1 & -a_1 & 0 \\ -a_1 & 1 + a_1^2 + \dfrac{a_3^2}{(2+a_3)^2} & \dfrac{-2a_3}{(2+a_3)^2} \\ 0 & \dfrac{-2a_3}{(2+a_3)^2} & \dfrac{4}{(2+a_3)^2} \end{bmatrix}.$$

It follows that

$$\hat{v} \cdot \mathbf{c} \cdot \hat{v} = \frac{1}{2} \begin{bmatrix} 1 & 0 & 1 \end{bmatrix} \begin{bmatrix} 1 & -a_1 & 0 \\ -a_1 & 1 + a_1^2 + \dfrac{a_3^2}{(2+a_3)^2} & \dfrac{-2a_3}{(2+a_3)^2} \\ 0 & \dfrac{-2a_3}{(2+a_3)^2} & \dfrac{4}{(2+a_3)^2} \end{bmatrix} \begin{bmatrix} 1 \\ 0 \\ 1 \end{bmatrix}$$

$$= \frac{8 + 4a_3 + a_3^2}{2(2+a_3)^2}.$$

Substituting this result into the expression (3.3.11) for the longitudinal strain yields

$$\eta(\hat{v}) = \frac{\sqrt{2}(2+a_3)}{\sqrt{8 + 4a_3 + a_3^2}} - 1.$$

Notice that the longitudinal strain in this example is not the same as the longitudinal strain found in Example 3.6, even though the material point in each is the same and $\hat{v} = \hat{\Upsilon}$. This is because a motion generally maps a reference material direction $\hat{\Upsilon}$ into a different current material direction. The longitudinal strains found here and in Example 3.6 are the changes in length per unit reference length at the same material point on different (nontangential) material lines.

Shear strain

Recall that at a material point P, the angle Θ formed by two reference material directions $\hat{\mathbf{\Upsilon}}'$ and $\hat{\mathbf{\Upsilon}}''$ and the angle θ formed by the corresponding current material directions $\hat{v}'$ and $\hat{v}''$ (see Figure 3.4) are given by

$$\cos\Theta = \hat{\mathbf{\Upsilon}}' \cdot \hat{\mathbf{\Upsilon}}'', \quad \cos\theta = \hat{v}' \cdot \hat{v}''. \tag{3.3.12}$$

The difference in the angles is a measure of the shear strain at that point.

It follows from the relation (3.3.8) between corresponding reference and current material directions that

$$\hat{\mathbf{\Upsilon}}' = \lambda(\hat{v}')\mathbf{F}^{-1} \cdot \hat{v}', \quad \hat{\mathbf{\Upsilon}}'' = \lambda(\hat{v}'')\mathbf{F}^{-1} \cdot \hat{v}''. \tag{3.3.13}$$

Therefore,

$$\begin{aligned}
\cos\Theta &= \lambda(\hat{v}')\lambda(\hat{v}'')(\mathbf{F}^{-1} \cdot \hat{v}') \cdot (\mathbf{F}^{-1} \cdot \hat{v}'') \\
&= \lambda(\hat{v}')\lambda(\hat{v}'')\hat{v}' \cdot (\mathbf{F}^{-T} \cdot \mathbf{F}^{-1}) \cdot \hat{v}'', \tag{3.3.14}
\end{aligned}$$

or, using the definition (3.3.3) for the Cauchy strain tensor,

$$\Theta = \cos^{-1}[\lambda(\hat{v}')\lambda(\hat{v}'')(\hat{v}' \cdot \mathbf{c} \cdot \hat{v}'')]. \tag{3.3.15a}$$

In terms of the Almansi-Hamel strain tensor, this expression is given by

$$\Theta = \cos^{-1}[\lambda(\hat{v}')\lambda(\hat{v}'')(\hat{v}' \cdot \hat{v}'' - 2\hat{v}' \cdot \mathbf{e} \cdot \hat{v}'')]. \tag{3.3.15b}$$

Given two current material directions $\hat{v}'$ and $\hat{v}''$ and the value of either the Cauchy or Almansi-Hamel strain tensor at a material point point, the angle Θ formed by the two corresponding reference material directions is given by (3.3.15) and the angle θ formed by the two current material directions is given by (3.3.12). The difference in these angles can thus be determined. Note that there is no meaningful spatial formulation of the orthogonal shear strain since it presumes that $\Theta = \pi/2$.

Dilatation

Recall that the dilatation (3.2.32), the change in volume per unit reference volume, is given by $e = J - 1$, where $J \equiv \det \mathbf{F}$ is the Jacobian of the motion. The dilatation can be given in terms of the spatial formulation strain tensors by noting that

$$J \equiv \det \mathbf{F} = \det(\mathbf{V} \cdot \mathbf{R}) = \det \mathbf{V} \det \mathbf{R} \tag{3.3.16}$$

and recalling that the rotation tensor must be *proper* orthogonal, $\det \mathbf{R} = +1$, if the volume of an arbitrary material volume is positive in both the reference and current configurations. Thus,

$$e = \det \mathbf{V} - 1 = \sqrt{\det \mathbf{B}} - 1 = \frac{1}{\sqrt{\det \mathbf{c}}} - 1 = \frac{1}{\sqrt{\det(\mathbf{I} - 2\mathbf{e})}} - 1 \,.$$

$$(3.3.17)$$

Conservation of mass

The condition that mass be conserved was previously shown to lead to the continuity condition

$$\rho_0 = J\rho \,, \tag{3.3.18}$$

where ρ is the current mass density at a material point, ρ_0 is the mass density at the material point at time $t = t_0$, and J is the Jacobian of the motion from the configuration at time $t = t_0$ to the current configuration. Thus, ρ_0 is independent of time and, consequently,

$$\frac{\mathrm{D}}{\mathrm{D}t}(J\rho) = 0 \,. \tag{3.3.19}$$

It can be shown that

$$\frac{\mathrm{D}J}{\mathrm{D}t} = J \operatorname{div} \mathbf{v} \,, \tag{3.3.20}$$

where $\mathbf{v}$ is the material velocity. Thus, using the chain rule, the continuity condition can alternatively be given as

$$\frac{\mathrm{D}\rho}{\mathrm{D}t} + \rho \operatorname{div} \mathbf{v} = 0 \,, \qquad \frac{\mathrm{D}\rho}{\mathrm{D}t} + \rho \frac{\partial v_i}{\partial x_i} = 0 \tag{3.3.21}$$

or, equivalently, using the definition (3.1.28) of the material time derivative,

$$\frac{\partial \rho}{\partial t} + \operatorname{div}(\rho \mathbf{v}) = 0 \,, \qquad \frac{\partial \rho}{\partial t} + \frac{\partial}{\partial x_i}(\rho v_i) = 0 \,. \tag{3.3.22}$$

The continuity conditions (3.3.21) and (3.3.22) are independent of ρ_0; that is, they involve only quantities defined over the current configuration of a body and their time rates of change. They can, in fact, be derived without introducing a reference configuration.

There is an important consequence of the conservation of mass related to the material time derivative of volume integrals in the current configuration. Suppose that the tensor $\mathbf{T}(P, t)$ represents the amount of some physical quantity *per unit mass* associated with a material point P at time *t*. Then, $\rho(P,t)\mathbf{T}(P,t)$ is the amount of the physical quantity *per unit volume* and

$$\mathbf{\Phi}_{\mathbb{B}'}(t) = \int_{\mathcal{R}'_t} \rho\mathbf{T}\,dv \tag{3.3.23}$$

is the amount of the physical quantity at time t associated with a material volume $\mathbb{B}'$ that occupies the region $\mathcal{R}'$ in the reference configuration and $\mathcal{R}'_t$ in the current configuration. Using (3.2.35) and (3.2.45), $\mathbf{\Phi}_{\mathbb{B}'}(t)$ can be rewritten as an integral over $\mathcal{R}'$:

$$\mathbf{\Phi}_{\mathbb{B}'}(t) = \int_{\mathcal{R}'} \rho\mathbf{T}J\,dV = \int_{\mathcal{R}'} \rho_0\mathbf{T}\,dV \ . \tag{3.3.24}$$

Since $\mathbf{\Phi}_{\mathbb{B}'}(t)$ is a quantity that is associated with a fixed set of material points, its time rate of change is a material time derivative:

$$\frac{d}{dt}\mathbf{\Phi}_{\mathbb{B}'}(t) = \frac{\mathbf{D}}{\mathbf{D}t}\int_{\mathcal{R}'_t} \rho\mathbf{T}\,dv = \frac{\mathbf{D}}{\mathbf{D}t}\int_{\mathcal{R}'} \rho_0\mathbf{T}\,dV \ . \tag{3.3.25}$$

However, since $\mathcal{R}'$ and ρ_0 are independent of time,

$$\frac{\mathbf{D}}{\mathbf{D}t}\int_{\mathcal{R}'} \rho_0\mathbf{T}\,dV = \int_{\mathcal{R}'} \rho_0\frac{\mathbf{D}\mathbf{T}}{\mathbf{D}t}\,dV = \int_{\mathcal{R}'_t} \rho\frac{\mathbf{D}\mathbf{T}}{\mathbf{D}t}\,dv \ . \tag{3.3.26}$$

Thus,

$$\boxed{\frac{\mathbf{D}}{\mathbf{D}t}\int_{\mathcal{R}'_t} \rho\mathbf{T}\,dv = \int_{\mathcal{R}'_t} \rho\frac{\mathbf{D}\mathbf{T}}{\mathbf{D}t}\,dv \ .} \tag{3.3.27}$$

3.3.2 Strain-Displacement Relations

The Almansi-Hamel strain tensor can be expressed in terms of the displacement field in a body as well as in terms of the inverse deformation gradient tensor. Recall that the positions of a material point in the reference and current configurations can be related by the displacement such that $\mathbf{X} = \mathbf{x} - \mathbf{u}$. Therefore, the inverse deformation gradient tensor $\mathbf{F}^{-1} \equiv \text{grad}\,\mathbf{X}$ can be given as

$$\mathbf{F}^{-1} = \mathbf{I} - \text{grad}\,\mathbf{u}\ , \quad F_{Ai}^{-1} = \delta_{Ai} - \frac{\partial u_A}{\partial x_i}\ . \tag{3.3.28}$$

The Almansi-Hamel strain tensor (3.3.6) can thus be written as

$$\mathbf{e} = \frac{1}{2}(\mathbf{I} - \mathbf{F}^{-\mathsf{T}} \cdot \mathbf{F}^{-1}) = \frac{1}{2}\left[\mathbf{I} - (\mathbf{I} - \text{grad}\,\mathbf{u})^{\mathsf{T}} \cdot (\mathbf{I} - \text{grad}\,\mathbf{u})\right]\ , \tag{3.3.29}$$

which, after simplifying, gives

$$
\mathbf{e} = \frac{1}{2}\left[\operatorname{grad}\mathbf{u} + (\operatorname{grad}\mathbf{u})^{\mathsf{T}} - (\operatorname{grad}\mathbf{u})^{\mathsf{T}}\cdot(\operatorname{grad}\mathbf{u})\right] ,
$$
$$
e_{ij} = \frac{1}{2}\left(\frac{\partial u_i}{\partial x_j} + \frac{\partial u_j}{\partial x_i} - \frac{\partial u_k}{\partial x_i}\frac{\partial u_k}{\partial x_j}\right) .
$$

(3.3.30)

3.4   Kinematic Linearization

Let the *norm*, or "length," of a second-order tensor $\mathbf{A}$ be a scalar measure of its magnitude defined by

$$
\|\mathbf{A}\| \equiv \sqrt{\mathbf{A}:\mathbf{A}} , \quad \|\mathbf{A}\| \equiv \sqrt{A_{ij}A_{ij}} ,
\tag{3.4.1}
$$

where A_{ij} are the scalar components of $\mathbf{A}$ in an orthonormal basis $\{\hat{\mathbf{e}}_i\}$. In an orthonormal basis, the norm of $\mathbf{A}$ is the square root of the sum of the squares of each of its scalar components. One says that this norm is positive definite, since $\|\mathbf{A}\| \geq 0$ with $\|\mathbf{A}\| = 0$ if and only if $\mathbf{A} = \mathbf{0}$. It can easily be shown that

$$
|A_{ij}| \leq \|\mathbf{A}\| ,
\tag{3.4.2}
$$

the scalar components of $\mathbf{A}$ have absolute values that are each less then or equal to its norm. The absolute value of a scalar component of $\mathbf{A}$ is equal to the norm of $\mathbf{A}$ if and only if that is the only nonzero scalar component. Conversely, it can also be shown that

$$
\|\mathbf{A}\| \leq 3\,|A_{ij}|_{\max} ,
\tag{3.4.3}
$$

where $|A_{ij}|_{\max}$ is the largest of all absolute values of scalar components of $\mathbf{A}$. The equality in (3.4.3) holds if and only if all the scalar components of $\mathbf{A}$ have the same absolute values.

One says that the material gradient of the displacement field $\operatorname{Grad}\mathbf{u}$ is "infinitesimal" at a material point if, at that material point, $\|\operatorname{Grad}\mathbf{u}\| \ll 1$. Then, there exists a positive scalar $\epsilon \ll 1$ such that

$$
\|\operatorname{Grad}\mathbf{u}\| = \mathcal{O}(\epsilon) ,
\tag{3.4.4}
$$

where the *order relation* $\mathcal{O}(\epsilon^n)$ (read "terms of order ϵ^n and higher") has the following meaning (Cole 1968):

$$
\lim_{\epsilon\to 0}\left|\frac{1}{\epsilon^n}\mathcal{O}(\epsilon^n)\right| < \infty .
\tag{3.4.5}
$$

Note that, if $f = \mathcal{O}(\epsilon^m)$ and $g = \mathcal{O}(\epsilon^n)$, where $m < n$, then $f \ll 1$, $g \ll f$, $fg = \mathcal{O}(\epsilon^{m+n})$, $g/f = \mathcal{O}(\epsilon^{n-m})$, $f \pm g = \mathcal{O}(\epsilon^m)$, and $f \pm g \approx f$ in the sense that the error

$$\frac{(f \pm g) - f}{f \pm g} = \mathcal{O}(\epsilon^{n-m}) \tag{3.4.6}$$

is much less than 1.

It follows from (3.4.2) and (3.4.3) that, in a Cartesian coordinate system, the material gradient of the displacement field is infinitesimal if and only if each of its scalar components satisfies $|\partial u_i / \partial X_A| \ll 1$. It follows that

$$\frac{\partial u_i}{\partial X_A} = \mathcal{O}_{iA}(\epsilon) , \tag{3.4.7}$$

where the nine order relations $\mathcal{O}_{iA}(\epsilon)$ each have the same meaning as $\mathcal{O}(\epsilon)$.

Using the expression (3.2.46) for the deformation gradient tensor in terms of the material gradient of the displacement field, it follows that at a point where the material gradient of the displacement field is infinitesimal,

$$F_{iA} = \delta_{iA} + \mathcal{O}_{iA}(\epsilon) , \tag{3.4.8}$$

and, by applying the chain rule, one finds that for a tensor field $\mathbf{T}$,

$$\frac{\partial \mathbf{T}}{\partial X_A} = \frac{\partial x_i}{\partial X_A} \frac{\partial \mathbf{T}}{\partial x_i} = F_{iA} \frac{\partial \mathbf{T}}{\partial x_i} = \frac{\partial \mathbf{T}}{\partial x_A} + \mathcal{O}_{iA}(\epsilon) \frac{\partial \mathbf{T}}{\partial x_i} . \tag{3.4.9}$$

The terms with $\mathcal{O}_{iA}(\epsilon)$ in (3.4.9) are negligible and $\operatorname{Grad} \mathbf{T} \approx \operatorname{grad} \mathbf{T}$; the material and spatial gradients of a tensor are approximately the same at a material point where the material gradient of the displacement field is infinitesimal. This means, of course, that the spatial gradient of the displacement field is also infinitesimal and one need only say that the displacement gradient is infinitesimal. The displacement itself need not necessarily be small, even at material points where the displacement gradient is infinitesimal, so the reference and current positions of such a material point are *not* approximately the same, in general.

Recall from (3.1.33) that $d\mathbf{x} = \mathbf{F} \cdot d\mathbf{X}$, where $d\mathbf{X} = \hat{\mathbf{\Upsilon}} \, dS$ and $d\mathbf{x} = \hat{v} \, ds$ are differential vectors in the reference and current configurations. The reference material direction $\hat{\mathbf{\Upsilon}}$ is mapped into the current material direction $\hat{v}$ and dS and ds are corresponding differential arc lengths. It follows from (3.4.8) that, at a material point where the displacement gradient is infinitesimal,

$$dx_i = dX_i + \mathcal{O}_{iA}(\epsilon) dX_A \tag{3.4.10}$$

and, consequently, that $ds = [1 + \mathcal{O}(\epsilon)] \, dS$ and

$$\hat{v}_i = \hat{\Upsilon}_i + \mathcal{O}_i(\epsilon) . \tag{3.4.11}$$

At a material point where the displacement gradient is infinitesimal, the difference between corresponding reference and current material directions is negligible, $\hat{\upsilon} \approx \hat{\Upsilon}$. In other words, material rotations are small at a material point where the displacement gradient is infinitesimal.

From the expressions (3.2.48) and (3.3.30) for the Green-St. Venant and Almansi-Hamel strain tensors, one can see that when the displacement gradient is infinitesimal, then $E_{AB} = \mathcal{O}_{AB}(\epsilon)$ and $e_{ij} = \mathcal{O}_{ij}(\epsilon)$ and

$$E_{AB} = \frac{1}{2}\left(\frac{\partial u_A}{\partial X_B} + \frac{\partial u_B}{\partial X_A}\right) + \mathcal{O}_{AB}(\epsilon^2)\,, \qquad (3.4.12)$$

$$e_{ij} = \frac{1}{2}\left(\frac{\partial u_i}{\partial x_j} + \frac{\partial u_j}{\partial x_i}\right) + \mathcal{O}_{ij}(\epsilon^2)\,. \qquad (3.4.13)$$

Thus, given (3.4.9), the Almansi-Hamel and Green-St. Venant strain tensors at a material point are approximately equal, $\mathbf{e} \approx \mathbf{E}$, if the displacement gradient is infinitesimal at that point.

A body is said to experience *infinitesimal deformation* if the displacement gradient is infinitesimal at every material point in the body. Then, the difference between material and spatial gradients and between material directions in the reference and current configurations are negligible everywhere in the body.

3.4.1   Infinitesimal Strain and Rotation Tensors

For the rest of this chapter, it is presumed that all quantities are given in terms of their material descriptions, that is, as fields over the reference configuration $\mathcal{R}$ of a body. Accordingly, given this understanding, the comma convention can be reinstated since there is no longer any ambiguity about whether differentiation is with respect to X_A or x_i—one understands that $\phi_{,i} = \partial\phi/\partial X_i$. In addition, the original notation for the gradient, divergence, and curl will be used where it is understood that they represent their material forms so that, for instance, $\nabla\phi = \text{Grad}\,\phi$. Finally, no distinction between uppercase and lowercase indices is needed for the linearized theory to follow, so lowercase indices will typically be used in accordance with the generally accepted practice.

The *infinitesimal strain tensor* is defined by

$$\boxed{\varepsilon \equiv \frac{1}{2}\left[\nabla\mathbf{u} + (\nabla\mathbf{u})^{\mathsf{T}}\right]\,, \quad \varepsilon_{ij} \equiv \frac{1}{2}(u_{i,j} + u_{j,i})\,,} \qquad (3.4.14)$$

and the *infinitesimal rotation tensor* is defined by

$$\boxed{\omega \equiv \frac{1}{2}\left[\nabla\mathbf{u} - (\nabla\mathbf{u})^{\mathsf{T}}\right]\,, \quad \omega_{ij} \equiv \frac{1}{2}(u_{i,j} - u_{j,i})\,.} \qquad (3.4.15)$$

These definitions can always be used to determine infinitesimal strain and rotation tensor fields. However, it will be shown that they only have physical significance for infinitesimal deformations.

It follows from the definitions (3.4.14) and (3.4.15) that, for infinitesimal deformations, $\varepsilon_{ij} = \mathcal{O}_{ij}(\epsilon)$ and $\omega_{ij} = \mathcal{O}_{ij}(\epsilon)$. Also, from the expressions (3.2.48) and (3.3.30) for the Green-St. Venant and Almansi-Hamel strain tensors, $E_{ij} = \varepsilon_{ij} + \mathcal{O}_{ij}(\epsilon^2)$ and $e_{ij} = \varepsilon_{ij} + \mathcal{O}_{ij}(\epsilon^2)$. Thus, for infinitesimal deformation, the scalar components of the infinitesimal strain and rotation tensors satisfy $|\varepsilon_{ij}| \ll 1$ and $|\omega_{ij}| \ll 1$ and the infinitesimal, Green-St. Venant, and Almansi-Hamel strain tensors at a material point are approximately equal, $\boldsymbol{\varepsilon} \approx \mathbf{E} \approx \mathbf{e}$. Note that the Green-St. Venant and Almansi-Hamel strain tensors are quadratic in Grad $\mathbf{u}$ and grad $\mathbf{u}$, whereas the infinitesimal strain tensor is linear in $\boldsymbol{\nabla}\mathbf{u}$. Thus, the use of the infinitesimal strain tensor to (approximately) characterize the infinitesimal deformation of a body is a *linearization*. The linearity of (3.4.14) means that if $\boldsymbol{\varepsilon}^{(1)}$ is the strain corresponding to $\mathbf{u}^{(1)}$ and $\boldsymbol{\varepsilon}^{(2)}$ is the strain corresponding to $\mathbf{u}^{(2)}$, then $\boldsymbol{\varepsilon}^{(1)} + \boldsymbol{\varepsilon}^{(2)}$ is the strain corresponding to $\mathbf{u}^{(1)} + \mathbf{u}^{(2)}$. This superposition property will prove to be extremely convenient.

By construction, $\boldsymbol{\varepsilon}$ is a symmetric second-order tensor and $\boldsymbol{\omega}$ is a skew-symmetric second-order tensor. Note that $\boldsymbol{\nabla}\mathbf{u} = \boldsymbol{\varepsilon} + \boldsymbol{\omega}$. The infinitesimal strain and rotation tensors are said to be the symmetric and skew-symmetric parts of the displacement gradient tensor and $\|\boldsymbol{\nabla}\mathbf{u}\| \ll 1$ if and only if $\|\boldsymbol{\varepsilon}\| \ll 1$ *and* $\|\boldsymbol{\omega}\| \ll 1$.

Since $\boldsymbol{\omega}$ is skew-symmetric, it has only three independent scalar components, which can be used to define the scalar components of a vector $\boldsymbol{\theta}$, called the *axial vector of $\boldsymbol{\omega}$*, as follows:

$$\omega_{ij} = -e_{ijk}\theta_k \;, \quad [\boldsymbol{\omega}] = \begin{bmatrix} 0 & -\theta_3 & \theta_2 \\ \theta_3 & 0 & -\theta_1 \\ -\theta_2 & \theta_1 & 0 \end{bmatrix} . \tag{3.4.16}$$

Every skew-symmetric second-order tensor has such an axial vector, which completely characterizes the tensor. The axial vector $\boldsymbol{\theta}$ of the infinitesimal rotation tensor is also called the *infinitesimal rotation vector*. By multiplying (3.4.16) by e_{rij},

$$e_{rij}\omega_{ij} = -e_{ijr}e_{ijk}\theta_k = -2\delta_{rk}\theta_k = -2\theta_r \;, \tag{3.4.17}$$

it is seen that the scalar components of $\boldsymbol{\theta}$ can alternatively be expressed as

$$\theta_i = -\frac{1}{2}e_{ijk}\omega_{jk} \;. \tag{3.4.18}$$

By substituting (3.4.15) into this expression,

$$\theta_i = -\frac{1}{4}e_{ijk}(u_{j,k} - u_{k,j}) = \frac{1}{4}e_{ijk}(u_{k,j} + u_{k,j}) = \frac{1}{2}e_{ijk}u_{k,j} \;, \tag{3.4.19}$$

one finds that the infinitesimal rotation vector $\boldsymbol{\theta}$ can be given in terms of the displacement field by

$$\boxed{\boldsymbol{\theta} = \frac{1}{2}\nabla \times \mathbf{u}, \quad \theta_i = \frac{1}{2}e_{ijk}u_{k,j} \,.}$$
(3.4.20)

Noting that $e_{ijk}u_{k,ji} = 0$, it follows that

$$\nabla \cdot \boldsymbol{\theta} = 0, \quad \theta_{i,i} = 0 \,;$$
(3.4.21)

the infinitesimal rotation vector field is divergence-free.

Example 3.10
Find the infinitesimal strain tensor field for the motion given in Example 3.1. Examine the conditions under which the infinitesimal strain tensor approximates the Green-St. Venant and Almansi-Hamel strain tensors at every point in the body, that is, the conditions for infinitesimal deformation of the body.

Solution. Using the results from Example 3.3 for the displacement field, the material gradient of the displacement field is

$$[\nabla\mathbf{u}] = \begin{bmatrix} 0 & 2a_1 X_2 & 0 \\ 0 & 0 & 0 \\ 0 & a_3 X_3 & a_3 X_2 \end{bmatrix},$$

and the infinitesimal strain tensor is therefore

$$[\varepsilon] = \frac{1}{2}\left([\nabla\mathbf{u}] + [\nabla\mathbf{u}]^{\mathsf{T}}\right) = \begin{bmatrix} 0 & a_1 X_2 & 0 \\ a_1 X_2 & 0 & \frac{1}{2}a_3 X_3 \\ 0 & \frac{1}{2}a_3 X_3 & a_3 X_2 \end{bmatrix}.$$

This infinitesimal strain field approximates the Almansi-Hamel and Green-St. Venant strain tensors if the absolute values of the scalar components of the material gradient of the displacement field are each much less than 1. From the material gradient of the displacement field given above and recalling that a_1, a_2, and a_3 are positive and that $0 \le X_A \le 1$ for every material point in the body, this infinitesimal deformation condition requires that $a_1 \ll 1$ and $a_3 \ll 1$. An examination of the results from Examples 3.5 and 3.8 shows then that if second-order and higher terms in $a_1 X_2$, $a_3 X_2$, and $a_3 X_3$ are neglected, then $\varepsilon \approx \mathbf{E} \approx \mathbf{e}$.

3.4.2 Physical Interpretation

The exact physical interpretations derived for the nonlinear strain tensors, such as the Green-St. Venant and Almansi-Hamel strain tensors, can be given *approximately* in terms of the infinitesimal strain tensor when the deformation is infinitesimal: $\|\text{Grad }\mathbf{u}\| \ll 1$.

Longitudinal strain

Consider the expression (3.2.13) for the longitudinal strain $H(\hat{\boldsymbol{\Upsilon}})$, the change in length per unit reference length, in the reference material direction $\hat{\boldsymbol{\Upsilon}}$. For infinitesimal deformation, $|2\hat{\boldsymbol{\Upsilon}} \cdot \mathbf{E} \cdot \hat{\boldsymbol{\Upsilon}}| \ll 1$, and using the Taylor series expansion $(1 + x)^{1/2} = 1 + \frac{1}{2}x + \mathcal{O}(x^2)$ for $-1 < x \leq 1$, it follows that

$$H(\hat{\boldsymbol{\Upsilon}}) = (1 + 2\hat{\boldsymbol{\Upsilon}} \cdot \mathbf{E} \cdot \hat{\boldsymbol{\Upsilon}})^{1/2} - 1 = \hat{\boldsymbol{\Upsilon}} \cdot \mathbf{E} \cdot \hat{\boldsymbol{\Upsilon}} + \mathcal{O}(\epsilon^2) \qquad (3.4.22)$$

and $H(\hat{\boldsymbol{\Upsilon}}) \ll 1$. Alternatively, consider the expression (3.3.11) for the longitudinal strain $\eta(\hat{\boldsymbol{v}})$ in the current material direction $\hat{\boldsymbol{v}}$. For infinitesimal deformation, $|2\hat{\boldsymbol{v}} \cdot \mathbf{e} \cdot \hat{\boldsymbol{v}}| \ll 1$, and using the Taylor series expansion $(1 - x)^{-1/2} = 1 + \frac{1}{2}x + \mathcal{O}(x^2)$ for $-1 \leq x < 1$, it follows that

$$\eta(\hat{\boldsymbol{v}}) = (1 - 2\hat{\boldsymbol{v}} \cdot \mathbf{e} \cdot \hat{\boldsymbol{v}})^{-1/2} - 1 = \hat{\boldsymbol{v}} \cdot \mathbf{e} \cdot \hat{\boldsymbol{v}} + \mathcal{O}(\epsilon^2) \qquad (3.4.23)$$

and $\eta(\hat{\boldsymbol{v}}) \ll 1$. Finally, recalling that $E_{ij} = \varepsilon_{ij} + \mathcal{O}_{ij}(\epsilon^2)$, $e_{ij} = \varepsilon_{ij} + \mathcal{O}_{ij}(\epsilon^2)$, and $\hat{\Upsilon}_i = \hat{v}_i + \mathcal{O}_i(\epsilon)$, one has that for infinitesimal deformation

$$\boxed{\begin{aligned} H(\hat{\boldsymbol{v}}) &\approx \eta(\hat{\boldsymbol{v}}) \approx \hat{\boldsymbol{v}} \cdot \boldsymbol{\varepsilon} \cdot \hat{\boldsymbol{v}} \,, \\ H(\hat{v}_1, \hat{v}_2, \hat{v}_3) &\approx \eta(\hat{v}_1, \hat{v}_2, \hat{v}_3) \approx \varepsilon_{ij} \hat{v}_i \hat{v}_j \,, \end{aligned}} \qquad (3.4.24)$$

where $\hat{\boldsymbol{v}}$ is a unit vector representing either the reference or current material direction (the distinction is negligible for infinitesimal deformation). Note that ε_{11} is the (approximate) longitudinal strain in the direction $\hat{\mathbf{e}}_1$, ε_{22} is the (approximate) longitudinal strain in the direction $\hat{\mathbf{e}}_2$, and ε_{33} is the (approximate) longitudinal strain in the direction $\hat{\mathbf{e}}_3$.

Orthogonal shear strain

The orthogonal shear strain, the reduction in angle between two orthogonal reference material directions, is found by linearizing the nonlinear strain relation (3.2.20). By applying reasoning similar to that for the longitudinal strain above, and noting the Taylor series expansion $\sin^{-1} x = x + \mathcal{O}(x^3)$ for $|x| < 1$, it can be shown that for infinitesimal deformation, $\gamma(\hat{\boldsymbol{v}}', \hat{\boldsymbol{v}}'') \ll 1$ and

$$\boxed{\begin{aligned} \gamma(\hat{\boldsymbol{v}}', \hat{\boldsymbol{v}}'') &\approx 2\hat{\boldsymbol{v}}' \cdot \boldsymbol{\varepsilon} \cdot \hat{\boldsymbol{v}}'' \,, \\ \gamma(\hat{v}_1', \hat{v}_2', \hat{v}_3', \hat{v}_1'', \hat{v}_2'', \hat{v}_3'') &\approx 2\varepsilon_{ij} \hat{v}_i' \hat{v}_j'' \,, \end{aligned}} \qquad (3.4.25)$$

where $\hat{\boldsymbol{v}}'$ and $\hat{\boldsymbol{v}}''$ are orthogonal unit vectors ($\hat{\boldsymbol{v}}' \cdot \hat{\boldsymbol{v}}'' = 0$) representing reference material directions and $\gamma(\hat{\boldsymbol{v}}', \hat{\boldsymbol{v}}'')$ is the reduction in angle between $\hat{\boldsymbol{v}}'$ and $\hat{\boldsymbol{v}}''$ due to the motion of the body. Note that ε_{12} is (approximately) *one-half* the reduction in angle between the $\hat{\mathbf{e}}_1$ and $\hat{\mathbf{e}}_2$ directions, and so on for ε_{23} and ε_{31}.

Dilatation

The change in volume per unit reference volume is found by linearizing either of the relations (3.2.34) or (3.3.17). To linearize (3.2.34), note first that for infinitesimal deformation, $\det(\mathbf{I} + 2\mathbf{E}) = 1 + 2\operatorname{tr}\mathbf{E} + \mathcal{O}(\epsilon^2)$. Therefore,

$$
\begin{aligned}
e &= \sqrt{\det(\mathbf{I} + 2\mathbf{E})} - 1 \\
  &= \sqrt{1 + 2\operatorname{tr}\mathbf{E} + \mathcal{O}(\epsilon^2)} - 1 \\
  &= \operatorname{tr}\mathbf{E} + \mathcal{O}(\epsilon^2)
\end{aligned}
\tag{3.4.26}
$$

and $e \ll 1$. Thus, since $E_{ij} = \varepsilon_{ij} + \mathcal{O}_{ij}(\epsilon^2)$, the dilatation for infinitesimal deformation is given approximately by

$$
\boxed{\,e \approx \operatorname{tr}\boldsymbol{\varepsilon} = \boldsymbol{\nabla}\cdot\mathbf{u}\,,\quad e \approx \varepsilon_{kk} = u_{k,k}\;.\,}
\tag{3.4.27}
$$

Conservation of mass

Since $F_{ij} = \delta_{ij} + \mathcal{O}_{ij}(\epsilon)$, the Jacobian $J \equiv \det\mathbf{F} = 1 + \mathcal{O}(\epsilon)$. Therefore, from the continuity condition (3.2.45), the mass density ρ_0 of a material point at time $t = t_0$ is related to the material point's current mass density ρ by $\rho_0 = \rho[1 + \mathcal{O}(\epsilon)]$, where the reference configuration is the configuration of the body at time $t = t_0$. In general, the continuity condition is an integral part of the formulation of problems in continuum mechanics. However, if the deformation is infinitesimal, the continuity condition just means that $\rho_0 \approx \rho$ and it need not be considered further.

Rigid-body rotation

Rigid-body motion is defined as motion in which the distances between material points remain constant; that is, the longitudinal strain in every direction at every material point is zero. It was previously seen that a motion of a material volume is a rigid-body motion if and only if the Green-St. Venant and Almansi-Hamel strain tensors $\mathbf{E}$ and $\mathbf{e}$, respectively, are zero everywhere in the material volume. Therefore, if an infinitesimal deformation is a rigid-body motion, then $\boldsymbol{\varepsilon} \approx \mathbf{0}$, the infinitesimal strain tensor is approximately zero everywhere in the material volume. Conversely, if at every point in a material volume $\boldsymbol{\varepsilon} = \mathbf{0}$ *and* $\|\boldsymbol{\omega}\| \ll 1$ so that the deformation is infinitesimal, then the motion of the material volume is approximately a rigid-body motion. Be aware, however, that for *finite* rigid-body rotations, the infinitesimal strain tensor $\boldsymbol{\varepsilon}$ *will not be approximately zero*. This means that if one blithely calculates $\boldsymbol{\varepsilon}$ in this case and uses it to determine the longitudinal, shear, and volume strains, the results may be large, giving the impression that there is significant deformation of the material volume where there is none. On the other hand, if one analyzes a motion of a material volume and determines that $\boldsymbol{\varepsilon} = \mathbf{0}$ everywhere in the material volume, but $\|\boldsymbol{\omega}\|$ is not much less than 1, then the motion is not an approximate rigid-body motion.

If $\mathbf{u}(\mathbf{X})$ is the displacement at a point $\mathbf{X}$ in the body, then the relative displacement at a point $\mathbf{X}+\Delta\mathbf{X}$ is $\Delta\mathbf{u} \equiv \mathbf{u}(\mathbf{X}+\Delta\mathbf{X}) - \mathbf{u}(\mathbf{X})$. This relative displacement can be given in terms of the displacement gradient by

$$\Delta\mathbf{u} = \int_{\mathbf{X}}^{\mathbf{X}+\Delta\mathbf{X}} \nabla\mathbf{u} \cdot d\mathbf{X} , \quad \Delta u_i = \int_{\mathbf{X}}^{\mathbf{X}+\Delta\mathbf{X}} u_{i,j} \, dX_j , \qquad (3.4.28)$$

where the path of integration in (3.4.28) is along any material line with end points at $\mathbf{X}$ and $\mathbf{X}+\Delta\mathbf{X}$. Let $\mathbb{B}'$ be a material volume within which $\boldsymbol{\varepsilon} = \mathbf{0}$ and, therefore, $\nabla\mathbf{u} = \boldsymbol{\omega}$. It follows from the relation (3.4.36) derived later that, since $\boldsymbol{\varepsilon} = \mathbf{0}$ everywhere in $\mathbb{B}'$, the infinitesimal rotation vector field $\boldsymbol{\theta}$ and, consequently, the infinitesimal rotation tensor field $\boldsymbol{\omega}$ are homogeneous in $\mathbb{B}'$. Thus, the relative displacement (3.4.28) between any two points $\mathbf{X}$ and $\mathbf{X}+\Delta\mathbf{X}$ in $\mathbb{B}'$ can in this case be rewritten as

$$\Delta\mathbf{u} = \boldsymbol{\omega} \cdot \Delta\mathbf{X} , \quad \Delta u_i = \omega_{ij} \, \Delta X_j , \qquad (3.4.29)$$

or, since $\omega_{ij} = -e_{ijk}\theta_k = e_{kji}\theta_k$, it can be rewritten as

$$\Delta\mathbf{u} = \boldsymbol{\theta} \times \Delta\mathbf{X} , \quad \Delta u_i = e_{kji}\theta_k \, \Delta X_j . \qquad (3.4.30)$$

Therefore, if $\|\boldsymbol{\omega}\| \ll 1$ in $\mathbb{B}'$, then the material volume experiences an approximate rigid-body motion and the rigid-body rotation is characterized via (3.4.30) by the infinitesimal rotation vector $\boldsymbol{\theta}$ (a rigid-body translation will have no effect on the relative displacement vector).

In order to provide a physical interpretation of the infinitesimal rotation vector $\boldsymbol{\theta}$, consider a rigid-body rotation through an angle $\theta = |\boldsymbol{\theta}|$ about an axis through the material point at $\mathbf{X}$ that is parallel to $\boldsymbol{\theta}$, as shown in Figure 3.5. It can be seen that the relative displacement $\Delta\mathbf{u}$ of a material point at $\mathbf{X}+\Delta\mathbf{X}$ has a magnitude of

$$|\Delta\mathbf{u}| = 2(|\Delta\mathbf{X}| \sin\phi) \sin\frac{\theta}{2} = \left(2\sin\frac{\theta}{2}\right) |\Delta\mathbf{X}| \sin\phi , \qquad (3.4.31)$$

where ϕ is the angle formed by $\boldsymbol{\theta}$ and $\Delta\mathbf{X}$. At the same time, $\boldsymbol{\theta} \times \Delta\mathbf{X}$ is a vector that forms an angle $\theta/2$ with $\Delta\mathbf{u}$ and has a magnitude of

$$|\boldsymbol{\theta} \times \Delta\mathbf{X}| = |\boldsymbol{\theta}||\Delta\mathbf{X}| \sin\phi = \theta|\Delta\mathbf{X}| \sin\phi . \qquad (3.4.32)$$

Thus, if $\theta \ll 1$, then $\Delta\mathbf{u} \approx \boldsymbol{\theta} \times \Delta\mathbf{X}$. If $\|\boldsymbol{\omega}\| \ll 1$ in $\mathbb{B}'$, then the material volume experiences an approximate rigid-body motion and the rigid-body rotation is a small rotation through an angle $|\boldsymbol{\theta}|$ about an axis that is parallel to $\boldsymbol{\theta}$. Recall that an orthogonal second-order tensor is required to characterize a finite rigid-body rotation—it is only for infinitesimal deformation that a rigid-body rotation can be approximately characterized by a vector.

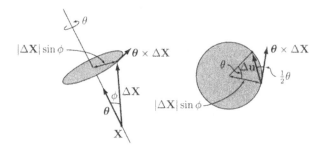

FIGURE 3.5. The infinitesimal rotation vector interpreted as a rigid-body rotation.

Example 3.11

For the motion given in Example 3.1, use the approximate relations (3.4.24) and (3.4.25) to find the longitudinal strain in the direction $\hat{v} = (\hat{e}_1 + \hat{e}_3)/\sqrt{2}$ and the shear strain between $\hat{v}' = (\hat{e}_1 + \hat{e}_3)/\sqrt{2}$ and $\hat{v}'' = \hat{e}_2$ at the material point whose reference position is given by the Cartesian coordinates $(X_1, X_2, X_3) = (1/2, 1/2, 1/2)$. Compare the results to the exact answers found in Examples 3.6 and 3.9.

Solution. Using the infinitesimal strain found in Example 3.10 with the coordinates $(X_1, X_2, X_3) = (1/2, 1/2, 1/2)$,

$$\hat{v} \cdot \varepsilon \cdot \hat{v} = \frac{1}{2} \begin{bmatrix} 1 & 0 & 1 \end{bmatrix} \begin{bmatrix} 0 & \frac{1}{2}a_1 & 0 \\ \frac{1}{2}a_1 & 0 & \frac{1}{4}a_3 \\ 0 & \frac{1}{4}a_3 & \frac{1}{2}a_3 \end{bmatrix} \begin{bmatrix} 1 \\ 0 \\ 1 \end{bmatrix} = \frac{1}{4}a_3 \ .$$

Therefore, the approximate longitudinal strain at this material point in the direction $\hat{v}$ is given by (3.4.24) as

$$H(\hat{v}) \approx \eta(\hat{v}) \approx \frac{1}{4}a_3 \ .$$

Similarly, one has that

$$\hat{v}' \cdot \varepsilon \cdot \hat{v}'' = \frac{1}{\sqrt{2}} \begin{bmatrix} 1 & 0 & 1 \end{bmatrix} \begin{bmatrix} 0 & \frac{1}{2}a_1 & 0 \\ \frac{1}{2}a_1 & 0 & \frac{1}{4}a_3 \\ 0 & \frac{1}{4}a_3 & \frac{1}{2}a_3 \end{bmatrix} \begin{bmatrix} 0 \\ 1 \\ 0 \end{bmatrix} = \frac{1}{2\sqrt{2}} \left(a_1 + \frac{1}{2}a_3 \right) \ .$$

so that the approximate shear strain at this material point between $\hat{v}'$ and $\hat{v}''$ is given by (3.4.25) as

$$\gamma(\hat{v}', \hat{v}'') \approx \frac{1}{\sqrt{2}} \left(a_1 + \frac{1}{2}a_3 \right) \ .$$

Recall that the infinitesimal deformation condition requires that $a_1 \ll 1$ and $a_3 \ll 1$.

The exact answers found in Example 3.6 can be expanded in a Taylor series to show that, for $a_1 \ll 1$ and $a_3 \ll 1$,

$$H(\hat{v}) = \frac{1}{4}a_3 + \frac{1}{32}a_3^2 + \mathcal{O}(\epsilon^3)$$

and

$$\gamma(\hat{v}', \hat{v}'') = \frac{1}{\sqrt{2}}a_1 + \frac{1}{2\sqrt{2}}a_3 - \frac{1}{4\sqrt{2}}a_1 a_3 + \frac{1}{8\sqrt{2}}a_3^2 + \mathcal{O}(\epsilon^3),$$

thus confirming that the results found by using the infinitesimal strain tensor and relations (3.4.24) and (3.4.25) do approximate the exact results. The exact answer found in Example 3.9 can also be expanded in a Taylor series to show that

$$\eta(\hat{v}) = \frac{1}{4}a_3 - \frac{3}{32}a_3^2 + \mathcal{O}(\epsilon^3),$$

Recall that this exact result was for the longitudinal strain in a different material direction than was used in Example 3.6, since no account was made for material rotation. Nevertheless, it can be seen that the two longitudinal strains are approximately the same for infinitesimal deformation. This is because corresponding reference and current material directions are approximately the same for infinitesimal deformation.

3.4.3 Principal Strains and Principal Directions of Strain

In the terminology of Section 2.3.6, the longitudinal strain (3.4.24) in the material direction $\hat{v}$ is approximately equal to the normal of the infinitesimal strain tensor in that direction (2.3.99), $H(\hat{v}) \approx \eta(\hat{v}) \approx N_\varepsilon(\hat{v})$, and the orthogonal shear strain (3.4.25) between the orthogonal directions $\hat{v}'$ and $\hat{v}''$ is approximately equal to *twice* the shear between those direction (2.3.110), $\gamma(\hat{v}', \hat{v}'') \approx 2S_\varepsilon(\hat{v}', \hat{v}'')$. It follows that the eigenvectors of the infinitesimal strain tensor are approximately the directions corresponding to extrema of the longitudinal strain and that the eigenvalues are the approximate values of these extrema. Also, the orthogonal shear strain between the direction of an eigenvector and every orthogonal direction is approximately zero.

- The eigenvalues of $\boldsymbol{\varepsilon}$, $\varepsilon^{(1)} \geq \varepsilon^{(2)} \geq \varepsilon^{(3)}$, are called the *principal strains*.

- The corresponding eigenvectors, $\hat{v}^{(1)}$, $\hat{v}^{(2)}$, and $\hat{v}^{(3)}$ are called the *principal directions of strain* and are chosen so as to form a right-handed system, $\hat{v}^{(1)} \times \hat{v}^{(2)} = \hat{v}^{(3)}$.

- Let the vector basis formed by the principal directions be denoted by

$$\{\hat{e}_i^*\}_\varepsilon = \begin{bmatrix} \hat{v}^{(1)} \\ \hat{v}^{(2)} \\ \hat{v}^{(3)} \end{bmatrix} . \tag{3.4.33}$$

Then, the components of strain in this vector basis are

$$[\varepsilon]^* = \begin{bmatrix} \varepsilon^{(1)} & 0 & 0 \\ 0 & \varepsilon^{(2)} & 0 \\ 0 & 0 & \varepsilon^{(3)} \end{bmatrix} . \tag{3.4.34}$$

- The maximum longitudinal strain $H_{max} = \eta_{max} \approx \varepsilon^{(1)}$ is in the direction of the eigenvector $\hat{v}^{(1)}$, the minimum longitudinal strain $H_{min} = \eta_{min} \approx \varepsilon^{(3)}$ is in the direction of the eigenvector $\hat{v}^{(3)}$, and the maximum orthogonal shear strain $|\gamma|_{max} \approx \varepsilon^{(1)} - \varepsilon^{(3)}$ is between directions $\pm(\hat{v}^{(1)} + \hat{v}^{(3)})/\sqrt{2}$ and $\pm(\hat{v}^{(1)} - \hat{v}^{(3)})/\sqrt{2}$ that bisect the eigenvectors $\hat{v}^{(1)}$ and $\hat{v}^{(3)}$.

Example 3.12
For the motion given in Example 3.1, with $a_1 = 0.05$ and $a_3 = 0.01$, determine the principal strains and principal directions of strain at the material point whose reference position is given by the Cartesian coordinates $(X_1, X_2, X_3) = (1/2, 1/2, 1/2)$.

Solution. Using the infinitesimal strain found in Example 3.10 with $a_1 = 0.05$ and $a_3 = 0.01$, one has that

$$[\varepsilon] = \begin{bmatrix} 0 & 0.025 & 0 \\ 0.025 & 0 & 0.0025 \\ 0 & 0.0025 & 0.0050 \end{bmatrix} .$$

By using a computer application to solve for the eigenvalues and eigenvectors of this matrix, one finds that the principal strains are

$$\varepsilon^{(1)} = 0.025 , \quad \varepsilon^{(2)} = 0.0049 , \quad \varepsilon^{(3)} = -0.025$$

and that the corresponding principal directions of strain are

$$\hat{v}^{(1)} = 0.702\hat{e}_1 + 0.707\hat{e}_2 + 0.088\hat{e}_3 ,$$
$$\hat{v}^{(2)} = -0.103\hat{e}_1 - 0.020\hat{e}_2 + 0.994\hat{e}_3 ,$$
$$\hat{v}^{(3)} = -0.704\hat{e}_1 + 0.707\hat{e}_2 - 0.059\hat{e}_3 .$$

Thus, the maximum longitudinal strain is an extension $\varepsilon^{(1)} = 0.025$ in the direction of $\hat{v}^{(1)}$ and the minimum longitudinal strain is a contraction $\varepsilon^{(3)} = -0.025$ in the direction of $\hat{v}^{(3)}$.

3.4.4 Compatibility

For any given displacement field **u** describing a one-to-one mapping of a reference configuration onto a current configuration, the Cartesian scalar components of the infinitesimal strain field (3.4.14) are uniquely given by the strain-displacement relations $\varepsilon_{ij} = \frac{1}{2}(u_{i,j} + u_{j,i})$. What if a strain field $\boldsymbol{\varepsilon}$ is given for which a corresponding displacement field **u** has not been presented? Solving the strain-displacement relations for **u** involves six independent equations and only three unknowns—the problem is overdetermined and it is possible that a continuous, single-valued displacement field does not exist. If an allowable displacement field does exist, then the corresponding strain field is said to be *compatible*. Otherwise, the strain field is incompatible and unacceptable unless some additional provisions are made (as, for instance, in the theory of fracture mechanics). What is required are *compatibility conditions* to be satisfied by a strain field, that are necessary and sufficient to guarantee that there is a continuous, single-valued displacement field corresponding to the strain field through the strain-displacement relations.

Necessary conditions

Take an arbitrary continuous, single-valued displacement field **u** and consider the corresponding strain field $\boldsymbol{\varepsilon}$, the Cartesian scalar components of which are given by $\varepsilon_{ij} = \frac{1}{2}(u_{i,j} + u_{j,i})$. By using these strain-displacement relations, the curl of the strain tensor is given in terms of the Cartesian scalar components of displacement by

$$\boldsymbol{\nabla} \times \boldsymbol{\varepsilon} = e_{ijk}\varepsilon_{rj,i}\,\hat{\mathbf{e}}_k\hat{\mathbf{e}}_r = \frac{1}{2}e_{ijk}(u_{r,ji} + u_{j,ri})\hat{\mathbf{e}}_k\hat{\mathbf{e}}_r \ . \tag{3.4.35}$$

However, since $u_{r,ji} = u_{r,ij}$, it follows that $e_{ijk}u_{r,ji} = 0$. Also, from (3.4.20), it follows that $\frac{1}{2}e_{ijk}u_{j,ri} = \theta_{k,r}$ where $\boldsymbol{\theta} = \theta_i\hat{\mathbf{e}}_i$ is the infinitesimal rotation vector. Thus,

$$\boldsymbol{\nabla} \times \boldsymbol{\varepsilon} = \boldsymbol{\nabla}\boldsymbol{\theta} \ ; \tag{3.4.36}$$

the curl of the infinitesimal strain tensor is equal to the gradient of the infinitesimal rotation vector. Taking the curl a second time and recalling from (2.4.42) that $\boldsymbol{\nabla} \times (\boldsymbol{\nabla}\boldsymbol{\theta}) = \mathbf{0}$, one obtains the necessary conditions

$$\boxed{\boldsymbol{\nabla} \times (\boldsymbol{\nabla} \times \boldsymbol{\varepsilon}) = \mathbf{0} \ , \quad e_{ikr}e_{jls}\varepsilon_{ij,kl} = 0} \tag{3.4.37}$$

for the existence of **u**. These conditions must necessarily be satisfied by the infinitesimal strain tensor if the displacement field exists; there is no solution to the strain-displacement equations if these conditions are not satisfied. It is shown below that these conditions are also sufficient—they guarantee the existence of an acceptable displacement field.

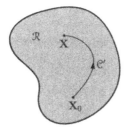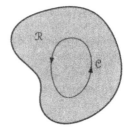

FIGURE 3.6. *Integration paths in the analysis of compatibility.*

Sufficient conditions

Assume that the necessary conditions (3.4.37) are satisfied everywhere in
the region $\mathcal{R}$ of a body. Does this guarantee the existence of a continuous, single-valued displacement field $\mathbf{u}$? If so, then they are also sufficient
conditions.

If the infinitesimal rotation vector field is known, then it can be expressed
in terms of its gradient field as

$$\boldsymbol{\theta}(\mathbf{X}) = \boldsymbol{\theta}(\mathbf{X}_0) + \int_{\mathbf{X}_0}^{\mathbf{X}} (\boldsymbol{\nabla}\boldsymbol{\theta}) \cdot d\mathbf{X} \ ,$$

$$\theta_i(\mathbf{X}) = \theta_i(\mathbf{X}_0) + \int_{\mathbf{X}_0}^{\mathbf{X}} \theta_{i,j} \ dX_j \ ,$$

$\qquad(3.4.38)$

where $\boldsymbol{\theta}(\mathbf{X}_0)$ is the infinitesimal rotation vector at some reference point
$\mathbf{X}_0$, which must be specified to fix the rigid-body rotation, and the line
integral is along an arbitrary path $\mathcal{C}'$, completely within the body, from
$\mathbf{X}_0$ to $\mathbf{X}$ (Figure 3.6). On the other hand, if an infinitesimal strain field
$\boldsymbol{\varepsilon}$ is specified, then the gradient field of the infinitesimal rotation vector is
given by (3.4.36), and (3.4.38) implies that the corresponding infinitesimal
rotation vector field can be expressed as

$$\boldsymbol{\theta}(\mathbf{X}) = \boldsymbol{\theta}(\mathbf{X}_0) + \int_{\mathbf{X}_0}^{\mathbf{X}} (\boldsymbol{\nabla} \times \boldsymbol{\varepsilon}) \cdot d\mathbf{X} \ . \qquad(3.4.39)$$

However, this expression for the infinitesimal rotation vector field corresponding to the infinitesimal strain field $\boldsymbol{\varepsilon}$ is uniquely determined if and
only if for every pair of points $\mathbf{X}_0$ and $\mathbf{X}$ in the body, the value of the
integral on the right-hand side is independent of the integration path between these points. This condition of *path independence* can alternatively
be stated as

$$\oint_{\mathcal{C}} (\boldsymbol{\nabla} \times \boldsymbol{\varepsilon}) \cdot d\mathbf{X} = \mathbf{0} \ , \qquad(3.4.40)$$

which must hold for all *closed curves* $\mathcal{C}$ that lie completely within the body (Figure 3.6). This can be seen by considering the path independence of (3.4.39) when $\mathbf{X} = \mathbf{X}_0$. However, from Stokes's theorem,

$$\oint_{\mathcal{C}} (\boldsymbol{\nabla} \times \boldsymbol{\varepsilon}) \cdot d\mathbf{X} = \int_{\mathcal{S}} \hat{\mathbf{N}} \cdot [\boldsymbol{\nabla} \times (\boldsymbol{\nabla} \times \boldsymbol{\varepsilon})] \, dA \,, \tag{3.4.41}$$

where $\mathcal{S}$ is any unbroken surface in the body such that $\partial \mathcal{S} = \mathcal{C}$ and $\hat{\mathbf{N}}$ is the unit normal to that surface. Thus, it is seen that if for *every* closed curve $\mathcal{C}$ in the body there exists *at least one* unbroken surface $\mathcal{S}$ that is bounded by $\mathcal{C}$, then the condition that $\boldsymbol{\nabla} \times (\boldsymbol{\nabla} \times \boldsymbol{\varepsilon}) = \mathbf{0}$ everywhere in the body guarantees the path independence of the integral on the right-hand side of (3.4.39) and therefore the existence of the infinitesimal rotation vector field $\boldsymbol{\theta}$.

Bodies in which every closed curve $\mathcal{C}$ is the boundary for at least one unbroken surface $\mathcal{S}$ are said to be *simply connected*. In a simply connected body, every closed curve can be "shrunken" to a point without passing outside of the body. For example, a hollow sphere is simply connected, but a torus (a doughnut) is not.

If a strain field $\boldsymbol{\varepsilon}$ is specified that satisfies $\boldsymbol{\nabla} \times (\boldsymbol{\nabla} \times \boldsymbol{\varepsilon}) = \mathbf{0}$ everywhere in the body, it has been shown that the infinitesimal rotation vector field $\boldsymbol{\theta}$ is uniquely determined by (3.4.39). Then, the infinitesimal rotation tensor field $\boldsymbol{\omega}$ is uniquely determined by (3.4.16), the gradient of the displacement field is uniquely given by $\boldsymbol{\nabla} \mathbf{u} = \boldsymbol{\varepsilon} + \boldsymbol{\omega}$, and the displacement field can be found by integrating $\boldsymbol{\nabla} \mathbf{u}$:

$$\mathbf{u}(\mathbf{X}) = \mathbf{u}(\mathbf{X}_0) + \int_{\mathbf{X}_0}^{\mathbf{X}} (\boldsymbol{\varepsilon} + \boldsymbol{\omega}) \cdot d\mathbf{X} \,,$$

$$u_i(\mathbf{X}) = u_i(\mathbf{X}_0) + \int_{\mathbf{X}_0}^{\mathbf{X}} (\varepsilon_{ij} + \omega_{ij}) \, dX_j \,. \tag{3.4.42}$$

The displacement field $\mathbf{u}$ is uniquely determined by (3.4.42) if and only if the integral on the right-hand side is path independent:

$$\oint_{\mathcal{C}} (\boldsymbol{\varepsilon} + \boldsymbol{\omega}) \cdot d\mathbf{X} = \mathbf{0} \,. \tag{3.4.43}$$

However, from Stokes's theorem and assuming the body is simply connected,

$$\oint_{\mathcal{C}} (\boldsymbol{\varepsilon} + \boldsymbol{\omega}) \cdot d\mathbf{X} = \int_{\mathcal{S}} \mathbf{N} \cdot [\boldsymbol{\nabla} \times (\boldsymbol{\varepsilon} + \boldsymbol{\omega})] \, dA \,. \tag{3.4.44}$$

It can be shown that $\nabla \times \boldsymbol{\omega} = -\nabla\boldsymbol{\theta}$ (see Problem 3.14) and recall from (3.4.36) that $\nabla \times \boldsymbol{\varepsilon} = \nabla\boldsymbol{\theta}$. Therefore, $\nabla \times (\boldsymbol{\varepsilon} + \boldsymbol{\omega}) = \mathbf{0}$, the integral is path independent, and the displacement field is uniquely given by (3.4.42).

Given a strain field $\boldsymbol{\varepsilon}$, the condition that $\nabla \times (\nabla \times \boldsymbol{\varepsilon}) = \mathbf{0}$ everywhere is sufficient in a simply connected body to guarantee the existence of a corresponding infinitesimal rotation vector field $\boldsymbol{\theta}$, given by (3.4.39). The gradient of the displacement field can then be determined and the displacement field $\mathbf{u}$ found by evaluating (3.4.42).

The conditions $\nabla \times (\nabla \times \boldsymbol{\varepsilon}) = \mathbf{0}$ are known as the *compatibility conditions* and they are necessary and sufficient for the existence of a corresponding displacement field in a simply connected body. In a Cartesian coordinate system, the indicial notation form of the compatibility conditions (3.4.37) are symmetric in the free indices r and s, so there are only six independent compatibility conditions. In explicit form, these are

$$
\begin{aligned}
\varepsilon_{11,22} + \varepsilon_{22,11} - 2\varepsilon_{12,12} &= 0 \,, \\
\varepsilon_{22,33} + \varepsilon_{33,22} - 2\varepsilon_{23,23} &= 0 \,, \\
\varepsilon_{33,11} + \varepsilon_{11,33} - 2\varepsilon_{31,31} &= 0 \,, \\
(\varepsilon_{12,3} - \varepsilon_{23,1} + \varepsilon_{31,2})_{,1} - \varepsilon_{11,23} &= 0 \,, \\
(\varepsilon_{23,1} - \varepsilon_{31,2} + \varepsilon_{12,3})_{,2} - \varepsilon_{22,31} &= 0 \,, \\
(\varepsilon_{31,2} - \varepsilon_{12,3} + \varepsilon_{23,1})_{,3} - \varepsilon_{33,12} &= 0 \,.
\end{aligned}
\tag{3.4.45}
$$

Whenever, in the course of solving a problem for instance, one arrives at an expression for the infinitesimal strain field in a body without first obtaining an expression for the corresponding displacement field, then it must be required that the strain satisfy these compatibility conditions. However, if the strain field is derived from a given displacement field, then this requirement is trivially satisfied and need not be explicitly considered.

3.5   Cylindrical and Spherical Coordinates

3.5.1   Cylindrical Coordinates

The scalar components of displacement and strain in cylindrical coordinates are related to those in Cartesian coordinates by the transformation matrix $[L]$ given by (2.5.12):

$$
u'_i = \ell_{ij} u_j \,, \qquad\qquad [\mathbf{u}]' = [L][\mathbf{u}] \,, \tag{3.5.1}
$$

$$
\varepsilon'_{ij} = \ell_{im}\ell_{jn}\varepsilon_{mn} \,, \qquad [\boldsymbol{\varepsilon}]' = [L][\boldsymbol{\varepsilon}][L]^{\mathsf{T}} \,, \tag{3.5.2}
$$

where the primed components are in cylindrical coordinates and their indices 1, 2, and 3 are understood to represent r, θ, and z, respectively. In dyadic notation, the displacement and strain are written as

$$\mathbf{u} = u_r \hat{\mathbf{e}}_r + u_\theta \hat{\mathbf{e}}_\theta + u_z \hat{\mathbf{e}}_z \; , \tag{3.5.3}$$

$$\boldsymbol{\varepsilon} = \varepsilon_{rr} \hat{\mathbf{e}}_r \hat{\mathbf{e}}_r + \varepsilon_{\theta\theta} \hat{\mathbf{e}}_\theta \hat{\mathbf{e}}_\theta + \varepsilon_{zz} \hat{\mathbf{e}}_z \hat{\mathbf{e}}_z + \varepsilon_{r\theta} (\hat{\mathbf{e}}_r \hat{\mathbf{e}}_\theta + \hat{\mathbf{e}}_\theta \hat{\mathbf{e}}_r) + \cdots \; . \tag{3.5.4}$$

Tensor notation equations hold in all coordinate systems, so the strain-displacement relation $\boldsymbol{\varepsilon} = \frac{1}{2}[\nabla \mathbf{u} + (\nabla \mathbf{u})^\mathsf{T}]$ is valid for cylindrical coordinates, where the gradient of a vector is given by (2.5.19). Thus,

$$
\begin{aligned}
\varepsilon_{rr} &= \frac{\partial u_r}{\partial r} \; , \quad \varepsilon_{\theta\theta} = \frac{1}{r}\left(\frac{\partial u_\theta}{\partial \theta} + u_r\right) \; , \quad \varepsilon_{zz} = \frac{\partial u_z}{\partial z} \; , \\[2mm]
\varepsilon_{r\theta} &= \frac{1}{2}\left(\frac{1}{r}\frac{\partial u_r}{\partial \theta} + \frac{\partial u_\theta}{\partial r} - \frac{u_\theta}{r}\right) \; , \\[2mm]
\varepsilon_{\theta z} &= \frac{1}{2}\left(\frac{\partial u_\theta}{\partial z} + \frac{1}{r}\frac{\partial u_z}{\partial \theta}\right) \; , \\[2mm]
\varepsilon_{zr} &= \frac{1}{2}\left(\frac{\partial u_r}{\partial z} + \frac{\partial u_z}{\partial r}\right) \; .
\end{aligned}
\tag{3.5.5}
$$

Compatibility conditions can be expressed in cylindrical and spherical coordinates, but they are quite lengthy and ultimately not very useful.

3.5.2 Spherical Coordinates

The scalar components of displacement and strain in spherical coordinates are related to those in Cartesian coordinates by the transformation matrix $[L]$ given by (2.5.42):

$$u_i' = \ell_{ij} u_j \; , \qquad [\mathbf{u}]' = [L][\mathbf{u}] \; , \tag{3.5.6}$$

$$\varepsilon_{ij}' = \ell_{im}\ell_{jn}\varepsilon_{mn} \; , \qquad [\varepsilon]' = [L][\varepsilon][L]^\mathsf{T} \; , \tag{3.5.7}$$

where the primed components are in spherical coordinates and their indices 1, 2, and 3 are understood to represent R, θ, and ϕ, respectively. In dyadic notation, the displacement and strain are written as

$$\mathbf{u} = u_R \hat{\mathbf{e}}_R + u_\theta \hat{\mathbf{e}}_\theta + u_\phi \hat{\mathbf{e}}_\phi \; , \tag{3.5.8}$$

$$\boldsymbol{\varepsilon} = \varepsilon_{RR} \hat{\mathbf{e}}_R \hat{\mathbf{e}}_R + \varepsilon_{\theta\theta} \hat{\mathbf{e}}_\theta \hat{\mathbf{e}}_\theta + \varepsilon_{\phi\phi} \hat{\mathbf{e}}_\phi \hat{\mathbf{e}}_\phi + \varepsilon_{R\theta} (\hat{\mathbf{e}}_R \hat{\mathbf{e}}_\theta + \hat{\mathbf{e}}_\theta \hat{\mathbf{e}}_R) + \cdots \; . \tag{3.5.9}$$

Tensor notation equations hold in all coordinate systems, so the strain-displacement relation $\boldsymbol{\varepsilon} = \frac{1}{2}[\nabla \mathbf{u} + (\nabla \mathbf{u})^\mathsf{T}]$ is valid for spherical coordinates, where the gradient of a vector is given by (2.5.48). Thus,

$$\varepsilon_{RR} = \frac{\partial u_R}{\partial R} , \quad \varepsilon_{\theta\theta} = \frac{1}{R}\left(\frac{\partial u_\theta}{\partial \theta} + u_R\right),$$

$$\varepsilon_{\phi\phi} = \frac{1}{R\sin\theta}\left(\frac{\partial u_\phi}{\partial \phi} + u_R\sin\theta + u_\theta\cos\theta\right),$$

$$\varepsilon_{R\theta} = \frac{1}{2}\left(\frac{1}{R}\frac{\partial u_R}{\partial \theta} + \frac{\partial u_\theta}{\partial R} - \frac{u_\theta}{R}\right), \qquad (3.5.10)$$

$$\varepsilon_{\theta\phi} = \frac{1}{2R}\left(\frac{1}{\sin\theta}\frac{\partial u_\theta}{\partial \phi} + \frac{\partial u_\phi}{\partial \theta} - u_\phi\cot\theta\right),$$

$$\varepsilon_{\phi R} = \frac{1}{2}\left(\frac{1}{R\sin\theta}\frac{\partial u_R}{\partial \phi} + \frac{\partial u_\phi}{\partial R} - \frac{u_\phi}{R}\right).$$

Problems

3.1 Show that the material and spatial gradients of a tensor field $\mathbf{T}$ are related by $\mathrm{Grad}\,\mathbf{T} = (\mathrm{grad}\,\mathbf{T})\cdot\mathbf{F}$, where $\mathbf{F}$ is the deformation gradient tensor.

3.2 Rederive the continuity condition (3.3.22) by equating the rate of change of the mass contained within an arbitrary *fixed* region of space and the rate of flow of mass across the region's boundary. (**Hint:** note that the rate of mass per unit area flowing across a fixed surface with unit normal $\hat{\mathbf{n}}$ is $\rho\mathbf{v}\cdot\hat{\mathbf{n}}$ and use the divergence theorem.)

3.3 Consider a body which, in the reference configuration $\mathcal{R}$, occupies the unit cube $(0 \leq X_1 \leq 1,\, 0 \leq X_2 \leq 1,\, 0 \leq X_3 \leq 1)$. The mapping $x_i = \chi_i(X_1, X_2, X_3)$ for the Cartesian coordinates of a material point in the current configuration in terms of its Cartesian coordinates in the reference configuration is given by

$$x_1 = X_1 + \kappa X_2 , \quad x_2 = X_2 , \quad x_3 = X_3 ,$$

where κ is a constant. This homogeneous motion is known as a *simple shear*.

(a) Give a geometric interpretation of this motion by providing a sketch of the body's current configuration.

(b) Show that there is no change in volume of the body during this motion, that is, that the motion is isochoric.

(c) Determine the stretches in each of the reference material directions $\hat{e}_1$, $\hat{e}_2$, $(\hat{e}_1 + \hat{e}_2)/\sqrt{2}$, and $(\hat{e}_1 - \hat{e}_2)/\sqrt{2}$ by using (3.2.10a). Verify your results by comparison with the geometric interpretation of (a).

(d) Determine the orthogonal shear strain between the reference material directions $\hat{e}_1$ and $\hat{e}_2$ and between the reference material directions $(\hat{e}_1 + \hat{e}_2)/\sqrt{2}$ and $(\hat{e}_1 - \hat{e}_2)/\sqrt{2}$, by using (3.2.20a). Verify your results by comparison with the geometric interpretation of (a).

(e) Determine the principal stretches and the principal reference directions of stretch in the special case where $\kappa = 0.4$.

3.4 Consider a body which, in the reference configuration $\mathcal{R}$, occupies the unit cube $(0 \leq X_1 \leq 1, 0 \leq X_2 \leq 1, 0 \leq X_3 \leq 1)$. The mapping $x_i = \chi_i(X_1, X_2, X_3)$ for the Cartesian coordinates of a material point in the current configuration in terms of its Cartesian coordinates in the reference configuration is given by

$$x_1 = \kappa X_1 , \quad x_2 = \frac{1}{\kappa} X_2 , \quad x_3 = X_3 ,$$

where $\kappa \neq 0$ is a constant. This homogeneous motion is known as a *pure shear*.

(a) Give a geometric interpretation of this motion by providing a sketch of the body's current configuration.

(b) Show that there is no change in volume of the body during this motion, that is, that the motion is isochoric.

(c) Determine the stretches in each of the reference material directions $\hat{e}_1$, $\hat{e}_2$, $(\hat{e}_1 + \hat{e}_2)/\sqrt{2}$, and $(\hat{e}_1 - \hat{e}_2)/\sqrt{2}$ by using (3.2.10a). Verify your results by comparison with the geometric interpretation of (a).

(d) Determine the orthogonal shear strain between the reference material directions $\hat{e}_1$ and $\hat{e}_2$ and between the reference material directions $(\hat{e}_1 + \hat{e}_2)/\sqrt{2}$ and $(\hat{e}_1 - \hat{e}_2)/\sqrt{2}$, by using (3.2.20a). Verify your results by comparison with the geometric interpretation of (a).

(e) Determine the principal stretches and the principal reference directions of stretch.

3.5 Consider a body which, in the reference configuration $\mathcal{R}$, occupies the unit cube $(0 \leq X_1 \leq 1, 0 \leq X_2 \leq 1, 0 \leq X_3 \leq 1)$. The mapping $x_i = \chi_i(X_1, X_2, X_3)$ for the Cartesian coordinates of a material point in the current configuration in terms of its Cartesian coordinates in the reference configuration is given by

$$x_1 = X_1 + \kappa X_2 , \quad x_2 = X_2 + \kappa X_3 , \quad x_3 = X_3 ,$$

where κ is a constant.

(a) Give a geometric interpretation of this homogeneous motion by providing a sketch of the body's current configuration.

(b) Show that there is no change in volume of the body during this motion, that is, that the motion is isochoric.

(c) Consider a material line $\mathbb{C}$ whose reference configuration lies along a straight line $\mathcal{C}$ in the plane $X_1 = 0$ that forms an angle α with the X_3-axis. Since $\mathcal{C}$ is a straight line, it has a constant unit tangent vector $\hat{\mathbf{\Upsilon}} = \sin\alpha\,\hat{\mathbf{e}}_2 + \cos\alpha\,\hat{\mathbf{e}}_3$. If S is the length of $\mathbb{C}$ in the reference configuration, show that the length of $\mathbb{C}$ in the current configuration is

$$s = S\sqrt{1 + \kappa^2 + \kappa\sin 2\alpha} \ .$$

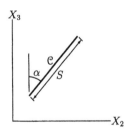

3.6 Consider a body which, in the reference configuration $\mathcal{R}$, occupies the rectangular parallelepiped $(0 \le X_1 \le a, \ 0 \le X_2 \le b, \ 0 \le X_3 \le c)$. The mapping $x_i = \chi_i(X_1, X_2, X_3)$ for the Cartesian coordinates of a material point in the current configuration in terms of its Cartesian coordinates in the reference configuration is given by

$$x_1 = \alpha X_1 , \quad x_2 = \beta X_2 , \quad x_3 = \beta X_3 ,$$

where α and β are constants.

(a) Give a geometric interpretation of this homogeneous motion by providing a sketch of the body's current configuration.

(b) Assuming that the body actually occupies its reference configuration at some point in time, what restrictions (if any) must be placed on the possible values of α and β?

(c) What relationship between α and β corresponds to isochoric deformation, that is, will result in conservation of volume?

(d) Determine the Cartesian scalar components of the infinitesimal strain tensor.

(e) Under what conditions can the infinitesimal deformation approximations by applied?

(f) Assuming infinitesimal deformation, what relationship between α and β corresponds to approximate isochoric deformation based on the infinitesimal strain tensor? Show that this approximate result is consistent with the exact result found above.

3.7 Consider the mapping $x_i = \chi_i(X_1, X_2, X_3)$ for the Cartesian coordinates of a material point in the current configuration in terms of its Cartesian coordinates in the reference configuration given by

$$x_1 = X_1 , \quad x_2 = X_2 + \kappa X_3 , \quad x_3 = X_3 + \kappa X_2 ,$$

where κ is a constant. Determine the material and spatial descriptions of the displacement field for this motion.

3.8 For each of the following displacement fields sketch, in the $x_1 x_2$-plane, the deformed configuration of the body which, in the $X_1 X_2$-plane of the reference configuration, occupies the unit square ($0 \leq X_1 \leq 1$, $0 \leq X_2 \leq 1$). In each case, κ is a constant.

(a) $\mathbf{u} = \kappa X_2 \hat{\mathbf{e}}_1 + \kappa X_1 \hat{\mathbf{e}}_2$

(b) $\mathbf{u} = -\kappa X_2 \hat{\mathbf{e}}_1 + \kappa X_1 \hat{\mathbf{e}}_2$

(c) $\mathbf{u} = \kappa X_1^2 \hat{\mathbf{e}}_2$

3.9 Consider the following displacement field, where κ is a constant:

$$\mathbf{u} = \kappa X_2 \hat{\mathbf{e}}_1 + \kappa X_1 \hat{\mathbf{e}}_2 .$$

(a) Determine the Green-St. Venant strain tensor $\mathbf{E}$, the infinitesimal strain tensor ε, the infinitesimal rotation tensor $\boldsymbol{\omega}$, and the infinitesimal rotation vector $\boldsymbol{\theta}$. Discuss the significance of the difference between $\mathbf{E}$ and ε in terms of the magnitude of κ.

(b) Determine the exact longitudinal strain in the reference material direction $\hat{\mathbf{e}}_1$.

(c) Determine the approximate longitudinal strain in the direction $\hat{\mathbf{e}}_1$ based on the infinitesimal strain tensor ε.

(d) Interpret the results from (b) and (c) geometrically by considering the deformation of a material line element that lies along the X_1-axis in the reference configuration.

3.10 The infinitesimal strain tensor at a point in a body experiencing infinitesimal deformation is

$$\varepsilon = (5\hat{e}_1\hat{e}_1 - 3\hat{e}_2\hat{e}_2 + \hat{e}_3\hat{e}_3) \times 10^{-3} .$$

(a) What are the principal strains and principal directions of strain at this point in the body (be sure to follow the proper labeling and sign conventions)?

(b) Find the orthogonal shear strain at this point in the body between the two orthogonal directions $\hat{v}' = (\hat{e}_1 + \hat{e}_2 + \hat{e}_3)/\sqrt{3}$ and $\hat{v}'' = (-\hat{e}_1 + \hat{e}_3)/\sqrt{2}$.

3.11 Plane elastic waves are characterized by a time-dependent displacement field of the form

$$\mathbf{u} = f(\mathbf{X} \cdot \hat{\mathbf{p}} - ct)\hat{\mathbf{d}} ,$$

where $\hat{\mathbf{p}}$ and $\hat{\mathbf{d}}$ are constant unit vectors, $\mathbf{X}$ is the reference position vector, t is time, c is the constant wave speed, and f is an arbitrary scalar-valued function. Assuming infinitesimal deformation, determine the relations between $\hat{\mathbf{p}}$ and $\hat{\mathbf{d}}$ necessary for

(a) approximate isochoric motion $(e = \nabla \cdot \mathbf{u} = 0)$

(b) approximate irrotational motion $(\boldsymbol{\theta} = \frac{1}{2}\nabla \times \mathbf{u} = \mathbf{0})$.

3.12 For the displacement field whose material description in cylindrical coordinates is given by

$$\mathbf{u} = ar\hat{e}_r + brz\hat{e}_\theta + c\sin\theta\hat{e}_z ,$$

where a, b, and c are constants, determine the scalar components of the infinitesimal strain field in cylindrical coordinates. What restrictions on the values of a, b, c, r, and z are necessary to ensure that the approximate physical interpretations of this strain field are valid?

3.13 Consider a displacement field whose material description, in spherical coordinates, is $\mathbf{u} = a^3 R^{-2} \ln(R/a)\sin\theta\cos\phi\hat{e}_R$. Assuming infinitesimal deformation, determine the approximate dilatation field.

3.14 Show that $\nabla \times \boldsymbol{\omega} = -\nabla\boldsymbol{\theta}$, where $\boldsymbol{\omega}$ is the infinitesimal rotation tensor and $\boldsymbol{\theta}$ is the corresponding infinitesimal rotation vector.

3.15 Consider an infinitesimal strain field for a simply connected body whose Cartesian scalar components are given by $\varepsilon_{31} = \varepsilon_{32} = \varepsilon_{33} = 0$ and

$$\varepsilon_{11} = AX_2^2, \quad \varepsilon_{22} = AX_1^2, \quad \varepsilon_{12} = BX_1 X_2,$$

where A and B are constants.

(a) Determine the relation between A and B required for there to exist a continuous, single-valued displacement field that corresponds to this strain field.

(b) After applying this relation between A and B, determine the most general form of the corresponding displacement field.

(c) Determine the specific corresponding displacement field that is fixed at the origin, so that $\mathbf{u} = \mathbf{0}$ and $\boldsymbol{\omega} = \mathbf{0}$ when $\mathbf{X} = \mathbf{0}$.

3.16 Consider an infinitesimal strain field whose Cartesian scalar components are given by $\varepsilon_{31} = \varepsilon_{32} = \varepsilon_{33} = 0$ and

$$\varepsilon_{11} = BX_1 X_2, \quad \varepsilon_{22} = -\nu B X_1 X_2, \quad \varepsilon_{12} = \frac{B}{2}(1+\nu)(h^2 - X_2^2),$$

where B, ν, and h are constants.

(a) Verify that the compatibility conditions are satisfied by this strain field.

(b) Determine the most general form of the corresponding displacement field.

(c) Show that the specific corresponding displacement field that satisfies $\mathbf{u} = \mathbf{0}$ and $\boldsymbol{\omega} = \mathbf{0}$ when $\mathbf{X} = L\hat{\mathbf{e}}_1$ has scalar components

$$u_1 = \frac{B}{2}X_1^2 X_2 - \frac{B}{3}\left(1 + \frac{\nu}{2}\right)X_2^3 + \frac{B}{2}\left[(1+\nu)h^2 - L^2\right]X_2,$$

$$u_2 = -\frac{\nu B}{2}X_1 X_2^2 - \frac{B}{6}(X_1^3 - L^3)$$

$$+ \frac{B}{2}\left[(1+\nu)h^2 + L^2\right](X_1 - L),$$

$$u_3 = 0.$$

Chapter 4

Forces and Stress

Material volumes interact by exerting forces on one another and it is through the actions of these forces that motion of a material volume occurs. Forces can be categorized either as *body forces* that act at a distance, such as the gravitational forces of attraction that exist between any two material volumes, or as *surface forces* that two material volumes exert on one another when they are in contact. An important concept in the analysis of forces is that of the *free body* and the identification of forces as either internal or external *relative to a free body*. Any material volume can be identified as a free body. Then, any forces exerted by one portion of the free body on another portion of the free body are said to be internal relative to the free body. Forces exerted on the free body by other material volumes that are not part of the free body are external relative to the free body. Whether a force is internal or external will depend on what material volume one chooses to be the free body. Whereas external body forces act, in general, on all the material points of a free body, external surface forces only act on material points on the free body's boundary.

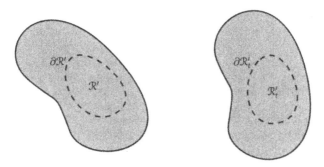

FIGURE 4.1. *Reference and current configurations of an arbitrary material volume.*

The internal forces in a body are related to the body's deformation. The nature of this relationship will depend on what material (steel, aluminum, mahogany, etc.) constitutes the body. The study of this *constitutive behavior* will be discussed in Chapter 5. Of concern here is the manner in which internal forces are described and how they relate to the motion of the body, independent of the body's material makeup. By considering arbitrary material volumes as free bodies, it will be seen that the internal forces at a material point in a body can be characterized by any one of several second-order *stress tensors* and that there are local *equations of motion* that the stress tensor fields must satisfy. The characterization of internal forces is simplified when deformations are assumed to be infinitesimal.

4.1   Stress Tensors: Referential Formulation

In this section, the current forces within a body will be related to the body's reference configuration. In problems involving the deformation of solids, this is often the natural choice, since one knows *a priori* the geometry of the reference configuration of a body.

Consider a body $\mathbb{B}$ bounded by the material surface $\partial \mathbb{B}$ and let $\mathbb{B}' \subset \mathbb{B}$ be an arbitrary portion of the body. The material volume $\mathbb{B}'$ occupies the region $\mathcal{R}'$ with boundary $\partial \mathcal{R}'$ in the reference configuration and the region $\mathcal{R}'_t$ with boundary $\partial \mathcal{R}'_t$ in the current configuration (Figure 4.1). With this material volume taken to be a free body, the current external forces acting on $\mathbb{B}'$ can be classified as either *surface forces* acting on the boundary $\partial \mathbb{B}'$ of $\mathbb{B}'$ or *body forces* which act on the internal mass of $\mathbb{B}'$. Surface forces are due to contact with the rest of the body or, at points where the boundaries of $\mathbb{B}'$ and $\mathbb{B}$ coincide, they are due to contact with another body. Body forces are due to remote influences such as the gravitational attraction of the Earth.

4.1.1   Surface Forces

Consider a material point $P \in \partial \mathbb{B}'$ on the boundary of the material volume $\mathbb{B}'$ and let the material surface $\Delta \mathbb{S}$ be an arbitrary portion of the boundary that includes the material point, so that $P \in \Delta \mathbb{S} \subset \partial \mathbb{B}'$. At time t, there is a distribution of surface forces acting on the current configuration of $\Delta \mathbb{S}$ that is statically equivalent to a resultant force $\Delta \mathbf{F}^S$ acting at the current position $\mathbf{x}(P, t)$ of P and a resultant couple $\Delta \mathbf{M}^S$ (Figure 4.2). Note that surface forces acting on the boundary of the body $\mathbb{B}$ are just a special case, in which $\Delta \mathbb{S} \subset \partial \mathbb{B}$, and that no exceptional provisions for this special case are required.

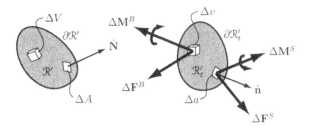

FIGURE 4.2. Free-body diagram showing surface and body forces.

In the reference configuration, the area of $\Delta\mathbb{S}$ is ΔA and the *reference unit outward normal vector* to $\partial\mathbb{B}'$ at P is $\hat{\mathbf{N}}$. In the limit as the material surface $\Delta\mathbb{S}$ is reduced to the material point P, the area $\Delta A \to 0$. One can make the following assumptions:

(a) Assume that in the limit as the material surface $\Delta\mathbb{S}$ is reduced to the material point P,

$$\lim_{\Delta\mathbb{S}\to P} \frac{\Delta\mathbf{M}^S}{\Delta A} = \mathbf{0}. \tag{4.1.1}$$

(b) Assume that in the limit as the material surface $\Delta\mathbb{S}$ is reduced to the material point P, the ratio $\Delta\mathbf{F}^S/\Delta A$ tends to a definite limit,

$$\mathbf{t}^{(\mathrm{N})} \equiv \lim_{\Delta\mathbb{S}\to P} \frac{\Delta\mathbf{F}^S}{\Delta A}, \tag{4.1.2}$$

known as the *nominal traction vector*. This nominal traction vector represents the *current* force per unit *reference* area acting at the material point P on the material surface $\partial\mathbb{B}'$.

(c) Consider two arbitrary material volumes $\mathbb{B}'$ and $\mathbb{B}''$ with boundaries $\partial\mathbb{B}'$ and $\partial\mathbb{B}''$, respectively, that both have the material point P on their boundaries. Assume that if $\partial\mathbb{B}'$ and $\partial\mathbb{B}''$ also have the same reference unit outward normal vector $\hat{\mathbf{N}}$ at P, then the nominal traction vectors acting at the material point P on $\partial\mathbb{B}'$ and $\partial\mathbb{B}''$ are the same. Thus, the material description of the nominal traction vector acting at a point P on a material surface, at time t, can be given as

$$\mathbf{t}^{(\mathrm{N})} = \mathbf{t}^{(\mathrm{N})}(\mathbf{X}, \hat{\mathbf{N}}, t), \tag{4.1.3}$$

where $\mathbf{X}$ is the reference position of P and $\hat{\mathbf{N}}$ is the unit outward normal vector to the reference configuration of the surface at P. The nominal traction vector acting at a point of the boundary of a material volume does not depend on the curvature of the boundary at that point, for instance.

It follows from assumptions (b) and (c) that the resultant force of the surface forces acting on a material volume $\mathbb{B}'$ at time t is given in terms of the reference configuration of $\mathbb{B}'$ by

$$\mathbf{F}^S(t) = \int_{\partial \mathcal{R}'} \mathbf{t}^{(\mathrm{N})}(\mathbf{X}, \hat{\mathbf{N}}, t) \, dA = \int_{\partial \mathcal{R}'} \mathbf{t}^{(\mathrm{N})} \, dA \,, \qquad (4.1.4)$$

where $\partial \mathcal{R}'$ is the boundary of the region occupied by the reference configuration of $\mathbb{B}'$ and $\hat{\mathbf{N}}$ is the unit outward normal at the point $\mathbf{X}$ on $\partial \mathcal{R}'$. It follows from assumptions (a)–(c) that the resultant moment about the origin O of the surface forces acting on a material volume $\mathbb{B}'$ at time t is given in terms of the reference configuration of $\mathbb{B}'$ by

$$\mathbf{M}_O^S(t) = \int_{\partial \mathcal{R}'} \chi(\mathbf{X}, t) \times \mathbf{t}^{(\mathrm{N})}(\mathbf{X}, \hat{\mathbf{N}}, t) \, dA = \int_{\partial \mathcal{R}'} \mathbf{x} \times \mathbf{t}^{(\mathrm{N})} \, dA \,, \qquad (4.1.5)$$

where $\mathbf{x} = \chi(\mathbf{X}, t)$ is the current position vector, relative to the origin O, of the material point whose reference position is $\mathbf{X}$.

4.1.2   Body Forces

Consider a material point Q within the material volume $\mathbb{B}'$ and let $\Delta\mathbb{B}$ be an arbitrary portion of the material volume that contains the material point, so that $Q \in \Delta\mathbb{B} \subset \mathbb{B}'$. At time t, there is a distribution of body forces acting on the current configuration of $\Delta\mathbb{B}$ that is statically equivalent to a resultant force $\Delta\mathbf{F}^B$ acting at the current position $\mathbf{x}(Q, t)$ of Q and a resultant couple $\Delta\mathbf{M}^B$ (Figure 4.2).

In the reference configuration, the volume of $\Delta\mathbb{B}$ is ΔV, and in the limit as this material volume reduces to the material point Q, the volume $\Delta V \to 0$. One can make the following assumptions:

(d) Assume that in the limit as the material volume $\Delta\mathbb{B}$ is reduced to the material point Q:

$$\lim_{\Delta\mathbb{B} \to Q} \frac{\Delta\mathbf{M}^B}{\Delta V} = \mathbf{0} \,. \qquad (4.1.6)$$

(e) Assume that in the limit as the material volume $\Delta\mathbb{B}$ is reduced to the material point Q, the ratio $\Delta\mathbf{F}^B / \Delta V$ tends to a definite limit:

$$\rho_0 \mathbf{b} \equiv \lim_{\Delta\mathbb{B} \to Q} \frac{\Delta\mathbf{F}^B}{\Delta V} \,, \qquad (4.1.7)$$

where ρ_0 is the mass density at the material point Q when the body is in its reference configuration and $\rho_0 \mathbf{b}$ is the *current* body force per unit *reference* volume acting at the material point Q. Thus, the *body force vector* $\mathbf{b}$ is the current body force per unit mass and its material description $\mathbf{b}(\mathbf{X}, t)$ is as a function of reference position $\mathbf{X}$ and time t.

It follows from assumption (e) that the resultant force of the body forces acting on a material volume $\mathbb{B}'$ at time t is given in terms of the reference configuration of $\mathbb{B}'$ by

$$\mathbf{F}^B(t) = \int_{\mathcal{R}'} \rho_0(\mathbf{X})\mathbf{b}(\mathbf{X}, t)\, dV = \int_{\mathcal{R}'} \rho_0 \mathbf{b}\, dV , \qquad (4.1.8)$$

where $\mathcal{R}'$ is the region occupied by the reference configuration of $\mathbb{B}'$. It follows from assumptions (d) and (e) that the resultant moment about the origin O of the body forces acting on a material volume $\mathbb{B}'$ at time t is given in terms of the reference configuration of $\mathbb{B}'$ by

$$\mathbf{M}_O^B(t) = \int_{\mathcal{R}'} \chi(\mathbf{X}, t) \times [\rho_0(\mathbf{X})\mathbf{b}(\mathbf{X}, t)]\, dV = \int_{\mathcal{R}'} \rho_0 \mathbf{x} \times \mathbf{b}\, dV , \qquad (4.1.9)$$

where $\mathbf{x} = \chi(\mathbf{X}, t)$ is the current position vector, relative to the origin O, of the material point whose reference position is $\mathbf{X}$.

The assumptions (a) and (d) that $\Delta \mathbf{M}^S / \Delta A$ and $\Delta \mathbf{M}^B / \Delta V$ vanish in the limit is, in some sense, arbitrary. The assumptions (a)–(e) are justified by the resulting model's success in predicting the behavior of deformable bodies. However, there are a few specialized applications where it is desirable to relax assumptions (a) and (d). The result, known as couple-stress theory [see, for example, Toupin (1964)], will not be addressed here.

4.1.3 Balance of Linear and Angular Momentum

Consider an arbitrary material volume $\mathbb{B}'$, occupying the region $\mathcal{R}'$ with boundary $\partial \mathcal{R}'$ in its reference configuration. The current linear momentum per unit mass at a material point is defined to be the material velocity $\mathbf{v}$, so that the *current* total linear momentum of the material volume $\mathbb{B}'$ is given as an integral over the region $\mathcal{R}'$ of its *reference* configuration by

$$\int_{\mathcal{R}'} \rho_0(\mathbf{X})\mathbf{v}(\mathbf{X}, t)\, dV = \int_{\mathcal{R}'} \rho_0 \mathbf{v}\, dV . \qquad (4.1.10)$$

It is *postulated*, as a generalization of the principle of conservation of linear momentum for systems of particles, that the resultant force of the external forces acting on $\mathbb{B}'$ is equal to the rate of change of the total linear momentum of $\mathbb{B}'$,

$$\int_{\partial\mathcal{R}'} \mathbf{t}^{(\mathrm{N})}\, dA + \int_{\mathcal{R}'} \rho_0 \mathbf{b}\, dV = \frac{\mathrm{D}}{\mathrm{D}t} \int_{\mathcal{R}'} \rho_0 \mathbf{v}\, dV \ . \qquad (4.1.11)$$

However, since $\mathcal{R}'$ and ρ_0 are independent of time,

$$\frac{\mathrm{D}}{\mathrm{D}t} \int_{\mathcal{R}'} \rho_0 \mathbf{v}\, dV = \int_{\mathcal{R}'} \rho_0 \frac{\mathrm{D}\mathbf{v}}{\mathrm{D}t}\, dV \ . \qquad (4.1.12)$$

Thus, the postulated *balance of linear momentum* requires that

$$\boxed{\begin{aligned} \int_{\partial\mathcal{R}'} \mathbf{t}^{(\mathrm{N})}\, dA + \int_{\mathcal{R}'} \rho_0 \mathbf{b}\, dV &= \int_{\mathcal{R}'} \rho_0 \mathbf{a}\, dV \ , \\[2mm] \int_{\partial\mathcal{R}'} t_i^{(\mathrm{N})}\, dA + \int_{\mathcal{R}'} \rho_0 b_i\, dV &= \int_{\mathcal{R}'} \rho_0 a_i\, dV \ , \end{aligned}} \qquad (4.1.13)$$

where $\mathbf{a} \equiv \mathrm{D}\mathbf{v}/\mathrm{D}t$ is the material acceleration.

The current angular momentum per unit mass at a material point, relative to the origin, is defined to be $\mathbf{x} \times \mathbf{v}$, where $\mathbf{x}$ is the material point's current position vector. The total *current* angular momentum of the material volume $\mathbb{B}'$ is therefore given as an integral over the region $\mathcal{R}'$ of its *reference* configuration by

$$\int_{\mathcal{R}'} \rho_0(\mathbf{X})[\chi(\mathbf{X},t) \times \mathbf{v}(\mathbf{X},t)]\, dV = \int_{\mathcal{R}'} \rho_0 \mathbf{x} \times \mathbf{v}\, dV \ . \qquad (4.1.14)$$

It is *postulated*, as a generalization of the principle of conservation of angular momentum for systems of particles, that the resultant moment about the origin of the external forces acting on $\mathbb{B}'$ is equal to the rate of change of the total angular momentum of $\mathbb{B}'$:

$$\int_{\partial\mathcal{R}'} \mathbf{x} \times \mathbf{t}^{(\mathrm{N})}\, dA + \int_{\mathcal{R}'} \rho_0 \mathbf{x} \times \mathbf{b}\, dV = \frac{\mathrm{D}}{\mathrm{D}t} \int_{\mathcal{R}'} \rho_0 \mathbf{x} \times \mathbf{v}\, dV \ . \qquad (4.1.15)$$

However, since $\mathcal{R}'$ and ρ_0 are independent of time, $\mathrm{D}\mathbf{x}/\mathrm{D}t = \mathbf{v}$, and $\mathbf{v} \times \mathbf{v} = 0$,

$$\frac{\mathrm{D}}{\mathrm{D}t} \int_{\mathcal{R}'} \rho_0 \mathbf{x} \times \mathbf{v}\, dV = \int_{\mathcal{R}'} \rho_0 \frac{\mathrm{D}}{\mathrm{D}t}(\mathbf{x} \times \mathbf{v})\, dV = \int_{\mathcal{R}'} \rho_0 \mathbf{x} \times \frac{\mathrm{D}\mathbf{v}}{\mathrm{D}t}\, dV \ . \qquad (4.1.16)$$

Thus, the postulated *balance of angular momentum* requires that

$$\int_{\partial \mathcal{R}'} \mathbf{x} \times \mathbf{t}^{(\mathrm{N})} \, dA + \int_{\mathcal{R}'} \rho_0 \mathbf{x} \times \mathbf{b} \, dV = \int_{\mathcal{R}'} \rho_0 \mathbf{x} \times \mathbf{a} \, dV \ ,$$

$$\int_{\partial \mathcal{R}'} e_{ijk} x_j t_k^{(\mathrm{N})} \, dA + \int_{\mathcal{R}'} \rho_0 e_{ijk} x_j b_k \, dV = \int_{\mathcal{R}'} \rho_0 e_{ijk} x_j a_k \, dV \ ,$$

(4.1.17)

where the origin O of the current position vector $\mathbf{x}$ is arbitrary.[1]

4.1.4 Generalization of Newton's Third Law

Consider two arbitrary, distinct material volumes $\mathbb{B}_1$ and $\mathbb{B}_2$, so that their intersection $\mathbb{B}_1 \cap \mathbb{B}_2 = \emptyset$ is the empty set, and let the union of these material volumes be $\mathbb{B}' = \mathbb{B}_1 \cup \mathbb{B}_2$. Thus, there are no material points that are in both $\mathbb{B}_1$ and $\mathbb{B}_2$, and a material point is in $\mathbb{B}'$ if and only if it is in either $\mathbb{B}_1$ or $\mathbb{B}_2$. If the material volumes $\mathbb{B}_1$ and $\mathbb{B}_2$ are in contact, let the material surface $\mathcal{S}_{12}$ be their common boundary so that a material point in $\mathcal{S}_{12}$ is on the boundary of both $\mathbb{B}_1$ and $\mathbb{B}_2$. Let the material surface $\mathcal{S}_1$ be the set of material points that are on the boundary of $\mathbb{B}_1$ but not on the boundary of $\mathbb{B}_2$ and let the material surface $\mathcal{S}_2$ be the set of material points that are on the boundary of $\mathbb{B}_2$ but not on the boundary of $\mathbb{B}_1$. Thus, the boundary of $\mathbb{B}_1$ is $\partial \mathbb{B}_1 = \mathcal{S}_1 \cup \mathcal{S}_{12}$, the boundary of $\mathbb{B}_2$ is $\partial \mathbb{B}_2 = \mathcal{S}_2 \cup \mathcal{S}_{12}$, and the boundary of $\mathbb{B}'$ is $\partial \mathbb{B}' = \mathcal{S}_1 \cup \mathcal{S}_2$, where $\mathcal{S}_1 \cap \mathcal{S}_2 = \emptyset$. If $\mathbb{B}_1$ and $\mathbb{B}_2$ are not in contact, then their common boundary is just the empty set, $\mathcal{S}_{12} = \emptyset$.

The balance of linear momentum (4.1.13) for $\mathbb{B}_1$ can be expressed as

$$\int_{\mathcal{S}_1 \cup \mathcal{S}_{12}} \mathbf{t}^{(\mathrm{N})}(\mathbf{X}, \hat{\mathbf{N}}_1, t) \, dA + \int_{\mathcal{R}_1} \rho_0(\mathbf{X})[\mathbf{b}(\mathbf{X}, t) + \mathbf{b}_{2 \to 1}(\mathbf{X}, t)] \, dV$$

$$= \int_{\mathcal{R}_1} \rho_0(\mathbf{X}) \mathbf{a}(\mathbf{X}, t) \, dV \ , \quad (4.1.18)$$

where $\mathcal{R}_1$ is the region occupied by the reference configuration of $\mathbb{B}_1$, $\mathcal{S}_1$ and $\mathcal{S}_{12}$ are the surfaces occupied by the reference configurations of $\mathcal{S}_1$

---

[1] The balance statements for linear and angular momentum presume an inertial frame of reference. An inertial frame of reference can be defined as one in which linear and angular momentum are balanced. This may seem at first like a self-fulfilling definition, but it can be shown that once one such frame of reference has been identified, any other frame of reference that is neither rotating nor accelerating with respect to the former is also inertial. A frame of reference "fixed at the center of the Universe" is sometimes given as a universal inertial frame of reference. A frame of reference fixed to the surface of the Earth, although only approximately inertial, is much more useful. Any errors accrued from using such a frame of reference will typically be negligible.

and S_{12}, $\hat{\mathbf{N}}_1$ is the unit outward normal at a point $\mathbf{X}$ on the boundary $\partial \mathcal{R}_1 = S_1 \cup S_{12}$ of $\mathcal{R}_1$, $\mathbf{b}$ is the body force exerted by material outside of $\mathbb{B}' = \mathbb{B}_1 \cup \mathbb{B}_2$, and $\mathbf{b}_{2\to1}$ is the body force exerted by the material in $\mathbb{B}_2$, so that $\mathbf{b} + \mathbf{b}_{2\to1}$ is the total body force exerted by material outside of $\mathbb{B}_1$. Similarly, the balance of linear momentum for $\mathbb{B}_2$ can be expressed as

$$\int_{S_2 \cup S_{12}} \mathbf{t}^{(N)}(\mathbf{X}, \hat{\mathbf{N}}_2, t)\, dA + \int_{\mathcal{R}_2} \rho_0(\mathbf{X})[\mathbf{b}(\mathbf{X}, t) + \mathbf{b}_{1\to2}(\mathbf{X}, t)]\, dV$$

$$= \int_{\mathcal{R}_2} \rho_0(\mathbf{X})\mathbf{a}(\mathbf{X}, t)\, dV\ , \quad (4.1.19)$$

where $\mathcal{R}_2$ is the region occupied by the reference configuration of $\mathbb{B}_2$, S_2 is the surface occupied by the reference configuration of S_2, $\hat{\mathbf{N}}_2$ is the unit outward normal at a point $\mathbf{X}$ on the boundary $\partial \mathcal{R}_2 = S_2 \cup S_{12}$ of $\mathcal{R}_2$, and $\mathbf{b}_{1\to2}$ is the body force exerted by the material in $\mathbb{B}_1$. By adding (4.1.18) and (4.1.19) and noting that $\mathcal{R}' = \mathcal{R}_1 \cup \mathcal{R}_2$ is the region occupied by the reference configuration of $\mathbb{B}'$, $\partial \mathcal{R}' = \partial S_1 \cup \partial S_2$ is the boundary of $\mathcal{R}'$, $\hat{\mathbf{N}} = \hat{\mathbf{N}}_1$ is the unit outward normal to $\partial \mathcal{R}'$ at a point $\mathbf{X}$ on ∂S_1, $\hat{\mathbf{N}} = \hat{\mathbf{N}}_2$ is the unit outward normal to $\partial \mathcal{R}'$ at a point $\mathbf{X}$ on ∂S_2, and $\mathbf{b}$ is the body force exerted by material outside of $\mathbb{B}'$, it follows from the balance of linear momentum (4.1.13) for $\mathbb{B}'$ that

$$\int_{S_{12}} [\mathbf{t}^{(N)}(\mathbf{X}, \hat{\mathbf{N}}_1, t) + \mathbf{t}^{(N)}(\mathbf{X}, \hat{\mathbf{N}}_2, t)]\, dA$$

$$+ \int_{\mathcal{R}_1} \rho_0(\mathbf{X})\mathbf{b}_{2\to1}(\mathbf{X}, t)\, dV + \int_{\mathcal{R}_2} \rho_0(\mathbf{X})\mathbf{b}_{1\to2}(\mathbf{X}, t)\, dV = 0\ . \quad (4.1.20)$$

Finally, since $\mathbb{B}_1$ and $\mathbb{B}_2$ are arbitrary and noting that $\hat{\mathbf{N}}_2 = -\hat{\mathbf{N}}_1$ at any point on S_{12}, it follows that the resultant body force exerted by $\mathbb{B}_2$ on $\mathbb{B}_1$ is equal in magnitude and opposite in direction to the resultant body force exerted by $\mathbb{B}_1$ on $\mathbb{B}_2$,

$$\int_{\mathcal{R}_1} \rho_0(\mathbf{X})\mathbf{b}_{2\to1}(\mathbf{X}, t)\, dV = -\int_{\mathcal{R}_2} \rho_0(\mathbf{X})\mathbf{b}_{1\to2}(\mathbf{X}, t)\, dV\ , \quad (4.1.21)$$

and that the traction acting on one side of a material surface is equal in magnitude and opposite in direction to the traction acting on the opposite side,

$$\boxed{\mathbf{t}^{(N)}(\mathbf{X}, -\hat{\mathbf{N}}, t) = -\mathbf{t}^{(N)}(\mathbf{X}, \hat{\mathbf{N}}, t)\ .} \quad (4.1.22)$$

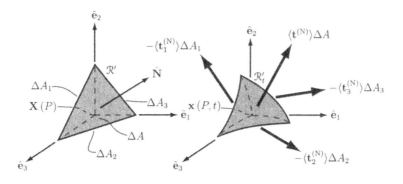

FIGURE 4.3. Reference configuration as a Cauchy tetrahedron.

This is a generalization of Newton's third law of mechanics which states that the force exerted by one particle on a second is equal in magnitude and opposite in direction to the force exerted by the second particle on the first.

An analogous treatment of the balance of angular momentum (4.1.17) can be performed to show that the equal and opposite forces in (4.1.21) must also be colinear.

4.1.5   First Piola-Kirchhoff Stress

How can the dependence of the nominal traction vector $\mathbf{t}^{(N)}(\mathbf{X}, \hat{\mathbf{N}}, t)$ on the unit normal $\hat{\mathbf{N}}$ be described? Recall that $\mathbf{t}^{(N)}$ is the current force per unit reference area acting on a material surface whose unit outward normal in the reference configuration is $\hat{\mathbf{N}}$. To answer this question, consider a material volume $\mathbb{B}'$ that occupies the *Cauchy tetrahedron* shown in Figure 4.3 when in its reference configuration. The volume of the region $\mathcal{R}'$ that is the Cauchy tetrahedron is ΔV. The boundary $\partial \mathcal{R}'$ of the Cauchy tetrahedron is composed of four plane faces. Three of the faces, with areas ΔA_1, ΔA_2, and ΔA_3, are parallel to the coordinate planes defined by the vector basis $\{\hat{\mathbf{e}}_i\}$ and intersect at the material point P. Their unit outward normals are $-\hat{\mathbf{e}}_1$, $-\hat{\mathbf{e}}_2$, and $-\hat{\mathbf{e}}_3$, respectively. The fourth face has area ΔA and unit outward normal $\hat{\mathbf{N}}$.

Let the current nominal traction vector acting on a material surface whose unit outward normal in the reference configuration is $\hat{\mathbf{e}}_J$, at a material point on the surface whose position in the reference configuration is $\mathbf{X}$, be denoted by $\mathbf{t}_J^{(N)}(\mathbf{X}, t) \equiv \mathbf{t}^{(N)}(\mathbf{X}, \hat{\mathbf{e}}_J, t)$. It follows from (4.1.22) that the nominal traction fields on the ΔA_J faces of the tetrahedron can be given as $\mathbf{t}^{(N)}(\mathbf{X}, -\hat{\mathbf{e}}_J, t) = -\mathbf{t}_J^{(N)}(\mathbf{X}, t)$. The nominal traction field on the ΔA face is $\mathbf{t}^{(N)}(\mathbf{X}, \hat{\mathbf{N}}, t)$.

Consider the balance of linear momentum (4.1.13) of this material volume $\mathbb{B}'$ whose reference configuration occupies the Cauchy tetrahedron $\mathcal{R}'$. The value of the surface integral can be given as

$$\int_{\partial\mathcal{R}'} \mathbf{t}^{(N)} \, dA = \langle \mathbf{t}^{(N)} \rangle \Delta A - \langle \mathbf{t}_J^{(N)} \rangle \Delta A_J \, , \qquad (4.1.23)$$

where $\langle \mathbf{t}^{(N)} \rangle$ is the average nominal traction on the face with unit outward normal $\hat{\mathbf{N}}$ and $-\langle \mathbf{t}_J^{(N)} \rangle$ is the average nominal traction on the face with unit outward normal $-\hat{\mathbf{e}}_J$, each at time t. Similarly, the volume integrals can be given as

$$\int_{\mathcal{R}'} \rho_0 \mathbf{b} \, dV = \langle \rho_0 \mathbf{b} \rangle \, \Delta V \, , \qquad (4.1.24)$$

$$\int_{\mathcal{R}'} \rho_0 \mathbf{a} \, dV = \langle \rho_0 \mathbf{a} \rangle \, \Delta V \, , \qquad (4.1.25)$$

where $\langle \rho_0 \mathbf{b} \rangle$ and $\langle \rho_0 \mathbf{a} \rangle$ are the averages, over the reference configuration of the material volume $\mathbb{B}'$, of the body force and inertial force, respectively, at time t. Using (4.1.23)–(4.1.25) and noting that $\Delta A_J = \hat{N}_J \Delta A$, where $\hat{N}_J = \hat{\mathbf{N}} \cdot \hat{\mathbf{e}}_J$ are the scalar components of the unit outward normal to the ΔA face, the balance of linear momentum (4.1.13) can then be rewritten as

$$\langle \mathbf{t}^{(N)} \rangle - \langle \mathbf{t}_J^{(N)} \rangle \hat{N}_J + (\langle \rho_0 \mathbf{b} \rangle - \langle \rho_0 \mathbf{a} \rangle) \frac{\Delta V}{\Delta A} = \mathbf{0} \, . \qquad (4.1.26)$$

By the mean-value theorem of calculus, in the limit as the tetrahedron shrinks to the material point P the average quantities approach their exact values at the point P. Also, $\Delta V / \Delta A \to 0$, so that, in this limit, the balance of linear momentum becomes

$$\mathbf{t}^{(N)}(\mathbf{X}, \hat{\mathbf{N}}, t) = \hat{N}_J \mathbf{t}_J^{(N)}(\mathbf{X}, t) \, , \qquad (4.1.27)$$

where $\mathbf{X}$ is the reference position of the material point P. The scalar components of the nominal traction vector, in the vector basis $\{\hat{\mathbf{e}}_i\}$, are then

$$t_i^{(N)} = \mathbf{t}^{(N)} \cdot \hat{\mathbf{e}}_i = \hat{N}_J \mathbf{t}_J^{(N)} \cdot \hat{\mathbf{e}}_i = \hat{N}_J (\mathbf{t}_J^{(N)} \cdot \hat{\mathbf{e}}_i) \, . \qquad (4.1.28)$$

Thus, as a consequence of the balance of linear momentum, the nominal traction vector at a material point, acting on a material surface which in its reference configuration has unit normal $\hat{\mathbf{N}}$, is completely determined by

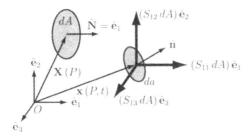

FIGURE 4.4.   Physical interpretation for scalar components of the first Piola-Kirchhoff stress.

the nominal traction vectors at the same material point acting on material surfaces which, in the reference configuration, are parallel to the coordinate planes. In other words, if, at a material point, the nominal traction vectors acting on the three material surfaces that are parallel to the coordinate planes in the reference configuration are known, then (4.1.27) gives the nominal traction vector acting on any other material surface.

Define the scalar components, in the vector basis $\{\hat{\mathbf{e}}_i\}$, of the *first Piola-Kirchhoff stress* to be $S_{Ji} \equiv \mathbf{t}_J^{(N)} \cdot \hat{\mathbf{e}}_i$. At a material point and instant in time, S_{Ji} is the ith scalar component of the nominal traction vector acting on the material surface whose unit outward normal in the reference configuration is $\hat{\mathbf{e}}_J$. To illustrate this physical interpretation, consider a material point P and a differential material surface $d\mathbb{S}$ that contains P and has a unit normal in the reference configuration $\hat{\mathbf{N}} = \hat{\mathbf{e}}_1$ (Figure 4.4). Let the area of $d\mathbb{S}$ in its reference configuration be dA. Then, the current force acting at $\mathbf{x}(P,t)$ on $d\mathbb{S}$ has scalar components $S_{11}\,dA$, $S_{12}\,dA$, and $S_{13}\,dA$ in the vector basis $\{\hat{\mathbf{e}}_i\}$. Note that $(S_{11}\,dA)\hat{\mathbf{e}}_1$ is *not* normal to the current configuration of the material surface and $(S_{12}\,dA)\hat{\mathbf{e}}_2$ and $(S_{13}\,dA)\hat{\mathbf{e}}_3$ are *not* tangent to the surface.

From (4.1.28), the nominal traction vector—the *current* force per unit *reference* area—acting at a point on a material surface, whose unit outward normal in the *reference* configuration is $\hat{\mathbf{N}}$, is given in terms of the S_{Ji} by

$$\boxed{\mathbf{t}^{(N)} = \hat{\mathbf{N}} \cdot \mathbf{S}, \quad t_i^{(N)} = \hat{N}_J S_{Ji}.} \qquad (4.1.29)$$

This is the referential formulation of the *traction-stress relation*. Since $\mathbf{t}^{(N)}$ and $\hat{\mathbf{N}}$ are both vectors, it follows from (4.1.29) that $\mathbf{S} = S_{Ji}\hat{\mathbf{e}}_J\hat{\mathbf{e}}_i$ is a (two-point) second-order tensor called the *first Piola-Kirchhoff stress tensor*. The material description of the first Piola-Kirchhoff stress tensor, $\mathbf{S}(\mathbf{X},t)$, is as a function of reference position $\mathbf{X}$ and time t.

The first Piola-Kirchhoff stress tensor completely characterizes the inter-

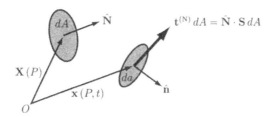

FIGURE 4.5.   Physical interpretation for the first Piola-Kirchhoff stress tensor.

nal forces at a material point—it can be used to find the current force per unit reference area acting on any surface that passes through the material point. Consider a material point P and a differential material surface $d\mathbb{S}$ that contains P and, in the reference configuration, has a unit normal $\hat{\mathbf{N}}$ and area dA (Figure 4.5). The unit normal $\hat{\mathbf{n}}$ to $d\mathbb{S}$ in the current configuration is, in general, not parallel to $\hat{\mathbf{N}}$ and the area da of $d\mathbb{S}$ in the current configuration is generally not equal to dA. Then, from (4.1.29), the current force acting at $\mathbf{x}(P,t)$ on $d\mathbb{S}$ is $\mathbf{t}^{(\mathrm{N})}\, dA = \hat{\mathbf{N}} \cdot \mathbf{S}\, dA$.

4.1.6   Equations of Motion

Consider an arbitrary material volume $\mathbb{B}'$, occupying in its reference configuration the region $\mathcal{R}'$ with boundary $\partial\mathcal{R}'$. Substituting the traction-stress relation (4.1.29) into the referential formulation of the balance of linear momentum (4.1.13) for $\mathbb{B}'$ gives

$$\int_{\partial\mathcal{R}'} \hat{N}_J S_{Ji}\, dA + \int_{\mathcal{R}'} \rho_0 b_i\, dV = \int_{\mathcal{R}'} \rho_0 a_i\, dV \ . \qquad (4.1.30)$$

By applying the divergence theorem, the integral over $\partial\mathcal{R}'$ can be rewritten as

$$\int_{\partial\mathcal{R}'} \hat{N}_J S_{Ji}\, dA = \int_{\mathcal{R}'} \frac{\partial S_{Ji}}{\partial X_J}\, dV \ . \qquad (4.1.31)$$

Thus, by combining these expressions and rearranging terms, the condition for balance of linear momentum can be rewritten as

$$\int_{\mathcal{R}'} \left(\frac{\partial S_{Ji}}{\partial X_J} + \rho_0 b_i - \rho_0 a_i \right) dV = 0 \ . \qquad (4.1.32)$$

This is satisfied for all possible $\mathcal{R}'$ if and only if the integrand vanishes everywhere, giving the *linear equation of motion* for the first Piola-Kirchhoff stress tensor:

$$\text{Div } \mathbf{S}^\mathsf{T} + \rho_0 \mathbf{b} = \rho_0 \mathbf{a} \ , \qquad \frac{\partial S_{Ji}}{\partial X_J} + \rho_0 b_i = \rho_0 a_i \ . \tag{4.1.33}$$

This equation, which must be satisfied everywhere within a body, represents the *local* condition for balance of linear momentum.

Substituting the traction-stress relation (4.1.29) into the referential formulation of the balance of angular momentum (4.1.17) for $\mathbb{B}'$ gives

$$\int_{\partial \mathcal{R}'} e_{ijk} x_j \hat{N}_M S_{Mk} \, dA + \int_{\mathcal{R}'} \rho_0 e_{ijk} x_j b_k \, dV = \int_{\mathcal{R}'} \rho_0 e_{ijk} x_j a_k \, dV \ . \tag{4.1.34}$$

Applying the divergence theorem to the integral over $\partial \mathcal{R}'$ yields

$$\int_{\partial \mathcal{R}'} e_{ijk} x_j \hat{N}_M S_{Mk} \, dA = \int_{\mathcal{R}'} \frac{\partial}{\partial X_M} (e_{ijk} x_j S_{Mk}) \, dV$$

$$= \int_{\mathcal{R}'} e_{ijk} \left(\frac{\partial x_j}{\partial X_M} S_{Mk} + x_j \frac{\partial S_{Mk}}{\partial X_M} \right) dV \ . \tag{4.1.35}$$

Recalling that $F_{jM} \equiv \partial x_j / \partial X_M$ are the scalar components of the deformation gradient tensor, these expressions can be combined to give

$$\int_{\mathcal{R}'} e_{ijk} \left[x_j \left(\frac{\partial S_{Mk}}{\partial X_M} + \rho_0 b_k - \rho_0 a_k \right) + F_{jM} S_{Mk} \right] dV = 0 \ . \tag{4.1.36}$$

However, the local condition for balance of linear momentum (4.1.33) requires that $\partial S_{Mk}/\partial X_M + \rho_0 b_k - \rho_0 a_k = 0$, so that the remaining condition for balance of angular momentum is

$$\int_{\mathcal{R}'} e_{ijk} F_{jM} S_{Mk} \, dV = 0 \ . \tag{4.1.37}$$

This condition is satisfied for all possible $\mathcal{R}'$ if and only if the integrand vanishes everywhere, giving $e_{ijk} F_{jM} S_{Mk} = 0$, which implies the *angular equation of motion* for the first Piola-Kirchhoff stress tensor:

$$\mathbf{F} \cdot \mathbf{S} = \mathbf{S}^\mathsf{T} \cdot \mathbf{F}^\mathsf{T} \ , \qquad F_{jM} S_{Mk} = F_{kM} S_{Mj} \ . \tag{4.1.38}$$

The equations of motion (4.1.33) and (4.1.38) are necessary and sufficient local conditions for the balance of linear and angular momentums.

Note from (4.1.38) that a first Piola-Kirchhoff stress tensor that satisfies the linear and angular equations of motion is not symmetric, but rather

$$\mathbf{S}^\mathsf{T} = \mathbf{F} \cdot \mathbf{S} \cdot \mathbf{F}^{-\mathsf{T}} \ , \qquad S_{Ji} = F_{iA} S_{Ak} F_{Jk}^{-1} \ . \tag{4.1.39}$$

It is often useful to have a symmetric referential measure of stress. One such measure is the *second Piola-Kirchhoff stress tensor*, defined in terms of the first Piola-Kirchhoff stress tensor and the inverse deformation gradient tensor as

$$\widetilde{\mathbf{S}} \equiv \mathbf{S} \cdot \mathbf{F}^{-\mathsf{T}}, \quad \widetilde{S}_{AB} \equiv S_{Ai} F_{Bi}^{-1} \,. \tag{4.1.40}$$

To see that the second Piola-Kirchhoff stress tensor is indeed symmetric, note first from the definition (4.1.40) that $\widetilde{\mathbf{S}}^{\mathsf{T}} = \mathbf{F}^{-1} \cdot \mathbf{S}^{\mathsf{T}}$. Thus, it follows from (4.1.39) that

$$\widetilde{\mathbf{S}}^{\mathsf{T}} = \mathbf{F}^{-1} \cdot \mathbf{F} \cdot \mathbf{S} \cdot \mathbf{F}^{-\mathsf{T}} = \mathbf{S} \cdot \mathbf{F}^{-\mathsf{T}} = \widetilde{\mathbf{S}} \,. \tag{4.1.41}$$

Unfortunately, the second Piola-Kirchhoff stress tensor does not have a convenient physical interpretation. However, the traction-stress relation (4.1.29) and the linear equation of motion (4.1.33) can be expressed in terms of the second Piola-Kirchhoff stress tensor by making the substitutions $\mathbf{S} = \widetilde{\mathbf{S}} \cdot \mathbf{F}^{\mathsf{T}}$ and $\mathbf{S}^{\mathsf{T}} = \mathbf{F} \cdot \widetilde{\mathbf{S}}$, where the angular equation of motion requires that $\widetilde{\mathbf{S}}^{\mathsf{T}} = \widetilde{\mathbf{S}}$.

4.2   Stress Tensors: Spatial Formulation

In the referential formulation, areas and unit normals to material surfaces were given with respect to the reference configuration. In this section, a different measure of stress will be derived—the Cauchy stress tensor—that relates the forces acting on surfaces to areas and unit normals in the current configuration. The reader should compare the steps in this derivation with those of the referential formulation, which it closely mirrors. This spatial formulation is required, for instance, in the study of fluids where there is no natural reference configuration.

4.2.1   Surface Forces

Consider a material point $P \in \partial \mathbb{B}'$ on the boundary of the material volume $\mathbb{B}'$ and let the material surface $\Delta \mathbb{S}$ be an arbitrary portion of the boundary that includes the material point, so that $P \in \Delta \mathbb{S} \subset \partial \mathbb{B}'$. At time t, there is a distribution of surface forces acting on the current configuration of $\Delta \mathbb{S}$ that is statically equivalent to a resultant force $\Delta \mathbf{F}^S$ acting at the current position $\mathbf{x}(P, t)$ of P and a resultant couple $\Delta \mathbf{M}^S$ (Figure 4.2). The resultant force $\Delta \mathbf{F}^S$ and the resultant couple $\Delta \mathbf{M}^S$ are the same as those considered in the referential formulation.

In the current configuration, the area of $\Delta\mathbb{S}$ is Δa and the *current unit outward normal vector* to $\partial\mathbb{B}'$ at P is $\hat{\mathbf{n}}$. In the limit as the material surface $\Delta\mathbb{S}$ is reduced to the material point P, the area $\Delta a \to 0$. One can make the following assumptions:

(a) Assume that in the limit as the material surface $\Delta\mathbb{S}$ is reduced to the material point P,

$$\lim_{\Delta\mathbb{S}\to P} \frac{\Delta\mathbf{M}^S}{\Delta a} = \mathbf{0} . \tag{4.2.1}$$

(b) Assume that in the limit as the material surface $\Delta\mathbb{S}$ is reduced to the material point P, the ratio $\Delta\mathbf{F}^S/\Delta a$ tends to a definite limit,

$$\mathbf{t} \equiv \lim_{\Delta\mathbb{S}\to P} \frac{\Delta\mathbf{F}^S}{\Delta a} , \tag{4.2.2}$$

known as the *(true) traction vector*. This traction vector represents the current force per unit area acting at the material point P on the material surface $\partial\mathbb{B}'$.

(c) Consider two arbitrary material volumes $\mathbb{B}'$ and $\mathbb{B}''$, with boundaries $\partial\mathbb{B}'$ and $\partial\mathbb{B}''$, that both have the material point P on their boundaries. Assume that if $\partial\mathbb{B}'$ and $\partial\mathbb{B}''$ also have the same current unit outward normal vector $\hat{\mathbf{n}}$ at P, then the traction vectors acting at the material point P on $\partial\mathbb{B}'$ and $\partial\mathbb{B}''$ are the same. Thus, the spatial description of the traction vector acting at a point P on a material surface, at time t, can be given as

$$\mathbf{t} = \mathbf{t}(\mathbf{x}, \hat{\mathbf{n}}, t) , \tag{4.2.3}$$

where $\mathbf{x}$ is the current position of P and $\hat{\mathbf{n}}$ is the unit outward normal vector to the current configuration of the surface at P.

It follows from assumptions (b) and (c) that the resultant force of the surface forces acting on a material volume $\mathbb{B}'$ at time t is given in terms of the current configuration of $\mathbb{B}'$ by

$$\mathbf{F}^S(t) = \int_{\partial\mathcal{R}'_t} \mathbf{t}(\mathbf{x}, \hat{\mathbf{n}}, t)\, da = \int_{\partial\mathcal{R}'_t} \mathbf{t}\, da , \tag{4.2.4}$$

where $\partial\mathcal{R}'_t$ is the boundary of the region occupied by the current configuration of $\mathbb{B}'$ and $\hat{\mathbf{n}}$ is the unit outward normal at the point $\mathbf{x}$ on $\partial\mathcal{R}'_t$. It follows from assumptions (a)–(c) that the resultant moment about the origin O of the surface forces acting on a material volume $\mathbb{B}'$ at time t is given in terms of the current configuration of $\mathbb{B}'$ by

$$\mathbf{M}_O^S(t) = \int_{\partial \mathcal{R}_t'} \mathbf{x} \times \mathbf{t}(\mathbf{x}, \hat{\mathbf{n}}, t) \, da = \int_{\partial \mathcal{R}_t'} \mathbf{x} \times \mathbf{t} \, da , \qquad (4.2.5)$$

where $\mathbf{x}$ is the current position vector relative to the origin O.

4.2.2 Body Forces

Consider a material point Q within the material volume $\mathbb{B}'$ and let $\Delta\mathbb{B}$ be an arbitrary portion of the material volume that contains the material point, so that $Q \in \Delta\mathbb{B} \subset \mathbb{B}'$. At time t, there is a distribution of body forces acting on the current configuration of $\Delta\mathbb{B}$ that is statically equivalent to a resultant force $\Delta\mathbf{F}^B$ acting at the current position $\mathbf{x}(Q, t)$ of Q and a resultant couple $\Delta\mathbf{M}^B$ (Figure 4.2). Again, the resultant force $\Delta\mathbf{F}^B$ and the resultant couple $\Delta\mathbf{M}^B$ are the same as those considered in the referential formulation.

In the current configuration, the volume of $\Delta\mathbb{B}$ is Δv, and in the limit as this material volume reduces to the material point Q, the volume $\Delta v \to 0$. One can make the following assumptions:

(d) Assume that in the limit as the material volume $\Delta\mathbb{B}$ is reduced to the material point Q, the ratio $\Delta\mathbf{M}^B/\Delta v$ vanishes:

$$\lim_{\Delta\mathbb{B} \to Q} \frac{\Delta\mathbf{M}^B}{\Delta v} = \mathbf{0} . \qquad (4.2.6)$$

(e) Assume that in the limit as the material volume $\Delta\mathbb{B}$ is reduced to the material point Q, the ratio $\Delta\mathbf{F}^B/\Delta v$ tends to a definite limit:

$$\rho\mathbf{b} \equiv \lim_{\Delta\mathbb{B} \to Q} \frac{\Delta\mathbf{F}^B}{\Delta v} , \qquad (4.2.7)$$

where ρ is the current mass density at the material point Q and $\rho\mathbf{b}$ is the current body force per unit volume acting at the material point Q. The body force vector $\mathbf{b}$, the current body force per unit mass, is the same as that defined in the referential formulation and its spatial description $\mathbf{b}(\mathbf{x}, t)$ is as a function of current position $\mathbf{x}$ and time t.

It follows from assumption (e) that the resultant force of the body forces acting on a material volume $\mathbb{B}'$ at time t is given in terms of the current configuration of $\mathbb{B}'$ by

$$\mathbf{F}^B(t) = \int_{\mathcal{R}_t'} \rho(\mathbf{x}, t)\mathbf{b}(\mathbf{x}, t) \, dv = \int_{\mathcal{R}_t'} \rho\mathbf{b} \, dv , \qquad (4.2.8)$$

where $\mathcal{R}'_t$ is the region occupied by the current configuration of $\mathbb{B}'$. It follows from assumptions (d) and (e) that the resultant moment about the origin O of the body forces acting on a material volume $\mathbb{B}'$ at time t is given in terms of the current configuration of $\mathbb{B}'$ by

$$\mathbf{M}_O^B(t) = \int_{\mathcal{R}'_t} \mathbf{x} \times [\rho(\mathbf{x}, t)\mathbf{b}(\mathbf{x}, t)]\, dv = \int_{\mathcal{R}'_t} \rho \mathbf{x} \times \mathbf{b}\, dv \,, \qquad (4.2.9)$$

where $\mathbf{x}$ is the current position vector relative to the origin O.

4.2.3 Balance of Linear and Angular Momentum

Consider an arbitrary material volume $\mathbb{B}'$, occupying the region $\mathcal{R}'_t$ with boundary $\partial\mathcal{R}'_t$ in its current configuration. The current linear momentum per unit mass at a material point is defined to be the material velocity $\mathbf{v}$, so that the total linear momentum of the material volume $\mathbb{B}'$ is given as an integral over the region $\mathcal{R}'_t$ by

$$\int_{\mathcal{R}'_t} \rho(\mathbf{x}, t)\mathbf{v}(\mathbf{x}, t)\, dv = \int_{\mathcal{R}'_t} \rho \mathbf{v}\, dv \,. \qquad (4.2.10)$$

It is *postulated* that the resultant force of the external forces acting on $\mathbb{B}'$ is equal to the rate of change of the total linear momentum of $\mathbb{B}'$,

$$\int_{\partial\mathcal{R}'_t} \mathbf{t}\, da + \int_{\mathcal{R}'_t} \rho\mathbf{b}\, dv = \frac{D}{Dt} \int_{\mathcal{R}'_t} \rho\mathbf{v}\, dv \,. \qquad (4.2.11)$$

However, applying the result (3.3.27),

$$\frac{D}{Dt} \int_{\mathcal{R}'_t} \rho\mathbf{v}\, dv = \int_{\mathcal{R}'_t} \rho\frac{D\mathbf{v}}{Dt}\, dv \,. \qquad (4.2.12)$$

Thus, the postulated *balance of linear momentum* requires that

$$\boxed{\begin{aligned} \int_{\partial\mathcal{R}'_t} \mathbf{t}\, da + \int_{\mathcal{R}'_t} \rho\mathbf{b}\, dv &= \int_{\mathcal{R}'_t} \rho\mathbf{a}\, dv \,, \\ \int_{\partial\mathcal{R}'_t} t_i\, da + \int_{\mathcal{R}'_t} \rho b_i\, dv &= \int_{\mathcal{R}'_t} \rho a_i\, dv \,, \end{aligned}} \qquad (4.2.13)$$

where $\mathbf{a} \equiv D\mathbf{v}/Dt$ is the material acceleration.

The current angular momentum per unit mass at a material point, relative to the origin, is defined to be $\mathbf{x} \times \mathbf{v}$, where $\mathbf{x}$ is the material point's current position vector. The total current angular momentum of the material volume $\mathbb{B}'$ is therefore given as an integral over the region $\mathcal{R}'_t$ by

$$\int_{\mathcal{R}'_t} \rho(\mathbf{x}, t)[\mathbf{x} \times \mathbf{v}(\mathbf{x}, t)]\, dv = \int_{\mathcal{R}'_t} \rho \mathbf{x} \times \mathbf{v}\, dv \ . \tag{4.2.14}$$

It is *postulated*, as a generalization of the principle of conservation of angular momentum for systems of particles, that the resultant moment about the origin of the external forces acting on $\mathbb{B}'$ is equal to the rate of change of the total angular momentum of $\mathbb{B}'$:

$$\int_{\partial \mathcal{R}'_t} \mathbf{x} \times \mathbf{t}\, da + \int_{\mathcal{R}'_t} \rho \mathbf{x} \times \mathbf{b}\, dv = \frac{\mathrm{D}}{\mathrm{D}t} \int_{\mathcal{R}'_t} \rho \mathbf{x} \times \mathbf{v}\, dv \ . \tag{4.2.15}$$

However, applying the result (3.3.27) and since $\mathrm{D}\mathbf{x}/\mathrm{D}t = \mathbf{v}$ and $\mathbf{v} \times \mathbf{v} = 0$,

$$\frac{\mathrm{D}}{\mathrm{D}t} \int_{\mathcal{R}'_t} \rho \mathbf{x} \times \mathbf{v}\, dv = \int_{\mathcal{R}'_t} \rho \frac{\mathrm{D}}{\mathrm{D}t}(\mathbf{x} \times \mathbf{v})\, dv = \int_{\mathcal{R}'_t} \rho \mathbf{x} \times \frac{\mathrm{D}\mathbf{v}}{\mathrm{D}t}\, dv \ . \tag{4.2.16}$$

Thus, the postulated *balance of angular momentum* requires that

$$
\boxed{
\begin{aligned}
\int_{\partial \mathcal{R}'_t} \mathbf{x} \times \mathbf{t}\, da + \int_{\mathcal{R}'_t} \rho \mathbf{x} \times \mathbf{b}\, dv &= \int_{\mathcal{R}'_t} \rho \mathbf{x} \times \mathbf{a}\, dv \ , \\
\int_{\partial \mathcal{R}'_t} e_{ijk} x_j t_k\, da + \int_{\mathcal{R}'_t} \rho e_{ijk} x_j b_k\, dv &= \int_{\mathcal{R}'_t} \rho e_{ijk} x_j a_k\, dv \ ,
\end{aligned}
}
\tag{4.2.17}
$$

where the origin O of the current position vector $\mathbf{x}$ is arbitrary.

4.2.4   Generalization of Newton's Third Law

In a manner similar to that used in the referential formulation, it can be shown that the postulated balance of linear momentum implies that the traction acting on one side of a material surface is equal in magnitude and opposite in direction to the traction acting on the opposite side,

$$\boxed{\mathbf{t}(\mathbf{x}, -\hat{\mathbf{n}}, t) = -\mathbf{t}(\mathbf{x}, \hat{\mathbf{n}}, t) \ ,} \tag{4.2.18}$$

and that the resultant body force exerted by one material volume on a second material volume is also equal and opposite to that exerted by the second material volume on the first. The balance of angular momentum implies that these equal and opposite forces are colinear.

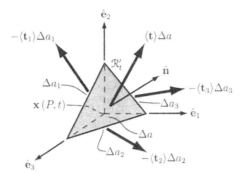

FIGURE 4.6. *Current configuration as a Cauchy tetrahedron.*

4.2.5 Cauchy Stress

To determine the relation between the traction vector $\mathbf{t}(\mathbf{x}, \hat{\mathbf{n}}, t)$ and the current unit normal $\hat{\mathbf{n}}$ to a material surface, consider a material volume $\mathbb{B}'$ whose *current* configuration occupies the Cauchy tetrahedron shown in Figure 4.6. The volume of the region $\mathcal{R}'_t$ that is the the Cauchy tetrahedron is Δv. The boundary $\partial \mathcal{R}'_t$ of the Cauchy tetrahedron is composed of four plane faces. Three of the faces, with areas Δa_1, Δa_2, and Δa_3, are parallel to the coordinate planes defined by the vector basis $\{\hat{\mathbf{e}}_i\}$ and intersect at the material point P. Their unit outward normals are $-\hat{\mathbf{e}}_1$, $-\hat{\mathbf{e}}_2$, and $-\hat{\mathbf{e}}_3$, respectively. The fourth face has area Δa and unit outward normal $\hat{\mathbf{n}}$.

Let the traction vector acting on a material surface whose current unit outward normal is $\hat{\mathbf{e}}_i$, at a material point on the surface whose current position is $\mathbf{x}$, be denoted by $\mathbf{t}_i(\mathbf{x}, t) \equiv \mathbf{t}(\mathbf{x}, \hat{\mathbf{e}}_i, t)$. It follows from (4.2.18) that the traction fields on the Δa_i faces of the tetrahedron can be given as $\mathbf{t}(\mathbf{x}, -\hat{\mathbf{e}}_i, t) = -\mathbf{t}_i(\mathbf{x}, t)$. The traction field on the Δa face is $\mathbf{t}(\mathbf{x}, \hat{\mathbf{n}}, t)$.

Consider the balance of linear momentum (4.2.13) of this material volume $\mathbb{B}'$ whose current configuration occupies the Cauchy tetrahedron $\mathcal{R}'_t$. The value of the surface integral can be given as

$$\int_{\partial \mathcal{R}'} \mathbf{t}\, da = \langle \mathbf{t} \rangle \Delta a - \langle \mathbf{t}_i \rangle \Delta a_i , \qquad (4.2.19)$$

where $\langle \mathbf{t} \rangle$ is the average traction on the face with unit outward normal $\hat{\mathbf{n}}$ and $-\langle \mathbf{t}_i \rangle$ is the average traction on the face with unit outward normal

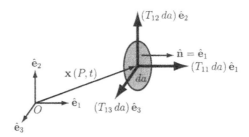

FIGURE 4.7. *Physical interpretation for scalar components of the Cauchy stress.*

$-\hat{\mathbf{e}}_i$, each at time t. Similarly, the volume integrals can be given as

$$\int_{\mathcal{R}'} \rho \mathbf{b}\, dv = \langle \rho \mathbf{b}\rangle\, \Delta v\ ,\tag{4.2.20}$$

$$\int_{\mathcal{R}'} \rho \mathbf{a}\, dv = \langle \rho \mathbf{a}\rangle\, \Delta v\ ,\tag{4.2.21}$$

where $\langle \rho \mathbf{b}\rangle$ and $\langle \rho \mathbf{a}\rangle$ are the averages, over the current configuration of the material volume $\mathbb{B}'$, of the body force and inertial force, respectively, at time t. Using (4.2.19)–(4.2.21) and noting that $\Delta a_i = \hat{n}_i \Delta a$, where $\hat{n}_i = \hat{\mathbf{n}}\cdot\hat{\mathbf{e}}_i$ are the scalar components of the unit outward normal to the Δa face, the balance of linear momentum (4.2.13) can then be rewritten as

$$\langle \mathbf{t}\rangle - \langle \mathbf{t}_j\rangle\hat{n}_j + (\langle \rho \mathbf{b}\rangle - \langle \rho \mathbf{a}\rangle)\frac{\Delta v}{\Delta a} = \mathbf{0}\ .\tag{4.2.22}$$

By the mean-value theorem of calculus, in the limit as the tetrahedron shrinks to the material point P, the average quantities approach their exact values at the point P. Also, $\Delta v/\Delta a \to 0$, so that, in this limit, the balance of linear momentum becomes

$$\mathbf{t}(\mathbf{x},\hat{\mathbf{n}},t) = \hat{n}_j \mathbf{t}_j(\mathbf{x},t)\ ,\tag{4.2.23}$$

where $\mathbf{x}$ is the current position of the material point P. The scalar components in the vector basis $\{\hat{\mathbf{e}}_i\}$ of the traction vector are then

$$t_i = \mathbf{t}\cdot\hat{\mathbf{e}}_i = \hat{n}_j \mathbf{t}_j\cdot\hat{\mathbf{e}}_i = \hat{n}_j(\mathbf{t}_j\cdot\hat{\mathbf{e}}_i)\ .\tag{4.2.24}$$

This gives the traction vector at a material point acting on a material surface with current unit normal $\hat{\mathbf{n}}$, in terms of the traction vectors at the same material point acting on the material surfaces which, in the current configuration, are parallel to the coordinate planes.

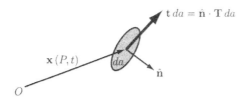

FIGURE 4.8. Physical interpretation for the Cauchy stress tensor.

Define the scalar components, in the vector basis $\{\hat{\mathbf{e}}_i\}$, of the *Cauchy stress* as $T_{ji} \equiv \mathbf{t}_j \cdot \hat{\mathbf{e}}_i$. At a material point and instant in time, T_{ji} is the ith scalar component of the traction vector acting on the material surface whose current unit outward normal is $\hat{\mathbf{e}}_j$. To illustrate this physical interpretation, consider a material point P and a differential material surface $d\mathbb{S}$ that contains P and has a current unit outward normal $\hat{\mathbf{n}} = \hat{\mathbf{e}}_1$ (Figure 4.7). Let the current area of $d\mathbb{S}$ be da. The current force acting at $\mathbf{x}(P, t)$ on $d\mathbb{S}$ has scalar components $T_{11}\,da$, $T_{12}\,da$, and $T_{13}\,da$ in the vector basis $\{\hat{\mathbf{e}}_i\}$. Note that, in contrast to the situation involving the first Piola-Kirchhoff stress, $(T_{11}\,da)\hat{\mathbf{e}}_1$ *is* normal to the current configuration of the material surface and $(T_{12}\,da)\hat{\mathbf{e}}_2$ and $(T_{13}\,da)\hat{\mathbf{e}}_3$ *are* tangent to the surface.

From (4.2.24), the traction vector acting at a point on a material surface, whose current unit outward normal is $\hat{\mathbf{n}}$, is given in terms of the T_{ji} by

$$\boxed{\mathbf{t} = \hat{\mathbf{n}} \cdot \mathbf{T}, \quad t_i = \hat{n}_j T_{ji}\,.} \qquad (4.2.25)$$

This is the spatial formulation of the *traction-stress relation*. Since $\mathbf{t}$ and $\hat{\mathbf{n}}$ are both vectors, it follows from (4.2.25) that $\mathbf{T} = T_{ij}\hat{\mathbf{e}}_i\hat{\mathbf{e}}_j$ is a second-order tensor called the *Cauchy stress tensor*. The spatial description of the Cauchy stress tensor, $\mathbf{T}(\mathbf{x}, t)$, is as a function of current position $\mathbf{x}$ and time t.

Like the first and second Piola-Kirchhoff stress tensors, the Cauchy stress tensor also completely characterizes the internal forces at a material point— it can be used to find the current force per unit area acting on any surface that passes through the material point. Consider a material point P and a differential material surface $d\mathbb{S}$ that contains P and, in the current configuration, has a unit outward normal $\hat{\mathbf{n}}$ and area da (Figure 4.8). Then, from (4.2.25), the current force acting at $\mathbf{x}(P, t)$ on $d\mathbb{S}$ is $\mathbf{t}\,da = \hat{\mathbf{n}} \cdot \mathbf{T}\,da$.

4.2.6 Equations of Motion

By considering an arbitrary material volume $\mathbb{B}'$ occupying in its current configuration the region $\mathcal{R}'_t$ with boundary $\partial\mathcal{R}'_t$, substituting the traction-stress relation (4.2.25) into the spatial formulation of the balance of linear momentum (4.2.13), and applying the divergence theorem, it can be shown that the local condition for the balance of linear momentum in terms of the Cauchy stress tensor is

$$\operatorname{div} \mathbf{T}^{\mathsf{T}} + \rho \mathbf{b} = \rho \mathbf{a} , \quad \frac{\partial T_{ji}}{\partial x_j} + \rho b_i = \rho a_i . \qquad (4.2.26)$$

This *linear equation of motion* for the Cauchy stress tensor must be satisfied everywhere within a body. Likewise, it can be shown that the local condition for the balance of angular momentum (4.2.17) in terms of the Cauchy stress tensor is

$$\mathbf{T} = \mathbf{T}^{\mathsf{T}} , \quad T_{ij} = T_{ji} . \qquad (4.2.27)$$

This *angular equation of motion* for the Cauchy stress tensor just requires that the Cauchy stress tensor be symmetric. The equations of motion (4.2.26) and (4.2.27) are necessary and sufficient local conditions for the balance of linear and angular momentums.

4.2.7 Relations Between Stress Tensors

Since both the first Piola-Kirchhoff stress tensor and the Cauchy stress tensor completely characterize the internal (surface) forces at a material point, it stands to reason that there should be some relation between the two. To derive this relation, consider an arbitrary material volume $\mathcal{B}'$ that occupies the region $\mathcal{R}'$ with boundary $\partial\mathcal{R}'$ in the reference configuration and the region $\mathcal{R}'_t$ with boundary $\partial\mathcal{R}'_t$ in the current configuration. The current resultant force acting on $\mathcal{B}'$ due to the distribution of surface forces on its boundary is given in terms of the nominal and true traction vectors by

$$\mathbf{F}^S(t) = \int_{\partial\mathcal{R}'} \mathbf{t}^{(\mathrm{N})} \, dA = \int_{\partial\mathcal{R}'_t} \mathbf{t} \, da . \qquad (4.2.28)$$

Using the traction-stress relations (4.1.29) and (4.2.25), this equality can be rewritten in terms of the first Piola-Kirchhoff and Cauchy stress tensors as

$$\int_{\partial\mathcal{R}'} \hat{\mathbf{N}} \cdot \mathbf{S} \, dA = \int_{\partial\mathcal{R}'_t} \hat{\mathbf{n}} \cdot \mathbf{T} \, da . \qquad (4.2.29)$$

However, the integral over the current configuration $\partial\mathcal{R}'_t$ on the right-hand side of (4.2.29) can be rewritten as an integral over the reference configuration $\partial \mathcal{R}'$ using the result (3.2.42):

$$\int_{\partial \mathcal{R}'_t} \hat{\mathbf{n}} \cdot \mathbf{T} \, da = \int_{\partial \mathcal{R}'_t} \mathbf{T}^\mathsf{T} \cdot \hat{\mathbf{n}} \, da$$

$$= \int_{\partial \mathcal{R}'} \mathbf{T}^\mathsf{T} \cdot \mathbf{F}^{-\mathsf{T}} \cdot \hat{\mathbf{N}} J \, dA$$

$$= \int_{\partial \mathcal{R}'} \hat{\mathbf{N}} \cdot \mathbf{F}^{-1} \cdot \mathbf{T} J \, dA , \qquad (4.2.30)$$

where $\mathbf{F}^{-1}$ is the inverse deformation gradient tensor and $J = \det \mathbf{F}$ is the Jacobian of the motion. Therefore,

$$\int_{\partial \mathcal{R}'} \hat{\mathbf{N}} \cdot \mathbf{S} \, dA = \int_{\partial \mathcal{R}'} \hat{\mathbf{N}} \cdot \mathbf{F}^{-1} \cdot \mathbf{T} J \, dA . \qquad (4.2.31)$$

Since $\mathbb{B}'$ (and therefore $\partial \mathcal{R}'$ and $\hat{\mathbf{N}}$) is arbitrary, it follows that

$$\boxed{\mathbf{S} = J\mathbf{F}^{-1} \cdot \mathbf{T} , \quad S_{Aj} = J F_{Ai}^{-1} T_{ij}} \qquad (4.2.32)$$

and that

$$\boxed{\mathbf{T} = \frac{1}{J}\mathbf{F} \cdot \mathbf{S} , \quad T_{ij} = \frac{1}{J}F_{iA}S_{Aj} .} \qquad (4.2.33)$$

In terms of the second Piola-Kirchhoff stress tensor, $\widetilde{\mathbf{S}} \equiv \mathbf{S} \cdot \mathbf{F}^{-\mathsf{T}}$,

$$\boxed{\widetilde{\mathbf{S}} = J\mathbf{F}^{-1} \cdot \mathbf{T} \cdot \mathbf{F}^{-\mathsf{T}} , \quad \widetilde{S}_{AB} = J F_{Ai}^{-1} T_{ij} F_{Bj}^{-1}} \qquad (4.2.34)$$

and

$$\boxed{\mathbf{T} = \frac{1}{J}\mathbf{F} \cdot \widetilde{\mathbf{S}} \cdot \mathbf{F}^\mathsf{T} , \quad T_{ij} = \frac{1}{J}F_{iA}\widetilde{S}_{AB}F_{jB} .} \qquad (4.2.35)$$

4.3 Kinematic Linearization

Recall that for infinitesimal deformation, in which $\|\mathrm{Grad}\,\mathbf{u}\| \ll 1$, there exists a positive scalar $\epsilon \ll 1$ such that $\|\mathrm{Grad}\,\mathbf{u}\| = \mathcal{O}(\epsilon)$. Thus, for infinitesimal deformation, the scalar components of the displacement gradient tensor are such that $F_{iA} = \delta_{iA} + \mathcal{O}_{iA}(\epsilon)$, the scalar components of the inverse displacement gradient tensor are such that $F_{Ai}^{-1} = \delta_{Ai} + \mathcal{O}_{Ai}(\epsilon)$, and the Jacobian of the deformation is such that $J = 1 + \mathcal{O}(\epsilon)$. It follows from the angular equation of motion (4.1.38) for the first Piola-Kirchhoff stress tensor $\mathbf{S}$ that

$$S_{Ji} = S_{iJ} + \mathcal{O}_{iA}(\epsilon)S_{AJ} + \mathcal{O}_{Jk}(\epsilon)S_{ik} , \qquad (4.3.1)$$

from the definition (4.1.40) of the second Piola-Kirchhoff stress tensor $\widetilde{\mathbf{S}}$ that

$$\widetilde{S}_{AB} = S_{AB} + \mathcal{O}_{Bi}(\epsilon)S_{Ai} , \qquad (4.3.2)$$

and from the relation (4.2.33) between the Cauchy stress tensor $\mathbf{T}$ and the first Piola-Kirchhoff stress tensor that

$$T_{ij} = S_{ij} + \mathcal{O}_{iA}(\epsilon)S_{Aj} . \qquad (4.3.3)$$

Also, from Nanson's formula (3.2.41) relating the unit normals and differential areas of a differential material surface in the reference and current configurations,

$$\hat{n}_i da = \left[\hat{N}_i + \mathcal{O}_{Ai}(\epsilon)\hat{N}_A \right] dA . \qquad (4.3.4)$$

Thus, for infinitesimal deformation, if it is assumed that the terms above involving $\mathcal{O}(\epsilon)$ are negligible, the angular equation of motion requires the first Piola-Kirchhoff stress tensor to be approximately symmetric,

$$\mathbf{S} \approx \mathbf{S}^\mathsf{T} , \quad S_{ij} \approx S_{ji} , \qquad (4.3.5)$$

the three measures of stress to be approximately the same,

$$\mathbf{S} \approx \widetilde{\mathbf{S}} \approx \mathbf{T} , \quad S_{ij} \approx \widetilde{S}_{ij} \approx T_{ij} , \qquad (4.3.6)$$

and the areas and unit normals of material surfaces to be approximately the same in the reference and current configurations,

$$\hat{n} da \approx \hat{N} dA , \quad \hat{n}_i da \approx \hat{N}_i dA . \qquad (4.3.7)$$

The latter result means that there is approximately no difference between the nominal and true traction vectors.

For the rest of this chapter, it is assumed that deformation is infinitesimal and that all quantities are given as fields over the reference configuration $\mathcal{R}$ of a body (i.e., in terms of their material descriptions). The comma convention will be used to denote differentiation of the material description of a quantity with respect to Cartesian coordinate variables so that, for instance, $\sigma_{ji,j} = \partial\sigma_{ji}/\partial X_j$.

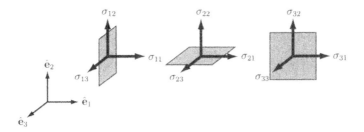

FIGURE 4.9.   Scalar components of stress.

Since the three stress tensors defined previously are approximately equivalent during infinitesimal deformation, a single *(kinematically infinitesimal) stress tensor,*

$$\boxed{\boldsymbol{\sigma} = \boldsymbol{\sigma}(\mathbf{X}, t) , \quad \sigma_{ij} = \sigma_{ij}(X_1, X_2, X_3, t) ,} \qquad (4.3.8)$$

will be used. This notation conforms with the convention in much of the literature on linearized elasticity [see, for instance, (Fung 1965)] and will also serve to reinforce the understanding that one is assuming infinitesimal deformation. The *traction-stress relation* is then

$$\boxed{\mathbf{t} = \hat{\mathbf{n}} \cdot \boldsymbol{\sigma} , \quad t_i = \hat{n}_j \sigma_{ji} ,} \qquad (4.3.9)$$

where the traction vector $\mathbf{t}$ is the force per unit area and $\hat{\mathbf{n}}$ is the unit normal to the surface. No distinction is made between either the reference and current areas or the reference and current unit normals, since they are approximately the same. Thus, σ_{ij} is the jth scalar component of the traction acting on a surface with unit outward normal $\hat{\mathbf{e}}_i$ (Figure 4.9).

The *linear equation of motion* for the stress tensor field $\boldsymbol{\sigma}$ is

$$\boxed{\boldsymbol{\nabla} \cdot \boldsymbol{\sigma}^{\mathsf{T}} + \mathbf{f} = \rho \mathbf{a} , \quad \sigma_{ji,j} + f_i = \rho a_i ,} \qquad (4.3.10)$$

where $\mathbf{f} = \rho \mathbf{b}$ is the body force per unit volume and ρ is the mass density. Again, no distinction is made between the reference and current volumes or between the initial and current mass densities. The *angular equation of motion* is

$$\boxed{\boldsymbol{\sigma} = \boldsymbol{\sigma}^{\mathsf{T}} , \quad \sigma_{ij} = \sigma_{ji} .} \qquad (4.3.11)$$

Typically, it will be implicitly assumed that the stress tensor is symmetric (recall that the infinitesimal strain tensor $\boldsymbol{\varepsilon}$ is symmetric by construction) and therefore satisfies the angular equation of motion. The remaining linear equation of motion is then just referred to as the *equation of motion* or, if the material acceleration field is negligible ($\mathbf{a} \approx \mathbf{0}$), as the *equilibrium equation*.

Because the stress tensor is symmetric, the traction-stress relation can equivalently be written as

$$\mathbf{t} = \boldsymbol{\sigma} \cdot \hat{\mathbf{n}} , \quad t_i = \sigma_{ij} \hat{n}_j \qquad (4.3.12)$$

and the equation of motion as

$$\nabla \cdot \boldsymbol{\sigma} + \mathbf{f} = \rho \mathbf{a} , \quad \sigma_{ij,j} + f_i = \rho a_i . \qquad (4.3.13)$$

In the literature, they are sometimes presented in this form.

In an inertial frame of reference, the inertial force $-\rho \mathbf{a}$ is indistinguishable from a body force. For the sake of conciseness, from now on *equations will be derived with* $\mathbf{a} = \mathbf{0}$. If the acceleration in a problem is not negligible, it is taken into account by treating the inertial force as a body force, that is, by making the substitution

$$\mathbf{f} \to \mathbf{f} - \rho \mathbf{a} . \qquad (4.3.14)$$

If the acceleration is not mentioned in a problem, it is understood to be negligible.

Example 4.1
Consider a body in which the stress field is given by

$$[\boldsymbol{\sigma}] = \begin{bmatrix} a_1 X_1^2 & 0 & a_1 X_1 X_2 \\ 0 & a_2 X_3 & a_2 X_2 \\ a_1 X_1 X_2 & a_2 X_2 & a_2 X_3 \end{bmatrix} .$$

Find the body force field that is required for the body to be in equilibrium.

Solution. Note first that the given stress field is symmetric everywhere in the body, as required. It is crucial to remember that this must always be the case. Rewriting the equilibrium equations as $f_i = -\sigma_{ji,j}$, one finds that the body force field must be

$$[\mathbf{f}] = \begin{bmatrix} -2a_1 X_1 \\ 0 \\ -a_1 X_2 - 2a_2 \end{bmatrix}$$

or, written alternatively in dyadic notation,

$$\mathbf{f} = -2a_1 X_1 \hat{\mathbf{e}}_1 - (a_1 X_2 + 2a_2) \hat{\mathbf{e}}_3 .$$

This example is instructional, but one should be aware that the option of specifying the body force field will usually not be available. The body force field is usually determined by factors (such as gravity) that are outside of one's control.

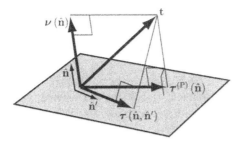

FIGURE 4.10. Normal, shear, and projected shear tractions.

4.3.1 Normal and Shear Tractions

Consider the traction vector $\mathbf{t}$ acting at a point on a surface whose unit outward normal is $\hat{\mathbf{n}}$. The traction is related to the stress at the point by the traction-stress relation $\mathbf{t} = \hat{\mathbf{n}} \cdot \boldsymbol{\sigma}$. What are the components of the traction vector normal to the surface, in a given direction tangent to the surface, and projected onto the surface (Figure 4.10)?

Normal traction

The *normal traction* $\nu(\hat{\mathbf{n}}) = \mathbf{t} \cdot \hat{\mathbf{n}}$ acting on a surface with unit outward normal $\hat{\mathbf{n}}$ is the orthogonal projection of the traction vector $\mathbf{t}$ acting on the surface onto the direction of the unit normal. Since $\mathbf{t} = \hat{\mathbf{n}} \cdot \boldsymbol{\sigma}$, it follows that

$$\nu(\hat{\mathbf{n}}) = \hat{\mathbf{n}} \cdot \boldsymbol{\sigma} \cdot \hat{\mathbf{n}} \;, \quad \nu(\hat{n}_1, \hat{n}_2, \hat{n}_3) = \sigma_{ij} \hat{n}_i \hat{n}_j \;. \tag{4.3.15}$$

Therefore, the corresponding *normal traction vector* $\boldsymbol{\nu}(\hat{\mathbf{n}}) = \nu(\hat{\mathbf{n}})\hat{\mathbf{n}}$ is

$$\boldsymbol{\nu}(\hat{\mathbf{n}}) = (\hat{\mathbf{n}} \cdot \boldsymbol{\sigma} \cdot \hat{\mathbf{n}})\hat{\mathbf{n}} \;, \quad \nu_k(\hat{n}_1, \hat{n}_2, \hat{n}_3) = \sigma_{ij} \hat{n}_i \hat{n}_j \hat{n}_k \;, \tag{4.3.16}$$

where the scalar components $\nu_k = \boldsymbol{\nu} \cdot \hat{\mathbf{e}}_k$ of the normal traction vector should not be confused with the normal traction $\nu = |\boldsymbol{\nu}|$. The normal traction is referred to as *tensile* if $\nu > 0$ and *compressive* if $\nu < 0$.

Shear traction

The *shear traction* $\tau(\hat{\mathbf{n}}, \hat{\mathbf{n}}') = \mathbf{t} \cdot \hat{\mathbf{n}}'$ acting on a surface with unit outward normal $\hat{\mathbf{n}}$, in the direction of a unit vector $\hat{\mathbf{n}}'$ that is tangent to the surface $(\hat{\mathbf{n}} \cdot \hat{\mathbf{n}}' = 0)$, is the orthogonal projection of the traction vector $\mathbf{t}$ acting on the surface onto the direction of $\hat{\mathbf{n}}'$. Since $\mathbf{t} = \hat{\mathbf{n}} \cdot \boldsymbol{\sigma}$, it follows that

$$\tau(\hat{\mathbf{n}}, \hat{\mathbf{n}}') = \hat{\mathbf{n}} \cdot \boldsymbol{\sigma} \cdot \hat{\mathbf{n}}' \;, \quad \tau(\hat{n}_1, \hat{n}_2, \hat{n}_3, \hat{n}_1', \hat{n}_2', \hat{n}_3') = \sigma_{ij} \hat{n}_i \hat{n}_j' \;. \tag{4.3.17}$$

The corresponding *shear traction vector* $\boldsymbol{\tau}(\hat{\mathbf{n}}, \hat{\mathbf{n}}') = \tau(\hat{\mathbf{n}}, \hat{\mathbf{n}}')\hat{\mathbf{n}}'$ is therefore

$$\boldsymbol{\tau}(\hat{\mathbf{n}}, \hat{\mathbf{n}}') = (\hat{\mathbf{n}} \cdot \boldsymbol{\sigma} \cdot \hat{\mathbf{n}}')\hat{\mathbf{n}}' , \quad \tau_k(\hat{n}_1, \hat{n}_2, \hat{n}_3, \hat{n}_1', \hat{n}_2', \hat{n}_3') = \sigma_{ij}\hat{n}_i\hat{n}_j'\hat{n}_k' ,$$
$$(4.3.18)$$

where the scalar components $\tau_k = \boldsymbol{\tau} \cdot \hat{\mathbf{e}}_k$ of the shear traction vector should not be confused with the shear traction $\tau = |\boldsymbol{\tau}|$. Since the stress tensor is symmetric, it is seen that $\tau(\hat{\mathbf{n}}, \hat{\mathbf{n}}') = \tau(\hat{\mathbf{n}}', \hat{\mathbf{n}})$ which means that the shear traction on the surface with unit normal $\hat{\mathbf{n}}$ in the direction $\hat{\mathbf{n}}'$ is equal to the shear traction on the surface with unit normal $\hat{\mathbf{n}}'$ in the direction $\hat{\mathbf{n}}$.

Projected shear traction

Among the shear tractions in all possible directions $\hat{\mathbf{n}}'$ tangent to a surface with unit outward normal $\hat{\mathbf{n}}$, the *projected shear traction* $\tau^{(\mathrm{P})}(\hat{\mathbf{n}}) = |\tau(\hat{\mathbf{n}}, \hat{\mathbf{n}}')|_{\max}$ is the maximum shear traction acting on this surface and is the orthogonal projection of the traction vector onto the surface. The corresponding *projected shear traction vector* $\boldsymbol{\tau}^{(\mathrm{P})}(\hat{\mathbf{n}})$ lies in the plane defined by the traction vector $\mathbf{t}$ and the normal traction vector $\boldsymbol{\nu}$ (Figure 4.10) and is given through vector addition by

$$\boldsymbol{\tau}^{(\mathrm{P})}(\hat{\mathbf{n}}) = \mathbf{t} - \boldsymbol{\nu}(\hat{\mathbf{n}}) . \qquad (4.3.19)$$

The normal traction vector and the projected shear traction vector are orthogonal so that, from the Pythagorean theorem, the projected shear traction is

$$\tau^{(\mathrm{P})}(\hat{\mathbf{n}}) = \sqrt{|\mathbf{t}|^2 - |\boldsymbol{\nu}(\hat{\mathbf{n}})|^2} . \qquad (4.3.20)$$

Do not confuse the projected shear traction $\tau^{(\mathrm{P})} = |\boldsymbol{\tau}^{(\mathrm{P})}|$ with the scalar components $\tau_k^{(\mathrm{P})} = \boldsymbol{\tau}^{(\mathrm{P})} \cdot \hat{\mathbf{e}}_k$ of the projected shear traction vector. Note also that $\tau^{(\mathrm{P})} \geq 0$.

Example 4.2

It is known that, at a given point on the surface of a body, the unit outward normal to the surface is $\hat{\mathbf{n}} = (\hat{\mathbf{e}}_1 + \hat{\mathbf{e}}_2 - \hat{\mathbf{e}}_3)/\sqrt{3}$ and the traction vector acting at that point on the surface is $\mathbf{t} = (P\hat{\mathbf{e}}_1 + 2P\hat{\mathbf{e}}_2)/\sqrt{5}$. Find the normal traction vector $\boldsymbol{\nu}$ and the projected shear traction vector $\boldsymbol{\tau}^{(\mathrm{P})}$ at this point on the surface of the body. What conditions must be satisfied by the components of the stress field at this point?

Solution. The normal traction is given by

$$\nu = \mathbf{t} \cdot \hat{\mathbf{n}} = \sqrt{\frac{3}{5}} P .$$

Note that the normal traction is tensile if $P > 0$ and compressive if $P < 0$. The normal traction vector $\boldsymbol{\nu} = \nu \hat{\mathbf{n}}$ is

$$\boldsymbol{\nu} = \sqrt{\frac{3}{5}} P \hat{\mathbf{n}} = \frac{P}{\sqrt{5}} \hat{\mathbf{e}}_1 + \frac{P}{\sqrt{5}} \hat{\mathbf{e}}_2 - \frac{P}{\sqrt{5}} \hat{\mathbf{e}}_3$$

and the projected shear traction vector is, therefore,

$$\boldsymbol{\tau}^{(P)} = \mathbf{t} - \boldsymbol{\nu} = \frac{P}{\sqrt{5}} \hat{\mathbf{e}}_2 + \frac{P}{\sqrt{5}} \hat{\mathbf{e}}_3 \ .$$

Since the traction and the unit normal on the surface of the body at the given point are specified, the traction-stress relations $\hat{\mathbf{n}} \cdot \boldsymbol{\sigma} = \mathbf{t}$ give the conditions

$$\frac{1}{\sqrt{3}} \sigma_{11} + \frac{1}{\sqrt{3}} \sigma_{21} - \frac{1}{\sqrt{3}} \sigma_{31} = \frac{P}{\sqrt{5}} \ ,$$

$$\frac{1}{\sqrt{3}} \sigma_{12} + \frac{1}{\sqrt{3}} \sigma_{22} - \frac{1}{\sqrt{3}} \sigma_{32} = \frac{2P}{\sqrt{5}} \ ,$$

$$\frac{1}{\sqrt{3}} \sigma_{13} + \frac{1}{\sqrt{3}} \sigma_{23} - \frac{1}{\sqrt{3}} \sigma_{33} = 0 \ ,$$

which the scalar components of the stress field must satisfy at this point. These three equations are known as traction boundary conditions. If the traction on a given surface is specified, then the components of stress must satisfy the corresponding boundary conditions, given by the traction-stress relations. Note that the three boundary conditions in this example involve all six components of stress. It will usually be to one's advantage to define a vector basis aligned with the unit normal to the surface, so that the number of stress components involved in the traction boundary conditions reduces to three.

4.3.2 Principal Stresses and Principal Directions of Stress

In the terminology of Section 2.3.6 on eigenvalue problems, the normal traction (4.3.15) is the normal (2.3.99) of the stress tensor in the direction $\hat{\mathbf{n}}$, $\nu = N_\sigma(\hat{\mathbf{n}})$, and the shear traction (4.3.17) is the shear (2.3.110) of the stress tensor between the orthogonal directions $\hat{\mathbf{n}}$ and $\hat{\mathbf{n}}'$, $\tau = S_\sigma(\hat{\mathbf{n}}, \hat{\mathbf{n}}')$. It follows that the eigenvectors of the stress tensor are the unit outward normals to surfaces corresponding to extrema of the normal traction ν and that the eigenvalues are the values of these extrema. Also, the shear traction in every direction tangent to these surfaces is zero; that is, the traction vector is normal to the surfaces whose unit outward normals are eigenvectors of the stress tensor.

- The eigenvalues of the stress tensor, $\sigma^{(1)} \geq \sigma^{(2)} \geq \sigma^{(3)}$, are called the *principal stresses*.

- The corresponding eigenvectors, $\hat{n}^{(1)}$, $\hat{n}^{(2)}$, and $\hat{n}^{(3)}$ are called the *principal directions of stress* and are chosen so as to form a right-handed system, $\hat{n}^{(1)} \times \hat{n}^{(2)} = \hat{n}^{(3)}$.

- Let the vector basis formed by the principal directions be denoted by

$$\{\hat{e}_i^*\}_\sigma = \begin{bmatrix} \hat{n}^{(1)} \\ \hat{n}^{(2)} \\ \hat{n}^{(3)} \end{bmatrix} . \tag{4.3.21}$$

Then, the components of stress in this vector basis are

$$[\sigma]^* = \begin{bmatrix} \sigma^{(1)} & 0 & 0 \\ 0 & \sigma^{(2)} & 0 \\ 0 & 0 & \sigma^{(3)} \end{bmatrix} . \tag{4.3.22}$$

- The maximum normal traction $\nu_{max} = \sigma^{(1)}$ acts on the surface whose unit normal is $\hat{n}^{(1)}$, the minimum normal traction $\nu_{min} = \sigma^{(3)}$ acts on the surface whose unit normal is $\hat{n}^{(3)}$, and the maximum shear traction $|\tau|_{max} = |\tau^{(P)}|_{max} = \frac{1}{2}(\sigma^{(1)} - \sigma^{(3)})$ acts on the surface whose unit outward normal is $\pm(\hat{n}^{(1)} + \hat{n}^{(3)})/\sqrt{2}$ in the direction $\pm(\hat{n}^{(1)} - \hat{n}^{(3)})/\sqrt{2}$ and also acts on the surface whose unit outward normal is $\pm(\hat{n}^{(1)} - \hat{n}^{(3)})/\sqrt{2}$ in the direction $\pm(\hat{n}^{(1)} + \hat{n}^{(3)})/\sqrt{2}$.

Example 4.3
A body composed of a brittle material is to be subjected to a homogeneous state of stress[2] given by

$$[\sigma] = \begin{bmatrix} 50 & 20 & 100 \\ 20 & 200 & -50 \\ 100 & -50 & 0 \end{bmatrix} \text{ (MPa)} .$$

The material has an experimentally determined uniaxial failure stress of $\sigma_f = 225$ MPa. Using the maximum principal stress criterion, predict whether or not the body will fail.

Solution. Solving the characteristic value problem numerically (there are several commercial software packages on the market and some modern calculators that make quick work of this), one finds that the principal stresses are

$$\sigma^{(1)} = 211.91 \text{ MPa} , \quad \sigma^{(2)} = 125.51 \text{ MPa} , \quad \sigma^{(3)} = -87.418 \text{ MPa}$$

---

[2] A homogeneous state of stress is one in which the stress is independent of position, that is, the same at all points in a body.

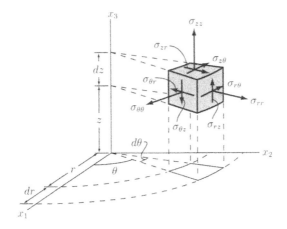

FIGURE 4.11. *Components of stress in cylindrical coordinates.*

and the corresponding principal directions of stress are

$$[\hat{\mathbf{n}}^{(1)}] = \begin{bmatrix} 0.0304 \\ -0.9695 \\ 0.2431 \end{bmatrix}, \quad [\hat{\mathbf{n}}^{(2)}] = \begin{bmatrix} -0.8025 \\ -0.1686 \\ -0.5723 \end{bmatrix}, \quad [\hat{\mathbf{n}}^{(3)}] = \begin{bmatrix} 0.5958 \\ -0.1777 \\ -0.7824 \end{bmatrix}.$$

Since $\sigma^{(1)} < \sigma_\mathrm{f}$, the maximum principle stress criterion predicts that the body will not fail.

4.4 Cylindrical and Spherical Coordinates

4.4.1 Cylindrical Coordinates

The scalar components of traction and stress in cylindrical coordinates are related to those in Cartesian coordinates by the transformation tensor (2.5.12):

$$t'_i = \ell_{ij} t_j \,, \qquad\qquad [\mathbf{t}]' = [L][\mathbf{t}] \,, \qquad\qquad (4.4.1)$$

$$\sigma'_{ij} = \ell_{im} \ell_{jn} \sigma_{mn} \,, \qquad [\boldsymbol{\sigma}]' = [L][\boldsymbol{\sigma}][L]^\mathsf{T} \,, \qquad\qquad (4.4.2)$$

where the primed components are in cylindrical coordinates and their indices 1, 2, and 3 are understood to represent r, θ, and z, respectively. The traction components on the coordinates faces are shown in Figure 4.11.

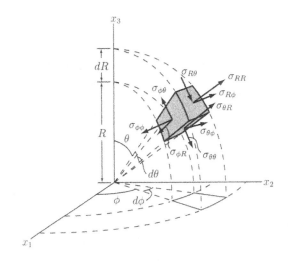

FIGURE 4.12. Components of stress in spherical coordinates.

Tensor notation equations hold in all coordinate systems, so the traction-stress relation $\mathbf{t} = \hat{\mathbf{n}} \cdot \boldsymbol{\sigma}$ in cylindrical coordinates gives

$$\begin{aligned}
t_r &= \sigma_{rr}\hat{n}_r + \sigma_{\theta r}\hat{n}_\theta + \sigma_{zr}\hat{n}_z \,, \\
t_\theta &= \sigma_{r\theta}\hat{n}_r + \sigma_{\theta\theta}\hat{n}_\theta + \sigma_{z\theta}\hat{n}_z \,, \\
t_z &= \sigma_{rz}\hat{n}_r + \sigma_{\theta z}\hat{n}_\theta + \sigma_{zz}\hat{n}_z \,,
\end{aligned}$$

(4.4.3)

and the equilibrium equation $\boldsymbol{\nabla} \cdot \boldsymbol{\sigma}^\mathsf{T} + \mathbf{f} = \mathbf{0}$, with the divergence of a second-order tensor in cylindrical coordinates given by (2.5.25), gives

$$\begin{aligned}
\frac{\partial \sigma_{rr}}{\partial r} + \frac{1}{r}\frac{\partial \sigma_{\theta r}}{\partial \theta} + \frac{\partial \sigma_{zr}}{\partial z} + \frac{1}{r}(\sigma_{rr} - \sigma_{\theta\theta}) + f_r &= 0 \,, \\
\frac{\partial \sigma_{r\theta}}{\partial r} + \frac{1}{r}\frac{\partial \sigma_{\theta\theta}}{\partial \theta} + \frac{\partial \sigma_{z\theta}}{\partial z} + \frac{2}{r}\sigma_{r\theta} + f_\theta &= 0 \,, \\
\frac{\partial \sigma_{rz}}{\partial r} + \frac{1}{r}\frac{\partial \sigma_{\theta z}}{\partial \theta} + \frac{\partial \sigma_{zz}}{\partial z} + \frac{1}{r}\sigma_{rz} + f_z &= 0 \,.
\end{aligned}$$

(4.4.4)

4.4.2 Spherical Coordinates

The scalar components of traction and stress in spherical coordinates are related to those in Cartesian coordinates by the transformation tensor (2.5.42):

$$t'_i = \ell_{ij}t_j \,, \qquad\qquad [\mathbf{t}]' = [L][\mathbf{t}] \,,$$

(4.4.5)

$$\sigma'_{ij} = \ell_{im}\ell_{jn}\sigma_{mn} \,, \quad [\boldsymbol{\sigma}]' = [L][\boldsymbol{\sigma}][L]^\mathsf{T} \,,$$

(4.4.6)

where the primed components are in spherical coordinates and their indices 1, 2, and 3 are understood to represent R, θ, and ϕ, respectively. The traction components on the coordinates faces are shown in Figure 4.12. Tensor notation equations hold in all coordinate systems, so the traction-stress relation $\mathbf{t} = \hat{\mathbf{n}} \cdot \boldsymbol{\sigma}$ in spherical coordinates gives

$$
\begin{aligned}
t_R &= \sigma_{RR}\hat{n}_R + \sigma_{\theta R}\hat{n}_\theta + \sigma_{\phi R}\hat{n}_\phi , \\
t_\theta &= \sigma_{R\theta}\hat{n}_R + \sigma_{\theta\theta}\hat{n}_\theta + \sigma_{\phi\theta}\hat{n}_\phi , \\
t_\phi &= \sigma_{R\phi}\hat{n}_R + \sigma_{\theta\phi}\hat{n}_\theta + \sigma_{\phi\phi}\hat{n}_\phi
\end{aligned}
\tag{4.4.7}
$$

and the equilibrium equations $\nabla \cdot \boldsymbol{\sigma}^{\mathsf{T}} + \mathbf{f} = \mathbf{0}$, with the divergence of a second-order tensor in spherical coordinates given by (2.5.55), gives

$$
\begin{aligned}
&\frac{\partial \sigma_{RR}}{\partial R} + \frac{1}{R}\frac{\partial \sigma_{\theta R}}{\partial \theta} + \frac{1}{R\sin\theta}\frac{\partial \sigma_{\phi R}}{\partial \phi} \\
&\qquad + \frac{1}{R}(2\sigma_{RR} - \sigma_{\theta\theta} - \sigma_{\phi\phi} + \sigma_{\theta R}\cot\theta) + f_R = 0 , \\
&\frac{\partial \sigma_{R\theta}}{\partial R} + \frac{1}{R}\frac{\partial \sigma_{\theta\theta}}{\partial \theta} + \frac{1}{R\sin\theta}\frac{\partial \sigma_{\phi\theta}}{\partial \phi} \\
&\qquad + \frac{1}{R}[(\sigma_{\theta\theta} - \sigma_{\phi\phi})\cot\theta + 3\sigma_{R\theta}] + f_\theta = 0 , \\
&\frac{\partial \sigma_{R\phi}}{\partial R} + \frac{1}{R}\frac{\partial \sigma_{\theta\phi}}{\partial \theta} + \frac{1}{R\sin\theta}\frac{\partial \sigma_{\phi\phi}}{\partial \phi} \\
&\qquad + \frac{1}{R}(2\sigma_{\theta\phi}\cot\theta + 3\sigma_{R\phi}) + f_\phi = 0 .
\end{aligned}
\tag{4.4.8}
$$

Problems

4.1 Derive the equations of motion (4.2.26) and (4.2.27) that are necessary and sufficient local conditions for the Cauchy stress tensor to satisfy the balance of linear and angular momentum.

4.2 Consider a homogeneous (kinematically infinitesimal) stress field with matrix of scalar components in the vector basis $\{\hat{\mathbf{e}}_i\}$ given by

$$
[\sigma] = \begin{bmatrix} 3 & 1 & 1 \\ 1 & 0 & 2 \\ 1 & 2 & 0 \end{bmatrix} \text{ (MPa)} ,
$$

where the units of force are Newtons (N) and the units of stress are Pascals (1 MPa $= 10^6$ Pa $= 10^6$ N/m^2).

(a) Determine the traction and the normal and projected shear tractions acting on a surface with unit outward normal $\hat{n} = (\hat{e}_2 + \hat{e}_3)/\sqrt{2}$.

(b) Determine the principal stresses and the principal directions of stress.

4.3 Consider a homogeneous first Piola-Kirchhoff stress field with known scalar components $S_{21} = A$, $S_{22} = B$, and $S_{11} = S_{33} = S_{31} = S_{32} = 0$, and suppose the accompanying motion is a simple shear where the mapping $x_i = \chi_i(X_1, X_2, X_3)$ for the Cartesian coordinates of a material point in the current configuration in terms of its Cartesian coordinates in the reference configuration is given by

$$x_1 = X_1 + \kappa X_2 , \quad x_2 = X_2 , \quad x_3 = X_3 ,$$

where κ is a constant.

(a) Determine the remaining scalar components S_{12}, S_{23}, and S_{13} of the first Piola-Kirchhoff stress field that are required to satisfy the angular equation of motion (4.1.38).

(b) Determine the corresponding second Piola-Kirchhoff and Cauchy stress fields and show that all three stress fields are approximately the same for infinitesimal deformation.

(c) For a body which, in the reference configuration $\mathcal{R}$, occupies the unit cube $(0 \leq X_1 \leq 1, 0 \leq X_2 \leq 1, 0 \leq X_3 \leq 1)$, determine the resultant force vectors acting on each of the cubic faces and illustrate your results on a sketch of the current configuration of the body.

4.4 Consider a (kinematically infinitesimal) stress field whose matrix of scalar components in the vector basis $\{\hat{e}_i\}$ is

$$[\sigma] = \begin{bmatrix} 6X_1X_3^2 & 0 & -2X_3^3 \\ 0 & 1 & 2 \\ -2X_3^3 & 2 & 3X_1^2 \end{bmatrix} \text{ (MPa)} ,$$

where the Cartesian coordinate variables X_i are in meters (m).

(a) Assuming body forces are negligible, does this stress field satisfy the equilibrium equations?

(b) Determine the traction vector acting at a point $\mathbf{X} = 2\hat{e}_1 + 3\hat{e}_2 + 2\hat{e}_3$ (m) on the plane $2X_1 + X_2 - X_3 = 5$ m. Note that the unit normal to a plane defined by $a_i X_i = b$ is

$$\hat{n} = \pm \frac{a_i \hat{e}_i}{\sqrt{a_j a_j}} .$$

(c) Determine the normal and projected shear tractions acting at this point on this plane.

(d) Determine the principal stresses and principal directions of stress at this point.

4.5 Consider a (kinematically infinitesimal) stress field whose matrix of scalar components in the vector basis $\{\hat{e}_i\}$ is

$$[\sigma] = \begin{bmatrix} 1 & 0 & 2X_2 \\ 0 & 1 & 4X_1 \\ 2X_2 & 4X_1 & 1 \end{bmatrix} \times 10^3 \text{ (psi)} ,$$

where the Cartesian coordinate variables X_i are in inches (in.) and the units of stress are pounds per square inch (psi).

(a) Assuming body forces are negligible, does this stress field satisfy the equilibrium equations?

(b) Determine the traction vector acting at a point $\mathbf{X} = \hat{e}_1 + 2\hat{e}_2 + 3\hat{e}_3$ (in.) on the plane $X_1 + X_2 + X_3 = 6$ in.

(c) Determine the normal and projected shear tractions acting at this point on this plane.

(d) Determine the principal stresses and principal directions of stress at this point.

4.6 A homogeneous (kinematically infinitesimal) stress field in a body is given, in the vector basis $\{\hat{e}_i\}$, by the matrix of scalar components

$$[\sigma] = \begin{bmatrix} \sigma_{11} & 2 & 1 \\ 2 & 0 & 2 \\ 1 & 2 & 0 \end{bmatrix} \text{ (MPa)} .$$

Determine a value of σ_{11} such that there will be traction-free planes in the body and determine the unit normal $\hat{n}$ to the traction-free planes.

4.7 A plane with unit normal equally inclined to the three principal directions of stress is known as an *octahedral plane*. For a given state of stress, there are eight such planes. Show that the normal and projected shear tractions on an octahedral plane are given by

$$\nu_{\text{oct}} = \frac{1}{3}(\sigma^{(1)} + \sigma^{(2)} + \sigma^{(3)}) = \frac{1}{3}I_\sigma ,$$

$$\tau_{\text{oct}}^{(P)} = \frac{1}{3}\sqrt{(\sigma^{(1)} - \sigma^{(2)})^2 + (\sigma^{(2)} - \sigma^{(3)})^2 + (\sigma^{(3)} - \sigma^{(1)})^2}$$

$$= \frac{1}{3}\sqrt{2I_\sigma^2 - 6II_\sigma} ,$$

where $\sigma^{(1)}$, $\sigma^{(2)}$, and $\sigma^{(3)}$ are the principal stresses and I_σ and II_σ are the first two principal invariants of stress.

4.8 According to the plane stress theory of linearized elasticity (to be discussed later), the stress field in a semi-infinite plate occupying the region $(X_1 \geq 0, -\infty < X_2 < \infty, 0 \leq X_3 \leq d)$ with a concentrated normal edge load of magnitude $P = pd$, as shown below, has Cartesian scalar components $\sigma_{31} = \sigma_{32} = \sigma_{33} = 0$ and

$$\sigma_{11} = -\frac{2p\cos^3\theta}{\pi r},$$

$$\sigma_{22} = -\frac{2p\sin^2\theta\cos\theta}{\pi r},$$

$$\sigma_{12} = -\frac{2p\sin\theta\cos^2\theta}{\pi r},$$

where r and θ are cylindrical coordinates.

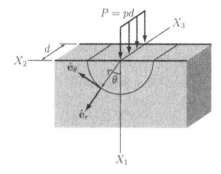

(a) Determine the scalar components of stress in the cylindrical coordinate vector basis.

(b) Derive an expression for the traction vector acting at points on the semicylindrical surface $r = a$.

(c) Show that the traction distribution on the semicylindrical surface $r = a$ is in equilibrium with the edge load, for arbitrary a. To do this, you will need to calculate the resultant forces and moments due to the traction distribution by integrating over the surface.

4.9 If the (kinematically infinitesimal) stress field in a body has a matrix of scalar components in the vector basis $\{\hat{e}_i\}$ given by

$$[\sigma] = B \begin{bmatrix} X_1^2 X_2 & (b^2 - X_2^2)X_1 & 0 \\ (b^2 - X_2^2)X_1 & \frac{1}{3}(X_2^2 - 3b^2)X_2 & 0 \\ 0 & 0 & 2bX_3^2 \end{bmatrix},$$

where B and b are constants, find the body force field necessary for the body to be in equilibrium.

Chapter 5

Constitutive Equations

The equations of motion, given for instance in a referential formulation by (4.1.33) and (4.1.38), are generally insufficient to determine the stress field in a body. There are nine scalar components of stress and only six scalar equations of motion. The equations of motion do not directly relate the deformation of a body to the external forces applied to the body either. What is lacking are a statement of the particular boundary conditions of a problem (discussed in Chapter 6) and a consideration of the mechanical properties associated with the particular material that constitutes the body, relating deformation and stress in the form of *constitutive equations*.

Constitutive equations are *mathematical models* for the behavior of real materials. The literature on the subject is vast and encompasses constitutive models for solids and fluids, for elasticity, viscoelasticity, and plasticity, for Newtonian fluids and non-Newtonian fluids, and so forth. Although the possibilities are endless, any worthwhile constitutive model must be in reasonable agreement with physical observations. Of considerable importance in this regard is the requirement that constitutive models satisfy the principle of material frame-indifference, as briefly discussed below. For more on this and other aspects of the general theory of constitutive models, see, for example, Malvern (1969), Ogden (1984), or Truesdell and Noll (1992).

5.1 Elasticity

Assume that the current stress at a material point is uniquely determined by the *local* history of the deformation at that material point. A material for which this is true is said to be a *simple material*. Thus, for example, the current value of the Cauchy stress tensor $\mathbf{T}$ at a material point can be expressed as a function of the history of the deformation gradient tensor $\mathbf{F}$ at that point. The available experimental evidence indicates that nearly all real materials may be modeled as simple materials.

An *elastic material* is defined as one in which the change in stress at a
material point in a body, between two arbitrary configurations of the body,
is independent of the time taken in going from one configuration to the
other and of the path followed in the space of all possible configurations.
This means that, in an elastic material, the current stress at a material
point is uniquely determined by the current local deformation at that material point. Thus, for example, in an elastic material, the Cauchy stress
tensor $\mathbf{T}$ can be given as a symmetric, second-order tensor-valued function
of the deformation gradient tensor $\mathbf{F}$,

$$\mathbf{T} = \mathbf{f}[\mathbf{F}(\mathbf{x}, t), \mathbf{x}] \,, \tag{5.1.1}$$

where the dependence of the mapping $\mathbf{f}(\cdot, \mathbf{x})$ on the current position $\mathbf{x}$ allows for the possibility that material properties may be different at different
material points in a body. If the material properties are the same at every
material point in a body, so that the constitutive relation (5.1.1) does not
depend explicitly on position,

$$\mathbf{T} = \mathbf{f}(\mathbf{F}) \,, \tag{5.1.2}$$

then the body is said to be *materially homogeneous*. Material homogeneity
will be assumed from now on, unless stated otherwise.

Whether or not any materials are truly elastic is a matter of debate,
but many materials can be effectively *modeled* as elastic materials under
certain circumstances. For instance, metallic materials such as steel can be
modeled as elastic materials so long as the relative crystal lattice positions
of their constituent molecules do not change (or at least most of them do
not change). If these positions do change (through the motion of dislocations), then *plastic deformation* is said to have occurred and the stress at a
material point will depend on the history of the deformation. The former
state of affairs is sometimes referred to as *elastic deformation*. Other materials such as rubber, wood, concrete, ceramic, and fiber composites are
often modeled as elastic materials.

5.1.1   Material Frame-Indifference

Suppose an observer records an event as occurring at a place $\mathbf{x}$ in (a three-
dimensional Euclidean point) space and at time t. Such an observation
always occurs with respect to the particular observer's own *frame of reference*. For instance, if one thinks of an observer as a person, then associated
with that person are an intrinsic measure of the place at which an event
occurs, such as given by the directions left, right, up, and down and the
distance from the person, and an intrinsic measure of the time at which
an event occurs, such as that given by a personal watch. Thus, a second
observer, with a different frame of reference, will observe the same event

as occurring at a different place $\mathbf{x}^*$ and at a different time t^*. The terms observer and frame of reference are treated here as being synonymous.

The concept of frames of reference should not be confused with that of coordinate systems, as they are not the same thing at all. A given observer is free to define any coordinate system as may be convenient to characterize the place $\mathbf{x}$ at which an event is seen to occur, with the observer at the origin say, but this coordinate system is not intrinsic to the observer. Invariance with respect to changes in an observer's choice of coordinate system was previously seen to be the basis of tensor analysis. Questions of invariance with respect to changes of observer (i.e., frame-indifference) are briefly discussed below.

If relativistic effects are negligible (all relevant velocities are much less than the speed of light) and it is assumed that all observers record the same distance between any two simultaneous events and the same time interval between any two sequential events, then it can be shown that the most general transformation between the observations $\{\mathbf{x}, t\}$ and $\{\mathbf{x}^*, t^*\}$ of the same event by two different observers is given by

$$\mathbf{x}^* = \mathbf{Q}(t) \cdot \mathbf{x} + \mathbf{c}(t) , \quad t^* = t - a , \tag{5.1.3}$$

where $\mathbf{Q}(t)$ is a proper orthogonal tensor-valued function of time characterizing the relative rotation of one observer with respect to the other, $\mathbf{c}(t)$ is a vector-valued function of time characterizing the relative translation of one observer with respect to the other, and a is a constant time difference between the observers' measures of time. The time difference a has no bearing on what follows, so one may just as well set $a = 0$. Also, for notational simplicity, from now on the explicit dependence of $\mathbf{Q}$ and $\mathbf{c}$ on time is understood but will not be indicated.

Consider two simultaneous events recorded as $\{\mathbf{x}, t\}$ and $\{\mathbf{x}_0, t\}$ by one observer and as $\{\mathbf{x}^*, t\}$ and $\{\mathbf{x}_0^*, t\}$ by a second observer. The relative position vector of the place of the first event relative to the place of the second event is seen by the first observer as $\mathbf{r} = \mathbf{x} - \mathbf{x}_0$ and is seen by the second observer as $\mathbf{r}^* = \mathbf{x}^* - \mathbf{x}_0^*$, where it follows from (5.1.3) that

$$\mathbf{r}^* = \mathbf{Q} \cdot \mathbf{r} . \tag{5.1.4}$$

Note that since $\mathbf{Q}$ is orthogonal,

$$\mathbf{r}^* \cdot \mathbf{r}^* = (\mathbf{Q} \cdot \mathbf{r}) \cdot (\mathbf{Q} \cdot \mathbf{r}) = \mathbf{r} \cdot (\mathbf{Q}^\mathsf{T} \cdot \mathbf{Q}) \cdot \mathbf{r} = \mathbf{r} \cdot \mathbf{r} , \tag{5.1.5}$$

thus showing that $|\mathbf{r}^*| = |\mathbf{r}|$; the transformation (5.1.4) is analogous to a rigid-body rotation. Vectors which transform under changes of observer like the relative position vector in (5.1.4) are said to be *frame-indifferent*, or *objective*, vectors.

Consider two frame-indifferent vectors, viewed by one observer as $\mathbf{a}$ and $\mathbf{b}$, that are related by a second-order tensor viewed by the same observer as $\mathbf{A}$ such that $\mathbf{a} = \mathbf{A} \cdot \mathbf{b}$. Then, a second observer views the two vectors as $\mathbf{a}^* = \mathbf{Q} \cdot \mathbf{a}$ and $\mathbf{b}^* = \mathbf{Q} \cdot \mathbf{b}$ and views the second-order tensor as $\mathbf{A}^*$, such that $\mathbf{a}^* = \mathbf{A}^* \cdot \mathbf{b}^*$. To find the transformation from $\mathbf{A}$ to $\mathbf{A}^*$, note that

$$\mathbf{a}^* = \mathbf{Q} \cdot \mathbf{a} = \mathbf{Q} \cdot \mathbf{A} \cdot \mathbf{b} = \mathbf{Q} \cdot \mathbf{A} \cdot \mathbf{Q}^\mathsf{T} \cdot \mathbf{b}^* \tag{5.1.6}$$

so that

$$\mathbf{A}^* = \mathbf{Q} \cdot \mathbf{A} \cdot \mathbf{Q}^\mathsf{T} . \tag{5.1.7}$$

Second-order tensors that transform under changes of observer according to (5.1.7) are said to be frame-indifferent second-order tensors.

Consider next the motion of a body as seen by two observers. Assume that each observer specifies the same reference configuration of the body, as seen in their own frame of reference, so that the reference position $\mathbf{X}$ refers to the same material point for each observer. Then, the mappings $\mathbf{x} = \chi(\mathbf{X}, t)$ and $\mathbf{x}^* = \chi^*(\mathbf{X}, t)$ from the reference configuration to the current configuration of the body in the two frames of reference are different and are related by (5.1.3). It follows that the material velocities $\mathbf{v} = D\mathbf{x}/Dt$ and $\mathbf{v}^* = D\mathbf{x}^*/Dt$ as seen by the two observers are related by

$$\mathbf{v}^* = \mathbf{Q} \cdot \mathbf{v} + \frac{D\mathbf{Q}}{Dt} \cdot \mathbf{x} + \frac{D\mathbf{c}}{Dt} , \tag{5.1.8}$$

and the material accelerations $\mathbf{a} = D\mathbf{v}/Dt$ and $\mathbf{a}^* = D\mathbf{v}^*/Dt$ as seen by the two observers are related by

$$\mathbf{a}^* = \mathbf{Q} \cdot \mathbf{a} + 2\frac{D\mathbf{Q}}{Dt} \cdot \mathbf{v} + \frac{D^2\mathbf{Q}}{Dt^2} \cdot \mathbf{x} + \frac{D^2\mathbf{c}}{Dt^2} . \tag{5.1.9}$$

Thus, it is seen that the material velocity and acceleration vectors are not frame-indifferent. For instance, the velocity of a material point may be zero in one frame of reference and not in another. Finally, noting that

$$\mathrm{Grad}\,\mathbf{x}^* = \mathrm{Grad}(\mathbf{Q} \cdot \mathbf{x} + \mathbf{c}) = \mathbf{Q} \cdot \mathrm{Grad}\,\mathbf{x} , \tag{5.1.10}$$

it follows that the deformation gradient tensors $\mathbf{F} = \mathrm{Grad}\,\mathbf{x}$ and $\mathbf{F}^* = \mathrm{Grad}\,\mathbf{x}^*$ as seen by the two observers are related by

$$\mathbf{F}^* = \mathbf{Q} \cdot \mathbf{F} . \tag{5.1.11}$$

Thus, the deformation gradient tensor is not a frame-indifferent second-order tensor, but rather it transforms like a frame-indifferent vector. Note that the Jacobians of the motion as seen by the two observers are the same: $J^* = \det \mathbf{F}^* = \det \mathbf{F} = J$.

The unit outward normal vector $\hat{\mathbf{n}}$ to the current configuration of a material surface is a frame-indifferent vector and the true traction vector

t acting at a point on a material surface is assumed to also be frame-indifferent. Thus, since the Cauchy stress tensor **T** is related to these two frame-indifferent vectors by $\mathbf{t} = \hat{\mathbf{n}} \cdot \mathbf{T}$, it follows that the Cauchy stress tensor is a frame-indifferent second-order tensor that transforms under changes of observer according to (5.1.7).

The *principle of material frame-indifference* says that if a constitutive equation is true for one observer, then it must be true for all other observers. A frequently given example of this principle [see Truesdell and Noll (1992) or Ogden (1984)] involves the relation between force and extension in a simple spring and its use in measuring centrifugal force. A demonstration of the validity of the equations relating angular velocity to centrifugal force can be performed wherein a spring is first calibrated in a frame of reference at rest relative to the surface of the Earth and is then used to attach a known mass to an anchor in the center of a horizontal table. When the table is given a constant angular velocity about an axis through the anchor point, the centrifugal force acting on the mass can be determined by recording the extension of the spring and using the previously determined relation between the force and the spring's extension. It is implicitly assumed in this demonstration that the force/extension relation for the spring is the same when it is rotated as when it is stationary. This is an example of material frame-indifference.

For a constitutive equation such as (5.1.2), the principle of material frame-indifference says that if this equation is satisfied for the stress **T** and deformation gradient tensor **F** as seen by one observer, then it must be satisfied by these same quantities as they are seen by all other observers, so that for arbitrary proper orthogonal **Q**,

$$\mathbf{T}^* = \mathbf{f}(\mathbf{F}^*) , \tag{5.1.12}$$

where $\mathbf{T}^* = \mathbf{Q} \cdot \mathbf{T} \cdot \mathbf{Q}^\mathsf{T}$ and $\mathbf{F}^* = \mathbf{Q} \cdot \mathbf{F}$. It follows that the constitutive function $\mathbf{f}(\mathbf{F})$ in (5.1.2) will satisfy the principle of material frame-indifference if and only if

$$\mathbf{f}(\mathbf{Q} \cdot \mathbf{F}) = \mathbf{Q} \cdot \mathbf{f}(\mathbf{F}) \cdot \mathbf{Q}^\mathsf{T} \tag{5.1.13}$$

for all proper orthogonal **Q** and arbitrary deformation gradients **F**. Thus, the principle of material frame-indifference imposes restrictions on the possible forms of $\mathbf{f}(\mathbf{F})$.

There are many equivalent forms of the constitutive equation (5.1.2) involving various measures of stress and strain, and each form will have restrictions imposed upon it by the principle of material frame-indifference that are related to (5.1.13). For example, a form of the constitutive equation that will be useful in the following is the second Piola-Kirchhoff stress tensor $\widetilde{\mathbf{S}}$ given as a symmetric, second-order tensor-valued function of the

Green-St. Venant strain tensor $\mathbf{E}$,

$$\widetilde{\mathbf{S}} = \mathbf{g}(\mathbf{E}) . \tag{5.1.14}$$

To see what restrictions are imposed on $\mathbf{g}(\mathbf{E})$ by the principle of material frame-indifference, one needs first to determine how $\widetilde{\mathbf{S}}$ and $\mathbf{E}$ transform under changes of observer. Since $\mathbf{E} = \frac{1}{2}(\mathbf{F}^\mathsf{T} \cdot \mathbf{F} - \mathbf{I})$ and noting from (5.1.11) that

$$\mathbf{F}^{*\mathsf{T}} \cdot \mathbf{F}^* = \mathbf{F}^\mathsf{T} \cdot \mathbf{Q}^\mathsf{T} \cdot \mathbf{Q} \cdot \mathbf{F} = \mathbf{F}^\mathsf{T} \cdot \mathbf{F} , \tag{5.1.15}$$

it follows that

$$\mathbf{E}^* = \frac{1}{2}(\mathbf{F}^{*\mathsf{T}} \cdot \mathbf{F}^* - \mathbf{I}) = \mathbf{E} . \tag{5.1.16}$$

It can similarly be shown that $\widetilde{\mathbf{S}}^* = \widetilde{\mathbf{S}}$. Thus, the second Piola-Kirchhoff stress tensor and the Green-St. Venant strain tensor appear the same to all observers. Hence, the principle of material frame-indifference imposes no restrictions on the possible forms of $\mathbf{g}(\mathbf{E})$.

The constitutive equation (5.1.14) for elastic materials can be restricted by assuming that the *strain energy density* $\mathcal{U}$, defined as a function $\mathcal{U} = \mathcal{U}(\mathbf{E})$ of the Green-St. Venant strain tensor by

$$\mathcal{U}(\mathbf{E}) \equiv \int_{\mathbf{0}}^{\mathbf{E}} \widetilde{\mathbf{S}} : d\mathbf{E} , \quad \mathcal{U}(\mathbf{E}) \equiv \int_{\mathbf{0}}^{\mathbf{E}} \widetilde{S}_{ij} \, dE_{ij} , \tag{5.1.17}$$

is independent of the integration path followed in strain space. Then the material is said to be *hyperelastic* and it follows from the definition (5.1.17) that the constitutive equation (5.1.14) is given in an orthonormal basis by

$$\widetilde{S}_{ij} = \frac{\partial \mathcal{U}(\mathbf{E})}{\partial E_{ij}} . \tag{5.1.18}$$

It will subsequently be shown (Chapter 10) that, given certain assumptions regarding mechanical and thermal processes, the strain energy density (5.1.17) can be interpreted as the stored mechanical energy per unit volume at a material point.

5.1.2 Kinematic Linearization

Recall that for infinitesimal deformation ($\|\mathrm{Grad}\,\mathbf{u}\| \ll 1$), one can impose the kinematic linearizations $\mathbf{E} \approx \boldsymbol{\varepsilon}$ and $\widetilde{\mathbf{S}} \approx \boldsymbol{\sigma}$, where $\boldsymbol{\varepsilon}$ is the infinitesimal strain tensor and $\boldsymbol{\sigma}$ is the (kinematically infinitesimal) stress tensor. It follows that for infinitesimal deformation of an elastic material, the constitutive equation (5.1.14) reduces to

$$\boldsymbol{\sigma} = \mathbf{h}(\boldsymbol{\varepsilon}) , \tag{5.1.19}$$

where $\mathbf{h}(\boldsymbol{\varepsilon})$ is a symmetric, second-order tensor-valued function. For infinitesimal deformation of a hyperelastic material, (5.1.17) and (5.1.18) reduce to

$$\mathcal{U}(\boldsymbol{\varepsilon}) \equiv \int_0^{\boldsymbol{\varepsilon}} \boldsymbol{\sigma} : d\boldsymbol{\varepsilon}\ , \quad \mathcal{U}(\boldsymbol{\varepsilon}) \equiv \int_0^{\boldsymbol{\varepsilon}} \sigma_{ij}\, d\varepsilon_{ij}\ , \qquad (5.1.20)$$

and

$$\sigma_{ij} = \frac{\partial \mathcal{U}(\boldsymbol{\varepsilon})}{\partial \varepsilon_{ij}}\ . \qquad (5.1.21)$$

If a body is not materially homogeneous, then the functions $\mathbf{h}$ and $\mathcal{U}$ will also have the reference position $\mathbf{X}$ as an independent variable.

5.2 Constitutive Linearization

In general, the constitutive equation (5.1.19) for infinitesimal deformation of an elastic material is nonlinear—the stress need not be proportional to the strain. However, if the components of stress are assumed to be linear functions of the components of strain, then the most general form of the constitutive equation is[1]

$$\boldsymbol{\sigma} = \mathbf{C} : \boldsymbol{\varepsilon}\ , \quad \sigma_{ij} = \mathsf{C}_{ijkl}\varepsilon_{kl}\ , \qquad (5.2.1)$$

where the scalar components of the fourth-order *stiffness tensor* $\mathbf{C}$ are material parameters, the values of which may depend on the reference position $\mathbf{X}$ if the body is not materially homogeneous. This constitutive *model* for infinitesimal deformation of a *linear elastic material* is sometimes referred to as the *generalized Hooke's law*.

There are in general $3^4 = 81$ scalar components of a fourth-order tensor, but the number of *independent* scalar components of the stiffness tensor is considerably less due to symmetry.

- Recall that the stress tensor must be symmetric, $\sigma_{ij} = \sigma_{ji}$, in order to satisfy the angular equation of motion. It follows then, from (5.2.1), that the angular equation of motion will be satisfied if and only if the stiffness tensor is symmetric in its first two indices, $\mathsf{C}_{ijkl} = \mathsf{C}_{jikl}$, thereby reducing the number of independent stiffness components to $6 \times 3 \times 3 = 54$.

---

[1]It is tacitly assumed that stresses vanish in the reference configuration (in which $\boldsymbol{\varepsilon} = \mathbf{0}$ by definition). If required, a residual stress state $\boldsymbol{\sigma}_0$ may be superposed on (5.2.1), giving $\boldsymbol{\sigma} = \mathbf{C} : \boldsymbol{\varepsilon} + \boldsymbol{\sigma}_0$.

- The stiffness tensor can be uniquely decomposed into components
  that are symmetric and skew-symmetric in its last two indices as
  follows:

$$C_{ijkl} = \frac{1}{2}(C_{ijkl} + C_{ijlk}) + \frac{1}{2}(C_{ijkl} - C_{ijlk}) . \tag{5.2.2}$$

However, since the strain tensor is symmetric by construction, $\varepsilon_{ij} = \varepsilon_{ji}$, only the symmetric part of the stiffness tensor plays any role in the constitutive equation (5.2.1):

$$C_{ijkl}\varepsilon_{kl} = \frac{1}{2}(C_{ijkl}\varepsilon_{kl} + C_{ijlk}\varepsilon_{lk}) = \frac{1}{2}(C_{ijkl} + C_{ijlk})\varepsilon_{kl} . \tag{5.2.3}$$

Thus, the skew-symmetric component of the stiffness tensor is arbitrary and, without loss of generality, one may set it equal to zero and require that the stiffness tensor be symmetric in its last two indices, so that $C_{ijkl} = C_{ijlk}$. This reduces the number of independent stiffness components to $6 \times 6 = 36$.

- If the linear elastic material is also hyperelastic, then it follows from
  (5.1.21) and (5.2.1) that

$$\frac{\partial \mathcal{U}(\varepsilon)}{\partial \varepsilon_{ij}} = C_{ijrs}\varepsilon_{rs} . \tag{5.2.4}$$

Differentiating (5.2.4) with respect to ε_{kl} yields

$$\frac{\partial^2 \mathcal{U}(\varepsilon)}{\partial \varepsilon_{ij}\partial \varepsilon_{kl}} = C_{ijkl} . \tag{5.2.5}$$

The order of differentiation in (5.2.5) is arbitrary, so the stiffness tensor for infinitesimal deformation of a linear hyperelastic material must have the symmetry $C_{ijkl} = C_{klij}$, finally reducing the number of independent stiffness components to $\frac{1}{2}(6 \times 6 + 6) = 21$. Conversely, it can similarly be shown that a linear elastic material is also hyperelastic if $C_{ijkl} = C_{klij}$. Materials will be assumed to be hyperelastic from now on.

There are, in general, 21 independent material parameters required to describe the constitutive behavior for infinitesimal deformation of a linear hyperelastic material, referred to from now on as *linear elastic deformation*. If the material that constitutes a body has inherent symmetry, as defined below, then it will be seen that the number of independent material parameters is further reduced.

The strain energy density for linear elastic deformation is

$$\mathcal{U}(\varepsilon) = \frac{1}{2}\varepsilon : \mathbf{C} : \varepsilon , \quad \mathcal{U}(\varepsilon) = \frac{1}{2}C_{ijkl}\varepsilon_{ij}\varepsilon_{kl} . \tag{5.2.6}$$

This can be seen by differentiating (5.2.6) and recovering (5.1.21):

$$
\begin{aligned}
\frac{\partial \mathcal{U}(\varepsilon)}{\partial \varepsilon_{rs}} &= \frac{1}{2} C_{ijkl} \varepsilon_{kl} \delta_{ir} \delta_{js} + \frac{1}{2} C_{ijkl} \varepsilon_{ij} \delta_{kr} \delta_{ls} \\
&= \frac{1}{2} C_{rskl} \varepsilon_{kl} + \frac{1}{2} C_{ijrs} \varepsilon_{ij} \\
&= \frac{1}{2} C_{rskl} \varepsilon_{kl} + \frac{1}{2} C_{rsij} \varepsilon_{ij} \\
&= \sigma_{rs} .
\end{aligned}
\tag{5.2.7}
$$

By combining (5.2.1) and (5.2.6), the strain energy density for linear elastic deformation can alternatively be expressed as

$$
\mathcal{U} = \frac{1}{2} \boldsymbol{\sigma} : \boldsymbol{\varepsilon} , \quad \mathcal{U} = \frac{1}{2} \sigma_{ij} \varepsilon_{ij} ,
\tag{5.2.8}
$$

in terms of both the stress and strain.

The constitutive equation (5.2.1) for linear elastic deformation can be inverted to obtain a constitutive equation for the stress as a function of the strain:

$$
\boldsymbol{\varepsilon} = \mathbf{S} : \boldsymbol{\sigma} , \quad \varepsilon_{ij} = S_{ijkl} \sigma_{kl} ,
\tag{5.2.9}
$$

where the fourth-order tensor **S** called the *compliance tensor*. The scalar components of **S** are determined by the 21 independent components of the stiffness tensor **C**, so there are also 21 independent scalar components of the compliance tensor. By combining (5.2.8) and (5.2.9), it is seen that the strain energy density for linear elastic deformation can be given as a function of the stress by

$$
\mathcal{U}(\boldsymbol{\sigma}) = \frac{1}{2} \boldsymbol{\sigma} : \mathbf{S} : \boldsymbol{\sigma} , \quad \mathcal{U}(\boldsymbol{\sigma}) = \frac{1}{2} S_{ijkl} \sigma_{ij} \sigma_{kl} .
\tag{5.2.10}
$$

The unfortunate convention of denoting the **S**tiffness tensor as **C** and the **C**ompliance tensor as **S** is historic and so will be followed here.

5.3 Material Symmetry

Consider two spherical, stress-free reference configurations of the same arbitrarily small material volume, where the reference configurations are related by a rigid-body rotation plus, possibly, an inversion. The mapping (transformation) from one reference configuration to the other is completely characterized by an orthogonal transformation tensor (that is improper if there is an inversion). If, through mechanical testing, it is not possible to distinguish one reference configuration from the other, then the material is said to be *symmetric* with respect to that transformation.

The orthogonal transformation from one reference configuration to another can equivalently be viewed as a change in the orthonormal basis, given by an orthogonal transformation matrix $[L]$, used to define the scalar components of the stiffness tensor. The material at a point is symmetric with respect to the transformation $[L]$ if and only if the scalar components of the stiffness tensor at that point are invariant with respect to this transformation so that

$$C_{pqrs} = \ell_{pi}\ell_{qj}\ell_{rk}\ell_{sl}C_{ijkl} \,. \tag{5.3.1}$$

A necessary outcome of material symmetry is a reduction in the number of independent stiffness components.

5.3.1 Engineering Notation

To aid in the discussion of material symmetry, one can introduce the following *engineering notation*:

$$\bar{\sigma}_1 \equiv \sigma_{11}, \quad \bar{\sigma}_2 \equiv \sigma_{22}, \quad \bar{\sigma}_3 \equiv \sigma_{33}, \quad \bar{\sigma}_4 \equiv \sigma_{23}, \quad \bar{\sigma}_5 \equiv \sigma_{31}, \quad \bar{\sigma}_6 \equiv \sigma_{12},$$
$$\bar{\varepsilon}_1 \equiv \varepsilon_{11}, \quad \bar{\varepsilon}_2 \equiv \varepsilon_{22}, \quad \bar{\varepsilon}_3 \equiv \varepsilon_{33}, \quad \bar{\varepsilon}_4 \equiv 2\varepsilon_{23}, \quad \bar{\varepsilon}_5 \equiv 2\varepsilon_{31}, \quad \bar{\varepsilon}_6 \equiv 2\varepsilon_{12}. \tag{5.3.2}$$

Consider the expansion of the index notation equation (5.2.1) for σ_{11}:

$$\begin{aligned}
\sigma_{11} = {}& C_{1111}\varepsilon_{11} + C_{1112}\varepsilon_{12} + C_{1113}\varepsilon_{13} + C_{1121}\varepsilon_{21} + C_{1122}\varepsilon_{22} \\
& + C_{1123}\varepsilon_{23} + C_{1131}\varepsilon_{31} + C_{1132}\varepsilon_{32} + C_{1133}\varepsilon_{33} \,.
\end{aligned} \tag{5.3.3}$$

Since $\varepsilon_{ij} = \varepsilon_{ji}$ and $C_{ijkl} = C_{ijlk}$, this can be rewritten as

$$\begin{aligned}
\sigma_{11} = {}& C_{1111}\varepsilon_{11} + C_{1122}\varepsilon_{22} + C_{1133}\varepsilon_{33} \\
& + 2C_{1123}\varepsilon_{23} + 2C_{1131}\varepsilon_{31} + 2C_{1112}\varepsilon_{12} \,.
\end{aligned} \tag{5.3.4}$$

Using the engineering notation (5.3.2), this becomes

$$\bar{\sigma}_1 = C_{1111}\bar{\varepsilon}_1 + C_{1122}\bar{\varepsilon}_2 + C_{1133}\bar{\varepsilon}_3 + C_{1123}\bar{\varepsilon}_4 + C_{1131}\bar{\varepsilon}_5 + C_{1112}\bar{\varepsilon}_6 \,. \tag{5.3.5}$$

By expanding the other index notation equations for components of stress as functions of components of strain, it is clear that in engineering notation, the constitutive equation can be expressed as

$$\bar{\sigma}_i = \bar{C}_{ij}\bar{\varepsilon}_j \quad (i,j = 1,\dots,6) \,, \tag{5.3.6}$$

where the range and summation conventions still apply, with the indices ranging over the values 1 through 6 and the engineering notation stiffness quantities are given by

$$
\begin{bmatrix}
\bar{C}_{11} & \bar{C}_{12} & \bar{C}_{13} & \bar{C}_{14} & \bar{C}_{15} & \bar{C}_{16} \\
\bar{C}_{21} & \bar{C}_{22} & \bar{C}_{23} & \bar{C}_{24} & \bar{C}_{25} & \bar{C}_{26} \\
\bar{C}_{31} & \bar{C}_{32} & \bar{C}_{33} & \bar{C}_{34} & \bar{C}_{35} & \bar{C}_{36} \\
\bar{C}_{41} & \bar{C}_{42} & \bar{C}_{43} & \bar{C}_{44} & \bar{C}_{45} & \bar{C}_{46} \\
\bar{C}_{51} & \bar{C}_{52} & \bar{C}_{53} & \bar{C}_{54} & \bar{C}_{55} & \bar{C}_{56} \\
\bar{C}_{61} & \bar{C}_{62} & \bar{C}_{63} & \bar{C}_{64} & \bar{C}_{65} & \bar{C}_{66}
\end{bmatrix}
$$

$$
\equiv
\begin{bmatrix}
C_{1111} & C_{1122} & C_{1133} & C_{1123} & C_{1131} & C_{1112} \\
C_{2211} & C_{2222} & C_{2233} & C_{2223} & C_{2231} & C_{2212} \\
C_{3311} & C_{3322} & C_{3333} & C_{3323} & C_{3331} & C_{3312} \\
C_{2311} & C_{2322} & C_{2333} & C_{2323} & C_{2331} & C_{2312} \\
C_{3111} & C_{3122} & C_{3133} & C_{3123} & C_{3131} & C_{3112} \\
C_{1211} & C_{1222} & C_{1233} & C_{1223} & C_{1231} & C_{1212}
\end{bmatrix} .
\tag{5.3.7}
$$

It follows from the previously established symmetry of the stiffness tensor $\mathbf{C}$ that the engineering notation stiffness matrix is symmetric, $\bar{C}_{ij} = \bar{C}_{ji}$.

The scalar components of strain are given in terms of the scalar components of stress using engineering notation by

$$
\bar{\varepsilon}_i = \bar{S}_{ij}\bar{\sigma}_j \quad (i, j = 1, \ldots, 6) ,
\tag{5.3.8}
$$

where it follows from a similar expansion of the indicial notation equations, that the engineering notation compliance quantities $\bar{S}_{ij}$ are related to the scalar components of the compliance tensor $\mathbf{S}$ by

$$
\begin{bmatrix}
\bar{S}_{11} & \bar{S}_{12} & \bar{S}_{13} & \bar{S}_{14} & \bar{S}_{15} & \bar{S}_{16} \\
\bar{S}_{21} & \bar{S}_{22} & \bar{S}_{23} & \bar{S}_{24} & \bar{S}_{25} & \bar{S}_{26} \\
\bar{S}_{31} & \bar{S}_{32} & \bar{S}_{33} & \bar{S}_{34} & \bar{S}_{35} & \bar{S}_{36} \\
\bar{S}_{41} & \bar{S}_{42} & \bar{S}_{43} & \bar{S}_{44} & \bar{S}_{45} & \bar{S}_{46} \\
\bar{S}_{51} & \bar{S}_{52} & \bar{S}_{53} & \bar{S}_{54} & \bar{S}_{55} & \bar{S}_{56} \\
\bar{S}_{61} & \bar{S}_{62} & \bar{S}_{63} & \bar{S}_{64} & \bar{S}_{65} & \bar{S}_{66}
\end{bmatrix}
$$

$$
\equiv
\begin{bmatrix}
S_{1111} & S_{1122} & S_{1133} & 2S_{1123} & 2S_{1131} & 2S_{1112} \\
S_{2211} & S_{2222} & S_{2233} & 2S_{2223} & 2S_{2231} & 2S_{2212} \\
S_{3311} & S_{3322} & S_{3333} & 2S_{3323} & 2S_{3331} & 2S_{3312} \\
2S_{2311} & 2S_{2322} & 2S_{2333} & 4S_{2323} & 4S_{2331} & 4S_{2312} \\
2S_{3111} & 2S_{3122} & 2S_{3133} & 4S_{3123} & 4S_{3131} & 4S_{3112} \\
2S_{1211} & 2S_{1222} & 2S_{1233} & 4S_{1223} & 4S_{1231} & 4S_{1212}
\end{bmatrix} .
\tag{5.3.9}
$$

The engineering notation components of the compliance can also be found by inverting the 6×6 matrix of engineering notation stiffness quantities:

$$
[\bar{S}] = [\bar{C}]^{-1} .
\tag{5.3.10}
$$

The engineering notation quantities $\bar{\sigma}_i$, $\bar{\varepsilon}_i$, and $\bar{C}_{ij}$ depend on the vector basis $\{\hat{e}_i\}$, but *do not* represent the scalar components of tensors in so much as their values in different vector bases are not directly related by the tensor transformation rule. To determine their values in different vector bases, the tensor transformation rule must be applied to the scalar components of σ, ε, and C and then the definitions (5.3.2) and (5.3.7) of the engineering notation components can be applied to the results.

By expanding the expression $\mathcal{U} = \frac{1}{2}\sigma_{ij}\varepsilon_{ij}$ for the strain energy density and recalling that $\sigma_{ij} = \sigma_{ji}$ and $\varepsilon_{ij} = \varepsilon_{ji}$, it follows from the definition (5.3.2) for the engineering notation that

$$\mathcal{U} = \frac{1}{2}\bar{\sigma}_i\bar{\varepsilon}_i \quad (i,j = 1,\dots,6) . \tag{5.3.11}$$

Thus, using (5.3.6), the strain energy density is given as a function of the engineering notation components of strain by

$$\mathcal{U}(\bar{\varepsilon}) = \frac{1}{2}\bar{C}_{ij}\bar{\varepsilon}_i\bar{\varepsilon}_j \quad (i,j = 1,\dots,6) , \tag{5.3.12}$$

and, using (5.3.8), as a function of the engineering notation components of stress by

$$\mathcal{U}(\bar{\sigma}) = \frac{1}{2}\bar{S}_{ij}\bar{\sigma}_i\bar{\sigma}_j \quad (i,j = 1,\dots,6) . \tag{5.3.13}$$

It follows that

$$\bar{\sigma}_i = \frac{\partial \mathcal{U}(\bar{\varepsilon})}{\partial \bar{\varepsilon}_i} \quad (i = 1,\dots,6) \tag{5.3.14}$$

and

$$\bar{\varepsilon}_i = \frac{\partial \mathcal{U}(\bar{\sigma})}{\partial \bar{\sigma}_i} \quad (i = 1,\dots,6) . \tag{5.3.15}$$

The constraints on the scalar components of the stiffness tensor due to material symmetry with respect to a transformation $[L]$, as specified by (5.3.1), can be deduced more concisely by considering the strain energy density (5.3.12). Since the strain energy density is a scalar, it follows that for arbitrary strain ε,

$$\bar{C}_{ij}\bar{\varepsilon}_i\bar{\varepsilon}_j = \bar{C}'_{ij}\bar{\varepsilon}'_i\bar{\varepsilon}'_j \quad (i,j = 1,\dots,6) , \tag{5.3.16}$$

where $\bar{C}'_{ij}$ and $\bar{\varepsilon}'_i$ are the engineering notation components of the stiffness and strain, respectively, in the vector basis $\{\hat{e}'_i\} = [L]\{\hat{e}_i\}$. However, if the material is symmetric with respect to $[L]$, then $\bar{C}'_{ij} = \bar{C}_{ij}$ and

$$\bar{C}_{ij}(\bar{\varepsilon}_i\bar{\varepsilon}_j - \bar{\varepsilon}'_i\bar{\varepsilon}'_j) = 0 \quad (i,j = 1,\dots,6) . \tag{5.3.17}$$

Requiring (5.3.17) to hold for arbitrary strain ε will result in the desired constraints on the scalar components of the stiffness tensor.

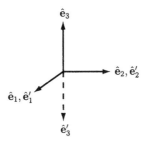

FIGURE 5.1. *Invariant transformation for a monoclinic material.*

5.3.2 Monoclinic Materials

Consider two reference configurations in which one configuration is the mirror image across a plane of the other. If the material at a point is symmetric with respect to this transformation, then the material is said to be *monoclinic* (i.e., it has one *plane of symmetry*) at that point in the body. Let the plane containing $\hat{e}_1$ and $\hat{e}_2$ be this plane of symmetry. Then, the material at the point is symmetric with respect to the following transformation (Figure 5.1):

$$[L] = \begin{bmatrix} 1 & 0 & 0 \\ 0 & 1 & 0 \\ 0 & 0 & -1 \end{bmatrix}. \tag{5.3.18}$$

Note that this is not a proper orthogonal transformation, because it involves an inversion; one vector basis is right-handed and the other is left-handed.

The scalar components of the strain tensor obey the tensor transformation rule, so

$$\begin{aligned}
\begin{bmatrix} \varepsilon'_{11} & \varepsilon'_{12} & \varepsilon'_{13} \\ \varepsilon'_{21} & \varepsilon'_{22} & \varepsilon'_{23} \\ \varepsilon'_{31} & \varepsilon'_{32} & \varepsilon'_{33} \end{bmatrix} &= \begin{bmatrix} 1 & 0 & 0 \\ 0 & 1 & 0 \\ 0 & 0 & -1 \end{bmatrix} \begin{bmatrix} \varepsilon_{11} & \varepsilon_{12} & \varepsilon_{13} \\ \varepsilon_{21} & \varepsilon_{22} & \varepsilon_{23} \\ \varepsilon_{31} & \varepsilon_{32} & \varepsilon_{33} \end{bmatrix} \begin{bmatrix} 1 & 0 & 0 \\ 0 & 1 & 0 \\ 0 & 0 & -1 \end{bmatrix} \\
&= \begin{bmatrix} \varepsilon_{11} & \varepsilon_{12} & -\varepsilon_{13} \\ \varepsilon_{21} & \varepsilon_{22} & -\varepsilon_{23} \\ -\varepsilon_{31} & -\varepsilon_{32} & \varepsilon_{33} \end{bmatrix}. \tag{5.3.19}
\end{aligned}$$

Therefore, in engineering notation,

$$\bar{\varepsilon}'_1 = \bar{\varepsilon}_1, \quad \bar{\varepsilon}'_2 = \bar{\varepsilon}_2, \quad \bar{\varepsilon}'_3 = \bar{\varepsilon}_3, \quad \bar{\varepsilon}'_4 = -\bar{\varepsilon}_4, \quad \bar{\varepsilon}'_5 = -\bar{\varepsilon}_5, \quad \bar{\varepsilon}'_6 = \bar{\varepsilon}_6. \tag{5.3.20}$$

Substituting (5.3.20) into (5.3.17) and keeping only the nonzero terms, one obtains the requirement

$$2\bar{C}_{14}\bar{\varepsilon}_1\bar{\varepsilon}_4 + 2\bar{C}_{15}\bar{\varepsilon}_1\bar{\varepsilon}_5 + 2\bar{C}_{24}\bar{\varepsilon}_2\bar{\varepsilon}_4 + 2\bar{C}_{25}\bar{\varepsilon}_2\bar{\varepsilon}_5$$
$$+ 2\bar{C}_{34}\bar{\varepsilon}_3\bar{\varepsilon}_4 + 2\bar{C}_{35}\bar{\varepsilon}_3\bar{\varepsilon}_5 + 2\bar{C}_{46}\bar{\varepsilon}_4\bar{\varepsilon}_6 + 2\bar{C}_{56}\bar{\varepsilon}_5\bar{\varepsilon}_6 = 0 . \quad (5.3.21)$$

Since the strains are arbitrary, it follows that $\bar{C}_{14} = \bar{C}_{15} = \bar{C}_{24} = \bar{C}_{25} = \bar{C}_{34} = \bar{C}_{35} = \bar{C}_{46} = \bar{C}_{56} = 0$ and the engineering notation stiffness matrix reduces to the following form for a monoclinic material when the base vectors $\hat{e}_1$ and $\hat{e}_2$ are in the plane of symmetry:

$$\begin{bmatrix} \bar{\sigma}_1 \\ \bar{\sigma}_2 \\ \bar{\sigma}_3 \\ \bar{\sigma}_4 \\ \bar{\sigma}_5 \\ \bar{\sigma}_6 \end{bmatrix} = \begin{bmatrix} \bar{C}_{11} & \bar{C}_{12} & \bar{C}_{13} & 0 & 0 & \bar{C}_{16} \\ \bar{C}_{12} & \bar{C}_{22} & \bar{C}_{23} & 0 & 0 & \bar{C}_{26} \\ \bar{C}_{13} & \bar{C}_{23} & \bar{C}_{33} & 0 & 0 & \bar{C}_{36} \\ 0 & 0 & 0 & \bar{C}_{44} & \bar{C}_{45} & 0 \\ 0 & 0 & 0 & \bar{C}_{45} & \bar{C}_{55} & 0 \\ \bar{C}_{16} & \bar{C}_{26} & \bar{C}_{36} & 0 & 0 & \bar{C}_{66} \end{bmatrix} \begin{bmatrix} \bar{\varepsilon}_1 \\ \bar{\varepsilon}_2 \\ \bar{\varepsilon}_3 \\ \bar{\varepsilon}_4 \\ \bar{\varepsilon}_5 \\ \bar{\varepsilon}_6 \end{bmatrix} . \quad (5.3.22)$$

Thus, there are 13 independent material parameters that need to be specified to characterize the elastic response of a monoclinic material. In vector bases not aligned with the plane of symmetry, there will *not*, in general, be elements of the stiffness matrix that are zero, but all of the elements are related to only 13 independent quantities.

5.3.3   Orthotropic Materials

A material is *orthotropic* at a point if it has three orthogonal planes of symmetry. Wood and laminated aligned fiber composites are good examples of orthotropic materials. Let the base vectors be normal to the planes of symmetry. Then, the stiffness components are invariant with respect to each of the following transformations:

$$[L^{(1)}] = \begin{bmatrix} 1 & 0 & 0 \\ 0 & 1 & 0 \\ 0 & 0 & -1 \end{bmatrix} , \quad [L^{(2)}] = \begin{bmatrix} -1 & 0 & 0 \\ 0 & 1 & 0 \\ 0 & 0 & 1 \end{bmatrix} ,$$
$$[L^{(3)}] = \begin{bmatrix} 1 & 0 & 0 \\ 0 & -1 & 0 \\ 0 & 0 & 1 \end{bmatrix} . \quad (5.3.23)$$

Clearly, an orthotropic material is also monoclinic (although a monoclinic material need not be orthotropic), and from the monoclinic result, invariance with respect to $[L^{(1)}]$ means that $\bar{C}_{14} = \bar{C}_{15} = \bar{C}_{24} = \bar{C}_{25} = \bar{C}_{34} = \bar{C}_{35} = \bar{C}_{46} = \bar{C}_{56} = 0$.

The result for invariance with respect to $[L^{(2)}]$ (i.e., symmetry with respect to the plane containing $\hat{e}_2$ and $\hat{e}_3$) can be inferred from the previous results for $[L^{(1)}]$, symmetry with respect to the plane containing $\hat{e}_1$ and $\hat{e}_2$. The tensor indices are cycled forward, $(1, 2, 3) \rightarrow (2, 3, 1)$, so that $11 \rightarrow 22$, $22 \rightarrow 33$, $33 \rightarrow 11$, $23 \rightarrow 31$, $31 \rightarrow 12$, and $12 \rightarrow 23$ and the engineering notation indices cycle like $1 \rightarrow 2$, $2 \rightarrow 3$, $3 \rightarrow 1$, $4 \rightarrow 5$, $5 \rightarrow 6$, and $6 \rightarrow 4$. Thus, invariance with respect to $[L^{(2)}]$ means that $\bar{C}_{25} = \bar{C}_{26} = \bar{C}_{35} = \bar{C}_{36} = \bar{C}_{15} = \bar{C}_{16} = \bar{C}_{54} = \bar{C}_{64} = 0$. The results for invariance with respect to $[L^{(3)}]$ can similarly be inferred, but no further reduction in the number of independent elastic constants is obtained—symmetry with respect to two orthogonal planes implies symmetry with respect to the third.

For an orthotropic material, the engineering notation stiffness matrix, in a vector basis with base vectors normal to the planes of symmetry, reduces to the following:

$$\begin{bmatrix} \bar{\sigma}_1 \\ \bar{\sigma}_2 \\ \bar{\sigma}_3 \\ \bar{\sigma}_4 \\ \bar{\sigma}_5 \\ \bar{\sigma}_6 \end{bmatrix} = \begin{bmatrix} \bar{C}_{11} & \bar{C}_{12} & \bar{C}_{13} & 0 & 0 & 0 \\ \bar{C}_{12} & \bar{C}_{22} & \bar{C}_{23} & 0 & 0 & 0 \\ \bar{C}_{13} & \bar{C}_{23} & \bar{C}_{33} & 0 & 0 & 0 \\ 0 & 0 & 0 & \bar{C}_{44} & 0 & 0 \\ 0 & 0 & 0 & 0 & \bar{C}_{55} & 0 \\ 0 & 0 & 0 & 0 & 0 & \bar{C}_{66} \end{bmatrix} \begin{bmatrix} \bar{\varepsilon}_1 \\ \bar{\varepsilon}_2 \\ \bar{\varepsilon}_3 \\ \bar{\varepsilon}_4 \\ \bar{\varepsilon}_5 \\ \bar{\varepsilon}_6 \end{bmatrix}. \tag{5.3.24}$$

Thus, nine independent material parameters must be specified to characterize the elastic response of an orthotropic material. In vector bases not aligned with the planes of symmetry, there will *not*, in general, be elements of the stiffness matrix that are zero, but they can always be related to nine independent quantities.

5.3.4 Transversely Isotropic Materials

A material is *transversely isotropic* at a point if it is symmetric with respect to an arbitrary rotation about a given axis, called the *axis of symmetry*. Aligned fiber composites are good examples of transversely isotropic materials, where the axis of symmetry is parallel to the fibers. Let the base vector $\hat{e}_3$ lie along the axis of symmetry. Then, the elastic constants are invariant with respect to the following transformation:

$$[L] = \begin{bmatrix} \cos\theta & \sin\theta & 0 \\ -\sin\theta & \cos\theta & 0 \\ 0 & 0 & 1 \end{bmatrix}, \tag{5.3.25}$$

where θ is arbitrary.

By applying the tensor transformation rule,

$$
\begin{bmatrix}
\varepsilon'_{11} & \varepsilon'_{12} & \varepsilon'_{13} \\
\varepsilon'_{21} & \varepsilon'_{22} & \varepsilon'_{23} \\
\varepsilon'_{31} & \varepsilon'_{32} & \varepsilon'_{33}
\end{bmatrix}
$$
$$
=
\begin{bmatrix}
\cos\theta & \sin\theta & 0 \\
-\sin\theta & \cos\theta & 0 \\
0 & 0 & 1
\end{bmatrix}
\begin{bmatrix}
\varepsilon_{11} & \varepsilon_{12} & \varepsilon_{13} \\
\varepsilon_{21} & \varepsilon_{22} & \varepsilon_{23} \\
\varepsilon_{31} & \varepsilon_{32} & \varepsilon_{33}
\end{bmatrix}
\begin{bmatrix}
\cos\theta & -\sin\theta & 0 \\
\sin\theta & \cos\theta & 0 \\
0 & 0 & 1
\end{bmatrix}, \quad (5.3.26)
$$

and converting to engineering notation (5.3.2), one finds that

$$
\begin{aligned}
\bar{\varepsilon}'_1 &= \bar{\varepsilon}_1 \cos^2\theta + \bar{\varepsilon}_2 \sin^2\theta + \bar{\varepsilon}_6 \sin\theta\cos\theta \;, \\
\bar{\varepsilon}'_2 &= \bar{\varepsilon}_1 \sin^2\theta + \bar{\varepsilon}_2 \cos^2\theta - \bar{\varepsilon}_6 \sin\theta\cos\theta \;, \\
\bar{\varepsilon}'_3 &= \bar{\varepsilon}_3 \;, \\
\bar{\varepsilon}'_4 &= \bar{\varepsilon}_4 \cos\theta - \bar{\varepsilon}_5 \sin\theta \;, \\
\bar{\varepsilon}'_5 &= \bar{\varepsilon}_4 \sin\theta + \bar{\varepsilon}_5 \cos\theta \;, \\
\bar{\varepsilon}'_6 &= -2(\bar{\varepsilon}_1 - \bar{\varepsilon}_2)\sin\theta\cos\theta + \bar{\varepsilon}_6(\cos^2\theta - \sin^2\theta) \;.
\end{aligned}
\qquad (5.3.27)
$$

One finds the constraints on the scalar components of the stiffness tensor by substituting (5.3.27) into the condition (5.3.17) and requiring that it hold for arbitrary strain. This is easier if one considers the cases for a couple of specific values of θ first.

Consider the case where $\theta = \pi$ for which, from (5.3.27),

$$
\bar{\varepsilon}'_1 = \bar{\varepsilon}_1 \;, \quad \bar{\varepsilon}'_2 = \bar{\varepsilon}_2 \;, \quad \bar{\varepsilon}'_3 = \bar{\varepsilon}_3 \;, \quad \bar{\varepsilon}'_4 = -\bar{\varepsilon}_4 \;, \quad \bar{\varepsilon}'_5 = -\bar{\varepsilon}_5 \;, \quad \bar{\varepsilon}'_6 = \bar{\varepsilon}_6 \;.
$$
$$
(5.3.28)
$$

This is identical to the monoclinic case from which it follows that $\bar{C}_{14} = \bar{C}_{15} = \bar{C}_{24} = \bar{C}_{25} = \bar{C}_{34} = \bar{C}_{35} = \bar{C}_{46} = \bar{C}_{56} = 0$.

Next consider the case where $\theta = \pi/2$ for which

$$
\bar{\varepsilon}'_1 = \bar{\varepsilon}_2 \;, \quad \bar{\varepsilon}'_2 = \bar{\varepsilon}_1 \;, \quad \bar{\varepsilon}'_3 = \bar{\varepsilon}_3 \;, \quad \bar{\varepsilon}'_4 = -\bar{\varepsilon}_5 \;, \quad \bar{\varepsilon}'_5 = \bar{\varepsilon}_4 \;, \quad \bar{\varepsilon}'_6 = -\bar{\varepsilon}_6 \;.
$$
$$
(5.3.29)
$$

Substituting this into (5.3.17), using the results obtained by considering $\theta = \pi$, and keeping only the nonzero terms, one obtains the requirement,

$$
\begin{aligned}
(\bar{C}_{11} - \bar{C}_{22})(\bar{\varepsilon}_1^2 - \bar{\varepsilon}_2^2) &+ 2(\bar{C}_{13} - \bar{C}_{23})(\bar{\varepsilon}_1 - \bar{\varepsilon}_2)\bar{\varepsilon}_3 \\
&+ 2(\bar{C}_{16} + \bar{C}_{26})(\bar{\varepsilon}_1 + \bar{\varepsilon}_2)\bar{\varepsilon}_6 + 4\bar{C}_{36}\bar{\varepsilon}_3\bar{\varepsilon}_6 \\
&+ (\bar{C}_{44} - \bar{C}_{55})(\bar{\varepsilon}_4^2 - \bar{\varepsilon}_5^2) + 4\bar{C}_{45}\bar{\varepsilon}_4\bar{\varepsilon}_5 = 0 \;. \quad (5.3.30)
\end{aligned}
$$

Since the strain is arbitrary, it follows that $\bar{C}_{22} = \bar{C}_{11}$, $\bar{C}_{23} = \bar{C}_{13}$, $\bar{C}_{26} = -\bar{C}_{16}$, $\bar{C}_{55} = \bar{C}_{44}$, and $\bar{C}_{36} = \bar{C}_{45} = 0$

Finally, by substituting (5.3.27) into the material symmetry condition (5.3.17), using the results obtained by considering the special cases of $\theta = \pi$ and $\theta = \pi/2$, and simplifying, one finds after some manipulation the condition

$$\left[\frac{1}{2}(\bar{C}_{11} - \bar{C}_{12}) - \bar{C}_{66}\right]\left\{\left[(\bar{\varepsilon}_1 - \bar{\varepsilon}_2)^2 - \bar{\varepsilon}_6^2\right]\sin 2\theta - 2(\bar{\varepsilon}_1 - \bar{\varepsilon}_2)\bar{\varepsilon}_6 \cos 2\theta\right\}$$

$$+ 2\bar{C}_{16}\left\{\left[(\bar{\varepsilon}_1 - \bar{\varepsilon}_2)^2 - \bar{\varepsilon}_6^2\right]\cos 2\theta + 2(\bar{\varepsilon}_1 - \bar{\varepsilon}_2)\bar{\varepsilon}_6 \sin 2\theta\right\} = 0 , \quad (5.3.31)$$

which can only be true for arbitrary strain ε and angle θ if $\bar{C}_{16} = 0$ and $\bar{C}_{66} = \frac{1}{2}(\bar{C}_{11} - \bar{C}_{12})$.

It has been shown then that, in a vector basis in which the base vector $\hat{e}_3$ is along the axis of symmetry, the engineering notation stiffness matrix for a transversely isotropic material reduces to the following:

$$\begin{bmatrix} \bar{\sigma}_1 \\ \bar{\sigma}_2 \\ \bar{\sigma}_3 \\ \bar{\sigma}_4 \\ \bar{\sigma}_5 \\ \bar{\sigma}_6 \end{bmatrix} = \begin{bmatrix} \bar{C}_{11} & \bar{C}_{12} & \bar{C}_{13} & 0 & 0 & 0 \\ \bar{C}_{12} & \bar{C}_{11} & \bar{C}_{13} & 0 & 0 & 0 \\ \bar{C}_{13} & \bar{C}_{13} & \bar{C}_{33} & 0 & 0 & 0 \\ 0 & 0 & 0 & \bar{C}_{44} & 0 & 0 \\ 0 & 0 & 0 & 0 & \bar{C}_{44} & 0 \\ 0 & 0 & 0 & 0 & 0 & \frac{1}{2}(\bar{C}_{11} - \bar{C}_{12}) \end{bmatrix} \begin{bmatrix} \bar{\varepsilon}_1 \\ \bar{\varepsilon}_2 \\ \bar{\varepsilon}_3 \\ \bar{\varepsilon}_4 \\ \bar{\varepsilon}_5 \\ \bar{\varepsilon}_6 \end{bmatrix} . \quad (5.3.32)$$

There are five independent elastic constants for a transversely isotropic material. In vector bases not aligned with the axis of rotational symmetry, there will *not*, in general, be elements of the stiffness matrix that are zero, but they can always be related to five independent quantities.

5.3.5 Principal Directions of Stress and Strain

Suppose that a material has symmetry properties such that, in a vector basis $\{\hat{e}_i\}$, the components of the stiffness $\bar{C}_{14} = \bar{C}_{15} = \bar{C}_{16} = \bar{C}_{24} = \bar{C}_{25} = \bar{C}_{26} = \bar{C}_{34} = \bar{C}_{35} = \bar{C}_{36} = 0$. In $\{\hat{e}_i\}$, the components of stress are related to the components of strain by

$$\begin{bmatrix} \bar{\sigma}_1 \\ \bar{\sigma}_2 \\ \bar{\sigma}_3 \\ \bar{\sigma}_4 \\ \bar{\sigma}_5 \\ \bar{\sigma}_6 \end{bmatrix} = \begin{bmatrix} \bar{C}_{11} & \bar{C}_{12} & \bar{C}_{13} & 0 & 0 & 0 \\ \bar{C}_{12} & \bar{C}_{22} & \bar{C}_{23} & 0 & 0 & 0 \\ \bar{C}_{13} & \bar{C}_{23} & \bar{C}_{33} & 0 & 0 & 0 \\ 0 & 0 & 0 & \bar{C}_{44} & \bar{C}_{45} & \bar{C}_{46} \\ 0 & 0 & 0 & \bar{C}_{45} & \bar{C}_{55} & \bar{C}_{56} \\ 0 & 0 & 0 & \bar{C}_{46} & \bar{C}_{56} & \bar{C}_{66} \end{bmatrix} \begin{bmatrix} \bar{\varepsilon}_1 \\ \bar{\varepsilon}_2 \\ \bar{\varepsilon}_3 \\ \bar{\varepsilon}_4 \\ \bar{\varepsilon}_5 \\ \bar{\varepsilon}_6 \end{bmatrix} . \quad (5.3.33)$$

Orthotropic and transversely isotropic materials have this symmetry property, and the corresponding vector bases $\{\hat{e}_i\}$ are the vector bases aligned with the planes of symmetry and the axis of symmetry, respectively. In other vector bases, there will, in general, be no zero terms in the matrix of stiffness components. In $\{\hat{e}_i\}$ then, vanishing shear strain components, $\bar{\varepsilon}_4 = \bar{\varepsilon}_5 = \bar{\varepsilon}_6 = 0$, imply vanishing shear stress components, $\bar{\sigma}_4 = \bar{\sigma}_5 = \bar{\sigma}_6 = 0$, and vice versa. Thus, if for a given state of stress in such a material the principal directions of stress coincide with the vector basis $\{\hat{e}_i\}$, then so do the principal directions of strain. In general, the principal directions of stress and strain are different.

Example 5.1
Suppose the stiffness components for a sample of Douglas fir wood are known to be

$$[\bar{C}] = \begin{bmatrix} 11.0 & 1.05 & 0.961 & 0 & 0 & 0 \\ 1.05 & 3.21 & 0.513 & 0 & 0 & 0 \\ 0.961 & 0.513 & 2.86 & 0 & 0 & 0 \\ 0 & 0 & 0 & 3.15 & 0 & 0 \\ 0 & 0 & 0 & 0 & 3.35 & 0 \\ 0 & 0 & 0 & 0 & 0 & 4.10 \end{bmatrix} \text{ (GPa)},$$

where $\hat{e}_1$ is along the grain of the wood and $\hat{e}_3$ is perpendicular to the growth rings. Find the components of strain corresponding to a state of stress of $\sigma_{11} = \sigma_{22} = \sigma_{33} = 10$ MPa and $\sigma_{12} = \sigma_{23} = \sigma_{31} = 5$ MPa.

Solution. In matrix form, the components of strain are given by

$$[\bar{\varepsilon}] = [\bar{S}][\bar{\sigma}] ,$$

where the compliance matrix is the inverse of the stiffness matrix,

$$[\bar{S}] = [\bar{C}]^{-1} .$$

Thus,

$$[\bar{S}] = \begin{bmatrix} 95.9 & -27.0 & -27.4 & 0 & 0 & 0 \\ -27.0 & 328 & -49.8 & 0 & 0 & 0 \\ -27.4 & -49.8 & 368 & 0 & 0 & 0 \\ 0 & 0 & 0 & 317 & 0 & 0 \\ 0 & 0 & 0 & 0 & 299 & 0 \\ 0 & 0 & 0 & 0 & 0 & 244 \end{bmatrix} \times 10^{-3} \text{ (GPa}^{-1})$$

and the components of strain in engineering notation are

$$[\bar{\varepsilon}] = \begin{bmatrix} 0.415 \\ 2.52 \\ 2.91 \\ 1.59 \\ 1.49 \\ 1.22 \end{bmatrix} \times 10^{-3} .$$

From the definition of the engineering notation (5.3.2), the scalar components of the strain tensor are therefore

$$[\varepsilon] = \begin{bmatrix} 0.415 & 0.610 & 0.745 \\ 0.610 & 2.52 & 0.795 \\ 0.745 & 0.795 & 2.91 \end{bmatrix} \times 10^{-3} .$$

5.4 Isotropic Materials

The material at a point in a body is *isotropic* if and only if it is symmetric with respect to *all* orthogonal transformations. Thus, for an isotropic material, the scalar components of the stiffness tensor are invariant with respect to all orthogonal transformations of the vector basis; that is, the stiffness tensor $\mathbf{C}$ is isotropic. From (2.3.67), all isotropic fourth-order tensors have scalar components of the form

$$C_{ijkl} = \lambda \delta_{ij} \delta_{kl} + \mu \delta_{ik} \delta_{jl} + \kappa \delta_{il} \delta_{jk} . \tag{5.4.1}$$

Recall that, in addition to being isotropic, the stiffness tensor must have the symmetries $C_{ijkl} = C_{klij}$, $C_{ijkl} = C_{jikl}$, and $C_{ijkl} = C_{jilk}$. The first of these is identically satisfied by (5.4.1), thus establishing that an isotropic linear elastic material is necessarily hyperelastic, but the later two symmetries are satisfied if and only if $\kappa = \mu$. Thus, for linear elastic deformation of an isotropic material, there are only two independent elastic constants, λ and μ, which are called the *Lamé constants*.

Using (5.4.1) with $\kappa = \mu$, the components of stress are given in terms of the components of strain by

$$\sigma_{ij} = C_{ijkl} \varepsilon_{kl} = \lambda \delta_{ij} \delta_{kl} \varepsilon_{kl} + \mu (\delta_{ik} \delta_{jl} + \delta_{il} \delta_{jk}) \varepsilon_{kl} . \tag{5.4.2}$$

By simplifying this expression, it is seen that the constitutive equation for an isotropic material can be given as

$$\boxed{\sigma = 2\mu\varepsilon + \lambda(\operatorname{tr}\varepsilon)\mathbf{I} , \qquad \sigma_{ij} = 2\mu\varepsilon_{ij} + \lambda\varepsilon_{kk}\delta_{ij} .} \tag{5.4.3}$$

Since the stress and strain tensors are symmetric, there are six independent scalar constitutive equations. As shown in Section 5.4.1, the isotropic constitutive equation (5.4.3) can be rewritten in terms of other elastic constants that are related to the Lamé constants λ and μ. The different forms of (5.4.3) will be convenient in different applications.

If follows from (5.4.3) that, in all vector bases, $\sigma_{12} = \sigma_{23} = \sigma_{31} = 0$ if and only if $\varepsilon_{12} = \varepsilon_{23} = \varepsilon_{31} = 0$. Thus, the principal directions of stress and strain always coincide in an isotropic material.

5.4.1   Equivalent Forms of the Constitutive Equation

By contracting (5.4.3), one obtains the constitutive relation between the trace of the stress tensor and the trace of the strain tensor:

$$\boxed{\operatorname{tr}\boldsymbol{\sigma} = (3\lambda + 2\mu)\operatorname{tr}\boldsymbol{\varepsilon}\,, \quad \sigma_{ii} = (3\lambda + 2\mu)\varepsilon_{ii}\,.} \tag{5.4.4}$$

The *mean stress* $\tilde{\sigma}$ is a scalar defined as the average of the normal components of the stress tensor (i.e., one-third the trace of the stress tensor):

$$\tilde{\sigma} \equiv \frac{1}{3}\operatorname{tr}\boldsymbol{\sigma}\,, \quad \tilde{\sigma} \equiv \frac{1}{3}\sigma_{ii}\,. \tag{5.4.5}$$

Recall that $\operatorname{tr}\boldsymbol{\sigma}$ and therefore the mean stress $\tilde{\sigma}$ are invariant with respect to changes in vector basis. Recall also that the dilatation equals the trace of the strain tensor, $e = \operatorname{tr}\boldsymbol{\varepsilon}$. Thus, the change in volume per unit volume at a point in a body is a function only of the mean stress at that point. This relation is used to define the *bulk modulus*, $K \equiv \tilde{\sigma}/e$, so that in terms of the Lamé constants, the bulk modulus is given by

$$K = \lambda + \frac{2}{3}\mu\,, \tag{5.4.6}$$

and (5.4.4) can be rewritten as

$$\boxed{\tilde{\sigma} = Ke\,.} \tag{5.4.7}$$

Hydrostatic pressure[2] is a special state of stress in which $\boldsymbol{\sigma} = -p\mathbf{I}$. Thus, the mean stress in this special case is $\tilde{\sigma} = -p$ and the dilatation is given by $e = -p/K$.

The *deviatoric stress* $\mathring{\boldsymbol{\sigma}}$ and *deviatoric strain* $\mathring{\boldsymbol{\varepsilon}}$ are defined by

$$\mathring{\boldsymbol{\sigma}} \equiv \boldsymbol{\sigma} - \frac{1}{3}(\operatorname{tr}\boldsymbol{\sigma})\mathbf{I}\,, \quad \mathring{\sigma}_{ij} \equiv \sigma_{ij} - \frac{1}{3}\sigma_{kk}\delta_{ij}\,, \tag{5.4.8}$$

$$\mathring{\boldsymbol{\varepsilon}} \equiv \boldsymbol{\varepsilon} - \frac{1}{3}(\operatorname{tr}\boldsymbol{\varepsilon})\mathbf{I}\,, \quad \mathring{\varepsilon}_{ij} \equiv \varepsilon_{ij} - \frac{1}{3}\varepsilon_{kk}\delta_{ij}\,. \tag{5.4.9}$$

Using (5.4.3)–(5.4.9), it can be shown that the deviatoric stress is a function of the deviatoric strain, which is a measure of distortion independent of volume change:

$$\boxed{\mathring{\boldsymbol{\sigma}} = 2\mu\mathring{\boldsymbol{\varepsilon}}\,, \quad \mathring{\sigma}_{ij} = 2\mu\mathring{\varepsilon}_{ij}\,.} \tag{5.4.10}$$

Note from their definitions that $\operatorname{tr}\mathring{\boldsymbol{\sigma}} = 0$ and $\operatorname{tr}\mathring{\boldsymbol{\varepsilon}} = 0$. It follows that (5.4.10) constitute only five independent scalar equations. Together, (5.4.4)

---

[2]Historically, the hydrostatic pressure p is defined to be positive in compression and negative in tension.

and (5.4.10) comprise six independent scalar constitutive equations which are equivalent to (5.4.3).

The constitutive equation (5.4.3) gives the components of stress as a function of the components of strain. It will often be useful to express the constitutive equation alternatively in terms of the components of strain as a function of the components of stress; (5.4.3) and (5.4.4) can be combined to give

$$
\begin{aligned}
\varepsilon &= \frac{1}{2\mu}\sigma - \frac{\lambda}{2\mu(3\lambda + 2\mu)}(\text{tr}\,\sigma)\mathbf{I}\,, \\
\varepsilon_{ij} &= \frac{1}{2\mu}\sigma_{ij} - \frac{\lambda}{2\mu(3\lambda + 2\mu)}\sigma_{kk}\delta_{ij}\,.
\end{aligned}
\tag{5.4.11}
$$

In the special case of *uniaxial tension*, $\sigma_{11} = \sigma$ and $\sigma_{22} = \sigma_{33} = \sigma_{12} = \sigma_{23} = \sigma_{31} = 0$. From (5.4.11), the corresponding components of strain are $\varepsilon_{12} = \varepsilon_{23} = \varepsilon_{31} = 0$ and

$$
\varepsilon_{11} = \frac{1}{\mu}\left(\frac{\lambda + \mu}{3\lambda + 2\mu}\right)\sigma\,, \quad \varepsilon_{22} = \varepsilon_{33} = \frac{-\lambda}{2\mu(3\lambda + 2\mu)}\sigma\,.
\tag{5.4.12}
$$

In such a state of uniaxial tension, *Young's modulus* is defined as $E \equiv \sigma/\varepsilon_{11}$ and *Poisson's ratio* is defined as $\nu \equiv -\varepsilon_{22}/\varepsilon_{11} = -\varepsilon_{33}/\varepsilon_{11}$ so that, from (5.4.12), E and ν are given in terms of the Lamé constants by

$$
E = \frac{\mu(3\lambda + 2\mu)}{\lambda + \mu}\,, \quad \nu = \frac{\lambda}{2(\lambda + \mu)}\,.
\tag{5.4.13}
$$

By solving (5.4.13) for the Lamé constants λ and μ, the constitutive equations (5.4.3) and (5.4.11) can be rewritten in terms of Young's modulus and Poisson's ratio as

$$
\begin{aligned}
\sigma &= \frac{E}{1+\nu}\varepsilon + \frac{\nu E}{(1+\nu)(1-2\nu)}(\text{tr}\,\varepsilon)\mathbf{I}\,, \\
\sigma_{ij} &= \frac{E}{1+\nu}\varepsilon_{ij} + \frac{\nu E}{(1+\nu)(1-2\nu)}\varepsilon_{kk}\delta_{ij}
\end{aligned}
\tag{5.4.14}
$$

and

$$
\begin{aligned}
\varepsilon &= \frac{1}{E}[(1+\nu)\sigma - \nu(\text{tr}\,\sigma)\mathbf{I}]\,, \\
\varepsilon_{ij} &= \frac{1}{E}[(1+\nu)\sigma_{ij} - \nu\sigma_{kk}\delta_{ij}]\,.
\end{aligned}
\tag{5.4.15}
$$

In the special case of *pure shear*, $\sigma_{12} = \tau$ and $\sigma_{11} = \sigma_{22} = \sigma_{33} = \sigma_{23} = \sigma_{31} = 0$. From (5.4.11), the corresponding components of strain are

$\varepsilon_{11} = \varepsilon_{22} = \varepsilon_{33} = \varepsilon_{23} = \varepsilon_{31} = 0$ and

$$\varepsilon_{12} = \frac{1}{2\mu}\tau . \qquad (5.4.16)$$

In a state of pure shear, the *shear modulus* is defined as $G \equiv \tau/\gamma$, where $\gamma = 2\varepsilon_{12}$ is the orthogonal shear strain; in terms of the Lamé constants,

$$G = \mu . \qquad (5.4.17)$$

Any one of the five isotropic elastic moduli λ, $\mu = G$, E, ν, and K can be expressed as a function of any two of the others (see Table 5.1).

5.4.2   Restrictions on Values of the Elastic Moduli

It follows from the expression (5.2.8) for the strain energy density in terms of stress and strain for linear elastic deformation and from the constitutive equation (5.4.3) that the strain energy density for an isotropic material can be given as a function of strain by

$$\mathcal{U}(\varepsilon) = \mu\varepsilon : \varepsilon + \frac{1}{2}\lambda(\operatorname{tr}\varepsilon)^2 ,$$
$$\mathcal{U}(\varepsilon) = \mu\varepsilon_{ij}\varepsilon_{ij} + \frac{1}{2}\lambda(\varepsilon_{kk})^2 . \qquad (5.4.18)$$

Similarly, from the constitutive equation (5.4.15), the strain energy density for an isotropic material can be given as a function of stress by

$$\mathcal{U}(\sigma) = \frac{1}{2E}\left[(1+\nu)\sigma : \sigma - \nu(\operatorname{tr}\sigma)^2\right] ,$$
$$\mathcal{U}(\sigma) = \frac{1}{2E}\left[(1+\nu)\sigma_{ij}\sigma_{ij} - \nu(\sigma_{kk})^2\right] . \qquad (5.4.19)$$

The strain energy density is assumed to be positive definite, $\mathcal{U} \geq 0$ with $\mathcal{U} = 0$ if and only if $\varepsilon = \mathbf{0}$ (see Chapter 10). This has implications for the allowable values of the elastic moduli.

For a state of uniaxial tension, $\sigma_{11} = \sigma$ and $\sigma_{22} = \sigma_{33} = \sigma_{12} = \sigma_{23} = \sigma_{31} = 0$ so that $\sigma_{ij}\sigma_{ij} = \sigma^2$ and $\sigma_{kk} = \sigma$, and from (5.4.19), the strain energy density is

$$\mathcal{U} = \frac{1}{2E}\sigma^2 . \qquad (5.4.20)$$

Thus, in order for $\mathcal{U}$ to be positive definite, Young's modulus must be positive,

$$E > 0 . \qquad (5.4.21)$$

Table 5.1 Relation between each of the isotropic elastic moduli and any two other moduli

	$\lambda =$	$\mu = G =$	$E =$	$\nu =$	$K =$
λ,μ	λ	μ	$\dfrac{\mu(3\lambda+2\mu)}{\lambda+\mu}$	$\dfrac{\lambda}{2(\lambda+\mu)}$	$\lambda+\dfrac{2}{3}\mu$
λ,E	λ	$\dfrac{E-3\lambda+r}{4}$	E	$\dfrac{2\lambda}{E+\lambda+r}$	$\dfrac{E+3\lambda+r}{6}$
λ,ν	λ	$\dfrac{\lambda(1-2\nu)}{2\nu}$	$\dfrac{\lambda(1+\nu)(1-2\nu)}{\nu}$	ν	$\dfrac{\lambda(1+\nu)}{3\nu}$
λ,K	λ	$\dfrac{3}{2}(K-\lambda)$	$\dfrac{9K(K-\lambda)}{3K-\lambda}$	$\dfrac{\lambda}{3K-\lambda}$	K
μ,E	$\dfrac{\mu(E-2\mu)}{3\mu-E}$	μ	E	$\dfrac{E-2\mu}{2\mu}$	$\dfrac{\mu E}{3(3\mu-E)}$
μ,ν	$\dfrac{2\mu\nu}{1-2\nu}$	μ	$2\mu(1+\nu)$	ν	$\dfrac{2\mu(1+\nu)}{3(1-2\nu)}$
μ,K	$K-\dfrac{2}{3}\mu$	μ	$\dfrac{9K\mu}{3K+\mu}$	$\dfrac{3K-2\mu}{6K+2\mu}$	K
E,ν	$\dfrac{E\nu}{(1+\nu)(1-2\nu)}$	$\dfrac{E}{2(1+\nu)}$	E	ν	$\dfrac{E}{3(1-2\nu)}$
E,K	$\dfrac{3K(3K-E)}{9K-E}$	$\dfrac{3KE}{9K-E}$	E	$\dfrac{3K-E}{6K}$	K
ν,K	$\dfrac{3K\nu}{1+\nu}$	$\dfrac{3K(1-2\nu)}{2(1+\nu)}$	$3K(1-2\nu)$	ν	K

Note: $r = \sqrt{E^2 + 9\lambda^2 + 2E\lambda}$

For pure shear, $\sigma_{12} = \tau$ and $\sigma_{11} = \sigma_{22} = \sigma_{33} = \sigma_{23} = \sigma_{31} = 0$ so that $\sigma_{ij}\sigma_{ij} = 2\tau^2$ and $\sigma_{kk} = 0$. It follows from (5.4.19) that the strain energy density for a state of pure shear is

$$ \mathcal{U} = \frac{1+\nu}{E}\tau^2 = \frac{1}{2\mu}\tau^2 \tag{5.4.22} $$

and therefore that the Lamé constant μ must be positive,

$$ \mu > 0 . \tag{5.4.23} $$

For hydrostatic pressure, $\sigma_{11} = \sigma_{22} = \sigma_{33} = -p$ and $\sigma_{12} = \sigma_{23} = \sigma_{31} = 0$ so that $\sigma_{ij}\sigma_{ij} = 3p^2$ and $\sigma_{kk} = -3p$, and from (5.4.19), the strain energy density is

$$ \mathcal{U} = \frac{3(1-2\nu)}{2E}p^2 = \frac{1}{2K}p^2 . \tag{5.4.24} $$

Therefore, the bulk modulus must also be positive,

$$ K > 0 . \tag{5.4.25} $$

These inequalities are consistent with physical intuition. For instance, as a result of (5.4.21), if one applies uniaxial tension to a bar, then the bar must increase in length. Having the bar to contract would violate the assumption that the strain energy density is positive definite.

Consider now the relations from Table 5.1 for Young's modulus, $E = 2\mu(1 + \nu)$ and $E = 3K(1 - 2\nu)$. Given the inequalities (5.4.21), (5.4.23), and (5.4.25), it follows from the assumption that the strain energy density is positive definite that the value of Poisson's ratio is limited to

$$ -1 < \nu < \frac{1}{2} . \tag{5.4.26} $$

Negative values of Poisson's ratio correspond to materials that expand laterally under uniaxial tension. Experience with real engineering materials shows that this does not occur and therefore that it is reasonable to assume

$$ 0 < \nu < \frac{1}{2} . \tag{5.4.27} $$

Then, based on the relation

$$ \lambda = \frac{E\nu}{(1+\nu)(1-2\nu)} \tag{5.4.28} $$

from Table 5.1 and the inequalities (5.4.21) and (5.4.27), it follows that the Lamé constant λ must be positive,

$$ \lambda > 0 . \tag{5.4.29} $$

Example 5.2

Find the Lamé constants and bulk modulus for (a) carbon steel ($E = 207$ GPa, $\nu = 0.292$) and for (b) aluminum ($E = 71.0$ GPa, $\nu = 0.334$).

Solution. Using the relations in Table 5.1, the answer for carbon steel is

$$\lambda = \frac{(207 \text{ GPa})(0.292)}{(1 + 0.292)[1 - 2(0.292)]} = 112 \text{ GPa},$$

$$\mu = \frac{207 \text{ GPa}}{2(1 + 0.292)} = 80.1 \text{ GPa},$$

$$K = \frac{207 \text{ GPa}}{3[1 - 2(0.292)]} = 166 \text{ GPa}$$

and the answer for aluminum is

$$\lambda = \frac{(71.0 \text{ GPa})(0.334)}{(1 + 0.334)[1 - 2(0.334)]} = 53.5 \text{ GPa},$$

$$\mu = \frac{71.0 \text{ GPa}}{2(1 + 0.334)} = 26.6 \text{ GPa},$$

$$K = \frac{71.0 \text{ GPa}}{3[1 - 2(0.334)]} = 71.3 \text{ GPa}.$$

Example 5.3

A deep-sea fisherman uses a 1.00-cm^3 lead weight ($E = 36.5$ GPa, $\nu = 0.425$) as a sinker for his bait. Calculate the volume of the sinker at a depth of 100 m. The density of seawater is $\rho = 1025$ kg/m^3 and the gravitational constant is $g = 9.81$ m/s^2. How deep would the sinker have to go to experience a 1% change in volume?

Solution. The pressure at a depth d is $p = \rho g d$, so that at $d = 100$ m, the pressure is $p = 1.006$ MPa. The bulk modulus for lead is

$$K = \frac{E}{3(1 - 2\nu)} = 81.1 \text{ GPa}.$$

Therefore, the dilatation for the lead sinker at a depth of 100 m is

$$e = -\frac{p}{K} = -1.24 \times 10^{-5}.$$

The change in volume is $\Delta V = eV_0 = -1.24 \times 10^{-11}$ m^3, much less than can be discerned given the number of significant digits. The change in volume is negligible.

In order to achieve a 1% change in volume ($e = -0.01$), the pressure would have to be

$$p = -eK = 811 \text{ MPa}.$$

The depth of seawater required to attain this pressure is

$$d = \frac{p}{\rho g} = 80.7 \text{ km} .$$

This is far deeper than the deepest part of the Earth's oceans.

5.5   Cylindrical and Spherical Coordinates

The constitutive equations are unaltered in cylindrical and spherical coordinates. For index notation equations, the indices 1, 2, and 3 are understood to represent r, θ, and z, respectively, in cylindrical coordinates and R, θ, and ϕ, respectively, in spherical coordinates. For example, using (5.4.3), one has that, in cylindrical coordinates,

$$\sigma_{\theta\theta} = 2\mu\varepsilon_{\theta\theta} + \lambda(\varepsilon_{rr} + \varepsilon_{\theta\theta} + \varepsilon_{zz}) . \qquad (5.5.1)$$

Likewise, the equations for the strain energy density are unaltered; for example, in spherical coordinates the expanded form of (5.2.8) is

$$\mathcal{U} = \frac{1}{2}(\sigma_{RR}\varepsilon_{RR} + \sigma_{\theta\theta}\varepsilon_{\theta\theta} + \sigma_{\phi\phi}\varepsilon_{\phi\phi} + 2\sigma_{R\theta}\varepsilon_{R\theta} + 2\sigma_{\theta\phi}\varepsilon_{\theta\phi} + 2\sigma_{\phi R}\varepsilon_{\phi R}) .$$
$$(5.5.2)$$

Problems

5.1 Consider two reference configurations that are related by a rigid-body, 90° rotation about an axis. If a material is symmetric with respect to this transformation, then the axis is said to be a *fourfold axis of symmetry* of the material at that point. A *cubic material* is a material with three mutually orthogonal fourfold axes of symmetry. Using a vector basis in which the base vectors are parallel to the fourfold axes of symmetry, show that there are only three independent elastic constants for a cubic material and give the engineering notation stiffness matrix for a cubic material in a form analogous to those given for a monoclinic material (5.3.22), an orthotropic material (5.3.24), and a transversely isotropic material (5.3.32). **Hint:** a 90° rotation about an axis parallel to $\hat{e}_1$ is given by the transformation

$$[L] = \begin{bmatrix} 1 & 0 & 0 \\ 0 & 0 & 1 \\ 0 & -1 & 0 \end{bmatrix} .$$

5.2 Give the engineering notation stiffness matrix for an isotropic material in a form analogous to those given for a monoclinic material (5.3.22), an orthotropic material (5.3.24), and a transversely isotropic material (5.3.32). Present your result in terms of the Lamé constants.

5.3 Give the engineering notation compliance matrix for an isotropic material in a form analogous to those given for a monoclinic material (5.3.22), an orthotropic material (5.3.24), and a transversely isotropic material (5.3.32). Present your result in terms of Young's modulus and Poisson's ratio.

5.4 Consider a homogeneous stress field given with respect to the vector basis $\{\hat{e}_i\}$ by the following matrix of scalar components:

$$[\sigma] = \begin{bmatrix} A \cos^2 \phi & A \sin \phi \cos \phi & 0 \\ A \sin \phi \cos \phi & A \sin^2 \phi & 0 \\ 0 & 0 & 0 \end{bmatrix} ,$$

where A is a constant.

(a) Determine the principal stresses and the principal directions of stress (let $\hat{e}_3$ be one of the principal directions of stress).

(b) Suppose that the body experiencing this homogeneous stress field is orthotropic, with the planes of symmetry aligned with the vector basis $\{\hat{e}_i\}$, so that the engineering notation compliance matrix is of the form

$$\begin{bmatrix} \bar{\varepsilon}_1 \\ \bar{\varepsilon}_2 \\ \bar{\varepsilon}_3 \\ \bar{\varepsilon}_4 \\ \bar{\varepsilon}_5 \\ \bar{\varepsilon}_6 \end{bmatrix} = \begin{bmatrix} \bar{S}_{11} & \bar{S}_{12} & \bar{S}_{13} & 0 & 0 & 0 \\ \bar{S}_{12} & \bar{S}_{22} & \bar{S}_{23} & 0 & 0 & 0 \\ \bar{S}_{13} & \bar{S}_{23} & \bar{S}_{33} & 0 & 0 & 0 \\ 0 & 0 & 0 & \bar{S}_{44} & 0 & 0 \\ 0 & 0 & 0 & 0 & \bar{S}_{55} & 0 \\ 0 & 0 & 0 & 0 & 0 & \bar{S}_{66} \end{bmatrix} \begin{bmatrix} \bar{\sigma}_1 \\ \bar{\sigma}_2 \\ \bar{\sigma}_3 \\ \bar{\sigma}_4 \\ \bar{\sigma}_5 \\ \bar{\sigma}_6 \end{bmatrix} .$$

Determine the corresponding scalar components of the strain tensor.

(c) Determine the principal directions of strain and show that they are the same as the principal directions of stress if either $\phi = 0$ or the scalar components of the compliance are those of an isotropic material (see Problem 5.3).

5.5 Given only that, for an isotropic material,

$$K = \lambda + \frac{2}{3}\mu, \quad E = \frac{\mu(3\lambda + 2\mu)}{\lambda + \mu}, \quad \nu = \frac{\lambda}{2(\lambda + \mu)},$$

verify that

(a) $\mu = \dfrac{E - 3\lambda + r}{4}$ (b) $K = \dfrac{E + 3\lambda + r}{6}$,

where $r = \sqrt{E^2 + 9\lambda^2 + 2E\lambda}$.

5.6 Consider an isotropic carbon steel ($E = 207$ GPa, $\nu = 0.292$) rod in uniaxial tension such that $\sigma = 200\hat{e}_1\hat{e}_1$ (MPa). Determine the dilatation in the rod.

5.7 For an isotropic material with $E = 71.0$ GPa and $G = 26.6$ GPa, determine the stress tensor and the strain energy density at a point in a body if the strain tensor at that point is given with respect to the vector basis $\{\hat{e}_i\}$ by the following matrix of scalar components:

$$[\varepsilon] = \begin{bmatrix} 200 & 100 & 0 \\ 100 & 300 & 400 \\ 0 & 400 & 0 \end{bmatrix} \times 10^{-6}.$$

5.8 For an isotropic material with $E = 71.0$ GPa and $G = 26.6$ GPa, determine the strain tensor and the strain energy density at a point in a body if the stress tensor at that point is given with respect to the vector basis $\{\hat{e}_i\}$ by the following matrix of scalar components:

$$[\sigma] = \begin{bmatrix} 20 & -4.0 & 5.0 \\ -4.0 & 0 & 10 \\ 5.0 & 10 & 15 \end{bmatrix} \text{ (MPa)}.$$

Chapter 6

Linearized Elasticity Problems

Problems in the linearized theory of elasticity will require finding the solution of a set of coupled partial differential equations known as field equations, over a specified domain that is the region occupied by the reference configuration of the body, subject to prescribed boundary conditions on the boundary of the domain. In other words, linearized elasticity problems are *boundary value problems*.

Unless stated otherwise, it is assumed from now on that all deformations are infinitesimal, that all materials are linear elastic and isotropic, and that all bodies are materially homogeneous. It is understood that tensor fields, such as the displacement and the stress, are fields given over the reference configuration of a body. If desired, the current configuration of a body can be constructed from the displacement field given over the reference configuration, once it is known. In general, the unknowns to be solved for in a linearized elasticity problem are the displacement, strain, and stress fields. In this chapter, it will be seen how linearized elasticity problems are formulated. Some general results that will be useful in solving these problems will be presented, different approaches to finding solutions are briefly discussed, and a couple of representative solutions are derived.

6.1 Field Equations

In general, linearized elasticity problems require one to determine 15 unknown scalar fields whose common domain is the region occupied by the reference configuration of a given body; there are 3 scalar components of the displacement field $\mathbf{u}$, 6 independent scalar components of the strain field $\boldsymbol{\varepsilon}$, and 6 independent scalar components of the stress field $\boldsymbol{\sigma}$ (recall that $\boldsymbol{\varepsilon}$ is symmetric by construction and that the local balance of angular momentum requires that $\boldsymbol{\sigma}$ be symmetric). These 15 scalar fields must satisfy the following 15 scalar *field equations*; 6 strain-displacement equations (3.4.14),

$$\varepsilon = \frac{1}{2}\left[\nabla u + (\nabla u)^T\right] , \quad \varepsilon_{ij} = \frac{1}{2}(u_{i,j} + u_{j,i}) ,$$

3 equilibrium equations (4.3.10),

$$\nabla \cdot \sigma^T + f = 0 , \quad \sigma_{ji,j} + f_i = 0 ,$$

and 6 constitutive equations (5.4.3),

$$\sigma = 2\mu\varepsilon + \lambda(\operatorname{tr}\varepsilon)I , \quad \sigma_{ij} = 2\mu\varepsilon_{ij} + \lambda\varepsilon_{kk}\delta_{ij} .$$

Note that these field equations are the same for all linearized elasticity problems—it is the boundary conditions, as discussed below, that distinguish one problem from another.

In addition to the 15 field equations given above, there are also 6 scalar compatibility equations (3.4.37),[1]

$$\nabla \times (\nabla \times \varepsilon) = 0 , \quad e_{ikr}e_{jls}\varepsilon_{ij,kl} = 0 .$$

Recall that these compatibility equations are necessary and sufficient conditions on the strain field to ensure the existence of a corresponding displacement field in a simply connected body. However, so long as the displacement field is one of the unknowns that one is trying to determine, the compatibility equations need not be explicitly considered—if a possible solution to a problem includes a continuous, single-valued displacement field, then there is no question of compatibility. It is only when one seeks a solution to a problem in the form of stress and/or strain fields, without solving explicitly for the displacement field, that the compatibility equations (3.4.37) must be included as field equations.

The field equations of linearized elasticity can be combined in various ways to eliminate unknowns and thus arrive at reduced forms of the field equations, involving a reduced number of equations and unknowns. The Lamé-Navier equations and Michell's equations, discussed next, are two particular reduced forms of the field equations.

6.1.1   Lamé-Navier Equations

The 15 scalar field equations can be combined to eliminate the stress and strain fields, thus leaving 3 scalar field equations for the 3 scalar displacement fields. Combining first the strain-displacement equations (3.4.14) and

---

[1] Actually, it can be shown that the six scalar compatibility equations given by (3.4.37) are not independent—the number of *independent* compatibility equations is actually three (Malvern 1969).

the constitutive equations (5.4.3) to eliminate the strain field, it follows that the scalar components of the stress field are given directly in terms of those of the displacement field by

$$\sigma_{ij} = \mu(u_{i,j} + u_{j,i}) + \lambda u_{k,k}\, \delta_{ij} \ . \tag{6.1.1}$$

Then, substituting (6.1.1) into the equilibrium equations (4.3.10) yields the equilibrium equations in terms of the scalar components of the displacement field

$$\begin{aligned}
0 &= \sigma_{ji,j} + f_i \\
&= \mu(u_{j,ij} + u_{i,jj}) + \lambda u_{k,kj}\, \delta_{ji} + f_i \\
&= \mu u_{k,ki} + \mu u_{i,jj} + \lambda u_{k,ki} + f_i \ ,
\end{aligned} \tag{6.1.2}$$

which can be rewritten as

$$\boxed{\begin{aligned}
(\lambda + \mu)\boldsymbol{\nabla}(\boldsymbol{\nabla} \cdot \mathbf{u}) + \mu\nabla^2 \mathbf{u} + \mathbf{f} &= \mathbf{0} \ , \\
(\lambda + \mu)u_{k,ki} + \mu u_{i,jj} + f_i &= 0 \ .
\end{aligned}} \tag{6.1.3}$$

These *Lamé-Navier equations*, which represent the equilibrium equation expressed in terms of the displacement field, provide three scalar field equations for three unknown scalar displacement fields. For any displacement field satisfying (6.1.3), the corresponding strain field is given by the strain-displacement equations (3.4.14) and the corresponding stress field, given either by the constitutive equations (5.4.3) or by the stress-displacement equations (6.1.1), will satisfy the equilibrium equations. The Lamé-Navier equations are the field equations to use when the displacement field is the only unknown that one wishes to determine. In practice, however, boundary conditions as discussed in Section 6.2 will very often require one to determine the stress field as well, which somewhat limits the utility of the Lamé-Navier equations.

Consider the special case in which the body force field is homogeneous. Such would be the case, for instance, in a body with uniform density in a uniform gravitational field. Taking the divergence of the Lamé-Navier equations (i.e., differentiating with respect to X_i) in this case yields

$$(\lambda + \mu)u_{k,kii} + \mu u_{i,jji} = (\lambda + 2\mu)u_{k,kii} = 0 \ . \tag{6.1.4}$$

Noting that $u_{k,k} = \varepsilon_{kk}$, it follows that

$$\nabla^2(\mathrm{tr}\,\varepsilon) = 0 \ , \quad \varepsilon_{kk,ii} = 0 \ . \tag{6.1.5}$$

Taking the Laplacian of the Lamé-Navier equations (i.e., differentiating twice with respect to X_m) yields

$$(\lambda + \mu)u_{k,kimm} + \mu u_{i,jjmm} = (\lambda + \mu)\varepsilon_{kk,mmi} + \mu u_{i,jjmm} = 0 \ . \tag{6.1.6}$$

However, from (6.1.5), $\varepsilon_{kk,mm} = 0$ so that

$$\nabla^4 \mathbf{u} = \mathbf{0} , \qquad u_{i,jjmm} = 0 . \tag{6.1.7}$$

Thus, it is interesting to note that if the body force field is homogeneous, then the dilatation field $e = \text{tr}\,\boldsymbol{\varepsilon}$ is harmonic (6.1.5) and the displacement field $\mathbf{u}$ is biharmonic (6.1.7). These are necessary but *insufficient* conditions for equilibrium—the Lamé-Navier equations (6.1.3) must still be satisfied.

6.1.2   Michell's and Beltrami's Equations

As previously noted, any set of field equations that does not involve the displacement field as an unknown must include the compatibility equations (3.4.37). To obtain a set of field equations for the stress field alone, the compatibility equations can be combined with the constitutive equations to eliminate the strain field, leaving an expression for compatibility in terms of the stress field. This result along with the equilibrium equations provide a set of six independent equations for the six unknown scalar components of the stress field.

By substituting the constitutive equations

$$\varepsilon_{ij} = \frac{1}{E}[(1+\nu)\sigma_{ij} - \nu\sigma_{kk}\delta_{ij}] \tag{6.1.8}$$

into the compatibility equations

$$e_{ikr}e_{jls}\varepsilon_{ij,kl} = 0 , \tag{6.1.9}$$

one obtains the compatibility equations in terms of the stress field:

$$
\begin{aligned}
0 &= e_{ikr}e_{jls}[(1+\nu)\sigma_{ij,kl} - \nu\sigma_{mm,kl}\,\delta_{ij}] \\
&= (1+\nu)e_{ikr}e_{jls}\sigma_{ij,kl} - \nu(\delta_{kl}\delta_{rs} - \delta_{ks}\delta_{rl})\sigma_{mm,kl} \\
&= (1+\nu)e_{ikr}e_{jls}\sigma_{ij,kl} - \nu(\sigma_{mm,kk}\,\delta_{rs} - \sigma_{mm,sr}) .
\end{aligned} \tag{6.1.10}
$$

It follows from (2.2.25) that the outer product of two permutation tensors in the first term of (6.1.10) can be rewritten in terms of Kronecker deltas as

$$
\begin{aligned}
e_{ikr}e_{jls} &= \delta_{ij}\delta_{kl}\delta_{rs} + \delta_{il}\delta_{ks}\delta_{rj} + \delta_{is}\delta_{kj}\delta_{rl} \\
&\quad - \delta_{ij}\delta_{ks}\delta_{rl} - \delta_{il}\delta_{kj}\delta_{rs} - \delta_{is}\delta_{kl}\delta_{rj} .
\end{aligned} \tag{6.1.11}
$$

Thus, substituting (6.1.11) into (6.1.10) and simplifying, the compatibility equations in terms of the stress field reduce to

$$\sigma_{ii,jj}\,\delta_{rs} - \sigma_{ii,rs} - (1+\nu)(\sigma_{ij,ij}\,\delta_{rs} + \sigma_{rs,ii} - \sigma_{is,ir} - \sigma_{ir,is}) = 0 . \tag{6.1.12}$$

Taking the contraction $r \to s$ of (6.1.12) yields

$$2\sigma_{ii,jj} - (1+\nu)(\sigma_{ij,ij} + \sigma_{ii,jj}) = 0 , \qquad (6.1.13)$$

so that, after simplifying, one obtains the necessary condition for compatibility,

$$\sigma_{ii,jj} = \frac{1+\nu}{1-\nu}\sigma_{ij,ij} . \qquad (6.1.14)$$

Thus, by substituting (6.1.14) back into (6.1.12), the compatibility equations in terms of stress can now be written as

$$\sigma_{rs,ii} + \frac{1}{1+\nu}\sigma_{ii,rs} = \frac{\nu}{1-\nu}\sigma_{ij,ij}\,\delta_{rs} + \sigma_{is,ir} + \sigma_{ir,is} . \qquad (6.1.15)$$

Finally, assuming that the stress field satisfies the equilibrium equation, it follows that $\sigma_{ij,ik} = -f_{j,k}$ and $\sigma_{ij,ij} = -f_{j,j}$. Therefore, from (6.1.14), a necessary condition for an equilibrated stress field to satisfy compatibility is

$$\nabla^2(\operatorname{tr}\boldsymbol{\sigma}) = -\frac{1+\nu}{1-\nu}\boldsymbol{\nabla}\cdot\mathbf{f} , \quad \sigma_{ii,jj} = -\frac{1+\nu}{1-\nu}f_{j,j} , \qquad (6.1.16)$$

and, from (6.1.15), a *necessary and sufficient condition* is

$$\boxed{\begin{aligned} &\nabla^2\boldsymbol{\sigma} + \frac{1}{1+\nu}\boldsymbol{\nabla}\left[\boldsymbol{\nabla}(\operatorname{tr}\boldsymbol{\sigma})\right] = -\frac{\nu}{1-\nu}(\boldsymbol{\nabla}\cdot\mathbf{f})\mathbf{I} - \left[\boldsymbol{\nabla}\mathbf{f} + (\boldsymbol{\nabla}\mathbf{f})^{\mathsf{T}}\right] , \\ &\sigma_{ij,kk} + \frac{1}{1+\nu}\sigma_{kk,ij} = -\frac{\nu}{1-\nu}f_{k,k}\,\delta_{ij} - (f_{i,j} + f_{j,i}) . \end{aligned}} \qquad (6.1.17)$$

These are *Michell's equations* and represent the compatibility equations in terms of stress. Michell's equations and the equilibrium equations constitute a set of field equations for the stress field. For any stress field satisfying these field equations, the corresponding strain field can easily be found from the constitutive equations and the corresponding displacement field can then be found to within an arbitrary rigid-body motion by integrating the strain-displacement equations.

If the body force field $\mathbf{f}$ is homogeneous, then the necessary condition (6.1.16) for compatibility reduces to

$$\nabla^2(\operatorname{tr}\boldsymbol{\sigma}) = 0 , \quad \sigma_{ii,jj} = 0 , \qquad (6.1.18)$$

and the necessary and sufficient Michell's equations (6.1.17) reduce to *Beltrami's equations*:

$$\boxed{\nabla^2\boldsymbol{\sigma} + \frac{1}{1+\nu}\boldsymbol{\nabla}\left[\boldsymbol{\nabla}(\operatorname{tr}\boldsymbol{\sigma})\right] = \mathbf{0} , \quad \sigma_{ij,kk} + \frac{1}{1+\nu}\sigma_{kk,ij} = 0 .} \qquad (6.1.19)$$

Thus, Beltrami's equation represent the compatibility equations in terms of the stress field in the special (but common) case that the body force field is homogeneous.

By taking the Laplacian of (6.1.19), it follows that another necessary condition for compatibility is

$$\nabla^4 \boldsymbol{\sigma} + \frac{1}{1+\nu} \nabla \left\{ \nabla \left[\nabla^2 (\operatorname{tr} \boldsymbol{\sigma}) \right] \right\} = \mathbf{0} \,. \tag{6.1.20}$$

However, from the previous necessary condition (6.1.18), the second term is zero, so that this necessary condition reduces to

$$\nabla^4 \boldsymbol{\sigma} = \mathbf{0} \,, \quad \sigma_{ij,kkmm} = 0 \,. \tag{6.1.21}$$

Thus, it is interesting to note that if the body force field is homogeneous, then necessary (but *insufficient*) conditions for compatibility are that the mean stress field $\tilde{\sigma} = \frac{1}{3} \operatorname{tr} \boldsymbol{\sigma}$ be harmonic (6.1.18) and the stress field $\boldsymbol{\sigma}$ be biharmonic (6.1.21).

6.2 Boundary Conditions

The problem of the linearized theory of elasticity is to find displacement, strain, and stress fields that satisfy the field equations everywhere within the region $\mathcal{R}$ occupied by (the reference configuration of) a body while meeting specified conditions, known as *boundary conditions*, on the boundary $\partial \mathcal{R}$ of the body. This is the boundary value problem of linearized elasticity.

At a point on the boundary of a body, either the motion of the point is constrained in some known way, a known force is exerted, or there is some known combination of kinematic constraint and applied force. Mathematically, these are expressed as the boundary conditions discussed below. In order for a boundary value problem of elasticity to be wellposed, suitable boundary conditions must be specified for *every* material point on the boundary of a body.

Consider a body occupying a region with boundary $\partial \mathcal{R}$, and let $\mathbf{X}$ be a point on the boundary ($\mathbf{X} \in \partial \mathcal{R}$). For a *displacement boundary condition* at $\mathbf{X}$, the displacement vector at $\mathbf{X}$ is completely specified. Thus, if $\partial \mathcal{R}^u \subset \partial \mathcal{R}$ is the portion of the boundary for which the displacement vector is completely prescribed (Figure 6.1), then the displacement boundary condition for the body can be expressed as

$$\boxed{\mathbf{u} = \tilde{\mathbf{u}} \,, \quad u_i = \tilde{u}_i \quad \forall \mathbf{X} \in \partial \mathcal{R}^u \,,} \tag{6.2.1}$$

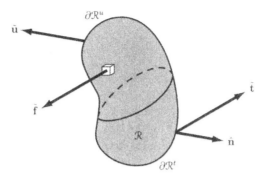

FIGURE 6.1. Specification of boundary conditions.

where $\tilde{\mathbf{u}}(\mathbf{X}, t)$ is a given function which, along with $\partial \mathcal{R}^u$, depends on the problem to be solved. The notation "$\forall \mathbf{X} \in \partial \mathcal{R}^u$" just means "for all $\mathbf{X}$ in $\partial \mathcal{R}^u$." For a *traction boundary condition* at $\mathbf{X}$, the traction vector at $\mathbf{X}$ is completely specified. Thus, if $\partial \mathcal{R}^t \subset \partial \mathcal{R}$ is the portion of the boundary for which the traction vector is completely prescribed (Figure 6.1) and using the traction-stress relation $\mathbf{t} = \hat{\mathbf{n}} \cdot \boldsymbol{\sigma}$, then the traction boundary condition for the body can be expressed as

$$\boxed{\hat{\mathbf{n}} \cdot \boldsymbol{\sigma} = \tilde{\mathbf{t}}\,, \quad \sigma_{ji}\hat{n}_j = \tilde{t}_i \quad \forall \mathbf{X} \in \partial \mathcal{R}^t\,,} \tag{6.2.2}$$

where $\tilde{\mathbf{t}}(\mathbf{X}, t)$ is another given function that depends on the problem to be solved and $\hat{\mathbf{n}}(\mathbf{X})$ is the unit outward normal at each point $\mathbf{X} \in \partial \mathcal{R}^t$. Note that the stress tensor $\boldsymbol{\sigma}$ at points on the surface $\partial \mathcal{R}^t$ is only partially determined by the specification of tractions—there are six independent components of stress but only three components of the specified traction.

Although it is not possible to completely prescribe both the displacement vector and the traction vector at the same material point on the boundary of a body, it is possible to prescribe part of the displacement and a complimentary part of the traction, resulting in a *mixed boundary condition*. In one form of mixed boundary condition, the scalar component of the displacement in the direction of a unit vector $\hat{\boldsymbol{\mu}}$ and the orthogonal projection of the traction onto the plane with unit normal $\hat{\boldsymbol{\mu}}$ are prescribed. Thus, if $\partial \mathcal{R}^{m1} \subset \partial \mathcal{R}$ is the portion of the boundary for which this type of mixed boundary condition is prescribed (not shown in Figure 6.1), then it can be expressed as

$$\boxed{\mathbf{u} \cdot \hat{\boldsymbol{\mu}} = \tilde{u}\,, \quad u_i \hat{\mu}_i = \tilde{u} \quad \forall \mathbf{X} \in \partial \mathcal{R}^{m1}\,,} \tag{6.2.3a}$$

where the scalar-valued function $\tilde{u}(\mathbf{X}, t)$ is given and

$$
\left.\begin{aligned}
\hat{\mathbf{n}} \cdot \boldsymbol{\sigma} - (\hat{\mathbf{n}} \cdot \boldsymbol{\sigma} \cdot \hat{\boldsymbol{\mu}})\hat{\boldsymbol{\mu}} &= \tilde{\mathbf{t}} \\
\sigma_{ji}\hat{n}_j - (\sigma_{jk}\hat{n}_j\hat{\mu}_k)\hat{\mu}_i &= \tilde{t}_i
\end{aligned}\right\} \quad \forall \mathbf{X} \in \partial \mathcal{R}^{m1} ,
\tag{6.2.3b}
$$

where the given vector-valued function $\tilde{\mathbf{t}}(\mathbf{X}, t)$ must be orthogonal to the unit vector $\hat{\boldsymbol{\mu}}$ (i.e., $\tilde{\mathbf{t}} \cdot \hat{\boldsymbol{\mu}} = 0$). For example, if $\hat{\mathbf{n}} = \hat{\mathbf{e}}_1$ and $\hat{\boldsymbol{\mu}} = \hat{\mathbf{e}}_1$, then the mixed boundary conditions (6.2.3) become $u_1 = \tilde{u}$, $\sigma_{12} = \tilde{t}_2$, and $\sigma_{13} = \tilde{t}_3$, where $\tilde{u}(\mathbf{X}, t)$, $\tilde{t}_2(\mathbf{X}, t)$, and $\tilde{t}_3(\mathbf{X}, t)$ are given scalar-valued functions of position and time. In the other form of mixed boundary condition, the scalar component of the traction in the direction of a unit vector $\hat{\boldsymbol{\mu}}$ and the orthogonal projection of the displacement onto the plane with unit normal $\hat{\boldsymbol{\mu}}$ are prescribed. Thus, if $\partial \mathcal{R}^{m2} \subset \partial \mathcal{R}$ is the portion of the boundary for which this type of mixed boundary condition is prescribed (not shown in Figure 6.1), then it can be expressed as

$$
\hat{\mathbf{n}} \cdot \boldsymbol{\sigma} \cdot \hat{\boldsymbol{\mu}} = \tilde{t} , \quad \hat{n}_i \sigma_{ij} \hat{\mu}_j = \tilde{t} \quad \forall \mathbf{X} \in \partial \mathcal{R}^{m2} ,
\tag{6.2.4a}
$$

where the scalar-valued function $\tilde{t}(\mathbf{X}, t)$ is given and

$$
\left.\begin{aligned}
\mathbf{u} - (\mathbf{u} \cdot \hat{\boldsymbol{\mu}})\hat{\boldsymbol{\mu}} &= \tilde{\mathbf{u}} \\
u_i - (u_j\hat{\mu}_j)\hat{\mu}_i &= \tilde{u}_i
\end{aligned}\right\} \quad \forall \mathbf{X} \in \partial \mathcal{R}^{m2} ,
\tag{6.2.4b}
$$

where the given vector-valued function $\tilde{\mathbf{u}}(\mathbf{X}, t)$ must be orthogonal to the unit vector $\hat{\boldsymbol{\mu}}$ (i.e. $\tilde{\mathbf{u}} \cdot \hat{\boldsymbol{\mu}} = 0$).

Suitable boundary conditions must be specified at *every* point on the boundary of a body. Therefore, every point on the boundary $\partial \mathcal{R}$ of the body is in one and only one of the four subsets $\partial \mathcal{R}^u$, $\partial \mathcal{R}^t$, $\partial \mathcal{R}^{m1}$, and $\partial \mathcal{R}^{m2}$. In other words, these material surfaces cover the boundary of the body without overlapping. If, as is frequently the case, there are no mixed boundary conditions so that $\partial \mathcal{R}^{m1} = \emptyset$ and $\partial \mathcal{R}^{m2} = \emptyset$ (see Figure 6.1), then a mathematical expression of this statement is $\partial \mathcal{R}^u \cup \partial \mathcal{R}^t = \partial \mathcal{R}$ and $\partial \mathcal{R}^u \cap \partial \mathcal{R}^t = \emptyset$.

In addition to a complete statement of the boundary conditions, the body force field must also be specified:

$$
\mathbf{f} = \tilde{\mathbf{f}} , \quad f_i = \tilde{f}_i \quad \forall \mathbf{X} \in \mathcal{R} ,
\tag{6.2.5}
$$

where $\tilde{\mathbf{f}}(\mathbf{X}, t)$ is a given vector-valued function. Often, the body force field is negligible so that $\tilde{\mathbf{f}} \approx 0$—if no mention of the body force field is made, then it is implicitly assumed to be negligible. It is the boundary conditions (6.2.1)–(6.2.4) and body force field (6.2.5) which define a problem.

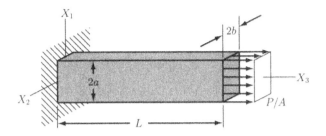

FIGURE 6.2. Prismatic bar in tension.

Example 6.1

Consider the prismatic bar shown in Figure 6.2. The bar has a rectangular cross section of dimensions $2a \times 2b$ and is of length L. One end of the bar is fixed to a rigid support and the other end has a tensile load P uniformly distributed over its area, $A = 4ab$. Using the coordinate system shown, where the X_3-axis is coincident with the centroidal axis of the bar, give the boundary conditions for this problem.

Solution. The surface of the bar at $X_3 = 0$ has a displacement boundary condition. Specifically, the prescribed displacement on this surface is zero, $\tilde{\mathbf{u}} = \mathbf{0}$, since the wall to which it is attached is assumed to be rigid. This is known as a *fixed-displacement boundary condition*. Therefore,

$$u_i = 0 \quad \text{on } X_3 = 0 .$$

The lateral surfaces of the bar, formed by the planes $X_1 = \pm a$ and $X_2 = \pm b$ and the end surface at $X_3 = L$ have traction boundary conditions. The prescribed tractions on the lateral surfaces are zero, $\tilde{\mathbf{t}} = \mathbf{0}$ (sometimes referred to as a *traction-free boundary condition*). On the top and bottom lateral surfaces ($X_1 = \pm a$), the unit outward normal is $\hat{\mathbf{n}} = \pm \hat{\mathbf{e}}_1$ so that the traction-stress relation $\mathbf{t} = \hat{\mathbf{n}} \cdot \boldsymbol{\sigma}$ gives $t_i = \pm \sigma_{1i}$. Thus,

$$\sigma_{11} = \sigma_{12} = \sigma_{13} = 0 \quad \text{on } X_1 = \pm a .$$

Similarly, on the front and back surfaces ($X_2 = \pm b$), the unit outward normal is $\hat{\mathbf{n}} = \pm \hat{\mathbf{e}}_2$, the traction-stress relation gives $t_i = \pm \sigma_{2i}$, and

$$\sigma_{21} = \sigma_{22} = \sigma_{23} = 0 \quad \text{on } X_2 = \pm b .$$

Finally, the prescribed traction on the loaded end of the bar ($X_3 = L$) is $\tilde{\mathbf{t}} = (P/A)\hat{\mathbf{e}}_3$. The unit outward normal to this surface is $\hat{\mathbf{n}} = \hat{\mathbf{e}}_3$, the traction-stress relation gives $t_i = \sigma_{3i}$, and

$$\sigma_{31} = \sigma_{32} = 0 , \quad \sigma_{33} = \frac{P}{A} \quad \text{on } X_3 = L .$$

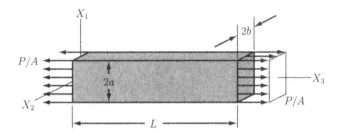

FIGURE 6.3. Prismatic bar in uniaxial tension.

This is a complete statement of the boundary conditions for this problem.

Unfortunately, once boundary conditions have been set it is not generally a simple task to find the solution to the corresponding boundary value problem. Throughout the remainder of this text, methods will be developed to help accomplish this for different types of problems.

Example 6.2

Consider the prismatic bar shown in Figure 6.3. The bar has a rectangular cross section of dimensions $2a \times 2b$ and is of length L. The ends of the bar have a tensile load P uniformly distributed over their area, $A = 4ab$. What is the stress field in the bar?

Solution. The bar has traction-only boundary conditions. The lateral surfaces is traction-free:

$$\sigma_{11} = \sigma_{12} = \sigma_{13} = 0 \quad \text{on } X_1 = \pm a \,,$$
$$\sigma_{21} = \sigma_{22} = \sigma_{23} = 0 \quad \text{on } X_2 = \pm b$$

and the end surfaces have a uniform normal traction:

$$\sigma_{31} = \sigma_{32} = 0 \,, \quad \sigma_{33} = \frac{P}{A} \quad \text{on } X_3 = 0 \text{ and } X_3 = L \,.$$

The homogeneous stress field $\sigma_{33} = P/A$, all other components of stress equal to zero, satisfies these boundary conditions. It also satisfies the equilibrium equations (4.3.10) and Beltrami's equations (6.1.19). Therefore, this is a solution stress field for the problem. It will be shown shortly that this solution is unique; that is, there is no other possible solution to the equilibrium equations and Beltrami's equations that also satisfies these boundary conditions.

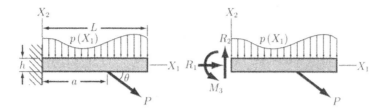

FIGURE 6.4. Cantilever beam problem and its approximation.

6.2.1   Saint-Venant's Principle

It has been shown how boundary value problems in elasticity are precisely formulated through the specification of boundary conditions. It is these boundary conditions which differentiate one problem from another. However, it will often be found that the exact solution to a particular boundary value problem in elasticity is elusive. However, suppose that the solution to a similar boundary value problem, wherein the body is the same but the conditions on part of the boundary have been replaced by statically equivalent conditions, can be found. *Saint-Venant's principle* says that the effect of this change in boundary conditions is local; that is, sufficiently far from the part of the boundary where the conditions are different, the solutions for the two problems will be approximately the same. How far is "sufficiently far" is rather ambiguous, but this will usually mean farther than the largest characteristic length scale of the portion of the boundary where the conditions are different.

Saint-Venant's principle will prove to be an extremely valuable concept. It offers hope in finding an approximate solution to a problem where none may exist for finding an exact solution.

Example 6.3
Consider Examples 6.1 and 6.2. The boundary conditions for these problems are statically equivalent. Therefore, from Saint-Venant's principle, the solution found in Example 6.2 is the approximate solution for the problem in Example 6.1 for positions in the bar sufficiently far, relative to the dimensions a and b, from the surface $X_3 = 0$.

Example 6.4
Consider the cantilever beam shown on the left in Figure 6.4. A distributed load $p(X_1)$ acts in the $-X_2$-direction on the top of the beam while a concentrated load P acts a distance a from the left end of the beam and forms an angle θ with the X_1-direction. The beam has height h, length L, and unit thickness. There is a zero-displacement boundary condition at the left end of the beam, $\mathbf{u} = \mathbf{0}$ on $X_1 = 0$.

Solving for the exact stress, strain, and displacements fields that satisfy
these boundary conditions and the equations of elasticity is a difficult,
maybe impossible, problem. Consider instead the beam shown on the right
in Figure 6.4. Noting that on the left end surface of the beam $\hat{\mathbf{n}} = -\hat{\mathbf{e}}_1$,
the resultant forces and moment shown are related to the tractions on this
surface and the components of stress by

$$R_1 = \int_{-h/2}^{h/2} t_1 \, dX_2 = \int_{-h/2}^{h/2} -\sigma_{11}(0, X_2, X_3) \, dX_2 \ ,$$

$$R_2 = \int_{-h/2}^{h/2} t_2 \, dX_2 = \int_{-h/2}^{h/2} -\sigma_{12}(0, X_2, X_3) \, dX_2 \ ,$$

$$M_3 = \int_{-h/2}^{h/2} -X_2 t_1 \, dX_2 = \int_{-h/2}^{h/2} X_2 \sigma_{11}(0, X_2, X_3) \, dX_2 \ .$$

These resultant forces and moment are statically equivalent to the boundary conditions in the original problem if

$$R_1 = -P \cos\theta \ ,$$

$$R_2 = P \sin\theta + \int_0^L p \, dX_1 \ ,$$

$$M_3 = Pa \sin\theta + \int_0^L p X_1 \, dX_1 \ .$$

These conditions do not uniquely define a traction distribution on the left
end of the beam in the second problem. Thus, there are an infinite number
of possible solution stress fields which satisfy these conditions, the boundary
conditions on the other surfaces of the beam, and the equations of elasticity.
However, Saint-Venant's principle states that $\boldsymbol{\sigma}(\mathbf{X})$, $\boldsymbol{\varepsilon}(\mathbf{X})$, and $\mathbf{u}(\mathbf{X})$ for
any of these solutions will be approximately equal to those for the original
problem for values of $\mathbf{X}$ sufficiently far from the surface $X_1 = 0$. Sufficiently
far in this case is with respect to the dimension h. It will be seen later that
this second problem is much more tractable.

6.3   Useful Consequences of Linearity

Leading up to this point, considerable effort has gone into developing a
linearized theory of elasticity. It is here that this linearization starts to pay
dividends. Of the consequences of linearity discussed below, perhaps the
most important are the principle of superposition and the closely related

fact that there can be only one solution for any given boundary value problem in linear elasticity; that is, solutions are unique.[2]

6.3.1 Superposition Principle

Consider a body occupying the region $\mathcal{R}$ with boundary $\partial\mathcal{R} = \partial\mathcal{R}^u \cup \partial\mathcal{R}^t$ (for simplicity, it is assumed that there are no mixed boundary conditions, although having mixed boundary conditions does not change the basic result). At a point on $\partial\mathcal{R}$, the unit outward normal is $\hat{\mathbf{n}}$. Suppose that for one set of boundary conditions,

$$
\begin{aligned}
\mathbf{u} &= \tilde{\mathbf{u}}^{(1)} && \forall \mathbf{X} \in \partial\mathcal{R}^u \,, \\
\hat{\mathbf{n}} \cdot \boldsymbol{\sigma} &= \tilde{\mathbf{t}}^{(1)} && \forall \mathbf{X} \in \partial\mathcal{R}^t \,, \\
\mathbf{f} &= \tilde{\mathbf{f}}^{(1)} && \forall \mathbf{X} \in \mathcal{R} \,,
\end{aligned}
\tag{6.3.1}
$$

the solution is known to be $\mathbf{u} = \mathbf{u}^{(1)}(\mathbf{X}, t)$, $\boldsymbol{\varepsilon} = \boldsymbol{\varepsilon}^{(1)}(\mathbf{X}, t)$, and $\boldsymbol{\sigma} = \boldsymbol{\sigma}^{(1)}(\mathbf{X}, t)$. Also suppose that for a second set of boundary conditions,

$$
\begin{aligned}
\mathbf{u} &= \tilde{\mathbf{u}}^{(2)} && \forall \mathbf{X} \in \partial\mathcal{R}^u \,, \\
\hat{\mathbf{n}} \cdot \boldsymbol{\sigma} &= \tilde{\mathbf{t}}^{(2)} && \forall \mathbf{X} \in \partial\mathcal{R}^t \,, \\
\mathbf{f} &= \tilde{\mathbf{f}}^{(2)} && \forall \mathbf{X} \in \mathcal{R} \,,
\end{aligned}
\tag{6.3.2}
$$

the solution is known to be $\mathbf{u} = \mathbf{u}^{(2)}(\mathbf{X}, t)$, $\boldsymbol{\varepsilon} = \boldsymbol{\varepsilon}^{(2)}(\mathbf{X}, t)$, and $\boldsymbol{\sigma} = \boldsymbol{\sigma}^{(2)}(\mathbf{X}, t)$. $\mathcal{R}$, $\partial\mathcal{R}^u$, and $\partial\mathcal{R}^t$ are the same in both cases.

Now consider the *superposition* of the above boundary conditions:

$$
\begin{aligned}
\mathbf{u} &= \tilde{\mathbf{u}}^{(1)} + \tilde{\mathbf{u}}^{(2)} && \forall \mathbf{X} \in \partial\mathcal{R}^u \,, \\
\hat{\mathbf{n}} \cdot \boldsymbol{\sigma} &= \tilde{\mathbf{t}}^{(1)} + \tilde{\mathbf{t}}^{(2)} && \forall \mathbf{X} \in \partial\mathcal{R}^t \,, \\
\mathbf{f} &= \tilde{\mathbf{f}}^{(1)} + \tilde{\mathbf{f}}^{(2)} && \forall \mathbf{X} \in \mathcal{R} \,.
\end{aligned}
\tag{6.3.3}
$$

Because the equations of elasticity are linear, the sum of any two solutions of the equations is itself a solution. Thus, *assuming that the prescribed traction and body force fields are independent of the deformation*, it follows immediately that the solution to this third boundary value problem [i.e., the displacement, strain, and stress fields that satisfy the equations of elasticity and the boundary conditions (6.3.3)] are $\mathbf{u} = \mathbf{u}^{(1)}(\mathbf{X}, t) + \mathbf{u}^{(2)}(\mathbf{X}, t)$, $\boldsymbol{\varepsilon} = \boldsymbol{\varepsilon}^{(1)}(\mathbf{X}, t) + \boldsymbol{\varepsilon}^{(2)}(\mathbf{X}, t)$, and $\boldsymbol{\sigma} = \boldsymbol{\sigma}^{(1)}(\mathbf{X}, t) + \boldsymbol{\sigma}^{(2)}(\mathbf{X}, t)$. This is the *superposition principle*.

---

[2]The more difficult question of whether or not a solution to a particular boundary value problem exists will not be considered here. Suffice it to say that the existence of solutions has been established under very general conditions (Sokolnikoff 1956).

6.3.2 Uniqueness of Solutions

Suppose that, for a given set of boundary conditions, it is possible to obtain two distinct solutions to the equations of linearized elasticity that satisfy the boundary conditions, say $\mathbf{u} = \mathbf{u}^{(1)}(\mathbf{X}, t)$, $\boldsymbol{\varepsilon} = \boldsymbol{\varepsilon}^{(1)}(\mathbf{X}, t)$, and $\boldsymbol{\sigma} = \boldsymbol{\sigma}^{(1)}(\mathbf{X}, t)$ and $\mathbf{u} = \mathbf{u}^{(2)}(\mathbf{X}, t)$, $\boldsymbol{\varepsilon} = \boldsymbol{\varepsilon}^{(2)}(\mathbf{X}, t)$, and $\boldsymbol{\sigma} = \boldsymbol{\sigma}^{(2)}(\mathbf{X}, t)$. The superposition principle above works equally well if, instead of addition of boundary conditions and solutions, one considers subtraction. In particular, in the present situation where one has two cases in which the boundary conditions in each case are the same but the solutions are different, a solution to the equations of elasticity that satisfies the boundary conditions formed by subtracting one set of boundary conditions from the other,

$$\mathbf{u} = \tilde{\mathbf{u}}^{(1)} - \tilde{\mathbf{u}}^{(2)} = \mathbf{0} \quad \forall \mathbf{X} \in \partial \mathcal{R}^u \,,$$
$$\hat{\mathbf{n}} \cdot \boldsymbol{\sigma} = \tilde{\mathbf{t}}^{(1)} - \tilde{\mathbf{t}}^{(2)} = \mathbf{0} \quad \forall \mathbf{X} \in \partial \mathcal{R}^t \,, \qquad (6.3.4)$$
$$\mathbf{f} = \tilde{\mathbf{f}}^{(1)} - \tilde{\mathbf{f}}^{(2)} = \mathbf{0} \quad \forall \mathbf{X} \in \mathcal{R} \,,$$

is found by subtracting one solution from the other. Thus, from the superposition principle, a solution corresponding to the boundary conditions (6.3.4) is given by the difference in the two distinct solutions for the original problem; $\mathbf{u} = \Delta\mathbf{u}(\mathbf{X}, t) \equiv \mathbf{u}^{(1)}(\mathbf{X}, t) - \mathbf{u}^{(2)}(\mathbf{X}, t)$, $\boldsymbol{\varepsilon} = \Delta\boldsymbol{\varepsilon}(\mathbf{X}, t) \equiv \boldsymbol{\varepsilon}^{(1)}(\mathbf{X}, t) - \boldsymbol{\varepsilon}^{(2)}(\mathbf{X}, t)$, and $\boldsymbol{\sigma} = \Delta\boldsymbol{\sigma}(\mathbf{X}, t) \equiv \boldsymbol{\sigma}^{(1)}(\mathbf{X}, t) - \boldsymbol{\sigma}^{(2)}(\mathbf{X}, t)$ are a solution to (6.3.4). However, in (6.3.4), clearly no work is done on the body by external forces. It will be shown (see Chapter 10) that this implies that the strain energy density vanishes ($\mathcal{U} = 0$) everywhere in the body. Since $\mathcal{U}$ is assumed to be a positive definite function of strain, it follows that $\Delta\boldsymbol{\varepsilon}(\mathbf{X}, t) = \mathbf{0}$ and therefore that $\Delta\boldsymbol{\sigma}(\mathbf{X}, t) = \mathbf{0}$. Since the difference between any two solution stress and strain fields must be zero, the solution stress and strain fields are unique; there cannot be two distinct solutions to any particular problem. The displacement field $\Delta\mathbf{u}(\mathbf{X}, t)$ corresponding to $\Delta\boldsymbol{\varepsilon}(\mathbf{X}, t) = \mathbf{0}$ is a rigid-body motion and so the solution displacement field is unique to within a rigid-body motion (which may be determined by the displacement boundary condition).

In using the superposition principle, it has been assumed that the prescribed traction and body force fields are independent of the deformation. If this assumption is violated, then there is no guarantee of uniqueness.

Example 6.5
Consider an arbitrary body occupying the region $\mathcal{R}$ with boundary $\partial \mathcal{R}$ (Figure 6.5). At a point on $\partial \mathcal{R}$, the unit outward normal is $\hat{\mathbf{n}}$. The body is subject to a uniform hydrostatic tension of magnitude A and body forces are negligible:

$$\hat{\mathbf{n}} \cdot \boldsymbol{\sigma} = A\hat{\mathbf{n}} \,, \quad \sigma_{ji}\hat{n}_j = An_i \quad \forall \mathbf{X} \in \partial \mathcal{R} \,,$$
$$\mathbf{f} = \mathbf{0} \,, \qquad f_i = 0 \qquad \forall \mathbf{X} \in \mathcal{R} \,.$$

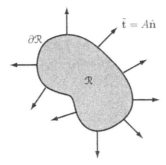

FIGURE 6.5. Body in hydrostatic tension.

Find the solution stress, strain, and displacement fields.

Solution. One can guess that the solution stress field for this problem is

$$\boldsymbol{\sigma} = A\mathbf{I} , \quad \sigma_{ij} = A\delta_{ij} .$$

This stress field satisfies the boundary conditions: $\sigma_{ji}\hat{n}_j = A\delta_{ji}\hat{n}_j = An_i$. It also satisfies equilibrium ($\sigma_{ji,j} = 0$) and Beltrami's equations (6.1.19). This then is the solution stress field since it satisfies the equations of elasticity and the boundary conditions. Furthermore, since solutions of linearized elasticity problems are unique, it is the only possible solution. The mean stress in the body is $\tilde{\sigma} = A$ and the deviatoric stress is zero. On every internal surface, the normal traction is A and the shear traction is zero.

Solving the constitutive relation (5.4.15) for the strain leads to

$$\boldsymbol{\varepsilon} = \frac{A}{3K}\mathbf{I} , \quad \varepsilon_{ij} = \frac{A}{3K}\delta_{ij} ,$$

where K is the bulk modulus. The dilatation is $e = A/K$ and the deviatoric strain is zero. The body expands (or contracts if $A < 0$) without changing shape. To within a rigid-body motion, the solution to the strain-displacement equations is

$$\mathbf{u} = \frac{A}{3K}\mathbf{X} , \quad u_i = \frac{A}{3K}X_i .$$

This displacement field corresponds to a body which has its rigid-body motion fixed such that there is neither translation nor rotation of the material point at the origin: $\mathbf{u}|_{\mathbf{X}=\mathbf{0}} = \mathbf{0}$ and $\boldsymbol{\omega}|_{\mathbf{X}=\mathbf{0}} = \mathbf{0}$.

6.3.3   Clapeyron's Theorem

Recall that the strain energy density for linear elastic deformation is given by $\mathcal{U} = \frac{1}{2}\sigma_{ij}\varepsilon_{ij}$. The total strain energy in a body occupying the region $\mathcal{R}$ with boundary $\partial\mathcal{R}$ is therefore

$$\int_{\mathcal{R}} \mathcal{U}\, dV = \frac{1}{2}\int_{\mathcal{R}} \sigma_{ij}\varepsilon_{ij}\, dV \ . \tag{6.3.5}$$

Since $\sigma_{ij} = \sigma_{ji}$, it follows from the strain-displacement equations that $\sigma_{ij}\varepsilon_{ij} = \sigma_{ji}u_{i,j}$. Therefore,

$$\int_{\mathcal{R}} \mathcal{U}\, dV = \frac{1}{2}\int_{\mathcal{R}} \sigma_{ji}u_{i,j}\, dV$$

$$= \frac{1}{2}\int_{\mathcal{R}} (\sigma_{ji}u_i)_{,j}\, dV - \frac{1}{2}\int_{\mathcal{R}} \sigma_{ji,j}\, u_i\, dV \ . \tag{6.3.6}$$

Finally, applying the divergence theorem, the traction-stress relations, and the equilibrium equations, it follows that

$$\int_{\mathcal{R}} \mathcal{U}\, dV = \frac{1}{2}\int_{\partial\mathcal{R}} \sigma_{ji}u_i\hat{n}_j\, dA + \frac{1}{2}\int_{\mathcal{R}} f_i u_i\, dV$$

$$= \frac{1}{2}\int_{\partial\mathcal{R}} t_i u_i\, dA + \frac{1}{2}\int_{\mathcal{R}} f_i u_i\, dV \ . \tag{6.3.7}$$

Thus, the total strain energy in a body is given by

$$\boxed{\begin{aligned} \int_{\mathcal{R}} \mathcal{U}\, dV &= \frac{1}{2}\int_{\partial\mathcal{R}} \mathbf{t}\cdot\mathbf{u}\, dA + \frac{1}{2}\int_{\mathcal{R}} \mathbf{f}\cdot\mathbf{u}\, dV \ , \\ \int_{\mathcal{R}} \mathcal{U}\, dV &= \frac{1}{2}\int_{\partial\mathcal{R}} t_i u_i\, dA + \frac{1}{2}\int_{\mathcal{R}} f_i u_i\, dV \ . \end{aligned}} \tag{6.3.8}$$

The right-hand side of (6.3.8) represents the work done by the external forces during linear elastic deformation. *Clapeyron's theorem* says that, in linearized elasticity, the total strain energy in a body is equal to the work done by the external forces.

Example 6.6

Consider the cantilever beam of variable cross section shown in Figure 6.6. A transverse load F acts on the end of the beam, which is displaced an amount u in the direction of the force. For linearized elasticity, the load-displacement curve is linear and it can be seen that the work done by the

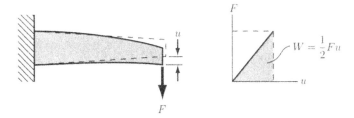

FIGURE 6.6. *Cantilever beam of variable cross section.*

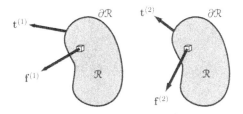

FIGURE 6.7. *Two equilibrium states.*

load, the area under the load-displacement curve, is $\frac{1}{2}Fu$. Equating the work done with the total strain energy of deformation in the beam, it is seen that the total strain energy in the beam is also $\frac{1}{2}Fu$.

6.3.4 Betti-Rayleigh Reciprocity Relations

Consider two equilibrium states of a linearized elastic body, as shown in Figure 6.7. Let Ω_{12} be the work that would be done by the first system of external forces $\mathbf{t}^{(1)}$ and $\mathbf{f}^{(1)}$ in acting through the displacements $\mathbf{u}^{(2)}$ of the second system:

$$
2\Omega_{12} = \int_{\partial \mathcal{R}} t_i^{(1)} u_i^{(2)} \, dA + \int_{\mathcal{R}} f_i^{(1)} u_i^{(2)} \, dV \ . \tag{6.3.9}
$$

From the traction-stress relation, (6.3.9) can be rewritten as

$$
2\Omega_{12} = \int_{\partial \mathcal{R}} \sigma_{ji}^{(1)} \hat{n}_j u_i^{(2)} \, dA + \int_{\mathcal{R}} f_i^{(1)} u_i^{(2)} \, dV \ , \tag{6.3.10}
$$

and applying the divergence theorem,

$$2\Omega_{12} = \int_{\mathcal{R}} (\sigma_{ji}^{(1)} u_i^{(2)})_{,j} \, dV + \int_{\mathcal{R}} f_i^{(1)} u_i^{(2)} \, dV$$

$$= \int_{\mathcal{R}} \left[\sigma_{ji}^{(1)} u_i^{(2)}{}_{,j} + (\sigma_{ji}^{(1)}{}_{,j} + f_i^{(1)}) u_i^{(2)} \right] dV . \qquad (6.3.11)$$

However, $\sigma_{ji}^{(1)} u_i^{(2)}{}_{,j} = \sigma_{ij}^{(1)} \varepsilon_{ij}^{(2)}$ and, from equilibrium, $\sigma_{ji}^{(1)}{}_{,j} + f_i^{(1)} = 0$. Therefore,

$$2\Omega_{12} = \int_{\mathcal{R}} \sigma_{ij}^{(1)} \varepsilon_{ij}^{(2)} \, dV . \qquad (6.3.12)$$

Finally, since $\sigma_{ij}^{(1)} = C_{ijkl} \varepsilon_{kl}^{(1)}$, (6.3.12) can be rewritten as

$$2\Omega_{12} = \int_{\mathcal{R}} C_{ijkl} \varepsilon_{kl}^{(1)} \varepsilon_{ij}^{(2)} \, dV . \qquad (6.3.13)$$

Since $C_{ijkl} = C_{klij}$, it is seen in (6.3.13) that $\Omega_{12} = \Omega_{21}$—the work that would be done by the first system of external forces in acting through the displacements of the second system is equal to the work that would be done by the second system of external forces in acting through the displacements of the first system:

$$\boxed{\begin{aligned} \int_{\partial \mathcal{R}} \mathbf{t}^{(1)} \cdot \mathbf{u}^{(2)} \, dA + \int_{\mathcal{R}} \mathbf{f}^{(1)} \cdot \mathbf{u}^{(2)} \, dV &= \int_{\partial \mathcal{R}} \mathbf{t}^{(2)} \cdot \mathbf{u}^{(1)} \, dA + \int_{\mathcal{R}} \mathbf{f}^{(2)} \cdot \mathbf{u}^{(1)} \, dV , \\ \int_{\partial \mathcal{R}} t_i^{(1)} u_i^{(2)} \, dA + \int_{\mathcal{R}} f_i^{(1)} u_i^{(2)} \, dV &= \int_{\partial \mathcal{R}} t_i^{(2)} u_i^{(1)} \, dA + \int_{\mathcal{R}} f_i^{(2)} u_i^{(1)} \, dV . \end{aligned}}$$

$$(6.3.14)$$

Also, from (6.3.12), it is seen that the strain energy that would be associated with the stress field of the first system and the strain field of the second system is equal to that associated with the stress field of the second system and the strain field of the first system:

$$\boxed{\begin{aligned} \int_{\mathcal{R}} \boldsymbol{\sigma}^{(1)} : \boldsymbol{\varepsilon}^{(2)} \, dV &= \int_{\mathcal{R}} \boldsymbol{\sigma}^{(2)} : \boldsymbol{\varepsilon}^{(1)} \, dV , \\ \int_{\mathcal{R}} \sigma_{ij}^{(1)} \varepsilon_{ij}^{(2)} \, dV &= \int_{\mathcal{R}} \sigma_{ij}^{(2)} \varepsilon_{ij}^{(1)} \, dV . \end{aligned}}$$

$$(6.3.15)$$

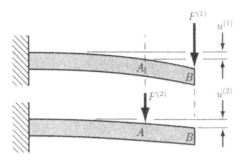

FIGURE 6.8. *Example of reciprocity.*

It is also equal to the work that would be done by the external forces of
the second system acting through the displacements of the first system:

$$
\int_{\mathcal{R}} \sigma^{(1)} : \varepsilon^{(2)} \, dV = \int_{\partial\mathcal{R}} \mathbf{t}^{(2)} \cdot \mathbf{u}^{(1)} \, dA + \int_{\mathcal{R}} \mathbf{f}^{(2)} \cdot \mathbf{u}^{(1)} \, dV ,
$$

$$
\int_{\mathcal{R}} \sigma_{ij}^{(1)} \varepsilon_{ij}^{(2)} \, dV = \int_{\partial\mathcal{R}} t_i^{(2)} u_i^{(1)} \, dA + \int_{\mathcal{R}} f_i^{(2)} u_i^{(1)} \, dV .
$$

(6.3.16)

The results (6.3.14)–(6.3.16) are known collectively as the *Betti-Rayleigh
reciprocity relations*.

Example 6.7
Consider a cantilever beam under two different loading conditions as shown
in Figure 6.8. For the first system, a vertical force $F^{(1)}$ is applied at B
and the corresponding vertical displacement at A is $u^{(1)}$. For the second
system, a vertical force $F^{(2)}$ is applied at A and the corresponding vertical
displacement at B is $u^{(2)}$. For linear elastic deformation of the beam, the
Betti-Rayleigh reciprocity relations state that $F^{(1)} u^{(2)} = F^{(2)} u^{(1)}$

Example 6.8
Consider an arbitrary body occupying the region $\mathcal{R}$ with boundary $\partial\mathcal{R}$. At
a point on $\partial\mathcal{R}$, the unit outward normal is $\hat{\mathbf{n}}$. Find an expression for the
change in volume of the body due to the traction distribution $\mathbf{t}$ on the
boundary $\partial\mathcal{R}$ and the body force field $\mathbf{f}$.

Solution. An expression for the change in volume can be found by applying the Betti-Rayleigh reciprocity relations. Let equilibrium state (1) be
that corresponding to a hydrostatic tension of magnitude A. Then, from

the results of Example 6.5,

$$u_i^{(1)} = \frac{A}{3K} X_i , \quad \sigma_{ij}^{(1)} = A\delta_{ij} ,$$

where K is the bulk modulus. Let equilibrium state (2) be the that for which the change in volume is sought:

$$t_i^{(2)} = t_i , \quad f_i^{(2)} = f_i , \quad \varepsilon_{ij}^{(2)} = \varepsilon_{ij} .$$

Then, from the Betti-Rayleigh reciprocity relation (6.3.16),

$$\int_{\mathcal{R}} A\delta_{ij}\varepsilon_{ij} \, dV = \int_{\partial\mathcal{R}} t_i \frac{A}{3K} X_i \, dA + \int_{\mathcal{R}} f_i \frac{A}{3K} X_i \, dV .$$

Simplifying,

$$\int_{\mathcal{R}} \varepsilon_{ii} \, dV = \frac{1}{3K} \left(\int_{\partial\mathcal{R}} t_i X_i \, dA + \int_{\mathcal{R}} f_i X_i \, dV \right) .$$

Recall that the dilatation $e = \varepsilon_{ii}$ is the change in volume per unit volume and, therefore, the integral on the left-hand side is the total change ΔV in volume. Thus,

$$\Delta V = \frac{1}{3K} \left(\int_{\partial\mathcal{R}} t_i X_i \, dA + \int_{\mathcal{R}} f_i X_i \, dV \right) .$$

6.4   Solution Methods

The solution of elasticity problems is often a matter of experience and intuition. In general, it will be a nontrivial problem to find a solution to the equations of elasticity which satisfies specified boundary conditions. One will often have to be satisfied with approximate solutions as described by the principle of Saint-Venant. Methods for finding solutions, whether exact or approximate, include the following:

- The *inverse method* wherein one finds stress, strain, and displacement fields that satisfy the equations of elasticity and then tries to find a problem with boundary conditions to which the fields correspond

- The *semi-inverse method* in which an educated guess is made as to the form of part of the solution, leaving enough freedom in the assumptions to hopefully satisfy the equations of elasticity and the boundary conditions

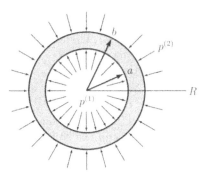

FIGURE 6.9. Spherical pressure vessel.

- The *method of potentials* in which potential functions are defined that incorporate some of the equations of elasticity, thereby simplifying the task of finding solutions to the remaining equations

- *Variational methods* which make use of certain minimization principles

- *Complex variable methods* that make use of the special properties of analytic functions of a complex variable

Other methods of interest include the integral transform method and numerical methods such as the finite element method.

6.4.1 Example: Spherical Pressure Vessel

Consider the spherical pressure vessel under internal and external pressure shown in Figure 6.9. The pressure vessel has inner radius a and outer radius b and is subject to internal and external hydrostatic pressures $p^{(1)}$ and $p^{(2)}$, respectively. Define a spherical coordinate system (R, θ, ϕ) with origin at the center of the pressure vessel; the body occupies the region $a < R < b$. Assume that body forces are negligible: $\mathbf{f} = \mathbf{0}$.

The first step in solving any elasticity problem is to establish the boundary conditions. For this problem, there are traction boundary conditions only. On the inner surface $(R = a)$, the prescribed traction is $\tilde{\mathbf{t}} = p^{(1)}\hat{\mathbf{e}}_R$ and the unit outward normal is $\hat{\mathbf{n}} = -\hat{\mathbf{e}}_R$. Therefore, from the traction-stress relations in spherical coordinates (4.4.7), $t_R = -\sigma_{RR}$, $t_\theta = -\sigma_{R\theta}$, $t_\phi = -\sigma_{R\phi}$, and

$$\sigma_{R\theta} = \sigma_{R\phi} = 0 , \quad \sigma_{RR} = -p^{(1)} \quad \text{on } R = a . \qquad (6.4.1)$$

On the outer surface $(R = b)$, the prescribed traction is $\tilde{\mathbf{t}} = -p^{(2)}\hat{\mathbf{e}}_R$ and the unit outward normal is $\hat{\mathbf{n}} = \hat{\mathbf{e}}_R$. Therefore, from the traction-stress relations, $t_R = \sigma_{RR}$, $t_\theta = \sigma_{R\theta}$, $t_\phi = \sigma_{R\phi}$, and

$$\sigma_{R\theta} = \sigma_{R\phi} = 0 , \quad \sigma_{RR} = -p^{(2)} \quad \text{on } R = b . \tag{6.4.2}$$

The problem is to find the stress field $\boldsymbol{\sigma}$ and the displacement field $\mathbf{u}$ that satisfy these boundary conditions and the equations of elasticity.

One way to solve this problem is to use the semi-inverse method. Assume that the solution displacement field exhibits the same spherical symmetry as the stated problem. Then, the components of displacement can be expressed in the form

$$u_R = U(R) , \quad u_\theta = u_\phi = 0 , \tag{6.4.3}$$

where $U(R)$ is an unknown function. If a solution can be found based on this assumption then, because linearized elasticity solutions are unique, one knows that the assumption is correct. If a solution can not be found, then the assumption must be abandoned.

Using the strain-displacement relations in spherical coordinates (3.5.10), the strains are

$$\varepsilon_{RR} = U'(R) , \quad \varepsilon_{\theta\theta} = \varepsilon_{\phi\phi} = \frac{1}{R}U(R) ,$$

$$\varepsilon_{R\theta} = \varepsilon_{\theta\phi} = \varepsilon_{\phi R} = 0 , \tag{6.4.4}$$

$$\varepsilon_{kk} = \varepsilon_{RR} + \varepsilon_{\theta\theta} + \varepsilon_{\phi\phi} = U'(R) + \frac{2}{R}U(R) .$$

The stresses are given by $\sigma_{ij} = 2\mu\varepsilon_{ij} + \lambda\varepsilon_{kk}\delta_{ij}$, where the index values $(1, 2, 3)$ are understood to represent (R, θ, ϕ), respectively;

$$\sigma_{RR} = (2\mu + \lambda)U'(R) + \frac{2\lambda}{R}U(R) ,$$

$$\sigma_{\theta\theta} = \sigma_{\phi\phi} = \lambda U'(R) + \frac{2}{R}(\mu + \lambda)U(R) , \tag{6.4.5}$$

$$\sigma_{R\theta} = \sigma_{\theta\phi} = \sigma_{\phi R} = 0 .$$

In order for the stress field given by (6.4.5) to be the solution to the problem, it must satisfy the equilibrium equations (4.4.8) and the boundary conditions (6.4.1) and (6.4.2).

The second and third equilibrium equations are identically satisfied and the first reduces to

$$\frac{\partial \sigma_{RR}}{\partial R} + \frac{1}{R}(2\sigma_{RR} - \sigma_{\theta\theta} - \sigma_{\phi\phi}) = 0 . \tag{6.4.6}$$

Substituting (6.4.5) into (6.4.6) gives the equilibrium condition

$$(2\mu + \lambda)U''(R) + \frac{2\lambda}{R}U'(R) - \frac{2\lambda}{R^2}U(R) + \frac{1}{R}\Big[2(2\mu + \lambda)U'(R)$$

$$+ \frac{4\lambda}{R}U(R) - 2\lambda U'(R) - \frac{4}{R}(\mu + \lambda)U(R)\Big] = 0 , \quad (6.4.7)$$

which reduces to

$$R^2 U''(R) + 2RU'(R) - 2U(R) = 0 . \tag{6.4.8}$$

The linear, ordinary differential equation (6.4.8) is of the Euler-Cauchy type and can be transformed into an ordinary differential equation with constant coefficients by the change of independent variable $R = e^\xi$, $\xi = \ln R$. Using the chain rule, this change of variable leads to the relations

$$U'(R) \equiv \frac{dU}{dR} = \frac{d\xi}{dR}\frac{dU}{d\xi} = \frac{1}{R}\frac{dU}{d\xi} ,$$

$$U''(R) \equiv \frac{d^2 U}{dR^2} = \frac{d}{dR}\left(\frac{1}{R}\frac{dU}{d\xi}\right) = \frac{1}{R^2}\left(\frac{d^2 U}{d\xi^2} - \frac{dU}{d\xi}\right) . \tag{6.4.9}$$

It follows that (6.4.8) is transformed by the change of variable into

$$\frac{d^2 U}{d\xi^2} + \frac{dU}{d\xi} - 2U = 0 . \tag{6.4.10}$$

Looking for solutions of the form $U = e^{m\xi}$, one finds that the general solution to (6.4.10) is

$$U(\xi) = Ae^\xi + Be^{-2\xi} , \tag{6.4.11}$$

or changing back to the original independent variable,

$$U(R) = AR + \frac{B}{R^2} , \tag{6.4.12}$$

where A and B are constants to be determined by the boundary conditions (6.4.1) and (6.4.2).

It is clear that the boundary conditions involving $\sigma_{R\theta}$ and $\sigma_{R\phi}$ are satisfied by the stress field (6.4.5). The remaining boundary conditions involve σ_{RR}, which can be written in terms of the unknown constants A and B by combining (6.4.12) and (6.4.5):

$$\sigma_{RR} = (2\mu + \lambda)(A - \frac{2B}{R^3}) + 2\lambda(A + \frac{B}{R^3})$$

$$= (2\mu + 3\lambda)A - \frac{4\mu B}{R^3} . \tag{6.4.13}$$

In terms of A and B then, the boundary conditions are

$$(2\mu + 3\lambda)A - \frac{4\mu B}{a^3} = -p^{(1)},$$

$$(2\mu + 3\lambda)A - \frac{4\mu B}{b^3} = -p^{(2)}.$$

(6.4.14)

Solving (6.4.14) for the two unknowns A and B,

$$A = \frac{1}{2\mu + 3\lambda}\left(\frac{a^3 p^{(1)} - b^3 p^{(2)}}{b^3 - a^3}\right),$$

$$B = \frac{a^3 b^3}{4\mu}\left(\frac{p^{(1)} - p^{(2)}}{b^3 - a^3}\right).$$

(6.4.15)

Thus, substituting (6.4.15) into (6.4.3) and (6.4.5), the exact solution to the linearized elasticity problem of a spherical vessel with internal and external pressure is

$$\boxed{\begin{aligned} u_R &= \frac{R}{2\mu + 3\lambda}\left(\frac{a^3 p^{(1)} - b^3 p^{(2)}}{b^3 - a^3}\right) + \frac{a^3 b^3}{4\mu R^2}\left(\frac{p^{(1)} - p^{(2)}}{b^3 - a^3}\right), \\ u_\theta &= u_\phi = 0, \end{aligned}}$$

(6.4.16)

$$\boxed{\begin{aligned} \sigma_{RR} &= \left(\frac{a^3 p^{(1)} - b^3 p^{(2)}}{b^3 - a^3}\right) - \frac{a^3 b^3}{R^3}\left(\frac{p^{(1)} - p^{(2)}}{b^3 - a^3}\right), \\ \sigma_{\theta\theta} &= \sigma_{\phi\phi} = \left(\frac{a^3 p^{(1)} - b^3 p^{(2)}}{b^3 - a^3}\right) + \frac{a^3 b^3}{2R^3}\left(\frac{p^{(1)} - p^{(2)}}{b^3 - a^3}\right), \\ \sigma_{R\theta} &= \sigma_{\theta\phi} = \sigma_{\phi R} = 0. \end{aligned}}$$

(6.4.17)

Since the off-diagonal elements of the matrix of scalar components stress are zero, the diagonal elements σ_{RR} and $\sigma_{\theta\theta} = \sigma_{\phi\phi}$ are the principal stresses.

It is a good idea to examine the behavior of solutions such as this under different limiting conditions. This serves both as a means of checking the accuracy of one's results and as a way to obtain solutions for related problems.

Consider first the special case when the internal and external pressures are equal: $p^{(1)} = p^{(2)} = p$. Then, the solution stress field reduces to

$$\sigma_{RR} = \sigma_{\theta\theta} = \sigma_{\phi\phi} = -p,$$

(6.4.18)

which is just pure hydrostatic pressure. This is consistent with the result derived in Example 6.5.

If the wall of the spherical shell is thin with thickness $t \equiv b - a$, $t \ll a$, then

$$b^3 - a^3 = (b - a)(b^2 + ab + a^2) \approx 3ta^2 \ ,$$
$$a^3 p^{(1)} - b^3 p^{(2)} \approx a^3 (p^{(1)} - p^{(2)}) \ ,$$
$$b^3 \approx a^3 \ ,$$
$$R^3 \approx a^3 \ ,$$

(6.4.19)

which leads to

$$\sigma_{\theta\theta} = \sigma_{\phi\phi} \approx \frac{a(p^{(1)} - p^{(2)})}{2t} \ .$$

(6.4.20)

This is the thin-walled pressure vessel approximation commonly given in texts on mechanics of materials.

In the limit as $b \to \infty$, the problem reduces to that of a pressurized spherical void in an infinite medium under remote hydrostatic pressure. The solution to this problem then is found by taking (6.4.17) in the limit as $b \to \infty$:

$$\sigma_{RR} = -p^{(2)} - (p^{(1)} - p^{(2)})\frac{a^3}{R^3} \ ,$$
$$\sigma_{\theta\theta} = \sigma_{\phi\phi} = -p^{(2)} + \frac{1}{2}(p^{(1)} - p^{(2)})\frac{a^3}{R^3} \ .$$

(6.4.21)

Far away from the void, as $R \to \infty$, the state of stress is one of pure hydrostatic pressure: $\sigma_{RR} = \sigma_{\theta\theta} = \sigma_{\phi\phi} = -p^{(2)}$. If additionally the void is unpressurized (i.e., $p^{(1)} = 0$), then

$$\sigma_{RR} = -p^{(2)}\left(1 - \frac{a^3}{R^3}\right) \ ,$$
$$\sigma_{\theta\theta} = \sigma_{\phi\phi} = -p^{(2)}\left(1 + \frac{a^3}{2R^3}\right) \ .$$

(6.4.22)

At $R = a$, $\sigma_{RR} = 0$ and $\sigma_{\theta\theta} = \sigma_{\phi\phi} = -3p^{(2)}/2$. Far away from the void, the maximum absolute value of the principal stresses is $|p^{(2)}|$, whereas at $R = a$, it is $3|p^{(2)}|/2$. The ratio of the later over the former is the *stress concentration factor*:

$$\text{s.c.f.} = \frac{3}{2} \ ,$$

(6.4.23)

for an unpressurized spherical void in an infinite medium.

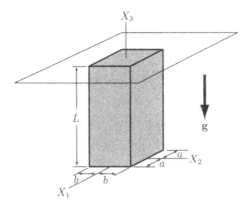

FIGURE 6.10. Rectangular beam hanging under its own weight.

6.4.2   Example: Hanging Beam

Consider a beam with dimensions $2a \times 2b \times L$ and density ρ in a gravitational field $\mathbf{g} = -g\hat{\mathbf{e}}_3$ (Figure 6.10). The top surface is attached to a rigid support. A Cartesian coordinate system is defined as shown, with the body occupying the region given by $-a < X_1 < a$, $-b < X_2 < b$, and $0 < X_3 < L$. Define the specific weight $\gamma \equiv \rho g$ so that the prescribed body force field can be expressed as $\tilde{\mathbf{f}} = -\gamma\hat{\mathbf{e}}_3$. Find the stress and displacement fields in the beam.

The displacement on the top surface is fixed ($\tilde{\mathbf{u}} = \mathbf{0}$), so that

$$u_i = 0 \quad \text{on } X_3 = L \ . \tag{6.4.24}$$

The sides and bottom of the beam are traction-free: $\tilde{\mathbf{t}} = \mathbf{0}$. Thus, on the bottom surface where the unit outward normal is $\hat{\mathbf{n}} = -\hat{\mathbf{e}}_3$,

$$\sigma_{31} = \sigma_{32} = \sigma_{33} = 0 \quad \text{on } X_3 = 0 \ , \tag{6.4.25}$$

on the side surfaces $X_1 = \pm a$ where the unit outward normal is $\hat{\mathbf{n}} = \pm\hat{\mathbf{e}}_1$,

$$\sigma_{11} = \sigma_{12} = \sigma_{13} = 0 \quad \text{on } X_1 = \pm a \ , \tag{6.4.26}$$

and on the side surfaces $X_2 = \pm b$ where the unit outward normal is $\hat{\mathbf{n}} = \pm\hat{\mathbf{e}}_2$,

$$\sigma_{21} = \sigma_{22} = \sigma_{23} = 0 \quad \text{on } X_2 = \pm b \ . \tag{6.4.27}$$

The problem is to find the stress field $\boldsymbol{\sigma}(\mathbf{X})$ and the displacement field $\mathbf{u}(\mathbf{X})$ that satisfy these boundary conditions and the equations of elasticity.

Again using the semi-inverse method, assume that the solution stress field can be expressed in the form

$$\sigma_{33} = h(X_3) , \quad \sigma_{11} = \sigma_{22} = \sigma_{12} = \sigma_{23} = \sigma_{31} = 0 . \tag{6.4.28}$$

This assumption satisfies the traction boundary conditions (6.4.25)–(6.4.27) if

$$h(0) = 0 . \tag{6.4.29}$$

Two of the equilibrium equations are identically satisfied by the stress field (6.4.28) and the third reduces to $\sigma_{33,3} = -f_3$. Substituting for σ_{33} and f_3, the equilibrium condition is

$$h'(X_3) = \gamma . \tag{6.4.30}$$

Solving (6.4.30) and enforcing (6.4.29) leads to $h(X_3) = \gamma X_3$. Beltrami's equations for compatibility in terms of stress (6.1.19) are satisfied by $\sigma_{33} = \gamma X_3$. Therefore, the stress field

$$\boxed{\sigma_{33} = \gamma X_3 , \quad \sigma_{11} = \sigma_{22} = \sigma_{12} = \sigma_{23} = \sigma_{31} = 0} \tag{6.4.31}$$

satisfies the traction boundary conditions (6.4.25)–(6.4.27) and the equations of elasticity.

What about the displacement boundary condition (6.4.24)? From the constitutive relations $\varepsilon_{ij} = [(1+\nu)\sigma_{ij} - \nu\sigma_{kk}\delta_{ij}]/E$, the strains corresponding to (6.4.31) are

$$\varepsilon_{11} = \varepsilon_{22} = -\frac{\nu}{E}\gamma X_3 , \quad \varepsilon_{33} = \frac{1}{E}\gamma X_3 , \quad \varepsilon_{12} = \varepsilon_{23} = \varepsilon_{31} = 0 . \tag{6.4.32}$$

In order to find the corresponding displacement field, it is necessary to solve the coupled set of partial differential equations represented by the strain-displacement relations $\varepsilon_{ij} = \frac{1}{2}(u_{i,j} + u_{j,i})$. Rewriting (6.4.32) in terms of the displacements,

$$u_{1,1} = u_{2,2} = -\frac{\nu}{E}\gamma X_3 , \tag{6.4.33}$$

$$u_{3,3} = \frac{1}{E}\gamma X_3 , \tag{6.4.34}$$

$$u_{1,2} + u_{2,1} = u_{2,3} + u_{3,2} = u_{3,1} + u_{1,3} = 0 . \tag{6.4.35}$$

Integrating (6.4.34),

$$u_3 = \frac{\gamma}{2E}X_3^2 + \eta(X_1, X_2) , \tag{6.4.36}$$

where η is an as yet unknown function of X_1 and X_2. From (6.4.35), $u_{1,3} = -u_{3,1}$. Thus, differentiating (6.4.36) with respect to X_1,

$$u_{1,3} = -\eta_{,1}(X_1, X_2) . \tag{6.4.37}$$

Now, integrating (6.4.37),

$$u_1 = -X_3 \eta_{,1}(X_1, X_2) + \xi(X_1, X_2) , \tag{6.4.38}$$

where ξ is another as yet unknown function of X_1 and X_2. Through a similar process, it is found that

$$u_2 = -X_3 \eta_{,2}(X_1, X_2) + \zeta(X_1, X_2) . \tag{6.4.39}$$

Keep in mind from now on that ξ, ζ, and η are functions of X_1 and X_2 only.

Differentiating (6.4.38) with respect to X_1 and comparing the result with (6.4.33),

$$-X_3 \eta_{,11} + \xi_{,1} = -\frac{\nu\gamma}{E} X_3 . \tag{6.4.40}$$

Since (6.4.40) holds for all values of X_3 in the body, it follows that

$$\eta_{,11} = \frac{\nu\gamma}{E} , \tag{6.4.41}$$

$$\xi_{,1} = 0 . \tag{6.4.42}$$

Similarly, it can be shown that

$$\eta_{,22} = \frac{\nu\gamma}{E} \tag{6.4.43}$$

$$\zeta_{,2} = 0 . \tag{6.4.44}$$

Substituting (6.4.38) and (6.4.39) into the first of (6.4.35),

$$-X_3 \eta_{,12} + \xi_{,2} - X_3 \eta_{,21} + \zeta_{,1} = 0 . \tag{6.4.45}$$

It follows from (6.4.45) that

$$\eta_{,12} = 0 , \tag{6.4.46}$$

$$\xi_{,2} + \zeta_{,1} = 0 . \tag{6.4.47}$$

All that remains now is to integrate (6.4.41)–(6.4.44), (6.4.46), and (6.4.47) for the functions ξ, ζ, and η.

Integrating (6.4.42) and (6.4.44), $\xi = f(X_2)$ and $\zeta = g(X_1)$. However, from (6.4.47), $f'(X_2) = -g'(X_1) = A$, where A is a constant. Thus,

$$\xi = AX_2 + B ,$$
$$\zeta = -AX_1 + C , \tag{6.4.48}$$

where B and C are constants. Integrating (6.4.46) alternately with respect to X_1 and X_2,

$$\eta_{,1} = \psi(X_1), \quad \eta_{,2} = \phi(X_2), \quad (6.4.49)$$

which from (6.4.41) and (6.4.43) reduce to

$$\eta_{,1} = \frac{\nu\gamma}{E}X_1 + D, \quad \eta_{,2} = \frac{\nu\gamma}{E}X_2 + F, \quad (6.4.50)$$

where D and F are constants. Integrating each of (6.4.50),

$$\eta = \frac{\nu\gamma}{2E}X_1^2 + DX_1 + \alpha(X_2), \quad \eta = \frac{\nu\gamma}{2E}X_2^2 + FX_2 + \beta(X_1). \quad (6.4.51)$$

Comparing each of (6.4.51), it follows that

$$\alpha(X_2) = \frac{\nu\gamma}{2E}X_2^2 + FX_2 + H, \quad \beta(X_1) = \frac{\nu\gamma}{2E}X_1^2 + DX_1 + H, \quad (6.4.52)$$

so that

$$\eta = \frac{\nu\gamma}{2E}(X_1^2 + X_2^2) + DX_1 + FX_2 + H, \quad (6.4.53)$$

where H is another constant.

Substituting (6.4.48) and (6.4.53) for ξ, ζ, and η in (6.4.36), (6.4.38), and (6.4.39),

$$u_1 = -\frac{\nu\gamma}{E}X_1X_3 - DX_3 + AX_2 + B,$$

$$u_2 = -\frac{\nu\gamma}{E}X_2X_3 - FX_3 - AX_1 + C, \quad (6.4.54)$$

$$u_3 = \frac{\gamma}{2E}\left[X_3^2 + \nu(X_1^2 + X_2^2)\right] + DX_1 + FX_2 + H.$$

This is the displacement field corresponding to the stress field (6.4.31), to within an arbitrary rigid-body motion. The constants B, C, and H correspond to a rigid-body translation and the constants A, D, and F to a rigid-body rotation. *There is no combination of these constants that will satisfy the displacement boundary condition* (6.4.24). However, the tractions on the top surface,

$$t_1 = t_2 = 0 \quad t_3 = \gamma L \quad \text{on } X_3 = L, \quad (6.4.55)$$

are in equilibrium with the weight $4ab\gamma L$ of the beam and are thus statically equivalent to the boundary condition (6.4.24). Saint-Venant's principle tells us then that the stress field (6.4.31) is approximately correct for points in the beam sufficiently far, with respect to a and b, from the top surface of the beam. This is probably as good an answer as can be found without resorting to numerical methods such as finite element analysis.

To fix the rigid-body motion in the (approximate) solution displacement field, let the point at the center of the top surface $(X_1, X_2, X_3) = (0, 0, L)$ be fixed, $\mathbf{u} = \mathbf{0}$, and have zero rotation, $\boldsymbol{\omega} = \mathbf{0}$, where recall that the scalar components of the infinitesimal rotation tensor are $\omega_{ij} = \frac{1}{2}(u_{i,j} - u_{j,i})$. This, in conjunction with (6.4.35) leads to the condition that

$$\left. \begin{array}{c} u_1 = u_2 = u_3 = 0 \\ u_{1,2} = u_{2,1} = u_{2,3} = u_{3,2} = u_{3,1} = u_{1,3} = 0 \end{array} \right\} \text{ at } (0,0,L), \qquad (6.4.56)$$

from which it follows that $A = B = C = D = F = 0$ and $H = -\gamma L^2/2E$. Therefore, the solution displacement field is

$$\boxed{ \begin{aligned} u_1 &= -\frac{\nu\gamma}{E} X_1 X_2 \,, \\ u_2 &= -\frac{\nu\gamma}{E} X_2 X_3 \,, \\ u_3 &= \frac{\gamma}{2E} \left[X_3^2 - L^2 + \nu(X_1^2 + X_2^2) \right] \,. \end{aligned} } \qquad (6.4.57)$$

Consideration of this displacement field shows that the plane sides remain plane and that the top and bottom are deformed into paraboloids of revolution; the deformed beam is a truncated pyramid bounded at the top and bottom by paraboloids of revolution.

Problems

6.1 A circular cylindrical material sample of radius a and height h is placed into a servo-hydraulic testing machine as shown below. The two surfaces of the testing machine that come into contact with the sample at A and B are called *platens*. These platens are made of a very stiff alloy that may be assumed to be rigid. The platen at B is held stationary and an active control circuit is integrated into the hydraulic actuator so that the vertical motion of the platen at A can be prescribed. Suppose that the platen at A is first adjusted until it is just in contact with the sample, and then an additional downward motion δ of the platen is prescribed. Define a coordinate system and give a complete statement of the boundary conditions for the sample, assuming

(a) that the sample is ideally bonded to the platens at A and B

(b) that the platens at A and B are frictionless (e.g., some sort of ideal lubricant has been applied).

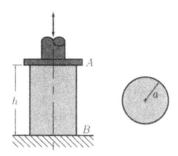

Note: do *not* attempt to solve the corresponding boundary value problem.

6.2 Consider the infinitely long cylindrical body, with semicircular cross section, illustrated below. In Cartesian coordinates, the body occupies the region $X_1^2 + X_2^2 \leq a^2$, $X_2 \geq 0$, $-\infty < X_3 < \infty$. In cylindrical coordinates, it occupies the region $r \leq a$, $0 \leq \theta \leq \pi$, $-\infty < z < \infty$. A uniform vertical traction of magnitude q acts on the semicircular surface of the body and the base is fixed to a rigid support.

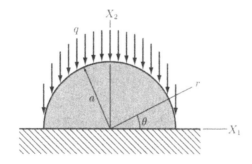

(a) Give a complete statement of the boundary conditions for this problem using the Cartesian coordinate system.

(b) Give a complete statement of the boundary conditions for this problem using the cylindrical coordinate system.

Note: do *not* attempt to solve the corresponding boundary value problem.

6.3 Consider the linear elastic deformation of a homogeneous sphere of radius a due to the mutual gravitational attraction of its parts. In a spherical coordinate system, with the origin at the center of the sphere, the body force field within such a self-gravitating sphere is

$$\mathbf{f} = -\frac{\rho g R}{a}\,\hat{\mathbf{e}}_R\,,$$

where ρ is the mass density and g is the gravitational acceleration at the surface of the sphere. Assuming the surface of the sphere is traction-free, determine the displacement field in the sphere.

6.4 Consider a spherical body, with Lamé constants λ and μ, that has embedded within it a concentric spherical *inclusion*. Initially, when the radius of the inclusion is a, the spherical body is stress-free and its outer radius is b. Suppose some event occurs that triggers a transformation in the inclusion that causes its radius to increase by a known amount κa. (This is *not* what the change in radius of the inclusion would be if it were isolated, but the actual change in radius of the inclusion while it is embedded in the spherical body.) Determine the change in the outer radius of the spherical body caused by this expansion of the inclusion.

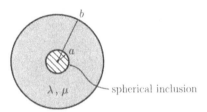

6.5 Consider a spherical body of radius b with a concentric, rigid spherical inclusion of radius a, as shown below. An external hydrostatic pressure of magnitude p is applied to the surface of the body.

    (a) Determine the displacement and stress fields in the body.

    (b) In the limiting case of a rigid spherical inclusion in an infinite body under remote hydrostatic pressure (i.e., as $b \to \infty$), show that the stress concentration factor for the rigid inclusion is

$$\text{s.c.f.} = 1 + \frac{4\mu}{3K}\,,$$

where μ is the shear modulus and K is the bulk modulus.

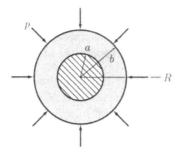

6.6 Consider a composite sphere composed of a solid spherical core of radius a, with Lamé constants μ_1 and λ_1, and an outer spherical shell of outer radius b, with Lamé constants μ_2 and λ_2. The core and shell are concentric and a spherical coordinate system is defined with the origin at their center. Assume that the core and shell are *ideally bonded* at their *interface* at $R = a$, which means that the displacement field **u** is continuous across the interface.

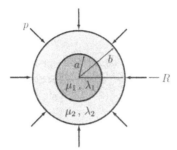

(a) Recall the generalization of Newton's third law, which says that the traction vector exerted by the shell on the core is the negative of the traction vector exerted by the core on the shell. Note also that, at a point on the interface, the unit outward normal to the core is the negative of the unit outward normal to the shell. What consequences does this have for continuity of the scalar components of stress (in spherical coordinates) across the interface?

(b) The composite sphere is subject to an external, hydrostatic pressure of magnitude p. Assuming that the spherical symmetry of the problem is preserved in the solution, use the semi-inverse method to find the displacement and stress fields in the composite sphere. Note that, given the symmetry assumption, the

displacement must vanish at the center of the core. The answer will involve separate expressions for the core and shell that
satisfy the continuity conditions at the interface.

6.7 A hollow circular cylinder has inner radius a, outer radius b, and
length L. The outer surface of the hollow cylinder is fixed and its
inner surface is ideally bonded to a rigid cylindrical core of radius a
and length L, as shown below. Suppose that an axial force $\mathbf{F} = F\hat{e}_z$
is applied to the rigid core along its centroidal axis.

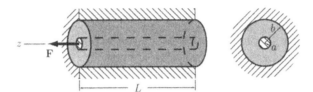

(a) Find the resulting (rigid-body) axial displacement of the core
and the work done by the applied force $\mathbf{F}$, by assuming a displacement field in the hollow cylinder of the form $u_r = u_\theta = 0$
and $u_z = U(r)$.

(b) Is this the exact solution to the problem, or is it a solution only
in the sense of Saint-Venant's principle? Explain.

Chapter 7

Two-Dimensional Problems

There is a class of problems in elasticity which, due to geometry and boundary conditions, have solution stress, strain, and displacement fields that are independent of one of the coordinate variables. The equations of elasticity for this class of problem reduce to a simplified form of the general equations. Recognizing that a body with given boundary conditions is a member of this class, in advance of solving the attendant boundary value problem, enables one to take advantage of these simplified equations. Members of this class of problem are subdivided into three types known as antiplane strain, plane strain, and plane stress, as discussed below. It will be seen that the reduced form of the equations of elasticity for the later two are functionally equivalent.

In the following analyses, Cartesian and cylindrical coordinate systems will be defined such that X_3 and z, respectively, are the coordinates for which the elastic fields are independent. In such a Cartesian coordinate system, a necessary condition for this class of problems is independence of a body's cross-sectional geometry with respect to X_3. The unit outward normal $\hat{n}$ at any point on a body's lateral surface must lie in a plane parallel to the X_1X_2-plane and be independent of X_3 (Figure 7.1), so that

$$\hat{n}_1 = \hat{n}_1(X_1, X_2), \quad \hat{n}_2 = \hat{n}_2(X_1, X_2), \quad \hat{n}_3 = 0. \tag{7.0.1}$$

The conditions in cylindrical coordinates are analogous, with r, θ, and z taking the places of X_1, X_2, and X_3.

Since there is no dependence on X_3, only the projection of the body onto the X_1X_2-plane will need to be considered. The region of the X_1X_2-plane occupied by the projection is $\mathcal{S}$ with boundary $\partial\mathcal{S}$. Two-dimensional problems in elasticity will involve finding stress, strain, and displacement fields that satisfy certain reduced elasticity equations at every point in the two-dimensional region $\mathcal{S}$ while satisfying boundary conditions on $\partial\mathcal{S}$.

When using indicial notation in two-dimensional problems, it will be understood from now on that, in contrast to Latin indices which range

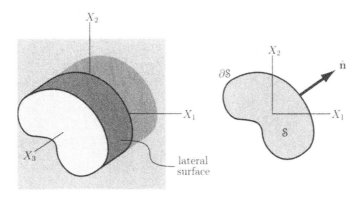

FIGURE 7.1.   Two-dimensional geometry.

over the values $\{1,2,3\}$, Greek indices range over the values $\{1,2\}$. For example,

$$u_{\alpha\beta,\beta} = d_\alpha \quad \Longleftrightarrow \quad \begin{cases} u_{11,1} + u_{12,2} = d_1 \\ u_{21,1} + u_{22,2} = d_2 \end{cases} . \qquad (7.0.2)$$

7.1   Antiplane Strain

A state of *antiplane strain* is defined as one in which

$$\boxed{u_\alpha = 0\,, \quad u_3 = u_3(X_1, X_2)\,.} \qquad (7.1.1)$$

If one is able to ascertain that this will indeed be the case for a particular problem, then it follows from (7.1.1) that the scalar components of strain, $\varepsilon_{ij} = \frac{1}{2}(u_{i,j} + u_{j,i})$, must have the form $\varepsilon_{11} = \varepsilon_{22} = \varepsilon_{33} = \varepsilon_{12} = 0$ and $\varepsilon_{\alpha3} = \varepsilon_{\alpha3}(X_1, X_2)$, where

$$\varepsilon_{\alpha3} = \frac{1}{2} u_{3,\alpha}\,. \qquad (7.1.2)$$

Therefore, the scalar components of stress, $\sigma_{ij} = 2\mu\varepsilon_{ij} + \lambda\varepsilon_{kk}\delta_{ij}$, must have the form $\sigma_{11} = \sigma_{22} = \sigma_{33} = \sigma_{12} = 0$ and $\sigma_{\alpha3} = \sigma_{\alpha3}(X_1, X_2)$, where

$$\sigma_{\alpha3} = \mu u_{3,\alpha}\,. \qquad (7.1.3)$$

Note that four of the six compatibility equations (3.4.45) are identically satisfied and the two remaining equations reduce to

$$\varepsilon_{23,11} - \varepsilon_{13,21} = 0 \, ,$$
$$\varepsilon_{23,12} - \varepsilon_{13,22} = 0 \, . \tag{7.1.4}$$

From (7.1.3), it is seen that for antiplane strain the equilibrium equations $\sigma_{ji,j} + f_i = 0$ reduce to

$$\sigma_{\alpha 3, \alpha} + f_3 = 0 \tag{7.1.5}$$

and the requirement that $f_1 = f_2 = 0$ and $f_3 = f_3(X_1, X_2)$. This lone equilibrium condition can be expressed in terms of the displacement field u_3 using (7.1.3) to give the *antiplane strain equilibrium condition,*

$$\boxed{\mu \nabla^2 u_3 + f_3 = 0 \, ,} \tag{7.1.6}$$

where, since there is no dependence on X_3, $\nabla^2 u_3 = u_{3,\alpha\alpha}$.

Let ∂S^u and ∂S^t be the portions of the boundary ∂S on which the displacement and traction are prescribed such that if there are no mixed boundary conditions, $\partial S^u \cup \partial S^t = \partial S$. Then, a *displacement boundary condition for antiplane strain* will have the form

$$\boxed{u_3 = \tilde{u}_3 \quad \forall (X_1, X_2) \in \partial S^u \, ,} \tag{7.1.7}$$

where $\tilde{u}_3(X_1, X_2)$ is a given function. For the traction boundary condition, note first that since $\hat{n}_3 = 0$ on ∂S^t, the traction-stress relations $t_i = \sigma_{ji}\hat{n}_j$ for the lateral surface reduce to the single relation

$$t_3 = \sigma_{\alpha 3}\hat{n}_\alpha \tag{7.1.8}$$

and the requirement that $t_1 = t_2 = 0$ and $t_3 = t_3(X_1, X_2)$. Therefore, using the traction-stress relation (7.1.8), a traction boundary condition on ∂S^t will have the form

$$\sigma_{\alpha 3}\hat{n}_\alpha = \tilde{t}_3 \quad \forall (X_1, X_2) \in \partial S^t \, , \tag{7.1.9}$$

where $\tilde{t}_3(X_1, X_2)$ is prescribed. Using (7.1.3) to express this traction boundary condition in terms of the displacement field u_3 gives the *traction boundary condition for antiplane strain:*

$$\boxed{\mu u_{3,\alpha}\hat{n}_\alpha = \tilde{t}_3 \quad \forall (X_1, X_2) \in \partial S^t \, .} \tag{7.1.10}$$

For antiplane strain, the boundary value problem of linearized elasticity reduces to that of finding a scalar displacement field $u_3(X_1, X_2)$ that satisfies the equilibrium condition (7.1.6) everywhere within the two-dimensional region S of the body while simultaneously satisfying the boundary conditions (7.1.7) and (7.1.10) on the boundary ∂S of the two-dimensional region.

7.1.1   Cylindrical Coordinates

In cylindrical coordinates, a state of antiplane strain is defined as one in which

$$u_r = u_\theta = 0 , \quad u_z = u_z(r, \theta) . \qquad (7.1.11)$$

Therefore, the cylindrical coordinate strain-displacement relations (3.5.5) reduce to $\varepsilon_{rr} = \varepsilon_{\theta\theta} = \varepsilon_{zz} = \varepsilon_{r\theta} = 0$ and

$$\varepsilon_{\theta z}(r, \theta) = \frac{1}{2r} \frac{\partial u_z}{\partial \theta} , \quad \varepsilon_{rz}(r, \theta) = \frac{1}{2} \frac{\partial u_z}{\partial r} , \qquad (7.1.12)$$

the constitutive equations give $\sigma_{rr} = \sigma_{\theta\theta} = \sigma_{zz} = \sigma_{r\theta} = 0$ and

$$\sigma_{\theta z}(r, \theta) = \frac{\mu}{r} \frac{\partial u_z}{\partial \theta} , \quad \sigma_{rz}(r, \theta) = \mu \frac{\partial u_z}{\partial r} , \qquad (7.1.13)$$

and the equilibrium equations (4.4.4) reduce to $f_r = f_\theta = 0$ and

$$\frac{\partial \sigma_{rz}}{\partial r} + \frac{1}{r} \frac{\partial \sigma_{\theta z}}{\partial \theta} + \frac{1}{r} \sigma_{rz} + f_z = 0 . \qquad (7.1.14)$$

By substituting (7.1.13) into (7.1.14), this equilibrium equation can alternatively be given in terms of the displacement field u_z, showing that the cylindrical coordinate form of the antiplane strain equilibrium condition (7.1.6) is

$$\mu \nabla^2 u_z + f_z = 0 , \qquad (7.1.15)$$

where

$$\nabla^2(\ ) = \frac{\partial^2(\ )}{\partial r^2} + \frac{1}{r} \frac{\partial(\ )}{\partial r} + \frac{1}{r^2} \frac{\partial^2(\ )}{\partial \theta^2} . \qquad (7.1.16)$$

Similarly, it follows that the displacement boundary condition for antiplane strain is

$$u_z = \tilde{u}_z \quad \forall (r, \theta) \in \partial S^u \qquad (7.1.17)$$

and the traction boundary condition for antiplane strain is

$$\mu \hat{n}_r \frac{\partial u_z}{\partial r} + \frac{\mu \hat{n}_\theta}{r} \frac{\partial u_z}{\partial \theta} = \tilde{t}_z \quad \forall (r, \theta) \in \partial S^t , \qquad (7.1.18)$$

where $\tilde{u}_z(r, \theta)$ and $\tilde{t}_z(r, \theta)$ are given functions.

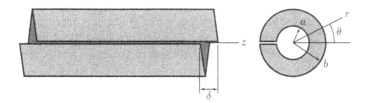

FIGURE 7.2. *Screw dislocation.*

7.1.2 Example: Screw Dislocation

Consider a long, hollow circular cylinder with inner and outer radii a and b, respectively, and let a cylindrical coordinate system be defined with the z-axis coincident with the centroidal axis of the cylinder. Now imagine that a cut is made along the length of the cylinder, creating two new boundary surfaces at $\theta = \pm\pi$, and that these new boundary surfaces are displaced axially an amount δ relative to each other (Figure 7.2). All other boundary surfaces are traction-free. Assume that body forces are negligible.

As usual, one should begin with a clear statement of the boundary conditions. On the surfaces $\theta = \pm\pi$ corresponding to the cut made along the length of the cylinder,

$$u_r = u_\theta = 0, \quad u_z = \pm\frac{\delta}{2} \quad \text{on } \theta = \pm\pi . \tag{7.1.19}$$

The inner and outer surfaces, $r = a$ and $r = b$, are traction-free ($\tilde{\mathbf{t}} = \mathbf{0}$) with unit outward normals $\hat{\mathbf{n}} = -\hat{\mathbf{e}}_r$ and $\hat{\mathbf{n}} = \hat{\mathbf{e}}_r$, respectively, so that

$$\sigma_{rr} = \sigma_{r\theta} = \sigma_{rz} = 0 \quad \text{on } r = a \text{ and } r = b . \tag{7.1.20}$$

The end surfaces, at $z = 0$ and $z = L$ say, are also traction-free and have unit outward normals $\hat{\mathbf{n}} = -\hat{\mathbf{e}}_z$ and $\hat{\mathbf{n}} = \hat{\mathbf{e}}_z$. Therefore,

$$\sigma_{zr} = \sigma_{z\theta} = \sigma_{zz} = 0 \quad \text{on } z = 0 \text{ and } z = L . \tag{7.1.21}$$

These are the boundary conditions for the full three-dimensional problem. In the study of crystal lattice defects in materials science, this problem corresponds to what is known as a screw dislocation.

Assume that the solution to this problem is a state of antiplane strain (i.e., that $u_r = u_\theta = 0$ everywhere and u_z is independent of z). Furthermore, assume that u_z is independent of r:

$$u_z = U(\theta) . \tag{7.1.22}$$

Recall that making *a priori* assumptions like this is known as the semi-inverse method. One makes these assumptions on the basis of physical intuition and if one is then able to solve the boundary value problem, the assumptions are correct. If not, the assumptions need to be reconsidered.

Substituting (7.1.22) into the equilibrium condition (7.1.15) gives

$$U''(\theta) = 0 . \tag{7.1.23}$$

The general solution to this equation is

$$U(\theta) = A\theta + B , \tag{7.1.24}$$

where A and B are constants to be determined by the boundary conditions. Substituting (7.1.24) into (7.1.19) and solving for A and B,

$$A = \frac{\delta}{2\pi} , \quad B = 0 . \tag{7.1.25}$$

It follows from (7.1.22), (7.1.24), and (7.1.13) that

$$u_z = \frac{\delta\theta}{2\pi} \tag{7.1.26}$$

and

$$\sigma_{rz} = 0 , \quad \sigma_{\theta z} = \frac{\mu\delta}{2\pi r} . \tag{7.1.27}$$

Going back to the full three-dimensional boundary conditions, the traction-free inner and outer surface condition (7.1.20) is satisfied, but the traction-free condition on the end surfaces (7.1.21) is not. The traction distribution on the end surfaces that arise from the antiplane strain assumption are statically equivalent to a torque applied to the ends of the cylinder. If the original problem does not have boundary conditions on the ends of the cylinder equivalent to this torque, rather than the stated conditions (7.1.21), then the solution is *not one of antiplane strain*—the cylinder will twist.

7.2 Plane Strain

A state of *plane strain* is defined as one in which

$$\boxed{u_\alpha = u_\alpha(X_1, X_2) , \quad u_3 = 0 .} \tag{7.2.1}$$

It is seen from the definition (7.2.1) that the scalar components of strain, $\varepsilon_{ij} = \frac{1}{2}(u_{i,j} + u_{j,i})$, must in a state of plane strain have the form $\varepsilon_{13} = \varepsilon_{23} = \varepsilon_{33} = 0$ and $\varepsilon_{\alpha\beta} = \varepsilon_{\alpha\beta}(X_1, X_2)$, where

$$\varepsilon_{\alpha\beta} = \frac{1}{2}(u_{\alpha,\beta} + u_{\beta,\alpha}) . \tag{7.2.2}$$

Noting that $\varepsilon_{kk} = \varepsilon_{11} + \varepsilon_{22} = \varepsilon_{\gamma\gamma}$, it follows that the scalar components of stress, $\sigma_{ij} = 2\mu\varepsilon_{ij} + \lambda\varepsilon_{kk}\delta_{ij}$, must correspondingly have the form $\sigma_{13} = \sigma_{23} = 0$, $\sigma_{\alpha\beta} = \sigma_{\alpha\beta}(X_1, X_2)$, and $\sigma_{33} = \sigma_{33}(X_1, X_2)$, where $\sigma_{33} = \lambda\varepsilon_{\gamma\gamma}$ and

$$\boxed{\sigma_{\alpha\beta} = 2\mu\varepsilon_{\alpha\beta} + \lambda\varepsilon_{\gamma\gamma}\delta_{\alpha\beta} \,.} \tag{7.2.3}$$

Observe that, even though the scalar component of stress σ_{33} is *not* identically zero in a state of plane strain, its value can be given in terms of the values of σ_{11} and σ_{22}. Since $\delta_{\alpha\alpha} = \delta_{11} + \delta_{22} = 2$, it follows from (7.2.3) and $\sigma_{33} = \lambda\varepsilon_{\gamma\gamma}$ that

$$\sigma_{\alpha\alpha} = 2(\mu + \lambda)\varepsilon_{\alpha\alpha} = \frac{2(\mu + \lambda)}{\lambda}\sigma_{33} = \frac{1}{\nu}\sigma_{33} \,. \tag{7.2.4}$$

Therefore, once σ_{11} and σ_{22} are known, the component of stress σ_{33} is just given by

$$\boxed{\sigma_{33} = \nu\sigma_{\alpha\alpha} \,.} \tag{7.2.5}$$

The formulation of plane strain boundary value problems will not explicitly involve σ_{33}—it will be eliminated from the plane strain equations of elasticity and boundary conditions. After a plane strain boundary value problem has been solved (so that σ_{11}, σ_{22}, and σ_{12} are known), then σ_{33} can be determined from (7.2.5).

To obtain the plane strain form of the expression

$$\varepsilon_{ij} = \frac{1}{E}[(1 + \nu)\sigma_{ij} - \nu\sigma_{kk}\delta_{ij}] \tag{7.2.6}$$

for the scalar components of strain in terms of the scalar components of stress, note first that $\sigma_{kk} = \sigma_{\gamma\gamma} + \sigma_{33} = (1 + \nu)\sigma_{\gamma\gamma}$. It follows therefore that

$$\boxed{\varepsilon_{\alpha\beta} = \frac{1 + \nu}{E}(\sigma_{\alpha\beta} - \nu\sigma_{\gamma\gamma}\delta_{\alpha\beta}) \,.} \tag{7.2.7}$$

The expressions (7.2.3) and (7.2.7) relating the scalar components of strain ε_{11}, ε_{22}, and ε_{12} to the scalar components of stress σ_{11}, σ_{22}, and σ_{12} are the *plane strain constitutive equations*.

From (7.2.3), one can see that the equilibrium equations $\sigma_{ji,j} + f_i = 0$ reduce to the two scalar *plane strain equilibrium equations*,

$$\boxed{\sigma_{\beta\alpha,\beta} + f_\alpha = 0 \,,} \tag{7.2.8}$$

and the requirement that $f_3 = 0$ and $f_\alpha = f_\alpha(X_1, X_2)$. Similarly, five of the six compatibility equations (3.4.45) are identically satisfied and the remaining equation reduces to

$$\varepsilon_{11,22} + \varepsilon_{22,11} - 2\varepsilon_{12,12} = 0 \,. \tag{7.2.9}$$

Using the constitutive equations (7.2.7), this lone remaining compatibility condition can be expressed in terms of stress:

$$(1 - \nu)(\sigma_{11,22} + \sigma_{22,11}) - \nu(\sigma_{11,11} + \sigma_{22,22}) = 2\sigma_{12,12} . \qquad (7.2.10)$$

However, differentiating the equilibrium equations (7.2.8) with respect to X_1 and X_2 gives, respectively, the relations $\sigma_{11,11} + \sigma_{12,12} + f_{1,1} = 0$ and $\sigma_{12,12} + \sigma_{22,22} + f_{2,2} = 0$, which, when adding together, yields a necessary condition for equilibrium:

$$2\sigma_{12,12} = -(\sigma_{11,11} + \sigma_{22,22}) - (f_{1,1} + f_{2,2}) . \qquad (7.2.11)$$

Thus, substituting (7.2.11) into (7.2.10), the *plane strain compatibility condition* can be expressed in terms of stress as

$$\boxed{\nabla^2(\sigma_{\alpha\alpha}) = \frac{-1}{1 - \nu} f_{\alpha,\alpha} ,} \qquad (7.2.12)$$

where $\nabla^2(\sigma_{\alpha\alpha}) = \sigma_{\alpha\alpha,\beta\beta}$.

Let ∂S^u and ∂S^t be the portions of the boundary ∂S on which displacements and tractions are prescribed such that if there are no mixed boundary conditions, $\partial S^u \cup \partial S^t = \partial S$. Then, a *displacement boundary condition for plane strain* will have the form

$$\boxed{u_\alpha = \tilde{u}_\alpha \quad \forall (X_1, X_2) \in \partial S^u ,} \qquad (7.2.13)$$

where $\tilde{u}_\alpha(X_1, X_2)$ are given functions. For the traction boundary condition, note first that since $\hat{n}_3 = 0$ on ∂S^t, the traction-stress relations $t_i = \sigma_{ji}\hat{n}_j$ for the lateral surface reduce to

$$t_\alpha = \sigma_{\beta\alpha}\hat{n}_\beta \qquad (7.2.14)$$

and the requirement that $t_3 = 0$ and $t_\alpha = t_\alpha(X_1, X_2)$. Therefore, a *traction boundary condition for plane strain* will have the form

$$\boxed{\sigma_{\beta\alpha}\hat{n}_\beta = \tilde{t}_\alpha \quad \forall (X_1, X_2) \in \partial S^t ,} \qquad (7.2.15)$$

where $\tilde{t}_\alpha(X_1, X_2)$ are given functions.

7.2.1 Cylindrical Coordinates

In cylindrical coordinates, a state of plane strain is defined as one in which

$$\boxed{u_r = u_r(r, \theta) , \quad u_\theta = u_\theta(r, \theta) , \quad u_z = 0 .} \qquad (7.2.16)$$

Therefore, the cylindrical coordinates strain-displacement relations (3.5.5) reduce to $\varepsilon_{rz} = \varepsilon_{\theta z} = \varepsilon_{zz} = 0$ and

$$\varepsilon_{rr}(r,\theta) = \frac{\partial u_r}{\partial r} \, ,$$

$$\varepsilon_{\theta\theta}(r,\theta) = \frac{1}{r} \left(\frac{\partial u_\theta}{\partial \theta} + u_r \right) , \tag{7.2.17}$$

$$\varepsilon_{r\theta}(r,\theta) = \frac{1}{2} \left(\frac{1}{r} \frac{\partial u_r}{\partial \theta} + \frac{\partial u_\theta}{\partial r} - \frac{u_\theta}{r} \right) .$$

The constitutive equations (7.2.3), (7.2.5), and (7.2.7) are unchanged except that the indices 1, 2, and 3 are replaced by r, θ, and z, respectively. The equilibrium equations (4.4.4) reduce to $f_z = 0$ and

$$\boxed{\begin{aligned} \frac{\partial \sigma_{rr}}{\partial r} + \frac{1}{r} \frac{\partial \sigma_{r\theta}}{\partial \theta} + \frac{1}{r}(\sigma_{rr} - \sigma_{\theta\theta}) + f_r = 0 \, , \\ \frac{\partial \sigma_{r\theta}}{\partial r} + \frac{1}{r} \frac{\partial \sigma_{\theta\theta}}{\partial \theta} + \frac{2}{r}\sigma_{r\theta} + f_\theta = 0 \end{aligned}} \tag{7.2.18}$$

and the compatibility condition in terms of stress is

$$\boxed{\nabla^2(\sigma_{rr} + \sigma_{\theta\theta}) = \frac{-1}{1-\nu} \left(\frac{\partial f_r}{\partial r} + \frac{1}{r} \frac{\partial f_\theta}{\partial \theta} + \frac{f_r}{r} \right) ,} \tag{7.2.19}$$

where

$$\nabla^2(\ ) = \frac{\partial^2(\ )}{\partial r^2} + \frac{1}{r} \frac{\partial(\ )}{\partial r} + \frac{1}{r^2} \frac{\partial^2(\ )}{\partial \theta^2} . \tag{7.2.20}$$

The boundary conditions (7.2.13) and (7.2.15) are the same in cylindrical coordinates except that the indices 1 and 2 are replaced by r and θ.

7.2.2 Example: Cylindrical Pressure Vessel

Consider the cylindrical pressure vessel under internal and external pressures shown in Figure 7.3. The pressure vessel has inner radius a and outer radius b and is under internal pressure $p^{(1)}$ and external pressure $p^{(2)}$. Define a cylindrical coordinate system (r, θ, z) with z along the centroidal axis of the pressure vessel. The body occupies the region $a \le r \le b$, $-L/2 \le z \le L/2$. The cylinder is held in rigid supports, so that the displacements are zero at $z = \pm L/2$. Assume that body forces are negligible. For cross sections sufficiently far from the ends, it is clear that $u_z = 0$ and that u_r and u_θ are independent of z—a state of plane strain exists.

As usual, the first order of business is to clearly establish the boundary conditions. For this plane strain problem, there are traction boundary

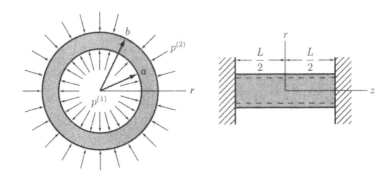

FIGURE 7.3. Cylindrical pressure vessel.

conditions only. On the inner surface of the pressure vessel, the traction vector is $\tilde{\mathbf{t}} = p^{(1)}\hat{\mathbf{e}}_r$ and the unit outward normal is $\hat{\mathbf{n}} = -\hat{\mathbf{e}}_r$. Therefore, using the traction-stress relations in cylindrical coordinates (7.2.15),

$$\sigma_{rr} = -p^{(1)}, \quad \sigma_{r\theta} = 0 \quad \text{on } r = a . \tag{7.2.21}$$

On the outer surface, the traction vector is $\tilde{\mathbf{t}} = -p^{(2)}\hat{\mathbf{e}}_r$ and the unit outward normal is $\hat{\mathbf{n}} = \hat{\mathbf{e}}_r$. Therefore,

$$\sigma_{rr} = -p^{(2)}, \quad \sigma_{r\theta} = 0 \quad \text{on } r = b . \tag{7.2.22}$$

The problem is to find the stress field $\boldsymbol{\sigma}$ and the displacement field $\mathbf{u}$ for the pressure vessel (sufficiently far from the ends).

Using the semi-inverse method, assume that the solution displacement field exhibits the same cylindrical symmetry as the stated problem. Then, the scalar components of displacement can be expressed in the form

$$u_r = U(r) , \quad u_\theta = 0 , \tag{7.2.23}$$

where $U(r)$ is an unknown function. Using the strain-displacement relations in cylindrical coordinates (7.2.17), the scalar components of strain are

$$\varepsilon_{rr} = U'(r) , \quad \varepsilon_{\theta\theta} = \frac{1}{r}U(r) , \quad \varepsilon_{r\theta} = 0 , \tag{7.2.24}$$

and from (7.2.3), the scalar components of stress are

$$\sigma_{rr} = (2\mu + \lambda)U'(r) + \frac{\lambda}{r}U(r) ,$$

$$\sigma_{\theta\theta} = \frac{1}{r}(2\mu + \lambda)U(r) + \lambda U'(r) , \tag{7.2.25}$$

$$\sigma_{r\theta} = 0 .$$

In order for the stress field (7.2.25) to be the solution to the problem, it must satisfy the equilibrium equations (7.2.18) and the boundary conditions (7.2.21) and (7.2.22).

Substituting (7.2.25) into the equilibrium equations (7.2.18), it is seen that the second equilibrium equation is identically satisfied and the first gives

$$(2\mu + \lambda)U''(r) + \frac{\lambda}{r}U'(r) - \frac{\lambda}{r^2}U(r) + \frac{2\mu}{r}U'(r) - \frac{2\mu}{r^2}U(r) = 0 , \quad (7.2.26)$$

which reduces to

$$r^2U''(r) + rU'(r) - U(r) = 0 . \quad (7.2.27)$$

This is an Euler-Cauchy equation which has a general solution of the form

$$U(r) = Br + \frac{C}{r} , \quad (7.2.28)$$

where B and C are constants to be determined by the boundary conditions (7.2.21) and (7.2.22).

It is clear that the boundary conditions involving $\sigma_{r\theta}$ are satisfied by the stress field (7.2.25). The remaining boundary conditions involve σ_{rr}, which can be written in terms of the unknown constants B and C by combining (7.2.28) and (7.2.25):

$$\sigma_{rr} = (2\mu + \lambda)\left(B - \frac{C}{r^2}\right) + \lambda\left(B + \frac{C}{r^2}\right)$$
$$= 2(\mu + \lambda)B - 2\mu\frac{C}{r^2} . \quad (7.2.29)$$

In terms of B and C then, the boundary conditions are

$$2(\mu + \lambda)B - \frac{2\mu}{a^2}C = -p^{(1)} ,$$
$$2(\mu + \lambda)B - \frac{2\mu}{b^2}C = -p^{(2)} . \quad (7.2.30)$$

Solving (7.2.30) for the two unknowns B and C,

$$B = \frac{1}{2(\mu + \lambda)}\left(\frac{a^2p^{(1)} - b^2p^{(2)}}{b^2 - a^2}\right) ,$$
$$C = \frac{a^2b^2}{2\mu}\left(\frac{p^{(1)} - p^{(2)}}{b^2 - a^2}\right) . \quad (7.2.31)$$

Thus, substituting (7.2.31) into (7.2.23) and (7.2.25) and using (7.2.5) to obtain the axial component of stress σ_{zz}, the solution displacement field is

$$u_r = \frac{r}{2(\mu + \lambda)} \left(\frac{a^2 p^{(1)} - b^2 p^{(2)}}{b^2 - a^2} \right) + \frac{a^2 b^2}{2\mu r} \left(\frac{p^{(1)} - p^{(2)}}{b^2 - a^2} \right) ,$$

$$u_\theta = 0$$

(7.2.32)

and the solution stress field is

$$\sigma_{rr} = \left(\frac{a^2 p^{(1)} - b^2 p^{(2)}}{b^2 - a^2} \right) - \frac{a^2 b^2}{r^2} \left(\frac{p^{(1)} - p^{(2)}}{b^2 - a^2} \right) ,$$

$$\sigma_{\theta\theta} = \left(\frac{a^2 p^{(1)} - b^2 p^{(2)}}{b^2 - a^2} \right) + \frac{a^2 b^2}{r^2} \left(\frac{p^{(1)} - p^{(2)}}{b^2 - a^2} \right) ,$$

$$\sigma_{zz} = 2\nu \left(\frac{a^2 p^{(1)} - b^2 p^{(2)}}{b^2 - a^2} \right) ,$$

$$\sigma_{r\theta} = 0 .$$

(7.2.33)

Remember that this solution is only valid for cross sections sufficiently far from the ends of the pressure vessel.

This solution can be generalized by superposing a uniaxial stress field. Consider a problem where, instead of rigid supports at $z = \pm L/2$, there are end caps, such as on an actual cylindrical pressure vessel. The internal pressure has a resultant force on each end cap of $F^{(1)} = \pi a^2 p^{(1)}$ along the z-direction and the external pressure has a resultant force of $F^{(2)} = \pi b^2 p^{(2)}$. The cross-sectional area of the vessel wall is $A = \pi(b^2 - a^2)$. For equilibrium with the pressure on the end caps, therefore, the axial stress σ_{zz} should be

$$\sigma_{zz} = \frac{F^{(1)} - F^{(2)}}{A} = \frac{a^2 p^{(1)} - b^2 p^{(2)}}{b^2 - a^2} .$$

(7.2.34)

To meet this condition, a uniaxial stress field given by

$$\sigma_{zz} = (1 - 2\nu) \frac{a^2 p^{(1)} - b^2 p^{(2)}}{b^2 - a^2} ,$$

(7.2.35)

along with the corresponding components of strain and displacement, can be superposed on the solution (7.2.33).

If the cylindrical vessel is thin-walled with thickness $t \equiv b - a$, $t \ll a$, then

$$b^2 - a^2 = (b - a)(b + a) \approx 2ta ,$$

$$a^2 p^{(1)} - b^2 p^{(2)} \approx a^2 (p^{(1)} - p^{(2)}) ,$$

$$b^2 \approx a^2 ,$$

$$r^2 \approx a^2$$

(7.2.36)

and it follows from (7.2.33) that

$$\sigma_{\theta\theta} \approx \frac{a(p^{(1)} - p^{(2)})}{t} . \tag{7.2.37}$$

This is the thin-walled cylindrical pressure vessel approximation from mechanics of materials.

In the limit as $b \to \infty$, the problem reduces to that of a pressurized cylindrical hole in an infinite body under remote "plane strain pressure" (as $r \to \infty$, $\sigma_{rr} = \sigma_{\theta\theta} = -p^{(2)}$ and $\sigma_{zz} = -2\nu p^{(2)}$). Letting $b \to \infty$ in (7.2.33),

$$\sigma_{rr} = -p^{(2)} - (p^{(1)} - p^{(2)})\frac{a^2}{r^2} ,$$

$$\sigma_{\theta\theta} = -p^{(2)} + (p^{(1)} - p^{(2)})\frac{a^2}{r^2} , \tag{7.2.38}$$

$$\sigma_{zz} = -2\nu p^{(2)} .$$

If the hole is unpressurized ($p^{(1)} = 0$), then

$$\sigma_{rr} = -p^{(2)} \left(1 - \frac{a^2}{r^2}\right) ,$$

$$\sigma_{\theta\theta} = -p^{(2)} \left(1 + \frac{a^2}{r^2}\right) , \tag{7.2.39}$$

$$\sigma_{zz} = -2\nu p^{(2)} .$$

Far from the hole ($r \to \infty$), the greatest principal stress magnitude is $|p^{(2)}|$, whereas at the hole ($r = a$), the greatest principal stress magnitude is $2|p^{(2)}|$. Thus, a cylindrical hole in an infinite body under far-field plane strain pressure has a stress concentration factor of 2.

7.3 Plane Stress

A state of *plane stress* is defined as one in which

$$\boxed{\sigma_{\alpha\beta} = \sigma_{\alpha\beta}(X_1, X_2) , \quad \sigma_{13} = \sigma_{23} = \sigma_{33} = 0 .} \tag{7.3.1}$$

Noting that $\sigma_{kk} = \sigma_{11} + \sigma_{22} = \sigma_{\gamma\gamma}$, it follows that in a state of plane stress, the scalar components of strain, $\varepsilon_{ij} = [(1 + \nu)\sigma_{ij} - \nu\sigma_{kk}\delta_{ij}]/E$, must have the form $\varepsilon_{13} = \varepsilon_{23} = 0$, $\varepsilon_{\alpha\beta} = \varepsilon_{\alpha\beta}(X_1, X_2)$, and $\varepsilon_{33} = \varepsilon_{33}(X_1, X_2)$, where

$\varepsilon_{33} = -\nu\sigma_{\gamma\gamma}/E$ and

$$\varepsilon_{\alpha\beta} = \frac{1}{E}\left[(1+\nu)\sigma_{\alpha\beta} - \nu\sigma_{\gamma\gamma}\delta_{\alpha\beta}\right] . \qquad (7.3.2)$$

In a manner analogous to the situation involving σ_{33} in a state of *plane strain*, in a state of *plane stress* the scalar component of strain ε_{33} can be given in terms of ε_{11} and ε_{22}, so that it need not appear explicitly in the formulation of plane stress problems. Noting that

$$\varepsilon_{\alpha\alpha} = \frac{1}{E}\left[(1+\nu)\sigma_{\alpha\alpha} - \nu\sigma_{\gamma\gamma}\delta_{\alpha\alpha}\right] = \frac{1-\nu}{E}\sigma_{\alpha\alpha} = -\frac{1-\nu}{\nu}\varepsilon_{33} , \qquad (7.3.3)$$

it is seen that the component of strain ε_{33} is given in terms of ε_{11} and ε_{22} by

$$\varepsilon_{33} = \frac{-\nu}{1-\nu}\varepsilon_{\alpha\alpha} = \frac{-\lambda}{\lambda+2\mu}\varepsilon_{\alpha\alpha} . \qquad (7.3.4)$$

After a plane stress boundary value problem has been solved (so that ε_{11}, ε_{22}, and ε_{12} are known), then ε_{33} can be determined from (7.3.4).

To obtain the plane stress form of the relation $\sigma_{ij} = 2\mu\varepsilon_{ij} + \lambda\varepsilon_{kk}\delta_{ij}$ for the scalar components of stress in terms of the scalar components of strain, note first that

$$\varepsilon_{kk} = \varepsilon_{\gamma\gamma} + \varepsilon_{33} = \frac{2\mu}{\lambda+2\mu}\varepsilon_{\gamma\gamma} . \qquad (7.3.5)$$

It follows that

$$\sigma_{\alpha\beta} = 2\mu\varepsilon_{\alpha\beta} + \frac{2\mu\lambda}{\lambda+2\mu}\varepsilon_{\gamma\gamma}\delta_{\alpha\beta} . \qquad (7.3.6)$$

The expressions (7.3.2) and (7.3.6) relating the scalar components of strain ε_{11}, ε_{22}, and ε_{12} to the scalar components of stress σ_{11}, σ_{22}, and σ_{12} are the *plane stress constitutive equations*.

From the definition (7.3.1), one can see that the equilibrium equations $\sigma_{ji,j} + f_i = 0$ reduce to the two scalar *plane stress equilibrium equations*,

$$\sigma_{\beta\alpha,\beta} + f_\alpha = 0 , \qquad (7.3.7)$$

and the requirement that $f_3 = 0$ and $f_\alpha = f_\alpha(X_1, X_2)$.

For a state of plane stress, only two of the six compatibility equations (3.4.45) are identically satisfied. The four remaining equations reduce to

$$\varepsilon_{11,22} + \varepsilon_{22,11} - 2\varepsilon_{12,12} = 0 ,$$
$$\varepsilon_{33,11} = 0 ,$$
$$\varepsilon_{33,22} = 0 , \qquad (7.3.8)$$
$$\varepsilon_{33,12} = 0 .$$

The last three of these are only satisfied if $\varepsilon_{33}(X_1, X_2)$ has the form $\varepsilon_{33} = a + bX_1 + cX_2$, where a, b, and c are constants. This requirement is usually ignored, however, resulting in an approximate theory that is *only acceptable for thin plates*—the dimension of the body in the X_3-direction must be small relative to the in-plane dimensions. This is related to the fact that, in thin plates, σ_{33} is only approximately equal to zero (this is shown in the discussion of generalized plane stress; Section 7.3.2). Thus, *implicit in the use of the plane stress formulation of the equations of elasticity is the assumption that σ_{33} is sufficiently small to be neglected, which is only true for thin plates.*

Using the constitutive relations (7.3.2) and the equations of motion (7.3.7) as was done in the derivation of the plane strain equations, the first compatibility condition (7.3.8) can be written as

$$\boxed{\nabla^2(\sigma_{\alpha\alpha}) = -(1+\nu)f_{\alpha,\alpha} ,} \qquad (7.3.9)$$

where $\nabla^2(\ ) = (\ )_{,\beta\beta}$.

Let $\partial \mathcal{S}^u$ and $\partial \mathcal{S}^t$ be the portions of the boundary $\partial \mathcal{S}$ on which displacements and tractions are prescribed such that if there are no mixed boundary conditions, $\partial \mathcal{S}^u \cup \partial \mathcal{S}^t = \partial \mathcal{S}$. Then, a *displacement boundary condition for plane stress* will have the form

$$\boxed{u_\alpha = \tilde{u}_\alpha \quad \forall (X_1, X_2) \in \partial \mathcal{S}^u ,} \qquad (7.3.10)$$

where $\tilde{u}_\alpha(X_1, X_2)$ are given functions. For the traction boundary condition, note first that since $\hat{n}_3 = 0$ on $\partial \mathcal{S}^t$, the traction-stress relations $t_i = \sigma_{ji}\hat{n}_j$ for the lateral surface reduce to

$$t_\alpha = \sigma_{\beta\alpha}\hat{n}_\beta \qquad (7.3.11)$$

and the requirement that $t_3 = 0$ and $t_\alpha = t_\alpha(X_1, X_2)$. Therefore, a *traction boundary condition for plane stress* will have the form

$$\boxed{\sigma_{\beta\alpha}\hat{n}_\beta = \tilde{t}_\alpha \quad \forall (X_1, X_2) \in \partial \mathcal{S}^t ,} \qquad (7.3.12)$$

where $\tilde{t}_\alpha(X_1, X_2)$ are given functions.

7.3.1 Cylindrical Coordinates

In cylindrical coordinates, a state of plane stress is defined as one in which $\sigma_{zz} = \sigma_{rz} = \sigma_{\theta z} = 0$ and

$$\boxed{\sigma_{rr} = \sigma_{rr}(r, \theta) , \quad \sigma_{\theta\theta} = \sigma_{\theta\theta}(r, \theta) , \quad \sigma_{r\theta} = \sigma_{r\theta}(r, \theta) .} \qquad (7.3.13)$$

The strain-displacement relations (3.5.5) reduce to $\varepsilon_{rz} = \varepsilon_{\theta z} = 0$ and

$$\varepsilon_{rr}(r, \theta) = \frac{\partial u_r}{\partial r} \,,$$

$$\varepsilon_{\theta\theta}(r, \theta) = \frac{1}{r} \left(\frac{\partial u_\theta}{\partial \theta} + u_r \right) \,, \tag{7.3.14}$$

$$\varepsilon_{r\theta}(r, \theta) = \frac{1}{2} \left(\frac{1}{r} \frac{\partial u_r}{\partial \theta} + \frac{\partial u_\theta}{\partial r} - \frac{u_\theta}{r} \right) \,.$$

The constitutive equations (7.3.2), (7.3.4), and (7.3.6) are unchanged except that the indices 1, 2, and 3 are replaced by r, θ, and z, respectively. The equilibrium equations (4.4.4) reduce to $f_z = 0$ and

$$\boxed{\begin{aligned} \frac{\partial \sigma_{rr}}{\partial r} + \frac{1}{r} \frac{\partial \sigma_{r\theta}}{\partial \theta} + \frac{1}{r}(\sigma_{rr} - \sigma_{\theta\theta}) + f_r = 0 \,, \\ \frac{\partial \sigma_{r\theta}}{\partial r} + \frac{1}{r} \frac{\partial \sigma_{\theta\theta}}{\partial \theta} + \frac{2}{r}\sigma_{r\theta} + f_\theta = 0 \end{aligned}} \tag{7.3.15}$$

and the compatibility condition in terms of stress is

$$\boxed{\nabla^2(\sigma_{rr} + \sigma_{\theta\theta}) = -(1 + \nu) \left(\frac{\partial f_r}{\partial r} + \frac{1}{r} \frac{\partial f_\theta}{\partial \theta} + \frac{f_r}{r} \right) \,,} \tag{7.3.16}$$

where

$$\nabla^2(\ ) = \frac{\partial^2(\ )}{\partial r^2} + \frac{1}{r} \frac{\partial(\ )}{\partial r} + \frac{1}{r^2} \frac{\partial^2(\ )}{\partial \theta^2} \,. \tag{7.3.17}$$

The boundary conditions (7.3.10) and (7.3.12) are the same in cylindrical coordinates except that the indices 1 and 2 are replaced by r and θ.

7.3.2   Generalized Plane Stress

An alternative derivation of the plane stress equations of elasticity involves the consideration of quantities *averaged* through the thickness of a thin plate, leading to the *generalized plane stress* formulation. Consider a plate of uniform thickness $2h$ with its middle surface (mid-plane) in the $X_1 X_2$-plane (Figure 7.4). Assume that h is much less than the smallest characteristic mid-plane dimension of the plate. The top and bottom surfaces of the plate are at $X_3 = \pm h$ and have unit outward normals $\hat{\mathbf{n}} = \pm \hat{\mathbf{e}}_3$, respectively. The edge surface of the plate is perpendicular to the top and bottom surfaces, so that $\hat{n}_3 = 0$ everywhere along the edge.

The top and bottom surfaces of the plate are traction-free, whereas, along the edge surface, prescribed tractions and displacements are parallel to the $X_1 X_2$-plane so that $\tilde{t}_3 = 0$ and $\tilde{u}_3 = 0$. The body force is also

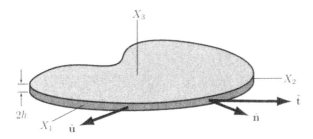

FIGURE 7.4. *Thin plate in a state of plane stress.*

parallel to the X_1X_2-plane so that $f_3 = 0$. The remaining components of the prescribed edge displacements and tractions, $\tilde{u}_\alpha$ and $\tilde{t}_\alpha$, and body force f_α are assumed to be even functions of X_3:

$$\tilde{u}_\alpha(X_1, X_2, -X_3) = \tilde{u}_\alpha(X_1, X_2, X_3) \,,$$
$$\tilde{t}_\alpha(X_1, X_2, -X_3) = \tilde{t}_\alpha(X_1, X_2, X_3) \,, \tag{7.3.18}$$
$$f_\alpha(X_1, X_2, -X_3) = f_\alpha(X_1, X_2, X_3) \,.$$

It follows from symmetry that, within the plate, u_α are even functions of X_3 and u_3 is an odd function of X_3. Consequently, $\varepsilon_{\alpha\beta}$, ε_{33}, $\sigma_{\alpha\beta}$, and σ_{33} are even functions of X_3 and $\varepsilon_{\alpha3}$ and $\sigma_{\alpha3}$ are odd functions of X_3.

Let $\langle\psi\rangle$ be the through-thickness average of the scalar field ψ:

$$\langle\psi\rangle \equiv \frac{1}{2h}\int_{-h}^{h} \psi(X_1, X_2, X_3)\,dX_3 \,. \tag{7.3.19}$$

Thus, the average quantity $\langle\psi\rangle$ is a function of X_1 and X_2. Since u_3, $\varepsilon_{\alpha3}$, and $\sigma_{\alpha3}$ are odd functions of X_3, their through-thickness averages vanish:

$$\langle u_3\rangle = 0 \,, \quad \langle\varepsilon_{13}\rangle = \langle\varepsilon_{23}\rangle = 0 \,, \quad \langle\sigma_{13}\rangle = \langle\sigma_{23}\rangle = 0 \,. \tag{7.3.20}$$

Also, given that the prescribed body force field and edge tractions and displacements are assumed to be parallel to the X_1X_2-plane, it follows trivially that $\langle f_3\rangle = 0$, $\langle\tilde{t}_3\rangle = 0$, and $\langle\tilde{u}_3\rangle = 0$.

Since the top and bottom surfaces are traction-free, it follows from the traction-stress relations that

$$\sigma_{13} = \sigma_{23} = \sigma_{33} = 0 \quad \text{on } X_3 = \pm h \,. \tag{7.3.21}$$

Therefore,

$$\sigma_{13,1} = \sigma_{23,2} = 0 \quad \text{on } X_3 = \pm h \,, \tag{7.3.22}$$

and from the equilibrium equation $\sigma_{13,1} + \sigma_{23,2} + \sigma_{33,3} + f_3 = 0$,

$$\sigma_{33,3} = -f_3 - \sigma_{13,1} - \sigma_{23,2} = 0 \quad \text{on } X_3 = \pm h . \tag{7.3.23}$$

Expanding σ_{33} in a Taylor series in X_3 about $X_3 = +h$,

$$\sigma_{33}(X_1, X_2, X_3) = \sigma_{33}(X_1, X_2, h) + \sigma_{33,3}(X_1, X_2, h)(X_3 - h)$$
$$+ \frac{1}{2}\sigma_{33,33}(X_1, X_2, h)(X_3 - h)^2 + \cdots , \tag{7.3.24}$$

so that, from (7.3.21) and (7.3.23),

$$\sigma_{33}(X_1, X_2, X_3) = \frac{1}{2}\sigma_{33,33}(X_1, X_2, h)(X_3 - h)^2 + \cdots . \tag{7.3.25}$$

Since $-h \leq X_3 \leq h$, it follows that $|X_3 - h| \leq 2h$ and, consequently, that the scalar component of stress $\sigma_{33} = \mathcal{O}(h^2)$, which is small for thin plates. Therefore,

$$\sigma_{33} \approx 0 , \quad \langle\sigma_{33}\rangle \approx 0 . \tag{7.3.26}$$

Recall that this was an implicit assumption in the regular plane stress formulation of the equations of elasticity.

It can easily be seen now that each of the equations in the regular plane stress formulation has an analog in the generalized plane stress formulation involving through-thickness averaged quantities. For example, the generalized form of the definition (7.3.1) is

$$\langle\sigma_{\alpha\beta}\rangle = \langle\sigma_{\alpha\beta}\rangle(X_1, X_2) , \quad \langle\sigma_{13}\rangle = \langle\sigma_{23}\rangle = 0 , \quad \langle\sigma_{33}\rangle \approx 0 . \tag{7.3.27}$$

Similarly, by integrating each term of the equilibrium equations through the thickness of the plate, dividing by $2h$, and reversing the order of integration and differentiation of the components of stress, one obtains a necessary condition for equilibrium:

$$\langle\sigma_{\beta\alpha}\rangle_{,\beta} + \langle f_\alpha\rangle = 0 , \tag{7.3.28}$$

which is a generalization of the plane stress equilibrium equation (7.3.7). Each of the other plane stress equations of elasticity can also be seen to have an analogous generalized form. It is found that the generalized plane stress equations are identical to their regular plane stress counterparts. This generalized plane stress formulation represents only an alternative *conceptual* approach to plane stress elasticity—there is no mathematical difference between generalized and regular plane stress.

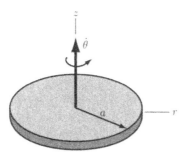

FIGURE 7.5. *Spinning thin disk.*

7.3.3 Example: Spinning Thin Disk

Consider the thin disk of radius a, spinning about its axis of symmetry with constant angular velocity $\dot{\theta}$, shown in Figure 7.5. Define a cylindrical coordinate system that rotates with the disk, with the z-axis coincident with the axis of symmetry. The material acceleration vector for points on the disk is $\mathbf{a} = -r\dot{\theta}^2\hat{\mathbf{e}}_r$. In this reference frame, the inertial force is indistinguishable from a body force:

$$\mathbf{f} = \rho r\dot{\theta}^2\hat{\mathbf{e}}_r \ . \tag{7.3.29}$$

Assume that other body forces are negligible. The disk is in a state of plane stress. Find the stress and displacement fields in the disk.

This plane stress problem has traction boundary conditions only. Using the traction-stress relations (7.3.12) in cylindrical coordinates, the traction boundary conditions in terms of stress are as follows. On the edge surface of the disk, the traction vector is $\tilde{\mathbf{t}} = \mathbf{0}$ and the unit outward normal is $\hat{\mathbf{n}} = \hat{\mathbf{e}}_r$ so that

$$\sigma_{rr} = \sigma_{r\theta} = 0 \quad \text{on } r = a \ . \tag{7.3.30}$$

Using the semi-inverse method, assume that the solution displacement field exhibits the same cylindrical symmetry as the stated problem. Then, the components of the displacement can be expressed in the form

$$u_r = U(r), \quad u_\theta = 0 , \tag{7.3.31}$$

where $U(r)$ is an unknown function. Using the strain-displacement relations (7.3.14), the strains are

$$\varepsilon_{rr} = U'(r), \quad \varepsilon_{\theta\theta} = \frac{1}{r}U(r), \quad \varepsilon_{r\theta} = 0 , \tag{7.3.32}$$

and from (7.3.6), the stresses are

$$
\sigma_{rr} = \left(2\mu + \frac{2\mu\lambda}{\lambda + 2\mu}\right) U'(r) + \frac{2\mu\lambda}{\lambda + 2\mu} \frac{U(r)}{r} ,
$$

$$
\sigma_{\theta\theta} = \left(2\mu + \frac{2\mu\lambda}{\lambda + 2\mu}\right) \frac{U(r)}{r} + \frac{2\mu\lambda}{\lambda + 2\mu} U'(r) , \tag{7.3.33}
$$

$$
\sigma_{r\theta} = 0 .
$$

In order for the stress field (7.3.33) to be the solution to the problem, it must satisfy the equilibrium equations (7.3.15) and the boundary conditions (7.3.30).

The second equilibrium equation (7.3.15) is identically satisfied and the first reduces to

$$
r^2 U''(r) + r U'(r) - U(r) = -\left[\frac{\lambda + 2\mu}{4\mu(\lambda + \mu)}\right] \rho\dot{\theta}^2 r^3 . \tag{7.3.34}
$$

This is an Euler-Cauchy equation whose complete solution is

$$
U(r) = Br + \frac{C}{r} - \left[\frac{\lambda + 2\mu}{32\mu(\lambda + \mu)}\right] \rho\dot{\theta}^2 r^3 . \tag{7.3.35}
$$

However, in order for the displacement to be bounded at $r = 0$, it follows that $C = 0$. Only B remains to be determined by the boundary conditions.

It is clear that the boundary condition involving $\sigma_{r\theta}$ is satisfied by the stress field (7.3.33). The remaining boundary condition involves σ_{rr}, which can be written in terms of the unknown constant B by combining (7.3.35) and (7.3.33):

$$
\sigma_{rr} = \frac{2\mu(3\lambda + 2\mu)}{\lambda + 2\mu} B - \frac{7\lambda + 6\mu}{16(\lambda + \mu)} \rho\dot{\theta}^2 r^2 . \tag{7.3.36}
$$

From the boundary condition $\sigma_{rr}(a, \theta) = 0$,

$$
B = \frac{(\lambda + 2\mu)(7\lambda + 6\mu)}{32\mu(\lambda + \mu)(3\lambda + 2\mu)} \rho\dot{\theta}^2 a^2 . \tag{7.3.37}
$$

Thus, substituting (7.3.37) into (7.3.31) and (7.3.33), the solution displacement field is

$$
\boxed{
\begin{aligned}
u_r &= \frac{\lambda + 2\mu}{32\mu(\lambda + \mu)} \left[\left(\frac{7\lambda + 6\mu}{3\lambda + 2\mu}\right) a^2 - r^2\right] \rho\dot{\theta}^2 r , \\
u_\theta &= 0
\end{aligned}
}
\tag{7.3.38}
$$

and the solution stress field is

$$
\boxed{
\begin{aligned}
\sigma_{rr} &= \frac{7\lambda + 6\mu}{16(\lambda + \mu)}(a^2 - r^2)\rho\dot{\theta}^2 , \\
\sigma_{\theta\theta} &= \frac{7\lambda + 6\mu}{16(\lambda + \mu)}\left[a^2 - \left(\frac{5\lambda + 2\mu}{7\lambda + 6\mu}\right)r^2\right]\rho\dot{\theta}^2 , \\
\sigma_{r\theta} &= 0 .
\end{aligned}
}
\tag{7.3.39}
$$

Example 7.1

Compare the angular velocity necessary to cause plastic yielding in a 1-m-radius spinning disk made of cold-drawn 1040 steel ($\lambda = 112$ GPa, $\mu = 80.1$ GPa, $\sigma_Y = 489$ MPa, $\rho = 7800$ kg/m^3) with that for one made of T6 aluminum ($\lambda = 53.5$ GPa, $\mu = 26.6$ GPa, $\sigma_Y = 413$ MPa, $\rho = 2710$ kg/m^3). Use the maximum shear stress yield criterion.

Solution. It is seen from the solution for a spinning disk (7.3.39) that the shear components of stress, in the cylindrical coordinate system defined in the problem, are zero. This then is the principal basis and σ_{rr}, $\sigma_{\theta\theta}$, and σ_{zz} are the principal stresses. Furthermore, an examination of the solution shows that $\sigma_{\theta\theta} \geq \sigma_{rr} \geq \sigma_{zz} = 0$, so that, at any given point in the disk, the maximum shear stress is

$$
|\tau^{(P)}|_{\max} = \frac{1}{2}(\sigma^{(1)} - \sigma^{(3)}) = \frac{1}{2}\sigma_{\theta\theta} .
$$

The greatest value of $\sigma_{\theta\theta}$ and therefore the greatest value of $|\tau^{(P)}|_{\max}$ will occur where $r = 0$.

The maximum shear stress criterion for plastic yielding holds that yielding will occur when $|\tau^{(P)}|_{\max} = \frac{1}{2}\sigma_Y$, where σ_Y is the uniaxial yield stress. Thus, plastic yielding will first occur at the center of the disk when $\sigma_{\theta\theta}|_{r=0} = \sigma_Y$. Solving for the corresponding angular velocity $\dot{\theta}_Y$,

$$
\dot{\theta}_Y = \sqrt{\frac{16(\lambda + \mu)\sigma_Y}{(7\lambda + 6\mu)a^2\rho}} .
$$

Substituting the material properties for the steel and aluminum alloys, one finds that $\dot{\theta}_Y = 390$ rad/s $= 62.1$ rev./s for cold-drawn 1040 steel and that $\dot{\theta}_Y = 605$ rad/s $= 96.2$ rev./s for T6 aluminum.

It is seen that the aluminum alloy disk can be taken to a higher angular velocity before yielding than the steel alloy disk. It is interesting to note, however, that for the same disk thickness, the kinetic energy stored in each disk at its respective maximum angular velocity (neglecting friction) is higher for the steel alloy disk:

$$
\text{K.E.} = \frac{1}{2}J_{zz}\dot{\theta}^2 = \frac{1}{4}\pi a^4 t\rho\dot{\theta}^2 ;
$$

so that for a disk thickness $t = 0.03$ m, the steel alloy has a maximum kinetic energy before yielding of $\text{K.E.}_{\text{max}} = 28.0$ MJ, whereas for the aluminum alloy, $\text{K.E.}_{\text{max}} = 23.4$ MJ. As a flywheel, for instance, the steel alloy may prove a better material (pending the evaluation of other considerations, of course).

7.3.4   Equivalence of Plane Stress and Plane Strain

Comparing the equations of plane strain and plane stress elasticity [e.g., the plane strain constitutive equations (7.2.3) and (7.2.7) and the corresponding plane stress equations (7.3.6) and (7.3.2)], it can be seen that each of the plane strain equations can be transformed into its corresponding plane stress equation, and vice versa, by a simple change in the material parameters as given in Table 7.1. Thus, there is no functional difference between the plane strain and plane stress equations of elasticity. This means that the solution to a plane strain problem can be converted to the solution of the corresponding plane stress problem, and vice versa, via the same change in the elastic constants (Table 7.1).

Table 7.1   Table for converting between plane strain and plane stress solutions

	$E \rightarrow$	$\nu \rightarrow$	$\lambda \rightarrow$	$\mu \rightarrow$
Plane stress → Plane strain	$\dfrac{E}{1 - \nu^2}$	$\dfrac{\nu}{1 - \nu}$	$\dfrac{2\mu\lambda}{2\mu - \lambda}$	μ
Plane strain → Plane stress	$\dfrac{(1 + 2\nu)E}{(1 + \nu)^2}$	$\dfrac{\nu}{1 + \nu}$	$\dfrac{2\mu\lambda}{2\mu + \lambda}$	μ

Example 7.2

Consider a thin circular washer that has been press-fit onto a rod and into a cylinder so that there is a radial pressure $p^{(1)}$ between the rod and the washer and a radial pressure $p^{(2)}$ between the washer and the outer cylinder. This is the plane stress analog of the cylindrical pressure vessel (Figure 7.3). Find the solution stress and displacement fields.

Solution. The solution stress and displacement fields are the same as already derived for the cylindrical pressure vessel (7.2.32) and (7.2.33), except with the substitution of material constants specified in Table 7.1.

Therefore,

$$u_r = \frac{r}{2\mu}\left(\frac{2\mu+\lambda}{2\mu+3\lambda}\right)\left(\frac{a^2 p^{(1)} - b^2 p^{(2)}}{b^2 - a_2}\right) + \frac{a^2 b^2}{2\mu r}\left(\frac{p^{(1)} - p^{(2)}}{b^2 - a^2}\right),$$

$$\varepsilon_{zz} = \frac{-\nu}{E}(\sigma_{rr} + \sigma_{\theta\theta}),$$

$$\sigma_{zz} = 0.$$

The remaining displacement and stress components are independent of material properties and thus remain unchanged from the cylindrical pressure vessel solution.

The equivalence of plane strain and plane stress can be exploited to give a single plane strain/stress formulation of the equations of elasticity. By using the relations between isotropic elastic moduli given in Table 5.1 and defining the elastic parameter α such that

$$\alpha \equiv \begin{cases} 1 - \nu & \text{for plane strain} \\ \dfrac{1}{1+\nu} & \text{for plane stress,} \end{cases} \tag{7.3.40}$$

it can been shown that the plane strain constitutive equations (7.2.3) and (7.2.7) and the corresponding plane stress equations (7.3.6) and (7.3.2) can be represented as *plane strain/stress constitutive equations* given by

$$\sigma_{\alpha\beta} = 2\mu\left[\varepsilon_{\alpha\beta} + \left(\frac{1-\alpha}{2\alpha - 1}\right)\varepsilon_{\gamma\gamma}\delta_{\alpha\beta}\right], \tag{7.3.41}$$

$$\varepsilon_{\alpha\beta} = \frac{1}{2\mu}[\sigma_{\alpha\beta} - (1-\alpha)\sigma_{\gamma\gamma}\delta_{\alpha\beta}]. \tag{7.3.42}$$

Similarly, the plane strain compatibility condition (7.2.12) and the corresponding plane stress condition (7.3.9) can be represented by the *plane strain/stress compatibility condition*

$$\nabla^2(\sigma_{\gamma\gamma}) = -\frac{1}{\alpha}f_{\gamma,\gamma}. \tag{7.3.43}$$

The plane strain and plane stress equilibrium conditions are precisely the same, so a *plane strain/stress equilibrium condition* would just be

$$\sigma_{\beta\alpha,\beta} + f_\alpha = 0. \tag{7.3.44}$$

The unknowns in the two-dimensional plane strain/stress boundary value problem are the displacement components u_1 and u_2, the strain components ε_{11}, ε_{22}, and ε_{12}, and the stress components σ_{11}, σ_{22}, and σ_{12}. Once a solution has been found in terms of the material parameter α, a distinction can be made between plane strain and plane stress by applying the definition (7.3.40) and recalling that

$$
\varepsilon_{33} = \begin{cases} 0 & \text{for plane strain} \\ \dfrac{-\nu}{1-\nu}\varepsilon_{\alpha\alpha} & \text{for plane stress} \end{cases} \tag{7.3.45}
$$

and that

$$
\sigma_{33} = \begin{cases} \nu\sigma_{\alpha\alpha} & \text{for plane strain} \\ 0 & \text{for plane stress.} \end{cases} \tag{7.3.46}
$$

Of course, in both cases, $\varepsilon_{13} = \varepsilon_{23} = 0$ and $\sigma_{13} = \sigma_{23} = 0$.

7.4   Airy Stress Function

Recall that for plane strain/stress problems, the equilibrium equations (in expanded form) are

$$
\begin{aligned}
\sigma_{11,1} + \sigma_{21,2} + f_1 &= 0 \,, \\
\sigma_{12,1} + \sigma_{22,2} + f_2 &= 0 \,.
\end{aligned} \tag{7.4.1}
$$

Also, many of the body force fields that one encounters are *conservative*— they can be expressed in terms of a scalar *potential function* $\Omega(X_1, X_2)$ as $\mathbf{f} = -\nabla\Omega$, so that

$$
f_1 = -\Omega_{,1} \,, \quad f_2 = -\Omega_{,2} \,. \tag{7.4.2}
$$

For example, $\Omega = \rho g X_1$ is a potential function for the body force field $\mathbf{f} = -\rho g \hat{\mathbf{e}}_1$ due to a uniform gravitational field acting in the $-\hat{\mathbf{e}}_1$-direction in a homogeneous body.

Combining (7.4.1) and (7.4.2) yields the plane strain/stress equilibrium equations in terms of the body force potential:

$$
\begin{aligned}
(\sigma_{11} - \Omega)_{,1} + \sigma_{21,2} &= 0 \,, \\
\sigma_{12,1} + (\sigma_{22} - \Omega)_{,2} &= 0 \,.
\end{aligned} \tag{7.4.3}
$$

Define the *Airy stress function* $\Phi(X_1, X_2)$ such that

$$
\boxed{\sigma_{11} = \Phi_{,22} + \Omega \,, \quad \sigma_{22} = \Phi_{,11} + \Omega \,, \quad \sigma_{12} = -\Phi_{,12} \,.} \tag{7.4.4}
$$

For plane strain/stress, a stress field has an Airy stress function representation (7.4.4) if and only if it satisfies the equilibrium equations—the equilibrium equations (7.4.3) provide necessary and sufficient conditions for the existence of an Airy stress function Φ, which when substituted into (7.4.4) yields the desired components of stress. In other words, every equilibrated stress field has an Airy stress function representation and every stress field that is derived from an Airy stress function via (7.4.4) is equilibrated. The latter can been seen by substituting the defining equations (7.4.4) for the Airy stress function into the equilibrium equations (7.4.3) and observing that the equilibrium equations are then identically satisfied.

The plane strain/stress compatibility condition (7.3.43) is given in expanded form by

$$\nabla^2(\sigma_{11} + \sigma_{22}) = -\frac{1}{\alpha}(f_{1,1} + f_{2,2}),\qquad(7.4.5)$$

where the material parameter α is defined by (7.3.40). Using (7.4.2) and (7.4.4) to rewrite the compatibility equation in terms of the Airy stress function and the body force potential gives

$$\Phi_{,1111} + 2\Phi_{,1122} + \Phi_{,2222} + 2\Omega_{,11} + 2\Omega_{,22} = \frac{1}{\alpha}(\Omega_{,11} + \Omega_{,22}),\qquad(7.4.6)$$

which simplifies to

$$\boxed{\nabla^4\Phi + \left(2 - \frac{1}{\alpha}\right)\nabla^2\Omega = 0,}\qquad(7.4.7)$$

where since there is no dependence on X_3,

$$\nabla^4\Phi = \Phi_{,1111} + 2\Phi_{,1122} + \Phi_{,2222}.\qquad(7.4.8)$$

If the body force field is homogeneous, this compatibility condition reduces to

$$\boxed{\nabla^4\Phi = 0.}\qquad(7.4.9)$$

The plane strain/stress problem has been reduced to finding solutions to (7.4.7) that satisfy the boundary conditions. Stress fields derived from an Airy stress function that satisfies (7.4.7) will be in equilibrium and correspond to compatible strain fields.

Example 7.3
Consider a second-order polynomial form of the Airy stress function:

$$\Phi = aX_1^2 + bX_1X_2 + cX_2^2,$$

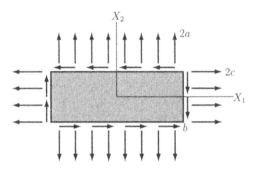

FIGURE 7.6.   Homogeneous stress field corresponding to a second-order polynomial form of the Airy stress function.

where a, b, and c are constants. Assume that the body force field is negligible. This function is biharmonic and therefore corresponds to components of stress that satisfy the plane strain/stress equations of elasticity. For what boundary conditions is this stress field a solution? This is an example of the *inverse method* of finding elasticity solutions.

Solution. The components of stress for this Airy stress function are

$$\sigma_{11} = \Phi_{,22} = 2c \ , \quad \sigma_{22} = \Phi_{,11} = 2a \ , \quad \sigma_{12} = -\Phi_{,12} = -b \ .$$

Thus, a second-order polynomial Airy stress function represents a homogeneous stress field. There are, in fact, an infinite number of problems for which this stress field is the solution; one can consider any plane geometry and, with the given distribution of unit outward normal to the surface and using the traction-stress relations, find the corresponding tractions. This is generally true of the inverse method. One possible example is a rectangular body with uniform tractions as shown in Figure 7.6.

Example 7.4
Consider a third-order polynomial form of the Airy stress function:

$$\Phi = aX_1^3 + bX_1^2 X_2 + cX_1 X_2^2 + dX_2^3 \ ,$$

where a, b, c, and d are constants. Assume that the body force field is negligible. This function is biharmonic and therefore corresponds to components of stress that satisfy the plane strain/stress equations of elasticity. For what boundary conditions is this stress field a solution?

Solution. The components of stress for this Airy stress function are

$$\sigma_{11} = 2cX_1 + 6dX_2 \ , \quad \sigma_{22} = 6aX_1 + 2bX_2 \ , \quad \sigma_{12} = -2bX_1 - 2cX_2 \ .$$

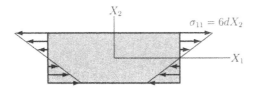

FIGURE 7.7. *Pure beam bending field corresponding to a third-order polynomial form of the Airy stress function.*

Again, there are an infinite number of possible problems for which this stress field is the solution. If $a = b = c = 0$, one such problem would be a thin beam (plane stress) in pure bending as shown in Figure 7.7.

Example 7.5
Consider a fourth-order polynomial form of the Airy stress function:

$$\Phi = aX_1^4 + bX_1^3X_2 + cX_1^2X_2^2 + dX_1X_2^3 + eX_2^4 \ ,$$

where a, b, c, d, and e are constants. Assume that the body force field is negligible. What conditions must the coefficients satisfy? What are the corresponding components of stress?

Solution. Applying the biharmonic operator to this Airy stress function,

$$\nabla^4\Phi = \Phi_{,1111} + 2\Phi_{,1122} + \Phi_{,2222} = 24a + 8c + 24e \ .$$

Then, since the Airy stress function must satisfy the compatibility condition $\nabla^4\Phi = 0$, the coefficients a, c, and e must satisfy

$$3a + c + 3e = 0 \ .$$

The corresponding components of stress are

$$\sigma_{11} = 2cX_1^2 + 6dX_1X_2 + 12eX_2^2 \ ,$$
$$\sigma_{22} = 12aX_1^2 + 6bX_1X_2 + 2cX_2^2 \ ,$$
$$\sigma_{12} = -3bX_1^2 - 4cX_1X_2 - 3dX_2^2 \ .$$

7.4.1 Displacements

If the body force field is negligible, then the displacement components can be expressed in the form

$$\boxed{2\mu u_1 = -\Phi_{,1} + \alpha\psi_{,2} \ , \quad 2\mu u_2 = -\Phi_{,2} + \alpha\psi_{,1} \ ,} \qquad (7.4.10)$$

where α is given by (7.3.40) and $\psi(X_1, X_2)$ is a potential function that satisfies the conditions

$$\nabla^2 \psi = 0 \,, \tag{7.4.11}$$

$$\psi_{,12} = \nabla^2 \Phi \,. \tag{7.4.12}$$

This is proved by showing that the stress field corresponding to the displacements (7.4.10) is that specified in the definition of the Airy stress function (7.4.4).

The plane strain/stress constitutive relations (7.3.41) for components of stress in terms of components of strain can be combined with the strain-displacement relations to give

$$
\begin{aligned}
\sigma_{11} &= 2\mu u_{1,1} + \frac{2\mu(1-\alpha)}{2\alpha-1}(u_{1,1} + u_{2,2}) \,, \\
\sigma_{22} &= 2\mu u_{2,2} + \frac{2\mu(1-\alpha)}{2\alpha-1}(u_{1,1} + u_{2,2}) \,, \\
\sigma_{12} &= \mu(u_{1,2} + u_{2,1}) \,.
\end{aligned}
\tag{7.4.13}
$$

Using (7.4.10) and (7.4.12),

$$
\begin{aligned}
2\mu(u_{1,1} + u_{2,2}) &= -\Phi_{,11} + \alpha\nabla^2\Phi - \Phi_{,22} + \alpha\nabla^2\Phi \\
&= (2\alpha - 1)\nabla^2\Phi \,.
\end{aligned}
\tag{7.4.14}
$$

Substituting this result into (7.4.13) and using (7.4.11) and (7.4.12),

$$
\begin{aligned}
\sigma_{11} &= -\Phi_{,11} + \alpha\nabla^2\Phi + (1-\alpha)\nabla^2\Phi \\
&= -\Phi_{,11} + \nabla^2\Phi \\
&= \Phi_{,22} \,, \\
\sigma_{22} &= -\Phi_{,22} + \alpha\nabla^2\Phi + (1-\alpha)\nabla^2\Phi \\
&= -\Phi_{,22} + \nabla^2\Phi \\
&= \Phi_{,11} \,, \\
\sigma_{12} &= \frac{1}{2}(-\Phi_{,12} + \alpha\psi_{,22} - \Phi_{,21} + \alpha\psi_{,11}) \\
&= -\Phi_{,12} + \frac{1}{2}\alpha\nabla^2\psi \\
&= -\Phi_{,12} \,.
\end{aligned}
$$

$$\tag{7.4.15}$$
$$\tag{7.4.16}$$
$$\tag{7.4.17}$$

Thus, the displacement field (7.4.10) is consistent with the definition of the Airy stress function (7.4.4).

7.4.2 Cylindrical Coordinates

If the body force field is conservative, that is, it can be expressed in terms of a potential function as $\mathbf{f} = -\nabla\Omega$,

$$f_r = -\frac{\partial\Omega}{\partial r}, \quad f_\theta = -\frac{1}{r}\frac{\partial\Omega}{\partial\theta}, \qquad (7.4.18)$$

then the equilibrium equations can be written as

$$\frac{\partial\sigma_{rr}}{\partial r} + \frac{1}{r}\frac{\partial\sigma_{r\theta}}{\partial\theta} + \frac{1}{r}(\sigma_{rr} - \sigma_{\theta\theta}) - \frac{\partial\Omega}{\partial r} = 0,$$
$$\frac{\partial\sigma_{r\theta}}{\partial r} + \frac{1}{r}\frac{\partial\sigma_{\theta\theta}}{\partial\theta} + \frac{2}{r}\sigma_{r\theta} - \frac{1}{r}\frac{\partial\Omega}{\partial\theta} = 0. \qquad (7.4.19)$$

Define the Airy stress function $\Phi(r,\theta)$ such that

$$\boxed{\begin{aligned} \sigma_{rr} &= \frac{1}{r}\frac{\partial\Phi}{\partial r} + \frac{1}{r^2}\frac{\partial^2\Phi}{\partial\theta^2} + \Omega, \\ \sigma_{\theta\theta} &= \frac{\partial^2\Phi}{\partial r^2} + \Omega, \\ \sigma_{r\theta} &= -\frac{\partial}{\partial r}\left(\frac{1}{r}\frac{\partial\Phi}{\partial\theta}\right). \end{aligned}} \qquad (7.4.20)$$

Substituting these defining equations for the Airy stress function into the equilibrium equations (7.4.19), it is seen that components of stress derived from an Airy stress function via (7.4.20) are guaranteed to be in equilibrium.

The tensor form of the compatibility condition is invariant, so that compatibility is satisfied if

$$\nabla^4\Phi + \left(2 - \frac{1}{\alpha}\right)\nabla^2\Omega = 0, \qquad (7.4.21)$$

where in cylindrical coordinates, since there is no dependence on z,

$$\nabla^2() = \frac{\partial^2()}{\partial r^2} + \frac{1}{r}\frac{\partial()}{\partial r} + \frac{1}{r^2}\frac{\partial^2()}{\partial\theta^2}, \qquad (7.4.22)$$
$$\nabla^4() = \nabla^2\left[\nabla^2()\right].$$

If the body force field is negligible, then the displacement components can be expressed in the form

$$\boxed{\begin{aligned} 2\mu u_r &= -\frac{\partial\Phi}{\partial r} + \alpha r\frac{\partial\psi}{\partial\theta}, \\ 2\mu u_\theta &= -\frac{1}{r}\frac{\partial\Phi}{\partial\theta} + \alpha r^2\frac{\partial\psi}{\partial r}, \end{aligned}} \qquad (7.4.23)$$

where $\psi(r, \theta)$ is a potential function that satisfies the conditions

$$\nabla^2 \psi = 0 , \qquad (7.4.24)$$

$$\frac{\partial}{\partial r}\left(r\frac{\partial \psi}{\partial \theta}\right) = \nabla^2 \Phi . \qquad (7.4.25)$$

Example 7.6
In cylindrical coordinates, some biharmonic functions that may be used as
Airy stress functions are

$$\Phi = C\theta ,$$
$$\Phi = Cr^2\theta ,$$
$$\Phi = Cr\theta \cos\theta ,$$
$$\Phi = Cr\theta \sin\theta ,$$

where C is a constant. It is left to the reader to verify that these functions
are biharmonic. Other useful cylindrical coordinate forms of Airy stress
function are

$$\Phi = f_n(r) \cos(n\theta) ,$$
$$\Phi = f_n(r) \sin(n\theta) ,$$

where in order for these functions to be biharmonic, it can be shown that
$f_n(r)$ must have the form

$$f_0(r) = a_0 r^2 + b_0 r^2 \ln r + c_0 + d_0 \ln r ,$$
$$f_1(r) = a_1 r^3 + b_1 r + c_1 r \ln r + d_1 r^{-1} ,$$
$$f_n(r) = a_n r^{n+2} + b_n r^n + c_n r^{-n+2} + d_n r^{-n} , \quad n > 1 ,$$

where a_n, b_n, c_n, and d_n are constants.

7.4.3   Example: Cantilevered Beam

Consider the cantilevered beam with transverse end load **P**, as shown in
Figure 7.8. The beam has length L, height h, and width w. Define a
Cartesian coordinate system as shown, with the X_1-axis along the centroidal axis of the beam. The end of the beam at $X_1 = L$ is fixed and the
end at $X_1 = 0$ has tractions applied to it that are statically equivalent to
a transverse load **P**. If the beam is thin (i.e., $w \ll h$), then σ_{33}, σ_{13}, and
σ_{23} are approximately zero and the beam is in a state of plane stress. If
$w \gg L$, then the beam is in a state of plane strain.

The beam has both traction and displacement boundary conditions. On
the top and bottom surfaces, the prescribed traction vector is $\tilde{\mathbf{t}} = \mathbf{0}$ and
the unit outward normal $\hat{\mathbf{n}} = \pm\hat{\mathbf{e}}_2$. Therefore,

$$\sigma_{22} = \sigma_{12} = 0 \quad \text{on } X_2 = \pm h/2 . \qquad (7.4.26)$$

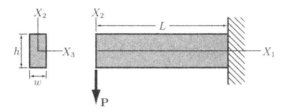

FIGURE 7.8. Plane stress cantilevered beam.

On the left vertical surface, there is a distribution of traction whose resultant is a force $\mathbf{P} = -P\hat{\mathbf{e}}_2$. Saint-Venant's principle states that, for points in the beam sufficiently far from this surface, the details of this distribution will have approximately no effect. Since the unit outward normal is $\hat{\mathbf{n}} = -\hat{\mathbf{e}}_1$, the traction-stress relations are $t_1 = -\sigma_{11}$ and $t_2 = -\sigma_{12}$ and the boundary conditions (in terms of the resultant force components and the resultant moment) are

$$\int_{-h/2}^{h/2} \sigma_{11}(0, X_2)\, dX_2 = 0 \,,$$

$$\int_{-h/2}^{h/2} X_2 \sigma_{11}(0, X_2)\, dX_2 = 0 \,, \tag{7.4.27}$$

$$\int_{-h/2}^{h/2} \sigma_{12}(0, X_2)\, dX_2 = \frac{P}{w} \,.$$

On the right vertical surface, the exact boundary condition is one of zero displacement,

$$\mathbf{u} = \mathbf{0} \quad \text{on } X_1 = L \,. \tag{7.4.28}$$

Using the semi-inverse method and taking elementary beam theory as inspiration, assume that the stress component σ_{11} (the *axial stress* in the parlance of mechanics of materials) is proportional to the bending moment in the beam $M(X_1) = -PX_1$:

$$\sigma_{11} = X_1 \eta''(X_2) \,. \tag{7.4.29}$$

In terms of the Airy stress function, $\sigma_{11} = \Phi_{,22}$ so that, integrating twice with respect to X_2, the Airy stress function must have the form

$$\Phi = X_1 \eta(X_2) + X_2 \xi(X_1) + \zeta(X_1) \tag{7.4.30}$$

and the remaining components of stress $\sigma_{22} = \Phi,_{11}$ and $\sigma_{12} = -\Phi,_{12}$ must have the form

$$\sigma_{22} = X_2\xi''(X_1) + \zeta''(X_1) , \tag{7.4.31}$$
$$\sigma_{12} = -\eta'(X_2) - \xi'(X_1) . \tag{7.4.32}$$

Since the components of stress are derived from the Airy stress function, equilibrium is guaranteed. If the initial assumption is correct, the unknown functions $\eta(X_2)$, $\xi(X_1)$, and $\zeta(X_1)$ will be determined by the boundary conditions (7.4.26)–(7.4.28) and the compatibility condition (7.4.9).

From the boundary condition $\sigma_{22}(X_1, \pm h/2) = 0$, it follows that

$$\frac{h}{2}\xi''(X_1) + \zeta''(X_1) = 0 ,$$
$$-\frac{h}{2}\xi''(X_1) + \zeta''(X_1) = 0 . \tag{7.4.33}$$

Therefore,

$$\xi''(X_1) = \zeta''(X_1) = 0 \tag{7.4.34}$$

and $\sigma_{22}(X_1, X_2) = 0$. Integrating (7.4.34),

$$\xi(X_1) = AX_1 ,$$
$$\zeta(X_1) = 0 , \tag{7.4.35}$$

where, without loss of generality, constants of integration that do not contribute to the components of stress have been set equal to zero.

Next, consider the compatibility condition $\nabla^4\Phi = 0$. Using (7.4.30) and (7.4.35), this condition becomes

$$X_1\eta^{(IV)}(X_2) = 0 , \tag{7.4.36}$$

so that the function $\eta(X_2)$ must have the form

$$\eta(X_2) = BX_2^3 + CX_2^2 + DX_2 , \tag{7.4.37}$$

where, again, irrelevant constants of integration have been set equal to zero. The Airy stress function can now be given in the form

$$\Phi = BX_1X_2^3 + CX_1X_2^2 + EX_1X_2 , \tag{7.4.38}$$

where $E \equiv A + D$.

Using (7.4.38), the boundary condition $\sigma_{12}(X_1, \pm h/2) = 0$ requires that

$$-\frac{3}{4}Bh^2 - Ch - E = 0 ,$$
$$-\frac{3}{4}Bh^2 + Ch - E = 0 . \tag{7.4.39}$$

Thus, $C = 0$ and $E = -\frac{3}{4}Bh^2$ and the components of stress in terms of the unknown constant B are

$$\sigma_{11} = 6BX_1X_2 \,,$$
$$\sigma_{22} = 0 \,,$$
$$\sigma_{12} = 3B\left[\left(\frac{h}{2}\right)^2 - X_2^2\right] \,. \tag{7.4.40}$$

Since $\sigma_{11}(0, X_2) = 0$, the first two of the boundary conditions (7.4.27) are satisfied. The third condition is

$$\frac{P}{w} = \int_{-h/2}^{h/2} 3B\left[\left(\frac{h}{2}\right)^2 - X_2^2\right] dX_2$$
$$= \frac{1}{2}Bh^3 \,, \tag{7.4.41}$$

from which it follows that $B = (2P)/(wh^3)$. Thus, the components of stress are

$$\boxed{\begin{aligned} \sigma_{11} &= \frac{P}{I_3}X_1X_2 \,, \\ \sigma_{22} &= 0 \,, \\ \sigma_{12} &= \frac{P}{2I_3}\left[\left(\frac{h}{2}\right)^2 - X_2^2\right] \end{aligned}} \tag{7.4.42}$$

and the Airy stress function is

$$\boxed{\Phi = \frac{P}{6I_3}\left(X_1X_2^3 - \frac{3}{4}h^2X_1X_2\right) \,,} \tag{7.4.43}$$

where $I_3 = \frac{1}{12}wh^3$ is the moment of inertia of the cross section about the X_3-axis. These are the same stress components predicted by elementary beam theory, where the bending moment is $M = -PX_1$ and the shear force is $V = P$.

All of the traction boundary conditions have been satisfied, but it remains to be seen if the displacement boundary condition $\mathbf{u}(L, X_2) = \mathbf{0}$ is satisfied. Recall that the displacement components can be given in terms of the Airy stress function by

$$2\mu u_1 = -\Phi_{,1} + \alpha\psi_{,2} \,,$$
$$2\mu u_2 = -\Phi_{,2} + \alpha\psi_{,1} \,, \tag{7.4.44}$$

where $\nabla^2 \psi = 0$ and $\psi_{,12} = \nabla^2 \Phi$. Using (7.4.43),

$$\psi_{,12} = \frac{P}{I_3} X_1 X_2 , \tag{7.4.45}$$

and integrating,

$$\psi = \frac{P}{4I_3} X_1^2 X_2^2 + f(X_1) + g(X_2) . \tag{7.4.46}$$

Relation (7.4.46) must satisfy $\nabla^2 \psi = 0$,

$$\frac{P}{2I_3} (X_1^2 + X_2^2) + f''(X_1) + g''(X_2) = 0 , \tag{7.4.47}$$

which is only possible if

$$f''(X_1) = -\frac{P}{2I_3} X_1^2 + G ,$$
$$g''(X_2) = -\frac{P}{2I_3} X_2^2 - G , \tag{7.4.48}$$

where G is a constant. Integrating (7.4.48),

$$f(X_1) = \frac{-P}{24I_3} X_1^4 + \frac{1}{2} G X_1^2 + H X_1 ,$$
$$g(X_2) = \frac{-P}{24I_3} X_2^4 - \frac{1}{2} G X_2^2 + J X_2 , \tag{7.4.49}$$

where constants of integration that do not contribute to the displacement components have been set equal to zero. Substituting (7.4.49) into (7.4.46),

$$\psi = \frac{P}{4I_3} X_1^2 X_2^2 - \frac{P}{24I_3} (X_1^4 + X_2^4) + \frac{1}{2} G(X_1^2 - X_2^2) + H X_1 + J X_2 , \tag{7.4.50}$$

and the displacement components (7.4.44) are

$$2\mu u_1 = -\frac{P(1+\alpha)}{6I_3} X_2^3 + \frac{P\alpha}{2I_3} X_1^2 X_2 + \left(\frac{Ph^2}{8I_3} - \alpha G \right) X_2 + \alpha J ,$$
$$2\mu u_2 = -\frac{P\alpha}{6I_3} X_1^3 - \frac{P(1-\alpha)}{2I_3} X_1 X_2^2 + \left(\frac{Ph^2}{8I_3} + \alpha G \right) X_1 + \alpha H . \tag{7.4.51}$$

The constants G, H, and J correspond to a rigid-body motion. To fix the rigid-body motion, let $\mathbf{u}(L,0) = \mathbf{0}$ and $\theta_3(L,0) = 0$. The rotation in the $X_1 X_2$-plane is given by

$$4\mu\theta_3 = 2\mu(u_{2,1} - u_{1,2})$$
$$= \frac{P\alpha}{I_3} (X_2^2 - X_1^2) + 2\alpha G . \tag{7.4.52}$$

Setting $\theta_3(L,0) = 0$,

$$G = \frac{PL^2}{2I_3} , \qquad (7.4.53)$$

and setting $\mathbf{u}(L,0) = \mathbf{0}$,

$$H = -\frac{PL^3}{3I_3} - \frac{Ph^2L}{8\alpha I_3} , \qquad (7.4.54)$$

$$J = 0 .$$

Assuming plane stress so that $\alpha = (1+\nu)^{-1}$ and using the relation $2\mu = E/(1+\nu)$, the displacement components are

$$
\begin{array}{|l|}
\hline
u_1 = \dfrac{-P}{2EI_3}(L^2 - X_1^2)X_2 - \dfrac{P(2+\nu)}{6EI_3}X_2^3 + \dfrac{P(1+\nu)h^2}{8EI_3}X_2 , \\[3mm]
u_2 = \dfrac{-PL^3}{6EI_3}\left[2 - \dfrac{3X_1}{L}\left(1 - \dfrac{\nu X_2^2}{L^2}\right) + \dfrac{X_1^3}{L^3} + \dfrac{3h^2}{4L^2}(1+\nu)\left(1 - \dfrac{X_1}{L}\right)\right] . \\
\hline
\end{array}
$$

$$(7.4.55)$$

Note that u_1 is not a linear function of X_2—plane sections *do not remain plane*. It is clear that $\mathbf{u}(L, X_2) \neq \mathbf{0}$. However, by Saint-Venant's principle, the solution is still approximately correct for points in the beam sufficiently far from the fixed end.

The deflection of the centroidal axis of the beam is

$$u_2(X_1,0) = \frac{-PL^3}{6EI_3}\left[2 - \frac{3X_1}{L} + \frac{X_1^3}{L^3} + \frac{3h^2}{4L^2}(1+\nu)\left(1 - \frac{X_1}{L}\right)\right]. \quad (7.4.56)$$

In the limit as $h/L \to 0$, this approaches the prediction of simple beam theory ($v'' = -M/EI_3$). The difference is due to the contribution of shear to the deflection, which is not accounted for in simple beam theory. The maximum deflection is

$$u_2(0,0) = \frac{-PL^3}{3EI_3} - \frac{Ph^2L}{16\mu I_3} . \qquad (7.4.57)$$

Simple beam theory predicts a maximum deflection of $(-PL^3)/(3EI_3)$. Thus, it is seen that simple beam theory does not correctly predict the beam deflection in this case, but gives acceptable results when $L \gg h$.

7.4.4 Example: Dam Problem

Consider the rectangular object shown in Figure 7.9. The object has width a and height h. The surface $X_2 = 0$ is fixed and the surface $X_1 = 0$ has

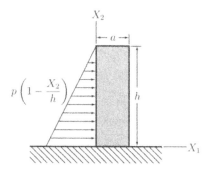

FIGURE 7.9. Plane strain dam.

a linear pressure distribution meant to represent the hydrostatic pressure due to a body of water. The remaining surfaces are traction-free. If the X_3 dimension of the object is large ($\gg h$), then the object is in a state of plane strain. This problem is an idealization of a dam with a rectangular cross section that is retaining a body of water.

The dam has both traction and displacement boundary conditions. On the surface $X_1 = 0$, the traction vector is $\tilde{\mathbf{t}} = [p(h - X_2)/h]\hat{\mathbf{e}}_1$ and $\hat{\mathbf{n}} = -\hat{\mathbf{e}}_1$. Therefore,

$$\sigma_{11} = \frac{p}{h}(X_2 - h) , \quad \sigma_{12} = 0 \quad \text{on } X_1 = 0 . \tag{7.4.58}$$

On the surface $X_1 = a$, the traction vector is $\tilde{\mathbf{t}} = \mathbf{0}$ and $\hat{\mathbf{n}} = \hat{\mathbf{e}}_1$. Therefore,

$$\sigma_{11} = \sigma_{12} = 0 \quad \text{on } X_1 = a . \tag{7.4.59}$$

On the surface $X_2 = h$, the traction vector is $\tilde{\mathbf{t}} = \mathbf{0}$ and $\hat{\mathbf{n}} = \hat{\mathbf{e}}_2$. Therefore,

$$\sigma_{22} = \sigma_{12} = 0 \quad \text{on } X_2 = h . \tag{7.4.60}$$

On the surface $X_2 = 0$, the boundary condition is one of zero displacement:

$$\mathbf{u} = \mathbf{0} \quad \text{on } X_2 = 0 . \tag{7.4.61}$$

Using the semi-inverse method, assume that the stress component σ_{11} has the form

$$\sigma_{11} = \frac{p}{h}(X_2 - h)\eta(X_1) , \tag{7.4.62}$$

where

$$\eta(0) = 1 , \quad \eta(a) = 0 . \tag{7.4.63}$$

This assumption satisfies all the boundary conditions that involve σ_{11}. In terms of the Airy stress function, $\sigma_{11} = \Phi_{,22}$ so that, integrating twice with respect to X_2, the Airy stress function must have the form

$$\Phi = \frac{p}{6h}(X_2 - h)^3\eta(X_1) + X_2\xi(X_1) + \zeta(X_1) \qquad (7.4.64)$$

and the remaining components of stress $\sigma_{22} = \Phi_{,11}$ and $\sigma_{12} = -\Phi_{,12}$ must have the form

$$
\begin{aligned}
\sigma_{22} &= \frac{p}{6h}(X_2 - h)^3\eta''(X_1) + X_2\xi''(X_1) + \zeta''(X_1) \, , \\
\sigma_{12} &= -\frac{p}{2h}(X_2 - h)^2\eta'(X_1) - \xi'(X_1) \, .
\end{aligned}
\qquad (7.4.65)
$$

Since the components of stress are derived from the Airy stress function, equilibrium is guaranteed. If the initial assumption is correct, the unknown functions $\eta(X_1)$, $\xi(X_1)$, and $\zeta(X_1)$ will be determined by the boundary conditions and the compatibility condition (7.4.9).

From the boundary condition $\sigma_{22}(X_1, h) = 0$, it follows that

$$h\xi''(X_1) + \zeta''(X_1) = 0 \, . \qquad (7.4.66)$$

Therefore, by integrating this expression,

$$\zeta(X_1) = -h\xi(X_1) \, , \qquad (7.4.67)$$

where constants of integration that do not contribute to the components of stress have been set equal to zero, without loss of generality. Rewriting the Airy stress function,

$$\Phi = \frac{p}{6h}(X_2 - h)^3\eta(X_1) + (X_2 - h)\xi(X_1) \, . \qquad (7.4.68)$$

From the boundary conditions $\sigma_{12}(0, X_2) = 0$ and $\sigma_{12}(a, X_2) = 0$,

$$
\begin{aligned}
-\frac{p}{2h}(X_2 - h)^2\eta'(0) - \xi'(0) &= 0 \, , \\
-\frac{p}{2h}(X_2 - h)^2\eta'(a) - \xi'(a) &= 0 \, ,
\end{aligned}
\qquad (7.4.69)
$$

so that, for these conditions to be satisfied for all values of X_2 on these surfaces, it follows that

$$\eta'(0) = \eta'(a) = \xi'(0) = \xi'(a) = 0 \, . \qquad (7.4.70)$$

Next, consider the compatibility condition $\nabla^4\Phi = 0$. Using (7.4.68),

$$\frac{p}{6h}(X_2 - h)^3\eta^{(IV)}(X_1) + (X_2 - h)\left[\frac{2p}{h}\eta''(X_1) + \xi^{(IV)}(X_1)\right] = 0 \, ,$$

$$\qquad (7.4.71)$$

which can only hold for all points in the dam if

$$\eta^{(IV)}(X_1) = 0 \,,$$
$$\xi^{(IV)}(X_1) = -\frac{2p}{h}\eta''(X_1) \,. \tag{7.4.72}$$

Integrating the first of (7.4.72),

$$\eta(X_1) = AX_1^3 + BX_1^2 + CX_1 + D \,. \tag{7.4.73}$$

From (7.4.63), $\eta(0) = 1$ and $\eta(a) = 0$ so that

$$D = 1 \,, \quad Aa^3 + Ba^2 + Ca + 1 = 0 \,, \tag{7.4.74}$$

and from (7.4.70), $\eta'(0) = 0$ and $\eta'(a) = 0$ so that

$$C = 0 \,, \quad 3Aa^2 + 2Ba = 0 \,. \tag{7.4.75}$$

Solving (7.4.74) and (7.4.75) for A and B,

$$A = \frac{2}{a^3} \,, \quad B = -\frac{3}{a^2} \,, \tag{7.4.76}$$

and $\eta(X_1)$ is

$$\eta(X_1) = \frac{2}{a^3}X_1^3 - \frac{3}{a^2}X_1^2 + 1 \,. \tag{7.4.77}$$

Substituting (7.4.77) into the second of equations (7.4.72),

$$\xi^{(IV)}(X_1) = -\frac{12p}{a^3h}(2X_1 - a) \,. \tag{7.4.78}$$

Integrating three times with respect to X_1,

$$\xi'(X_1) = -\frac{p}{a^3h}(X_1^4 - 2aX_1^3 + EX_1^2 + FX_1 + G) \,. \tag{7.4.79}$$

From the boundary condition (7.4.70), $\xi'(0) = 0$ and $\xi'(a) = 0$ so that

$$G = 0 \,, \quad F = a^3 - Ea \,, \tag{7.4.80}$$

and rewriting (7.4.79),

$$\xi'(X_1) = -\frac{p}{a^3h}\left[X_1^4 - 2aX_1^3 + a^3X_1 + EX_1(X_1 - a)\right] \,. \tag{7.4.81}$$

Only one unknown constant, E, remains to be determined.

One traction boundary condition, $\sigma_{12}(X_1, h) = 0$, remains to be satisfied. However, from examination of (7.4.65) and (7.4.81), it is seen that this condition requires that $\xi'(X_1) = 0$, which cannot be satisfied; this boundary condition cannot be exactly satisfied given the original assumption (7.4.62). Instead, let one require that the resultant shear force on the surface $X_2 = h$ be zero:

$$\int_0^a \sigma_{12}(X_1, h)\, dX_1 = -\int_0^a \xi'(X_1)\, dX_1 = -\xi(a) + \xi(0) = 0. \quad (7.4.82)$$

Integrating (7.4.81), this condition becomes

$$\frac{1}{5}a^5 - \frac{1}{2}a^5 + \frac{1}{2}a^5 + Ea^2\left(\frac{1}{3}a - \frac{1}{2}a\right) = 0, \quad (7.4.83)$$

so that by solving for E,

$$E = \frac{6}{5}a^2. \quad (7.4.84)$$

Therefore, using (7.4.68), (7.4.77), (7.4.81), and (7.4.84), the components of stress in the dam are

$$\sigma_{11} = \frac{p}{h}(X_2 - h)\left(\frac{2X_1^3}{a^3} - \frac{3X_1^2}{a^2} + 1\right),$$

$$\sigma_{22} = \frac{p}{a^2 h}(X_2 - h)^3\left(\frac{2X_1}{a} - 1\right) - \frac{p}{h}(X_2 - h)\left(\frac{4X_1^3}{a^3} - \frac{6X_1^2}{a^2} + \frac{12X_1}{5a} - \frac{1}{5}\right),$$

$$\sigma_{12} = -\frac{3p}{ah}(X_2 - h)^2\left(\frac{X_1^2}{a^2} - \frac{X_1}{a}\right) + \frac{pa}{h}\left(\frac{X_1^4}{a^4} - \frac{2X_1^3}{a^3} + \frac{6X_1^2}{5a^2} - \frac{X_1}{5a}\right).$$

$$(7.4.85)$$

Note that if one were to solve for the displacement field in the dam, one would find that the boundary condition $\mathbf{u}(X_1, 0) = \mathbf{0}$ is not exactly satisfied. Owing to Saint-Venant's principle, this solution is valid only for points in the dam sufficiently far from the surfaces $X_2 = 0$ and $X_2 = h$.

7.4.5 Example: Axially Loaded Wedge

Consider the axially loaded wedge of infinite length shown in Figure 7.10. The edges of the wedge are, in cylindrical coordinates, at $\theta = \pm\beta$, the vertex is at the origin, and the base is at infinity. A *concentrated load* $\mathbf{P} = P\hat{\mathbf{e}}_1$ per unit wedge thickness acts at the vertex. "Concentrated load" is the generic term for an applied load acting at a point or line of zero area.

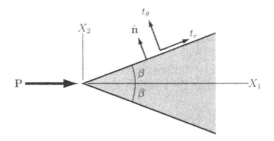

FIGURE 7.10. Axially loaded infinite wedge.

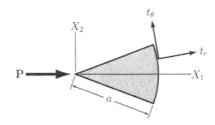

FIGURE 7.11. Global equilibrium of the axially loaded wedge.

Assume that this is either a plane strain or plane stress problem. Find the solution stress and displacement fields.

The edges of the wedge, $\theta = \pm\beta$, must be traction-free. The unit normals to these surfaces are $\hat{\mathbf{n}} = \pm\hat{\mathbf{e}}_\theta$. Therefore, the components of the traction on these surfaces are $t_r = \pm\sigma_{\theta r}$ and $t_\theta = \pm\sigma_{\theta\theta}$ (see Figure 7.10) and

$$\sigma_{\theta r} = \sigma_{\theta\theta} = 0 \quad \text{on } \theta = \pm\beta \, . \tag{7.4.86}$$

The concentrated load at the vertex of the wedge presents some difficulty in the specification of corresponding boundary conditions, both because the unit outward normal at the vertex is undefined and because a nonzero forces applied at a point with zero area implies an unbounded traction (force per unit area). To overcome this difficulty, recall that an arbitrary portion of the wedge (in particular, one that contains the vertex) must be in overall equilibrium. Consider the portion bounded on the right by the surface $r = a$, where a is arbitrary, as shown in Figure 7.11. The edges $\theta = \pm\beta$ are traction-free from the previous boundary condition. The unit normal to the surface $r = a$ is $\hat{\mathbf{n}} = \hat{\mathbf{e}}_r$ so that the tractions on this surface are $t_r = \sigma_{rr}$ and $t_\theta = \sigma_{r\theta}$. In order for this portion of the wedge to be in overall equilibrium,

$$\sum F_1 = P + \int_{-\beta}^{\beta} [\sigma_{rr}(a,\theta)\cos\theta - \sigma_{r\theta}(a,\theta)\sin\theta]a\,d\theta = 0 , \qquad (7.4.87)$$

$$\sum F_2 = \int_{-\beta}^{\beta} [\sigma_{rr}(a,\theta)\sin\theta + \sigma_{r\theta}(a,\theta)\cos\theta]a\,d\theta = 0 , \qquad (7.4.88)$$

$$\sum M_3 = \int_{-\beta}^{\beta} [a\sigma_{r\theta}(a,\theta)]a\,d\theta = 0 . \qquad (7.4.89)$$

The boundary conditions corresponding to the concentrated load are thus given indirectly by requiring that the solution satisfy (7.4.87)–(7.4.89) for *all* values of a.

Using the semi-inverse method, assume that $\sigma_{r\theta} = 0$ everywhere. This is consistent with the traction-free edge condition (7.4.86) and identically satisfies the condition (7.4.89). The boundary condition (7.4.86) then reduces to

$$\sigma_{\theta\theta} = 0 \quad \text{on } \theta = \pm\beta , \qquad (7.4.90)$$

and the equilibrium conditions (7.4.87)–(7.4.89) reduce to

$$\int_{-\beta}^{\beta} a\sigma_{rr}(a,\theta)\cos\theta\,d\theta = -P , \qquad (7.4.91)$$

$$\int_{-\beta}^{\beta} a\sigma_{rr}(a,\theta)\sin\theta\,d\theta = 0 . \qquad (7.4.92)$$

Using the definition (7.4.20) for the Airy stress function in cylindrical coordinates, the assumption $\sigma_{r\theta} = 0$ means that

$$\frac{\partial}{\partial r}\left(\frac{1}{r}\frac{\partial\Phi}{\partial\theta}\right) = 0 . \qquad (7.4.93)$$

The general solution to this equation is

$$\Phi = r\eta(\theta) + \xi(r) , \qquad (7.4.94)$$

where $\eta(\theta)$ and $\xi(r)$ are unknown functions that (assuming the assumption is correct) will be determined by the remaining boundary conditions (7.4.90)–(7.4.92) and the compatibility condition $\nabla^4\Phi = 0$.

With the Airy stress function given by (7.4.94), it follows from (7.4.20) that

$$\sigma_{\theta\theta} = \frac{\partial^2\Phi}{\partial r^2} = \xi''(r) . \qquad (7.4.95)$$

Thus, as a consequence of the assumption that $\sigma_{r\theta} = 0$ everywhere, $\sigma_{\theta\theta}$ is independent of θ. Comparing this with the boundary condition (7.4.90), one sees therefore that $\sigma_{\theta\theta} = 0$ everywhere, as well, and

$$\xi''(r) = 0 . \tag{7.4.96}$$

The general solution to this equation is

$$\xi(r) = c_1 r + c_2 \tag{7.4.97}$$

and therefore the Airy stress function (7.4.94) is

$$\Phi = r[\eta(\theta) + c_1] = r\zeta(\theta) , \tag{7.4.98}$$

where c_2 has been set equal to zero, with no loss of generality since only derivatives of Φ appear in the expressions for the components of stress.

Consider next the compatibility condition $\nabla^4 \Phi = 0$ (7.4.22), which with (7.4.98) leads to the condition

$$\zeta^{(IV)}(\theta) + 2\zeta''(\theta) + \zeta(\theta) = 0 . \tag{7.4.99}$$

The general solution to this equation is

$$\zeta(\theta) = A \sin\theta + B \cos\theta + C\theta \sin\theta + D\theta \cos\theta . \tag{7.4.100}$$

Therefore, the Airy stress function is

$$\Phi = r(A \sin\theta + B \cos\theta + C\theta \sin\theta + D\theta \cos\theta) , \tag{7.4.101}$$

where A, B, C, and D are constants to be determined by the remaining boundary conditions (7.4.91) and (7.4.92).

From (7.4.20) and (7.4.101), it follows that

$$\sigma_{rr} = \frac{1}{r}(2C \cos\theta - 2D \sin\theta) . \tag{7.4.102}$$

Note that the constants A and B do not appear in this expression and so, without loss of generality, they may be set equal to zero. Substituting (7.4.102) into the condition (7.4.92),

$$\begin{aligned}
0 &= \int_{-\beta}^{\beta} (2C \cos\theta - 2D \sin\theta) \sin\theta \, d\theta \\
&= 2C \int_{-\beta}^{\beta} \cos\theta \sin\theta \, d\theta - 2D \int_{-\beta}^{\beta} \sin^2\theta \, d\theta \\
&= C \sin^2\theta \Big|_{-\beta}^{\beta} - D\left(\theta - \frac{1}{2}\sin 2\theta\right)\Big|_{-\beta}^{\beta} \\
&= -D(2\beta - \sin 2\beta) .
\end{aligned} \tag{7.4.103}$$

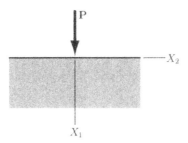

FIGURE 7.12. *Semi-infinite plate with a concentrated normal load.*

Since β is arbitrary, the boundary condition (7.4.92) is only satisfied if $D = 0$. Using this result and (7.4.102), the final boundary condition (7.4.91) becomes

$$-P = 2C \int_{-\beta}^{\beta} \cos^2 \theta \, d\theta$$

$$= C \left(\theta + \frac{1}{2} \sin 2\theta \right) \Big|_{-\beta}^{\beta}$$

$$= C(2\beta + \sin 2\beta) . \tag{7.4.104}$$

Thus, the Airy stress function for this problem is

$$\boxed{\Phi = Cr\theta \sin \theta ,} \tag{7.4.105}$$

where

$$\boxed{C = -\frac{P}{2\beta + \sin 2\beta} ,} \tag{7.4.106}$$

and the stress field in an axially loaded wedge has components given by

$$\boxed{\sigma_{rr} = -\frac{2P \cos \theta}{r(2\beta + \sin 2\beta)} , \quad \sigma_{\theta\theta} = \sigma_{r\theta} = 0 .} \tag{7.4.107}$$

An interesting special case of this solution occurs when $\beta = \pi/2$, as shown in Figure 7.12. In plane stress, this corresponds to an semi-infinite plate $(X_1 \geq 0)$ with a concentrated normal load per unit thickness P applied to the edge. The stress field for this problem has the following components:

$$\sigma_{rr} = -\frac{2P}{\pi r} \cos \theta , \quad \sigma_{\theta\theta} = \sigma_{r\theta} = 0 . \tag{7.4.108}$$

In plane strain, this would be the solution for a normal line load acting on the surface of a semi-infinite space.

The displacement field in the axially loaded wedge can be determined from the Airy stress function via (7.4.23), but one first has to determine ψ. From (7.4.25),

$$\frac{\partial}{\partial r}\left(r\frac{\partial\psi}{\partial\theta}\right) = \nabla^2\Phi = \frac{2C}{r}\cos\theta . \tag{7.4.109}$$

Solving for ψ,

$$\psi = \frac{2C}{r}\ln r \,\sin\theta + \frac{1}{r}\eta(\theta) + \xi(r) . \tag{7.4.110}$$

From (7.4.24), $\nabla^2\psi = 0$. Substituting (7.4.110) into this requirement,

$$\frac{1}{r^3}\eta''(\theta) + \frac{1}{r^3}\eta(\theta) + \xi''(r) + \frac{1}{r}\xi'(r) - \frac{4C}{r^3}\sin\theta = 0 . \tag{7.4.111}$$

This equation is satisfied for arbitrary r and θ if

$$\eta''(\theta) + \eta(\theta) = 4C\sin\theta + b ,$$
$$\xi''(r) + \frac{1}{r}\xi'(r) = -\frac{b}{r^3} , \tag{7.4.112}$$

where b is a constant. Solving for $\eta(\theta)$ and $\xi'(r)$,

$$\eta(\theta) = -2C\theta\cos\theta + d\cos\theta + e\sin\theta + b ,$$
$$\xi'(r) = \frac{f}{r} + \frac{b}{r^2} . \tag{7.4.113}$$

where d, e, and f are constants of integration. Therefore, from (7.4.23), the components of displacement are

$$2\mu u_r = 2\alpha C\ln r\,\cos\theta + (2\alpha - 1)C\theta\sin\theta + \alpha(e - 2C)\cos\theta - \alpha d\sin\theta ,$$
$$2\mu u_\theta = -2\alpha C\ln r\,\sin\theta + (2\alpha - 1)C\sin\theta + (2\alpha - 1)C\theta\cos\theta$$
$$- \alpha d\cos\theta - \alpha e\sin\theta + \alpha fr ; \tag{7.4.114}$$

recall that α is a material parameter, with different values for plane strain and plane stress, given by (7.3.40).

This is the solution displacement field to within an arbitrary rigid-body motion. To fix the rigid-body motion, let $u_\theta = 0$ when $\theta = 0$ and let $u_r = 0$ when $\theta = 0$ and $r = L$, where L is some arbitrary length. Then, $d = f = 0$ and $e = 2C - 2C\ln L$ and the components of displacement are

$$\boxed{\begin{aligned} u_r &= \frac{\alpha C}{\mu}\ln\left(\frac{r}{L}\right)\cos\theta + \frac{(2\alpha - 1)C}{2\mu}\theta\sin\theta , \\ u_\theta &= -\frac{\alpha C}{\mu}\ln\left(\frac{r}{L}\right)\sin\theta + \frac{(2\alpha - 1)C}{2\mu}\theta\cos\theta - \frac{C}{2\mu}\sin\theta . \end{aligned}} \tag{7.4.115}$$

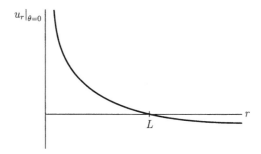

FIGURE 7.13. Displacement profile for the axially loaded wedge.

Notice that the displacements have a logarithmic singularity at $r = 0$ and $r = \infty$.

On $\theta = 0$, $u_\theta = 0$ and

$$u_r = \frac{\alpha C}{\mu} \ln\left(\frac{r}{L}\right) . \tag{7.4.116}$$

This displacement profile is sketched in Figure 7.13 (recall that $C < 0$ if $P > 0$). The logarithmic singularities at $r = 0$ and $r = \infty$ are accommodated by noting that, in reality, there cannot be a concentrated load acting at the point and the wedge cannot extend to infinity. The load at the tip of the wedge must be distributed over some finite area, the tip itself must have some nonzero radius of curvature, and the wedge must have some finite length, which can be taken to be the previously introduced length L. The solution stress and displacement fields given above are the solution of such a problem in the sense of Saint-Venant's principle. Then, the singularities in the expressions for the solution occur at values of r and θ outside the domain of the body and, therefore, do not present a problem.

It may also have occurred to the observant reader that, given the singularities in the solution, there may be some values of r and θ where the infinitesimal displacement gradient condition for the linearized equations of elasticity may be violated. The gradient of the displacement (7.4.115) is

$$\nabla \mathbf{u} = \frac{\alpha C}{\mu r} \cos\theta \hat{\mathbf{e}}_r \hat{\mathbf{e}}_r + \frac{(\alpha - 1)C}{\mu r} \cos\theta \hat{\mathbf{e}}_\theta \hat{\mathbf{e}}_\theta + \frac{\alpha C}{\mu r} \sin\theta (\hat{\mathbf{e}}_r \hat{\mathbf{e}}_\theta - \hat{\mathbf{e}}_\theta \hat{\mathbf{e}}_r) ,$$
$$\tag{7.4.117}$$

so the linearization condition is

$$\left|\frac{\alpha C}{\mu r}\right| \ll 1 \tag{7.4.118}$$

or, noting that $\alpha < 1$,

$$r \gg \frac{|C|}{\mu}.$$ (7.4.119)

Suppose $P = 1$ kN/m, $\beta = \pi/4$, and $\mu = 80.1$ GPa (carbon steel). Then, $|C| = 0.389$ kN/m and the linearized equations of elasticity are not valid unless $r \gg 5 \times 10^{-9}$ m. The above solution is not valid where this condition is not satisfied (i.e., near the wedge vertex). This problem is dealt with in the same way as the singularity.

Problems

7.1 The thin plate shown below is rectangular with length L and height h and has a circular hole of radius a in the middle. A uniform normal traction of magnitude σ_0 is applied to the ends of the plate as shown. Define a coordinate system and give a complete statement of the boundary conditions for this plane stress problem. **Note:** do *not* attempt to solve the corresponding boundary value problem.

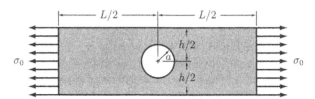

7.2 Consider the plane strain hollow circular shaft shown in cross section below. A cylindrical coordinate system is defined with its origin at the center of the cross section. The inner surface of the shaft at $r = a$ is ideally bonded to a fixed, rigid core and the outer boundary at $r = b$ is subject to a uniform shearing traction τ_0. Using the semi-inverse method, find the displacement and stress fields in the shaft.

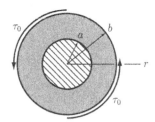

7.3 Consider a homogeneous solid circular cylinder of radius a and length L. The lateral surface of the cylinder ($r = a$) is ideally bonded to a rigid surface and gravity exerts a uniform body force field acting down: $\mathbf{f} = -\gamma \hat{\mathbf{e}}_z$ where γ is the weight per unit volume. The ends of the cylinder at $z = 0$ and $z = L$ are traction-free.

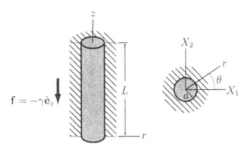

(a) Determine the displacement and stress fields in the cylinder by assuming antiplane strain and using the semi-inverse method.

(b) Is this an exact solution to the problem or is it a solution only in the sense of Saint-Venant's principle? Explain.

7.4 Consider the semi-infinite plate shown below. The uniform shear traction τ_0 acts on the portion of the boundary defined by $X_1 < 0$, $X_2 = 0$.

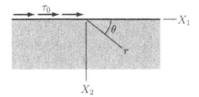

(a) Show that an Airy stress function of the form

$$\Phi = C\left[\frac{1}{2}X_2^2 \ln(X_1^2 + X_2^2) + X_1 X_2 \tan^{-1}\left(\frac{X_2}{X_1}\right) - X_2^2\right]$$

provides a solution to this problem and determine the value of the constant C and the stress field in the plate.

(b) By expressing the stress field in terms of the cylindrical coordinates r and θ, show that σ_{22} and σ_{12} are independent of r and that σ_{11} becomes unbounded as $r \to 0$.

7.5 The curved beam shown below is curved along a circular arc. The beam is fixed at the upper end and is subject at the lower end to a distribution of tractions statically equivalent to a force per unit thickness $\mathbf{P} = -P\hat{\mathbf{e}}_1$. Assume that the beam is in a state of plane strain/stress. Show that an Airy stress function of the form

$$\Phi = \left(Ar^3 + \frac{B}{r} + Cr\ln r \right) \sin\theta$$

provides an approximate solution to this problem (in the sense of Saint-Venant's principle) and solve for the values of the constants A, B, and C.

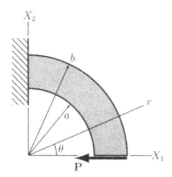

7.6 Show that an Airy stress function of the form

$$\Phi = Cr^2(\alpha + \theta - \sin\theta\cos\theta - \cos^2\theta\tan\alpha)$$

provides an approximate solution (in the sense of Saint-Venant's principle) for a cantilevered triangular beam with a uniform normal traction p applied to the upper surface, as shown below. Determine the value of the constant C in terms of the load p and the angle of the beam α.

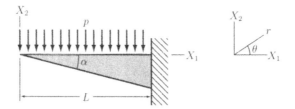

7.7 Consider the cantilevered beam shown below. A uniform shear traction of magnitude τ_0 acts on the top surface of the beam. Assuming plane strain/stress, determine the stress field in the beam (approximate in the sense of Saint-Venant's principle) by assuming a general form for σ_{12} that is consistent with the boundary conditions at the top and bottom surfaces of the beam.

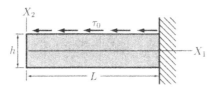

7.8 Consider the wedge of infinite length shown below. A concentrated load $\mathbf{P} = P\hat{\mathbf{e}}_2$ per unit wedge thickness acts at the vertex. Assuming plane strain/stress, determine the stress field in the wedge.

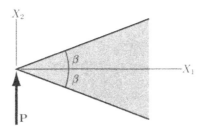

7.9 Consider the wedge of infinite length shown below. A concentrated moment $\mathbf{M} = M\hat{\mathbf{e}}_3$ per unit wedge thickness acts at the vertex. Assuming plane strain/stress, show that an Airy stress function of the form $\Phi = A\theta + B\sin 2\theta$, where A and B are constants, provides the solution to the problem and determine the stress field in the wedge.

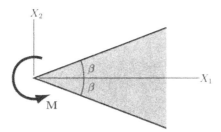

Chapter 8

Torsion of Noncircular Cylinders

Consider a cylindrical body of length L, the ends of which are subject to distributions of traction that are statically equivalent to equal and opposite torques $\pm\mathbf{T} = \pm T\hat{\mathbf{e}}_3$. The lateral surface of the cylinder is traction-free. It is assumed that there is an axis that passes through the center of twist of each cross section of the cylinder (i.e., an axis on which the displacement perpendicular to the axis is zero) and this is defined to be the X_3-axis (see Figure 8.1). Let S be the cross section of the cylinder and let ∂S be its boundary. The cross section S is, in general, noncircular.

For a given cross section, one wishes to know such things as the torsional rigidity of the cylinder (the ratio of the torque applied to the amount of twist per unit length) and the maximum shear stress. The solution to this problem was first found by Saint-Venant using the semi-inverse method. On the basis of the known solution for torsion of circular cylinders, Saint-Venant made certain assumptions concerning the nature of deformation of noncircular cylinders and showed that an exact solution to the appropriate boundary value problem could subsequently be found, thus validating the original assumptions.

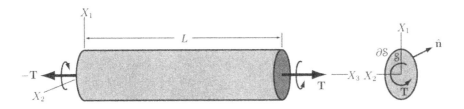

FIGURE 8.1. Torsion of a noncircular cylinder.

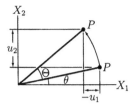

FIGURE 8.2. In-plane displacement during torsion.

8.1 Warping Function

The assumptions made by Saint-Venant in solving the torsion problem are, as the torque is applied to the ends of the cylinder, as follows:

- Each cross section's projection onto the X_1X_2-plane rotates, but remains undistorted.

- The amount each cross section rotates is proportional to its distance from the end of the cylinder; that is, the *twist* of each cross section is $\Theta = \alpha X_3$, where α is the *twist per unit length*.

- Each cross section's out-of-plane distortion is the same, and the magnitude of the distortion is proportional to the twist per unit length α.

According to the first two assumptions, the in-plane components of the displacement should be (Figure 8.2)

$$u_1 = r\cos(\Theta + \theta) - r\cos\theta = X_1(\cos\Theta - 1) - X_2\sin\Theta , \qquad (8.1.1)$$
$$u_2 = r\sin(\Theta + \theta) - r\sin\theta = X_1\sin\Theta + X_2(\cos\Theta - 1) . \qquad (8.1.2)$$

The third assumption states that the out-of-plane displacement of each cross section is the same, that is, that

$$u_3 = \alpha\psi(X_1, X_2) . \qquad (8.1.3)$$

$\psi(X_1, X_2)$ is called the *warping function* and describes the out-of-plane distortion of each cross section. If $\Theta = \alpha X_3 \ll 1$, then the displacement components are approximately given by the linearized relations

$$\boxed{u_1 \approx -\alpha X_2 X_3 , \quad u_2 \approx \alpha X_1 X_3 , \quad u_3 = \alpha\psi(X_1, X_2) .} \qquad (8.1.4)$$

One must try to satisfy the equations of elasticity and the boundary conditions based on (8.1.4). If this can be done, then the assumptions are valid and the solution to the problem (for small Θ) is found.

8.1.1 Stress

Given the assumed form of the displacement field (8.1.4), the components of strain $\varepsilon_{ij} = \frac{1}{2}(u_{i,j} + u_{j,i})$ are $\varepsilon_{11} = \varepsilon_{22} = \varepsilon_{33} = \varepsilon_{12} = 0$ and

$$\varepsilon_{31} = \frac{\alpha}{2}(\psi_{,1} - X_2), \quad \varepsilon_{32} = \frac{\alpha}{2}(\psi_{,2} + X_1) \tag{8.1.5}$$

and the components of stress $\sigma_{ij} = 2\mu\varepsilon_{ij} + \lambda\varepsilon_{kk}\delta_{ij}$ are $\sigma_{11} = \sigma_{22} = \sigma_{33} = \sigma_{12} = 0$ and

$$\boxed{\sigma_{31} = \mu\alpha(\psi_{,1} - X_2), \quad \sigma_{32} = \mu\alpha(\psi_{,2} + X_1) .} \tag{8.1.6}$$

At any point on a cross section of the cylinder, the unit outward normal is $\hat{n} = \hat{e}_3$. Thus, the normal traction on each cross section is zero,

$$\nu(\hat{e}_3) = \hat{e}_3 \cdot \boldsymbol{\sigma} \cdot \hat{e}_3 = \sigma_{33} = 0 , \tag{8.1.7}$$

the projected shear traction vector is

$$\tau^{(P)}(\hat{e}_3) = \mathbf{t} = \hat{e}_3 \cdot \boldsymbol{\sigma} = \sigma_{31}\hat{e}_1 + \sigma_{32}\hat{e}_2 , \tag{8.1.8}$$

and the projected shear traction acting at a point on the cross section is just the magnitude of the traction vector,

$$\boxed{\tau^{(P)}(\hat{e}_3) = \sqrt{\sigma_{31}^2 + \sigma_{32}^2} .} \tag{8.1.9}$$

In other words, the traction vector acting at a point on a cross section of the cylinder is tangent to the cross section.

8.1.2 Equilibrium

Assuming that the body force field is negligible, two of the equilibrium equations $\sigma_{ji,j} = 0$ are identically satisfied and the third reduces to

$$\sigma_{31,1} + \sigma_{32,2} = \mu\alpha\psi_{,11} + \mu\alpha\psi_{,22} = \mu\alpha\nabla^2\psi = 0 , \tag{8.1.10}$$

so that the equilibrium condition is

$$\boxed{\nabla^2\psi = 0 \quad \forall(X_1, X_2) \in \mathcal{S} .} \tag{8.1.11}$$

The warping function $\psi(X_1, X_2)$ must be harmonic in the region $\mathcal{S}$ of the cross section in order for the cylinder to be in equilibrium.

FIGURE 8.3. Unit normal to the lateral surface.

8.1.3   Boundary Conditions

The lateral surface of the cylinder is traction-free. In a view of the cross section (Figure 8.3), the unit normal to the lateral surface appears as an in-plane unit normal to the boundary $\partial \mathcal{S}$. This unit normal can be related to the parametric description of the boundary. Recall that the closed curve $\partial \mathcal{S}$ can be defined parametrically (2.4.53) by giving the position vector of points on the curve as a function of their distance s along the curve from some origin,

$$\mathbf{X} = \tilde{\mathbf{X}}(s), \quad X_\alpha = \tilde{X}_\alpha(s) \quad 0 \le s \le l, \qquad (8.1.12)$$

where l is the distance around $\partial \mathcal{S}$, so that $\tilde{\mathbf{X}}(0) = \tilde{\mathbf{X}}(l)$. In terms of this parametric description of the curve, the unit tangent vector to $\partial \mathcal{S}$ in the direction of increasing s is

$$\hat{v} = \frac{d\mathbf{X}}{ds}, \quad \hat{v}_\alpha = \frac{dX_\alpha}{ds}. \qquad (8.1.13)$$

By defining the direction of increasing s such that, as one progresses around $\partial \mathcal{S}$, the interior of $\mathcal{S}$ is always to one's left, it follows that the unit outward normal $\hat{n}$ to $\partial \mathcal{S}$ is related to $\hat{v}$ by $\hat{n} = \hat{v} \times \hat{e}_3$, so that

$$\hat{n} = \frac{dX_2}{ds}\hat{e}_1 - \frac{dX_1}{ds}\hat{e}_2. \qquad (8.1.14)$$

Expressing the unit normal in this way will prove useful in the examples that follow.

Since $\hat{n}_3 = 0$ and $\sigma_{11} = \sigma_{22} = \sigma_{12} = 0$, the conditions that $t_1 = t_2 = 0$ on the lateral surface are identically satisfied. The remaining traction-free condition $t_3 = 0$ becomes

$$\sigma_{31}\hat{n}_1 + \sigma_{32}\hat{n}_2 = \alpha\mu(\psi_{,1} - X_2)\hat{n}_1 + \alpha\mu(\psi_{,2} + X_1)\hat{n}_2 = 0. \qquad (8.1.15)$$

Thus, the warping function $\psi(X_1, X_2)$ must satisfy the condition

$$\boxed{(\psi_{,1} - X_2)\hat{n}_1 + (\psi_{,2} + X_1)\hat{n}_2 = 0 \quad \forall (X_1, X_2) \in \partial \mathcal{S},} \qquad (8.1.16a)$$

or alternatively using (8.1.14),

$$\boxed{(\psi_{,1} - X_2)\frac{dX_2}{ds} - (\psi_{,2} + X_1)\frac{dX_1}{ds} = 0 \quad \forall (X_1, X_2) \in \partial \mathcal{S}} \qquad (8.1.16b)$$

in order for the lateral surface to be traction-free.

The traction distribution on the end surfaces of the cylinder must be statically equivalent to the torque **T**. On the end surface $X_3 = L$, the unit outward normal is $\mathbf{n} = \hat{\mathbf{e}}_3$ so the tractions are $t_1 = \sigma_{31}$, $t_2 = \sigma_{32}$, and $t_3 = \sigma_{33} = 0$. Thus, the resultant force on the end surface in the X_1-direction is

$$F_1 = \int_{\mathcal{S}} \sigma_{31} \, dA = \mu\alpha \int_{\mathcal{S}} (\psi_{,1} - X_2) \, dA \,. \qquad (8.1.17)$$

Using (8.1.11), this can be rewritten as

$$F_1 = \mu\alpha \int_{\mathcal{S}} \{ [X_1(\psi_{,1} - X_2)]_{,1} + [X_1(\psi_{,2} + X_1)]_{,2} \} \, dA \,, \qquad (8.1.18)$$

so that, using the Green-Riemann formula (2.4.69),

$$F_1 = \mu\alpha \oint_{\partial \mathcal{S}} [X_1(\psi_{,1} - X_2) \, dX_2 - X_1(\psi_{,2} + X_1) \, dX_1] \,. \qquad (8.1.19)$$

Finally, from the boundary condition (8.1.16b), it follows that

$$F_1 = 0 \,. \qquad (8.1.20)$$

The resultant force on the end surface in the X_2-direction can similarly be shown to be zero. Finally, since $t_3 = 0$, the resultant force on the end surface in the X_3-direction is also zero.

The resultant moments on the end surface about the X_1 and X_2 axes are likewise zero:

$$M_1 = \int_{\mathcal{S}} X_2 \sigma_{33} \, dA = 0 \,, \quad M_2 = \int_{\mathcal{S}} -X_1 \sigma_{33} \, dA = 0 \,. \qquad (8.1.21)$$

The moment about the X_3-axis (Figure 8.4) is

$$M_3 = \int_{\mathcal{S}} (X_1 \sigma_{32} - X_2 \sigma_{31}) \, dA$$

$$= \mu\alpha \int_{\mathcal{S}} (X_1 \psi_{,2} + X_1^2 - X_2 \psi_{,1} + X_2^2) \, dA \,. \qquad (8.1.22)$$

FIGURE 8.4.   Tractions on the cross section of a cylinder in torsion.

However, it is required that $M_3 = T$, the magnitude of the applied torque. Thus, the twist per unit length α is related to the applied torque by

$$\alpha = \frac{T}{\mu \tilde{J}} , \tag{8.1.23}$$

where the *torsion constant* is

$$\tilde{J} \equiv \int_S (X_1^2 + X_2^2 + X_1\psi_{,2} - X_2\psi_{,1}) \, dA . \tag{8.1.24}$$

Note that the torsion constant is equal to the polar moment of inertia if the warping function is zero. The quantity $\mu\tilde{J}$ is known as the *torsional rigidity*; it relates the torque to the twist per unit length.

In summary, the torsion problem is reduced to finding a warping function $\psi(X_1, X_2)$ that is harmonic everywhere in the cross section S of the cylinder, (8.1.11), and satisfies the traction-free boundary condition on the lateral surface of the cylinder, (8.1.16). Compatibility is not an issue since the analysis begins with a displacement field. Note that this problem involves neither the applied torque nor the material properties of the cylinder; the warping function $\psi(X_1, X_2)$ and therefore the torsion constant (8.1.24) are geometric quantities—they depend only on the geometry of the cross section, not on the torque or the material properties. Once the warping function is known, the displacement and stress fields are given by (8.1.4) and (8.1.6) and the twist per unit length is given by (8.1.23).

8.1.4   Example: Circular Cylinder

As an example of using the inverse method, consider the warping function $\psi = 0$. This function is trivially harmonic, so it satisfies the equilibrium condition (8.1.11). The boundary condition (8.1.16b) in this case becomes

$$X_1\frac{dX_1}{ds} + X_2\frac{dX_2}{ds} = \frac{1}{2}\frac{d}{ds}(X_1^2 + X_2^2) = 0 \quad \forall(X_1, X_2) \in \partial S . \tag{8.1.25}$$

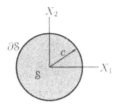

FIGURE 8.5. *Torsion of a circular cylinder.*

Integrating this equation gives the boundary condition

$$X_1^2 + X_2^2 = c^2 \quad \forall (X_1, X_2) \in \partial S , \qquad (8.1.26)$$

where c is a constant. This is the equation for a circle of radius c—for the warping function $\psi = 0$ to satisfy the traction-free boundary condition on the lateral surface of the cylinder, all the material points on ∂S must satisfy this equation; that is, this is the warping function for torsion of a circular cylinder (Figure 8.5). Only for a circular cylinder do plane cross sections remain plane.

The torsion constant in this case is the polar moment of inertia,

$$\tilde{J} = \int_S (X_1^2 + X_2^2)\, dA = \int_S r^2\, dA = \frac{1}{2}\pi c^4 , \qquad (8.1.27)$$

and the twist per unit length is

$$\alpha = \frac{T}{\mu \tilde{J}} = \frac{2T}{\mu \pi c^4} . \qquad (8.1.28)$$

The nonzero components of stress (8.1.6) are

$$\sigma_{31} = -\mu \alpha X_2 = -\frac{2T X_2}{\pi c^4} , \quad \sigma_{32} = \mu \alpha X_1 = \frac{2T X_1}{\pi c^4} \qquad (8.1.29)$$

and, assuming that $T \geq 0$, the projected shear traction at any point on a cross section (8.1.9) is

$$\tau^{(\mathrm{P})}(\hat{\mathbf{e}}_3) = \frac{2Tr}{\pi c^4} . \qquad (8.1.30)$$

Thus, the maximum shear traction on a cross section occurs at the boundary ∂S, where $r = c$:

$$\tau_{\max} = \frac{2T}{\pi c^3} . \qquad (8.1.31)$$

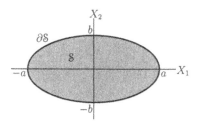

FIGURE 8.6.   Torsion of an elliptical cylinder.

8.1.5   Example: Elliptical Cylinder

As another example using the inverse method, consider the warping function $\psi = kX_1X_2$, where k is a constant. This function satisfies the equilibrium condition (8.1.11) and the boundary condition (8.1.16b) in this case becomes

$$(k-1)X_2\frac{dX_2}{ds} - (k+1)X_1\frac{dX_1}{ds}$$
$$= -\frac{1}{2}\frac{d}{ds}\left[(1+k)X_1^2 + (1-k)X_2^2\right] = 0 , \quad (8.1.32)$$

so that

$$\frac{d}{ds}\left(X_1^2 + \frac{1-k}{1+k}X_2^2\right) = 0 \quad \forall(X_1, X_2) \in \partial S . \quad (8.1.33)$$

Integrating this equation gives the boundary condition

$$X_1^2 + \frac{1-k}{1+k}X_2^2 = a^2 \quad \forall(X_1, X_2) \in \partial S , \quad (8.1.34)$$

where a is a constant. This is the equation for an ellipse (Figure 8.6) with semiaxes a and b:

$$\frac{X_1^2}{a^2} + \frac{X_2^2}{b^2} = 1 , \quad (8.1.35)$$

where

$$\frac{a^2}{b^2} = \frac{1-k}{1+k} . \quad (8.1.36)$$

Solving (8.1.36) for k,

$$k = -\frac{a^2 - b^2}{a^2 + b^2} . \quad (8.1.37)$$

Therefore,

$$\psi = -\left(\frac{a^2 - b^2}{a^2 + b^2}\right) X_1 X_2 \qquad (8.1.38)$$

is the warping function for torsion of an elliptical cylinder with semiaxes a and b as shown in Figure 8.6.

The torsion constant (8.1.24) in this case is

$$\tilde{J} = \int_S (X_1^2 + X_2^2 + kX_1^2 - kX_2^2)\, dA$$

$$= (1 + k) \int_S X_1^2\, dA + (1 - k) \int_S X_2^2\, dA$$

$$= \frac{2b^2}{a^2 + b^2} I_2 + \frac{2a^2}{a^2 + b^2} I_1 , \qquad (8.1.39)$$

where $I_1 = \pi a b^3 / 4$ and $I_2 = \pi a^3 b / 4$ are the moments of inertia about the X_1- and X_2-axes, respectively. Thus,

$$\tilde{J} = \frac{\pi a^3 b^3}{a^2 + b^2} \qquad (8.1.40)$$

and the twist per unit length (8.1.23) is

$$\alpha = \frac{(a^2 + b^2)T}{\mu \pi a^3 b^3} . \qquad (8.1.41)$$

It is interesting to note that the polar moment of inertia for the ellipse is $J = \pi a b(a^2 + b^2)/4$ so that

$$\tilde{J} = \left(\frac{2ab}{a^2 + b^2}\right)^2 J . \qquad (8.1.42)$$

Since $(a - b)^2 = a^2 + b^2 - 2ab \geq 0$, it follows that $a^2 + b^2 \geq 2ab$ and, consequently, that if $a \neq b$, then $\tilde{J} < J$. Thus, the torsional rigidity is less than would have been predicted if one had assumed that plane sections remain plane. This is always true for torsion of noncircular cylinders.

The nonzero scalar components of stress (8.1.6) are

$$\sigma_{31} = -\mu\alpha\left(1 + \frac{a^2 - b^2}{a^2 + b^2}\right) X_2 = -\frac{2\mu\alpha a^2 X_2}{a^2 + b^2} ,$$

$$\sigma_{32} = \mu\alpha\left(1 - \frac{a^2 - b^2}{a^2 + b^2}\right) X_1 = \frac{2\mu\alpha b^2 X_1}{a^2 + b^2} . \qquad (8.1.43)$$

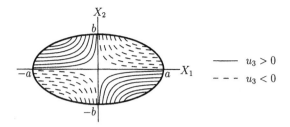

FIGURE 8.7. *Contours of constant out-of-plane displacement for torsion of an elliptical cylinder when $b < a$.*

The projected shear traction at any point on a cross section (8.1.9) is

$$\tau^{(P)}(\hat{\mathbf{e}}_3) = \frac{2\mu\alpha}{a^2 + b^2} \sqrt{b^4 X_1^2 + a^4 X_2^2} \,, \tag{8.1.44}$$

and if $b < a$, then it can be shown that the maximum shear occurs at $(X_1, X_2) = (0, \pm b)$ (i.e., at the point on the boundary ∂S nearest the centroid of the cross section) and therefore that the magnitude of the maximum shear is

$$\tau_{\max} = \frac{2\mu\alpha a^2 b}{a^2 + b^2} \,. \tag{8.1.45}$$

It can be proven, in fact, that for *any* torsion problem where ∂S is convex,[1] the maximum projected shear traction occurs at the point on ∂S nearest the centroid of S.

Using (8.1.38) and (8.1.41), the out-of-plane displacement (8.1.4) is

$$u_3 = -\frac{(a^2 - b^2)T}{\mu\pi a^3 b^3} X_1 X_2 \,. \tag{8.1.46}$$

Assuming $b < a$, contours of constant u_3 are sketched in Figure 8.7.

8.1.6   Example: Rectangular Cylinder

Consider torsion of a cylinder with a rectangular cross section $2a \times 2b$, as shown in Figure 8.8. Since $\hat{\mathbf{n}} = \pm\hat{\mathbf{e}}_1$ on the boundary surfaces $X_1 = \pm a$ and $\hat{\mathbf{n}} = \pm\hat{\mathbf{e}}_2$ on the boundary surfaces $X_2 = \pm b$, the boundary condition (8.1.16a) in this case reduces to

$$\begin{aligned} \psi_{,1} &= X_2 & \text{when } X_1 = \pm a \,, \\ \psi_{,2} &= -X_1 & \text{when } X_2 = \pm b \,. \end{aligned} \tag{8.1.47}$$

---

[1] The boundary ∂S is convex if for each and every pair of points on ∂S, the line segment connecting these points lies entirely within $S \cup \partial S$.

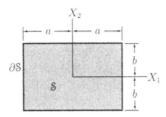

FIGURE 8.8. Torsion of a rectangular cylinder.

To solve this problem, one must find a warping function that satisfies the equilibrium condition (8.1.11), $\nabla^2 \psi = 0$, and these boundary conditions.

By defining a new function $\bar{\psi} \equiv X_1 X_2 - \psi$, the problem is transformed into one of finding $\bar{\psi}$ such that

$$\nabla^2 \bar{\psi} = 0 \quad \forall (X_1, X_2) \in \mathcal{S} \tag{8.1.48}$$

and

$$\bar{\psi}_{,1} = 0 \qquad \text{when } X_1 = \pm a , \tag{8.1.49}$$
$$\bar{\psi}_{,2} = 2X_1 \quad \text{when } X_2 = \pm b . \tag{8.1.50}$$

Assuming that the problem can be solved by using separation of variables, let

$$\bar{\psi}(X_1, X_2) = F(X_1)G(X_2) . \tag{8.1.51}$$

Substituting into the equilibrium condition (8.1.48) and collecting terms with the same independent variable, equilibrium is satisfied if

$$\frac{F''(X_1)}{F(X_1)} = -\frac{G''(X_2)}{G(X_2)} = \eta , \tag{8.1.52}$$

where η is an arbitrary constant.

Consider first the case where $\eta > 0$ and let $\eta = k^2$ where $k > 0$. Then, the complete solutions for $F(X_1)$ and $G(X_2)$ from (8.1.52) are

$$F(X_1) = C_1 \cosh kX_1 + C_2 \sinh kX_1 , \tag{8.1.53}$$
$$G(X_2) = C_3 \cos kX_2 + C_4 \sin kX_2 . \tag{8.1.54}$$

From (8.1.51), it is seen that the boundary condition (8.1.49) is satisfied only if $F'(\pm a) = 0$:

$$F'(a) = C_1 k \sinh ka + C_2 k \cosh ka = 0 ,$$
$$F'(-a) = -C_1 k \sinh ka + C_2 k \cosh ka = 0 . \tag{8.1.55}$$

Because $ka > 0$, $\sinh ka \neq 0$ and $\cosh ka \neq 0$. Therefore, the only solution to (8.1.55) is $C_1 = C_2 = 0$—the only separable solution to (8.1.48) corresponding to $\eta > 0$ that satisfies the boundary condition (8.1.49) is $\bar{\psi}(X_1, X_2) = 0$. This is a trivial solution and plays no role in the problem.

Next, consider the case $\eta = 0$. Then, the complete solutions for $F(X_1)$ and $G(X_2)$ from (8.1.52) are

$$F(X_1) = C_1 X_1 + C_2 , \tag{8.1.56}$$
$$G(X_2) = C_3 X_2 + C_4 . \tag{8.1.57}$$

Again, it follows that the boundary condition (8.1.49) is only satisfied if $C_1 = C_2 = 0$ and therefore that the only separable solution to (8.1.48) corresponding to $\eta = 0$ that satisfies the boundary condition (8.1.49) is $\bar{\psi}(X_1, X_2) = 0$. This is another trivial solution that plays no role in the problem.

Finally, consider the case where $\eta < 0$ and let $\eta = -k^2$ where $k > 0$. Then, the complete solutions for $F(X_1)$ and $G(X_2)$ from (8.1.52) are

$$F(X_1) = C_1 \cos kX_1 + C_2 \sin kX_1 , \tag{8.1.58}$$
$$G(X_2) = C_3 \cosh kX_2 + C_4 \sinh kX_2 . \tag{8.1.59}$$

From the boundary condition (8.1.50),

$$\begin{aligned} F(X_1)G'(b) &= 2X_1 , \\ F(X_1)G'(-b) &= 2X_1 . \end{aligned} \tag{8.1.60}$$

Since the right-hand sides of (8.1.60) are odd, $F(X_1)$ must also be an odd function. Therefore, $C_1 = 0$. Eliminating $2X_1$ from (8.1.60),

$$F(X_1)[G'(b) - G'(-b)] = 0 . \tag{8.1.61}$$

Since b is arbitrary, if follows that $G'(X_2)$ must be an even function and therefore that $C_3 = 0$. Thus, with $A = C_2 C_4$,

$$\bar{\psi}(X_1, X_2) = A \sin kX_1 \sinh kX_2 . \tag{8.1.62}$$

The boundary condition (8.1.50) still has to be satisfied, but consider first the boundary condition (8.1.49):

$$\bar{\psi}_{,1}(\pm a, X_2) = Ak \cos ka \sinh kX_2 = 0 . \tag{8.1.63}$$

The solution $A = 0$ is trivial, so it is required that $\cos ka = 0$:

$$k_n = \frac{(2n+1)\pi}{2a} , \quad n = 0, 1, 2, \ldots . \tag{8.1.64}$$

Note that negative values of n do not yield any different functions. Thus, every term of a series solution of the form

$$\bar{\psi}(X_1, X_2) = \sum_{n=0}^{\infty} A_n \sin k_n X_1 \sinh k_n X_2 \qquad (8.1.65)$$

satisfies the boundary condition (8.1.49). The coefficients A_n are determined by the boundary condition (8.1.50):

$$2X_1 = \bar{\psi}_{,2}(X_1, \pm b)$$

$$= \sum_{n=0}^{\infty} A_n k_n \sin k_n X_1 \cosh k_n b$$

$$= \sum_{n=0}^{\infty} B_n \sin k_n X_1 , \qquad (8.1.66)$$

where $B_n \equiv A_n k_n \cosh k_n b$. The $\sin k_n X_1$ form an orthogonal set on $(-a, a)$:

$$\int_{-a}^{a} \sin k_m X_1 \sin k_n X_1 \, dX_1 = \begin{cases} 0 & \text{if } m \neq n \\ a & \text{if } m = n. \end{cases} \qquad (8.1.67)$$

It follows, by multiplying both sides of (8.1.66) by $\sin k_m X_1$ and integrating on $(-a, a)$, that

$$B_m = \frac{1}{a} \int_{-a}^{a} 2X_1 \sin k_m X_1 \, dX_1$$

$$= \frac{2}{a} \left(\frac{1}{k_m^2} \sin k_m X_1 - \frac{X_1}{k_m} \cos k_m X_1 \right) \Big|_{-a}^{a}$$

$$= \frac{4}{a k_m^2} \sin k_m a$$

$$= \frac{16a}{(2m+1)^2 \pi^2} \sin \frac{(2m+1)\pi}{2} . \qquad (8.1.68)$$

Thus, since $\sin(2n+1)\pi/2 = (-1)^n$ and $A_n = B_n/(k_n \cosh k_n b)$,

$$A_n = \frac{(-1)^n 32a^2}{(2n+1)^3 \pi^3 \cosh k_n b} \qquad (8.1.69)$$

and the warping function $\psi = X_1 X_2 - \bar{\psi}$ is

$$\psi = X_1 X_2 - \frac{32a^2}{\pi^3} \sum_{n=0}^{\infty} \frac{(-1)^n \sin k_n X_1 \sinh k_n X_2}{(2n+1)^3 \cosh k_n b} , \qquad (8.1.70)$$

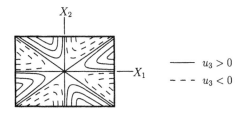

FIGURE 8.9. *Contours of constant out-of-plane displacement for torsion of a rectangular cylinder.*

where k_n is given by (8.1.64). The out-of-plane displacement $u_3 = \alpha\psi$ is sketched in Figure 8.9.

By evaluating (8.1.24) and using the identity $\sum_{n=0}^{\infty}(2n+1)^{-4} = \pi^4/96$, the torsion constant can be shown to be

$$\tilde{J} = ab^3\Omega_1\left(\frac{a}{b}\right) , \tag{8.1.71}$$

where

$$\Omega_1(x) = \frac{16x^2}{3}\left[1 - \frac{192x}{\pi^5}\sum_{n=0}^{\infty}\frac{1}{(2n+1)^5}\tanh\frac{(2n+1)\pi}{2x}\right] . \tag{8.1.72}$$

The boundary $\partial\mathcal{S}$ is convex, so the maximum projected shear traction occurs at the point on $\partial\mathcal{S}$ closest to the centroid. If $a > b$, this point is $(0, b)$ and the maximum projected shear traction can be shown from (8.1.9) to be

$$\tau_{\max} = \mu ab\Omega_2\left(\frac{a}{b}\right) = \frac{T}{ab^2}\left[\frac{\Omega_2(a/b)}{\Omega_1(a/b)}\right] , \tag{8.1.73}$$

where

$$\Omega_2(x) = \frac{16x}{\pi^2}\sum_{n=0}^{\infty}\frac{(-1)^n}{(2n+1)^2}\tanh\frac{(2n+1)\pi}{2x} . \tag{8.1.74}$$

Values of $\Omega_1(a/b)$ and $\Omega_2(a/b)$ are shown in Table 8.1 for various values of a/b.

8.2 Prandtl Stress Function

It has already been shown that for torsion of noncircular cylinders, $\sigma_{11} = \sigma_{22} = \sigma_{33} = \sigma_{12} = 0$ and the equations of equilibrium reduce to the single

Table 8.1 Table of values for torsion of a rectangular cylinder

a/b	$\Omega_1(a/b)$	$\Omega_2(a/b)$
1.0	2.249	1.351
1.2	2.658	1.518
1.5	3.132	1.695
2.0	3.659	1.860
2.5	3.990	1.936
3.0	4.213	1.971
4.0	4.493	1.994
5.0	4.661	1.999
10.0	4.997	2.000
∞	5.333	2.000

condition

$$\sigma_{31,1} + \sigma_{32,2} = 0 \ . \tag{8.2.1}$$

Accordingly, one defines the *Prandtl stress function* $\phi(X_1, X_2)$ by the relations

$$\boxed{\sigma_{31} = \phi_{,2} \ , \quad \sigma_{32} = -\phi_{,1} \ .} \tag{8.2.2}$$

It can easily be shown then that components of stress derived from the Prandtl stress function $\phi(X_1, X_2)$ via (8.2.2) are guaranteed to satisfy the equilibrium condition.

The compatibility conditions in terms of the components of stress with $\mathbf{f} = \mathbf{0}$ [i.e., Beltrami's equations (6.1.19)] reduce in this case to

$$\sigma_{3\alpha,\beta\beta} = 0 \ , \tag{8.2.3}$$

since $\sigma_{kk} = 0$ and the nonzero components of stress σ_{31} and σ_{32} are independent of X_3. Using (8.2.2), the compatibility conditions for ϕ are therefore

$$\phi_{,211} + \phi_{,222} = (\nabla^2\phi)_{,2} = 0 \ ,$$
$$\phi_{,111} + \phi_{,122} = (\nabla^2\phi)_{,1} = 0 \ , \tag{8.2.4}$$

from which it follows that, for compatibility,

$$\boxed{\nabla^2\phi = C \quad \forall(X_1, X_2) \in \mathcal{S} \ ,} \tag{8.2.5}$$

where C is a constant.

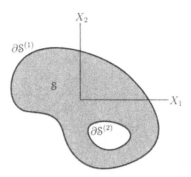

FIGURE 8.10. Multiply connected cross section.

Recall that the traction-free lateral boundary condition is

$$\sigma_{31}\hat{n}_1 + \sigma_{32}\hat{n}_2 = 0 \qquad (8.2.6)$$

and that the unit outward normal to the lateral surface $\hat{n}$ can be expressed in terms of the parametric equation for ∂S as

$$\hat{n} = \frac{dX_2}{ds}\hat{e}_1 - \frac{dX_1}{ds}\hat{e}_2 . \qquad (8.2.7)$$

From (8.2.2) then, the boundary condition in terms of the Prandtl stress function is

$$\phi_{,2}\frac{dX_2}{ds} + \phi_{,1}\frac{dX_1}{ds} = \frac{d\phi}{ds} = 0 \qquad (8.2.8)$$

so that

$$\boxed{\frac{d\phi}{ds} = 0 \quad \forall (X_1, X_2) \in \partial S .} \qquad (8.2.9)$$

In general, the cross section S may be multiply connected so that there are multiple distinct boundaries, $\partial S^{(1)}$ and $\partial S^{(2)}$ say (Figure 8.10). Then, ϕ must be constant on each, although the value of the constant may be different on different distinct boundaries. For example, for the cross section shown in Figure 8.10, the traction-free boundary conditions for $\partial S^{(1)}$ and $\partial S^{(2)}$ are

$$\begin{aligned}
\phi = C^{(1)} \quad &\forall (X_1, X_2) \in \partial S^{(1)} , \\
\phi = C^{(2)} \quad &\forall (X_1, X_2) \in \partial S^{(2)} ,
\end{aligned} \qquad (8.2.10)$$

where $C^{(1)}$ and $C^{(2)}$ are constants and, in general, $C^{(1)} \neq C^{(2)}$. Note from (8.2.2) that an arbitrary constant can be added to ϕ without affecting the components of stress. Thus, for example, $C^{(1)}$ can arbitrarily be set equal to zero, but then the value of $C^{(2)} \neq 0$ is no longer arbitrary. However, for a simply connected cross section (i.e., *for a solid cylinder*), one can arbitrarily set

$$\phi = 0 \quad \forall (X_1, X_2) \in \partial \mathcal{S} \quad \text{(solid cylinders)} . \tag{8.2.11}$$

It is important to remember that the boundary condition (8.2.11) is only valid for solid cylinders.

Comparing the definition of the Prandtl stress function (8.2.2) with the relations for components of stress in terms of the warping function (8.1.6), it is seen that

$$\psi_{,1} = \frac{1}{\mu\alpha}\phi_{,2} + X_2 , \tag{8.2.12}$$

$$\psi_{,2} = -\frac{1}{\mu\alpha}\phi_{,1} - X_1 . \tag{8.2.13}$$

By solving (8.2.12) and (8.2.13), the warping function ψ and thus the out-of-plane displacements can be found if ϕ is known.

The twist per unit length α can be found from ϕ by differentiating (8.2.12) and (8.2.13) with respect to X_2 and X_1, respectively, and eliminating $\psi_{,12}$ to get

$$\frac{1}{\mu\alpha}\phi_{,22} + 1 = -\frac{1}{\mu\alpha}\phi_{,11} - 1 . \tag{8.2.14}$$

Solving for α,

$$\alpha = -\frac{1}{2\mu}\nabla^2\phi . \tag{8.2.15}$$

Recall that $\nabla^2\phi$ must be a constant from the compatibility condition, so (8.2.15) relates that constant to the twist per unit length. This also means that the constant must be nonzero (or else the twist per unit length is zero).

Since the tractions on each cross section must be statically equivalent to the applied torque T,

$$T = \int_{\mathcal{S}} (X_1\sigma_{32} - X_2\sigma_{31}) \, dA , \tag{8.2.16}$$

it follows that, with the components of stress given by (8.2.2),

$$T = -\int_{\mathcal{S}} (X_1\phi_{,1} + X_2\phi_{,2}) \, dA . \tag{8.2.17}$$

This result can be rewritten as

$$T = - \int_S [(X_1\phi)_{,1} + (X_2\phi)_{,2}] \, dA + 2 \int_S \phi \, dA , \qquad (8.2.18)$$

which, upon applying the Green-Riemann theorem (2.4.69), yields

$$T = - \oint_{\partial S} (X_1\phi \, dX_2 - X_2\phi \, dX_1) + 2 \int_S \phi \, dA . \qquad (8.2.19)$$

In the *special case of a solid cylinder* where the boundary condition (8.2.11) has been enforced, the integrand of the first integral is zero everywhere on ∂S and

$$\boxed{T = 2 \int_S \phi \, dA \quad (\text{when } \phi = 0 \text{ on } \partial S) .} \qquad (8.2.20)$$

In summary, the torsion problem has been reduced to finding a Prandtl stress function $\phi(X_1, X_2)$ such that its Laplacian is a nonzero constant (8.2.5) and, in the case of solid cylinders, its value is zero on the boundary of the cross section (8.2.11). Then, the twist per unit length is related to the value of the Laplacian by (8.2.15) and the torque is given by (8.2.20). If desired, one can then find the torsional rigidity from (8.1.23):

$$\mu \tilde{J} = \frac{T}{\alpha} . \qquad (8.2.21)$$

One can make some additional useful observations. The normal traction on cross sections is zero, $\nu(\hat{\mathbf{e}}_3) = \sigma_{33} = 0$, and the projected shear traction vector is

$$\boldsymbol{\tau}^{(\mathrm{P})}(\hat{\mathbf{e}}_3) = \sigma_{31}\hat{\mathbf{e}}_1 + \sigma_{32}\hat{\mathbf{e}}_2 = \phi_{,2}\,\hat{\mathbf{e}}_1 - \phi_{,1}\,\hat{\mathbf{e}}_2 . \qquad (8.2.22)$$

Recall that

$$\boldsymbol{\nabla}\phi = \phi_{,1}\,\hat{\mathbf{e}}_1 + \phi_{,2}\,\hat{\mathbf{e}}_2 . \qquad (8.2.23)$$

Thus, it is seen that $\boldsymbol{\nabla}\phi \cdot \boldsymbol{\tau}^{(\mathrm{P})}(\hat{\mathbf{e}}_3) = 0$ and, since from elementary calculus $\boldsymbol{\nabla}\phi$ is normal to contours of constant ϕ, the projected shear traction at any point on the cross section is tangent to the contour of constant ϕ at that point (Figure 8.11). It also follows immediately from (8.2.22) and (8.2.23) that the projected shear traction is

$$\tau^{(\mathrm{P})}(\hat{\mathbf{e}}_3) = |\boldsymbol{\nabla}\phi| = \sqrt{(\phi_{,1})^2 + (\phi_{,2})^2} . \qquad (8.2.24)$$

The maximum gradient of ϕ and thus $\tau_{\max}$ occur where contour lines are closest together.

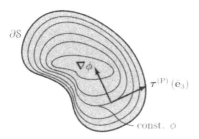

FIGURE 8.11. *Projected shear traction and contours of constant Prandtl stress function.*

8.2.1 Example: Elliptical Cylinder (Revisited)

Consider the elliptical cross section shown previously in Figure 8.6. The boundary $\partial \mathcal{S}$ is the ellipse with semiaxes a and b given by

$$\partial \mathcal{S} = \left\{ (X_1, X_2) \mid \frac{X_1^2}{a^2} + \frac{X_2^2}{b^2} = 1 \right\}. \qquad (8.2.25)$$

One can easily find a function based on this condition that is zero everywhere on $\partial \mathcal{S}$:

$$\phi = m \left(\frac{X_1^2}{a^2} + \frac{X_2^2}{b^2} - 1 \right), \qquad (8.2.26)$$

where m is a constant. However, in order to be a valid Prandtl stress function, the Laplacian of ϕ must be a nonzero constant. Taking the Laplacian of (8.2.26),

$$\nabla^2 \phi = 2m \left(\frac{1}{a^2} + \frac{1}{b^2} \right) = \frac{2m(a^2 + b^2)}{a^2 b^2}. \qquad (8.2.27)$$

Thus, it is seen that $\nabla^2 \phi$ is a constant. Therefore, (8.2.26) is the solution to the problem. This is serendipitous—be forewarned that this strategy for finding the Prandtl stress function only works for a few cross sections. The compatibility condition is usually much more difficult to satisfy.

From (8.2.15) and (8.2.27), the twist per unit length is related to m by

$$\alpha = -\frac{m(a^2 + b^2)}{\mu a^2 b^2}. \qquad (8.2.28)$$

Solving for m and substituting the result into (8.2.26), the Prandtl stress function can be rewritten as

$$\phi = -\frac{\mu \alpha a^2 b^2}{a^2 + b^2} \left(\frac{X_1^2}{a^2} + \frac{X_2^2}{b^2} - 1 \right). \qquad (8.2.29)$$

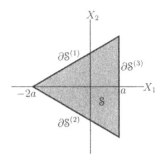

FIGURE 8.12. Equilateral triangular cross section.

Substituting (8.2.29) into (8.2.20) and evaluating the area integral, the torque is related to the twist per unit length by

$$T = 2 \int_S \phi \, dA = \frac{\mu \alpha \pi a^3 b^3}{a^2 + b^2} \,, \tag{8.2.30}$$

which is the same as previously found using the warping function. Likewise, the components of stress and the torsion constant given by (8.2.2), (8.2.21), and (8.2.29) are also the same as previously found.

8.2.2   Example: Equilateral Triangular Cylinder

Consider the cross section shown in Figure 8.12. The region S is an equilateral triangle with boundary composed of three straight-line segments, $\partial S = \partial S^{(1)} \cup \partial S^{(2)} \cup \partial S^{(3)}$. With the coordinate system defined as shown in Figure 8.12, these three straight-line segments are given by

$$\partial S^{(1)} = \left\{ (X_1, X_2) \mid X_1 - \sqrt{3} X_2 = -2a \quad \text{and} \quad -2a \le X_1 \le a \right\},$$

$$\partial S^{(2)} = \left\{ (X_1, X_2) \mid X_1 + \sqrt{3} X_2 = -2a \quad \text{and} \quad -2a \le X_1 \le a \right\},$$

$$\partial S^{(3)} = \left\{ (X_1, X_2) \mid X_1 = a \quad \text{and} \quad |X_2| \le \sqrt{3}\, a \right\}.$$

$$\tag{8.2.31}$$

As in the previous example, a function that is zero everywhere on ∂S is easily constructed from these conditions:

$$\phi = m(X_1 - \sqrt{3} X_2 + 2a)(X_1 + \sqrt{3} X_2 + 2a)(X_1 - a)$$
$$= m(X_1^3 + 3aX_1^2 + 3aX_2^2 - 3X_1 X_2^2 - 4a^3) \,. \tag{8.2.32}$$

Before one declares (8.2.32) to be the solution Prandtl stress function for the problem, remember that it needs to be established that the Laplacian of this function is a nonzero constant. It is easily shown that the Laplacian of (8.2.32) is

$$\nabla^2 \phi = 12ma \, , \tag{8.2.33}$$

which is indeed a nonzero constant. Thus, this is the solution to the problem. It is worth repeating that this approach to finding the Prandtl stress function only works for a few cross sections.

From (8.2.15) and (8.2.33), the twist per unit length is related to m by

$$\alpha = -\frac{6ma}{\mu} \, . \tag{8.2.34}$$

Solving for m and substituting the result into (8.2.32), the Prandtl stress function can be rewritten as

$$\phi = -\frac{\mu\alpha}{6a}(X_1^3 + 3aX_1^2 + 3aX_2^2 - 3X_1X_2^2 - 4a^3) \, . \tag{8.2.35}$$

Substituting (8.2.35) into (8.2.20) and evaluating the area integral, the torque is related to the twist per unit length by

$$T = 2\int_S \phi \, dA = \frac{27}{5\sqrt{3}}\mu\alpha a^4 \, . \tag{8.2.36}$$

Thus, the torsion constant (8.2.21) is

$$\tilde{J} = \frac{27}{5\sqrt{3}}a^4 \, . \tag{8.2.37}$$

The nonzero scalar components of stress (8.2.2) are

$$\sigma_{31} = \phi_{,2} = \frac{\mu\alpha}{a}(X_1 - a)X_2 \, ,$$
$$\sigma_{32} = -\phi_{,1} = \frac{\mu\alpha}{2a}(X_1^2 + 2aX_1 - X_2^2) \, . \tag{8.2.38}$$

The boundary ∂S is convex, so the maximum projected shear traction on the surface occurs at the points on ∂S nearest the centroid (i.e., at the midpoints of each of the three straight-line segments). Choosing the mid-point $(X_1, X_2) = (a, 0)$ of $\partial S^{(3)}$, (8.2.38) gives $\sigma_{31}(a, 0) = 0$ and, therefore, the maximum shear traction

$$\tau_{\max} = \sigma_{32}(a, 0) = \frac{3}{2}\mu\alpha a \, . \tag{8.2.39}$$

To find the out-of-plane displacements, one needs to solve (8.2.12),

$$\psi_{,1} = \frac{1}{\mu\alpha}\phi_{,2} + X_2 = \frac{1}{a}X_1X_2 \, , \tag{8.2.40}$$

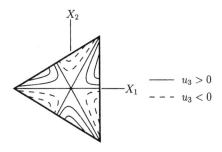

FIGURE 8.13.　Contours of constant out-of-plane displacement for the equilateral triangular cross section.

and (8.2.13),

$$\psi_{,2} = -\frac{1}{\mu\alpha}\phi_{,1} - X_1 = \frac{1}{2a}(X_1^2 - X_2^2) , \tag{8.2.41}$$

for the warping function ψ. Integrating (8.2.40) and (8.2.41),

$$\psi = \frac{1}{2a}X_1^2 X_2 + \eta(X_2) , \tag{8.2.42}$$

$$\psi = \frac{1}{2a}X_1^2 X_2 - \frac{1}{6a}X_2^3 + \xi(X_1) . \tag{8.2.43}$$

Comparing (8.2.42) and (8.2.43),

$$\eta(X_2) = -\frac{1}{6a}X_2^3 + C , \quad \xi(X_1) = C , \tag{8.2.44}$$

where C is a constant corresponding to a rigid-body translation in the X_3-direction. Setting $C = 0$, then the out-of-plane displacement is

$$u_3 = \alpha\psi = \frac{\alpha X_2}{6a}(3X_1^2 - X_2^2) . \tag{8.2.45}$$

Contours of constant u_3 are sketched in Figure 8.13.

Problems

8.1 Determine the Prandtl stress function for torsion of a hollow circular cylinder of inner radius a and outer radius b, in terms of the shear modulus μ and the twist per unit length α.

8.2 An axisymmetric composite cylinder is composed of a solid inner shaft, of radius a and shear modulus μ_1, and an outer sleeve of outer radius b and shear modulus μ_2. The shaft and sleeve are ideally bonded at their interface and the composite cylinder is subjected to an applied torque T.

(a) Determine the distribution of stress within the composite cylinder in terms of the twist per unit length α.

(b) Find an expression for the twist per unit length α in terms of the applied torque T.

(c) How much of the total torque T is carried by the outer sleeve?

8.3 Consider torsion of a cylinder with the rectangular cross section shown below. Explain why a function of the form

$$\phi = m\left(\frac{X_1^2}{a^2} - 1\right)\left(\frac{X_2^2}{b^2} - 1\right)$$

cannot be used as a Prandtl stress function for this cross section.

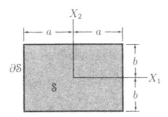

8.4 Consider the circular shaft with a circular keyway whose cross section is shown below.

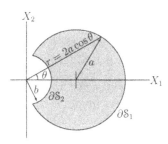

(a) Show that the Prandtl stress function,

$$\phi = m(r^2 - b^2)\left(\frac{2a\cos\theta}{r} - 1\right),$$

may be used to describe the torsion for this cross section and determine the value of the constant m in terms of the twist per unit length α.

(b) Assuming that the maximum shear traction τ_{max} on the cross section occurs at the point $(r,\theta) = (b,0)$ on the boundary, compute the value of τ_{max} in the limit as $b \to 0$ and compare it with the maximum shear traction for torsion of a solid circular cylinder of radius a (the two maximum shear tractions are different).

8.5 Consider the Prandtl stress function

$$\phi = m(a^2 - X_1^2 + b^2 X_2^2)(a^2 + b^2 X_1^2 - X_2^2),$$

where $|b| < 1$.

(a) What restrictions (if any) on the constants a, b, and m are required in order for this to be a valid Prandtl stress function?

(b) Determine the value of m in terms of the twist per unit length α.

(c) Determine and sketch the shape of the cross section of the cylinder for which this Prandtl stress function gives the torsion solution. Why was it necessary to have $|b| < 1$?

8.6 Consider the hollow cylinder shown below whose cross section is bounded by two concentric, similar ellipses $\partial \mathcal{S}_1$ and $\partial \mathcal{S}_2$:

$$\partial \mathcal{S}_1 = \left\{ (X_1, X_2) \mid \frac{X_1^2}{a^2} + \frac{X_2^2}{b^2} = k^2 \right\},$$

$$\partial \mathcal{S}_2 = \left\{ (X_1, X_2) \mid \frac{X_1^2}{a^2} + \frac{X_2^2}{b^2} = 1 \right\},$$

where a, b, and k are constants and $0 < k < 1$.

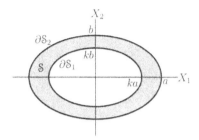

(a) Show that the Prandtl stress function,

$$\phi = m \left(\frac{X_1^2}{a^2} + \frac{X_2^2}{b^2} - 1 \right),$$

may be used to describe the torsion for this cross section and determine the value of the constant m in terms of the twist per unit length α.

(b) Find the torque T as a function of the twist per unit length and determine the torsion constant $\tilde{J}$. **Hint:** the moments of inertia for a region $\mathcal{S}$ with elliptic boundary

$$\partial \mathcal{S} = \left\{ (X_1, X_2) \mid \frac{X_1^2}{c^2} + \frac{X_2^2}{d^2} = 1 \right\}$$

are

$$I_1 \equiv \int_{\mathcal{S}} X_2^2 \, dA = \frac{\pi c d^3}{4}, \quad I_2 \equiv \int_{\mathcal{S}} X_1^2 \, dA = \frac{\pi c^3 d}{4}.$$

—

Chapter 9

Three-Dimensional Problems

Although the solution to some simple three-dimensional problems in linearized elasticity have been presented (e.g., spherical pressure vessels), the focus so far has been on special classes of problems for which the equations of elasticity could be reduced to a two-dimensional form. Some additional three-dimensional problems of interest will now be considered.

Recall that the problem of finding solutions to the equations of elasticity can be restated in terms of either finding a displacement field **u** that satisfies the Lamé-Navier equations or finding a stress field $\boldsymbol{\sigma}$ that satisfies the equations of equilibrium and Michell's equations. Such solutions are, of course, still subject to a particular problem's boundary conditions. In the method of potentials, potential functions are defined which facilitate the determination of solutions to these governing equations. Potential functions related to the displacement field are used to find solutions to the Lamé-Navier equations. These include the scalar and vector displacement potentials and the Papkovich functions. Potentials related to the stress field are used to help find solutions to Michell's equations and the equations of equilibrium. These include the Airy stress function for plane strain/stress problems, the Prandtl stress function for torsion problems, and the Maxwell and Morera stress functions for three-dimensional problems.

Only the scalar and vector displacement potentials and the Papkovich functions will be considered here. Complex variable stress functions for antiplane strain and plane strain/stress problems will be introduced later. See Fung (1965) for a discussion of Maxwell and Morera stress functions.

9.1 Field Theory Results

Recall that a tensor field is a tensor quantity that varies from point to point in space—it is a function of position. Presented below are some important results from the general theory of three-dimensional tensor fields.

The following field theory identities are easily proven and will be needed in the derivation of potential functions to follow. If ζ is a scalar field, $\mathbf{v}$ is a vector field, and $\mathbf{x}$ is the position vector of a point in space relative to some prescribed origin, then

$$\boldsymbol{\nabla}(\nabla^2\zeta) = \nabla^2(\boldsymbol{\nabla}\zeta) \,, \tag{9.1.1}$$

$$\boldsymbol{\nabla}(\mathbf{v}\cdot\mathbf{x}) = \mathbf{v} + \mathbf{x}\cdot(\boldsymbol{\nabla}\mathbf{v}) \,, \tag{9.1.2}$$

$$\boldsymbol{\nabla}\cdot(\boldsymbol{\nabla}\times\mathbf{v}) = 0 \,, \tag{9.1.3}$$

$$\boldsymbol{\nabla}\cdot(\boldsymbol{\nabla}\zeta) = \nabla^2\zeta \,, \tag{9.1.4}$$

$$\boldsymbol{\nabla}\cdot(\nabla^2\mathbf{v}) = \nabla^2(\boldsymbol{\nabla}\cdot\mathbf{v}) \,, \tag{9.1.5}$$

$$\boldsymbol{\nabla}\cdot(\zeta\mathbf{v}) = \zeta\boldsymbol{\nabla}\cdot\mathbf{v} + (\boldsymbol{\nabla}\zeta)\cdot\mathbf{v} \,, \tag{9.1.6}$$

$$\boldsymbol{\nabla}\times(\boldsymbol{\nabla}\zeta) = \mathbf{0} \,, \tag{9.1.7}$$

$$\boldsymbol{\nabla}\times(\nabla^2\mathbf{v}) = \nabla^2(\boldsymbol{\nabla}\times\mathbf{v}) \,, \tag{9.1.8}$$

$$\boldsymbol{\nabla}\times(\boldsymbol{\nabla}\times\mathbf{v}) = \boldsymbol{\nabla}(\boldsymbol{\nabla}\cdot\mathbf{v}) - \nabla^2\mathbf{v} \,, \tag{9.1.9}$$

$$\nabla^2(\zeta\mathbf{x}) = 2\boldsymbol{\nabla}\zeta + \mathbf{x}\nabla^2\zeta \,, \tag{9.1.10}$$

$$\nabla^2(\mathbf{v}\cdot\mathbf{x}) = 2\boldsymbol{\nabla}\cdot\mathbf{v} + \mathbf{x}\cdot\nabla^2\mathbf{v} \,. \tag{9.1.11}$$

9.1.1   Poisson's Equation

Consider the *Poisson equation*

$$\nabla^2\mathbf{F} = \mathbf{f} \,, \tag{9.1.12}$$

where $\mathbf{F}$ and $\mathbf{f}$ are tensor fields of the same order. Suppose that $\mathbf{f}$ is known and is analytic in some region $\mathcal{R}$.[1] Then, it is well known from the theory of Newtonian potentials (Courant and Hilbert 1937; Kellogg 1953) that a particular solution to (9.1.12) is

$$\mathbf{F}^{(p)}(\mathbf{x}) = -\frac{1}{4\pi}\int_{\mathcal{R}}\frac{\mathbf{f}(\mathbf{x}')}{|\mathbf{x}-\mathbf{x}'|}\,dV_{\mathbf{x}'} \,, \tag{9.1.13}$$

where $dV_{\mathbf{x}'}$ indicates that the variable of integration is $\mathbf{x}'$. Thus, a complete solution to (9.1.12) is

$$\mathbf{F}(\mathbf{x}) = \mathbf{F}^{(p)}(\mathbf{x}) + \mathbf{F}^{(h)}(\mathbf{x}) \,, \tag{9.1.14}$$

---

[1] A tensor field is said to be *analytic at a point* $\mathbf{x} = \mathbf{x}_0$ if its scalar components can be represented by power series in powers of $\mathbf{x} - \mathbf{x}_0$ with radius of convergence $R > 0$. If a tensor field is analytic at every point in a region $\mathcal{R}$, then the tensor field is said to be analytic in $\mathcal{R}$.

where the homogeneous solution $\mathbf{F}^{(h)}$ is a tensor field that is regular harmonic in $\mathcal{R}$, but is otherwise arbitrary.[2]

9.1.2 Three-Dimensional Dirac Delta Function

The *three-dimensional Dirac delta function* δ is a scalar field defined such that its value is zero for all nonzero values of its argument,

$$\delta(\mathbf{x}) = 0 \quad \forall \mathbf{x} \neq \mathbf{0} , \tag{9.1.15}$$

and for an arbitrary tensor field $\mathbf{T}$ and an open region $\mathcal{R}$ with boundary $\partial \mathcal{R}$,

$$\int_{\mathcal{R}} \mathbf{T}(\mathbf{x}')\delta(\mathbf{x}' - \mathbf{x}) \, dV_{\mathbf{x}'} = \begin{cases} \mathbf{T}(\mathbf{x}) & \mathbf{x} \in \mathcal{R} \\ \frac{1}{2}\mathbf{T}(\mathbf{x}) & \mathbf{x} \in \partial \mathcal{R} \\ \mathbf{0} & \mathbf{x} \notin (\mathcal{R} \cup \partial \mathcal{R}) . \end{cases} \tag{9.1.16}$$

Example 9.1
Find a particular solution, over the open region $\mathcal{R}$ with boundary $\partial \mathcal{R}$, for the Poisson equation

$$\nabla^2 \mathbf{F} = \mathbf{A}\delta(\mathbf{x} - \boldsymbol{\xi}) ,$$

where $\mathbf{A}$ is a constant tensor and $\boldsymbol{\xi}$ is the position vector of a fixed point in space.

Solution. From (9.1.13), a particular solution to this equation is given by

$$\mathbf{F}^{(p)}(\mathbf{x}) = -\frac{1}{4\pi} \int_{\mathcal{R}} \frac{\mathbf{A}\delta(\mathbf{x}' - \boldsymbol{\xi})}{|\mathbf{x} - \mathbf{x}'|} \, dV_{\mathbf{x}'} .$$

Thus, using (9.1.16), a particular solution is

$$\mathbf{F}^{(p)}(\mathbf{x}) = \begin{cases} -\mathbf{A}/(4\pi\rho) & \boldsymbol{\xi} \in \mathcal{R} \\ -\mathbf{A}/(8\pi\rho) & \boldsymbol{\xi} \in \partial \mathcal{R} \\ \mathbf{0} & \boldsymbol{\xi} \notin (\mathcal{R} \cup \partial \mathcal{R}) , \end{cases}$$

where $\rho \equiv |\mathbf{x} - \boldsymbol{\xi}| = \sqrt{(x_i - \xi_i)(x_i - \xi_i)}$ is the distance between $\mathbf{x}$ and $\boldsymbol{\xi}$.

---

[2] A tensor field $\mathbf{T}$ is *regular harmonic in a region* $\mathcal{R}$ if, at every point in $\mathcal{R}$, the scalar components of $\mathbf{T}$ are twice continuously differentiable and $\nabla^2 \mathbf{T} = \mathbf{0}$.

9.1.3 Helmholtz's Representation Theorem

If $\mathbf{u}$ is a vector field that is analytic in the region $\mathcal{R}$, then there exists a scalar field ζ and a vector field $\boldsymbol{\psi}$ such that, in the region $\mathcal{R}$,

$$\boxed{\mathbf{u} = \boldsymbol{\nabla}\zeta + \boldsymbol{\nabla} \times \boldsymbol{\psi} \,, \quad \boldsymbol{\nabla} \cdot \boldsymbol{\psi} = 0 \,.} \tag{9.1.17}$$

In this representation of $\mathbf{u}$, ζ is called a *scalar potential* and $\boldsymbol{\psi}$ is called a *vector potential*.

Helmholtz's representation theorem is proven by showing that ζ and $\boldsymbol{\psi}$ can be found. Define the vector field $\mathbf{w}$ such that in the region $\mathcal{R}$,

$$\nabla^2 \mathbf{w} = \mathbf{u} \,. \tag{9.1.18}$$

This is a Poisson equation for $\mathbf{w}$ and since $\mathbf{u}$ is analytic in $\mathcal{R}$, a particular solution $\mathbf{w}^{(p)}$ is given by (9.1.13). Thus, for any given $\mathbf{u}$, a corresponding $\mathbf{w} = \mathbf{w}^{(p)} + \mathbf{w}^{(h)}$ can be found, where $\mathbf{w}^{(h)}$ is a vector field that is regular harmonic in $\mathcal{R}$, but is otherwise arbitrary. Using the identity (9.1.9) to rewrite (9.1.18),

$$\mathbf{u} = \boldsymbol{\nabla}(\boldsymbol{\nabla} \cdot \mathbf{w}) - \boldsymbol{\nabla} \times (\boldsymbol{\nabla} \times \mathbf{w}) \,. \tag{9.1.19}$$

By defining the scalar potential $\zeta \equiv \boldsymbol{\nabla} \cdot \mathbf{w}$ and the vector potential $\boldsymbol{\psi} \equiv -\boldsymbol{\nabla} \times \mathbf{w}$ and noting from the identity (9.1.3) that $\boldsymbol{\nabla} \cdot \boldsymbol{\psi} = -\boldsymbol{\nabla} \cdot (\boldsymbol{\nabla} \times \mathbf{w}) = 0$, ζ and $\boldsymbol{\psi}$ are found and the Helmholtz representation theorem is proven. Note, however, that since the solution $\mathbf{w}$ to (9.1.18) includes an arbitrary homogeneous solution, the representation (9.1.17) is not unique—for any given $\mathbf{u}$, there are an infinite number of scalar and vector potential pairs.

It is useful to note that the divergence and curl of $\mathbf{u}$ can be expressed in terms of its scalar and vector potentials. Taking the divergence of (9.1.17) gives

$$\boldsymbol{\nabla} \cdot \mathbf{u} = \boldsymbol{\nabla} \cdot (\boldsymbol{\nabla}\zeta) + \boldsymbol{\nabla} \cdot (\boldsymbol{\nabla} \times \boldsymbol{\psi}) \,, \tag{9.1.20}$$

so that, by using the identities (9.1.3) and (9.1.4), one finds that

$$\boxed{\boldsymbol{\nabla} \cdot \mathbf{u} = \nabla^2 \zeta \,.} \tag{9.1.21}$$

Taking the curl of (9.1.17) gives

$$\boldsymbol{\nabla} \times \mathbf{u} = \boldsymbol{\nabla} \times (\boldsymbol{\nabla}\zeta) + \boldsymbol{\nabla} \times (\boldsymbol{\nabla} \times \boldsymbol{\psi}) \,, \tag{9.1.22}$$

so that, by using the field relations (9.1.7) and (9.1.9) and recalling that $\boldsymbol{\nabla} \cdot \boldsymbol{\psi} = 0$, one finds that

$$\boxed{\boldsymbol{\nabla} \times \mathbf{u} = -\nabla^2 \boldsymbol{\psi} \,.} \tag{9.1.23}$$

9.1.4 Green's Theorem

Consider the vector field $\mathbf{v} = \zeta_1(\nabla\zeta_2)$, where ζ_1 and ζ_2 are scalar fields, and note from (9.1.4) and (9.1.6) that

$$\nabla \cdot \mathbf{v} = \zeta_1(\nabla^2\zeta_2) + (\nabla\zeta_1) \cdot (\nabla\zeta_2) . \tag{9.1.24}$$

If $\mathcal{R}$ is a closed bounded region whose boundary $\partial\mathcal{R}$ is a regular surface, it follows from the divergence theorem that if ζ_1 is continuously differentiable and ζ_2 is two-times continuously differentiable (so that $\mathbf{v}$ and $\nabla \cdot \mathbf{v}$ are continuous) in $\mathcal{R}$, then

$$\int_{\mathcal{R}} [\zeta_1(\nabla^2\zeta_2) + (\nabla\zeta_1) \cdot (\nabla\zeta_2)] \, dV = \int_{\partial\mathcal{R}} \zeta_1(\nabla\zeta_2) \cdot \hat{\mathbf{n}} \, dA , \tag{9.1.25}$$

where $\hat{\mathbf{n}}$ is the unit outward normal to $\partial\mathcal{R}$. This is known as the *first form of Green's theorem.* By reversing the roles of ζ_1 and ζ_2 and subtracting the result from (9.1.25), one finds that if ζ_1 and ζ_2 are two-times continuously differentiable in $\mathcal{R}$, then

$$\int_{\mathcal{R}} [\zeta_1(\nabla^2\zeta_2) - \zeta_2(\nabla^2\zeta_1)] \, dV = \int_{\partial\mathcal{R}} [\zeta_1(\nabla\zeta_2) - \zeta_2(\nabla\zeta_1)] \cdot \hat{\mathbf{n}} \, dA , \tag{9.1.26}$$

which is known as *the second form of Green's theorem.*

If $\zeta_1 = \zeta_2 = \zeta$ is a regular harmonic function in $\mathcal{R}$, then it follows from (9.1.25) that

$$\int_{\mathcal{R}} |\nabla\zeta|^2 \, dV = \int_{\partial\mathcal{R}} \zeta(\nabla\zeta \cdot \hat{\mathbf{n}}) \, dA . \tag{9.1.27}$$

Recall that $\nabla\zeta \cdot \hat{\mathbf{n}}$ is the direction derivative of ζ in the direction of the unit outward normal $\hat{\mathbf{n}}$, also known as the *normal derivative of ζ relative to $\partial\mathcal{R}$* and represented as

$$\frac{\partial\zeta}{\partial n} \equiv \nabla\zeta \cdot \hat{\mathbf{n}} . \tag{9.1.28}$$

The result (9.1.27) leads to the following two consequences of Green's theorem:

1. If a scalar field ζ is regular harmonic in a closed bounded region $\mathcal{R}$, whose boundary $\partial\mathcal{R}$ is a regular closed surface, and the normal derivative $\partial\zeta/\partial n$ vanishes on the boundary $\partial\mathcal{R}$, then the right-hand side of (9.1.27) is zero. Since the left-hand-side integrand $|\nabla\zeta|^2$ is

positive definite, it follows that $\nabla \zeta = 0$ everywhere in $\mathcal{R}$—the scalar field ζ is homogeneous in $\mathcal{R}$:

$$\left. \begin{array}{ll} \nabla^2 \zeta = 0 & \forall \mathbf{x} \in \mathcal{R} \\ \dfrac{\partial \zeta}{\partial n} = 0 & \forall \mathbf{x} \in \partial \mathcal{R} \end{array} \right\} \quad \Rightarrow \quad \zeta = C \quad \forall \mathbf{x} \in \mathcal{R}, \qquad (9.1.29)$$

where C is a constant.

2. If a scalar field ζ is regular harmonic in a closed bounded region $\mathcal{R}$, whose boundary $\partial \mathcal{R}$ is a regular closed surface, and ζ vanishes on the boundary $\partial \mathcal{R}$, then the right-hand side of (9.1.27) is once again zero and the scalar field ζ must equal some constant C in $\mathcal{R}$ by the same reasoning as in (9.1.29). However, in this case, since ζ vanishes at the boundary, it must be that $C = 0$:

$$\left. \begin{array}{ll} \nabla^2 \zeta = 0 & \forall \mathbf{x} \in \mathcal{R} \\ \zeta = 0 & \forall \mathbf{x} \in \partial \mathcal{R} \end{array} \right\} \quad \Rightarrow \quad \zeta = 0 \quad \forall \mathbf{x} \in \mathcal{R}. \qquad (9.1.30)$$

9.2 Potentials in Elasticity

Displacement potentials are first defined using Helmholtz's representation theorem. This result is then used to derive the Papkovich representation for displacement fields. The Papkovich representation will prove to be convenient when solving some important three-dimensional problems in elasticity.

The deformation is assumed to be infinitesimal and it is understood that all quantities related to the deformation of a body are given in terms of their material descriptions (i.e., as fields over the reference configuration of the body).

9.2.1 Displacement Potentials

From Helmholtz's theorem, an analytic displacement field $\mathbf{u}(\mathbf{X}, t)$ can be represented by a *scalar displacement potential* $\zeta(\mathbf{X}, t)$ and a *vector displacement potential* $\boldsymbol{\psi}(\mathbf{X}, t)$ as

$$\boxed{\mathbf{u} = \nabla \zeta + \nabla \times \boldsymbol{\psi}; \quad \nabla \cdot \boldsymbol{\psi} = 0,} \qquad (9.2.1)$$

where $\mathbf{X}$ is the reference position of material points in the body and t is time. Note from (9.1.21) and (9.1.23) that the dilatation $e = \varepsilon_{kk} = \nabla \cdot \mathbf{u}$

and the infinitesimal rotation vector $\boldsymbol{\theta} = \frac{1}{2}\boldsymbol{\nabla} \times \mathbf{u}$ are related to ζ and $\boldsymbol{\psi}$ by

$$e = \nabla^2\zeta , \quad \boldsymbol{\theta} = -\frac{1}{2}\nabla^2\boldsymbol{\psi} . \tag{9.2.2}$$

Recall the Lamé-Navier equation for equilibrium in terms of displacement:

$$(\lambda + \mu)\boldsymbol{\nabla}(\boldsymbol{\nabla} \cdot \mathbf{u}) + \mu\nabla^2\mathbf{u} + \mathbf{f} = \mathbf{0} . \tag{6.1.3'}$$

Using the field relation (9.1.9), this can be rewritten as

$$(\lambda + 2\mu)\boldsymbol{\nabla}(\boldsymbol{\nabla} \cdot \mathbf{u}) - \mu\boldsymbol{\nabla} \times (\boldsymbol{\nabla} \times \mathbf{u}) + \mathbf{f} = \mathbf{0} . \tag{9.2.3}$$

It follows that since $\boldsymbol{\nabla} \cdot \mathbf{u} = \nabla^2\zeta$ and $\boldsymbol{\nabla} \times \mathbf{u} = -\nabla^2\boldsymbol{\psi}$, the equilibrium condition for the scalar and vector displacement potentials is

$$\boxed{(\lambda + 2\mu)\boldsymbol{\nabla}(\nabla^2\zeta) + \mu\boldsymbol{\nabla} \times (\nabla^2\boldsymbol{\psi}) + \mathbf{f} = \mathbf{0} .} \tag{9.2.4}$$

Any ζ and $\boldsymbol{\psi}$ which satisfy (9.2.4) and the condition $\boldsymbol{\nabla} \cdot \boldsymbol{\psi} = 0$ correspond, via (9.2.1), to a displacement field that satisfies the Lamé-Navier equilibrium equations. Furthermore, the representation is complete—scalar and vector potentials for an analytic displacement field can always be found, in principle—although it is not unique. It is interesting to note from (9.2.2) and (9.2.4) that if body forces are negligible, then any displacement field with homogeneous dilatation and infinitesimal rotation vector fields will satisfy the Lamé-Navier equilibrium equations.

9.2.2 Papkovich Representation

Unfortunately, although the representation (9.2.1) of the displacement field in terms of scalar and vector displacement potentials is quite elegant, one would generally find it difficult to solve the equilibrium condition (9.2.4) for scalar and vector displacement potentials corresponding to any particular body force field $\mathbf{f}$. It is desirable, therefore, to formulate a representation of the displacement field that facilitates the finding of solutions of the Lamé-Navier equations for a given body force field. The development of the Papkovich representation given below is guided by the knowledge that solutions of Poisson equations are known.

Using the identities (9.1.1) and (9.1.8), recalling that $\lambda = 2\mu\nu/(1 - 2\nu)$ and dividing by $2(1-\nu)$, the equilibrium condition (9.2.4) can be rewritten as

$$\nabla^2\left[\frac{\mu}{1 - 2\nu}\boldsymbol{\nabla}\zeta + \frac{\mu}{2(1 - \nu)}\boldsymbol{\nabla} \times \boldsymbol{\psi}\right] = -\frac{\mathbf{f}}{2(1 - \nu)} . \tag{9.2.5}$$

Thus, if one defines the *Papkovich vector strain function* as

$$\mathbf{B} \equiv -\frac{\mu}{1-2\nu}\nabla\zeta - \frac{\mu}{2(1-\nu)}\nabla\times\boldsymbol{\psi}\,, \tag{9.2.6}$$

it follows that

$$\nabla^2\mathbf{B} = \frac{\mathbf{f}}{2(1-\nu)}\,, \tag{9.2.7}$$

which is a Poisson equation that one can, in principle, solve for $\mathbf{B}$. However, the representation is not yet complete. Given a Papkovich vector strain function $\mathbf{B}$, there are too many unknowns in (9.2.6) to solve for the scalar and vector displacement potentials ζ and $\boldsymbol{\psi}$ and therefore the displacement field $\mathbf{u}$. An additional scalar quantity is needed.

Taking the divergence of (9.2.6) and using the identities (9.1.3) and (9.1.4), one finds that

$$\nabla\cdot\mathbf{B} = -\frac{\mu}{1-2\nu}\nabla^2\zeta\,. \tag{9.2.8}$$

Also, from the identity (9.1.11),

$$\mathbf{X}\cdot\nabla^2\mathbf{B} = \nabla^2(\mathbf{X}\cdot\mathbf{B}) - 2\nabla\cdot\mathbf{B}\,, \tag{9.2.9}$$

so that, using (9.2.7) and (9.2.8),

$$\frac{\mathbf{X}\cdot\mathbf{f}}{2(1-\nu)} = \nabla^2\left[\mathbf{X}\cdot\mathbf{B} + \frac{2\mu}{1-2\nu}\zeta\right]\,. \tag{9.2.10}$$

Thus, if one defines the *Papkovich scalar strain function* as

$$\beta \equiv -\mathbf{X}\cdot\mathbf{B} - \frac{2\mu}{1-2\nu}\zeta\,, \tag{9.2.11}$$

it follows that

$$\nabla^2\beta = -\frac{\mathbf{X}\cdot\mathbf{f}}{2(1-\nu)}\,, \tag{9.2.12}$$

which is another Poisson equation. It remains to be shown that the Papkovich scalar and vector strain functions provide a complete (although non-unique) representation of the displacement field (i.e., that the displacement field $\mathbf{u}$ can be expressed in terms of β and $\mathbf{B}$).

Substituting for $\nabla\times\boldsymbol{\psi}$ in the representation $\mathbf{u} = \nabla\zeta + \nabla\times\boldsymbol{\psi}$ using (9.2.6),

$$\mathbf{u} = -\frac{1}{1-2\nu}\nabla\zeta - \frac{2(1-\nu)}{\mu}\mathbf{B}\,. \tag{9.2.13}$$

Finally, substituting for ζ using (9.2.11), one finds the *Papkovich represen-tation* for the displacement field **u**,

$$\begin{aligned} 2\mu\mathbf{u} &= \boldsymbol{\nabla}(\mathbf{X}\cdot\mathbf{B}+\beta) - 4(1-\nu)\mathbf{B}\ , \\ 2\mu u_i &= (X_j B_j + \beta)_{,i} - 4(1-\nu)B_i\ , \end{aligned} \tag{9.2.14}$$

where the equilibrium conditions for the Papkovich vector and scalar strain functions are

$$\nabla^2\mathbf{B} = \frac{\mathbf{f}}{2(1-\nu)}\ , \qquad B_{i,jj} = \frac{f_i}{2(1-\nu)} \tag{9.2.15}$$

and

$$\nabla^2\beta = -\frac{\mathbf{X}\cdot\mathbf{f}}{2(1-\nu)}\ , \qquad \beta_{,ii} = -\frac{X_i f_i}{2(1-\nu)}\ . \tag{9.2.16}$$

The representation (9.2.14) of the displacement field in terms of the Pap-kovich scalar and vector strain functions is arguably not as elegant as the representation (9.2.1) in terms of the scalar and vector displacement poten-tials, but the Papkovich representation has the virtue of possessing equilib-rium conditions (9.2.15) and (9.2.16) that take the form of Poisson equa-tions, for which solutions can, in principle, be found.

To derive an expression for the stress field $\boldsymbol{\sigma}$ in terms of the Papkovich scalar and vector strain functions, first note from (9.2.14) that the displace-ment gradient is given by

$$2\mu\boldsymbol{\nabla}\mathbf{u} = \boldsymbol{\nabla}\boldsymbol{\nabla}(\mathbf{X}\cdot\mathbf{B}+\beta) - 4(1-\nu)\boldsymbol{\nabla}\mathbf{B}\ . \tag{9.2.17}$$

Since for any scalar ξ, $(\boldsymbol{\nabla}\boldsymbol{\nabla}\xi)^{\mathsf{T}} = \boldsymbol{\nabla}\boldsymbol{\nabla}\xi$, it follows that the strain field $\varepsilon = \frac{1}{2}[\boldsymbol{\nabla}\mathbf{u} + (\boldsymbol{\nabla}\mathbf{u})^{\mathsf{T}}]$ is given by

$$2\mu\varepsilon = \boldsymbol{\nabla}\boldsymbol{\nabla}(\mathbf{X}\cdot\mathbf{B}+\beta) - 2(1-\nu)\left[\boldsymbol{\nabla}\mathbf{B} + (\boldsymbol{\nabla}\mathbf{B})^{\mathsf{T}}\right]\ . \tag{9.2.18}$$

Using (9.1.4), the trace of the strain field $\operatorname{tr}\varepsilon = \boldsymbol{\nabla}\cdot\mathbf{u}$ is given by

$$2\mu\operatorname{tr}\varepsilon = \nabla^2(\mathbf{X}\cdot\mathbf{B}+\beta) - 4(1-\nu)\boldsymbol{\nabla}\cdot\mathbf{B}\ , \tag{9.2.19}$$

which, from (9.1.11), can be rewritten as

$$2\mu\operatorname{tr}\varepsilon = \mathbf{X}\cdot\nabla^2\mathbf{B} + \nabla^2\beta - 2(1-2\nu)\boldsymbol{\nabla}\cdot\mathbf{B}\ . \tag{9.2.20}$$

However, from the equilibrium conditions (9.2.15) and (9.2.16) it can be seen that $\mathbf{X}\cdot\nabla^2\mathbf{B}+\nabla^2\beta = 0$. Consequently, noting that $\lambda = 2\mu\nu/(1-2\nu)$,

$$\lambda\operatorname{tr}\varepsilon = -2\nu\boldsymbol{\nabla}\cdot\mathbf{B}\ . \tag{9.2.21}$$

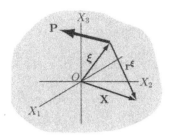

FIGURE 9.1. *Concentrated force* **P** *acting point* $\boldsymbol{\xi}$ *in an infinite body.*

Thus, the stress field $\boldsymbol{\sigma} = 2\mu\boldsymbol{\varepsilon} + \lambda(\operatorname{tr}\boldsymbol{\varepsilon})\mathbf{I}$ is given in terms of the Papkovich strain functions by

$$\boxed{\begin{aligned}
\boldsymbol{\sigma} &= \boldsymbol{\nabla}\boldsymbol{\nabla}(\mathbf{X}\cdot\mathbf{B} + \beta) - 2(1-\nu)\left[\boldsymbol{\nabla}\mathbf{B} + (\boldsymbol{\nabla}\mathbf{B})^{\mathsf{T}}\right] - 2\nu(\boldsymbol{\nabla}\cdot\mathbf{B})\mathbf{I}\,, \\
\sigma_{ij} &= (X_k B_k + \beta)_{,ij} - 2(1-\nu)(B_{i,j} + B_{j,i}) - 2\nu B_{k,k}\,\delta_{ij}\,.
\end{aligned}}$$

$$(9.2.22)$$

9.2.3 Kelvin's Problem

Consider an infinite body subjected to a concentrated body force **P** (units of force) acting at a point with position vector $\boldsymbol{\xi}$ relative to the origin O (Figure 9.1). The body force field is zero everywhere else. This means that for any region $\mathcal{R}'$ of the body with boundary $\partial\mathcal{R}'$, the resultant body force acting on $\mathcal{R}'$ is

$$\int_{\mathcal{R}'} \mathbf{f}\, dV = \begin{cases} \mathbf{P} & \boldsymbol{\xi}\in\mathcal{R}' \\ \frac{1}{2}\mathbf{P} & \boldsymbol{\xi}\in\partial\mathcal{R}' \\ \mathbf{0} & \boldsymbol{\xi}\notin(\mathcal{R}'\cup\partial\mathcal{R}')\,, \end{cases} \qquad (9.2.23)$$

where **f** is the body force field. Thus, the body force field can be expressed as a function of the position **X** relative to the origin O by

$$\mathbf{f} = \mathbf{P}\delta(\mathbf{X} - \boldsymbol{\xi})\,, \qquad (9.2.24)$$

where δ is the three-dimensional Dirac delta function. Assume also that

$$\lim_{|\mathbf{X}|\to\infty} \mathbf{u} = \mathbf{0}\,; \qquad (9.2.25)$$

the displacement field vanishes at infinity.

From the equilibrium conditions (9.2.15) and (9.2.16), the Papkovich vector and scalar strain functions must, in this case, satisfy

$$\nabla^2\mathbf{B} = \frac{\mathbf{P}}{2(1-\nu)}\delta(\mathbf{X} - \boldsymbol{\xi})\,, \qquad \nabla^2\beta = -\frac{\mathbf{X}\cdot\mathbf{P}}{2(1-\nu)}\delta(\mathbf{X} - \boldsymbol{\xi})\,. \qquad (9.2.26)$$

The particular solutions to these Poisson equations are

$$\mathbf{B}^{(p)}(\mathbf{X}) = -\frac{\mathbf{P}}{8\pi(1-\nu)} \int_{\mathcal{R}} \frac{\delta(\mathbf{X}'-\boldsymbol{\xi})}{|\mathbf{X}-\mathbf{X}'|} dV_{\mathbf{X}'} = -\frac{\mathbf{P}}{8\pi(1-\nu)\rho},$$

$$\beta^{(p)}(\mathbf{X}) = \frac{1}{8\pi(1-\nu)} \int_{\mathcal{R}} \frac{\mathbf{X}' \cdot \mathbf{P}\delta(\mathbf{X}'-\boldsymbol{\xi})}{|\mathbf{X}-\mathbf{X}'|} dV_{\mathbf{X}'} = \frac{\boldsymbol{\xi} \cdot \mathbf{P}}{8\pi(1-\nu)\rho},$$

$$\tag{9.2.27}$$

where $\mathcal{R}$ is the infinite domain of the body and

$$\rho \equiv |\mathbf{X}-\boldsymbol{\xi}| = \sqrt{(X_i - \xi_i)(X_i - \xi_i)} \tag{9.2.28}$$

is the distance from the point of application of the concentrated body force. The stress field corresponding to the particular solution (9.2.27) is in equilibrium with the body force field (9.2.24). If, in addition, the displacement field corresponding to (9.2.27) vanishes at infinity (as it will shortly be seen to do), then the boundary condition (9.2.25) is satisfied and homogeneous solutions of (9.2.26) need not be considered. Therefore, the solution to Kelvin's problem can be expressed in terms of Papkovich vector and scalar strain functions as

$$\boxed{\begin{aligned} \mathbf{B} &= -\frac{\mathbf{P}}{8\pi(1-\nu)\rho}, & B_i &= -\frac{P_i}{8\pi(1-\nu)\rho}, \\ \beta &= \frac{\boldsymbol{\xi}\cdot\mathbf{P}}{8\pi(1-\nu)\rho}, & \beta &= \frac{\xi_i P_i}{8\pi(1-\nu)\rho}, \end{aligned}} \tag{9.2.29}$$

where ρ is the scalar field given by (9.2.28).

Substituting (9.2.29) into (9.2.14), the displacement field for Kelvin's problem is given by

$$\begin{aligned} 2\mu\mathbf{u} &= -\frac{1}{8\pi(1-\nu)}\nabla\left[\frac{(\mathbf{X}-\boldsymbol{\xi})\cdot\mathbf{P}}{\rho}\right] + \frac{\mathbf{P}}{2\pi\rho} \\ &= -\frac{1}{8\pi(1-\nu)}\mathbf{P}\cdot\nabla\left(\frac{\mathbf{r}^{\boldsymbol{\xi}}}{\rho}\right) + \frac{\mathbf{P}}{2\pi\rho}, \end{aligned} \tag{9.2.30}$$

where

$$\mathbf{r}^{\boldsymbol{\xi}}(\mathbf{X}) \equiv \mathbf{X}-\boldsymbol{\xi} \tag{9.2.31}$$

is the position vector of the material point at $\mathbf{X}$ relative to the point of application $\boldsymbol{\xi}$ of the concentrated body force, and $|\mathbf{r}^{\boldsymbol{\xi}}(\mathbf{X})| = \rho$. By noting that

$$\nabla\left(\frac{\mathbf{r}^{\boldsymbol{\xi}}}{\rho}\right) = \frac{\mathbf{I}}{\rho} - \frac{\mathbf{r}^{\boldsymbol{\xi}}\mathbf{r}^{\boldsymbol{\xi}}}{\rho^3}, \tag{9.2.32}$$

the representation for the displacement field can be rewritten as

$$
\begin{aligned}
2\mu\mathbf{u} &= \frac{1}{8\pi(1-\nu)}\left[(3-4\nu)\frac{\mathbf{P}}{\rho} + \frac{(\mathbf{P}\cdot\mathbf{r}^\xi)\mathbf{r}^\xi}{\rho^3}\right] , \\
2\mu u_i &= \frac{1}{8\pi(1-\nu)}\left[(3-4\nu)\frac{P_i}{\rho} + \frac{P_j r_j^\xi r_i^\xi}{\rho^3}\right] .
\end{aligned}
\tag{9.2.33}
$$

The first term on the right-hand side of (9.2.33) gives the component of the displacement in the direction of the concentrated body force $\mathbf{P}$, and the second term gives the component in the direction of $\mathbf{r}^\xi$. The component of the displacement in the direction of $\mathbf{r}^\xi$ vanishes when $\mathbf{r}^\xi$ is orthogonal to $\mathbf{P}$—on the plane that is orthogonal to $\mathbf{P}$ and contains its point of application, the displacement is in the direction of $\mathbf{P}$. The magnitude of the displacement vanishes as ρ^{-1} as the distance ρ from the point of application of the concentrated body force goes to infinity.

Substituting (9.2.29) into (9.2.22), the stress field for Kelvin's problem is given by

$$
\begin{aligned}
\boldsymbol{\sigma} &= -\frac{1}{8\pi(1-\nu)}\boldsymbol{\nabla}\boldsymbol{\nabla}\left(\frac{\mathbf{r}^\xi\cdot\mathbf{P}}{\rho}\right) + \frac{1}{4\pi}\left\{\boldsymbol{\nabla}\left(\frac{\mathbf{P}}{\rho}\right) + \left[\boldsymbol{\nabla}\left(\frac{\mathbf{P}}{\rho}\right)\right]^\mathsf{T}\right\} \\
&\quad + \frac{\nu}{4\pi(1-\nu)}\left[\boldsymbol{\nabla}\cdot\left(\frac{\mathbf{P}}{\rho}\right)\right]\mathbf{I} \\
&= -\frac{1}{8\pi(1-\nu)}\mathbf{P}\cdot\boldsymbol{\nabla}\boldsymbol{\nabla}\left(\frac{\mathbf{r}^\xi}{\rho}\right) + \frac{1}{4\pi}\left\{\mathbf{P}\left[\boldsymbol{\nabla}\left(\frac{1}{\rho}\right)\right] + \left[\boldsymbol{\nabla}\left(\frac{1}{\rho}\right)\right]\mathbf{P}\right\} \\
&\quad + \frac{\nu}{4\pi(1-\nu)}\left[\mathbf{P}\cdot\boldsymbol{\nabla}\left(\frac{1}{\rho}\right)\right]\mathbf{I} .
\end{aligned}
\tag{9.2.34}
$$

Noting that

$$
\mathbf{P}\cdot\boldsymbol{\nabla}\boldsymbol{\nabla}\left(\frac{\mathbf{r}^\xi}{\rho}\right) = -\frac{\mathbf{P}\mathbf{r}^\xi + \mathbf{r}^\xi\mathbf{P} + (\mathbf{P}\cdot\mathbf{r}^\xi)\mathbf{I}}{\rho^3} + \frac{3\left(\mathbf{P}\cdot\mathbf{r}^\xi\right)\mathbf{r}^\xi\mathbf{r}^\xi}{\rho^5} ,
\tag{9.2.35}
$$

$$
\boldsymbol{\nabla}\left(\frac{1}{\rho}\right) = -\frac{\mathbf{r}^\xi}{\rho^3} ,
\tag{9.2.36}
$$

it follows that the stress field due to a concentrated body force can be expressed as

$$
\begin{aligned}
\boldsymbol{\sigma} &= \frac{1}{8\pi(1-\nu)}\left[(1-2\nu)\frac{(\mathbf{P}\cdot\mathbf{r}^\xi)\mathbf{I} - \mathbf{P}\mathbf{r}^\xi - \mathbf{r}^\xi\mathbf{P}}{\rho^3} - \frac{3(\mathbf{P}\cdot\mathbf{r}^\xi)\mathbf{r}^\xi\mathbf{r}^\xi}{\rho^5}\right] , \\
\sigma_{ij} &= \frac{1}{8\pi(1-\nu)}\left[(1-2\nu)\frac{P_k r_k^\xi\delta_{ij} - P_i r_j^\xi - r_i^\xi P_j}{\rho^3} - \frac{3P_k r_k^\xi r_i^\xi r_j^\xi}{\rho^5}\right] .
\end{aligned}
\tag{9.2.37}
$$

It is interesting to note that the mean stress $\tilde{\sigma} = \frac{1}{3} \operatorname{tr} \boldsymbol{\sigma} = \frac{1}{3}\sigma_{ii}$ is

$$\tilde{\sigma} = -\frac{1+\nu}{12\pi(1-\nu)} \left(\frac{\mathbf{P} \cdot \mathbf{r}^\xi}{\rho^3} \right) . \tag{9.2.38}$$

The mean stress is negative (compressive) when $\mathbf{P} \cdot \mathbf{r}^\xi > 0$, is positive (tensile) when $\mathbf{P} \cdot \mathbf{r}^\xi < 0$, and vanishes at points where $\mathbf{P} \cdot \mathbf{r}^\xi = 0$. Also, the magnitude of the stress vanishes as ρ^{-2} as the distance ρ from the point of application of the concentrated body force goes to infinity.

It will be seen later how this solution to Kelvin's problem can be used, along with the superposition principle and the Betti-Rayleigh reciprocity relations, to generate solutions to much more complicated and interesting problems. For this, the general expressions given above will be required. However, for a single concentrated body force, one can, without loss of generality, define a new cylindrical coordinate system such that the concentrated force acts at the origin and is directed along the z-axis. Then, $\boldsymbol{\xi} = \mathbf{0}$ so that

$$\mathbf{r}^\xi = \mathbf{X} = r\hat{\mathbf{e}}_r + z\hat{\mathbf{e}}_z \tag{9.2.39}$$

and

$$\mathbf{P} = P\hat{\mathbf{e}}_z , \tag{9.2.40}$$

where P is the magnitude of the concentrated body force.

By first noting that

$$\mathbf{P} \cdot \mathbf{r}^\xi = Pz , \tag{9.2.41}$$

it follows from (9.2.33) that the displacement field in this coordinate system can be expressed as

$$2\mu\mathbf{u} = \frac{P}{8\pi(1-\nu)} \left[(3 - 4\nu)\frac{\hat{\mathbf{e}}_z}{R} + \frac{z(r\hat{\mathbf{e}}_r + z\hat{\mathbf{e}}_z)}{R^3} \right] , \tag{9.2.42}$$

where

$$R = \rho|_{\xi=0} = \sqrt{r^2 + z^2} \tag{9.2.43}$$

is the distance from the origin. Thus, by equating $2\mu u_r$ and $2\mu u_z$ with the coefficients of $\hat{\mathbf{e}}_r$ and $\hat{\mathbf{e}}_z$, respectively, it is seen that the scalar components of the displacement field in this cylindrical coordinate system are given by

$$\begin{aligned} 2\mu u_r &= \frac{Prz}{8\pi(1-\nu)R^3} , \\ 2\mu u_z &= \frac{P}{8\pi(1-\nu)} \left(\frac{3-4\nu}{R} + \frac{z^2}{R^3} \right) . \end{aligned} \tag{9.2.44}$$

As could easily have been predicted, $u_\theta = 0$ and u_r and u_z are independent of θ—the displacement field is cylindrically symmetric about the z-axis. Also, u_r is an odd function of z that vanishes along the z-axis and on the plane $z = 0$, and u_z is an even function of z.

Since, in this coordinate system,

$$\mathbf{Pr}^\xi + \mathbf{r}^\xi \mathbf{P} = Pr(\hat{\mathbf{e}}_r \hat{\mathbf{e}}_z + \hat{\mathbf{e}}_z \hat{\mathbf{e}}_r) + 2Pz\hat{\mathbf{e}}_z \hat{\mathbf{e}}_z \ , \tag{9.2.45}$$

$$\mathbf{r}^\xi \mathbf{r}^\xi = r^2 \hat{\mathbf{e}}_r \hat{\mathbf{e}}_r + z^2 \hat{\mathbf{e}}_z \hat{\mathbf{e}}_z + rz(\hat{\mathbf{e}}_r \hat{\mathbf{e}}_z + \hat{\mathbf{e}}_z \hat{\mathbf{e}}_r) \ , \tag{9.2.46}$$

the stress field (9.2.37) can be expressed as

$$\boldsymbol{\sigma} = \frac{P}{8\pi(1-\nu)} \left\{ (1-2\nu) \frac{z(\hat{\mathbf{e}}_r \hat{\mathbf{e}}_r + \hat{\mathbf{e}}_\theta \hat{\mathbf{e}}_\theta - \hat{\mathbf{e}}_z \hat{\mathbf{e}}_z) - r(\hat{\mathbf{e}}_r \hat{\mathbf{e}}_z + \hat{\mathbf{e}}_z \hat{\mathbf{e}}_r)}{R^3} \right.$$
$$\left. - \frac{3z[r^2 \hat{\mathbf{e}}_r \hat{\mathbf{e}}_r + z^2 \hat{\mathbf{e}}_z \hat{\mathbf{e}}_z + rz(\hat{\mathbf{e}}_r \hat{\mathbf{e}}_z + \hat{\mathbf{e}}_z \hat{\mathbf{e}}_r)]}{R^5} \right\} \ . \tag{9.2.47}$$

By equating σ_{rr} with the coefficient of $\hat{\mathbf{e}}_r \hat{\mathbf{e}}_r$, $\sigma_{\theta\theta}$ with the coefficient of $\hat{\mathbf{e}}_\theta \hat{\mathbf{e}}_\theta$, and so forth, it follows that the scalar components of the stress field are given by

$$\sigma_{rr} = \frac{P}{8\pi(1-\nu)} \left[(1-2\nu) \frac{z}{R^3} - \frac{3r^2 z}{R^5} \right] \ , \tag{9.2.48}$$

$$\sigma_{\theta\theta} = \frac{(1-2\nu)Pz}{8\pi(1-\nu)R^3} \ , \tag{9.2.49}$$

$$\sigma_{zz} = -\frac{P}{8\pi(1-\nu)} \left[(1-2\nu) \frac{z}{R^3} + \frac{3z^3}{R^5} \right] \ , \tag{9.2.50}$$

$$\sigma_{rz} = \sigma_{zr} = -\frac{P}{8\pi(1-\nu)} \left[(1-2\nu) \frac{r}{R^3} + \frac{3rz^2}{R^5} \right] \ , \tag{9.2.51}$$

where, again consistent with cylindrical symmetry about the z-axis, $\sigma_{r\theta} = \sigma_{\theta z} = 0$ and σ_{rr}, $\sigma_{\theta\theta}$, σ_{zz}, and σ_{rz} are independent of θ. Note that σ_{rz} is an even function of z and σ_{rr}, $\sigma_{\theta\theta}$, and σ_{zz} are odd functions of z.

Example 9.2

Consider an infinite body with a concentrated body force $\mathbf{P}$. Show that a cylindrical portion of the body, with radius a and centroidal axis coincident with the line of action of $\mathbf{P}$, is in equilibrium (Figure 9.2).

Solution. Define a cylindrical coordinate system with the concentrated body force $\mathbf{P}$ acting at the origin and directed along the z-axis. Then, the boundary of the cylindrical portion of the body under consideration is the surface $r = a$ and the unit outward normal to the boundary is $\hat{\mathbf{e}}_r$. The scalar components of the traction vector acting on this boundary are

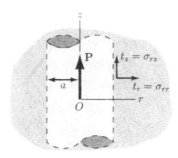

FIGURE 9.2. *Cylindrical portion of an infinite body with a concentrated body force **P** acting along the centroidal axis.*

therefore $t_r = \sigma_{rr}$, $t_\theta = \sigma_{r\theta}$, and $t_z = \sigma_{rz}$, where the scalar components of the stress are given by (9.2.48)–(9.2.51).

Since σ_{rr} and σ_{rz} are independent of θ and $\sigma_{r\theta} = 0$, the only nontrivial equilibrium condition is equilibrium of forces in the z-direction. The sum of forces in the z-direction acting on the cylindrical portion of the body is

$$\sum F_z = P + \int_{-\infty}^{\infty} \int_0^{2\pi} \sigma_{rz}(a, z)\, a\, d\theta\, dz$$

$$= P - \frac{P}{4(1 - \nu)} \int_{-\infty}^{\infty} \left[(1 - 2\nu)\frac{a^2}{R_a^3} + \frac{3a^2 z^2}{R_a^5} \right] dz \;,$$

where $R_a \equiv R|_{r=a} = \sqrt{a^2 + z^2}$. Noting that

$$\frac{d}{dz}\left(\frac{z}{R_a} \right) = \frac{a^2}{R_a^3}\;, \qquad \frac{d}{dz}\left(\frac{z^3}{R_a^3} \right) = \frac{3a^2 z^2}{R_a^5}\;,$$

it follows that

$$\sum F_z = P - \frac{P}{4(1 - \nu)}\left[(1 - 2\nu)\frac{z}{R_a} + \frac{z^3}{R_a^3} \right]\Bigg|_{-\infty}^{\infty}\;.$$

Finally, since

$$\lim_{z \to \infty} \frac{z}{R_a} = \lim_{z \to \infty} \frac{z^3}{R_a^3} = 1\;, \qquad \lim_{z \to -\infty} \frac{z}{R_a} = \lim_{z \to -\infty} \frac{z^3}{R_a^3} = -1\;,$$

one can see that

$$\sum F_z = P - \frac{P}{2(1 - \nu)}\left[(1 - 2\nu) + 1 \right] = 0\;;$$

equilibrium of the cylindrical portion of the body is satisfied.

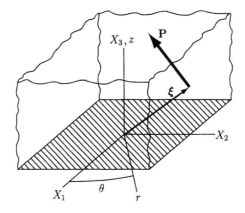

*FIGURE 9.3. Concentrated force **P** acting at a point **ξ** in a semi-infinite body.*

9.2.4   Mindlin's Problem

Define a cylindrical coordinate system and consider a semi-infinite body occupying the half-space region $\mathcal{R} = \{\, \mathbf{X} \mid z > 0 \,\}$ above the $z = 0$ plane, with boundary $\partial\mathcal{R} = \{\, \mathbf{X} \mid z = 0 \,\}$, and a concentrated body force $\mathbf{P}$ acting at a point $\boldsymbol{\xi} \in \mathcal{R}$ in the body (Figure 9.3). The body force field within the body can be expressed in terms of the three-dimensional Dirac delta function as

$$\mathbf{f} = \mathbf{P}\delta(\mathbf{X} - \boldsymbol{\xi}) \quad \forall \mathbf{X} \in \mathcal{R} . \tag{9.2.52}$$

The boundary $\partial\mathcal{R}$ is traction-free, and the unit outward normal to this boundary is $\hat{\mathbf{n}} = -\hat{\mathbf{e}}_z$, giving the boundary condition

$$\sigma_{zr} = \sigma_{z\theta} = \sigma_{zz} = 0 \quad \forall \mathbf{X} \in \partial\mathcal{R} . \tag{9.2.53}$$

Assume also that the displacement field in the body vanishes at infinity:

$$\lim_{\substack{|\mathbf{X}| \to \infty \\ z \geq 0}} \mathbf{u} = \mathbf{0} . \tag{9.2.54}$$

This is the general form of *Mindlin's problem* and the solution is given by Mindlin (1953).

To simplify matters here, consider only the *special axisymmetric case* of this problem (Figure 9.4) in which the concentrated body force's line of action is coincident with the z-axis,

$$\mathbf{P} = P\hat{\mathbf{e}}_z , \tag{9.2.55}$$

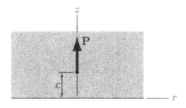

FIGURE 9.4. *Special axisymmetric case of Mindlin's problem, with the concentrated body force's line of action coincident with the z-axis.*

and its point of application is a distance c above the traction-free surface,

$$\boldsymbol{\xi} = c\hat{\mathbf{e}}_z \quad (c > 0) . \tag{9.2.56}$$

If one were to neglect the traction-free boundary condition (9.2.53), the solution to Kelvin's problem (9.2.29) would give Papkovich vector and scalar strain functions of the form $B_r = B_\theta = 0$ and

$$B_z = \frac{-P}{8\pi(1-\nu)\rho} , \quad \beta = \frac{cP}{8\pi(1-\nu)\rho} , \tag{9.2.57}$$

where

$$\rho = \sqrt{r^2 + (z-c)^2} . \tag{9.2.58}$$

Using the semi-inverse method, this suggests one look for a solution to the special axisymmetric case of Mindlin's problem of the form $B_r = B_\theta = 0$ and

$$B_z = \frac{P}{8\pi(1-\nu)}\left(\bar{B} - \frac{1}{\rho}\right) , \quad \beta = \frac{P}{8\pi(1-\nu)}\left(\bar{\beta} + \frac{c}{\rho}\right) , \tag{9.2.59}$$

where $\bar{B} = \bar{B}(r, z)$ and $\bar{\beta} = \bar{\beta}(r, z)$ are as yet to be determined scalar fields that are assumed to be independent of θ and can be thought of as correction factors required to satisfy the traction-free boundary condition (9.2.53).

When $B_r = B_\theta = 0$ and B_z and β are independent of θ, it follows from (9.2.14) that the scalar components of the displacement field in a cylindrical coordinate systems are given by $u_\theta = 0$ and

$$2\mu u_r = z\frac{\partial B_z}{\partial r} + \frac{\partial \beta}{\partial r} ,$$

$$2\mu u_z = z\frac{\partial B_z}{\partial z} - (3 - 4\nu)B_z + \frac{\partial \beta}{\partial z} . \tag{9.2.60}$$

Thus, the condition (9.2.54) that the displacement field vanish at infinity will, given the assumed form of the Papkovich vector and scalar strain functions (9.2.59), require that

$$\lim_{\substack{|\mathbf{X}| \to \infty \\ z \geq 0}} \bar{B} = 0 , \qquad \lim_{\substack{|\mathbf{X}| \to \infty \\ z \geq 0}} \frac{\partial \bar{\beta}}{\partial r} = 0 , \qquad \lim_{\substack{|\mathbf{X}| \to \infty \\ z \geq 0}} \frac{\partial \bar{\beta}}{\partial z} = 0 . \qquad (9.2.61)$$

In addition, it is clear from the equilibrium equations (9.2.15) and (9.2.16) and the analysis of Kelvin's problem that the proposed solution (9.2.59) is in equilibrium with the body force field (9.2.52) if and only if $\bar{B}$ and $\bar{\beta}$ are regular harmonic in the upper half-space:

$$\nabla^2 \bar{B} = 0 , \qquad \nabla^2 \bar{\beta} = 0 \quad \forall \mathbf{X} \in \mathcal{R} . \qquad (9.2.62)$$

If scalar fields $\bar{B}$ and $\bar{\beta}$ can be found that satisfy (9.2.61) and (9.2.62), such that the stress field corresponding to (9.2.59) satisfies the traction-free boundary condition (9.2.53), then the solution to the problem will be known.

When $B_r = B_\theta = 0$ and B_z and β are independent of θ, the scalar components of stress (9.2.22) are expressed in a cylindrical coordinate system by $\sigma_{r\theta} = \sigma_{\theta z} = 0$ and

$$
\begin{aligned}
\sigma_{rr} &= z\frac{\partial^2 B_z}{\partial r^2} + \frac{\partial^2 \beta}{\partial r^2} - 2\nu\frac{\partial B_z}{\partial z} , \\
\sigma_{\theta\theta} &= \frac{z}{r}\frac{\partial B_z}{\partial r} + \frac{1}{r}\frac{\partial \beta}{\partial r} - 2\nu\frac{\partial B_z}{\partial z} , \\
\sigma_{zz} &= z\frac{\partial^2 B_z}{\partial z^2} - 2(1-\nu)\frac{\partial B_z}{\partial z} + \frac{\partial^2 \beta}{\partial z^2} , \\
\sigma_{rz} &= \frac{\partial}{\partial r}\left[z\frac{\partial B_z}{\partial z} - (1-2\nu)B_z + \frac{\partial \beta}{\partial z} \right] .
\end{aligned}
\qquad (9.2.63)
$$

Thus, the boundary condition $\sigma_{z\theta} = 0$ on $\partial\mathcal{R}$ is trivially satisfied. From the boundary condition $\sigma_{zr} = 0$ on $\partial\mathcal{R}$, one obtains the condition

$$\sigma_{rz}\big|_{z=0} = \frac{\partial}{\partial r}\left[-(1-2\nu)B_z + \frac{\partial \beta}{\partial z} \right]\bigg|_{z=0} = 0 , \qquad (9.2.64)$$

which, along with (9.2.54), implies that

$$\frac{\partial \beta}{\partial z} - (1-2\nu)B_z = 0 \quad \forall \mathbf{X} \in \partial\mathcal{R} . \qquad (9.2.65)$$

From the boundary condition $\sigma_{zz} = 0$ on $\partial\mathcal{R}$,

$$\sigma_{zz}\big|_{z=0} = \left[-2(1-\nu)\frac{\partial B_z}{\partial z} + \frac{\partial^2 \beta}{\partial z^2} \right]\bigg|_{z=0} = 0 , \qquad (9.2.66)$$

from which one obtains the condition

$$\frac{\partial^2 \beta}{\partial z^2} - 2(1-\nu)\frac{\partial B_z}{\partial z} = 0 \quad \forall \mathbf{X} \in \partial \mathcal{R}. \tag{9.2.67}$$

Finally, substituting for B_z and β in (9.2.65) and (9.2.67) using the assumed forms of the Papkovich vector and scalar strain functions (9.2.59) gives

$$\left.\begin{array}{l} \dfrac{\partial \bar{\beta}}{\partial z} - (1-2\nu)\bar{B} + c\dfrac{\partial}{\partial z}\left(\dfrac{1}{\rho}\right) + (1-2\nu)\dfrac{1}{\rho} = 0 \\[2mm] \dfrac{\partial^2 \bar{\beta}}{\partial z^2} - 2(1-\nu)\dfrac{\partial \bar{B}}{\partial z} + c\dfrac{\partial^2}{\partial z^2}\left(\dfrac{1}{\rho}\right) + 2(1-\nu)\dfrac{\partial}{\partial z}\left(\dfrac{1}{\rho}\right) = 0 \end{array}\right\} \quad \forall \mathbf{X} \in \partial \mathcal{R}. \tag{9.2.68}$$

The problem now is reduced to finding scalar fields $\bar{B}$ and $\bar{\beta}$ that vanish at infinity (9.2.61), are regular harmonic in $\mathcal{R}$ (9.2.62) and satisfy the traction-free boundary conditions (9.2.68).

The approach taken is to define new functions involving $\bar{B}$ and $\bar{\beta}$ which vanish on $\partial \mathcal{R}$ as a result of (9.2.68), vanish at infinity as a result of (9.2.61), and are regular harmonic on $\mathcal{R}$. It then follows from the consequence (9.1.30) of Green's theorem that the functions must vanish identically in $\mathcal{R}$ and one can then solve for $\bar{B}$ and $\bar{\beta}$. In doing this, keep in mind that even though the scalar field $1/\rho$ (and its derivatives) is harmonic everywhere except $\mathbf{X} = c\hat{\mathbf{e}}_z$,

$$\nabla^2\left(\frac{1}{\rho}\right) = 0 \quad \forall \mathbf{X} \neq c\hat{\mathbf{e}}_z, \tag{9.2.69}$$

ρ (9.2.58) vanishes at $\mathbf{X} = c\hat{\mathbf{e}}_z$ so that $1/\rho$ and its derivatives are undefined at this point and are therefore *not* regular harmonic in $\mathcal{R}$.

To begin, define the scalar field

$$\bar{\rho} \equiv \sqrt{r^2 + (z+c)^2} \tag{9.2.70}$$

and note that $\bar{\rho}$ is nonzero everywhere in $\mathcal{R}$ and

$$\nabla^2\left(\frac{1}{\bar{\rho}}\right) = 0 \quad \forall \mathbf{X} \in \mathcal{R}. \tag{9.2.71}$$

Thus, $1/\bar{\rho}$ and its derivatives are regular harmonic in the upper half-space $\mathcal{R}$. Also, it is easily shown that

$$\begin{aligned} \left.\frac{1}{\bar{\rho}}\right|_{z=0} &= \left.\frac{1}{\rho}\right|_{z=0}, \\[2mm] \left.\frac{\partial}{\partial z}\left(\frac{1}{\bar{\rho}}\right)\right|_{z=0} &= -\left.\frac{\partial}{\partial z}\left(\frac{1}{\rho}\right)\right|_{z=0}, \\[2mm] \left.\frac{\partial^2}{\partial z^2}\left(\frac{1}{\bar{\rho}}\right)\right|_{z=0} &= \left.\frac{\partial^2}{\partial z^2}\left(\frac{1}{\rho}\right)\right|_{z=0}. \end{aligned} \tag{9.2.72}$$

With (9.2.72) in mind, one can then define the following scalar fields:

$$
\begin{aligned}
\chi_1 &\equiv \frac{\partial \bar{\beta}}{\partial z} - (1 - 2\nu)\bar{B} - c\frac{\partial}{\partial z}\left(\frac{1}{\bar{\rho}}\right) + (1 - 2\nu)\frac{1}{\bar{\rho}} \,, \\
\chi_2 &\equiv \frac{\partial^2 \bar{\beta}}{\partial z^2} - 2(1 - \nu)\frac{\partial \bar{B}}{\partial z} + c\frac{\partial^2}{\partial z^2}\left(\frac{1}{\bar{\rho}}\right) - 2(1 - \nu)\frac{\partial}{\partial z}\left(\frac{1}{\bar{\rho}}\right) \,,
\end{aligned}
\tag{9.2.73}
$$

so that the boundary conditions (9.2.68) and the identities (9.2.72) require that

$$
\chi_1 = 0 \,, \quad \chi_2 = 0 \quad \forall \mathbf{X} \in \partial \mathcal{R} \,.
\tag{9.2.74}
$$

In addition, the conditions (9.2.61) at infinity mean that

$$
\lim_{\substack{|\mathbf{X}| \to \infty \\ z \geq 0}} \chi_1 = 0 \,, \qquad \lim_{\substack{|\mathbf{X}| \to \infty \\ z \geq 0}} \chi_2 = 0
\tag{9.2.75}
$$

and the equilibrium conditions (9.2.62) mean that χ_1 and χ_2 are regular harmonic in $\mathcal{R}$:

$$
\nabla^2 \chi_1 = 0 \,, \quad \nabla^2 \chi_2 = 0 \quad \forall \mathbf{X} \in \mathcal{R} \,.
\tag{9.2.76}
$$

Thus, from the consequences (9.1.30) of Green's theorem, $\chi_1 = 0$ and $\chi_2 = 0$ everywhere in $\mathcal{R}$ so that

$$
\begin{aligned}
\frac{\partial \bar{\beta}}{\partial z} - (1 - 2\nu)\bar{B} &= c\frac{\partial}{\partial z}\left(\frac{1}{\bar{\rho}}\right) - (1 - 2\nu)\frac{1}{\bar{\rho}} \,, \\
\frac{\partial \bar{\beta}}{\partial z} - 2(1 - \nu)\bar{B} &= -c\frac{\partial}{\partial z}\left(\frac{1}{\bar{\rho}}\right) + 2(1 - \nu)\frac{1}{\bar{\rho}} \,,
\end{aligned}
\tag{9.2.77}
$$

where, in integrating χ_2 with respect to z, the resulting arbitrary additive function of r must be identically equal to zero in order for (9.2.61) and (9.2.62) to be satisfied.

Solving (9.2.77) for $\bar{B}$ and $\partial \bar{\beta}/\partial z$, one finds that

$$
\bar{B} = 2c\frac{\partial}{\partial z}\left(\frac{1}{\bar{\rho}}\right) - (3 - 4\nu)\frac{1}{\bar{\rho}} \,,
\tag{9.2.78}
$$

$$
\frac{\partial \bar{\beta}}{\partial z} = (3 - 4\nu)c\frac{\partial}{\partial z}\left(\frac{1}{\bar{\rho}}\right) - 4(1 - \nu)(1 - 2\nu)\frac{1}{\bar{\rho}} \,.
\tag{9.2.79}
$$

Integrating (9.2.79),

$$
\bar{\beta} = (3 - 4\nu)\frac{c}{\bar{\rho}} - 4(1 - \nu)(1 - 2\nu)\ln(z + c + \bar{\rho}) \,,
\tag{9.2.80}
$$

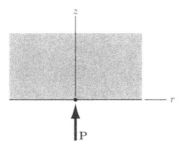

FIGURE 9.5.   Concentrated normal traction acting at the boundary of a
semi-infinite body.

where the constant of integration has been set equal to zero without loss of
generality, since from (9.2.59) and (9.2.60), it can be seen that it will have
no bearing on the displacement field. Therefore, substituting (9.2.78) and
(9.2.80) into the assumed forms (9.2.59), one finds that the Papkovich vector and scalar strain functions for the special axisymmetric case of Mindlin's
problem are given by $B_r = B_\theta = 0$ and

$$
B_z = -\frac{P}{8\pi(1-\nu)}\left[\frac{1}{\rho} + \frac{3-4\nu}{\bar{\rho}} + \frac{2c(z+c)}{\bar{\rho}^3}\right] ,
$$

$$
\beta = \frac{P}{8\pi(1-\nu)}\left[\frac{c}{\rho} + \frac{(3-4\nu)c}{\bar{\rho}} - 4(1-\nu)(1-2\nu)\ln(z+c+\bar{\rho})\right] ,
$$

(9.2.81)

where $\rho = \sqrt{r^2 + (z-c)^2}$ and $\bar{\rho} = \sqrt{r^2 + (z+c)^2}$. The corresponding scalar components of displacement and stress can be determined via
(9.2.60) and (9.2.63), although the resulting expressions are rather cum-
bersome.

   In the limit as $c \to 0$, the concentrated body force $\mathbf{P} = P\hat{\mathbf{e}}_z$ acting at a
point $\boldsymbol{\xi} = c\hat{\mathbf{e}}_z$ within the semi-infinite body becomes a concentrated normal
traction acting at a point on the boundary $\partial\mathcal{R}$ of the body (Figure 9.5).
This is known as *Boussinesq's problem* and its solution is given by (9.2.81)
with $c = 0$ so that $B_r = B_\theta = 0$ and

$$
B_z = -\frac{P}{2\pi R} , \quad \beta = -\frac{(1-2\nu)P}{2\pi}\ln(z+R) ,
$$

(9.2.82)

where $R = \rho|_{c=0} = \bar{\rho}|_{c=0} = \sqrt{r^2 + z^2}$ is the distance from the origin, which
is the point of application of the concentrated normal traction. It follows

from (9.2.60) that the displacement field for Boussinesq's problem is given by $u_\theta = 0$ and

$$
\begin{aligned}
u_r &= \frac{Pr}{4\pi\mu}\left[\frac{z}{R^3} - \frac{1-2\nu}{R(z+R)}\right], \\
u_z &= \frac{P}{4\pi\mu}\left[\frac{2(1-\nu)}{R} + \frac{z^2}{R^3}\right].
\end{aligned}
\tag{9.2.83}
$$

The stress field is given by (9.2.63) to be $\sigma_{r\theta} = \sigma_{\theta z} = 0$ and

$$
\begin{aligned}
\sigma_{rr} &= \frac{P}{2\pi}\left[\frac{1-2\nu}{R(z+R)} - \frac{3r^2 z}{R^5}\right], \\
\sigma_{\theta\theta} &= \frac{P(1-2\nu)}{2\pi}\left[\frac{z}{R^3} - \frac{1}{R(z+R)}\right], \\
\sigma_{zz} &= -\frac{3Pz^3}{2\pi R^5}, \\
\sigma_{rz} &= -\frac{3Prz^2}{2\pi R^5}.
\end{aligned}
\tag{9.2.84}
$$

It is interesting to note that the mean stress $\tilde{\sigma} = \frac{1}{3}(\sigma_{rr} + \sigma_{\theta\theta} + \sigma_{zz})$ for Boussinesq's problem is

$$
\tilde{\sigma} = -\frac{(1+\nu)Pz}{3\pi R^3}.
\tag{9.2.85}
$$

The companion to Boussinesq's problem, in which a semi-infinite body is subjected to a concentrated shear traction instead of a normal traction, is known as *Cerruti's problem*. For a discussion of this and related problems, see Johnson (1985) and Fung (1965).

9.2.5   Fundamental Problem

The solution to Kelvin's problem can be used in conjunction with the superposition principle and the Betti-Rayleigh reciprocity relations to derive solutions for some other problems with important physical significance. In particular, the displacement field due to a dislocation surface and the stress field in and around an inclusion of a dissimilar material will be considered shortly. In anticipation of this, a basic fundamental problem will be considered first.

From now on, let the previously derived solution to Kelvin's problem [i.e., the displacement field (9.2.33) and the stress field (9.2.37) due to a concentrated body force $\mathbf{P}$ acting at a point $\boldsymbol{\xi}$ in an infinite body] be denoted by $\overset{*}{\mathbf{u}}$ and $\overset{*}{\boldsymbol{\sigma}}$, respectively. Stated explicitly,

$$
\overset{*}{\mathbf{u}}(\mathbf{X};\mathbf{P},\boldsymbol{\xi}) \equiv \frac{1}{16\pi\mu(1-\nu)}\left[(3-4\nu)\frac{\mathbf{P}}{\rho} + \frac{(\mathbf{P}\cdot\mathbf{r}^\xi)\mathbf{r}^\xi}{\rho^3}\right]
\tag{9.2.86}
$$

and

$$\overset{*}{\sigma}(\mathbf{X};\mathbf{P},\boldsymbol{\xi}) \equiv \frac{1}{8\pi(1-\nu)}\left[(1-2\nu)\frac{(\mathbf{P}\cdot\mathbf{r}^{\xi})\mathbf{I} - \mathbf{P}\mathbf{r}^{\xi} - \mathbf{r}^{\xi}\mathbf{P}}{\rho^3} - \frac{3(\mathbf{P}\cdot\mathbf{r}^{\xi})\mathbf{r}^{\xi}\mathbf{r}^{\xi}}{\rho^5}\right],$$

$$(9.2.87)$$

where $\mathbf{r}^{\xi}(\mathbf{X}) \equiv \mathbf{X} - \boldsymbol{\xi}$ and $\rho(\mathbf{X}) \equiv |\mathbf{r}^{\xi}(\mathbf{X})|$. When necessary, the dependence of $\overset{*}{\mathbf{u}}$ and $\overset{*}{\sigma}$ on $\mathbf{X}$, $\mathbf{P}$, and $\boldsymbol{\xi}$ will be emphasized by writing $\overset{*}{\mathbf{u}}(\mathbf{X};\mathbf{P},\boldsymbol{\xi})$ and $\overset{*}{\sigma}(\mathbf{X};\mathbf{P},\boldsymbol{\xi})$.

Consider a body, subject to negligible body force, in which the displacement field $\mathbf{u}$ and stress field σ, corresponding to some unspecified boundary conditions, satisfy all of the equations of elasticity. Let $\mathcal{R}$ be an arbitrary region of the body with boundary $\partial\mathcal{R}$. Derive an expression that relates the traction distribution $\mathbf{t} = \hat{\mathbf{n}} \cdot \sigma$ and the displacement distribution on $\partial\mathcal{R}$, where $\hat{\mathbf{n}}$ is the unit outward normal to $\partial\mathcal{R}$.

This problem can be approached through the use of the Betti-Rayleigh reciprocity relation (6.3.14):

$$\int_{\partial\mathcal{R}} \mathbf{t}^{(1)} \cdot \mathbf{u}^{(2)}\, dA + \int_{\mathcal{R}} \mathbf{f}^{(1)} \cdot \mathbf{u}^{(2)}\, dV = \int_{\partial\mathcal{R}} \mathbf{t}^{(2)} \cdot \mathbf{u}^{(1)}\, dA + \int_{\mathcal{R}} \mathbf{f}^{(2)} \cdot \mathbf{u}^{(1)}\, dV .$$

$$(9.2.88)$$

Let equilibrium state (1) be that corresponding to the given displacement and stress fields, so that

$$\mathbf{t}^{(1)} = \mathbf{t}, \quad \mathbf{f}^{(1)} = \mathbf{0}, \quad \mathbf{u}^{(1)} = \mathbf{u}, \tag{9.2.89}$$

and let equilibrium state (2) be that given by the solution $\overset{*}{\mathbf{u}}(\mathbf{X};\mathbf{P},\boldsymbol{\xi})$ and $\overset{*}{\sigma}(\mathbf{X};\mathbf{P},\boldsymbol{\xi})$ to Kelvin's problem of a concentrated body force $\mathbf{P}$ acting at $\boldsymbol{\xi}$ in an *infinite* body, so that

$$\mathbf{t}^{(2)} = \overset{*}{\mathbf{t}}, \quad \mathbf{f}^{(2)} = \mathbf{P}\delta(\mathbf{X} - \boldsymbol{\xi}), \quad \mathbf{u}^{(2)} = \overset{*}{\mathbf{u}}, \tag{9.2.90}$$

where $\overset{*}{\mathbf{t}} \equiv \hat{\mathbf{n}} \cdot \overset{*}{\sigma}$. Do not let the fact that the fields $\overset{*}{\mathbf{u}}$ and $\overset{*}{\sigma}$ were derived for a problem with an infinite body and the current problem involves a body that may not be infinite confuse you. Regardless of whether or not $\boldsymbol{\xi} \in \mathcal{R}$, the system satisfies equilibrium everywhere in $\mathcal{R}$ and thus is suitable for use in the Betti-Rayleigh reciprocity relations.

The Betti-Rayleigh reciprocity relations thus give

$$\int_{\partial\mathcal{R}} \mathbf{t} \cdot \overset{*}{\mathbf{u}}\, dA_{\mathbf{X}} = \int_{\partial\mathcal{R}} \overset{*}{\mathbf{t}} \cdot \mathbf{u}\, dA_{\mathbf{X}} + \int_{\mathcal{R}} \mathbf{P} \cdot \mathbf{u}\delta(\mathbf{X} - \boldsymbol{\xi})\, dV_{\mathbf{X}} . \tag{9.2.91}$$

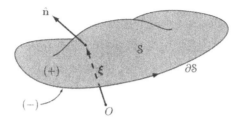

FIGURE 9.6. Dislocation surface $\mathcal{S}$ in an infinite body.

where the notation $dA_{\mathbf{X}}$ and $dV_{\mathbf{X}}$ is a reminder that $\mathbf{X}$ is the variable of integration. However, from the definition of the Dirac delta function,

$$\int_{\mathcal{R}} \mathbf{P} \cdot \mathbf{u}\delta(\mathbf{X} - \boldsymbol{\xi})\, dV_{\mathbf{X}} = \begin{cases} \mathbf{P} \cdot \mathbf{u}(\boldsymbol{\xi}) & \boldsymbol{\xi} \in \mathcal{R} \\ \frac{1}{2}\mathbf{P} \cdot \mathbf{u}(\boldsymbol{\xi}) & \boldsymbol{\xi} \in \partial\mathcal{R} \\ 0 & \boldsymbol{\xi} \notin (\mathcal{R} \cup \partial\mathcal{R}). \end{cases} \tag{9.2.92}$$

Therefore, since $\mathbf{t} = \hat{\mathbf{n}} \cdot \boldsymbol{\sigma}$ and $\overset{*}{\mathbf{t}} = \hat{\mathbf{n}} \cdot \overset{*}{\boldsymbol{\sigma}}$, where $\hat{\mathbf{n}}$ is the unit outward normal to $\partial\mathcal{R}$,

$$\boxed{\int_{\partial\mathcal{R}} \hat{\mathbf{n}} \cdot (\boldsymbol{\sigma} \cdot \overset{*}{\mathbf{u}} - \overset{*}{\boldsymbol{\sigma}} \cdot \mathbf{u})\, dA_{\mathbf{X}} = \begin{cases} \mathbf{P} \cdot \mathbf{u}(\boldsymbol{\xi}) & \boldsymbol{\xi} \in \mathcal{R} \\ \frac{1}{2}\mathbf{P} \cdot \mathbf{u}(\boldsymbol{\xi}) & \boldsymbol{\xi} \in \partial\mathcal{R} \\ 0 & \boldsymbol{\xi} \notin (\mathcal{R} \cup \partial\mathcal{R}). \end{cases}} \tag{9.2.93}$$

Keep in mind that one is free to choose values for $\mathbf{P}$ and $\boldsymbol{\xi}$ arbitrarily and that $\overset{*}{\mathbf{u}}(\mathbf{X}; \mathbf{P}, \boldsymbol{\xi})$ and $\overset{*}{\boldsymbol{\sigma}}(\mathbf{X}; \mathbf{P}, \boldsymbol{\xi})$ are known.

9.3   Dislocation Surface

Consider an infinite body in which the displacement field is continuous everywhere except across a prescribed open surface $\mathcal{S}$, across which there is a uniform displacement discontinuity (Figure 9.6). Imagine that the body has been cut along $\mathcal{S}$ and that along this cut, material has been added or removed and the two sides of the cut have been moved laterally relative to each other as necessary so that when the two sides are "welded" back together, one side has been displaced relative to the other by a vector amount $\mathbf{b}$ that is the same at every point on $\mathcal{S}$.

The surface $\mathcal{S}$ is called a *dislocation surface* and its boundary $\partial\mathcal{S}$ is called a *dislocation loop*. They are related to the crystallographic imperfections

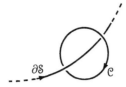

FIGURE 9.7.   Closed curve $\mathcal{C}$ used to define the Berger's vector.

known as *dislocations* that play an integral part in the plastic deformation of crystalline materials (Hirth and Lothe 1992).

Let the positive $(+)$ and negative $(-)$ sides of $\mathcal{S}$ be defined relative to the direction of positive progression around $\partial\mathcal{S}$ as shown in Figure 9.6—when facing the positive side of $\mathcal{S}$, positive progression around $\partial\mathcal{S}$ is counterclockwise. The unit normal $\hat{\mathbf{n}}$ to $\mathcal{S}$ at a point on $\boldsymbol{\xi} \in \mathcal{S}$ is positive as shown, emanating from the positive side of the surface.

The displacement discontinuity $\mathbf{b}$ across $\mathcal{S}$ is known as the *Berger's vector* and can be defined by

$$\mathbf{b} \equiv \oint_{\mathcal{C}} d\mathbf{u}\,, \tag{9.3.1}$$

where the closed curve $\mathcal{C}$ is any closed curve through which the dislocation loop passes once (and only once). The direction of integration around $\mathcal{C}$ is as shown in Figure 9.7. In other words,

$$\boxed{\mathbf{u}(\boldsymbol{\xi}^-) - \mathbf{u}(\boldsymbol{\xi}^+) = \mathbf{b} \quad \forall \boldsymbol{\xi} \in \mathcal{S}\,,} \tag{9.3.2}$$

where the Berger's vector $\mathbf{b}$ is a constant and $f(\boldsymbol{\xi}^-)$ and $f(\boldsymbol{\xi}^+)$ are the limiting values of $f(\mathbf{X})$ as $\mathbf{X} \to \boldsymbol{\xi}$ from the negative and positive sides of $\mathcal{S}$, respectively.

The stress field will also be discontinuous across $\mathcal{S}$, in general, but the traction exerted across $\mathcal{S}$ by the material on the positive side against the material on the negative side is equal and opposite to that exerted by the material on the negative side against the material on the positive side. Letting $\hat{\mathbf{n}}^+$ be the unit outward normal to $\mathcal{S}$ for the material on the positive side and $\hat{\mathbf{n}}^-$ be the unit outward normal to $\mathcal{S}$ for the material on the negative side, this traction continuity condition says that

$$\hat{\mathbf{n}}^+ \cdot \boldsymbol{\sigma}(\boldsymbol{\xi}^+) = -\hat{\mathbf{n}}^- \cdot \boldsymbol{\sigma}(\boldsymbol{\xi}^-) \quad \forall \boldsymbol{\xi} \in \mathcal{S}\,. \tag{9.3.3}$$

Thus, by noting that $\hat{\mathbf{n}}^+ = -\hat{\mathbf{n}}$ and $\hat{\mathbf{n}}^- = \hat{\mathbf{n}}$ (Figure 9.6), it is seen that the traction continuity condition across $\mathcal{S}$ requires that

$$\boxed{\hat{\mathbf{n}} \cdot [\boldsymbol{\sigma}(\boldsymbol{\xi}^+) - \boldsymbol{\sigma}(\boldsymbol{\xi}^-)] = \mathbf{0} \quad \forall \boldsymbol{\xi} \in \mathcal{S}\,.} \tag{9.3.4}$$

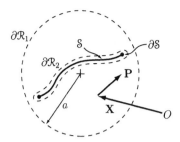

FIGURE 9.8.   Boundary $\partial \mathcal{R}' = \partial \mathcal{R}_1 \cup \partial \mathcal{R}_2$ of the region $\mathcal{R}'$.

Finally, assume that the displacement field vanishes at infinity,

$$\lim_{|\mathbf{X}| \to \infty} \mathbf{u} = \mathbf{0} \,, \tag{9.3.5}$$

which means that the stress field will vanish at infinity as well. The problem then is defined by the continuity conditions (9.3.2) and (9.3.4) and the condition (9.3.5) at infinity.

Consider a portion of the body occupying a region $\mathcal{R}'$ that is bounded by two surfaces: a spherical surface $\partial \mathcal{R}_1$ that is centered at a point in the vicinity of $\mathcal{S}$ and has radius a that is large enough that $\mathcal{S}$ is entirely within $\partial \mathcal{R}_1$ and a surface $\partial \mathcal{R}_2$ that closely envelopes the dislocation surface $\mathcal{S}$ (Figure 9.8). From the result (9.2.93) of the fundamental problem considered earlier (with the roles of $\mathbf{X}$ and $\boldsymbol{\xi}$ switched and the independent variables given explicitly), if $\mathbf{X} \in \mathcal{R}'$, then

$$\int_{\partial \mathcal{R}'} \hat{\mathbf{n}}'(\boldsymbol{\xi}) \cdot \left[\boldsymbol{\sigma}(\boldsymbol{\xi}) \cdot \overset{*}{\mathbf{u}}(\boldsymbol{\xi}; \mathbf{P}, \mathbf{X}) - \overset{*}{\boldsymbol{\sigma}}(\boldsymbol{\xi}; \mathbf{P}, \mathbf{X}) \cdot \mathbf{u}(\boldsymbol{\xi}) \right] dA_{\boldsymbol{\xi}} = \mathbf{P} \cdot \mathbf{u} \,, \tag{9.3.6}$$

where $\partial \mathcal{R}' = \partial \mathcal{R}_1 \cup \partial \mathcal{R}_2$ is the boundary of $\mathcal{R}'$, $\hat{\mathbf{n}}'(\boldsymbol{\xi})$ is the unit outward normal to $\partial \mathcal{R}'$ at the point $\boldsymbol{\xi} \in \partial \mathcal{R}'$, $\overset{*}{\mathbf{u}}(\boldsymbol{\xi}; \mathbf{P}, \mathbf{X})$ and $\overset{*}{\boldsymbol{\sigma}}(\boldsymbol{\xi}; \mathbf{P}, \mathbf{X})$ are the displacement and stress at a point $\boldsymbol{\xi}$ in an infinite body due to a concentrated body force $\mathbf{P}$ acting at $\mathbf{X}$, and $\mathbf{u}$ and $\boldsymbol{\sigma}$ are the displacement and stress fields due to the dislocation surface.

Let the radius a of the spherical surface $\partial \mathcal{R}_1$ go to infinity. Each of $\mathbf{u}$, $\boldsymbol{\sigma}$, $\overset{*}{\mathbf{u}}$, and $\overset{*}{\boldsymbol{\sigma}}$ vanishes in the limit, so

$$\lim_{a \to \infty} \int_{\partial \mathcal{R}_1} \hat{\mathbf{n}}'(\boldsymbol{\xi}) \cdot \left[\boldsymbol{\sigma}(\boldsymbol{\xi}) \cdot \overset{*}{\mathbf{u}}(\boldsymbol{\xi}; \mathbf{P}, \mathbf{X}) - \overset{*}{\boldsymbol{\sigma}}(\boldsymbol{\xi}; \mathbf{P}, \mathbf{X}) \cdot \mathbf{u}(\boldsymbol{\xi}) \right] dA_{\boldsymbol{\xi}} = 0 \tag{9.3.7}$$

and, therefore,

$$\int_{\partial \mathcal{R}_2} \hat{\mathbf{n}}'(\boldsymbol{\xi}) \cdot \left[\boldsymbol{\sigma}(\boldsymbol{\xi}) \cdot \overset{*}{\mathbf{u}}(\boldsymbol{\xi}; \mathbf{P}, \mathbf{X}) - \overset{*}{\boldsymbol{\sigma}}(\boldsymbol{\xi}; \mathbf{P}, \mathbf{X}) \cdot \mathbf{u}(\boldsymbol{\xi}) \right] dA_{\boldsymbol{\xi}} = \mathbf{P} \cdot \mathbf{u} , \quad (9.3.8)$$

Now, let the surface $\partial \mathcal{R}_2$ shrink until it conforms perfectly to the positive and negative sides of $\mathcal{S}$, so that

$$\int_{\partial \mathcal{R}_2} \psi(\boldsymbol{\xi}) \, dA_{\boldsymbol{\xi}} = \int_{\mathcal{S}} \psi(\boldsymbol{\xi}^+) \, dA_{\boldsymbol{\xi}} + \int_{\mathcal{S}} \psi(\boldsymbol{\xi}^-) \, dA_{\boldsymbol{\xi}}$$

$$= \int_{\mathcal{S}} [\psi(\boldsymbol{\xi}^+) + \psi(\boldsymbol{\xi}^-)] \, dA_{\boldsymbol{\xi}} . \quad (9.3.9)$$

Since the displacement field from Kelvin's problem is continuous across $\mathcal{S}$, $\overset{*}{\mathbf{u}}(\boldsymbol{\xi}^+; \mathbf{P}, \mathbf{X}) = \overset{*}{\mathbf{u}}(\boldsymbol{\xi}^-; \mathbf{P}, \mathbf{X}) = \overset{*}{\mathbf{u}}(\boldsymbol{\xi}; \mathbf{P}, \mathbf{X})$, and noting that $\hat{\mathbf{n}}'(\boldsymbol{\xi}^+) = -\hat{\mathbf{n}}(\boldsymbol{\xi})$ and $\hat{\mathbf{n}}'(\boldsymbol{\xi}^-) = \hat{\mathbf{n}}(\boldsymbol{\xi})$, where $\hat{\mathbf{n}}(\boldsymbol{\xi})$ is the unit normal to $\mathcal{S}$ at $\boldsymbol{\xi} \in \mathcal{S}$ as shown in Figure 9.6, it follows from the traction continuity condition (9.3.4) that

$$\int_{\partial \mathcal{R}_2} \hat{\mathbf{n}}'(\boldsymbol{\xi}) \cdot \boldsymbol{\sigma}(\boldsymbol{\xi}) \cdot \overset{*}{\mathbf{u}}(\boldsymbol{\xi}; \mathbf{P}, \mathbf{X}) \, dA_{\boldsymbol{\xi}}$$

$$= -\int_{\mathcal{S}} \hat{\mathbf{n}}(\boldsymbol{\xi}) \cdot [\boldsymbol{\sigma}(\boldsymbol{\xi}^+) - \boldsymbol{\sigma}(\boldsymbol{\xi}^-)] \cdot \overset{*}{\mathbf{u}}(\boldsymbol{\xi}; \mathbf{P}, \mathbf{X}) \, dA_{\boldsymbol{\xi}} = 0 . \quad (9.3.10)$$

Since the stress field from Kelvin's problem is continuous across $\mathcal{S}$,

$$\overset{*}{\boldsymbol{\sigma}}(\boldsymbol{\xi}^+; \mathbf{P}, \mathbf{X}) = \overset{*}{\boldsymbol{\sigma}}(\boldsymbol{\xi}^-; \mathbf{P}, \mathbf{X}) = \overset{*}{\boldsymbol{\sigma}}(\boldsymbol{\xi}; \mathbf{P}, \mathbf{X}) , \quad (9.3.11)$$

it follows that

$$\int_{\partial \mathcal{R}_2} \hat{\mathbf{n}}'(\boldsymbol{\xi}) \cdot \overset{*}{\boldsymbol{\sigma}}(\boldsymbol{\xi}; \mathbf{P}, \mathbf{X}) \cdot \mathbf{u}(\boldsymbol{\xi}) \, dA_{\boldsymbol{\xi}}$$

$$= \int_{\mathcal{S}} \hat{\mathbf{n}}(\boldsymbol{\xi}) \cdot \overset{*}{\boldsymbol{\sigma}}(\boldsymbol{\xi}; \mathbf{P}, \mathbf{X}) \cdot [\mathbf{u}(\boldsymbol{\xi}^-) - \mathbf{u}(\boldsymbol{\xi}^+)] \, dA_{\boldsymbol{\xi}} . \quad (9.3.12)$$

Finally, substituting (9.3.10) and (9.3.12) into (9.3.8) and using the definition (9.3.2) of the Berger's vector $\mathbf{b}$, it follows that

$$\mathbf{P} \cdot \mathbf{u} = -\int_{\mathcal{S}} \hat{\mathbf{n}}(\boldsymbol{\xi}) \cdot \overset{*}{\boldsymbol{\sigma}}(\boldsymbol{\xi}; \mathbf{P}, \mathbf{X}) \cdot \mathbf{b} \, dA_{\boldsymbol{\xi}} . \quad (9.3.13)$$

where $\mathbf{P}$ is arbitrary, except that $\mathbf{P} \neq \mathbf{0}$.

Substituting (9.2.87) for the stress $\overset{*}{\boldsymbol{\sigma}}$ in (9.3.13) (with the roles of $\mathbf{X}$ and $\boldsymbol{\xi}$ reversed) and factoring out the constant $\mathbf{P}$ gives

$$
\mathbf{P} \cdot \mathbf{u} = -\,\mathbf{P} \cdot \int_S \frac{1}{8\pi(1-\nu)} \left[(1-2\nu)\frac{\mathbf{r}^{\mathbf{X}}(\hat{\mathbf{n}} \cdot \mathbf{b}) - \hat{\mathbf{n}}(\mathbf{r}^{\mathbf{X}} \cdot \mathbf{b}) - \mathbf{b}(\hat{\mathbf{n}} \cdot \mathbf{r}^{\mathbf{X}})}{\rho^3} \right.
$$
$$
\left. -\,\frac{3\mathbf{r}^{\mathbf{X}}(\hat{\mathbf{n}} \cdot \mathbf{r}^{\mathbf{X}})(\mathbf{r}^{\mathbf{X}} \cdot \mathbf{b})}{\rho^5} \right] dA_{\boldsymbol{\xi}} \; ,
$$

$$\tag{9.3.14}$$

where

$$
\mathbf{r}^{\mathbf{X}}(\boldsymbol{\xi}) \equiv \boldsymbol{\xi} - \mathbf{X} \tag{9.3.15}
$$

is the relative position vector of the point $\boldsymbol{\xi}$ on the surface S with respect to the point $\mathbf{X}$ and

$$
\rho(\boldsymbol{\xi}) \equiv |\mathbf{r}^{\mathbf{X}}(\boldsymbol{\xi})| \; . \tag{9.3.16}
$$

Since $\mathbf{P}$ is arbitrary, it follows that the displacement field due to the dislocation surface is

$$
\mathbf{u} = \int_S \frac{1}{8\pi(1-\nu)} \left[(1-2\nu)\frac{\hat{\mathbf{n}}(\mathbf{r}^{\mathbf{X}} \cdot \mathbf{b}) + \mathbf{b}(\hat{\mathbf{n}} \cdot \mathbf{r}^{\mathbf{X}}) - \mathbf{r}^{\mathbf{X}}(\hat{\mathbf{n}} \cdot \mathbf{b})}{\rho^3} \right.
$$
$$
\left. +\,\frac{3\mathbf{r}^{\mathbf{X}}(\hat{\mathbf{n}} \cdot \mathbf{r}^{\mathbf{X}})(\mathbf{r}^{\mathbf{X}} \cdot \mathbf{b})}{\rho^5} \right] dA_{\boldsymbol{\xi}} \; ,
$$

$$\tag{9.3.17}$$

It is known from the problem statement that this displacement field should have a discontinuity across S equal to the Berger's vector $\mathbf{b}$. It should therefore be possible to isolate this discontinuous component of $\mathbf{u}$ from any remaining continuous component.

To rewrite (9.3.17) in terms of its continuous and discontinuous components, note first that for an arbitrary vector field $\mathbf{v}$ defined in some region containing a regular surface S that is bounded by a simple closed curve ∂S, Stoke's theorem (2.4.66) can be used to show that

$$
\oint_{\partial S} \mathbf{v} \times d\mathbf{x} = \oint_{\partial S} e_{piq} v_p \hat{\mathbf{e}}_q \, dx_i
$$
$$
= \int_S e_{ijk} e_{pjq} v_{p,i} \, \hat{\mathbf{e}}_q \hat{n}_k \, dA
$$
$$
= \int_S (\delta_{ip}\delta_{kq} - \delta_{iq}\delta_{kp}) v_{p,i} \, \hat{\mathbf{e}}_q \hat{n}_k \, dA
$$
$$
= \int_S (v_{i,i}\, \hat{n}_q - v_{p,q}\, \hat{n}_p) \hat{\mathbf{e}}_q \, dA \; , \tag{9.3.18}
$$

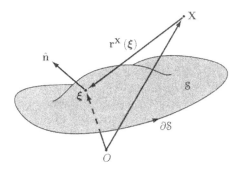

FIGURE 9.9. Relative position vector used in the solution for the displacement due to a dislocation vector.

where $\mathbf{x}$ is the position vector of a point in space. Thus, an alternative form of Stoke's theorem is

$$\int_S [\hat{\mathbf{n}}(\boldsymbol{\nabla} \cdot \mathbf{v}) - \hat{\mathbf{n}} \cdot (\boldsymbol{\nabla}\mathbf{v})]\, dA = \oint_{\partial S} \mathbf{v} \times d\mathbf{x}\,, \qquad (9.3.19)$$

where $\hat{\mathbf{n}}$ is the unit normal to the surface S.

Let $\boldsymbol{\nabla}$ and $\boldsymbol{\nabla}_{\boldsymbol{\xi}}$ denote differentiation with respect to $\mathbf{X}$ and $\boldsymbol{\xi}$, respectively, and recall that $\hat{\mathbf{n}}(\boldsymbol{\xi})$ is independent of $\mathbf{X}$. Then, it can be shown that the displacement field (9.3.17) can be rewritten as

$$\mathbf{u} = \int_S \left\{ \frac{1}{4\pi} \frac{\mathbf{b}(\mathbf{r}^{\mathbf{X}} \cdot \hat{\mathbf{n}})}{\rho^3} - \frac{1}{4\pi} \left[\hat{\mathbf{n}}\left(\boldsymbol{\nabla}_{\boldsymbol{\xi}} \cdot \frac{\hat{\mathbf{n}}}{\rho} \right) - \hat{\mathbf{n}} \cdot \left(\boldsymbol{\nabla}_{\boldsymbol{\xi}} \frac{\hat{\mathbf{n}}}{\rho} \right) \right] \right.$$
$$\left. + \frac{1}{8\pi(1-\nu)} \boldsymbol{\nabla} \left\{ \mathbf{b} \cdot \left[\hat{\mathbf{n}}\left(\boldsymbol{\nabla}_{\boldsymbol{\xi}} \cdot \frac{\mathbf{r}^{\mathbf{X}}}{\rho} \right) - \hat{\mathbf{n}} \cdot \left(\boldsymbol{\nabla}_{\boldsymbol{\xi}} \frac{\mathbf{r}^{\mathbf{X}}}{\rho} \right) \right] \right\} \right\} dA_{\boldsymbol{\xi}}\,.$$
$$(9.3.20)$$

Therefore, using the alternate form of Stoke's theorem (9.3.19),

$$\mathbf{u} = \frac{\mathbf{b}}{4\pi} \int_S \frac{\mathbf{r}^{\mathbf{X}} \cdot \hat{\mathbf{n}}}{\rho^3}\, dA_{\boldsymbol{\xi}} - \frac{1}{4\pi} \oint_{\partial S} \frac{\mathbf{b}}{\rho} \times d\boldsymbol{\xi} + \frac{1}{8\pi(1-\nu)} \boldsymbol{\nabla} \left(\mathbf{b} \cdot \oint_{\partial S} \frac{\mathbf{r}^{\mathbf{X}}}{\rho} \times d\boldsymbol{\xi} \right)\,.$$
$$(9.3.21)$$

Finally, by recalling that $\mathbf{u} \cdot (\mathbf{v} \times \mathbf{w}) = (\mathbf{u} \times \mathbf{v}) \cdot \mathbf{w}$, it follows that the displacement field due to a dislocation surface S with boundary ∂S and

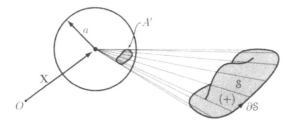

FIGURE 9.10. Spherical projection of S onto a sphere of radius a.

Berger's vector $\mathbf{b}$ can be expressed as

$$\mathbf{u} = -\frac{\mathbf{b}}{4\pi}\Omega(\mathbf{X}) - \frac{1}{4\pi}\oint_{\partial S}\frac{\mathbf{b}\times d\boldsymbol{\xi}}{\rho} + \frac{1}{8\pi(1-\nu)}\nabla\oint_{\partial S}\frac{(\mathbf{b}\times\mathbf{r}^{\mathbf{X}})\cdot d\boldsymbol{\xi}}{\rho},$$

$$u_i = -\frac{b_i}{4\pi}\Omega(\mathbf{X}) - \frac{1}{4\pi}\oint_{\partial S}\frac{e_{ijk}b_j\,d\xi_k}{\rho} + \frac{1}{8\pi(1-\nu)}\frac{\partial}{\partial X_i}\oint_{\partial S}\frac{e_{pqr}b_p r_q^{\mathbf{X}}\,d\xi_r}{\rho},$$

$$(9.3.22)$$

where

$$\Omega(\mathbf{X}) \equiv -\int_S \frac{\mathbf{r}^{\mathbf{X}}\cdot\hat{\mathbf{n}}}{\rho^3}\,dA_{\boldsymbol{\xi}}, \quad \Omega(\mathbf{X}) \equiv -\int_S \frac{r_i^{\mathbf{X}}\hat{n}_i}{\rho^3}\,dA_{\boldsymbol{\xi}}.$$

$$(9.3.23)$$

Recall that $\mathbf{r}^{\mathbf{X}}(\boldsymbol{\xi}) \equiv \boldsymbol{\xi} - \mathbf{X}$ is the relative position vector for a point $\boldsymbol{\xi}$ on the surface S relative to the point $\mathbf{X}$ (Figure 9.9) and that $\rho(\boldsymbol{\xi}) = |\mathbf{r}^{\mathbf{X}}(\boldsymbol{\xi})|$ is the distance between the two points.

The second and third terms in (9.3.22) are undefined for positions $\mathbf{X}$ on the boundary ∂S, since their integrands have singularities that are not integrable as $\boldsymbol{\xi} \to \mathbf{X}$ (and $\rho \to 0$). These terms are continuous everywhere else. The function $\Omega(\mathbf{X})$ in the first term is the *solid angle* at $\mathbf{X}$ of the surface S, which is discontinuous across S.

9.3.1 Solid Angle of a Surface

Consider a sphere of radius a centered at $\mathbf{X}$, with a sufficiently small so that the surface S is entirely outside of the sphere. Let rays join all of the points in S to the point at $\mathbf{X}$. The intersection of these rays with the spherical surface define the *spherical projection* of S onto the sphere, with the area of the projection corresponding to rays emanating from the (+) side of S being positive and the area corresponding to rays emanating from the

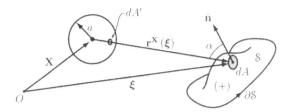

FIGURE 9.11. Spherical projection of an infinitesimal element dA of the surface S.

$(-)$ side being negative (Figure 9.10). Let A' be the total projected area. Then, the *solid angle* at $\mathbf{X}$ of the surface S is defined to be A'/a^2. This solid angle is independent of the radius a of the sphere, and the maximum possible solid angle is 4π (and the minimum possible solid angle is -4π).

The spherical projection dA' of an infinitesimal element dA of the surface S, at a point $\boldsymbol{\xi} \in S$, is

$$dA' = \frac{a^2}{\rho^2} \, dA \cos\alpha \, , \qquad (9.3.24)$$

where $\rho = |\mathbf{r}^{\mathbf{X}}|$ and α is the angle formed by the relative position vector $-\mathbf{r}^{\mathbf{X}}$ and the unit normal vector $\hat{\mathbf{n}}$ as shown in Figure 9.11, so that

$$\cos\alpha = \frac{-\mathbf{r}^{\mathbf{X}} \cdot \hat{\mathbf{n}}}{|\mathbf{r}^{\mathbf{X}}||\hat{\mathbf{n}}|} = -\frac{\mathbf{r}^{\mathbf{X}} \cdot \hat{\mathbf{n}}}{\rho} \, . \qquad (9.3.25)$$

Thus, the solid angle at $\mathbf{X}$ of the surface S is

$$\frac{A'}{a^2} = \int_S \frac{dA'}{a^2} = -\int_S \frac{\mathbf{r}^{\mathbf{X}} \cdot \hat{\mathbf{n}}}{\rho^3} \, dA \, . \qquad (9.3.26)$$

Comparing (9.3.23) and (9.3.26), it is seen that $\Omega(\mathbf{X})$ is the solid angle at $\mathbf{X}$ of the surface S.

A few moments of contemplation should convince one that if $\boldsymbol{\xi} \in S$ is a point on the surface, $\Omega(\boldsymbol{\xi}^+)$ is the limiting value of the solid angle $\Omega(\mathbf{X})$ as $\mathbf{X}$ approaches $\boldsymbol{\xi}$ from the positive side of S, and $\Omega(\boldsymbol{\xi}^-)$ is the limiting value when approached from the negative side, then

$$\Omega(\boldsymbol{\xi}^+) - \Omega(\boldsymbol{\xi}^-) = 4\pi \, . \qquad (9.3.27)$$

It follows from (9.3.22) that the discontinuity in the displacement field across S is

$$\mathbf{u}(\boldsymbol{\xi}^+) - \mathbf{u}(\boldsymbol{\xi}^-) = -\frac{\mathbf{b}}{4\pi}[\Omega(\boldsymbol{\xi}^+) - \Omega(\boldsymbol{\xi}^-)] = -\mathbf{b} \, , \qquad (9.3.28)$$

which is in agreement with the definition (9.3.2) of the Berger's vector $\mathbf{b}$.

9.4   Eshelby's Inclusion Problems

Consider an infinite body $\mathbb{B}$ of which a portion $\mathbb{B}^{\mathrm{incl}} \subset \mathbb{B}$ has some distinguishing characteristic that differentiates it from the rest of the body. The material volume $\mathbb{B}^{\mathrm{incl}}$ is referred to as an *inclusion* and the rest of the infinite body as the *matrix*. Let $\mathcal{R}^{\mathrm{incl}}$ be the region occupied by the reference configuration of the inclusion, with boundary $\partial \mathcal{R}^{\mathrm{incl}}$. Let $\mathcal{R}^{\mathrm{mtrx}}$ be the region occupied by the reference configuration of the matrix.

In the following, two related problems involving inclusions in an infinite matrix are addressed. Their solutions have applications in the study of mechanics of composites and fracture mechanics. The treatment of these problems follows that given by Eshelby (1957).

9.4.1   Transforming Inclusions

Recall that the constitutive model, $\boldsymbol{\sigma} = \mathbf{C} : \boldsymbol{\varepsilon}$, for infinitesimal deformation of a linear elastic material presumed that the strain tensor $\boldsymbol{\varepsilon}$ was determined relative to a stress-free reference configuration, so that a state of zero strain implied a state of zero stress. However, a material point can, under certain conditions, undergo a *transformation* that causes a change in its stress-free configuration. This transformation, which could for example be due to thermal expansion, crystallographic twinning, or a martensitic transformation, manifests itself in an *unconstrained* material point as a *transformation strain* $\boldsymbol{\varepsilon}^{(T)}$ that occurs at zero stress. The constitutive relation is then given by $\boldsymbol{\sigma} = \mathbf{C} : (\boldsymbol{\varepsilon} - \boldsymbol{\varepsilon}^{(T)})$, where $\boldsymbol{\varepsilon}$ is the strain relative to the untransformed configuration. If the material is *constrained*, by surrounding material for example, then the transformation will give rise to a strain that is not equal to the transformation strain and to a corresponding nonzero *residual stress*.

Suppose that an infinite homogeneous isotropic elastic matrix contains a homogeneous isotropic elastic inclusion that is indistinguishable from the matrix, except that the inclusion at some instant experiences a homogeneous transformation. Prior to the inclusion's transformation, both the matrix and the inclusion are stress-free. Also, the configuration of the body prior to the transformation is taken to be the reference configuration so that the matrix and inclusion are both strain-free before the inclusion's transformation as well.

If the inclusion were not constrained by the surrounding matrix, the transformation would result in a stress-free, *homogeneous* transformation strain $\boldsymbol{\varepsilon}^{(T)}$ within the inclusion. However, the constraint imposed by the surrounding elastic matrix will result in an elastic state different than that of an unconstrained inclusion—the inclusion will, in general, experience an *inhomogeneous* strain field that is not equal to the transformation strain together with a nonzero, inhomogeneous stress field. The surrounding matrix will also be deformed. Thus, given the region $\mathcal{R}^{\text{incl}}$ occupied by the inclusion in its reference configuration along with its homogeneous transformation strain $\boldsymbol{\varepsilon}^{(T)}$, the problem of the transforming inclusion is to find the stress, strain, and displacement fields in the inclusion and the surrounding matrix that are a result of the inclusion's transformation.

The solution to this problem is found via an application of the superposition principle that is conceptualized through the following sequence of imaginary operations:

1. Remove the inclusion from the rest of the infinite body prior to its transformation and then allow the transformation to occur without constraint. At this stage, the inclusion is stress-free and it has a homogeneous strain field equal to the transformation strain $\boldsymbol{\varepsilon}^{(T)}$.

2. Apply the necessary traction distribution to the boundary of the isolated inclusion required to return it to its pretransformation (i.e., reference) configuration. The necessary traction distribution corresponds to a homogeneous strain $-\boldsymbol{\varepsilon}^{(T)}$ and thus to a homogeneous stress $-\boldsymbol{\sigma}^{(T)}$, where the *transformation stress* is defined as

$$\boldsymbol{\sigma}^{(T)} \equiv 2\mu\boldsymbol{\varepsilon}^{(T)} + \lambda(\operatorname{tr}\boldsymbol{\varepsilon}^{(T)})\mathbf{I} . \tag{9.4.1}$$

Therefore, the traction distribution is

$$\mathbf{t} = -\hat{\mathbf{n}} \cdot \boldsymbol{\sigma}^{(T)} \quad \forall \mathbf{X} \in \partial\mathcal{R}^{\text{incl}} , \tag{9.4.2}$$

where $\hat{\mathbf{n}}$ is the unit outward normal to $\partial\mathcal{R}^{\text{incl}}$. Note that the Lamé constants μ and λ for the inclusion are the same as those for the matrix.

3. Put the inclusion back into the hole from which it came and rebond the interface between the inclusion and the matrix, all the while maintaining the traction distribution (9.4.2) on the interface. At this juncture, the displacement and strain fields in both the matrix and the inclusion are identically zero:

$$\left. \begin{array}{l} \mathbf{u} = \mathbf{0} \\ \boldsymbol{\varepsilon} = \mathbf{0} \end{array} \right\} \quad \forall \mathbf{X} \in \mathcal{R}^{\text{incl}} \cup \mathcal{R}^{\text{mtrx}} . \tag{9.4.3}$$

On the other hand, while the stress field in the matrix is identically zero, the inclusion is in a state of homogeneous, nonzero stress:

$$\boldsymbol{\sigma} = \begin{cases} -\boldsymbol{\sigma}^{(T)} & \forall \mathbf{X} \in \mathcal{R}^{\text{incl}} \\ \mathbf{0} & \forall \mathbf{X} \in \mathcal{R}^{\text{mtrx}} , \end{cases} \tag{9.4.4}$$

where $\boldsymbol{\sigma}^{(T)}$ is defined by (9.4.1).

4. Finally, remove the traction distribution (9.4.2) that is preventing the inclusion from leaving its reference configuration or, in what amounts to the same thing, superpose an equal and opposite traction distribution to cancel out (9.4.2). Such a traction distribution is equivalent to a distribution over $\partial \mathcal{R}^{\,\text{incl}}$ of concentrated body forces $d\mathbf{P} = -\mathbf{t}\,dA = \hat{\mathbf{n}} \cdot \boldsymbol{\sigma}^{(T)}\,dA$, where dA is a differential element of the surface $\partial \mathcal{R}^{\,\text{incl}}$. Let the *constrained displacement* $\mathbf{u}^{(C)}$, the *constrained strain* $\boldsymbol{\varepsilon}^{(C)}$, and the *constrained stress* $\boldsymbol{\sigma}^{(C)}$ represent the displacement, strain, and stress fields in an infinite body due to such a distribution of concentrated body forces. Kelvin's solution will be used below to determine these fields. Once they are known, it follows from the superposition principle that the displacement and strain fields in both the inclusion and the matrix are given by $\mathbf{u}^{(C)}$ and $\boldsymbol{\varepsilon}^{(C)}$,

$$\left.\begin{array}{l} \mathbf{u} = \mathbf{u}^{(C)}(\mathbf{X}) \\[4pt] \boldsymbol{\varepsilon} = \boldsymbol{\varepsilon}^{(C)}(\mathbf{X}) \end{array}\right\} \quad \forall \mathbf{X} \in \mathcal{R}^{\,\text{incl}} \cup \mathcal{R}^{\,\text{mtrx}} , \qquad (9.4.5)$$

whereas the stress field requires different representations for the inclusion and the matrix,

$$\boldsymbol{\sigma} = \begin{cases} \boldsymbol{\sigma}^{(C)}(\mathbf{X}) - \boldsymbol{\sigma}^{(T)} & \forall \mathbf{X} \in \mathcal{R}^{\,\text{incl}} \\[4pt] \boldsymbol{\sigma}^{(C)}(\mathbf{X}) & \forall \mathbf{X} \in \mathcal{R}^{\,\text{mtrx}} . \end{cases} \qquad (9.4.6)$$

Note that, although the traction and displacement fields satisfy the necessary continuity conditions across $\partial \mathcal{R}^{\,\text{incl}}$, the strain and stress fields will generally by discontinuous across this interface between the inclusion and the matrix.

Attention is now turned to determining the "constrained" fields $\mathbf{u}^{(C)}$, $\boldsymbol{\varepsilon}^{(C)}$, and $\boldsymbol{\sigma}^{(C)}$ due to a distribution over $\partial \mathcal{R}^{\,\text{incl}}$ of concentrated body forces $d\mathbf{P} = \hat{\mathbf{n}} \cdot \boldsymbol{\sigma}^{(T)}\,dA$.

From the solution to Kelvin's problem (9.2.86), the displacement field $d\mathbf{u}(\mathbf{X})$ due to a concentrated body force $d\mathbf{P}$ acting at $\boldsymbol{\xi}$ is

$$d\mathbf{u} = \overset{*}{\mathbf{u}}(\mathbf{X}; d\mathbf{P}, \boldsymbol{\xi}) = \frac{1}{16\pi\mu(1-\nu)} \left[(3 - 4\nu)\frac{\mathbf{I}}{\rho} + \frac{\mathbf{r}^{\xi}\mathbf{r}^{\xi}}{\rho^3} \right] \cdot d\mathbf{P} , \qquad (9.4.7)$$

where $\mathbf{r}^{\xi} \equiv \mathbf{X} - \boldsymbol{\xi}$ and $\rho \equiv |\mathbf{r}^{\xi}|$. By noting that

$$\nabla\nabla\rho = \frac{\mathbf{I}}{\rho} - \frac{\mathbf{r}^{\xi}\mathbf{r}^{\xi}}{\rho^3} , \qquad (9.4.8)$$

the displacement field (9.4.7) can be rewritten as

$$d\mathbf{u} = \left[\frac{1}{4\pi\mu} \frac{\mathbf{I}}{\rho} - \frac{1}{16\pi\mu(1-\nu)} \boldsymbol{\nabla}\boldsymbol{\nabla}\rho \right] \cdot d\mathbf{P} \, . \tag{9.4.9}$$

It follows that the displacement field $\mathbf{u}^{(C)}$ due to a distribution over $\partial\mathcal{R}^{\mathrm{incl}}$ of concentrated body forces $d\mathbf{P} = \hat{\mathbf{n}} \cdot \boldsymbol{\sigma}^{(T)} \, dA$ is

$$\mathbf{u}^{(C)} = \int\limits_{\partial\mathcal{R}^{\mathrm{incl}}} \left[\frac{1}{4\pi\mu} \frac{\mathbf{I}}{\rho} - \frac{1}{16\pi\mu(1-\nu)} \boldsymbol{\nabla}\boldsymbol{\nabla}\rho \right] \cdot \boldsymbol{\sigma}^{(T)} \cdot \hat{\mathbf{n}} \, dA_{\boldsymbol{\xi}} \, , \tag{9.4.10}$$

where $\hat{\mathbf{n}}$ is the unit outward normal to $\partial\mathcal{R}^{\mathrm{incl}}$ and the notation $dA_{\boldsymbol{\xi}}$ is used to emphasize that $\boldsymbol{\xi}$ is the variable of integration. By applying the divergence theorem, it can be shown after some manipulation that (9.4.10) can be rewritten as

$$\boxed{\begin{aligned} \mathbf{u}^{(C)} &= \frac{1}{16\pi\mu(1-\nu)} \boldsymbol{\sigma}^{(T)} : (\boldsymbol{\nabla}\boldsymbol{\nabla}\boldsymbol{\nabla}\psi) - \frac{1}{4\pi\mu} \boldsymbol{\sigma}^{(T)} \cdot \boldsymbol{\nabla}\phi \, , \\ u_i^{(C)} &= \frac{1}{16\pi\mu(1-\nu)} \sigma_{pq}^{(T)} \psi_{,pqi} - \frac{1}{4\pi\mu} \sigma_{ik}^{(T)} \phi_{,k} \, , \end{aligned}} \tag{9.4.11}$$

where

$$\boxed{\psi(\mathbf{X}) \equiv \int\limits_{\mathcal{R}^{\mathrm{incl}}} \rho \, dV_{\boldsymbol{\xi}} \, , \quad \phi(\mathbf{X}) \equiv \int\limits_{\mathcal{R}^{\mathrm{incl}}} \frac{1}{\rho} \, dV_{\boldsymbol{\xi}} \, .} \tag{9.4.12}$$

Thus, in principle, the problem of a transforming inclusion is solved. The dependence of the solution on the geometry of the inclusion (i.e., on the region $\mathcal{R}^{\mathrm{incl}}$ occupied by the inclusion's reference configuration) is completely determined by the scalar fields ψ and ϕ defined in (9.4.12). Then, the constrained strain field,

$$\boldsymbol{\varepsilon}^{(C)} = \frac{1}{2} \left[\boldsymbol{\nabla}\mathbf{u}^{(C)} + (\boldsymbol{\nabla}\mathbf{u}^{(C)})^{\mathsf{T}} \right] \, , \tag{9.4.13}$$

and the constrained stress field,

$$\boldsymbol{\sigma}^{(C)} = 2\mu\boldsymbol{\varepsilon}^{(C)} + \lambda(\mathrm{tr}\,\boldsymbol{\varepsilon}^{(C)})\mathbf{I} \, , \tag{9.4.14}$$

are easily determined from the constrained displacement field (9.4.11) and the solution is given by (9.4.5) and (9.4.6).

The catch, of course, is in evaluating the volume integrals (9.4.12). However, some useful information about the solution can be obtained simply by noting from classical potential theory (Kellogg 1953) that

$$\nabla^2 \phi = \begin{cases} -4\pi & \forall \mathbf{X} \in \mathcal{R}^{\mathrm{incl}} \\ 0 & \forall \mathbf{X} \in \mathcal{R}^{\mathrm{mtrx}} \, , \end{cases} \tag{9.4.15}$$

and, since $\nabla^2 \rho = 2/\rho$, that

$$\nabla^2 \psi = 2\phi \, . \tag{9.4.16}$$

Thus, for instance, it follows from (9.4.11) and (9.4.16) that the dilatation field $e^{(C)} = \nabla \cdot \mathbf{u}^{(C)}$ can be expressed in terms of ϕ only:

$$e^{(C)} = -\frac{1-2\nu}{8\pi\mu(1-\nu)}\sigma^{(T)} : (\nabla\nabla\phi) \, , \quad e^{(C)} = -\frac{1-2\nu}{8\pi\mu(1-\nu)}\sigma_{ij}^{(T)}\phi_{,ij} \, . \tag{9.4.17}$$

One does not need to know ψ to determine the dilatation field throughout the inclusion and the matrix.

Similarly, in the special case in which the *transformation strain is a pure dilatation* $e^{(T)}$ so that $\boldsymbol{\varepsilon}^{(T)} = \frac{1}{3}e^{(T)}\mathbf{I}$, the solution can also be expressed in terms of ϕ only. In this case, the transformation stress is

$$\boldsymbol{\sigma}^{(T)} = Ke^{(T)}\mathbf{I} = \frac{2\mu(1+\nu)}{3(1-2\nu)}e^{(T)}\mathbf{I} \tag{9.4.18}$$

and it follows from (9.4.11) and (9.4.16) that the constrained displacement field is

$$\mathbf{u}^{(C)} = -\frac{1+\nu}{12\pi(1-\nu)}e^{(T)}\nabla\phi \, , \quad u_i^{(C)} = -\frac{1+\nu}{12\pi(1-\nu)}e^{(T)}\phi_{,i} \, . \tag{9.4.19}$$

Noting that the gradient of this displacement field is symmetric, one then finds that the constrained strain field due to a purely dilatational transformation strain $\boldsymbol{\varepsilon}^{(T)} = \frac{1}{3}e^{(T)}\mathbf{I}$ is given by

$$\boldsymbol{\varepsilon}^{(C)} = -\frac{1+\nu}{12\pi(1-\nu)}e^{(T)}\nabla\nabla\phi \, , \quad \varepsilon_{ij}^{(C)} = -\frac{1+\nu}{12\pi(1-\nu)}e^{(T)}\phi_{,ij} \tag{9.4.20}$$

and that the corresponding rotation tensor is identically zero everywhere.

Finally, it follows from (9.4.15) and either (9.4.17) or (9.4.20) that the dilatation field due to a purely dilatational transformation strain $\boldsymbol{\varepsilon}^{(T)} = \frac{1}{3}e^{(T)}\mathbf{I}$ is

$$e^{(C)} = \begin{cases} \dfrac{1+\nu}{3(1-\nu)}e^{(T)} & \forall \mathbf{X} \in \mathcal{R}^{\text{incl}} \\[2mm] 0 & \forall \mathbf{X} \in \mathcal{R}^{\text{mtrx}} \, . \end{cases} \tag{9.4.21}$$

The dilatation field in this case is identically zero throughout the matrix and is homogeneous in the inclusion, where it is completely determined by the transformation dilatation $e^{(T)}$ and Poisson's ratio ν. The dilatation in the inclusion is equal to the transformation dilatation if the material is incompressible (i.e., when $\nu = 1/2$). In the other extreme, when $\nu = 0$, the dilatation in the inclusion is equal to one-third the transformation dilatation.

Consistent with the assumption that the inclusion and matrix are ideally bonded across their interface, the displacement field (9.4.11) is continuous everywhere and the stress field (9.4.6) is such that the traction vector at a point on the interface exerted by the inclusion on the matrix is equal in magnitude and opposite in direction to that exerted by the matrix on the inclusion. However, the strain and stress fields are discontinuous across the interface and the nature of these discontinuities can be determined from known properties of the scalar fields ψ and ϕ (9.4.12).

If $\boldsymbol{\zeta} \in \partial \mathcal{R}^{\text{incl}}$ is a point on the interface between the inclusion and the matrix, and $f(\boldsymbol{\zeta}^{\text{i}})$ and $f(\boldsymbol{\zeta}^{\text{m}})$ are the limiting values of $f(\mathbf{X})$ as $\mathbf{X}$ approaches $\boldsymbol{\zeta}$ through the inclusion and through the matrix, respectively, then it is known from classical potential theory (Eshelby 1957) that

$$\boldsymbol{\nabla}\boldsymbol{\nabla}\phi(\boldsymbol{\zeta}^{\text{m}}) - \boldsymbol{\nabla}\boldsymbol{\nabla}\phi(\boldsymbol{\zeta}^{\text{i}}) = 4\pi\hat{\mathbf{n}}\hat{\mathbf{n}} \ ,$$
$$\phi_{,ij}(\boldsymbol{\zeta}^{\text{m}}) - \phi_{,ij}(\boldsymbol{\zeta}^{\text{i}}) = 4\pi\hat{n}_i\hat{n}_j \tag{9.4.22}$$

and

$$\boldsymbol{\nabla}\boldsymbol{\nabla}\boldsymbol{\nabla}\boldsymbol{\nabla}\psi(\boldsymbol{\zeta}^{\text{m}}) - \boldsymbol{\nabla}\boldsymbol{\nabla}\boldsymbol{\nabla}\boldsymbol{\nabla}\psi(\boldsymbol{\zeta}^{\text{i}}) = 8\pi\hat{\mathbf{n}}\hat{\mathbf{n}}\hat{\mathbf{n}}\hat{\mathbf{n}} \ ,$$
$$\psi_{,ijkl}(\boldsymbol{\zeta}^{\text{m}}) - \psi_{,ijkl}(\boldsymbol{\zeta}^{\text{i}}) = 8\pi\hat{n}_i\hat{n}_j\hat{n}_k\hat{n}_l \ , \tag{9.4.23}$$

where $\hat{\mathbf{n}}$ is the unit outward normal to $\partial \mathcal{R}^{\text{incl}}$ at $\boldsymbol{\zeta}$. It follows then from (9.4.11) and (9.4.13) that the discontinuity in the constrained strain field across the interface is given by

$$\boldsymbol{\varepsilon}^{(C)}(\boldsymbol{\zeta}^{\text{m}}) - \boldsymbol{\varepsilon}^{(C)}(\boldsymbol{\zeta}^{\text{i}}) = \frac{\hat{\mathbf{n}} \cdot \boldsymbol{\sigma}^{(T)} \cdot \hat{\mathbf{n}}}{2\mu(1-\nu)}\hat{\mathbf{n}}\hat{\mathbf{n}} - \frac{1}{2\mu}\left[(\boldsymbol{\sigma}^{(T)} \cdot \hat{\mathbf{n}})\hat{\mathbf{n}} + \hat{\mathbf{n}}(\boldsymbol{\sigma}^{(T)} \cdot \hat{\mathbf{n}})\right] \ , \tag{9.4.24}$$

and, from (9.4.14), that the discontinuity in the constrained stress field is given by

$$\boldsymbol{\sigma}^{(C)}(\boldsymbol{\zeta}^{\text{m}}) - \boldsymbol{\sigma}^{(C)}(\boldsymbol{\zeta}^{\text{i}}) = \frac{\hat{\mathbf{n}} \cdot \boldsymbol{\sigma}^{(T)} \cdot \hat{\mathbf{n}}}{1-\nu}(\hat{\mathbf{n}}\hat{\mathbf{n}} - \nu\mathbf{I}) - (\boldsymbol{\sigma}^{(T)} \cdot \hat{\mathbf{n}})\hat{\mathbf{n}} - \hat{\mathbf{n}}(\boldsymbol{\sigma}^{(T)} \cdot \hat{\mathbf{n}}). \tag{9.4.25}$$

From (9.4.6), the discontinuity in the *actual* stress field is

$$\boldsymbol{\sigma}(\boldsymbol{\zeta}^{\text{m}}) - \boldsymbol{\sigma}(\boldsymbol{\zeta}^{\text{i}}) = \frac{\hat{\mathbf{n}} \cdot \boldsymbol{\sigma}^{(T)} \cdot \hat{\mathbf{n}}}{1-\nu}(\hat{\mathbf{n}}\hat{\mathbf{n}} - \nu\mathbf{I})$$
$$- (\boldsymbol{\sigma}^{(T)} \cdot \hat{\mathbf{n}})\hat{\mathbf{n}} - \hat{\mathbf{n}}(\boldsymbol{\sigma}^{(T)} \cdot \hat{\mathbf{n}}) + \boldsymbol{\sigma}^{(T)} \ . \tag{9.4.26}$$

Thus, for instance, if the stress field within the inclusion is known, then the stress at points in the matrix adjacent to the inclusion can be determined from (9.4.26). Note also that

$$\hat{\mathbf{n}} \cdot \left[\boldsymbol{\sigma}(\boldsymbol{\zeta}^{m}) - \boldsymbol{\sigma}(\boldsymbol{\zeta}^{i}) \right] = \mathbf{0} , \qquad (9.4.27)$$

which verifies that the tractions on either side of the interface $\partial \mathcal{R}^{incl}$ are equal in magnitude and opposite in direction.

One would expect the displacement, strain, and stress fields to vanish at points in the matrix sufficiently far from the transforming inclusion. That this is, in fact, the case is easily established by noting first from (9.4.12) that

$$\lim_{|\mathbf{X}| \to \infty} \frac{1}{|\mathbf{X}|} \psi(\mathbf{X}) = \lim_{|\mathbf{X}| \to \infty} |\mathbf{X}| \phi(\mathbf{X}) = V^{incl} , \qquad (9.4.28)$$

where V^{incl} is the volume of the inclusion. If follows from (9.4.5), (9.4.6), and (9.4.11) that, in the limit as $|\mathbf{X}| \to \infty$, the displacement field vanishes like $|\mathbf{X}|^{-2}$ and the strain and stress fields vanish like $|\mathbf{X}|^{-3}$.

Transforming, ellipsoidal inclusions

It will be shown that, perhaps surprisingly, the strain and stress fields within a transforming, *ellipsoidal* inclusion are homogeneous, regardless of the material properties and the transformation strain $\boldsymbol{\varepsilon}^{(T)}$. This, of course, is not the case for arbitrarily shaped inclusions. It is this characteristic of transforming, ellipsoidal inclusions that will later be exploited to study the problem of a dissimilar, ellipsoidal inclusion in an infinite matrix subject to a far-field, homogeneous state of stress.

Let the region $\mathcal{R}^{incl}$ occupied by the reference configuration of the inclusion be an ellipsoid with semiaxes a_1, a_2, and a_3 and define a Cartesian coordinate system (X_1, X_2, X_3) so that the coordinate axes are aligned with the semiaxes of the ellipsoid. Thus, $\mathcal{R}^{incl}$ can be interpreted as the open set of points given by

$$\mathcal{R}^{incl} \equiv \left\{ \mathbf{X} \mid \frac{X_1^2}{a_1^2} + \frac{X_2^2}{a_2^2} + \frac{X_3^2}{a_3^2} < 1 \right\} . \qquad (9.4.29)$$

For this geometry, the scalar fields ϕ and ψ defined in (9.4.12) have been studied and it has been shown that ϕ and $\boldsymbol{\nabla}\psi$ can be expressed in terms of elliptic integrals (Kellogg 1953; Eshelby 1957). Within the inclusion, it has been shown that

$$\left. \begin{aligned} \phi &= A - \mathbf{X} \cdot \mathbf{B} \cdot \mathbf{X} \\ \phi &= A - B_{ij} X_i X_j \end{aligned} \right\} \quad \forall \mathbf{X} \in \mathcal{R}^{incl} \qquad (9.4.30)$$

and

$$
\left.\begin{aligned}
\boldsymbol{\nabla}\psi &= (\mathbf{C} - \mathbf{D} : \mathbf{X}\mathbf{X}) \cdot \mathbf{X} \\
\psi_{,i} &= (C_{ij} - D_{ijkl}X_k X_l)X_j
\end{aligned}\right\} \quad \forall \mathbf{X} \in \mathcal{R}^{\mathrm{incl}} ,
$$

(9.4.31)

where the scalar A, the two second-order tensors $\mathbf{B}$ and $\mathbf{C}$, and the fourth-order tensor $\mathbf{D}$ are constants that are completely determined by the semi-axes a_1, a_2, and a_3. The scalar constant A is given by

$$
A = \pi a_1 a_2 a_3 \int_0^\infty \frac{ds}{g(s)} ,
$$

(9.4.32)

where

$$
g(s) \equiv \sqrt{(a_1^2 + s)(a_2^2 + s)(a_3^2 + s)} .
$$

(9.4.33)

The constants $\mathbf{B}$ and $\mathbf{C}$ are symmetric, second-order tensors which, in the Cartesian coordinate system (X_1, X_2, X_3) aligned with the semiaxes of the ellipsoid, have matrices of scalar components that are diagonal, so that

$$
B_{ij} = B^{(i)}\delta_{ij} \quad (\text{no sum } i) ,
$$

(9.4.34)

$$
C_{ij} = C^{(i)}\delta_{ij} \quad (\text{no sum } i) ,
$$

(9.4.35)

where

$$
B^{(i)} = \pi a_1 a_2 a_3 \int_0^\infty \frac{ds}{(a_i^2 + s)g(s)} ,
$$

(9.4.36)

$$
C^{(i)} = \pi a_1 a_2 a_3 \int_0^\infty \frac{s\, ds}{(a_i^2 + s)g(s)} .
$$

(9.4.37)

Finally, the constant, fourth-order tensor $\mathbf{D}$ has scalar components given in this coordinate system by

$$
D_{ijkl} = D^{(ik)}\delta_{ij}\delta_{kl} \quad (\text{no sum } i,\ k) ,
$$

(9.4.38)

where

$$
D^{(ij)} = \pi a_1 a_2 a_3 \int_0^\infty \frac{s\, ds}{(a_i^2 + s)(a_j^2 + s)g(s)} .
$$

(9.4.39)

Note that $\mathbf{D}$ has $D_{ijkl} = D_{jikl} = D_{ijlk} = D_{klij}$ symmetry (in any orthonormal basis) and that, in the Cartesian coordinate system (X_1, X_2, X_3) aligned with the semiaxes of the ellipsoid, it has only nine nonzero scalar components.

Substituting (9.4.30) and (9.4.31) into (9.4.11), one finds that the constrained displacement field within the ellipsoidal inclusion $\mathcal{R}^{\text{incl}}$ is a linear function of $\mathbf{X}$ given by

$$\boxed{\mathbf{u}^{(C)} = (\hat{\mathbf{U}} : \boldsymbol{\sigma}^{(T)}) \cdot \mathbf{X}, \quad u_i^{(C)} = \hat{U}_{ijpq}\sigma_{pq}^{(T)} X_j \quad \forall \mathbf{X} \in \mathcal{R}^{\text{incl}},} \tag{9.4.40}$$

where

$$\hat{U}_{ijpq} = \frac{1}{4\pi\mu}\left[\delta_{ip}B_{jq} + \delta_{iq}B_{jp} - \frac{1}{2(1-\nu)}(D_{ijpq} + D_{ipjq} + D_{iqjp}\right] \tag{9.4.41}$$

are the scalar components of a constant fourth-order tensor $\hat{\mathbf{U}}$ with $\hat{U}_{ijpq} = \hat{U}_{ijqp}$ symmetry. This displacement field within the inclusion can be rewritten in terms of the transformation strain (which will be convenient later) by substituting

$$\sigma_{pq}^{(T)} \equiv 2\mu\varepsilon_{pq}^{(T)} + \lambda\varepsilon_{rr}^{(T)}\delta_{pq}$$
$$= \mu\left(\delta_{pr}\delta_{qs} + \delta_{ps}\delta_{qr} + \frac{2\nu}{1-2\nu}\delta_{pq}\delta_{rs}\right)\varepsilon_{rs}^{(T)} \tag{9.4.42}$$

into (9.4.40) to get

$$\boxed{\mathbf{u}^{(C)} = (\mathbf{U} : \boldsymbol{\varepsilon}^{(T)}) \cdot \mathbf{X}, \quad u_i^{(C)} = U_{ijrs}\varepsilon_{rs}^{(T)} X_j \quad \forall \mathbf{X} \in \mathcal{R}^{\text{incl}},} \tag{9.4.43}$$

where

$$U_{ijrs} = \frac{1}{2\pi}\left\{\delta_{ir}B_{js} + \delta_{is}B_{jr} - \frac{1}{2(1-\nu)}(D_{ijrs} + D_{irjs} + D_{isjr})\right.$$
$$\left. + \frac{\nu}{1-2\nu}\left[2B_{ij} - \frac{1}{2(1-\nu)}(D_{ijpp} + 2D_{ipjp})\right]\delta_{rs}\right\}. \tag{9.4.44}$$

The constant fourth-order tensor $\mathbf{U}$ has the same symmetry as $\hat{\mathbf{U}}$ (i.e., $U_{ijrs} = U_{ijsr}$), indicating that it could have $3 \times 3 \times 6 = 54$ independent scalar components. However, (9.4.44) simplifies considerably in the Cartesian coordinate system (X_1, X_2, X_3) aligned with the semiaxes of the ellipsoid. In this coordinate system, $\mathbf{B}$ and $\mathbf{D}$ have scalar components given by (9.4.34) and (9.4.38), from which it follows that $U_{ijrs} = 0$ whenever $i = j$ and $r \neq s$ ($U_{1112} = U_{1123} = \cdots = U_{3331} = 0$), whenever $r = s$ and $i \neq j$ ($U_{1211} = U_{2311} = \cdots = U_{3133} = 0$), and whenever $i \neq j$ and $r \neq s$, and neither $ij = rs$ nor $ij = sr$ ($U_{1223} = U_{1231} = \cdots = U_{3112} = 0$). There remain only 15 nonzero scalar components of **U** and one has that

$$
\begin{bmatrix}
U_{11rs}\varepsilon_{rs}^{(T)} \\
U_{22rs}\varepsilon_{rs}^{(T)} \\
U_{33rs}\varepsilon_{rs}^{(T)} \\
U_{12rs}\varepsilon_{rs}^{(T)} \\
U_{23rs}\varepsilon_{rs}^{(T)} \\
U_{31rs}\varepsilon_{rs}^{(T)} \\
U_{21rs}\varepsilon_{rs}^{(T)} \\
U_{32rs}\varepsilon_{rs}^{(T)} \\
U_{13rs}\varepsilon_{rs}^{(T)}
\end{bmatrix}
=
\begin{bmatrix}
U_{1111} & U_{1122} & U_{1133} & 0 & 0 & 0 \\
U_{2211} & U_{2222} & U_{2233} & 0 & 0 & 0 \\
U_{3311} & U_{3322} & U_{3333} & 0 & 0 & 0 \\
0 & 0 & 0 & U_{1212} & 0 & 0 \\
0 & 0 & 0 & 0 & U_{2323} & 0 \\
0 & 0 & 0 & 0 & 0 & U_{3131} \\
0 & 0 & 0 & U_{2121} & 0 & 0 \\
0 & 0 & 0 & 0 & U_{3232} & 0 \\
0 & 0 & 0 & 0 & 0 & U_{1313}
\end{bmatrix}
\begin{bmatrix}
\varepsilon_{11}^{(T)} \\
\varepsilon_{22}^{(T)} \\
\varepsilon_{33}^{(T)} \\
2\varepsilon_{12}^{(T)} \\
2\varepsilon_{23}^{(T)} \\
2\varepsilon_{31}^{(T)}
\end{bmatrix},
$$

$$(9.4.45)$$

where three of the nonzero scalar components of **U** are given by

$$
\mathsf{U}_{iiii} = \frac{1}{4\pi(1-\nu)}\left[2(2-\nu)B^{(i)} - 3D^{(ii)}\right] \quad \text{(no sum } i) , \qquad (9.4.46)
$$

six more are given by

$$
\mathsf{U}_{iijj} = \frac{1}{4\pi(1-\nu)}\left[2\nu B^{(i)} - D^{(ij)}\right] \quad \text{(no sum } i, j \text{ and } i \neq j) , \qquad (9.4.47)
$$

and the final six are given by

$$
\mathsf{U}_{ijij} = \frac{1}{2\pi}\left[B^{(j)} - \frac{1}{2(1-\nu)}D^{(ij)}\right] \quad \text{(no sum } i, j \text{ and } i \neq j) . \qquad (9.4.48)
$$

Note that the identity (9.4.56) discussed below has been anticipated in deriving (9.4.46)–(9.4.48).

It follows from (9.4.43) that the constrained strain field (9.4.13) within the inclusion is homogeneous and can be expressed in terms of the transformation strain as

$$
\boxed{\varepsilon^{(C)} = \mathbf{E} : \varepsilon^{(T)} , \qquad \varepsilon_{ij}^{(C)} = \mathsf{E}_{ijrs}\varepsilon_{rs}^{(T)} \quad \forall \mathbf{X} \in \mathcal{R}^{\text{incl}} ,} \qquad (9.4.49)
$$

where $\mathsf{E}_{ijrs} = \frac{1}{2}(\mathsf{U}_{ijrs} + \mathsf{U}_{jirs})$. It follows from the symmetry of **U** that the fourth-order tensor **E** has $\mathsf{E}_{ijrs} = \mathsf{E}_{ijsr} = \mathsf{E}_{jirs}$ symmetry, although $\mathsf{E}_{ijrs} \neq \mathsf{E}_{rsij}$ in general. In addition, **E** simplifies considerably in the Cartesian coordinate system (X_1, X_2, X_3) aligned with the semiaxes of the ellipsoid. In this coordinate system, there is no coupling in (9.4.49) between normal and shear components of strain ($\mathsf{E}_{1112} = \mathsf{E}_{1123} = \mathsf{E}_{1211} = \cdots = 0$) and there is no coupling between different shear components ($\mathsf{E}_{1223} = \mathsf{E}_{1231} =$

$E_{2312} = \cdots = 0$). Explicitly,

$$
\begin{bmatrix}
\varepsilon_{11}^{(C)} \\
\varepsilon_{22}^{(C)} \\
\varepsilon_{33}^{(C)} \\
\varepsilon_{12}^{(C)} \\
\varepsilon_{23}^{(C)} \\
\varepsilon_{31}^{(C)}
\end{bmatrix}
=
\begin{bmatrix}
E_{1111} & E_{1122} & E_{1133} & 0 & 0 & 0 \\
E_{2211} & E_{2222} & E_{2233} & 0 & 0 & 0 \\
E_{3311} & E_{3322} & E_{3333} & 0 & 0 & 0 \\
0 & 0 & 0 & E_{1212} & 0 & 0 \\
0 & 0 & 0 & 0 & E_{2323} & 0 \\
0 & 0 & 0 & 0 & 0 & E_{3131}
\end{bmatrix}
\begin{bmatrix}
\varepsilon_{11}^{(T)} \\
\varepsilon_{22}^{(T)} \\
\varepsilon_{33}^{(T)} \\
2\varepsilon_{12}^{(T)} \\
2\varepsilon_{23}^{(T)} \\
2\varepsilon_{31}^{(T)}
\end{bmatrix},
$$

(9.4.50)

where it follows from (9.4.46)–(9.4.48) that three of the nonzero components of $\mathbf{E}$ are give by

$$
E_{iiii} = \frac{1}{4\pi(1-\nu)}\left[2(2-\nu)B^{(i)} - 3D^{(ii)}\right] \quad \text{(no sum } i\text{)}, \qquad (9.4.51)
$$

six more are given by

$$
E_{iijj} = \frac{1}{4\pi(1-\nu)}\left[2\nu B^{(i)} - D^{(ij)}\right] \quad \text{(no sum } i, j \text{ and } i \neq j\text{)}, \quad (9.4.52)
$$

and the final three nonzero components of $\mathbf{E}$ are given by

$$
E_{ijij} = \frac{1}{4\pi}\left[B^{(i)} + B^{(j)} - \frac{D^{(ij)}}{1-\nu}\right] \quad \text{(no sum } i, j \text{ and } i \neq j\text{)}. \quad (9.4.53)
$$

Since the constrained strain field within the inclusion is homogeneous, the constrained stress field given by (9.4.14) is also homogeneous. After $\mathbf{u}^{(C)}$, $\boldsymbol{\varepsilon}^{(C)}$, and $\boldsymbol{\sigma}^{(C)}$ have been determined, the solution for the transforming ellipsoidal inclusion is given by (9.4.5) and (9.4.6).

As it turns out, the coefficients A (9.4.32) and $\mathbf{C}$ (9.4.35) do not appear in the expressions for $\mathbf{u}^{(C)}$, $\boldsymbol{\varepsilon}^{(C)}$, and $\boldsymbol{\sigma}^{(C)}$ within the ellipsoidal inclusion— the second-order tensor $\mathbf{B}$ (9.4.34) and the fourth-order tensor $\mathbf{D}$ (9.4.38) are the only geometric quantities required to solve this problem. Noting from (9.4.39) that $D^{(ij)} = D^{(ji)}$, it is evident that there are nine scalar geometric parameters whose values need to be determined for any given set $\{a_1, a_2, a_3\}$ of semiaxes. Conveniently, however, these nine geometric parameters are not independent—once any two of the $B^{(i)}$'s are determined, the remaining $B^{(i)}$ and the six $D^{(ij)}$'s can easily be found via simple algebraic relations. Note first that

$$
\left(\frac{1}{a_1^2 + s} + \frac{1}{a_2^2 + s} + \frac{1}{a_3^2 + s}\right)\frac{1}{g(s)} = -2\frac{d}{ds}\left[\frac{1}{g(s)}\right]. \qquad (9.4.54)
$$

Then, it follows directly that

$$\boxed{B^{(1)} + B^{(2)} + B^{(3)} = 2\pi \, .} \tag{9.4.55}$$

Furthermore, by using (9.4.54) and integrating by parts, it can be shown that

$$\boxed{D^{(i1)} + D^{(i2)} + D^{(i3)} + 2D^{(ii)} = 2B^{(i)} \quad \text{(no sum } i) \, .} \tag{9.4.56}$$

Finally, by splitting the integrand of $D^{(ij)}$ into partial fractions, one finds that

$$\boxed{(a_i^2 - a_j^2)D^{(ij)} = a_i^2 B^{(i)} - a_j^2 B^{(j)} \quad \text{(no sum } i, j \text{ and } i \neq j) \, .} \tag{9.4.57}$$

Once any two of the $B^{(i)}$'s are known, the third is determined by (9.4.55). Then, the six remaining unknown geometric parameters $D^{(11)}$, $D^{(22)}$, $D^{(33)}$, $D^{(12)}$, $D^{(23)}$, and $D^{(31)}$ can be found via (9.4.56) and (9.4.57).

Consider the case where $a_1 > a_2 > a_3 \geq 0$. Then, one has that (Kellogg 1953; Eshelby 1957)

$$B^{(1)}(\hat{a}_2, \hat{a}_3) = \frac{2\pi \hat{a}_2 \hat{a}_3}{(1 - \hat{a}_2^2)\sqrt{1 - \hat{a}_3^2}} [F(k, \theta) - E(k, \theta)] \, ,$$

$$B^{(3)}(\hat{a}_2, \hat{a}_3) = \frac{2\pi \hat{a}_2 \hat{a}_3}{(\hat{a}_2^2 - \hat{a}_3^2)\sqrt{1 - \hat{a}_3^2}} \left[\frac{\hat{a}_2 \sqrt{1 - \hat{a}_3^2}}{\hat{a}_3} - E(k, \theta) \right] \, , \tag{9.4.58}$$

where $\hat{a}_2 \equiv a_2/a_1$ and $\hat{a}_3 \equiv a_3/a_1$ are dimensionless parameters that characterize the shape of the ellipsoidal inclusion independent of its size, and $F(k, \theta)$ and $E(k, \theta)$ are elliptic integrals of the first and second kind, respectively, with modulus and amplitude

$$k(\hat{a}_2, \hat{a}_3) = \sqrt{\frac{1 - \hat{a}_2^2}{1 - \hat{a}_3^2}} \, , \quad \theta(\hat{a}_3) = \sin^{-1}\sqrt{1 - \hat{a}_3^2} \, , \quad 0 < \theta \leq \frac{\pi}{2} \, . \tag{9.4.59}$$

Once $B^{(1)}$ and $B^{(3)}$ are known, one has from (9.4.55) that

$$B^{(2)}(\hat{a}_2, \hat{a}_3) = 2\pi - B^{(1)} - B^{(3)} \, . \tag{9.4.60}$$

Then, once the $B^{(i)}$'s are known, (9.4.57) gives $D^{(12)}$, $D^{(23)}$, and $D^{(31)}$:

$$D^{(12)}(\hat{a}_2, \hat{a}_3) = \frac{B^{(1)} - \hat{a}_2^2 B^{(2)}}{1 - \hat{a}_2^2} \, ,$$

$$D^{(23)}(\hat{a}_2, \hat{a}_3) = \frac{\hat{a}_2^2 B^{(2)} - \hat{a}_3^2 B^{(3)}}{\hat{a}_2^2 - \hat{a}_3^2} \, , \tag{9.4.61}$$

$$D^{(31)}(\hat{a}_2, \hat{a}_3) = \frac{\hat{a}_3^2 B^{(3)} - B^{(1)}}{\hat{a}_3^2 - 1} \, .$$

Finally, $D^{(11)}$, $D^{(22)}$, and $D^{(33)}$ are determined by (9.4.56):

$$D^{(11)}(\hat{a}_2, \hat{a}_3) = \frac{2}{3}B^{(1)} - \frac{1}{3}D^{(12)} - \frac{1}{3}D^{(31)} ,$$

$$D^{(22)}(\hat{a}_2, \hat{a}_3) = \frac{2}{3}B^{(2)} - \frac{1}{3}D^{(23)} - \frac{1}{3}D^{(12)} , \qquad (9.4.62)$$

$$D^{(33)}(\hat{a}_2, \hat{a}_3) = \frac{2}{3}B^{(3)} - \frac{1}{3}D^{(31)} - \frac{1}{3}D^{(23)} .$$

Now, for any given transformation strain $\boldsymbol{\varepsilon}^{(T)}$, the constrained strain field within the inclusion is given by (9.4.50)–(9.4.53) and (9.4.58)–(9.4.62). The constrained displacement field can be determined from (9.4.43), and the constrained stress field is just given by (9.4.14).

An *oblate spheroid* is an ellipsoid in which $a_1 = a_2 > a_3 \geq 0$. For an oblate spheroidal inclusion,

$$g(s) = (a_1^2 + s)\sqrt{a_3^2 + s} , \qquad (9.4.63)$$

and one has from (9.4.36) that

$$B^{(1)}(\hat{a}_3) = B^{(2)}(\hat{a}_3) = \frac{\pi\hat{a}_3}{(1 + \hat{a}_3^2)^{3/2}} \left(\cos^{-1}\hat{a}_3 - \hat{a}_3\sqrt{1 - \hat{a}_3^2} \right) , \qquad (9.4.64)$$

where $\hat{a}_3 \equiv a_3/a_1$ and $0 < \cos^{-1}\hat{a}_3 \leq \pi/2$. Once $B^{(1)} = B^{(2)}$ is known, one has from (9.4.55) that

$$B^{(3)}(\hat{a}_3) = 2(\pi - B^{(1)}) \qquad (9.4.65)$$

and then from (9.4.57) that

$$D^{(23)}(\hat{a}_3) = D^{(31)}(\hat{a}_3) = \frac{\hat{a}_3^2 B^{(3)} - B^{(1)}}{\hat{a}_3^2 - 1} \qquad (9.4.66)$$

and from (9.4.56) that

$$D^{(33)}(\hat{a}_3) = \frac{2}{3}(B^{(3)} - D^{(31)}) . \qquad (9.4.67)$$

Although (9.4.57) cannot be used to find $D^{(12)}$, the remaining geometric parameters can be found by noting from (9.4.39) that $D^{(11)} = D^{(22)} = D^{(12)}$, from which it follows from (9.4.56) that

$$D^{(11)}(\hat{a}_3) = D^{(22)}(\hat{a}_3) = D^{(12)}(\hat{a}_3) = \frac{1}{2}B^{(1)} - \frac{1}{4}D^{(31)} . \qquad (9.4.68)$$

The problem of a transforming *prolate spheroidal* inclusion, in which $a_1 > a_2 = a_3 > 0$, can be dealt with similarly.

The focus until now has been exclusively on determining $\mathbf{u}^{(C)}$, $\boldsymbol{\varepsilon}^{(C)}$, and $\boldsymbol{\sigma}^{(C)}$ within the ellipsoidal inclusion. Expressions for ϕ and $\nabla\psi$ in *the matrix* can also be derived, which are similar to (9.4.30) and (9.4.31) except that the coefficients A, $\mathbf{B}$, $\mathbf{C}$, and $\mathbf{D}$ are complicated functions of $\mathbf{X}$ (Eshelby 1957). These expressions can, in principle, be used to determine $\mathbf{u}^{(C)}$, $\boldsymbol{\varepsilon}^{(C)}$, and $\boldsymbol{\sigma}^{(C)}$ in the matrix, although $\boldsymbol{\varepsilon}^{(C)}$ and $\boldsymbol{\sigma}^{(C)}$ will not, of course, be homogeneous. However, the results are exceedingly complex and ultimately unenlightening. Consequently, they will not be pursued further except to recall that once the solution within the inclusion is known, the strain and stress at points in the matrix adjacent to the inclusion are given by the discontinuity results (9.4.24) and (9.4.25). Also, in the limit as $|\mathbf{X}| \to \infty$, the displacement field vanishes like $|\mathbf{X}|^{-2}$ and the strain and stress fields vanish like $|\mathbf{X}|^{-3}$. This limited information about the solution in the matrix will be sufficient for what follows.

Transforming, spherical inclusions

The simplest special case of the transforming, ellipsoidal inclusion is the spherical case in which $a_1 = a_2 = a_3 > 0$. By noting from (9.4.36) that $B^{(1)} = B^{(2)} = B^{(3)}$, in this case it follows immediately from (9.4.55) that $B^{(i)} = 2\pi/3$ and thus from (9.4.34) that

$$B_{ij} = \frac{2\pi}{3}\delta_{ij} \; . \tag{9.4.69}$$

Then, after noting from (9.4.39) that $D^{(11)} = D^{(12)} = D^{(13)} = \cdots = D^{(33)}$, it follows from (9.4.56) that $D^{(ij)} = 4\pi/15$ and thus from (9.4.38) that

$$D_{ijkl} = \frac{4\pi}{15}\delta_{ij}\delta_{kl} \; . \tag{9.4.70}$$

Substituting (9.4.69) and (9.4.70) into (9.4.44) leads to

$$U_{ijrs} = \frac{4-5\nu}{15(1-\nu)}(\delta_{ir}\delta_{js} + \delta_{jr}\delta_{is}) + \frac{5\nu-1}{15(1-\nu)}\delta_{ij}\delta_{rs} \; , \tag{9.4.71}$$

from which one gets

$$\left.\begin{aligned}
u_i^{(C)} &= \frac{2(4-5\nu)}{15(1-\nu)}\varepsilon_{ij}^{(T)}X_j + \frac{5\nu-1}{15(1-\nu)}\varepsilon_{kk}^{(T)}X_i \\
\mathbf{u}^{(C)} &= \frac{2(4-5\nu)}{15(1-\nu)}\boldsymbol{\varepsilon}^{(T)} \cdot \mathbf{X} + \frac{5\nu-1}{15(1-\nu)}(\mathrm{tr}\,\boldsymbol{\varepsilon}^{(T)})\mathbf{X}
\end{aligned}\right\} \quad \forall \mathbf{X} \in \mathcal{R}^{\mathrm{incl}} \; . \tag{9.4.72}$$

Since in this case $U_{ijrs} = U_{jirs}$, it follows that $\mathbf{E} = \mathbf{U}$ and

$$\left. \begin{aligned} \varepsilon_{ij}^{(C)} &= \frac{2(4 - 5\nu)}{15(1 - \nu)}\varepsilon_{ij}^{(T)} + \frac{5\nu - 1}{15(1 - \nu)}\varepsilon_{kk}^{(T)}\delta_{ij} \\ \varepsilon^{(C)} &= \frac{2(4 - 5\nu)}{15(1 - \nu)}\varepsilon^{(T)} + \frac{5\nu - 1}{15(1 - \nu)}(\operatorname{tr}\varepsilon^{(T)})\mathbf{I} \end{aligned} \right\} \quad \forall \mathbf{X} \in \mathcal{R}^{\text{incl}} . \qquad (9.4.73)$$

The constrained rotation field, $\boldsymbol{\omega}^{(C)} = \frac{1}{2}[\nabla \mathbf{u}^{(C)} - (\nabla \mathbf{u}^{(C)})^{\mathsf{T}}]$ is identically zero within the spherical inclusion.

Recall that a strain tensor ε can be decomposed into its dilatational component $e \equiv \operatorname{tr}\varepsilon$ and its deviatoric component $\grave{\varepsilon} \equiv \varepsilon - \frac{1}{3}\operatorname{tr}\varepsilon\mathbf{I}$, so that $\varepsilon = \grave{\varepsilon} + \frac{1}{3}e\mathbf{I}$. Then, it follows from (9.4.73) that

$$\boxed{e^{(C)} = \alpha e^{(T)} , \quad \grave{\varepsilon}^{(C)} = \beta \grave{\varepsilon}^{(T)} \quad \forall \mathbf{X} \in \mathcal{R}^{\text{incl}} ,} \qquad (9.4.74)$$

where

$$\alpha = \frac{1}{3}\left(\frac{1 + \nu}{1 - \nu}\right) , \quad \beta = \frac{2}{15}\left(\frac{4 - 5\nu}{1 - \nu}\right) \qquad (9.4.75)$$

are material parameters that depend only on Poisson's equation. Thus, it is seen that, for a spherical inclusion, the dilatational component of the constrained strain depends only on the dilatational component of the transformation strain, and the deviatoric component of the constrained strain depends only on the deviatoric component of the transformation strain. This convenient decoupling of dilatational and deviatoric components holds only in the spherical case—for a nonspherical ellipsoidal inclusion, $e^{(C)}$ and $\grave{\varepsilon}^{(C)}$ each depend on both $e^{(T)}$ and $\grave{\varepsilon}^{(T)}$.

Decomposing each stress tensor $\boldsymbol{\sigma}$ into its mean component $\tilde{\sigma} \equiv \frac{1}{3}\operatorname{tr}\boldsymbol{\sigma}$ and its deviatoric component $\grave{\boldsymbol{\sigma}} \equiv \boldsymbol{\sigma} - \frac{1}{3}\operatorname{tr}\boldsymbol{\sigma}$, so that $\boldsymbol{\sigma} = \grave{\boldsymbol{\sigma}} + \tilde{\sigma}\mathbf{I}$, and recalling that $\tilde{\sigma} = Ke$ and $\grave{\boldsymbol{\sigma}} = 2\mu\grave{\varepsilon}$, where $K = \lambda + \frac{2}{3}\mu$ is the bulk modulus, it is easily shown that

$$\boxed{\tilde{\sigma}^{(C)} = \alpha\tilde{\sigma}^{(T)} , \quad \grave{\boldsymbol{\sigma}}^{(C)} = \beta\grave{\boldsymbol{\sigma}}^{(T)} \quad \forall \mathbf{X} \in \mathcal{R}^{\text{incl}} .} \qquad (9.4.76)$$

From (9.4.6), the mean and deviatoric components of the actual stress in the inclusion are related to those of the transformation stress by

$$\boxed{\tilde{\sigma} = (\alpha - 1)\tilde{\sigma}^{(T)} , \quad \grave{\boldsymbol{\sigma}} = (\beta - 1)\grave{\boldsymbol{\sigma}}^{(T)} \quad \forall \mathbf{X} \in \mathcal{R}^{\text{incl}} .} \qquad (9.4.77)$$

Again, this decoupling of mean and deviatoric components is only possible for the spherical inclusion.

9.4.2 Dissimilar, Ellipsoidal Inclusions

Suppose that a homogeneous strain field ε^∞ is superposed onto the previously analyzed solution for a transforming ellipsoidal inclusion, whose reference configuration occupies the ellipsoidal region $\mathcal{R}^{\mathrm{incl}}$ and experiences a transformation strain $\varepsilon^{(T)}$. It follows from (9.4.5) that the resulting strain field within the inclusion is

$$\varepsilon = \varepsilon^{(C)} + \varepsilon^\infty \quad \forall X \in \mathcal{R}^{\mathrm{incl}} \tag{9.4.78}$$

and from (9.4.6) that the resulting stress field within the inclusion is given by

$$\sigma = \sigma^{(C)} - \sigma^{(T)} + \sigma^\infty \quad \forall X \in \mathcal{R}^{\mathrm{incl}}, \tag{9.4.79}$$

where $\sigma^\infty = 2\mu\varepsilon^\infty + \lambda(\mathrm{tr}\,\varepsilon^\infty)\mathbf{I}$ is the superposed homogeneous stress field and the constrained strain $\varepsilon^{(C)}$ and the constrained stress $\sigma^{(C)}$ are homogeneous fields that depend on the transformation strain $\varepsilon^{(T)}$. This stress field within the inclusion can be rewritten in terms of the strains as

$$\sigma = 2\mu(\varepsilon^{(C)} - \varepsilon^{(T)} + \varepsilon^\infty) + \lambda\,\mathrm{tr}(\varepsilon^{(C)} - \varepsilon^{(T)} + \varepsilon^\infty)\mathbf{I} \quad \forall X \in \mathcal{R}^{\mathrm{incl}}. \tag{9.4.80}$$

At points in the matrix sufficiently far from the inclusion, $\varepsilon^{(C)}$ and $\sigma^{(C)}$ vanish so that

$$\lim_{|\mathbf{X}|\to\infty} \varepsilon = \varepsilon^\infty, \quad \lim_{|\mathbf{X}|\to\infty} \sigma = \sigma^\infty, \tag{9.4.81}$$

and since σ^∞ is continuous across the interface $\partial\mathcal{R}^{\mathrm{incl}}$ between the inclusion and the matrix, the stress discontinuity across $\partial\mathcal{R}^{\mathrm{incl}}$ is still given by (9.4.26).

Now, consider another inclusion whose reference configuration occupies that same ellipsoidal region $\partial\mathcal{R}^{\mathrm{incl}}$, but, rather than experiencing a transformation $\varepsilon^{(T)}$, instead has elastic material properties μ^* and λ^* that are different than those of the matrix. Furthermore, assume that μ^* and λ^* are such that when this dissimilar inclusion experiences the homogeneous strain (9.4.78) experienced by the transforming inclusion, the corresponding stress,

$$\sigma = 2\mu^*(\varepsilon^{(C)} + \varepsilon^\infty) + \lambda^*\,\mathrm{tr}(\varepsilon^{(C)} + \varepsilon^\infty)\mathbf{I} \quad \forall X \in \mathcal{R}^{\mathrm{incl}}, \tag{9.4.82}$$

is the same as that experienced by the transforming inclusion (9.4.80), so that

$$2\mu(\varepsilon^{(C)} - \varepsilon^{(T)} + \varepsilon^\infty) + \lambda\,\mathrm{tr}(\varepsilon^{(C)} - \varepsilon^{(T)} + \varepsilon^\infty)\mathbf{I}$$
$$= 2\mu^*(\varepsilon^{(C)} + \varepsilon^\infty) + \lambda^*\,\mathrm{tr}(\varepsilon^{(C)} + \varepsilon^\infty)\mathbf{I}. \tag{9.4.83}$$

If the condition (9.4.83) is satisfied, then the above transforming inclusion can be replaced by this dissimilar inclusion and continuity of displacement and traction across $\partial \mathcal{R}^{\text{incl}}$ will be maintained.

By substituting the previously determined solution (9.4.49) for the constrained strain $\varepsilon^{(C)} = \mathbf{E} : \varepsilon^{(T)}$, one can rewrite the condition (9.4.83) as

$$2(\mu - \mu^*)\mathbf{E} : \varepsilon^{(T)} + (\lambda - \lambda^*)\operatorname{tr}(\mathbf{E} : \varepsilon^{(T)})\mathbf{I} - 2\mu\varepsilon^{(T)} - \lambda(\operatorname{tr}\varepsilon^{(T)})\mathbf{I}$$
$$= 2(\mu^* - \mu)\varepsilon^\infty + (\lambda^* - \lambda)(\operatorname{tr}\varepsilon^\infty)\mathbf{I} , \quad (9.4.84)$$

where $\mathbf{E}$ is a fourth-order tensor that is completely determined by the semiaxes a_1, a_2, and a_3 of the ellipsoidal inclusion and Poisson's ratio of the matrix. The condition (9.4.84) required for this approach to work constitutes six linearly independent scalar equations (corresponding to the six independent scalar components of stress) so that, given a far-field strain ε^∞ and a transformation strain $\varepsilon^{(T)}$, it will *not*, in general, be possible to find two material parameters μ^* and λ^* such that (9.4.84) is satisfied. However, when the far-field strain and the material properties of the matrix and of the dissimilar inclusion are specified, the six independent scalar components of the transformation strain are uniquely determined by (9.4.84). The result is the transformation strain required for the transforming inclusion to "mimic" the response of the dissimilar inclusion. In this way, the problem of a dissimilar, ellipsoidal inclusion embedded in an infinite matrix subject to a *far-field stress* σ^∞ is solved. Once the required transformation strain $\varepsilon^{(T)}$ is determined by solving (9.4.84), the strain and stress in the dissimilar inclusion are given by (9.4.78) and (9.4.82), sufficiently far from the inclusion the strain and stress are ε^∞ and σ^∞, and the discontinuity in stress across the interface $\partial \mathcal{R}^{\text{incl}}$ is given by (9.4.26).

The equation (9.4.84) to be solved for the required transformation strain $\varepsilon^{(T)}$ can be rewritten as

$$\mathbf{L} : \varepsilon^{(T)} = 2(\mu^* - \mu)\varepsilon^\infty + (\lambda^* - \lambda)e^\infty\mathbf{I} ,$$
$$\mathsf{L}_{ijkl}\varepsilon_{kl}^{(T)} = 2(\mu^* - \mu)\varepsilon_{ij}^\infty + (\lambda^* - \lambda)e^\infty\delta_{ij} , \quad (9.4.85)$$

where $e^\infty = \operatorname{tr}\varepsilon^\infty$ is the dilatational component of the *far-field strain* and

$$\mathsf{L}_{ijkl} = 2(\mu - \mu^*)\mathsf{E}_{ijkl} + (\lambda - \lambda^*)\mathsf{E}_{ppkl}\delta_{ij} - \mu(\delta_{ik}\delta_{jl} + \delta_{il}\delta_{jk}) - \lambda\delta_{ij}\delta_{kl} . \quad (9.4.86)$$

The fourth-order tensor $\mathbf{L}$ has $\mathsf{L}_{ijkl} = \mathsf{L}_{jikl} = \mathsf{L}_{ijlk}$ symmetry, although $\mathsf{L}_{ijkl} \neq \mathsf{L}_{klij}$. In addition, in the Cartesian coordinate system (X_1, X_2, X_3) aligned with the semiaxes of the ellipsoid, $\mathbf{L}$ has the same zero-valued components as $\mathbf{E}$ (i.e., $\mathsf{L}_{1112} = \mathsf{L}_{1123} = \mathsf{L}_{1211} = \cdots = 0$ and $\mathsf{L}_{1223} = \mathsf{L}_{1231} = \mathsf{L}_{2312} = \cdots = 0$). Therefore, in this coordinate system, the normal components of the required transformation strain are determined by a system of three simultaneous equations:

$$\begin{bmatrix} \mathsf{L}_{1111} & \mathsf{L}_{1122} & \mathsf{L}_{1133} \\ \mathsf{L}_{2211} & \mathsf{L}_{2222} & \mathsf{L}_{2233} \\ \mathsf{L}_{3311} & \mathsf{L}_{3322} & \mathsf{L}_{3333} \end{bmatrix} \begin{bmatrix} \varepsilon_{11}^{(T)} \\ \varepsilon_{22}^{(T)} \\ \varepsilon_{33}^{(T)} \end{bmatrix} = \begin{bmatrix} 2(\mu^* - \mu)\varepsilon_{11}^{\infty} + (\lambda^* - \lambda)e^{\infty} \\ 2(\mu^* - \mu)\varepsilon_{22}^{\infty} + (\lambda^* - \lambda)e^{\infty} \\ 2(\mu^* - \mu)\varepsilon_{33}^{\infty} + (\lambda^* - \lambda)e^{\infty} \end{bmatrix},$$

$$(9.4.87)$$

where L_{1111}, L_{2222}, and L_{3333} are given by

$$\mathsf{L}_{iiii} = 2(\mu - \mu^*)\mathsf{E}_{iiii} + (\lambda - \lambda^*)(\mathsf{E}_{11ii} + \mathsf{E}_{22ii} + \mathsf{E}_{33ii}) - 2\mu - \lambda$$

$$\text{(no sum } i) \quad (9.4.88)$$

and L_{1122}, L_{1133}, ..., L_{3322} are given by

$$\mathsf{L}_{iijj} = 2(\mu - \mu^*)\mathsf{E}_{iijj} + (\lambda - \lambda^*)(\mathsf{E}_{11ii} + \mathsf{E}_{22ii} + \mathsf{E}_{33ii}) - \lambda$$

$$\text{(no sum } i, j \text{ and } i \neq j). \quad (9.4.89)$$

The shear components of the required transformation strain, given by

$$\varepsilon_{ij}^{(T)} = \frac{\mu^* - \mu}{2(\mu - \mu^*)\mathsf{E}_{ijij} - \mu} \varepsilon_{ij}^{\infty} \quad \text{(no sum } i, j \text{ and } i \neq j), \quad (9.4.90)$$

are determined independently.

Once the required transformation strain $\varepsilon^{(T)}$ is found from (9.4.87) and (9.4.90), it follows from (9.4.78) and (9.4.83) that the homogeneous strain within the dissimilar, ellipsoidal inclusion is given in terms of its dilatational and deviatoric components by

$$\boxed{e = \frac{K}{K - K^*}e^{(T)}, \quad \grave{\varepsilon} = \frac{\mu}{\mu - \mu^*}\grave{\varepsilon}^{(T)} \quad \forall \mathbf{X} \in \mathcal{R}^{\text{incl}},} \quad (9.4.91)$$

where $K^* = \lambda^* + \frac{2}{3}\mu^*$ is the bulk modulus of the inclusion. Noting that $\tilde{\sigma} = K^*e$, $\grave{\sigma} = 2\mu^*\grave{\varepsilon}$, $\tilde{\sigma}^{(T)} = Ke^{(T)}$, and $\grave{\sigma}^{(T)} = 2\mu\grave{\varepsilon}^{(T)}$, it follows that the homogeneous stress within the dissimilar inclusion is given by

$$\boxed{\tilde{\sigma} = \frac{K^*}{K - K^*}\tilde{\sigma}^{(T)}, \quad \grave{\sigma} = \frac{\mu^*}{\mu - \mu^*}\grave{\sigma}^{(T)} \quad \forall \mathbf{X} \in \mathcal{R}^{\text{incl}}.} \quad (9.4.92)$$

Recall that $\varepsilon = \grave{\varepsilon} + \frac{1}{3}e\mathbf{I}$ and $\sigma = \grave{\sigma} + \tilde{\sigma}\mathbf{I}$. The displacement field within the inclusion is then given by

$$\boxed{\mathbf{u} = (\varepsilon + \omega) \cdot \mathbf{X}, \quad u_i = (\varepsilon_{ij} + \omega_{ij})X_j \quad \forall \mathbf{X} \in \mathcal{R}^{\text{incl}},} \quad (9.4.93)$$

where $\omega = -\omega^{\mathsf{T}}$ is an arbitrary, constant (infinitesimal) rotation tensor and it has been presumed that the displacement field vanishes at the origin.

For an ellipsoidal *void*, $\mu^* = \lambda^* = K^* = 0$ and the equation (9.4.49) for the required transformation strain reduces to $\varepsilon^{(T)} - \varepsilon^{(C)} = \varepsilon^{(T)} - \mathbf{E} : \varepsilon^{(T)} = \varepsilon^\infty$. Therefore, in the Cartesian coordinate system (X_1, X_2, X_3) aligned with the semiaxes of the ellipsoid, the normal components of the required transformation strain are determined by a system of three simultaneous equations:

$$
\begin{bmatrix}
1 - \mathsf{E}_{1111} & -\mathsf{E}_{1122} & -\mathsf{E}_{1133} \\
-\mathsf{E}_{2211} & 1 - \mathsf{E}_{2222} & -\mathsf{E}_{2233} \\
-\mathsf{E}_{3311} & -\mathsf{E}_{3322} & 1 - \mathsf{E}_{3333}
\end{bmatrix}
\begin{bmatrix}
\varepsilon_{11}^{(T)} \\
\varepsilon_{22}^{(T)} \\
\varepsilon_{33}^{(T)}
\end{bmatrix}
=
\begin{bmatrix}
\varepsilon_{11}^\infty \\
\varepsilon_{22}^\infty \\
\varepsilon_{33}^\infty
\end{bmatrix} ,
\tag{9.4.94}
$$

and the shear components are each given by

$$
\varepsilon_{ij}^{(T)} = \frac{\varepsilon_{ij}^\infty}{1 - 2\mathsf{E}_{ijij}} \quad \text{(no sum } i, j \text{ and } i \neq j) .
\tag{9.4.95}
$$

From (9.4.91), it follows that the homogeneous strain within the ellipsoidal void is $\varepsilon = \varepsilon^{(T)}$, which, along with (9.4.93), can be used to determine the displacement at any point on the boundary $\partial\mathcal{R}^{\text{incl}}$ of the void and thus its change in shape. It is seen from (9.4.92) that the stress within the void is, of course, zero. However, the transformation stress $\sigma^{(T)} = 2\mu\varepsilon^{(T)} + \lambda(\operatorname{tr}\varepsilon^{(T)})\mathbf{I}$ needed in (9.4.26) to determine the stress in the matrix at points adjacent to the void is nonzero.

For a *rigid* ellipsoidal inclusion, let μ^*, λ^*, and K^* go to infinity. In this case, the equation (9.4.49) for the required transformation strain reduces to $\varepsilon^{(C)} = \mathbf{E} : \varepsilon^{(T)} = -\varepsilon^\infty$. In the Cartesian coordinate system (X_1, X_2, X_3) aligned with the semiaxes of the ellipsoid, the normal components of the required transformation strain are determined by

$$
\begin{bmatrix}
\mathsf{E}_{1111} & \mathsf{E}_{1122} & \mathsf{E}_{1133} \\
\mathsf{E}_{2211} & \mathsf{E}_{2222} & \mathsf{E}_{2233} \\
\mathsf{E}_{3311} & \mathsf{E}_{3322} & \mathsf{E}_{3333}
\end{bmatrix}
\begin{bmatrix}
\varepsilon_{11}^{(T)} \\
\varepsilon_{22}^{(T)} \\
\varepsilon_{33}^{(T)}
\end{bmatrix}
=
\begin{bmatrix}
-\varepsilon_{11}^\infty \\
-\varepsilon_{22}^\infty \\
-\varepsilon_{33}^\infty
\end{bmatrix}
\tag{9.4.96}
$$

and the shear components are each given by

$$
\varepsilon_{ij}^{(T)} = -\frac{\varepsilon_{ij}^\infty}{2\mathsf{E}_{ijij}} \quad \text{(no sum } i, j \text{ and } i \neq j) .
\tag{9.4.97}
$$

The strain in the rigid inclusion is zero (9.4.91) and it follows from (9.4.92) that the homogeneous stress in the rigid inclusion is

$$
\sigma = -\sigma^{(T)} = -2\mu\varepsilon^{(T)} - \lambda(\operatorname{tr}\varepsilon^{(T)})\mathbf{I} ,
\tag{9.4.98}
$$

which can be used to determine the stress at points in the matrix adjacent to the rigid inclusion and the traction exerted by the matrix on the rigid inclusion across the interface $\partial\mathcal{R}^{\text{incl}}$.

Dissimilar, spherical inclusions

Because the dilatational and deviatoric components of the relation between the transformation and constrained strain fields is decoupled in the case of a spherical inclusion (9.4.74), the equations (9.4.87) and (9.4.90) for the required transformation strain are greatly simplified. By first decomposing the condition (9.4.83) into its dilatational and deviatoric components,

$$K(e^{(C)} - e^{(T)} + e^\infty) = K^*(e^{(C)} + e^\infty),$$
$$\mu(\dot{\varepsilon}^{(C)} - \dot{\varepsilon}^{(T)} + \dot{\varepsilon}^\infty) = \mu^*(\dot{\varepsilon}^{(C)} + \dot{\varepsilon}^\infty),$$

$$\text{(9.4.99)}$$

and after substituting (9.4.74) for $e^{(C)}$ and $\dot{\varepsilon}^{(C)}$ and solving for $e^{(T)}$ and $\dot{\varepsilon}^{(T)}$, one finds that

$$e^{(T)} = \frac{K - K^*}{K + \alpha(K^* - K)} e^\infty, \quad \dot{\varepsilon}^{(T)} = \frac{\mu - \mu^*}{\mu + \beta(\mu^* - \mu)} \dot{\varepsilon}^\infty, \quad \text{(9.4.100)}$$

where α and β are defined by (9.4.75). Therefore, using (9.4.91), the dilatational and deviatoric components of the strain within a dissimilar spherical inclusion embedded in an infinite matrix subjected to a far-field strain $\varepsilon^\infty = \dot{\varepsilon}^\infty + \frac{1}{3}e^\infty\mathbf{I}$ are

$$\left. \begin{aligned} e &= \frac{K}{K + \alpha(K^* - K)} e^\infty \\ \dot{\varepsilon} &= \frac{\mu}{\mu + \beta(\mu^* - \mu)} \dot{\varepsilon}^\infty \end{aligned} \right\} \quad \forall \mathbf{X} \in \mathcal{R}^{\text{incl}} . \quad \text{(9.4.101)}$$

It follows that the mean and deviatoric components of the stress within the dissimilar spherical inclusion in terms of the far-field stress $\sigma^\infty = \dot{\sigma}^\infty + \tilde{\sigma}^\infty\mathbf{I}$ are

$$\left. \begin{aligned} \tilde{\sigma} &= \frac{K^*}{K + \alpha(K^* - K)} \tilde{\sigma}^\infty \\ \dot{\sigma} &= \frac{\mu^*}{\mu + \beta(\mu^* - \mu)} \dot{\sigma}^\infty \end{aligned} \right\} \quad \forall \mathbf{X} \in \mathcal{R}^{\text{incl}} . \quad \text{(9.4.102)}$$

The displacement field within the spherical inclusion is still given by the relation (9.4.93).

Within a spherical *void* ($\mu^* = K^* = 0$), the stress (9.4.102) vanishes and the strain (9.4.101) is given by

$$\left. \begin{aligned} e &= \frac{1}{1 - \alpha} e^\infty = \frac{3}{2}\left(\frac{1 - \nu}{1 - 2\nu}\right) e^\infty \\ \dot{\varepsilon} &= \frac{1}{1 - \beta} \dot{\varepsilon}^\infty = \frac{15(1 - \nu)}{7 - 5\nu} \dot{\varepsilon}^\infty \end{aligned} \right\} \quad \forall \mathbf{X} \in \mathcal{R}^{\text{incl}} . \quad \text{(9.4.103)}$$

It is interesting to note that in the limiting case of an incompressible matrix ($\nu \to 1/2$), the dilatation in the void becomes unbounded, indicating that the spherical void becomes infinitely large. Within a *rigid* spherical inclusion ($K^* \to \infty$ and $\mu^* \to \infty$), the strain (9.4.101) vanishes and the stress (9.4.102) is given by

$$\left. \begin{aligned} \tilde{\sigma} &= \frac{1}{\alpha}\tilde{\sigma}^\infty = 3\left(\frac{1-\nu}{1+\nu}\right)\tilde{\sigma}^\infty \\ \grave{\sigma} &= \frac{1}{\beta}\grave{\sigma}^\infty = \frac{15}{2}\left(\frac{1-\nu}{4-5\nu}\right)\grave{\sigma}^\infty \end{aligned} \right\} \quad \forall \mathbf{X} \in \mathcal{R}^{\text{incl}} \ . \qquad (9.4.104)$$

Problems

9.1 Recall that the representation of a displacement field in terms of its scalar and vector displacement potentials is given by

$$\mathbf{u} = \nabla\zeta + \nabla \times \boldsymbol{\psi} \ ; \quad \nabla \cdot \boldsymbol{\psi} = 0 \ .$$

Show that the displacement field corresponding to the potentials ζ and ψ is the same as that corresponding to the potentials $\zeta + \nabla \cdot \mathbf{w}^{(h)}$ and $\psi - \nabla \times \mathbf{w}^{(h)}$, where $\mathbf{w}^{(h)}$ is an arbitrary vector field.

9.2 Consider an initially stress and strain-free infinite body and suppose that at a point $\mathbf{X} = \boldsymbol{\xi}$, an additional volume of material α is "inserted." This corresponds to a *concentrated dilatation*,

$$e = \nabla \cdot \mathbf{u} = \alpha\,\delta(\mathbf{X} - \boldsymbol{\xi}) \ ,$$

where there is no rotation and the displacement vanishes at infinity:

$$\boldsymbol{\theta} = \frac{1}{2}\nabla \times \mathbf{u} = \mathbf{0} \ , \quad \lim_{|\mathbf{X}|\to\infty} \mathbf{u} = \mathbf{0} \ .$$

Find the displacement field for this problem by considering the scalar and vector displacement potentials. For the special case where $\boldsymbol{\xi} = \mathbf{0}$, show that the displacement flux across any spherical surface centered at the origin is α.

9.3 Show that the displacement field corresponding to the following Papkovich strain functions,

$$\mathbf{B} = \mathbf{0} \ , \quad \beta = -\frac{\mu\alpha}{2\pi\rho} \ , \quad \rho \equiv |\mathbf{X} - \boldsymbol{\xi}| \ ,$$

is the same as that given in the solution to Problem 9.2. Use the definitions (9.2.6) and (9.2.11) for the Papkovich strain functions in terms of the scalar and vector displacement potentials and the expressions for ζ and ψ found in Problem 9.2 to derive alternative expressions for $\mathbf{B}$ and β. Explain the discrepancy in terms of the nonuniqueness of the scalar and vector displacement potential representation of displacement (and therefore of the Papkovich strain function representation), as examined in Problem 9.1.

9.4 Define a cylindrical coordinate system and consider a semi-infinite body occupying the half-space region $\mathcal{R} = \{\, \mathbf{X} \mid z > 0 \,\}$, with boundary $\partial\mathcal{R} = \{\, \mathbf{X} \mid z = 0 \,\}$. Find the solution for a concentrated dilatation of magnitude α acting at a point $\boldsymbol{\xi} = c\hat{\mathbf{e}}_z$ $(c > 0)$ in the body, where the boundary $\partial\mathcal{R}$ is traction-free and the displacement field vanishes at infinity, by assuming Papkovich strain functions of the form $B_r = B_\theta = 0$,

$$B_z = \frac{\mu\alpha}{2\pi}\bar{B}\,, \quad \beta = \frac{\mu\alpha}{2\pi}\left(\bar{\beta} - \frac{1}{\rho}\right)\,,$$

where $\rho \equiv |\mathbf{X} - \boldsymbol{\xi}|$.

9.5 Consider a dislocation surface $\mathcal{S} = \{\, \mathbf{X} \mid X_1 < 0, X_2 = 0 \,\}$ that occupies the portion of the X_1X_3-plane for which $X_1 < 0$, and has a Berger's vector $\mathbf{b} = b\hat{\mathbf{e}}_3$. Note the Berger's vector is parallel to the dislocation loop $\partial\mathcal{S} = \{\, \mathbf{X} \mid X_1 = X_2 = 0 \,\}$, which lies along the X_3-axis—this is known as a *screw dislocation*. Derive an expression for the displacement field due to this dislocation surface.

9.6 Beginning with (9.4.10), derive the expression (9.4.11) for the constrained displacement field $\mathbf{u}^{(C)}$ in the transforming inclusion problem.

9.7 Given an orthonormal basis $\{\hat{\mathbf{e}}_i\}$ show that at a point $\boldsymbol{\zeta}$ on the boundary of an inclusion where the unit outward normal is $\hat{\mathbf{n}} = \hat{\mathbf{e}}_1$, the discontinuity in the stress field across the interface (9.4.26) reduces to

$$\boldsymbol{\sigma}(\boldsymbol{\zeta}^{\mathrm{m}}) - \boldsymbol{\sigma}(\boldsymbol{\zeta}^{\mathrm{i}}) = \left(\sigma_{22}^{(T)} - \frac{\nu}{1-\nu}\sigma_{11}^{(T)}\right)\hat{\mathbf{e}}_2\hat{\mathbf{e}}_2$$
$$+ \left(\sigma_{33}^{(T)} - \frac{\nu}{1-\nu}\sigma_{11}^{(T)}\right)\hat{\mathbf{e}}_3\hat{\mathbf{e}}_3 + \sigma_{23}^{(T)}(\hat{\mathbf{e}}_2\hat{\mathbf{e}}_3 + \hat{\mathbf{e}}_3\hat{\mathbf{e}}_2)\,.$$

9.8 Consider a spherical inclusion with Lamé constants μ^* and λ^* embedded in a matrix with Lamé constants μ and λ and subject to a far-field shear $\boldsymbol{\sigma}^\infty = \tau(\hat{\mathbf{e}}_1\hat{\mathbf{e}}_3 + \hat{\mathbf{e}}_3\hat{\mathbf{e}}_1)$.

(a) Determine the strain and stress within the inclusion.

(b) Using a spherical coordinate system with the origin at the center of the inclusion, derive expressions for the scalar components of stress at points in the matrix adjacent to the inclusion.

(c) In the case of a spherical void, determine the maximum principal stress and the maximum shear stress that occur in the matrix adjacent to the void, and their corresponding locations.

9.9 Consider a spherical inclusion with Lamé constants μ^* and λ^* embedded in a matrix with Lamé constants μ and λ and subject to a far-field hydrostatic tension $\sigma^\infty = p\mathbf{I}$.

(a) Using a spherical coordinate system with the origin at the center of the inclusion, derive expressions for the scalar components of stress at points in the matrix adjacent to the inclusion.

(b) Show that, in the case of a spherical void, the maximum principal stress at points in the matrix adjacent to the inclusion is $3p/2$.

(c) Show that, in the case of a rigid spherical inclusion, the maximum principal stress at points in the matrix adjacent to the inclusion is $3(1-\nu)p/(1+\nu)$.

9.10 Consider a spherical void in a matrix with Lamé constants μ and λ and subject to a far-field uniaxial tension $\sigma^\infty = \sigma\hat{\mathbf{e}}_1\hat{\mathbf{e}}_1$.

(a) If the void is a unit sphere in the reference configuration, show that it is deformed into a prolate spherical void whose boundary $\partial\mathcal{R}_t^{\mathrm{incl}}$ can be given by

$$\partial\mathcal{R}_t^{\mathrm{incl}} = \left\{ \mathbf{x} \mid \frac{x_1^2}{a^2} + \frac{x_2^2 + x_3^2}{b^2} = 1 \right\},$$

where the semiaxes are given in terms of Young's modulus E and Poisson's ratio ν by

$$a = 1 + \frac{3(1-\nu)(9+5\nu)\sigma}{2(7-5\nu)E}, \quad b = 1 - \frac{3(1-\nu)(1+5\nu)\sigma}{2(7-5\nu)E}.$$

(b) Derive expressions for the stress in the matrix adjacent to the spherical void at the three points corresponding to $\hat{\mathbf{n}} = \hat{\mathbf{e}}_1$, $\hat{\mathbf{n}} = \hat{\mathbf{e}}_2$, and $\hat{\mathbf{n}} = \hat{\mathbf{e}}_3$. Use superposition to verify that the result is consistent with the result for far-field hydrostatic tension considered in Problem 9.9.

9.11 Recall that an oblate spheroid is an ellipsoid in which the semi-axes satisfy $a_1 = a_2 = a$ and $0 \leq a_3 < a$. An oblate spheroidal void becomes a *penny-shaped crack*, centered at the origin and lying in the X_1X_2-plane, when $a_3 \to 0$. The boundary $\partial \mathcal{R}^{\text{incl}}$ becomes two opposing *crack faces*—the portion of the boundary for which $X_3 > 0$ becomes the positive $(+)$ crack face and the portion of the boundary for which $X_3 < 0$ becomes the negative $(-)$ crack face. The portion of the boundary for which $X_1^2 + X_2^2 = a^2$ becomes the (circular) *crack front*. Let $(X_1, X_2, 0^+)$ and $(X_1, X_2, 0^-)$, where $X_1^2 + X_2^2 < a^2$, represent corresponding points on the positive and negative crack faces, respectively. The *crack opening displacement* $\boldsymbol{\delta}(X_1, X_2) \equiv \mathbf{u}(X_1, X_2, 0^+) - \mathbf{u}(X_1, X_2, 0^-)$ is the displacement discontinuity across the crack.

(a) If $\boldsymbol{\varepsilon}$ is the homogeneous strain field in the oblate spheroidal void and it is required that the displacement field within the void satisfy $\mathbf{u}|_{\mathbf{X}=\mathbf{0}} = \mathbf{0}$ and $u_3|_{X_3=0} = 0$, show that the crack opening displacement field for the corresponding penny-shaped crack has scalar components given by

$$\delta_1(X_1, X_2) = \lim_{\hat{a}_3 \to 0} 4\hat{a}_3 \varepsilon_{13} \sqrt{a^2 - X_1^2 - X_2^2},$$

$$\delta_2(X_1, X_2) = \lim_{\hat{a}_3 \to 0} 4\hat{a}_3 \varepsilon_{23} \sqrt{a^2 - X_1^2 - X_2^2},$$

$$\delta_3(X_1, X_2) = \lim_{\hat{a}_3 \to 0} 2\hat{a}_3 \varepsilon_{33} \sqrt{a^2 - X_1^2 - X_2^2},$$

where $\hat{a}_3 \equiv a_3/a$.

(b) Show that, in the limit as $\hat{a}_3 \to 0$, the strain $\boldsymbol{\varepsilon}$ within the oblate spheroidal void is given by the system of equations

$$\begin{bmatrix} 1 & 0 & \dfrac{\pi}{8}\left(\dfrac{1-2\nu}{1-\nu}\right) \\[2mm] 0 & 1 & \dfrac{\pi}{8}\left(\dfrac{1-2\nu}{1-\nu}\right) \\[2mm] -\dfrac{\nu}{1-\nu} & -\dfrac{\nu}{1-\nu} & \dfrac{\pi}{4}\left(\dfrac{1-2\nu}{1-\nu}\right) \end{bmatrix} \begin{bmatrix} \varepsilon_{11} \\[2mm] \varepsilon_{22} \\[2mm] \hat{a}_3 \varepsilon_{33} \end{bmatrix} = \begin{bmatrix} \varepsilon_{11}^{\infty} \\[2mm] \varepsilon_{22}^{\infty} \\[2mm] \varepsilon_{33}^{\infty} \end{bmatrix},$$

and

$$\varepsilon_{12} = \varepsilon_{12}^{\infty},$$

$$\hat{a}_3 \varepsilon_{23} = \frac{4}{\pi}\left(\frac{1-\nu}{2-\nu}\right)\varepsilon_{23}^{\infty},$$

$$\hat{a}_3 \varepsilon_{31} = \frac{4}{\pi}\left(\frac{1-\nu}{2-\nu}\right)\varepsilon_{31}^{\infty}.$$

(c) Find the crack opening displacement field for a penny-shaped crack of radius a that is centered at the origin and lies in the $X_1 X_2$-plane if the far-field stress in the matrix is $\boldsymbol{\sigma}^\infty = \sigma \hat{\mathbf{e}}_3 \hat{\mathbf{e}}_3 + \tau(\hat{\mathbf{e}}_1 \hat{\mathbf{e}}_3 + \hat{\mathbf{e}}_3 \hat{\mathbf{e}}_1)$.

Chapter 10

Variational Methods

Variational methods involve the use of minimization principles such as the principles of minimum potential energy and minimum complementary energy. These minimization principles can be used to derive governing equations and boundary conditions for specialized classes of problems in elasticity. Another important application of energy methods is in finding approximate solutions to elasticity problems, which involves making assumptions about the relative accuracy of different approximations based on the minimization of certain quantities to be defined. For instance, this is the basis of the numerical algorithm known as the finite element method.

The calculus of variations is an important tool in minimization problems, such as those that arise when using the principles of minimum potential energy and minimum complementary energy. It is appropriate then to begin with a brief review of the calculus of variations.

10.1   Calculus of Variations

Before introducing the calculus of variations, one can first motivate one's efforts by examining a representative physical problem that will require the calculus of variations to obtain a solution. Consider the surface of revolution formed by rotating a curve, defined in the xy-plane by $y = u(x)$, about the x-axis (see Figure 10.1). Given that the surface begins at x_0 with a radius of y_0 and ends at x_1 with a radius of y_1 [i.e., given the boundary conditions $u(x_0) = y_0$ and $u(x_1) = y_1$], one may ask which continuous function u minimizes the area of this surface of revolution?

To answer this question, one must first derive the relation between the area A of the surface of revolution and the function u. Such a relation, where the argument is a function, is called a *functional*. Let the differential area dA of the surface be a circumferential strip of width ds, where s is the

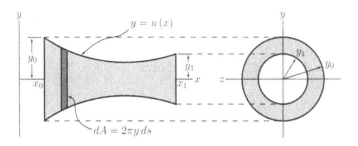

FIGURE 10.1. Surface of revolution.

arc length measured along the curve $y = u(x)$, so that dA is

$$dA = 2\pi y \, ds \; . \tag{10.1.1}$$

Using the Pythagorean theorem, the differential arc length ds is

$$\begin{aligned}
ds &= \sqrt{dx^2 + dy^2} \\
&= \sqrt{1 + (dy/dx)^2} \, dx \\
&= \sqrt{1 + (u')^2} \, dx \; ,
\end{aligned} \tag{10.1.2}$$

so that, with $y = u(x)$, the differential area is given as a function of x by

$$dA = 2\pi u \sqrt{1 + (u')^2} \, dx \; . \tag{10.1.3}$$

Thus, the total area of the surface of revolution is

$$A[u(x)] = \int_{x_0}^{x_1} 2\pi u \sqrt{1 + (u')^2} \, dx \; . \tag{10.1.4}$$

The relation $A[u(x)]$ is a *functional of the function* u, where the notation emphasizes that the argument of A is a function u with independent variable x. To minimize the area of the surface of revolution, one must find the continuous function u that minimizes the functional (10.1.4) subject to the boundary conditions $u(x_0) = y_0$ and $u(x_1) = y_1$. This is where the calculus of variations enters the picture.

Consider the more general problem of finding the continuous function u that minimizes the functional

$$I[u(x)] = \int_{x_0}^{x_1} F(x, u, u') \, dx \; . \tag{10.1.5}$$

The question of boundary conditions for u will be addressed below. The integrand F in (10.1.5) is a function of x in which $u(x)$ and $u'(x)$ appear.

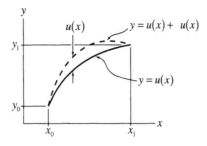

FIGURE 10.2.   *The variation of a function.*

The surface of revolution problem considered above is a special case of
(10.1.5), for which x does not appear explicitly in the integrand.

Suppose that the function u minimizes the functional (10.1.5) in some
neighborhood h of u, so that

$$I[u(x) + \delta u(x)] \geq I[u(x)] \tag{10.1.6}$$

for *all* continuous functions δu that satisfy

$$|\delta u(x)| < h \quad \forall x \in [x_0, x_1] \,, \tag{10.1.7}$$

where the equality in (10.1.6) holds only when $\delta u = 0$ everywhere on the
interval (see Figure 10.2). The function δu is known as the *variation of u.*

Now, let $\delta u = \epsilon \eta$, where η is an arbitrary continuous function and ϵ is a
parameter small enough that (10.1.7) is satisfied. One can then define the
function

$$\Phi(\epsilon) \equiv I[u(x) + \epsilon \eta(x)] = \int_{x_0}^{x_1} F(x, u + \epsilon \eta, u' + \epsilon \eta') \, dx \,. \tag{10.1.8}$$

Since u minimizes I, it follows that $\epsilon = 0$ minimizes Φ. A necessary (though
insufficient) condition for this is

$$\left. \frac{d\Phi}{d\epsilon} \right|_{\epsilon=0} = 0 \,. \tag{10.1.9}$$

Differentiating Φ with respect to ϵ and applying the chain rule,

$$\frac{d\Phi}{d\epsilon} = \int_{x_0}^{x_1} \left[\frac{\partial F(x, u + \epsilon \eta, u' + \epsilon \eta')}{\partial(u + \epsilon \eta)} \eta + \frac{\partial F(x, u + \epsilon \eta, u' + \epsilon \eta')}{\partial(u' + \epsilon \eta')} \eta' \right] dx \,. \tag{10.1.10}$$

Therefore,

$$\left. \frac{d\Phi}{d\epsilon} \right|_{\epsilon=0} = \int_{x_0}^{x_1} \left[\frac{\partial F(x, u, u')}{\partial u} \eta + \frac{\partial F(x, u, u')}{\partial u'} \eta' \right] dx \,. \tag{10.1.11}$$

By integrating the second term in the integrand by parts, (10.1.11) can be rewritten as

$$\frac{d\Phi}{d\epsilon}\bigg|_{\epsilon=0} = \frac{\partial F}{\partial u'}\eta\bigg|_{x_0}^{x_1} + \int_{x_0}^{x_1}\left[\frac{\partial F}{\partial u}\eta - \frac{d}{dx}\left(\frac{\partial F}{\partial u'}\right)\eta\right]dx . \qquad (10.1.12)$$

Therefore, a necessary condition for u to minimize the functional I is that it satisfy

$$\frac{\partial F}{\partial u'}\eta\bigg|_{x_0}^{x_1} + \int_{x_0}^{x_1}\left[\frac{\partial F}{\partial u} - \frac{d}{dx}\left(\frac{\partial F}{\partial u'}\right)\right]\eta\,dx = 0 \qquad (10.1.13)$$

for all continuous functions η. This is true if and only if u satisfies the ordinary differential equation

$$\boxed{\frac{\partial F}{\partial u} - \frac{d}{dx}\left(\frac{\partial F}{\partial u'}\right) = 0 \quad \forall x \in (x_0, x_1)} \qquad (10.1.14)$$

and the boundary conditions

$$\boxed{\text{either} \quad \frac{\partial F}{\partial u'}\bigg|_{x=x_0} = 0 \quad \text{or} \quad \eta(x_0) = 0} \qquad (10.1.15)$$

and

$$\boxed{\text{either} \quad \frac{\partial F}{\partial u'}\bigg|_{x=x_1} = 0 \quad \text{or} \quad \eta(x_1) = 0 .} \qquad (10.1.16)$$

If u does not satisfy (10.1.14)–(10.1.16), then it can be shown that it is always possible to find a continuous function η such that (10.1.13) is not satisfied. Using integration by parts to rewrite (10.1.11) as (10.1.12) was necessary in order to arrive at a condition (10.1.13) that involves only η and not its derivatives. Because η and its derivatives (specifically η') are not independent, it would be incorrect to conclude from (10.1.11) that $\partial F/\partial u = 0$ and $\partial F/\partial u' = 0$ on the interval (x_0, x_1) are necessary conditions for satisfying (10.1.9). The differential equation (10.1.14) is often referred to as the *Euler equation* for the functional I (10.1.5).

It is seen that the boundary condition (10.1.15) at x_0 that is necessary for u to minimize the functional I can be satisfied in one of two ways. Suppose first that the original minimization problem is augmented by the requirement that $u(x_0) = y_0$ where y_0 is a prescribed constant, as was the case for the motivating problem of minimizing the area of a surface of revolution. Then, since one is only considering functions that satisfy this boundary condition, it must be required in (10.1.6) that $u(x_0)+\delta u(x_0) = y_0$

or, equivalently, that only variations of u that satisfy $\delta u(x_0) = 0$ are to be considered. It follows from the relation $\delta u = \epsilon \eta$ that only functions η that satisfy $\eta(x_0) = 0$ are to be considered. Thus, in this case, the boundary condition (10.1.15) is automatically satisfied. This is known as an *imposed boundary condition*. Alternatively, if a boundary condition at x_0 is not imposed, then

$$\left. \frac{\partial F}{\partial u'} \right|_{x=x_0} = 0 , \tag{10.1.17}$$

is a necessary condition for u to minimize I. This is known as a *natural boundary condition*. An identical situation exists for the boundary condition (10.1.16) at x_1.

As already mentioned, the conditions (10.1.14)–(10.1.16) are necessary but insufficient conditions for minimizing the functional I. A function that satisfies these conditions may, in fact, maximize the functional or correspond to a saddle point. One also has to consider the possibility that such a solution may only represent a local, rather than a global or absolute, minimum. Independent of these questions, a function that satisfies the conditions (10.1.14)–(10.1.16) is said to render the functional I *stationary*, or to be a *stationary function* of the functional. To mathematically determine whether such a function minimizes (10.1.5) requires higher-order derivatives. In practice, when dealing with functionals derived from physical problems, minimization (or maximization) can often be established on physical grounds.

Returning now to the problem of finding the curve $y = u(x)$ that minimizes the area of the surface of revolution, that is, finding the function u that minimizes the functional $A[u(x)]$ given by (10.1.4), the integrand $F(x, u, u')$ is

$$F = 2\pi u \sqrt{1 + (u')^2} . \tag{10.1.18}$$

Since this problem has imposed boundary conditions at x_0 and x_1, one need only consider the Euler equation (10.1.14)—the necessary boundary conditions (10.1.15) and (10.1.16) are automatically satisfied. Substituting for F in (10.1.14) yields the following Euler equation for the functional A:

$$2\pi \sqrt{1 + (u')^2} - \frac{d}{dx} \left[\frac{2\pi u u'}{\sqrt{1 + (u')^2}} \right] = 0 . \tag{10.1.19}$$

One could, in principle, carry out the differentiation in x in the second term and solve the resultant second-order ordinary differential equation for u. The solution to (10.1.19) can more easily be determined, however, by using the argument of the differentiation in the second term as the basis for introducing the function

$$\xi(x) \equiv \frac{uu'}{\sqrt{1 + (u')^2}} \; , \tag{10.1.20}$$

and rewriting (10.1.19) as

$$2\pi \frac{uu'}{\xi} - 2\pi \xi' = 0 \; . \tag{10.1.21}$$

Separating variables (and canceling dx and π),

$$2u \, du = 2\xi \, d\xi \; , \tag{10.1.22}$$

and integrating yields

$$u^2 = \xi^2 + C^2 \; , \tag{10.1.23}$$

where C is a constant. Substituting (10.1.20) for ξ and solving for u' leads to the first-order ordinary differential equation

$$u' = \sqrt{(u/C)^2 - 1} \; . \tag{10.1.24}$$

This finally can be solved by separating variables,

$$\frac{du}{\sqrt{(u/C)^2 - 1}} = dx \; , \tag{10.1.25}$$

and integrating to get

$$C \cosh^{-1}(u/C) = x + D \; , \tag{10.1.26}$$

where D is another constant. Thus, the general solution to (10.1.19) is

$$u(x) = C \cosh\left(\frac{x + D}{C}\right) \; . \tag{10.1.27}$$

The constants of integration are determined by the boundary conditions $u(x_0) = y_0$ and $u(x_1) = y_1$. This is the only solution to Euler's equation in this case and physically one can see that this is the equation of the curve that minimizes the area of the surface of revolution [the maximum area is infinite, corresponding to curves $y = u(x)$ that are unbounded in the interval (x_0, x_1)].

10.1.1 Process of "Taking Variations"

The general functional (10.1.5) is just one of many possible general forms for functionals that may arise in minimization problems of the sort illustrated by the surface of revolution problem. For instance, functionals could have integrands that depend on higher-order derivatives of the independent function,

$$J[u(x)] = \int_{x_0}^{x_1} F(x, u, u', u'', \ldots, u^{(n)}) \, dx , \tag{10.1.28}$$

or on more than one independent variable,

$$K[u(x, y)] = \int_S F\left(x, y, u, \frac{\partial u}{\partial x}, \frac{\partial u}{\partial y}\right) dA . \tag{10.1.29}$$

In each case, the procedure outlined above can be used for minimization problems involving such functionals. In practice, however, a less formal approach is taken that one can refer to as the process of *taking variations*.

Consider again the general functional

$$I[u(x)] = \int_{x_0}^{x_1} F(x, u, u') \, dx . \tag{10.1.30}$$

For arbitrary, small variations δu of the continuous function u that satisfy

$$|\delta u(x)| \ll 1 , \quad |\delta u'(x)| \ll 1 \quad \forall x \in [x_0, x_1] , \tag{10.1.31}$$

the "variation" of the integrand F is

$$\begin{aligned}
\delta F &\equiv F(x, u + \delta u, u' + \delta u') - F(x, u, u') \\
&= F(x, u, u') + \frac{\partial F}{\partial u} \delta u + \frac{\partial F}{\partial u'} \delta u' - F(x, u, u') \\
&= \frac{\partial F}{\partial u} \delta u + \frac{\partial F}{\partial u'} \delta u' ,
\end{aligned} \tag{10.1.32}$$

and the corresponding "variation" of the functional I is

$$\begin{aligned}
\delta I &\equiv \int_{x_0}^{x_1} \delta F \, dx \\
&= \int_{x_0}^{x_1} \left(\frac{\partial F}{\partial u} \delta u + \frac{\partial F}{\partial u'} \delta u'\right) dx \\
&= \frac{\partial F}{\partial u'} \delta u \Big|_{x_0}^{x_1} + \int_{x_0}^{x_1} \left[\frac{\partial F}{\partial u} - \frac{d}{dx}\left(\frac{\partial F}{\partial u'}\right)\right] \delta u \, dx .
\end{aligned} \tag{10.1.33}$$

Assuming that a necessary condition for a continuous function u to minimize I is that it satisfy $\delta I = 0$, for every continuous function δu that satisfies (10.1.31), then (10.1.33) yields the same necessary conditions (10.1.14)– (10.1.16) as were derived previously.

This informal approach can be similarly applied to other functionals. Note the similarity to differentiation using the chain rule when taking variations of the integrand of a functional. One should bear in mind, however, that this informal process for applying the calculus of variations is not rigorously defined and that inconsistencies and ambiguities may on occasion crop up when it is applied. In such instances, one should revert to the formal process.

Example 10.1
Consider the functional

$$I[\mathbf{u}(\mathbf{x})] = \int_{\mathcal{R}} F(\mathbf{x}, \mathbf{u}, \nabla \mathbf{u}) \, dV \ ,$$

whose argument is the vector field $\mathbf{u}$ which is a function of the position vector $\mathbf{x}$. The domain of integration $\mathcal{R}$ is a region in space. Determine the Euler equation and associated boundary conditions that are necessary conditions for the vector field $\mathbf{u}$ to minimize this functional.

Solution. Using a Cartesian coordinate system in which $\mathbf{x} = x_i \hat{\mathbf{e}}_i$, $\mathbf{u} = u_i \hat{\mathbf{e}}_i$, and $\nabla \mathbf{u} = u_{i,j} \, \hat{\mathbf{e}}_i \hat{\mathbf{e}}_j$, the functional I can be expressed in terms of the scalar components of the vector field $\mathbf{u}$:

$$I[\mathbf{u}(\mathbf{x})] = \int_{\mathcal{R}} \tilde{F}(x_i, u_i, u_{i,j}) \, dV \ .$$

It follows that the variation of the functional is

$$\delta I = \int_{\mathcal{R}} \left(\frac{\partial \tilde{F}}{\partial u_i} \delta u_i + \frac{\partial \tilde{F}}{\partial u_{i,j}} \delta u_{i,j} \right) dV \ ,$$

where δu_i are three arbitrary scalar fields that may be taken to be the scalar components of an arbitrary vector field $\delta \mathbf{u}$. Since the partial derivatives $\delta u_{i,j}$ of these scalar fields are not independent, the variation of the functional needs to be expressed entirely in terms of δu_i. In the one-dimensional cases considered previously, this was accomplished using integration by parts. In this three-dimensional problem, this is accomplished by using the divergence theorem. By noting that

$$\int_{\mathcal{R}} \frac{\partial \tilde{F}}{\partial u_{i,j}} \delta u_{i,j} \, dV = \int_{\mathcal{R}} \frac{\partial}{\partial x_j} \left(\frac{\partial \tilde{F}}{\partial u_{i,j}} \delta u_i \right) dV - \int_{\mathcal{R}} \frac{\partial}{\partial x_j} \left(\frac{\partial \tilde{F}}{\partial u_{i,j}} \right) \delta u_i \, dV \ ,$$

it follows from the divergence theorem that

$$\int_{\mathcal{R}} \frac{\partial \tilde{F}}{\partial u_{i,j}} \delta u_{i,j} \, dV = \int_{\partial \mathcal{R}} \left(\frac{\partial \tilde{F}}{\partial u_{i,j}} \delta u_i \right) \hat{n}_j \, dA - \int_{\mathcal{R}} \frac{\partial}{\partial x_j} \left(\frac{\partial \tilde{F}}{\partial u_{i,j}} \right) \delta u_i \, dV ,$$

where $\partial \mathcal{R}$ is the boundary of $\mathcal{R}$, with unit outward normal $\hat{\mathbf{n}} = \hat{n}_i \hat{\mathbf{e}}_i$. Thus, the variation of the functional I can be rewritten as

$$\delta I = \int_{\partial \mathcal{R}} \left(\frac{\partial \tilde{F}}{\partial u_{i,j}} \hat{n}_j \right) \delta u_i \, dA + \int_{\mathcal{R}} \left[\frac{\partial \tilde{F}}{\partial u_i} - \frac{\partial}{\partial x_j} \left(\frac{\partial \tilde{F}}{\partial u_{i,j}} \right) \right] \delta u_i \, dV .$$

Assuming that a necessary condition for a continuous vector field $\mathbf{u}$ to minimize the functional I is that it satisfy $\delta I = 0$ for every (sufficiently small) variation $\delta \mathbf{u}$, it follows that the Euler equation

$$\frac{\partial \tilde{F}}{\partial u_i} - \frac{\partial}{\partial x_j} \left(\frac{\partial \tilde{F}}{\partial u_{i,j}} \right) = 0 \quad \forall \mathbf{x} \in \mathcal{R} ,$$

and the boundary condition

$$\text{either} \quad \frac{\partial \tilde{F}}{\partial u_{i,j}} \hat{n}_j = 0 \quad \text{or} \quad \delta u_i = 0 \quad \forall \mathbf{x} \in \partial \mathcal{R} ,$$

are necessary conditions in a Cartesian coordinate system. Compare these conditions with the necessary conditions (10.1.14)–(10.1.16) for the one-dimensional analog of this problem.

10.1.2 Using Lagrange Multipliers

At times, one will be faced with a problem requiring the minimization of some function or functional subject to some constraint. One way to deal with such problems is to use *Lagrange multipliers* to define a new function or functional, the minimization of which will encompass both the original minimization problem and the constraint. Lagrange multipliers were used previously in the derivation of the eigenvalue problem (Section 2.3.6). Lagrange multipliers will be required again in this section, so it is worthwhile examining their use a little more closely.

As a simple example, consider a parallelepiped with edge lengths a, b, and c (Figure 10.3). Suppose that the surface area of the parallelepiped is prescribed, $2ab + 2bc + 2ca = A_0$, and that, subject to this constraint, one is asked to find the values of the edge lengths a, b, and c that maximize the volume, $V = abc$. To solve this problem, define a new function

$$\tilde{W}(a, b, c, \lambda) = V - \lambda \left(ab + bc + ca - \frac{1}{2} A_0 \right) , \tag{10.1.34}$$

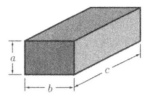

FIGURE 10.3. Parallelepiped.

where the Lagrange multiplier λ is an independent parameter. Values of a, b, c, and λ that correspond to extrema of $\tilde{W}$, which are found by setting the partial derivatives of $\tilde{W}$ with respect to the four independent variables equal to zero, correspond to extrema of V subject to the given constraint on the surface area. Setting the partial derivative of $\tilde{W}$ with respect to λ equal to zero,

$$\frac{\partial \tilde{W}}{\partial \lambda} = -ab - bc - ca + \frac{1}{2}A_0 = 0 , \qquad (10.1.35)$$

shows that the constraint on the surface area will be satisfied. Setting the partial derivatives of $\tilde{W}$ with respect to a, b, and c equal to zero,

$$\frac{\partial \tilde{W}}{\partial a} = bc - \lambda(b + c) = 0 ,$$

$$\frac{\partial \tilde{W}}{\partial b} = ac - \lambda(a + c) = 0 , \qquad (10.1.36)$$

$$\frac{\partial \tilde{W}}{\partial a} = ab - \lambda(a + b) = 0 ,$$

then yields $a = b = c = 2\lambda$. Thus, the parallelepiped with maximum volume for any given surface area is a cube. Substituting this result back into the constraint on the surface area gives $\lambda = \sqrt{A_0/24}$ and $a = b = c = \sqrt{A_0/6}$.

Note that the value A_0 of the prescribed surface area does not appear in the equations (10.1.36) for the values of a, b, and c that maximize the volume of the parallelepiped subject to the constraint on the surface area. Thus, in practice, to simplify subsequent computations, the function considered would typically be

$$W(a, b, c, \lambda) = V - \lambda(ab + bc + ca) . \qquad (10.1.37)$$

Setting the partial derivatives of W with respect to a, b, and c equal to zero gives the same results as (10.1.36). The resulting answer will generally involve the Lagrange multiplier, the value of which is not known, but this

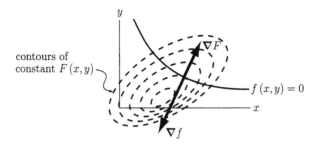

FIGURE 10.4. Contours of constant $F(x, y)$ and line $f(x, y) = 0$.

can be determined from the initial constraint. Setting the partial derivative of W with respect to λ equal to zero does *not* return the initial constraint on the surface area, but that does not matter since the constraint is known *a priori*.

The Lagrange multiplier in this example is a constant scalar parameter. It will be seen later that, in other instances, the Lagrange multiplier may be a tensor and it may be a function of the independent parameter(s).

Example 10.2

In the xy-plane, consider a field $F(x, y)$ and a line defined by $f(x, y) = 0$ (Figure 10.4). Find the points on the line $f(x, y) = 0$ where $F(x, y)$ is a minimum or maximum; that is, find the extrema of $F(x, y)$ subject to the constrain $f(x, y) = 0$.

Solution. Using the Lagrange multiplier λ, define the function

$$U(x, y, \lambda) = F(x, y) - \lambda f(x, y) .$$

Setting the partial derivatives of U with respect to x and y equal to zero,

$$\frac{\partial U}{\partial x} = \frac{\partial F}{\partial x} - \lambda \frac{\partial f}{\partial x} = 0 ,$$

$$\frac{\partial U}{\partial y} = \frac{\partial F}{\partial y} - \lambda \frac{\partial f}{\partial y} = 0 .$$

Thus, it is seen that the extrema of U correspond to solutions of

$$\nabla F = \lambda \nabla f .$$

Recall that the vector ∇F at a point in the xy-plane is normal to the contour of constant $F(x, y)$ at that point and, similarly, the vector ∇f at a point on the line $f(x, y) = 0$ is normal to the line at that point. Thus, at points on the line $f(x, y) = 0$ where $F(x, y)$ is a minimum or maximum, the line is tangent to contours of constant $F(x, y)$.

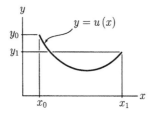

FIGURE 10.5. Hanging chain.

Example 10.3

Consider a uniform chain hanging between two points (x_0, y_0) and (x_1, y_1) with prescribed length L_0, where

$$L_0 > \sqrt{(x_1 - x_0)^2 + (y_1 - y_0)^2} \ .$$

The gravitational force acts in the negative y-direction (Figure 10.5). Formulate the calculus of variations problem for the curve $y = u(x)$ occupied by the chain in equilibrium.

Solution. The chain's potential energy is

$$\Pi = \int_{s_0}^{s_1} \gamma y \, ds \ ,$$

where γ is the weight per unit length of the chain and s is the distance measured along the length of the chain from s_0, corresponding to (x_0, y_0), to s_1, corresponding to (x_1, y_1). The total length of the chain is therefore

$$L = \int_{s_0}^{s_1} ds \ .$$

The equilibrium position of the chain minimizes the chain's potential energy, subject to the constrain $L = L_0$. Consequently, one needs to minimize

$$U = \Pi - \lambda L = \int_{s_0}^{s_1} (\gamma y - \lambda) \, ds \ ,$$

where λ is a Lagrange multiplier. Recall from the surface of revolution problem that, using $y = u(x)$, the differential arc length ds can be given as a function of x by

$$ds = \sqrt{1 + (u')^2} \, dx \ .$$

Thus, U can be rewritten as a functional of $u(x)$;

$$U[u(x)] = \int_{x_0}^{x_1} (\gamma u - \lambda) \sqrt{1 + (u')^2} \, dx \ .$$

Setting $\delta U = 0$ yields the solution for the curve $y = u(x)$ occupied by the chain in equilibrium. The answer involves λ, the value of which is determined by imposing the constraint on the total length of the chain.

10.2 Energy Theorems in Elasticity

In the derivations that follow, the kinematic linearization $\|\mathrm{Grad}\,\mathbf{u}\| \ll 1$ (which is implicit in the use of the infinitesimal strain tensor $\boldsymbol{\varepsilon}$) and hyperelasticity are assumed. However, the possibility of nonlinear constitutive relations $\boldsymbol{\sigma} = \mathbf{h}(\boldsymbol{\varepsilon})$ are considered.

10.2.1 Principle of Virtual Work

Consider a body occupying the region $\mathcal{R}$ with boundary $\partial\mathcal{R}$. At a point on $\partial\mathcal{R}$, the unit outward normal is $\hat{\mathbf{n}}$.

- Let $\boldsymbol{\sigma} = \boldsymbol{\sigma}^{(1)}(\mathbf{X})$ be *any* stress field defined on $\mathcal{R}$. The corresponding traction at a point on $\partial\mathcal{R}$ is $\mathbf{t}^{(1)} = \hat{\mathbf{n}} \cdot \boldsymbol{\sigma}^{(1)}$. Let the corresponding body force field be $\mathbf{f}^{(1)} = -\boldsymbol{\nabla} \cdot (\boldsymbol{\sigma}^{(1)})^{\mathsf{T}}$, so that the stress field is in equilibrium.

- Let $\mathbf{u} = \mathbf{u}^{(2)}(\mathbf{X})$ be *any* displacement field defined on $\mathcal{R}$, with corresponding strain field $\boldsymbol{\varepsilon}^{(2)} = \frac{1}{2}\left[\boldsymbol{\nabla}\mathbf{u}^{(2)} + (\boldsymbol{\nabla}\mathbf{u}^{(2)})^{\mathsf{T}}\right]$.

It is *not* required that the stress field $\boldsymbol{\sigma}^{(1)}$ satisfy compatibility nor is it required that the stress field be related to the strain field $\boldsymbol{\varepsilon}^{(2)}$ by any particular constitutive relation. The *principle of virtual work* says that

$$\boxed{\begin{aligned}\int_{\mathcal{R}} \boldsymbol{\sigma}^{(1)} : \boldsymbol{\varepsilon}^{(2)} \, dV &= \int_{\partial\mathcal{R}} \mathbf{t}^{(1)} \cdot \mathbf{u}^{(2)} \, dA + \int_{\mathcal{R}} \mathbf{f}^{(1)} \cdot \mathbf{u}^{(2)} \, dV \,, \\ \int_{\mathcal{R}} \sigma_{ij}^{(1)} \varepsilon_{ij}^{(2)} \, dV &= \int_{\partial\mathcal{R}} t_i^{(1)} u_i^{(2)} \, dA + \int_{\mathcal{R}} f_i^{(1)} u_i^{(2)} \, dV \,. \end{aligned}}$$

$$(10.2.1)$$

This principle of virtual work can easily be proven using indicial notation in a Cartesian coordinate system. Recall that if a tensor notation equation is true in one coordinate system, then it is true in all coordinate systems. First note that since the stress is symmetric,

$$\int_{\mathcal{R}} \sigma_{ij}{}^{(1)} \varepsilon_{ij}^{(2)} \, dV = \int_{\mathcal{R}} \sigma_{ji}^{(1)} u_i^{(2)}{}_{,j} \, dV \,, \qquad (10.2.2)$$

which can be rewritten as

$$\int_{\mathcal{R}} \sigma_{ij}{}^{(1)} \varepsilon_{ij}^{(2)} \, dV = \int_{\mathcal{R}} (\sigma_{ji}^{(1)} u_i^{(2)})_{,j} \, dV - \int_{\mathcal{R}} \sigma_{ji}^{(1)}{}_{,j} u_i^{(2)} \, dV \ . \qquad (10.2.3)$$

Then, by applying the divergence theorem to the first term on the right-hand side, one finds that

$$\int_{\mathcal{R}} \sigma_{ij}{}^{(1)} \varepsilon_{ij}^{(2)} \, dV = \int_{\partial\mathcal{R}} \sigma_{ji}^{(1)} u_i^{(2)} \hat{n}_j \, dA - \int_{\mathcal{R}} \sigma_{ji}^{(1)}{}_{,j} u_i^{(2)} \, dV \ . \qquad (10.2.4)$$

Finally, using the traction-stress relation and equilibrium, it follows that

$$\int_{\mathcal{R}} \sigma_{ij}{}^{(1)} \varepsilon_{ij}^{(2)} \, dV = \int_{\partial\mathcal{R}} t_i^{(1)} u_i^{(2)} \, dA + \int_{\mathcal{R}} f_i^{(1)} u_i^{(2)} \, dV \ , \qquad (10.2.5)$$

and the principle of virtual work is established.

10.2.2   Strain Energy Density

Recall that the deformation is hyperelastic if the strain energy density, defined by

$$\mathcal{U}(\varepsilon) \equiv \int_0^\varepsilon \boldsymbol{\sigma} : d\boldsymbol{\varepsilon} \ , \quad \mathcal{U}(\varepsilon) \equiv \int_0^\varepsilon \sigma_{ij} \, d\varepsilon_{ij} \ , \qquad (10.2.6)$$

is independent of the path followed in strain space. It follows then that the strain energy density is a potential function for the scalar components of stress,

$$\sigma_{ij} = \frac{\partial \mathcal{U}(\varepsilon)}{\partial \varepsilon_{ij}} \ . \qquad (10.2.7)$$

It can be shown (Fung 1965; Truesdell and Noll 1992) via continuum thermodynamics that the strain energy density $\mathcal{U}$ can be identified with the change in either the internal energy per unit volume if the deformation process is adiabatic, or the Helmholtz free energy per unit volume if the deformation process is isothermal.

Consider a body, with no external loads applied to it, in which the stress and strain are zero everywhere. If the body is in stable thermodynamic equilibrium (i.e., if no spontaneous changes in the state of the body will occur while it is isolated from the rest of the universe), then it is said to be in its *natural state*. It is assumed that if the body is deformed elastically by some system of applied loads, it will always return to its natural state when the loads vanish. This property, along with the Gibbs condition for stable thermodynamic equilibrium, implies that the strain energy density is *positive definite*, $\mathcal{U}(\varepsilon) \geq 0$ with $\mathcal{U}(\varepsilon) = 0$ if and only if $\varepsilon = \mathbf{0}$ (Fung 1965).

The work done on a body by external forces can be given as a functional of the displacement field $\mathbf{u}$:

$$
\mathcal{W}[\mathbf{u}(\mathbf{X})] = \int_{\partial \mathcal{R}} \int_0^{\mathbf{u}} \mathbf{t} \cdot d\mathbf{u} \, dA + \int_{\mathcal{R}} \int_0^{\mathbf{u}} \mathbf{f} \cdot d\mathbf{u} \, dV ,
$$

$$
\mathcal{W}[\mathbf{u}(\mathbf{X})] = \int_{\partial \mathcal{R}} \int_0^{\mathbf{u}} t_i \, du_i \, dA + \int_{\mathcal{R}} \int_0^{\mathbf{u}} f_i \, du_i \, dV .
$$

$$(10.2.8)$$

Noting that

$$
\frac{\partial}{\partial u_j} \int_0^{\mathbf{u}} t_i \, du_i = t_j , \qquad \frac{\partial}{\partial u_j} \int_0^{\mathbf{u}} f_i \, du_i = f_j , \qquad (10.2.9)
$$

it follows that the variation of $\mathcal{W}$, often called the *external virtual work*, is given by

$$
\delta \mathcal{W} = \int_{\partial \mathcal{R}} \mathbf{t} \cdot \delta \mathbf{u} \, dA + \int_{\mathcal{R}} \mathbf{f} \cdot \delta \mathbf{u} \, dV ,
$$

$$
\delta \mathcal{W} = \int_{\partial \mathcal{R}} t_i \delta u_i \, dA + \int_{\mathcal{R}} f_i \delta u_i \, dV .
$$

$$(10.2.10)$$

By applying the principle of virtual work (10.2.1) to (10.2.10) with $\mathbf{t}^{(1)} = \mathbf{t}$, $\mathbf{f}^{(1)} = \mathbf{f}$, $\boldsymbol{\sigma}^{(1)} = \boldsymbol{\sigma}$, $\mathbf{u}^{(2)} = \delta\mathbf{u}$, and

$$
\varepsilon^{(2)} = \delta\varepsilon = \frac{1}{2} \left\{ \nabla(\delta\mathbf{u}) + [\nabla(\delta\mathbf{u})]^{\mathsf{T}} \right\} , \qquad (10.2.11)
$$

the variation of $\mathcal{W}$ (10.2.10) can be rewritten as

$$
\delta \mathcal{W} = \int_{\mathcal{R}} \boldsymbol{\sigma} : \delta\varepsilon \, dV , \qquad \delta \mathcal{W} = \int_{\mathcal{R}} \sigma_{ij} \delta\varepsilon_{ij} \, dV . \qquad (10.2.12)
$$

The right-hand side of (10.2.12) is often called the *internal virtual work* and the principal of virtual work given as the equivalence of (10.2.10) and (10.2.12).

Assuming one has hyperelastic deformation and substituting the constitutive equation (10.2.7) into the expression (10.2.12), one finds that

$$
\delta \mathcal{W} = \int_{\mathcal{R}} \frac{\partial \mathcal{U}(\varepsilon)}{\partial \varepsilon_{ij}} \delta\varepsilon_{ij} \, dV . \qquad (10.2.13)
$$

It follows, by considering the calculus of variations in reverse and *assuming no change in the kinetic or gravitational energies of the body*, that the work done by external forces can be given as a functional of the strain field ε by

$$\mathcal{W}[\varepsilon(\mathbf{X})] = \int_{\mathcal{R}} \mathcal{U}(\varepsilon)\, dV \ . \tag{10.2.14}$$

Since the strain energy density is positive definite, it is seen from (10.2.14) that the work done on a body by external forces (10.2.8) is also positive definite, $\mathcal{W}[\varepsilon(\mathbf{X})] \geq 0$, where $\mathcal{W}[\varepsilon(\mathbf{X})] = 0$ if and only if $\varepsilon = \mathbf{0}$ everywhere in the body. Consider an arbitrary loading history in which external forces are applied to a body and then removed. There is zero net work done on the body at the end of this loading history—any work done at an intermediate stage is stored as "recoverable" strain energy.

10.2.3 Principle of Minimum Potential Energy

Consider a body occupying the region $\mathcal{R}$ with boundary $\partial \mathcal{R} = \partial \mathcal{R}^u \cup \partial \mathcal{R}^t$. At a point on $\partial \mathcal{R}$, the unit outward normal is $\hat{\mathbf{n}}$. The boundary conditions and body force field are prescribed:

$$\mathbf{u} = \tilde{\mathbf{u}} \quad \forall \mathbf{X} \in \partial \mathcal{R}^u \ , \tag{10.2.15}$$

$$\hat{\mathbf{n}} \cdot \boldsymbol{\sigma} = \tilde{\mathbf{t}} \quad \forall \mathbf{X} \in \partial \mathcal{R}^t \ , \tag{10.2.16}$$

$$\mathbf{f} = \tilde{\mathbf{f}} \quad \forall \mathbf{X} \in \mathcal{R} \ . \tag{10.2.17}$$

It has tacitly been assumed up to now that the prescribed external forces $\tilde{\mathbf{t}}$ and $\tilde{\mathbf{f}}$ were independent of the deformation. However, there are some interesting problems, those involving elastic foundations for example, where the prescribed traction and body force fields depend on the local displacement. In this event, assume that the prescribed traction and body force fields can be given in terms of scalar potentials $\psi[\mathbf{u}(\mathbf{X}), \mathbf{X}]$ and $\phi[\mathbf{u}(\mathbf{X}), \mathbf{X}]$ such that

$$\tilde{t}_i = -\frac{\partial \psi}{\partial u_i} \ , \quad \tilde{f}_i = -\frac{\partial \phi}{\partial u_i} \ . \tag{10.2.18}$$

The prescribed traction and body force fields are said to be *conservative* in this case. Then, the *potential energy functional* $\Pi[\mathbf{u}(\mathbf{X})]$ is a function of the displacement field defined as

$$\boxed{\Pi[\mathbf{u}(\mathbf{X})] \equiv \int_{\mathcal{R}} \mathcal{U}(\varepsilon)\, dV + \int_{\partial \mathcal{R}^t} \psi(\mathbf{u})\, dA + \int_{\mathcal{R}} \phi(\mathbf{u})\, dV \ , \tag{10.2.19}}$$

where ε it the strain field corresponding to the displacement field $\mathbf{u}$. The prescribed displacement field $\tilde{\mathbf{u}}$ in the boundary condition (10.2.15) is assumed to be completely predetermined.

In the majority of problems, the prescribed traction and body force fields *are* independent of the deformation. In this case, their potentials are linear functions of the displacement, $\psi = -\tilde{\mathbf{t}} \cdot \mathbf{u}$ and $\phi = -\tilde{\mathbf{f}} \cdot \mathbf{u}$, and the potential energy is given by

$$
\boxed{
\begin{aligned}
\Pi[\mathbf{u}(\mathbf{X})] &= \int_{\mathcal{R}} \mathcal{U}(\varepsilon)\, dV - \int_{\partial\mathcal{R}^t} \tilde{\mathbf{t}} \cdot \mathbf{u}\, dA - \int_{\mathcal{R}} \tilde{\mathbf{f}} \cdot \mathbf{u}\, dV \;, \\
\Pi[\mathbf{u}(\mathbf{X})] &= \int_{\mathcal{R}} \mathcal{U}(\varepsilon)\, dV - \int_{\partial\mathcal{R}^t} \tilde{t}_i u_i\, dA - \int_{\mathcal{R}} \tilde{f}_i u_i\, dV \;.
\end{aligned}
}
$$

$$(10.2.20)$$

A *kinematically admissible displacement field* is one that is continuously differentiable and satisfies the displacement boundary condition (10.2.15). An *actual displacement field* is kinematically admissible and corresponds, through the strain-displacement and constitutive relations, to a stress field that is in equilibrium with the prescribed body force field (10.2.17) and satisfies the traction boundary condition (10.2.16). It has previously been shown that if the constitutive relation is linear and the prescribed traction and body force fields are independent of the deformation, then there is only one actual displacement field; that is, the solution is unique. However, if the constitutive relation is nonlinear and/or the prescribed external loads are dependent on the deformation, one has to allow for the possibility of more than one actual solution.

Principle of minimum potential energy
Among all possible kinematically admissible displacement fields, the potential energy functional $\Pi[\mathbf{u}(\mathbf{X})]$ is rendered stationary by, and only by, those that are actual displacement fields. Actual displacement fields that are energetically stable correspond to local minima of Π. If the prescribed traction and body force fields are independent of the deformation and the constitutive relation is linear, then the actual displacement field makes Π an absolute minimum.

Taking the variation of (10.2.19) and using (10.2.7) and (10.2.18),

$$
\begin{aligned}
\delta\Pi &= \int_{\mathcal{R}} \frac{\partial \mathcal{U}}{\partial \varepsilon_{ij}} \delta\varepsilon_{ij}\, dV + \int_{\partial\mathcal{R}^t} \frac{\partial \psi}{\partial u_i} \delta u_i\, dA + \int_{\mathcal{R}} \frac{\partial \phi}{\partial u_i} \delta u_i\, dV \\
&= \int_{\mathcal{R}} \sigma_{ij} \delta\varepsilon_{ij}\, dV - \int_{\partial\mathcal{R}^t} \tilde{t}_i \delta u_i\, dA - \int_{\mathcal{R}} \tilde{f}_i \delta u_i\, dV \\
&= \int_{\mathcal{R}} \boldsymbol{\sigma} : \delta\boldsymbol{\varepsilon}\, dV - \int_{\partial\mathcal{R}^t} \tilde{\mathbf{t}} \cdot \delta\mathbf{u}\, dA - \int_{\mathcal{R}} \tilde{\mathbf{f}} \cdot \delta\mathbf{u}\, dV \;,
\end{aligned}
$$

$$(10.2.21)$$

where

$$\delta\varepsilon = \delta\left\{\frac{1}{2}\left[\boldsymbol{\nabla}\mathbf{u} + (\boldsymbol{\nabla}\mathbf{u})^{\mathsf{T}}\right]\right\} = \frac{1}{2}\left\{\boldsymbol{\nabla}(\delta\mathbf{u}) + [\boldsymbol{\nabla}(\delta\mathbf{u})]^{\mathsf{T}}\right\},$$

$$\delta\varepsilon_{ij} = \delta\left[\frac{1}{2}(u_{i,j} + u_{j,i})\right] = \frac{1}{2}\left[(\delta u_i)_{,j} + (\delta u_j)_{,i}\right] \tag{10.2.22}$$

can equivalently be thought of as either the variation of the strain field corresponding to the displacement field $\mathbf{u}$ or the strain field corresponding to the variation $\delta\mathbf{u}$ of the displacement field, and $\boldsymbol{\sigma}$ is the stress field corresponding to the kinematically admissible displacement field. Using the divergence theorem and the symmetry of $\boldsymbol{\sigma}$, the first integral on the right-hand side of (10.2.21) can be rewritten as

$$\int_{\mathcal{R}} \sigma_{ij}\delta\varepsilon_{ij}\, dV = \int_{\mathcal{R}} \sigma_{ji}\delta u_{i,j}\, dV$$

$$= \int_{\mathcal{R}} (\sigma_{ji}\delta u_i)_{,j}\, dV - \int_{\mathcal{R}} \sigma_{ji,j}\,\delta u_i\, dV$$

$$= \int_{\partial\mathcal{R}} \sigma_{ji}\delta u_i \hat{n}_j\, dA - \int_{\mathcal{R}} \sigma_{ji,j}\,\delta u_i\, dV$$

$$= \int_{\partial\mathcal{R}^t} \sigma_{ji}\hat{n}_j\delta u_i\, dA - \int_{\mathcal{R}} \sigma_{ji,j}\,\delta u_i\, dV, \tag{10.2.23}$$

where the fact that the variation $\delta\mathbf{u}$ vanishes on $\partial\mathcal{R}^u$, where $\mathbf{u}$ is prescribed, has been used to reduce the integral over the boundary $\partial\mathcal{R}$ to an integral over the portion of the boundary $\partial\mathcal{R}^t$. Substituting (10.2.23) into (10.2.21),

$$\delta\Pi = \int_{\partial\mathcal{R}^t} (\sigma_{ji}\hat{n}_j - \tilde{t}_i)\delta u_i\, dA - \int_{\mathcal{R}} (\sigma_{ji,j} + \tilde{f}_i)\delta u_i\, dV$$

$$= \int_{\partial\mathcal{R}^t} (\hat{\mathbf{n}}\cdot\boldsymbol{\sigma} - \tilde{\mathbf{t}})\cdot\delta\mathbf{u}\, dA - \int_{\mathcal{R}} (\boldsymbol{\nabla}\cdot\boldsymbol{\sigma}^{\mathsf{T}} + \tilde{\mathbf{f}})\cdot\delta\mathbf{u}\, dV \tag{10.2.24}$$

Since $\delta\mathbf{u}$ is an arbitrary vector field, $\delta\Pi = 0$ if and only if

$$\hat{\mathbf{n}}\cdot\boldsymbol{\sigma} = \tilde{\mathbf{t}} \quad \forall\mathbf{X}\in\partial\mathcal{R}^t, \tag{10.2.25}$$

$$\boldsymbol{\nabla}\cdot\boldsymbol{\sigma}^{\mathsf{T}} + \tilde{\mathbf{f}} = \mathbf{0} \quad \forall\mathbf{X}\in\mathcal{R}. \tag{10.2.26}$$

Thus, it is proven that a kinematically admissible displacement field renders $\Pi[\mathbf{u}(\mathbf{X})]$ stationary if and only if it is an actual displacement field.

Let $\mathbf{u}$ be an actual displacement field and $\mathbf{u} + \Delta\mathbf{u}$ be any other kinematically admissible displacement field. Let ε and $\varepsilon + \Delta\varepsilon$ be the strain fields corresponding to the actual displacement field and the other kinematically admissible displacement field, respectively. In particular, $\Delta\varepsilon$ is the strain field corresponding to $\Delta\mathbf{u}$. Then, *assuming that the prescribed traction and body force fields are independent of the deformation*, it follows from (10.2.20) that

$$\Pi[\mathbf{u}(\mathbf{X}) + \Delta\mathbf{u}(\mathbf{X})] - \Pi[\mathbf{u}(\mathbf{X})]$$
$$= \int_{\mathcal{R}} [\mathcal{U}(\varepsilon + \Delta\varepsilon) - \mathcal{U}(\varepsilon)]\, dV - \int_{\partial\mathcal{R}^t} \tilde{\mathbf{t}} \cdot \Delta\mathbf{u}\, dA - \int_{\mathcal{R}} \tilde{\mathbf{f}} \cdot \Delta\mathbf{u}\, dV . \quad (10.2.27)$$

Since $\mathbf{u}$ and $\mathbf{u} + \Delta\mathbf{u}$ are both kinematically admissible, it follows that $\Delta\mathbf{u} = \mathbf{0}$, $\forall\mathbf{X} \in \partial\mathcal{R}^u$. Thus, the integral over the portion $\partial\mathcal{R}^t$ of the boundary can be extended to the entire boundary $\partial\mathcal{R}$:

$$\Pi[\mathbf{u}(\mathbf{X}) + \Delta\mathbf{u}(\mathbf{X})] - \Pi[\mathbf{u}(\mathbf{X})]$$
$$= \int_{\mathcal{R}} [\mathcal{U}(\varepsilon + \Delta\varepsilon) - \mathcal{U}(\varepsilon)]\, dV - \int_{\partial\mathcal{R}} \mathbf{t} \cdot \Delta\mathbf{u}\, dA - \int_{\mathcal{R}} \tilde{\mathbf{f}} \cdot \Delta\mathbf{u}\, dV , \quad (10.2.28)$$

where $\mathbf{t}$ is the traction on $\partial\mathcal{R}$ corresponding to the actual displacement field $\mathbf{u}$, so that $\mathbf{t} = \tilde{\mathbf{t}}$, $\forall\mathbf{X} \in \partial\mathcal{R}^t$. Let σ be the stress field corresponding to the actual displacement field. Then, from the principle of virtual work (10.2.1) with $\mathbf{t}^{(1)} = \mathbf{t}$, $\mathbf{f}^{(1)} = \tilde{\mathbf{f}}$, $\sigma^{(1)} = \sigma$, $\mathbf{u}^{(2)} = \Delta\mathbf{u}$, and $\varepsilon^{(2)} = \Delta\varepsilon$,

$$\int_{\partial\mathcal{R}} \mathbf{t} \cdot \Delta\mathbf{u}\, dA + \int_{\mathcal{R}} \tilde{\mathbf{f}} \cdot \Delta\mathbf{u}\, dV = \int_{\mathcal{R}} \sigma : \Delta\varepsilon\, dV . \quad (10.2.29)$$

Thus, (10.2.28) can be rewritten as

$$\Pi[\mathbf{u}(\mathbf{X}) + \Delta\mathbf{u}(\mathbf{X})] - \Pi[\mathbf{u}(\mathbf{X})]$$
$$= \int_{\mathcal{R}} [\mathcal{U}(\varepsilon + \Delta\varepsilon) - \mathcal{U}(\varepsilon) - \sigma : \Delta\varepsilon]\, dV . \quad (10.2.30)$$

Recall that for a linear constitutive relation, the strain energy density is $\mathcal{U}(\varepsilon) = \frac{1}{2}\varepsilon : \mathbf{C} : \varepsilon$, where $\mathbf{C}$ is the fourth-order stiffness tensor. Therefore, it follows that for linear elastic deformation,

$$\mathcal{U}(\varepsilon + \Delta\varepsilon) = \mathcal{U}(\varepsilon) + \sigma : \Delta\varepsilon + \mathcal{U}(\Delta\varepsilon) . \quad (10.2.31)$$

Substituting (10.2.31) into (10.2.30), one finds that

$$\Pi[\mathbf{u}(\mathbf{X}) + \Delta\mathbf{u}(\mathbf{X})] - \Pi[\mathbf{u}(\mathbf{X})] = \int_{\mathcal{R}} \mathcal{U}(\Delta\varepsilon)\, dV . \quad (10.2.32)$$

Since the strain energy density is positive definite, it follows that

$$\Pi[\mathbf{u}(\mathbf{X}) + \Delta\mathbf{u}(\mathbf{X})] - \Pi[\mathbf{u}(\mathbf{X})] \geq 0 , \tag{10.2.33}$$

with the equality holding if and only if $\Delta\mathbf{u} = \mathbf{0}$, $\forall\mathbf{X} \in \mathcal{R}$. It is therefore proven that if the prescribed traction and body force fields are independent of the deformation and the constitutive relation is linear, then the actual displacement field makes $\Pi[\mathbf{u}(\mathbf{X})]$ an absolute minimum. Since there can be only one absolute minimum, this also establishes that the solution is unique.

If the prescribed traction and body force fields depend on the deformation and/or the constitutive relation is nonlinear, then there is no such assurance of the uniqueness of solutions. In fact, multiple solutions are possible, each rendering $\Pi[\mathbf{u}(\mathbf{X})]$ stationary with *stable solutions* assumed to correspond to local minima of Π. The existence of unstable solutions result in elastic buckling. It should be noted that buckling can also occur as a result of finite components of the displacement gradient (kinematic nonlinearity) or inelastic deformation (plastic buckling).

10.2.4   Complementary Strain Energy Density

Define the *complementary strain energy density* as

$$\mathcal{U}^c(\boldsymbol{\sigma}) \equiv \int_0^{\boldsymbol{\sigma}} \boldsymbol{\varepsilon} : d\boldsymbol{\sigma} , \quad \mathcal{U}^c(\boldsymbol{\sigma}) \equiv \int_0^{\boldsymbol{\sigma}} \varepsilon_{ij}\, d\sigma_{ij} . \tag{10.2.34}$$

Comparing (10.2.34) with the definition (10.2.6) of the strain energy density, it follows that

$$\mathcal{U} + \mathcal{U}^c = \boldsymbol{\sigma} : \boldsymbol{\varepsilon} , \quad \mathcal{U} + \mathcal{U}^c = \sigma_{ij}\varepsilon_{ij} . \tag{10.2.35}$$

Since $\mathcal{U}(\boldsymbol{\varepsilon})$ and $\boldsymbol{\sigma} : \boldsymbol{\varepsilon}$ are path independent, assuming hyperelasticity implies that $\mathcal{U}^c(\boldsymbol{\sigma})$ is also path independent. Thus,

$$\varepsilon_{ij} = \frac{\partial \mathcal{U}^c(\boldsymbol{\sigma})}{\partial \sigma_{ij}} . \tag{10.2.36}$$

It can also be shown that under the conditions for which the strain energy density is positive definite, the complementary strain energy density is also positive definite, $\mathcal{U}^c(\boldsymbol{\sigma}) \geq 0$ with $\mathcal{U}^c(\boldsymbol{\sigma}) = 0$ if and only if $\boldsymbol{\sigma} = \mathbf{0}$.

In the *special linear case*, recall that the components of strain can be given in terms of the components of stress by

$$\boldsymbol{\varepsilon} = \mathbf{S} : \boldsymbol{\sigma} , \quad \varepsilon_{ij} = S_{ijkl}\sigma_{kl} , \tag{10.2.37}$$

where S_{ijkl} are the scalar components of the fourth-order compliance tensor
S, of which only 21 are independent. It follows then that

$$\mathcal{U}^c(\boldsymbol{\sigma}) = \frac{1}{2}\boldsymbol{\sigma} : \mathbf{S} : \boldsymbol{\sigma}, \quad \mathcal{U}^c(\boldsymbol{\sigma}) = \frac{1}{2}S_{ijkl}\sigma_{ij}\sigma_{kl}, \tag{10.2.38}$$

and

$$\mathcal{U}^c = \frac{1}{2}\boldsymbol{\sigma} : \boldsymbol{\varepsilon}, \quad \mathcal{U}^c = \frac{1}{2}\sigma_{ij}\varepsilon_{kl}. \tag{10.2.39}$$

Thus, in the case of a linear constitutive relation,

$$\mathcal{U}^c = \mathcal{U}. \tag{10.2.40}$$

This is not generally true in the nonlinear case.

10.2.5 Principle of Minimum Complementary Energy

Consider a body occupying the region $\mathcal{R}$ with boundary $\partial\mathcal{R} = \partial\mathcal{R}^u \cup \partial\mathcal{R}^t$.
At a point on $\partial\mathcal{R}$, the unit outward normal is $\hat{\mathbf{n}}$. The boundary conditions
and body force field are prescribed:

$$\mathbf{u} = \tilde{\mathbf{u}} \quad \forall \mathbf{X} \in \partial\mathcal{R}^u, \tag{10.2.41}$$

$$\hat{\mathbf{n}} \cdot \boldsymbol{\sigma} = \tilde{\mathbf{t}} \quad \forall \mathbf{X} \in \partial\mathcal{R}^t, \tag{10.2.42}$$

$$\mathbf{f} = \tilde{\mathbf{f}} \quad \forall \mathbf{X} \in \mathcal{R}. \tag{10.2.43}$$

Define the *complementary energy functional* $\Pi^c[\boldsymbol{\sigma}(\mathbf{X})]$ as

$$\boxed{\begin{aligned}
\Pi^c[\boldsymbol{\sigma}(\mathbf{X})] &= \int_{\mathcal{R}} \mathcal{U}^c(\boldsymbol{\sigma})\, dV - \int_{\partial\mathcal{R}^u} \mathbf{t} \cdot \tilde{\mathbf{u}}\, dA, \\
\Pi^c[\boldsymbol{\sigma}(\mathbf{X})] &= \int_{\mathcal{R}} \mathcal{U}^c(\boldsymbol{\sigma})\, dV - \int_{\partial\mathcal{R}^u} t_i \tilde{u}_i\, dA,
\end{aligned}} \tag{10.2.44}$$

where $\mathbf{t} = \hat{\mathbf{n}} \cdot \boldsymbol{\sigma}$. It is assumed that the prescribed displacement field $\tilde{\mathbf{u}}$ is
independent of the local traction and body force fields.

A *statically admissible stress field* is one that is in equilibrium with the
prescribed body force field (10.2.43) and satisfies the traction boundary
condition (10.2.42). An *actual stress field* is statically admissible, satisfies
compatibility everywhere in the body, and corresponds to a displacement
field that satisfies the displacement boundary condition (10.2.41).

The value of the complementary energy is not independent of that of
the potential energy. Consider a body in which $\Pi = \Pi[\mathbf{u}(\mathbf{X})]$ and $\Pi^c =$

$\Pi^c[\boldsymbol{\sigma}(\mathbf{X})]$, where $\mathbf{u}$ and $\boldsymbol{\sigma}$ are the actual displacement and stress fields. From the definitions (10.2.20) and (10.2.44),

$$\Pi + \Pi^c = \int_{\mathcal{R}} (\mathcal{U} + \mathcal{U}^c)\, dV - \int_{\partial\mathcal{R}^t} \tilde{\mathbf{t}} \cdot \mathbf{u}\, dA - \int_{\partial\mathcal{R}^u} \mathbf{t} \cdot \tilde{\mathbf{u}}\, dA - \int_{\mathcal{R}} \tilde{\mathbf{f}} \cdot \mathbf{u}\, dV .$$

$$(10.2.45)$$

However, recalling that $\mathcal{U} + \mathcal{U}^c = \boldsymbol{\sigma} : \boldsymbol{\varepsilon}$ and combining the integrals over $\partial\mathcal{R}^t$ and $\partial\mathcal{R}^u$ into one integral over the entire boundary $\partial\mathcal{R} = \partial\mathcal{R}^t \cup \partial\mathcal{R}^u$, one has that

$$\Pi + \Pi^c = \int_{\mathcal{R}} \boldsymbol{\sigma} : \boldsymbol{\varepsilon}\, dV - \int_{\partial\mathcal{R}} \mathbf{t} \cdot \mathbf{u}\, dA - \int_{\mathcal{R}} \mathbf{f} \cdot \mathbf{u}\, dV . \qquad (10.2.46)$$

From the principle of virtual work (10.2.1), the right-hand side of this equation is zero. Therefore,

$$\Pi^c = -\Pi ; \qquad (10.2.47)$$

the complementary energy corresponding to the actual stress field is the negative of the potential energy corresponding to the actual displacement field.

Principle of minimum complementary energy
Among all statically admissible stress fields, the complementary energy functional $\Pi^c[\boldsymbol{\sigma}(\mathbf{X})]$ is rendered stationary by, and only by, those that are actual stress fields. Actual stress fields that are energetically stable correspond to local minima of Π^c. If the prescribed traction and body force fields are independent of the deformation and the constitutive relation is linear, then the actual stress field makes Π^c an absolute minimum.

To consider the stationarity of $\Pi^c[\boldsymbol{\sigma}(\mathbf{X})]$ subject to the constraint that $\boldsymbol{\sigma}$ be in equilibrium with the prescribed body force field, consider instead the functional

$$\tilde{\Pi}^c[\boldsymbol{\sigma}(\mathbf{X})] = \int_{\mathcal{R}} \mathcal{U}^c(\boldsymbol{\sigma})\, dV - \int_{\partial\mathcal{R}^u} \mathbf{t} \cdot \tilde{\mathbf{u}}\, dA + \int_{\mathcal{R}} (\boldsymbol{\nabla} \cdot \boldsymbol{\sigma}^\mathsf{T} + \tilde{\mathbf{f}}) \cdot \boldsymbol{\lambda}\, dV ,$$

$$\tilde{\Pi}^c[\boldsymbol{\sigma}(\mathbf{X})] = \int_{\mathcal{R}} \mathcal{U}^c(\boldsymbol{\sigma})\, dV - \int_{\partial\mathcal{R}^u} t_i \tilde{u}_i\, dA + \int_{\mathcal{R}} (\sigma_{ji,j} + \tilde{f}_i)\lambda_i\, dV .$$

$$(10.2.48)$$

The Lagrange multiplier $\boldsymbol{\lambda}$ is a vector field in this case. Stress fields that render (10.2.48) stationary are in equilibrium with the prescribed body force field $\tilde{\mathbf{f}}$ and render the complementary energy functional Π^c stationary. Taking the variation of (10.2.48), one has that

$$\delta\tilde{\Pi}^c = \int_{\mathcal{R}} \varepsilon_{ij}\delta\sigma_{ij}\,dV - \int_{\partial\mathcal{R}^u} \tilde{u}_i\delta t_i\,dA + \int_{\mathcal{R}} \lambda_i\delta\sigma_{ji,j}\,dV\,, \qquad (10.2.49)$$

where $\varepsilon_{ij} = \partial\mathcal{U}^c/\partial\sigma_{ij}$. However, the last term on the right-hand side of (10.2.49) can be rewritten using the divergence theorem:

$$\begin{aligned}
\int_{\mathcal{R}} \lambda_i\delta\sigma_{ji,j}\,dV &= \int_{\mathcal{R}} (\lambda_i\delta\sigma_{ji})_{,j}\,dV - \int_{\mathcal{R}} \lambda_{i,j}\,\delta\sigma_{ji}\,dV \\
&= \int_{\partial\mathcal{R}} \lambda_i\delta\sigma_{ji}\hat{n}_j\,dA - \int_{\mathcal{R}} \lambda_{i,j}\,\delta\sigma_{ji}\,dV \\
&= \int_{\partial\mathcal{R}^u} \lambda_i\delta t_i\,dA - \int_{\mathcal{R}} \lambda_{i,j}\,\delta\sigma_{ji}\,dV\,, \qquad (10.2.50)
\end{aligned}$$

where the fact that the variation of traction $\delta\mathbf{t}$ vanishes on $\partial\mathcal{R}^t$, where the traction is prescribed, has been used to reduce the integral over $\partial\mathcal{R}$ to an integral over $\partial\mathcal{R}^u$. Thus, substituting (10.2.50) into (10.2.49), the variation of (10.2.48) is given by

$$\begin{aligned}
\delta\tilde{\Pi}^c &= \int_{\mathcal{R}} (\varepsilon_{ij} - \lambda_{i,j})\delta\sigma_{ji}\,dV + \int_{\partial\mathcal{R}^u} (\lambda_i - \tilde{u}_i)\delta t_i\,dA \\
&= \int_{\mathcal{R}} (\boldsymbol{\varepsilon} - \boldsymbol{\nabla}\boldsymbol{\lambda}):\delta\boldsymbol{\sigma}^\mathsf{T}\,dV + \int_{\partial\mathcal{R}^u} (\boldsymbol{\lambda} - \tilde{\mathbf{u}})\cdot\delta\mathbf{t}\,dA\,. \qquad (10.2.51)
\end{aligned}$$

Aside from the requirement that $\delta\boldsymbol{\sigma}$ must be symmetric, $\delta\boldsymbol{\sigma}$ is arbitrary in $\mathcal{R}$ and $\delta\mathbf{t}$ is arbitrary on $\partial\mathcal{R}^u$. Thus, $\delta\tilde{\Pi}^c = 0$ if and only if

$$\boldsymbol{\varepsilon} = \frac{1}{2}\left[\boldsymbol{\nabla}\boldsymbol{\lambda} + (\boldsymbol{\nabla}\boldsymbol{\lambda})^\mathsf{T}\right] \quad \forall\mathbf{X}\in\mathcal{R}\,, \qquad (10.2.52)$$

$$\boldsymbol{\lambda} = \tilde{\mathbf{u}} \qquad\qquad\qquad \forall\mathbf{X}\in\partial\mathcal{R}^u\,, \qquad (10.2.53)$$

where the requirement that $\boldsymbol{\varepsilon}$ must be symmetric has been used in writing (10.2.52). From (10.2.52), it is seen that the strain field corresponding to the stress field $\boldsymbol{\sigma}$ must satisfy compatibility in order for the Lagrangian multiplier $\boldsymbol{\lambda}$ to exist. The vector field $\boldsymbol{\lambda}$ then is equal to the displacement field corresponding to $\boldsymbol{\sigma}$, which from (10.2.53) must satisfy the displacement boundary condition (10.2.41). Thus, a statically admissible stress field renders $\Pi^c[\boldsymbol{\sigma}(\mathbf{X})]$ stationary if and only if it is an actual stress field.

Let $\boldsymbol{\sigma}$ be an actual stress field and $\boldsymbol{\sigma} + \Delta\boldsymbol{\sigma}$ be any other statically admissible stress field. Let $\mathbf{t}$ and $\mathbf{t} + \Delta\mathbf{t}$ be the corresponding traction fields on $\partial\mathcal{R}$. In particular, $\Delta t = \hat{n} \cdot \Delta\sigma$. Then, it follows from (10.2.44) that

$$\Pi^c[\sigma(X) + \Delta\sigma(X)] - \Pi^c[\sigma(X)]$$
$$= \int_{\mathcal{R}} [\mathcal{U}^c(\sigma + \Delta\sigma) - \mathcal{U}^c(\sigma)]\, dV - \int_{\partial\mathcal{R}^u} \tilde{u} \cdot \Delta t\, dA . \quad (10.2.54)$$

Since σ and $\sigma + \Delta\sigma$ are both statically admissible and *assuming that the prescribed traction and body force fields are independent of the deformation*, it follows that $\Delta t = 0$, $\forall X \in \partial\mathcal{R}^t$, and $\Delta f = -\nabla \cdot (\Delta\sigma)^\mathsf{T} = 0$. Thus, the integral over $\partial\mathcal{R}^u$ can be extended to $\partial\mathcal{R}$,

$$\Pi^c[\sigma(X) + \Delta\sigma(X)] - \Pi^c[\sigma(X)]$$
$$= \int_{\mathcal{R}} [\mathcal{U}^c(\sigma + \Delta\sigma) - \mathcal{U}^c(\sigma)]\, dV - \int_{\partial\mathcal{R}} u \cdot \Delta t\, dA , \quad (10.2.55)$$

where u is the displacement field corresponding to the actual stress field so that $u = \tilde{u}$, $\forall X \in \partial\mathcal{R}^u$. Let ε be the strain field corresponding to the actual stress field. Then, from the principle of virtual work (10.2.1) with $t^{(1)} = \Delta t$, $f^{(1)} = \Delta f = 0$, $\sigma^{(1)} = \Delta\sigma$, $u^{(2)} = u$, and $\varepsilon^{(2)} = \varepsilon$, it follows that

$$\int_{\partial\mathcal{R}} u \cdot \Delta t\, dA = \int_{\mathcal{R}} \varepsilon : \Delta\sigma\, dV . \quad (10.2.56)$$

Thus, (10.2.55) can be rewritten as

$$\Pi^c[\sigma(X) + \Delta\sigma(X)] - \Pi^c[\sigma(X)]$$
$$= \int_{\mathcal{R}} [\mathcal{U}^c(\sigma + \Delta\sigma) - \mathcal{U}^c(\sigma) - \varepsilon : \Delta\sigma]\, dV . \quad (10.2.57)$$

By an argument similar to that used in proving the principle of minimum potential energy, it can be shown (see Problem 10.2) that if the constitutive relation is linear,

$$\Pi^c[\sigma(X) + \Delta\sigma(X)] - \Pi^c[\sigma(X)] = \int_{\mathcal{R}} \mathcal{U}^c(\Delta\sigma)\, dV . \quad (10.2.58)$$

Since the complementary strain energy density is positive definite,

$$\Pi^c[\sigma(X) + \Delta\sigma(X)] - \Pi^c[\sigma(X)] \geq 0 , \quad (10.2.59)$$

with the equality holding if and only if $\Delta\sigma = 0$, $\forall X \in \mathcal{R}$. Thus, it is proven that if the prescribed traction and body force fields are independent of the deformation and the constitutive relation is linear, the actual stress field makes $\Pi^c[\sigma(X)]$ an absolute minimum and the solution is unique.

10.3 Approximate Solutions

Assume that the prescribed traction and body force fields are independent of the deformation and let $\mathbf{u}^*$ be an arbitrary member of the set of all kinematically admissible displacement fields, that is, displacement fields that are continuously differentiable in the region $\mathcal{R}$ of a body and satisfy the displacement boundary condition. According to the principle of minimum potential energy, the actual displacement field $\mathbf{u}$ is the kinematically admissible displacement field that corresponds to the absolute minimum of the potential energy functional. One can propose that the difference in the value of the potential energy functional for the actual displacement field and any other kinematically admissible displacement field,

$$\Delta \Pi \equiv \Pi[\mathbf{u}^*(\mathbf{X})] - \Pi[\mathbf{u}(\mathbf{X})] \,, \tag{10.3.1}$$

be used as a measure of the error. According to the principle of minimum potential energy, $\Delta \Pi \geq 0$, where $\Delta \Pi = 0$ if and only if $\mathbf{u}^* = \mathbf{u}$.

Consider now a subset of kinematically admissible displacement fields, of which $\mathbf{u}^*_{\mathrm{approx}}$ is an arbitrary member, that are to be considered as possible approximate solutions for a particular problem. This subset may or may not include the actual displacement field. The supposition is that, among all members of this subset, the "best" approximation $\mathbf{u}_{\mathrm{approx}}$ of the actual displacement field minimizes the potential energy functional,

$$\Pi[\mathbf{u}_{\mathrm{approx}}(\mathbf{X})] \leq \Pi[\mathbf{u}^*_{\mathrm{approx}}(\mathbf{X})] \,, \tag{10.3.2}$$

and therefore minimizes the error,

$$\Delta \Pi = \Pi[\mathbf{u}^*_{\mathrm{approx}}(\mathbf{X})] - \Pi[\mathbf{u}(\mathbf{X})] \,. \tag{10.3.3}$$

This minimum error will equal zero if and only if the subset includes the actual displacement field, in which case the best approximate solution is the actual solution. Approximate solution displacement fields found in this way will, in general, satisfy neither equilibrium nor the traction boundary condition. They do, however, provide an upper bound on the potential energy of the system—the potential energy of the approximate solution exceeds that of the actual solution by an amount $\Delta \Pi$.

Alternatively, the principle of minimum complementary energy can be used to find approximate solutions. Let $\boldsymbol{\sigma}^*$ be an arbitrary member of the set of all statically admissible stress fields, that is, stress fields that are in equilibrium with the prescribed body force field and satisfy the traction boundary condition. One can propose that the difference in the value of the complementary energy functional for the actual stress field $\boldsymbol{\sigma}$ and any other statically admissible stress field,

$$\Delta\Pi^c \equiv \Pi^c[\boldsymbol{\sigma}^*(\mathbf{X})] - \Pi^c[\boldsymbol{\sigma}(\mathbf{X})] \,, \tag{10.3.4}$$

be used as a measure of the error. According to the principle of minimum complementary energy, $\Delta\Pi^c \geq 0$, where $\Delta\Pi^c = 0$ if and only if $\boldsymbol{\sigma}^* = \boldsymbol{\sigma}$.

Let $\boldsymbol{\sigma}^*_{\text{approx}}$ be an arbitrary member of a subset of statically admissible stress fields to be considered as possible approximate solutions of a particular problem. The supposition in this case is that, among all members of this subset, the "best" approximation $\boldsymbol{\sigma}_{\text{approx}}$ of the actual stress field minimizes the complementary energy functional,

$$\Pi^c[\boldsymbol{\sigma}_{\text{approx}}(\mathbf{X})] \leq \Pi^c[\boldsymbol{\sigma}^*_{\text{approx}}(\mathbf{X})] \,, \tag{10.3.5}$$

and therefore minimizes the error. This minimum error vanishes if and only if the subset includes the actual stress field, in which case the best approximate solution is the actual solution. Approximate solution stress fields found in this way will, in general, satisfy neither compatibility nor the displacement boundary condition. An approximate solution stress field provides an upper bound on the complementary energy of the system— the complementary energy of the approximate solution exceeds that of the actual solution by an amount $\Delta\Pi^c$.

It has be shown that an approximate solution that is kinematically admissible provides an upper bound on the potential energy of a system and that an approximate solution that is statically admissible provides an upper bound on the complementary energy. Furthermore, since the complementary energy is the negative of the potential energy, an approximate solution that is statically admissible also provides a *lower* bound on the potential energy. Of course, if an approximate solution is both kinematically and statically admissible, then it satisfies all of the boundary conditions and equations of elasticity and is, therefore, the actual solution. The upper and lower bounds on the potential energy in this case converge on the actual potential energy.

10.3.1   Rayleigh-Ritz Method

Consider a subset of kinematically admissible displacement fields given by

$$\mathbf{u}^*_{\text{approx}} = \mathbf{U}_0(\mathbf{X}) + a_1\mathbf{U}_1(\mathbf{X}) + a_2\mathbf{U}_2(\mathbf{X}) + \cdots + a_n\mathbf{U}_n(\mathbf{X}) \,, \tag{10.3.6}$$

where a_1, a_2, ..., a_n are arbitrary constants and $\mathbf{U}_0$, $\mathbf{U}_1$, ..., $\mathbf{U}_n$ are given continuously differentiable vector fields. To ensure that $\mathbf{u}^*_{\text{approx}}$ is kinematically admissible, it is required that

$$\left.\begin{aligned}\mathbf{U}_0 &= \tilde{\mathbf{u}}\\ \mathbf{U}_1 = \mathbf{U}_2 = \cdots = \mathbf{U}_n &= \mathbf{0}\end{aligned}\right\} \quad \forall \mathbf{X} \in \partial\mathcal{R}^u \,, \tag{10.3.7}$$

where $\tilde{\mathbf{u}}$ is the prescribed displacement distribution on the portion of the boundary $\partial \mathcal{R}^u$. The potential energy for this subset of kinematically admissible displacement fields, found by substituting (10.3.6) into the definition (10.2.20), is a function with independent variables $a_1, a_2, \ldots, a_n$:

$$\Pi = \Pi^*_{\text{approx}}(a_1, a_2, \ldots, a_n) . \tag{10.3.8}$$

The best approximate solution (10.3.6) minimizes (10.3.8) and, consequently, is the displacement field $\mathbf{u}^*_{\text{approx}}$ with constants $a_1, a_2, \ldots, a_n$ that satisfy the set of n algebraic equations given by

$$\frac{\partial \Pi}{\partial a_i} = 0 \quad (i = 1, 2, \ldots, n) . \tag{10.3.9}$$

This is the basis of the displacement-based finite element method for linearized elasticity problems, where the *shape functions* $\mathbf{U}_0, \mathbf{U}_1, \ldots, \mathbf{U}_n$ are determined by the finite element mesh and the constants $a_1, a_2, \ldots, a_n$ are the nodal displacements.

The Rayleigh-Ritz method can also be applied to the principle of minimum complementary energy. Consider a subset of statically admissible stress fields given by

$$\boldsymbol{\sigma}^*_{\text{approx}} = \boldsymbol{\Sigma}_0(\mathbf{X}) + b_1 \boldsymbol{\Sigma}_1(\mathbf{X}) + b_2 \boldsymbol{\Sigma}_2(\mathbf{X}) + \cdots + b_n \boldsymbol{\Sigma}_n(\mathbf{X}) , \tag{10.3.10}$$

where $b_1, b_2, \ldots, b_n$ are arbitrary constants and $\boldsymbol{\Sigma}_0, \boldsymbol{\Sigma}_1, \ldots, \boldsymbol{\Sigma}_n$ are given symmetric second-order tensor fields. To ensure that $\boldsymbol{\sigma}^*_{\text{approx}}$ is statically admissible, it is required that

$$\left. \begin{array}{r} \boldsymbol{\nabla} \cdot \boldsymbol{\Sigma}_0^{\mathsf{T}} + \tilde{\mathbf{f}} = \mathbf{0} \\ \boldsymbol{\nabla} \cdot \boldsymbol{\Sigma}_1^{\mathsf{T}} = \boldsymbol{\nabla} \cdot \boldsymbol{\Sigma}_2^{\mathsf{T}} = \cdots = \boldsymbol{\nabla} \cdot \boldsymbol{\Sigma}_n^{\mathsf{T}} = \mathbf{0} \end{array} \right\} \quad \forall \mathbf{X} \in \mathcal{R} \tag{10.3.11}$$

and

$$\left. \begin{array}{r} \hat{\mathbf{n}} \cdot \boldsymbol{\Sigma}_0 = \tilde{\mathbf{t}} \\ \hat{\mathbf{n}} \cdot \boldsymbol{\Sigma}_1 = \hat{\mathbf{n}} \cdot \boldsymbol{\Sigma}_2 = \cdots = \hat{\mathbf{n}} \cdot \boldsymbol{\Sigma}_n = \mathbf{0} \end{array} \right\} \quad \forall \mathbf{X} \in \partial \mathcal{R}^t , \tag{10.3.12}$$

where $\tilde{\mathbf{f}}$ is the prescribed body force field, $\tilde{\mathbf{t}}$ is the prescribed traction distribution, and $\hat{\mathbf{n}}$ is the unit outward normal at a point on $\partial \mathcal{R}^t$.

The complementary energy for the subset of statically admissible stress fields, found by substituting (10.3.10) into the definition (10.2.44), is a function with independent variables $b_1, b_2, \ldots, b_n$:

$$\Pi^c = \Pi^{c*}_{\text{approx}}(b_1, b_2, \ldots, b_n) . \tag{10.3.13}$$

The best approximate solution (10.3.10) minimizes (10.3.13) and, consequently, is the stress field $\boldsymbol{\sigma}^*_{\text{approx}}$ with constants $b_1, b_2, \ldots, b_n$ that satisfy

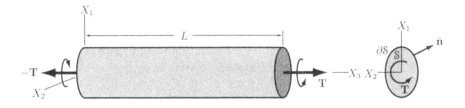

FIGURE 10.6. Torsion of a noncircular cylinder.

the set of n algebraic equations given by

$$\frac{\partial \Pi^c}{\partial b_i} = 0 \quad (i = 1, 2, \dots, n) . \tag{10.3.14}$$

This is the basis for a less frequently utilized stress-based finite element method.

10.3.2   Approximate Solutions to Torsion Problems

Consider a cylindrical body of length L, the ends of which are subject to traction distributions that are statically equivalent to equal and opposite torques $\pm \mathbf{T} = \pm T \hat{\mathbf{e}}_3$ (see Figure 10.6). The lateral surface of the cylinder is traction-free. It is assumed that there is an axis that passes through the center of twist of each cross section of the cylinder (i.e., an axis on which the displacement perpendicular to the axis is zero) and this is defined to be the X_3-axis. Let $\mathcal{S}$ be the cross section of the cylinder and let $\partial \mathcal{S}$ be its boundary. The cross section $\mathcal{S}$ is in general noncircular.

The traction boundary condition on the ends of the cylinder at $X_3 = 0$ and $X_3 = L$ is not uniquely prescribed—it is only required that the traction distribution on these surfaces be statically equivalent to the specified torque. However, it was shown in Chapter 8 that the displacement field in this problem can be given in terms of the warping function ψ by

$$u_1 = -\alpha X_2 X_3 , \quad u_2 = \alpha X_1 X_3 , \quad u_3 = \alpha \psi(X_1, X_2) , \tag{10.3.15}$$

where the twist per unit length α is a constant and the twist of the cylinder at $X_3 = 0$ is fixed at zero. The warping function is determined by the geometry of the cross section $\mathcal{S}$ and is independent of the material properties and the applied torque. Thus, the boundary conditions for this problem can alternatively be given as a traction-free condition $\tilde{\mathbf{t}} = \mathbf{0}$ on the lateral surface,

$$\hat{\mathbf{n}} \cdot \boldsymbol{\sigma} = \mathbf{0} \quad \forall (X_1, X_2) \in \partial \mathcal{S} , \tag{10.3.16}$$

where $\hat{\mathbf{n}}$ is the unit outward normal to the lateral surface, with displacement boundary conditions on the ends of the cylinder,

$$u_1 = u_2 = 0, \quad u_3 = \alpha\psi(X_1, X_2) \quad \text{on } X_3 = 0 \qquad (10.3.17)$$

and

$$u_1 = -\alpha L X_2, \quad u_2 = \alpha L X_1, \quad u_3 = \alpha\psi(X_1, X_2) \quad \text{on } X_3 = L. \qquad (10.3.18)$$

In this view of the problem, the twist per unit length α is prescribed rather than the torque.

Principle of minimum potential energy applied to torsion

Since the prescribed traction (10.3.16) and body force field are zero, the potential energy (10.2.20) in this case reduces to the total strain energy:

$$\Pi[\mathbf{u}(\mathbf{X})] = \int_{\mathcal{R}} \mathcal{U}(\varepsilon)\, dV. \qquad (10.3.19)$$

It follows from (10.3.15) that the components of strain for the torsion problem are $\varepsilon_{11} = \varepsilon_{22} = \varepsilon_{33} = \varepsilon_{12} = 0$ and

$$\varepsilon_{31} = \frac{\alpha}{2}(\psi_{,1} - X_2), \quad \varepsilon_{32} = \frac{\alpha}{2}(\psi_{,2} + X_1) \qquad (10.3.20)$$

and, assuming the material is isotropic, that the components of stress are $\sigma_{11} = \sigma_{22} = \sigma_{33} = \sigma_{12} = 0$ and

$$\sigma_{31} = \mu\alpha(\psi_{,1} - X_2), \quad \sigma_{32} = \mu\alpha(\psi_{,2} + X_1). \qquad (10.3.21)$$

Thus, the strain energy density $\mathcal{U} = \frac{1}{2}\sigma_{ij}\varepsilon_{ij}$ is

$$\mathcal{U} = \frac{1}{2}\mu\alpha^2 \left[(\psi_{,1} - X_2)^2 + (\psi_{,2} + X_1)^2\right]. \qquad (10.3.22)$$

Noting that $\mathcal{U}$ is independent of X_3, it follows that the potential energy per unit length $\bar{\Pi} \equiv \Pi/L$ is expressed as a functional of the warping function by

$$\boxed{\bar{\Pi}[\psi(X_1, X_2)] = \frac{1}{2}\mu\alpha^2 \int_{\mathcal{S}} \left[(\psi_{,1} - X_2)^2 + (\psi_{,2} + X_1)^2\right] dA.} \qquad (10.3.23)$$

All continuously differentiable warping functions are kinematically admissible, in that they correspond to continuously differentiable displacement fields that satisfy the displacement boundary conditions given by (10.3.17) and (10.3.18). The actual warping function corresponds via (10.3.21) to a stress field that satisfies equilibrium and the traction-free boundary condition on the lateral surface. According to the principle of minimum potential energy, the actual warping function will make $\bar{\Pi}(\psi)$ an absolute minimum.

The conditions necessary for the warping function to be the actual warping function can be determined by considering the stationarity of the potential energy functional (10.3.23). Taking the variation,

$$\delta\bar{\Pi} = \mu\alpha^2 \int_S [(\psi_{,1} - X_2)\delta\psi_{,1} + (\psi_{,2} + X_1)\delta\psi_{,2}]\,dA , \qquad (10.3.24)$$

which can be rewritten as

$$\delta\bar{\Pi} = \mu\alpha^2 \int_S \{[(\psi_{,1} - X_2)\delta\psi]_{,1} + [(\psi_{,2} + X_1)\delta\psi]_{,2}\}\,dA$$

$$- \mu\alpha^2 \int_S (\psi_{,11} + \psi_{,22})\delta\psi\,dA . \qquad (10.3.25)$$

Applying the Green-Riemann theorem to the first integral on the right-hand side and noting that $\psi_{,11} + \psi_{,22} = \nabla^2\psi$,

$$\delta\bar{\Pi} = \mu\alpha^2 \oint_{\partial S} \left[(\psi_{,1} - X_2)\frac{dX_2}{ds} - (\psi_{,2} + X_1)\frac{dX_1}{ds}\right]\delta\psi\,ds$$

$$- \mu\alpha^2 \int_S (\nabla^2\psi)\delta\psi\,dA , \qquad (10.3.26)$$

where s is the arc length around ∂S. Since $\delta\psi$ is arbitrary, $\delta\bar{\Pi} = 0$ if and only if

$$\nabla^2\psi = 0 \quad \forall(X_1, X_2) \in S , \qquad (10.3.27)$$

$$(\psi_{,1} - X_2)\frac{dX_2}{ds} - (\psi_{,2} + X_1)\frac{dX_1}{ds} = 0 \quad \forall(X_1, X_2) \in \partial S . \qquad (10.3.28)$$

These are the conditions for the warping function to satisfy equilibrium and the traction-free lateral surface boundary condition. They are exactly the same as those derived in Chapter 8.

The potential energy can be related to the torque applied to the cylinder. As previously noted, the potential energy for this problem is given by

$$\Pi = \int_{\mathcal{R}} \mathcal{U}\,dV = \frac{1}{2}\int_{\mathcal{R}} \sigma_{ij}\varepsilon_{ij}\,dV . \qquad (10.3.29)$$

Applying the principle of virtual work with $\mathbf{f} = \mathbf{0}$ and $\tilde{\mathbf{t}} = \mathbf{0}$ on the lateral surface $\partial \mathcal{R}^t$,

$$\Pi = \frac{1}{2} \int_{\partial \mathcal{R}^u} t_i u_i \, dA \, . \tag{10.3.30}$$

Recall that the portion of the boundary $\partial \mathcal{R}^u$ is composed of the ends of the cylinder at $X_3 = 0$ and $X_3 = L$. On the end $X_3 = 0$, $u_1 = u_2 = 0$ and $t_3 = 0$ so that the integrand above is zero on this portion of $\partial \mathcal{R}^u$. On the end $X_3 = L$, $u_1 = -\alpha L X_2$, $u_2 = \alpha L X_1$, and $t_3 = 0$. Thus,

$$\Pi = \frac{1}{2} \alpha L \int_S (X_1 t_2 - X_2 t_1) \, dA = \frac{1}{2} \alpha L T \, , \tag{10.3.31}$$

where the torque T is the resultant moment about the X_3-axis due to the traction distribution on the end of the cylinder. Noting that αL is the total twist of the cylinder, it is seen that the potential energy is equal to the work done by the torque. Recall that the torsion constant $\tilde{J}$ is a property of the geometry of the cylinder that relates the torque to the twist per unit length, $T = \mu \alpha \tilde{J}$. Thus, the torsion constant is given in terms of the normalized potential energy by

$$\boxed{\tilde{J} = \frac{2\bar{\Pi}}{\mu \alpha^2}} \, , \tag{10.3.32}$$

where $\bar{\Pi} \equiv \Pi / L$.

Let ψ_{approx} be a kinematically admissible warping function, that may or may not be the actual warping function ψ and let $\bar{\Pi}_{\mathrm{approx}}$ be the corresponding potential energy. According to the principle of minimum potential energy, if $\bar{\Pi}$ is the actual potential energy, then $\bar{\Pi}_{\mathrm{approx}} \geq \bar{\Pi}$ with $\bar{\Pi}_{\mathrm{approx}} = \bar{\Pi}$ if and only if $\psi_{\mathrm{approx}} = \psi$. It follows that if

$$\tilde{J}_{\mathrm{approx}} = \frac{2\bar{\Pi}_{\mathrm{approx}}}{\mu \alpha^2} \tag{10.3.33}$$

is the corresponding torsion constant and $\tilde{J}$ is the actual torsion constant, then $\tilde{J}_{\mathrm{approx}} \geq \tilde{J}$ with $\tilde{J}_{\mathrm{approx}} = \tilde{J}$ if and only if $\psi_{\mathrm{approx}} = \psi$. Therefore, approximate solutions to the torsion problem based on minimization of the potential energy will overestimate the torsion constant—they provide an upper bound on the torsion constant.

Example 10.4
Consider torsion of a cylinder with an arbitrary cross section S with boundary ∂S (Figure 10.7). Let the X_1- and X_2-axes be principal axes of inertia

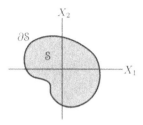

FIGURE 10.7. Torsion of a cylinder with arbitrary cross section.

of the area. Determine the best approximate warping function of the form

$$\psi = AX_1^2 + BX_1X_2 + CX_2^2$$

by minimizing the potential energy. Find the corresponding approximate torsion constant.

Solution. The potential energy per unit length (10.3.23) corresponding to this assumed form of the warping function is

$$\bar{\Pi} = \frac{1}{2}\mu\alpha^2 \int_S \left[(2AX_1 + BX_2 - X_2)^2 + (BX_1 + 2CX_2 + X_1)^2\right] dA$$

$$= \frac{1}{2}\mu\alpha^2 \int_S \left\{\left[4A^2 + (B+1)^2\right] X_1^2 + \left[4C^2 + (B-1)^2\right] X_2^2\right.$$

$$\left. + 4\left[A(B-1) + C(B+1)\right] X_1X_2\right\} dA .$$

The moments of inertia for S about the X_1- and X_2-axes are respectively

$$I_1 \equiv \int_S X_2^2 \, dA , \quad I_2 \equiv \int_S X_1^2 \, dA ,$$

and the product of inertia is

$$I_{12} \equiv \int_S X_1X_2 \, dA .$$

However, since the X_1- and X_2-axes are principal axes of inertia, $I_{12} = 0$. Thus, the potential energy per unit length can be expressed as

$$\bar{\Pi} = \frac{1}{2}\mu\alpha^2 \left\{\left[4A^2 + (B+1)^2\right] I_2 + \left[4C^2 + (B-1)^2\right] I_1\right\} .$$

The values of A, B, and C that minimize the potential energy satisfy

$$\frac{\partial \bar{\Pi}}{\partial A} = 4\mu\alpha^2 A I_2 = 0 \,,$$

$$\frac{\partial \bar{\Pi}}{\partial B} = \mu\alpha^2 [(B+1)I_2 + (B-1)I_1] = 0 \,,$$

$$\frac{\partial \bar{\Pi}}{\partial C} = 4\mu\alpha^2 C I_1 = 0 \,,$$

from which it follows that $A = C = 0$ and

$$B = \frac{I_1 - I_2}{I_1 + I_2} \,.$$

Thus, the best approximate warping function of the specified form, regardless of the cross section, is

$$\psi_{\text{approx}} = \frac{I_1 - I_2}{I_1 + I_2} X_1 X_2 \,.$$

The corresponding potential energy is

$$\bar{\Pi}_{\text{approx}} = 2\mu\alpha^2 \frac{I_1 I_2}{I_1 + I_2} \,,$$

so that, from (10.3.32), the approximate torsion constant is

$$\tilde{J}_{\text{approx}} = \frac{4 I_1 I_2}{I_1 + I_2} \,.$$

For cross sections in which $I_1 = I_2$ (e.g., for circular and square cross sections), the best approximate warping function of the specified form is $\psi_{\text{approx}} = 0$—the approximation in this case is equivalent to assuming that plane cross sections remain plane and undeformed. For circular cross sections, this yields the exact result. On the other hand, for a square cross section with edge length $2a$, $I_1 = I_2 = \frac{4}{3}a^4$ so that

$$\tilde{J}_{\text{approx}} = \frac{8}{3}a^4 = 2.667a^4 \,.$$

Comparing this with the exact result $\tilde{J} = 2.249a^4$ from Chapter 8, it is seen that there is an error of nearly 19%.

Principle of minimum complementary energy applied to torsion

Since for a cylinder in torsion, $\sigma_{11} = \sigma_{22} = \sigma_{33} = \sigma_{12} = 0$ and σ_{31} and σ_{32} are independent of X_3, the equilibrium equations reduce to the single equation

$$\sigma_{31,1} + \sigma_{32,2} = 0 \,. \tag{10.3.34}$$

Accordingly, the Prandtl stress function $\phi(X_1, X_2)$ is defined by the relations

$$\sigma_{31} = \phi_{,2} , \quad \sigma_{32} = -\phi_{,1} , \qquad (10.3.35)$$

so that components of stress derived from the Prandtl stress function are guaranteed to satisfy the equilibrium equation. It was shown in Chapter 8 that for cylinders with simply connected cross sections (i.e., for solid cylinders), the traction-free lateral surface boundary condition can be enforced by requiring

$$\phi = 0 \quad \forall (X_1, X_2) \in \partial \mathcal{S} , \qquad (10.3.36)$$

where $\partial \mathcal{S}$ is the boundary of the cross section.

The complementary energy functional,

$$\Pi^c[\boldsymbol{\sigma}(\mathbf{X})] = \int_{\mathcal{R}} \mathcal{U}^c(\boldsymbol{\sigma}) \, dV - \int_{\partial \mathcal{R}^u} t_i \tilde{u}_i \, dA , \qquad (10.3.37)$$

can be expressed in terms of the Prandtl stress function. Recall first that, assuming linear constitutive behavior, $\mathcal{U}^c = \mathcal{U}$ so that from (5.4.19), assuming the material is isotropic,

$$\mathcal{U}^c = \frac{1}{2E} \left[(1+\nu)\sigma_{ij}\sigma_{ij} - \nu(\sigma_{kk})^2 \right] = \frac{1+\nu}{E}(\sigma_{31}^2 + \sigma_{32}^2) = \frac{1}{2\mu}(\phi_{,1}^2 + \phi_{,2}^2) . \qquad (10.3.38)$$

For the second integral on the right-hand side of (10.3.37), it has already been shown in (10.3.30) and (10.3.31) that

$$\int_{\partial \mathcal{R}^u} t_i \tilde{u}_i \, dA = \alpha L T . \qquad (10.3.39)$$

Furthermore, it was shown in Chapter 8 that for solid cylinders with $\phi = 0$ on $\partial \mathcal{S}$,

$$T = 2 \int_{\mathcal{S}} \phi \, dA . \qquad (10.3.40)$$

Thus, noting that ϕ is independent of X_3, the complementary energy per unit length $\bar{\Pi}^c \equiv \Pi^c/L$ is expressed as a functional of the Prandtl stress function by

$$\boxed{\bar{\Pi}^c[\phi(X_1, X_2)] = \int_{\mathcal{S}} \left[\frac{1}{2\mu}(\phi_{,1}^2 + \phi_{,2}^2) - 2\alpha\phi \right] dA .} \qquad (10.3.41)$$

All twice continuously differentiable Prandtl stress functions that satisfy $\phi = 0$ on the boundary ∂S of the (simply connected) cross section of a solid cylinder are statically admissible, in that they correspond to equilibrated stress fields that satisfy the traction boundary condition (10.3.16). The actual Prandtl stress function satisfies compatibility and corresponds to a displacement field that satisfies the displacement boundary conditions (10.3.17) and (10.3.18). According to the principle of minimum complementary energy, the actual Prandtl stress function will make $\bar{\Pi}^c(\phi)$ an absolute minimum.

Taking the variation of $\bar{\Pi}^c(\phi)$,

$$\delta\bar{\Pi}^c = \int\limits_S \left[\frac{1}{\mu}(\phi_{,1}\,\delta\phi_{,1} + \phi_{,2}\,\delta\phi_{,2}) - 2\alpha\delta\phi \right] dA , \tag{10.3.42}$$

which can be rewritten as

$$\delta\bar{\Pi}^c = \frac{1}{\mu} \int\limits_S [(\phi_{,1}\,\delta\phi)_{,1} + (\phi_{,2}\,\delta\phi)_{,2}]\, dA - \int\limits_S \left[\frac{1}{\mu}(\phi_{,11} + \phi_{,22}) + 2\alpha \right]\delta\phi\, dA . \tag{10.3.43}$$

Applying the Green-Riemann theorem to the first integral on the right-hand side and noting that $\phi_{,11} + \phi_{,22} = \nabla^2\phi$,

$$\delta\bar{\Pi}^c = \frac{1}{\mu} \oint\limits_{\partial S} (\phi_{,1}\,\delta\phi\, dX_2 - \phi_{,2}\,\delta\phi\, dX_1) - \int\limits_S \left(\frac{1}{\mu}\nabla^2\phi + 2\alpha \right)\delta\phi\, dA . \tag{10.3.44}$$

However, enforcing the boundary condition (10.3.36) means that $\delta\phi = 0$ on ∂S and the first integral on the right-hand side vanishes. Thus,

$$\delta\bar{\Pi}^c = -\int\limits_S \left(\frac{1}{\mu}\nabla^2\phi + 2\alpha \right)\delta\phi\, dA . \tag{10.3.45}$$

Since $\delta\phi$ is otherwise arbitrary, $\delta\bar{\Pi}^c = 0$ if and only if

$$\nabla^2\phi = -2\mu\alpha \quad \forall(X_1, X_2) \in S . \tag{10.3.46}$$

This is precisely the compatibility condition that was derived in Chapter 8.

The complementary energy can be related to the torque T applied to the ends of the cylinder. Recall that the complementary energy is the negative of the potential energy, $\Pi^c = -\Pi$, and from (10.3.31) that $\Pi = \frac{1}{2}\alpha LT$. Thus,

$$\Pi^c = -\frac{1}{2}\alpha LT , \tag{10.3.47}$$

and the torsion constant $\tilde{J} = T/\mu\alpha$ is given in terms of the complementary energy per unit length by

$$\tilde{J} = -\frac{2\bar{\Pi}^c}{\mu\alpha^2} , \qquad (10.3.48)$$

where $\bar{\Pi}^c \equiv \Pi^c/L$.

Let ϕ_{approx} be a statically admissible Prandtl stress function, which may or may not be the actual Prandtl stress function ϕ, and let $\bar{\Pi}^c_{\text{approx}}$ be the corresponding complementary energy. According to the principle of minimum complementary energy, if $\bar{\Pi}^c$ is the actual complementary energy, then $\bar{\Pi}^c_{\text{approx}} \geq \bar{\Pi}^c$ with $\bar{\Pi}^c_{\text{approx}} = \bar{\Pi}^c$ if and only if $\phi_{\text{approx}} = \phi$. It follows that if

$$\tilde{J}_{\text{approx}} = -\frac{2\bar{\Pi}^c_{\text{approx}}}{\mu\alpha^2} \qquad (10.3.49)$$

is the corresponding torsion constant and $\tilde{J}$ is the actual torsion constant, then $\tilde{J}_{\text{approx}} \leq \tilde{J}$ with $\tilde{J}_{\text{approx}} = \tilde{J}$ if and only if $\phi_{\text{approx}} = \phi$. Therefore, approximate solutions to the torsion problem based on minimization of the complementary energy will underestimate the torsion constant—they provide a lower bound. Recall that approximate solutions based on minimization of the potential energy provide an upper bound on the torsion constant.

Example 10.5
Consider torsion of a cylinder with the cross section shown in Figure 10.8. The region S is bounded by the curves $X_1 = c$, $X_1 = d$, $X_2 = g(X_1)$, and $X_2 = -g(X_1)$, where $g(X_1)$ is a known, twice continuously differentiable function. Prandtl stress functions of the form

$$\phi = \left[X_2^2 - g^2(X_1)\right] f(X_1)$$

will be statically admissible provided $f(X_1)$ is twice continuously differentiable and

$$f(c) = f(d) = 0 .$$

Derive an expression for the best approximate Prandtl stress function of this form, by minimizing the complementary energy.

Solution. Noting first that for the given form of the Prandtl stress function,

$$\phi_{,1} = \left[X_2^2 - g^2(X_1)\right] f'(X_1) - 2g(X_1)g'(X_1)f(X_1) ,$$
$$\phi_{,2} = 2X_2 f(X_1) ,$$

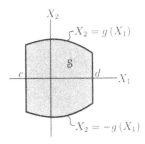

FIGURE 10.8. *Complementary-energy-based approximate torsion solutions.*

it follows from (10.3.41) that the complementary energy in this case is given by

$$\bar{\Pi}^c[f(X_1)] = \int_c^d \int_{-g(X_1)}^{g(X_1)} \left\{ \frac{1}{2\mu} \left[(X_2^2 - g^2)^2 (f')^2 - 4(X_2^2 - g^2)gg'ff' \right. \right.$$
$$\left. \left. + 4(gg'f)^2 + 4X_2^2 f^2 \right] - 2\alpha(X_2^2 - g^2)f \right\} dX_2\, dX_1,$$

where it must be remembered that f and g are functions of X_1. After some manipulation, this expression can be integrated in X_2 to give

$$\bar{\Pi}^c[f(X_1)] = \frac{16}{15\mu} \int_c^d \left[\frac{1}{2}g^5(f')^2 + \frac{5}{2}g^4g'ff' \right.$$
$$\left. + \frac{15}{4}g^3(g'f)^2 + \frac{5}{4}g^3f^2 + \frac{5}{2}\mu\alpha g^3 f \right] dX_1.$$

Since g is prescribed, one now has an expression for the complementary energy in terms of the unknown function f. Thus, the function f that renders $\bar{\Pi}^c[f(X_1)]$ stationary (i.e., that corresponds to $\delta\bar{\Pi}^c = 0$) gives the best approximate Prandtl stress function of the given form for this problem.

Taking the variation of $\bar{\Pi}^c[f(X_1)]$ gives

$$\delta\bar{\Pi}^c = \frac{16}{15\mu} \int_c^d \left[g^5 f'\, \delta f' + \frac{5}{2}g^4g'f\, \delta f' + \frac{5}{2}g^4g'f'\, \delta f \right.$$
$$\left. + \frac{15}{2}g^3(g')^2 f\, \delta f + \frac{5}{2}g^3 f\, \delta f + \frac{5}{2}\mu\alpha g^3\, \delta f \right] dX_1,$$

which can be rewritten as

$$\delta\bar{\Pi}^c = \frac{16}{15\mu} \int_c^d \left\{ \left(g^5 f' + \frac{5}{2}g^4 g' f \right) \delta f' \right.$$

$$\left. + \left[\frac{5}{2}g^4 g' f' + \frac{15}{2}g^3 (g')^2 f + \frac{5}{2}g^3 f + \frac{5}{2}\mu\alpha g^3 \right] \delta f \right\} dX_1 \; .$$

The variations δf and $\delta f'$ are not independent, so that it would be incorrect to say that requiring $\delta\bar{\Pi}^c = 0$ implies that the coefficients of δf and $\delta f'$ in the preceding expression must be zero. However, an expression for $\delta\bar{\Pi}^c$ given entirely in terms of δf is obtained by applying integration by parts to show that

$$\int_c^d \left(g^5 f' + \frac{5}{2}g^4 g' f \right) \delta f' \, dX_1$$

$$= \left(g^5 f' + \frac{5}{2}g^4 g' f \right) \delta f \Big|_c^d - \int_c^d \left(g^5 f' + \frac{5}{2}g^4 g' f \right)' \delta f \, dX_1 \; .$$

Recall that in order for the given form of the Prandtl stress function to be statically admissible, it is required that $f(c) = f(d) = 0$. This means that $\delta f(c) = \delta f(d) = 0$, so that

$$\int_c^d \left(g^5 f' + \frac{5}{2}g^4 g' f \right) \delta f' \, dX_1$$

$$= -\int_c^d \left[5g^4 g' f' + g^5 f'' + 10g^3 (g')^2 f + \frac{5}{2}g^4 g'' f + \frac{5}{2}g^4 g' f' \right] \delta f \, dX_1 \; .$$

Substituting this result back into the expression for $\delta\bar{\Pi}^c$ and simplifying gives

$$\delta\bar{\Pi}^c = \frac{16}{15\mu} \int_c^d \left\{ -g^5 f'' - 5g^4 g' f' \right.$$

$$\left. + \frac{5}{2} \left[-g^3 (g')^2 - g^4 g'' + g^3 \right] f + \frac{5}{2}\mu\alpha g^3 \right\} \delta f \, dX_1 \; ,$$

which can be rewritten as

$$\delta\bar{\Pi}^c = \frac{16}{15\mu} \int_c^d \frac{g^5}{4} \left[f'' - \frac{5}{g^2}(g^2 f)'' + \frac{10}{g^2} f + \frac{10\mu\alpha}{g^2} \right] \delta f \, dX_1 \; .$$

Thus, the function f that renders $\bar{\Pi}^c[f(X_1)]$ stationary satisfies the second-order linear ordinary differential equation with variable coefficients

$$f'' - \frac{5}{g^2}(g^2 f)'' + \frac{10}{g^2} f = -\frac{10\mu\alpha}{g^2} \; .$$

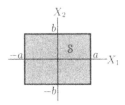

FIGURE 10.9. Approximate solution for torsion of a rectangular cylinder.

Given a cross section that is characterized by the function g, the best approximate Prandtl stress function of the specified form is obtained by solving this differential equation. The two constants of integration are determined by the static admissibility conditions $f(c) = f(d) = 0$.

Example 10.6
Consider, as a special case of the problem examined in Example 10.5, torsion of a cylinder with the rectangular cross section shown in Figure 10.9.

Solution. In this case, $c = -a$, $d = a$, and $g = b$ (a constant). Thus, the best approximate Prandtl stress function of the form

$$\phi = (X_2^2 - b^2)f(X_1) ,$$

is given by the function f that is the solution of the differential equation

$$f'' - \frac{5}{2b^2}f = \frac{5\mu\alpha}{2b^2}$$

that satisfies the static admissibility conditions $f(-a) = f(a) = 0$.
The general solution of the above differential equation is

$$f = A\sinh\left(\sqrt{\frac{5}{2}}\frac{X_1}{b}\right) + B\cosh\left(\sqrt{\frac{5}{2}}\frac{X_1}{b}\right) - \mu\alpha .$$

It follows from the conditions $f(-a) = f(a) = 0$ that

$$A = 0 , \quad B = \frac{\mu\alpha}{\cosh\left(\sqrt{\frac{5}{2}}\frac{a}{b}\right)} .$$

Thus, the best approximate Prandtl stress function of the given form is

$$\phi_{\text{approx}} = \mu\alpha(X_2^2 - b^2)\left[\frac{\cosh\left(\sqrt{\frac{5}{2}}\frac{X_1}{b}\right)}{\cosh\left(\sqrt{\frac{5}{2}}\frac{a}{b}\right)} - 1\right] .$$

Recall that for a solid cylinder in which $\phi = 0$, $\forall(X_1, X_2) \in \partial S$, the applied torque is related to the Prandtl stress function by $T = 2\int_S \phi \, dA$ and that the torsion constant is given by $\tilde{J} = T/\mu\alpha$. Therefore, the approximate torsion constant in this example is

$$
\begin{aligned}
\tilde{J}_{\text{approx}} &= \frac{2}{\mu\alpha} \int_S \phi_{\text{approx}} \, dA \\[2mm]
&= 2 \int_{-a}^{a} \int_{-b}^{b} (X_2^2 - b^2) \left[\frac{\cosh\left(\sqrt{\frac{5}{2}}\frac{X_1}{b}\right)}{\cosh\left(\sqrt{\frac{5}{2}}\frac{a}{b}\right)} - 1 \right] dX_2 \, dX_1 \\[2mm]
&= -\frac{8}{3} b^3 \int_{-a}^{a} \left[\frac{\cosh\left(\sqrt{\frac{5}{2}}\frac{X_1}{b}\right)}{\cosh\left(\sqrt{\frac{5}{2}}\frac{a}{b}\right)} - 1 \right] dX_1 \\[2mm]
&= \frac{8}{3} b^3 \left[X_1 - \sqrt{\frac{2}{5}} b \frac{\sinh\left(\sqrt{\frac{5}{2}}\frac{X_1}{b}\right)}{\cosh\left(\sqrt{\frac{5}{2}}\frac{a}{b}\right)} \right]\Bigg|_{-a}^{a} \\[2mm]
&= \frac{16}{3} b^3 \left[a - \sqrt{\frac{2}{5}} b \tanh\left(\sqrt{\frac{5}{2}}\frac{a}{b}\right) \right].
\end{aligned}
$$

The best approximate torsion constant for torsion of a cylinder with rectangular cross section, corresponding to the given form of the Prandtl stress function, is

$$
\tilde{J}_{\text{approx}} = \frac{16}{3} ab^3 \left[1 - \sqrt{\frac{2}{5}\frac{b}{a}} \tanh\left(\sqrt{\frac{5}{2}\frac{a}{b}}\right) \right].
$$

Let $\tilde{J}$ be the exact result from Chapter 8. If $a/b = 1$, then $\tilde{J}_{\text{approx}} = 2.234b^4$ and $\tilde{J} = 2.249b^4$, and the error is less than 0.7%. If $a/b = 5$, then $\tilde{J}_{\text{approx}} = 23.29b^4$ and $\tilde{J} = 23.31b^4$, and the error is less than 0.1%.

Problems

10.1 Find the curves $y = u(x)$ that render

$$
I[u(x)] = \int_{x_0}^{x_1} \sqrt{2a - b^2 u^2} \, ds, \quad ds = \sqrt{dx^2 + dy^2},
$$

stationary. a, b, x_0, and x_1 are prescribed constants.

10.2 Beginning with (10.2.57), derive the expression (10.2.58) for the difference in complementary energy between an actual stress field and any other statically admissible stress field.

10.3 By considering Prandtl stress functions of the form

$$\phi(r,\theta) = f(r)\cos\frac{\pi\theta}{2\beta} , \quad f(a) = 0 ,$$

and using the principle of minimum complementary energy, find an approximate expression of the torsion constant for a cylinder with a cross section in the form of a circular wedge, as shown below.

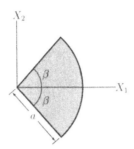

Note that, in cylindrical coordinates, the components of stress are related to the Prandtl stress function by $\sigma_{rr} = \sigma_{\theta\theta} = \sigma_{zz} = \sigma_{r\theta} = 0$ and

$$\sigma_{zr} = \frac{1}{r}\frac{\partial\phi}{\partial\theta} , \quad \sigma_{z\theta} = -\frac{\partial\phi}{\partial r} .$$

Part of the problem is to determine what conditions on $f(r)$ at $r = 0$ are appropriate. Finally, check the accuracy of the result for the special case of a semicircular cross section ($\beta = \pi/2$) by noting that the exact solution for the constant in this case is given by

$$\tilde{J}|_{\beta=\pi/2} = \left(\frac{\pi}{2} - \frac{4}{\pi}\right) a^4 .$$

10.4 Consider torsion of a cylinder with a cross section bounded by the following two curves:

$$X_2 = aF\left(\frac{X_1}{L}\right) , \quad X_2 = -bF\left(\frac{X_1}{L}\right) ,$$

where a and b are prescribed positive constants and $F(s)$ is a prescribed function that satisfies $F(0) = F(1) = 0$ and $F(s) > 0$ when

$0 < s < 1$. Use the principle of minimum complementary energy to find an approximate expression for the torsion constant by considering Prandtl stress functions of the form

$$\phi(X_1, X_2) = A\left[X_2 - aF\left(\frac{X_1}{L}\right)\right]\left[X_2 + bF\left(\frac{X_1}{L}\right)\right],$$

where A is a constant parameter. (**Hint:** simplify the answer as much as possible—by exploring different factorizations in powers of a and b, a relatively compact form of the answer can be found.) Show that, in the particular case when $F(s) = (1 - s)\sqrt{s}$, the best approximate torsion constant given the assumed form of the Prandtl stress function is

$$\tilde{J}_{\text{approx}} = \frac{32}{3465}\left\{\frac{(a+b)^3 L}{1 + (11/39)\left[(a^2 + ab + b^2)/L^2\right]}\right\}.$$

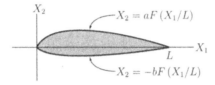

10.5 Assuming that plane cross sections remain plane, it can be shown that the potential energy functional for a beam in bending is expressible as

$$\Pi[y(x)] = \frac{1}{2}\int_0^L EI(y')^2\, dx - \int_0^L py\, dx$$
$$+ M_0 y'(0) - V_0 y(0) - M_L y'(L) + V_L y(L),$$

where x is the position along the length of the beam and $y(x)$ is the beam's deflection curve. The beam extends from $x = 0$ to $x = L$, along the length of the beam there is a prescribed transverse load distribution $p(x)$, at the end $x = 0$ there is a moment M_0 and a shear V_0, at the end $x = L$ there is a moment M_L and a shear V_L, and $E(x)$ and $I(x)$ are the prescribed Young's modulus of the material and moment of inertia of the cross section, respectively. Use the principle of minimum potential energy to derive the governing differential equation for $y(x)$ on $0 < x < L$ and the permissible boundary conditions at $x = 0$ and $x = L$.

10.6 It can be shown that the potential energy functional for a membrane stretched over a simply connected region S of the $X_1 X_2$-plane is expressible as

$$\Pi[w(X_1, X_2)] = \frac{1}{2} \int_S \eta \left[(w_{,1})^2 + (w_{,2})^2 \right] dA - \int_S pw \, dA,$$

where $w(X_1, X_2)$ is the deflection of the membrane, $p(X_1, X_2)$ is the prescribed transverse pressure distribution, and the constant η is the membrane stiffness. Use the principle of minimum potential energy to derive the governing differential equation for $w(X_1, X_2)$ on S and the permissible boundary conditions at the boundary ∂S of S.

Chapter 11

Complex Variable Methods

There are certain properties of complex-valued functions of a complex variable, discussed below, that prove to be very useful in the study of two-dimensional potential theory with applications to heat flow, inviscid fluids, and linearized elasticity, among others. The focus here is on the application of complex variable methods to antiplane strain and plane strain/stress problems in linearized elasticity.

Although some fundamental results in the theory of functions of a complex variable are presented here, it is assumed that the reader already has a basic familiarity with the subject. Many important results are presented below without formal proof. Many additional results from complex variable theory are not addressed at all, since they are not required for the purpose at hand. For the reader who desires additional background on the subject, there are any number of excellent texts devoted to the theory of functions of a complex variable, including those of Dettman (1965), Carrier et al. (1983), and Lang (1985). For more information on applications of the theory to linearized elasticity, see England (1971) and Muskhelishvili (1963).

11.1 Functions of a Complex Variable

11.1.1 Complex Numbers

A *complex number* u is expressed in its *standard form* as $u = a + ib$, where i is the *imaginary unit*, which has the defining property that $i^2 = -1$, and a and b are real numbers known as the *real part* and the *imaginary part* of u, respectively, which is denoted by writing $a = \text{Re}\{u\}$ and $b = \text{Im}\{u\}$. If the imaginary part of u vanishes (i.e., $b = 0$), then u is an ordinary real number. On the other hand, u is said to be *pure imaginary* if $a = 0$. A complex number u is zero if and only if both its real and imaginary parts are zero. The *complex conjugate* of u is defined as $\bar{u} \equiv a - ib$. In other words, $\mathrm{Re}\{\bar{u}\} = \mathrm{Re}\{u\}$ and $\mathrm{Im}\{\bar{u}\} = -\,\mathrm{Im}\{u\}$.

By definition, two complex numbers $u = a + ib$ and $v = c + id$ are *equal* if and only if their real parts are equal ($a = c$) and their imaginary parts are equal ($b = d$). The *sum* of two complex numbers u and v is defined to be another complex number whose real part is the sum of the real parts of u and v and whose imaginary part is the sum of the imaginary parts of u and v, so that

$$u + v \equiv (a + c) + i(b + d) \ . \tag{11.1.1}$$

Suppose that the usual distributive and associative rules for multiplication of real numbers can be applied to the standard forms of u and v, so that

$$(a + ib)(c + id) = ac + i^2 bd + iad + ibc \ . \tag{11.1.2}$$

The *product* of two complex numbers u and v is defined in such a way that it is consistent with this supposition. Specifically, recalling that $i^2 = -1$, the product is defined as

$$uv \equiv (ac - bd) + i(ad + bc) \ . \tag{11.1.3}$$

The *negative* of a complex number v is defined as $-v \equiv (-1)v = -c - id$ and the *difference* between the complex numbers u and v is

$$u - v \equiv u + (-v) = (a - c) + i(b - d) \ . \tag{11.1.4}$$

The *quotient* u/v ($v \neq 0$) of two complex numbers u and v is another complex number defined such that if $w = u/v$, then $u = vw$. Using (11.1.3) and solving for the necessary real and imaginary parts of w in terms of those of u and v leads to the definition

$$\frac{u}{v} \equiv \frac{ac + bd}{c^2 + d^2} + i\frac{bc - ad}{c^2 + d^2} \ . \tag{11.1.5}$$

It is useful to note that the quotient u/v can also be reduced to its standard form by multiplying the numerator and denominator by the complex conjugate of v:

$$\frac{u}{v} = \frac{u\bar{v}}{v\bar{v}} = \frac{(a + ib)(c - id)}{(c + id)(c - id)} = \frac{(ac + bd) + i(bc - ad)}{c^2 + d^2} \ . \tag{11.1.6}$$

This works because the product of a complex number and its complex conjugate is a real number, which is discussed further below. This algebra of complex numbers can be shown to satisfy the same associative, commutative, and distributive properties as the algebra of real numbers.

The real and imaginary parts of a complex number u can be expressed in terms of the sum and difference of the number and its complex conjugate:

$$\text{Re}\{u\} = \frac{1}{2}(u + \bar{u}), \quad \text{Im}\{u\} = -\frac{i}{2}(u - \bar{u}). \tag{11.1.7}$$

The *absolute value*, or *modulus*, of a complex number $u = a + ib$ is a non-negative real number defined as

$$|u| \equiv \sqrt{a^2 + b^2}. \tag{11.1.8}$$

The absolute value of zero is zero, $|0| = 0$. Note from (11.1.3) that the product of a complex number and its complex conjugate is equal to the square of its absolute value, $u\bar{u} = |u|^2$.

By defining two new real numbers r and θ such that $a = r\cos\theta$ and $b = r\sin\theta$, a nonzero complex number $u = a + ib$ ($u \neq 0$) can be expressed in a *trigonometric form* as

$$u = r\cos\theta + ir\sin\theta = r(\cos\theta + i\sin\theta). \tag{11.1.9}$$

Noting that $|u|^2 = r^2\cos^2\theta + r^2\sin^2\theta = r^2$ and taking the positive root, it follows that

$$r = |u|. \tag{11.1.10}$$

Thus, r is the absolute value of u, which is uniquely determined. The angle θ, however, is not uniquely determined by u. If $\theta = \theta_0$ satisfies $a = r\cos\theta$ and $b = r\sin\theta$, then so do $\theta = \theta_0 + 2k\pi$ for all $k = \ldots, -2, -1, 0, 1, 2, \ldots$. Each of these angles is called an *argument* of u, written

$$\theta = \arg\{u\}. \tag{11.1.11}$$

The argument of zero, $\arg\{0\}$, is undefined. Note that if θ is an argument of u, then $-\theta$ is an argument of its complex conjugate. The *principal argument* of u, written $\text{Arg}\{u\}$, is either defined to be the one argument that satisfies

$$0 \leq \text{Arg}\{u\} < 2\pi, \tag{11.1.12}$$

or it is defined to be the one argument that satisfies

$$-\pi < \text{Arg}\{u\} \leq \pi. \tag{11.1.13}$$

The most convenient choice for the definition of the principle argument will depend on the problem. In either case, once the definition is established, $\text{Arg}\{u\}$ is uniquely determined by u.

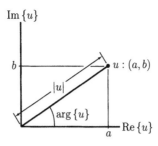

FIGURE 11.1. *Representation of a complex number as a point in the complex plane.*

The exponential of a complex number $v = c + id$ is defined by *Euler's formula*,

$$e^v = e^{c+id} = e^c e^{id} = e^c(\cos d + i \sin d) \ . \tag{11.1.14}$$

It will be seen later that Euler's formula preserves many of the familiar properties of the exponential function of real variables, particularly as relates to differentiation. For now, note that the exponential of a complex number reduces to the exponential of its real part when its imaginary part is zero. By setting $v = i\theta$, one finds that

$$e^{i\theta} = \cos \theta + i \sin \theta \ , \tag{11.1.15}$$

which when introduced into (11.1.9) leads to the *exponential form* of the complex number $u = a + ib$ $(u \neq 0)$,

$$u = re^{i\theta} \ , \tag{11.1.16}$$

where $r = |u|$ and $\theta = \arg\{u\}$. It follows that the exponential form of the complex conjugate of u is

$$\overline{u} = re^{-i\theta} \ . \tag{11.1.17}$$

The product of two nonzero complex numbers $u_1 = r_1 e^{i\theta_1}$ and $u_2 = r_2 e^{i\theta_2}$ is

$$u_1 u_2 = r_1 r_2 e^{i(\theta_1 + \theta_2)} \tag{11.1.18}$$

and their quotient is

$$\frac{u_1}{u_2} = \frac{r_1}{r_2} e^{i(\theta_1 - \theta_2)} \ . \tag{11.1.19}$$

These and other useful results are easily derived.

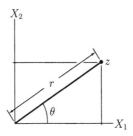

FIGURE 11.2. *One-to-one correspondence between values of $z = X_1 + iX_2$ and points in the X_1X_2-plane.*

Finally, all of the above results can be interpreted geometrically in the *complex plane* (or *Argand diagram*). The complex number $u = a + ib$ may be represented graphically as the point in the complex plane with Cartesian coordinates (a, b), as shown in Figure 11.1. The coordinate axes are called the *real axis* and the *imaginary axis*. The absolute value $r = |u|$ and the argument $\theta = \arg\{u\}$ are the polar coordinates of the point representing u.

11.1.2 Complex Variables

A one-to-one correspondence between the set of all points in the X_1X_2-plane and the set of all complex numbers can be constructed by defining the *complex variable* $z = X_1 + iX_2$. Each value of z corresponds to a unique point in the X_1X_2-plane and vice versa, as shown in Figure 11.2. The polar coordinates of the point in the plane are $r = |z|$ and $\theta = \arg\{z\}$, so that $z = re^{i\theta}$. Once this one-to-one correspondence is established, the X_1X_2-plane is often referred to as the *z-plane*.

It will be convenient to define an idealized *point at infinity*, denoted by ∞, such that

$$\lim_{|z|\to\infty} z = \infty . \tag{11.1.20}$$

The argument of the point at infinity, $\arg\{\infty\}$, is undefined. When the point at infinity is included, the complex plane is referred to as the *extended complex plane*.

Sets of complex numbers correspond to sets of points in the z-plane. For example,

$$\mathcal{S}_0 = \{z_1, z_2, z_3, z_4\} \tag{11.1.21}$$

is a set containing only four points (called elements of the set $\mathcal{S}_0$). The set of all points in the unextended, or finite, complex plane will be denoted by

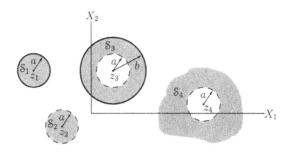

FIGURE 11.3. Some examples of sets in the z-plane.

$\mathfrak{C}$. The set of all points in the extended complex plane will be denoted by $\mathfrak{C}^\infty$. A *neighborhood* $\mathcal{N}_\epsilon$ of a point z_0 is a set of all the points satisfying $|z - z_0| < \epsilon$ for some $\epsilon > 0$ [i.e., the region of the z-plane within (but not including) a circle of radius ϵ centered at z_0], which is expressed by writing

$$\mathcal{N}_\epsilon(z_0) = \{\, z \in \mathfrak{C} \mid |z - z_0| < \epsilon \,\} \; . \qquad (11.1.22)$$

Some other examples illustrated in Figure 11.3 are the set S_1 corresponding to the region of the z-plane within and including a circle of radius a centered at z_1,

$$S_1 = \{\, z \in \mathfrak{C} \mid |z - z_1| \leq a \,\} \; ; \qquad (11.1.23)$$

the set S_2 corresponding to the region within, but not including, a circle of radius a centered at z_2,

$$S_2 = \{\, z \in \mathfrak{C} \mid |z - z_2| < a \,\} \; ; \qquad (11.1.24)$$

the set S_3 corresponding to the region bounded by circles of radius a and b that are centered at z_3, where the outer circle is included but the inner circle is not,

$$S_3 = \{\, z \in \mathfrak{C} \mid a < |z - z_3| \leq b \,\} \; ; \qquad (11.1.25)$$

and the set S_4 corresponding to the entire z-plane with the region within and on a circle of radius a centered at z_4 deleted,

$$S_4 = \{\, z \in \mathfrak{C} \mid |z - z_4| > a \,\} \; . \qquad (11.1.26)$$

Note that S_2 is a neighborhood of z_2.

Relative to a given set S, points in the z-plane are characterized as follows:

- A point z_0 is an *interior point* of a set S if and only if at least one neighborhood of z_0 consists entirely of points in S. An interior point is always an element of the set.

- A point z_0 is an *exterior point* of a set S if and only if at least one neighborhood of z_0 contains no points that are in S. An exterior point is never an element of the set.

- A point z_0 is a *boundary point* of a set S if and only if every neighborhood of z_0 contains both points that are in S and points that are not in S. Note that a boundary point may or may not be an element of the set.

- A point z_0 is a *limit point* of a set S if and only if every neighborhood of z_0 contains at least one point, other than z_0, that is in S. Note that a limit point may or may not be an element of the set. Every interior point is a limit point. A boundary point may or may not be a limit point. For example, all of the boundary points of the sets S_1, S_2, S_3, and S_4 are limit points, but the boundary points of S_0 are not.

Note that all the elements of the set S_0 are boundary points of that set. Sets are characterized as follows:

- A set S is *open* if and only if each of its elements is an interior point (i.e., if and only if S does not contain any of its boundary points). Thus, the sets S_2 and S_4 are open, as is every neighborhood of a point z_0.

- A set S is *closed* if and only if it contains all of its boundary points. The sets S_0 and S_1 are closed. Note that S_3 is neither open nor closed.

- A set S is *bounded* if and only if there exists a real number $d > 0$ such that $|z| < d$ for all z in S. The sets S_0, S_1, S_2, and S_3 are all bounded. The set S_4 is *unbounded*.

- A set S is *connected* if and only if every two points in S can be connected by a path in the z-plane, all of whose points are in S. The sets S_1, S_2, S_3, and S_4 are all connected. The set S_0 is not connected.

- A set S is *simply connected* if and only if it is connected and every simple closed curve which can be drawn in the interior of S contains only points of S in its interior. The sets S_1 and S_2 are simply connected. The sets S_3 and S_4 are *multiply connected*.

A *region* of the z-plane is a connected set, which may be open, closed, or neither open nor closed. Thus, each of the sets S_1, S_2, S_3, and S_4 is a region, although the set S_0 is not.

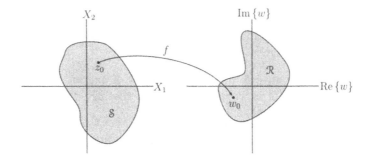

FIGURE 11.4. *Mapping of points in the z-plane into points in the w-plane.*

11.1.3   Functions of a Complex Variable

Let S be a set of points in the z-plane, and consider an operator f that assigns to each point $z \in S$ a unique complex value $w = u + iv$, so that z is the *independent variable* and w is the *dependent variable*. The operator f is said to be a *function of the complex variable* z defined in S. If $z_0 \in S$ and the function f assigns to z_0 the value w_0, then one writes that $w_0 = f(z_0)$ and says that z_0 is *mapped into* w_0 by the function or that w_0 is the *image* of z_0 under the function. In general, one writes that

$$w = f(z) \quad \forall z \in S \qquad (11.1.27)$$

and says that f maps S into the w-plane, the complex plane in which each value of w corresponds to a unique point with Cartesian coordinates (u, v) (Figure 11.4). The set S is the *domain* of the function and the set of all image points of S is the function's *range*. The function maps S *onto* the range $\mathcal{R}$, since every point in $\mathcal{R}$ is the image of *at least one* point in S. If each point in the range is the image of *precisely one* point in the domain, then the function is said to be a *one-to-one mapping*. If the function $w = f(z)$, $\forall z \in S$, is a one-to-one mapping of S onto $\mathcal{R}$, then an inverse function $z = g(w)$, $\forall w \in \mathcal{R}$, can be defined such that if $w_0 \in \mathcal{R}$ is the image of $z_0 \in S$ under f, then $z_0 \in S$ is the (uniquely defined) image of $w_0 \in \mathcal{R}$ under g. It follows in this case that $z = g[f(z)]$ and $w = f[g(w)]$.

Example 11.1

Consider the function $w = -z$, $\forall z \in \mathbb{C}$. This function maps each $z = X_1 + iX_2$ into $w = -X_1 - iX_2$. Each $z \in \mathbb{C}$ is mapped by the function into a point in the finite w-plane that is a reflection of z through the origin. If $z = re^{i\theta}$ ($z \neq 0$), then its image under the function is $w = re^{i(\theta + \pi)}$. Clearly, each $w \in \mathbb{C}$ is the image under the function of precisely one point in the domain—if $w = u + iv$, then it is the image under the function of

$z = -u - iv$. Therefore, the function is a one-to-one mapping of the finite z-plane onto the finite w-plane, with an inverse given by $z = -w$, $\forall w \in \mathfrak{C}$.

Example 11.2
Consider the function $w = |z|$, $\forall z \in \mathfrak{C}$. This function maps each $z = X_1 + iX_2$ into $w = \sqrt{X_1^2 + X_2^2}$, which is a point on the non-negative real axis, and each point on the non-negative real axis is the image under the function of at least one point in the domain. Therefore, the range of the function is the non-negative real axis. However, the mapping is not one-to-one—except for $w = 0$, every point in the range is the image under the function of infinitely many points in the domain—and the function does not have an inverse.

Example 11.3
Consider the function $w = \overline{z}$, $\forall z \in \mathfrak{C}$. This function maps each $z = X_1 + iX_2$ into $w = X_1 - iX_2$. Each $z \in \mathfrak{C}$ is mapped by the function into a point in the finite w-plane that is a reflection of z across the real axis. Each $w \in \mathfrak{C}$ is the image under the function of precisely one point in the domain—if $w = u + iv$, then it is the image under the function of $z = u - iv$. Therefore, the function is a one-to-one mapping of the finite z-plane onto the finite w-plane, with an inverse given by $z = \overline{w}$, $\forall w \in \mathfrak{C}$.

Example 11.4
Consider the function $w = 1/z$, $\forall z \in \mathcal{S}$, where the domain is the set $\mathcal{S} = \{ z \in \mathfrak{C} \mid z \neq 0 \}$ of all points in the finite z-plane with the origin deleted. It follows from (11.1.19) that each $z = re^{i\theta}$ in the domain is mapped by the function into the point $w = r^{-1}e^{-i\theta}$ in the w-plane. The function maps points on the unit circle (points that satisfy $|z| = 1$) into points on the unit circle, it maps points on the interior of the unit circle into points on the exterior of the unit circle, and its maps points on the exterior of the unit circle into points on the interior of the unit circle. Each $w \in \{ w \in \mathfrak{C} \mid w \neq 0 \}$ is the image of precisely one point in the domain—if $w = \rho e^{i\phi}$, then it is the image under the function of $z = \rho^{-1}e^{-i\phi}$. Therefore, the function is a one-to-one mapping of the finite z-plane, with the origin deleted, onto the finite w-plane, with the origin deleted, with an inverse given by $z = 1/w$, $\forall w \in \mathcal{S}$.

The domain of the function can be extended to include the origin if the image under the function of $z = 0$ is defined to be $w = \infty$, the idealized point at infinity. The domain can be further extended to include the point at infinity if the image of $z = \infty$ is defined to be $w = 0$. Then, the function $w = 1/z$, $\forall z \in \mathfrak{C}^\infty$, is a one-to-one mapping of the extended z-plane onto the extended w-plane, with an inverse given by $z = 1/w$, $\forall w \in \mathfrak{C}^\infty$.

Example 11.5

Consider the function $w = z^2$, $\forall z \in \mathbb{C}$. The function maps $z = 0$ into $w = 0$. If $z \neq 0$, and $|z| = r$ and $\arg\{z\} = \theta$, it follows from (11.1.18) that

$$z^2 = r^2 e^{i2\theta} .$$

The function maps a point $z = r e^{i\theta}$ ($z \neq 0$), that is a distance r from the origin in the z-plane with an argument θ, into the point $w = \rho e^{i\phi}$ that is a distance $\rho = r^2$ from the origin in the w-plane with an argument $\phi = 2\theta$. The function maps points on the unit circle (points that satisfy $|z| = 1$) into points on the unit circle, it maps points on the interior of the unit circle into points on the interior of the unit circle, and its maps points on the exterior of the unit circle into points on the exterior of the unit circle. Clearly, every point in the finite w-plane is an image under the function of at least one point in the function's domain $\mathbb{C}$. Therefore, the function maps the finite z-plane onto the finite w-plane—the range of the function is the finite w-plane.

The mapping is *not* one-to-one, however. To see this, consider a point w ($w \neq 0$) in the finite w-plane with absolute value $|w| = \rho$ and principal argument $\text{Arg}\{w\} = \phi$, defined such that $0 \leq \phi < 2\pi$. The absolute value of any point z_j in the finite z-plane that is mapped into w is uniquely given by

$$|z_j| = r_j = \rho^{\frac{1}{2}} .$$

Recall, however, that each $\arg\{w\} = \phi + 2k\pi$, $k = \ldots, -2, -1, 0, 1, 2, \ldots$, is an argument of w. Thus, determining the principal argument of a point z_j in the finite z-plane that is mapped into w [i.e., solving for $\text{Arg}\{z_j\} = \theta_j$ such that $\arg\{w\} = 2\theta_j$ and $0 \leq \theta_j < 2\pi$] yields two solutions:

$$\theta_1 = \frac{1}{2}\phi , \quad \theta_2 = \frac{1}{2}\phi + \pi .$$

It is seen, therefore, that each nonzero point $w = \rho e^{i\phi}$ in the finite w-plane is the image under the function of *two* distinct points in the finite z-plane given by

$$z_1 = \rho^{\frac{1}{2}} e^{\frac{1}{2}\phi} , \quad z_2 = \rho^{\frac{1}{2}} e^{\frac{1}{2}\phi + \pi} .$$

Since the function is not a one-to-one mapping, its inverse is not defined.

It is seen that the function $w = z^2$, $\forall z \in \mathbb{C}$, maps the finite z-plane onto the finite w-plane *twice*. This result generalizes to higher integer powers of z. A function $w = z^n$, $\forall z \in \mathbb{C}$, where n is a positive integer, can be seen to map the finite z-plane onto the finite w-plane n times. Each nonzero point in the finite w-plane is the image under the function of n distinct points in the finite z-plane.

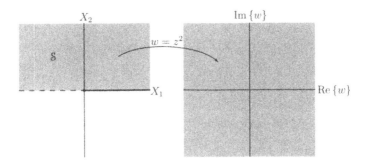

FIGURE 11.5. *A domain for $w = z^2$ that leads to a one-to-one mapping.*

Example 11.6

Consider the function $w = e^z$, $\forall z \in \mathbb{C}$. By applying Euler's formula (11.1.14), it is seen that the image w of $z = X_1 + iX_2$ under the function is

$$w = e^{X_1}(\cos X_2 + i \sin X_2)$$

so that

$$|w| = e^{\text{Re}\{z\}}, \quad \arg\{w\} = \text{Im}\{z\} .$$

Thus, the range $\mathcal{R}$ of the function is the finite w-plane with the origin deleted, $\mathcal{R} = \{ w \in \mathbb{C} \mid w \neq 0 \}$.

If a point z_0 is mapped by this function into w_0, then each point $z_0 + 2in\pi$, $n = \ldots, -2, -1, 0, 1, 2, \ldots$, will also be mapped into w_0. Thus, the function maps the finite z-plane onto the finite w-plane, with the origin deleted, an infinite number of times and the function is not one-to-one and does not have an inverse. In fact, each horizontal strip $\mathcal{S} = \{ z \in \mathbb{C} \mid \phi_0 \leq \text{Im}\{z\} < \phi_0 + 2\pi \}$ will be mapped by the function onto the range $\mathcal{R}$.

11.1.4 Branch Cuts and Branch Points

Functions that are not one-to-one, such as those considered in Examples 11.5 and 11.6, can be made one-to-one by *restricting their domains*. These restricted functions will then have inverses. Consider, for example, the function $w = z^2$, $\forall z \in \mathcal{S}$, where the domain

$$\mathcal{S} = \{ z \in \mathbb{C} \mid z = 0 \text{ or } 0 \leq \text{Arg}\{z\} < \pi \} \qquad (11.1.28)$$

is the set of all points in the finite z-plane whose imaginary part is positive (the "upperhalf-plane"), plus the points on the non-negative real axis (Figure 11.5). This is not the same function as defined in Example 11.5,

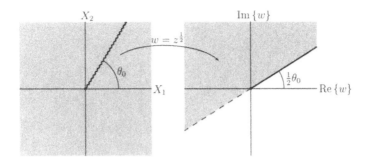

FIGURE 11.6. A domain for $w = z^{\frac{1}{2}}$ showing the branch cut and corresponding range.

because the domains are different. This function is a one-to-one mapping of S onto the finite w-plane. Its inverse $z = w^{\frac{1}{2}}$, $\forall w \in \mathfrak{C}$, is a one-to-one mapping of the finite w-plane onto S defined such that $w = 0$ is mapped into $z = 0$ and $w = \rho e^{i\phi}$ ($w \neq 0$) is mapped into $z = \rho^{\frac{1}{2}} e^{i\frac{\phi}{2}}$, where the argument of w is restricted to the interval $0 \leq \phi < 2\pi$. This restriction of the argument of w for the inverse function plays the same role as the restriction of the domain in the original function.

It is now clear that a function such as $w = z^{\frac{1}{2}}$, $\forall z \in \mathfrak{C}$, must include some suitable restriction on the values of $\arg\{z\}$ so that each point in the domain has a *uniquely defined* image under the function. For instance, in this function, one can choose to restrict the argument of $z = re^{i\theta}$ to the interval $\theta_0 \leq \theta < \theta_0 + 2\pi$. Then, with

$$w = z^{\frac{1}{2}} = r^{\frac{1}{2}} e^{i\frac{\theta}{2}} , \qquad (11.1.29)$$

it is clear that the range of the function will be

$$\mathcal{R} = \left\{ w \in \mathfrak{C} \mid \frac{\theta_0}{2} \leq \arg\{w\} < \frac{\theta_0}{2} + \pi \right\} . \qquad (11.1.30)$$

The domain and range of the function are shown in Figure 11.6. Notice that there is a discontinuity in the value of the function as one crosses a line in the domain that extends from the origin to infinity and forms an angle θ_0 with the positive real axis (shown as a wavy line in Figure 11.6). This line is known as a *branch cut*. To see this, consider a point in the domain $z_0 = r_0 e^{i\theta_0}$ that is on the branch cut and another point in the domain $z_0^- = r_0 e^{i\theta_0 + 2\pi - \epsilon}$, where $\epsilon > 0$ is a real number. The image under the function of z_0 is

$$w_0 = r_0^{\frac{1}{2}} e^{i\frac{\theta_0}{2}} , \qquad (11.1.31)$$

and the image under the function of z_0^- is

$$w_0^- = r_0^{\frac{1}{2}} e^{i\left(\frac{\theta_0}{2} + \pi - \frac{\epsilon}{2}\right)} . \tag{11.1.32}$$

In the limit as $\epsilon \to 0$, the point z_0^- approaches z_0 and

$$\lim_{\epsilon \to 0} w_0^- = r_0^{\frac{1}{2}} e^{i\left(\frac{\theta_0}{2} + \pi\right)} = w_0 e^{i\pi} = -w_0 . \tag{11.1.33}$$

Thus, as one crosses the branch cut in the domain, there is a discontinuity of magnitude π in the argument of the dependent variable of the function.

Suppose that, in the above function, one were to restrict the argument of $z = re^{i\theta}$ to the interval $\theta_0 + 2\pi \le \theta < \theta_0 + 4\pi$. Then, with

$$w = z^{\frac{1}{2}} = r^{\frac{1}{2}} e^{i\frac{\theta}{2}} , \tag{11.1.34}$$

it is clear that the range of the function will be

$$\mathcal{R}' = \left\{ w \in \mathbb{C} \ \bigg| \ \frac{\theta_0}{2} + \pi \le \arg\{w\} < \frac{\theta_0}{2} + 2\pi \right\} . \tag{11.1.35}$$

The branch cut in this case is the same as before, but the range $\mathcal{R}'$ is the complement of the previous range $\mathcal{R}$ (i.e., $\mathcal{R} \cup \mathcal{R}' = \mathbb{C}$). These two different functions are called the *branches* of the "function" $w = z^{\frac{1}{2}}$.[1]

The angle θ_0 used to define the branch cut is arbitrary. In fact, the branch cut need not be a straight line. The only requirement is that it be a continuous curve that begins at the origin and extends to infinity without intersecting itself. The discontinuity in the value of the function as one crosses the branch cut in the domain will be the same regardless. The point that is a terminus of all possible branch cuts, the origin in this case, is known as a *branch point*.

11.1.5 Differentiability and Analyticity

Prior to discussing differentiation of functions of a complex variable, the limit of such a function needs to be defined. Consider a function $w = f(z)$, $\forall z \in \mathcal{S}$, and let z_0 be a limit point of the domain $\mathcal{S}$. If there exists a complex number L such that for every real number $\epsilon > 0$, there exists a real number δ such that

$$|f(z) - L| < \epsilon \quad \forall z \in \{ z \in \mathcal{S} \mid 0 < |z - z_0| < \delta \} , \tag{11.1.36}$$

---

[1] The two branches of $w = z^{\frac{1}{2}}$ can be defined as a single function by introducing an abstract topological construct involving *Riemann sheets* as the domain. Essentially, it allows the domain to cover the z-plane twice. See Dettman (1965) for a discussion.

then L is the *limit of* $f(z)$ *as* z *approaches* z_0 in S, which is written as

$$\lim_{z \to z_0} f(z) = L . \tag{11.1.37}$$

In words, this says that L is the limit of $f(z)$ as z approaches z_0 through the domain S if and only if $f(z)$ can be kept arbitrarily close to L by restricting z to be in the intersection of S and some sufficiently small neighborhood of z_0. Thought of another way, the limit (11.1.37) is defined if and only if it approaches the same complex constant L regardless of the path followed in S as $z \to z_0$. The function does not need to be defined at z_0 in order for it to have a limit at z_0.

A function $w = f(z)$, $\forall z \in$ S, is *continuous* at a point $z_0 \in$ S if and only if $f(z_0) \neq \infty$ and it has a limit as z approaches z_0 that is equal to its value at z_0,

$$\lim_{z \to z_0} f(z) = f(z_0) . \tag{11.1.38}$$

The function is continuous in S if it is continuous at every point in S. If the function is continuous in S and the domain S is a closed set, then the function is bounded in S (i.e., its range is a bounded set).

Consider a function $w = f(z)$, $\forall z \in$ S, and let z_0 be an *interior* point of S where $f(z_0) \neq \infty$. The *derivative of* $w = f(z)$ *at* z_0 is defined by

$$\left. \frac{dw}{dz} \right|_{z=z_0} = f'(z_0) \equiv \lim_{h \to 0} \frac{f(z_0 + h) - f(z_0)}{h} , \tag{11.1.39}$$

provided the limit is not equal to infinity. If $f'(z_0)$ exists, then $f(z)$ is said to be *differentiable* at z_0. If $f(z)$ is differentiable at z_0, then it is continuous at z_0 and, conversely, if it is not continuous at z_0, then it is not differentiable at z_0. Note that the derivative (11.1.39) only exists if the limit exists; that is, it must have a value that is independent of the path followed in S as $z_0 + h \to z_0$. This restriction in the notion of derivative will be seen to have important consequences.

Example 11.7
Consider the function $f(z) = z^2$, $\forall z \in \mathbb{C}$. Evaluating the limit in the definition (11.1.39) of the derivative at an arbitrary interior point z_0,

$$\lim_{h \to 0} \frac{(z_0 + h)^2 - z_0^2}{h} = \lim_{h \to 0} (2z_0 + h) = 2z_0 ,$$

it is seen that the limit exists and is finite for all z_0, so that the function is differentiable everywhere in the finite z-plane and $f'(z) = 2z$. It can similarly be shown that the function $f(z) = z^n$, $\forall z \in \mathbb{C}$, where n is a positive integer is differentiable everywhere in the finite z-plane and that its derivative is $f'(z) = nz^{n-1}$.

Example 11.8

Consider the function $f(z) = \bar{z}, \forall z \in \mathfrak{C}$. The limit in the definition (11.1.39) of the derivative at an arbitrary interior point z_0 becomes

$$\lim_{h \to 0} \frac{(\bar{z}_0 + \bar{h}) - \bar{z}_0}{h} = \lim_{h \to 0} \frac{\bar{h}}{h}.$$

The limit, and therefore the derivative of the function at z_0, only exists if its value is independent of the path followed in the complex plane as $h \to 0$. However, if $h \to 0$ along the real axis, $\bar{h}/h = 1$, whereas if $h \to 0$ along the imaginary axis, $\bar{h}/h = -1$. Thus, the limit does not exist and the function is not differentiable at any point in the finite z-plane.

Example 11.9

Consider the function $f(z) = |z|^2, \forall z \in \mathfrak{C}$. The limit in the definition (11.1.39) of the derivative at an arbitrary interior point z_0 becomes

$$\lim_{h \to 0} \frac{|z_0 + h|^2 - |z_0|^2}{h} = \lim_{h \to 0} \frac{(z_0 + h)(\bar{z}_0 + \bar{h}) - z_0 \bar{z}_0}{h}$$

$$= \lim_{h \to 0} \left(\bar{z}_0 + z_0 \frac{\bar{h}}{h} + \bar{h} \right).$$

Only when $z_0 = 0$ is the value of this limit independent of the path followed in the complex plane as $h \to 0$. Thus, the function is only differentiable at the origin, at which point $f'(0) = 0$.

Example 11.10

Consider the function $f(z) = 1/z, \forall z \in \mathfrak{C}$. Since $f(0) = \infty$, the function is not differentiable at the origin. Evaluating the limit in the definition (11.1.39) of the derivative at an arbitrary interior point $z_0 \neq 0$,

$$\lim_{h \to 0} \frac{\frac{1}{z_0 + h} - \frac{1}{z_0}}{h} = \lim_{h \to 0} \frac{-1}{z_0^2 + z_0 h} = -\frac{1}{z_0^2},$$

it is seen that the limit exists and is finite for all $z_0 \neq 0$, so that the function is differentiable everywhere in the finite z-plane except the origin and $f'(z) = -1/z^2$. It can similarly be shown that the function $f(z) = 1/z^n = z^{-n}, \forall z \in \mathfrak{C}$, where n is a positive integer is differentiable everywhere in the finite z-plane except the origin and that its derivative is $f'(z) = -nz^{-n-1}$.

Example 11.11

Consider the function $w = f(z) = (z - \zeta)^{\frac{1}{3}}, \forall z \in \mathfrak{C}$, where ζ is a complex constant and the argument of $z - \zeta$ is restricted to the interval $\theta_0 \leq \arg\{z - \zeta\} < \theta_0 + 2\pi$. The inverse of this function, $z = w^3 + \zeta$, is the same for *all* values of θ_0 (i.e., for all three of the function's branches). By noting that

$$\frac{dz}{dw} = 3w^2 ,$$

it follows that the derivative of the function is given by

$$\frac{dw}{dz} = f'(z) = \frac{1}{3}w^{-2} = \frac{1}{3}(z - \zeta)^{-\frac{2}{3}} ,$$

where the argument of $z - \zeta$ is still restricted to the interval

$$\theta_0 \le \arg\{z - \zeta\} < \theta_0 + 2\pi .$$

The value of this expression is unbounded at $z = \zeta$, which is a branch point of both the original function and its derivative. Thus, the function is not differentiable at the point $z = \zeta$. The function is not differentiable at points on its branch cut since it is not continuous at these points. The function is differentiable everywhere else in the finite z-plane.

The usual differentiation formulas from real variable calculus apply to functions of a complex variable. In particular, if $f(z)$ and $g(z)$ are both differentiable at z, then

$$\frac{d}{dz}cf(z) = cf'(z) \quad c \text{ a constant} , \tag{11.1.40}$$

$$\frac{d}{dz}[f(z) \pm g(z)] = f'(z) \pm g'(z) , \tag{11.1.41}$$

$$\frac{d}{dz}[f(z)g(z)] = f(z)g'(z) + f'(z)g(z) , \tag{11.1.42}$$

$$\frac{d}{dz}\left[\frac{f(z)}{g(z)}\right] = \frac{g(z)f'(z) - g'(z)f(z)}{[g(z)]^2} \quad \text{where } g(z) \ne 0 , \tag{11.1.43}$$

$$\frac{d}{dz}[g(z)]^n = n[g(z)]^{n-1}g'(z) . \tag{11.1.44}$$

The formula (11.1.44) is a special case of the *chain rule*, which can be stated as follows. If $g(z)$ is differentiable at z and $f(\zeta)$ is differentiable at $\zeta = g(z)$, then

$$\frac{d}{dz}f[g(z)] = f'[g(z)]g'(z) . \tag{11.1.45}$$

These results can be used to show that a *polynomial*,

$$w = P(z) = a_0 + a_1 z + a_2 z^2 + \cdots + a_n z^n \quad \forall z \in \mathfrak{C} , \tag{11.1.46}$$

is differentiable everywhere in the finite z-plane and that a *rational function*,

$$w = \frac{P(z)}{Q(z)} = \frac{a_0 + a_1 z + a_2 z^2 + \cdots + a_n z^n}{b_0 + b_1 z + b_2 z^2 + \cdots + b_m z^m} \qquad \forall z \in \mathbb{C}, \qquad (11.1.47)$$

is differentiable everywhere in the finite z-plane except where $Q(z) = 0$.

Consider a function $w = f(z)$, $\forall z \in \mathcal{S}$, and let z_0 be an interior point of the domain $\mathcal{S}$. The function is *analytic* at z_0 if and only if it is differentiable at z_0, and at every point in some neighborhood of z_0. Then, z_0 is called a *regular point* of the function. If the function is *not* analytic at z_0, but every neighborhood of z_0 contains points at which the function *is* analytic, then z_0 is called a *singular point* of the function. A function that is analytic at every point in a region is said to be *analytic in the region*.

The function given in Example 11.7 is analytic everywhere in the finite z-plane. The function given in Example 11.8 is not analytic anywhere. The function given in Example 11.9 is differentiable at the origin, but there is no neighborhood of the origin in which the function is everywhere differentiable, so this function is also not analytic anywhere. The function given in Example 11.10 is differentiable everywhere except at the origin, so it is analytic everywhere except the origin and the origin is a singular point of the function.

The function given in Example 11.11 is differentiable everywhere except at its branch point, $z = \zeta$, and at points on its branch cut, which are all singular points of the function. Branch points and points on branch cuts will always be singular points. Recall, however, that the path followed by the branch cut can be altered by a change in the restriction on the argument of z that is part of the definition of the function. The location of the branch point is always the same. Thus, a *branch point singularity* of a function is in some sense more substantial than the singularities at points on a branch cut.

The definition of analyticity can be extended to the point at infinity. Suppose a function $w = f(z)$, $\forall z \in \mathcal{S}$, has both the origin and the point at infinity in its domain $\mathcal{S}$. Then, even though differentiability is not defined at the point at infinity, the function is said to be *analytic at infinity* if $f(1/\zeta)$ is analytic at $\zeta = 0$. For example, the function $f(z) = 1/z$, $\forall z \in \mathbb{C}^\infty$, is analytic at infinity since $f(1/\zeta) = \zeta$ is analytic at $\zeta = 0$.

The following important results are offered without proof:

- An analytic function $w = f(z)$ can always be expressed in terms of z alone, that is, such that X_1, X_2, $\bar{z}$, $|z|$, $\arg\{z\}$, and so forth do not appear explicitly.

- If a function $w = f(z)$ is analytic in a simply connected region $\mathcal{S}$, then at any interior point z_0 of $\mathcal{S}$ the derivatives of $f(z)$ of all orders exist and are analytic.

Cauchy-Riemann equations

Consider a function $w = f(z)$, $\forall z \in \mathcal{S}$, with independent variable $z = X_1 + iX_2$ and dependent variable $w = u + iv$. Since w is uniquely determined by z, which is uniquely determined by X_1 and X_2, the real and imaginary part of w must also be uniquely determined by X_1 and X_2 and one can write

$$w = f(z) = u(X_1, X_2) + iv(X_1, X_2) \quad \forall z \in \mathcal{S}, \tag{11.1.48}$$

where u and v are real-valued functions of the real variables X_1 and X_2. With this understanding, necessary and sufficient conditions for the function (11.1.48) to be analytic at an interior point $z_0 = \xi + i\eta$, are that u and v have continuous first partial derivatives in some neighborhood of z_0 and that

$$\boxed{u_{,1}(\xi, \eta) = v_{,2}(\xi, \eta), \quad u_{,2}(\xi, \eta) = -v_{,1}(\xi, \eta).} \tag{11.1.49}$$

The conditions (11.1.49) are the *Cauchy-Riemann equations*.

To prove that these conditions are necessary, assume that the function (11.1.48) is analytic at $z_0 = \xi + i\eta$. This means not only that the function is differentiable at z_0 but also that the function is differentiable at every point in some neighborhood of z_0, which implies the existence of continuous first partial derivatives of u and v in that neighborhood. Since the derivative of (11.1.48) exists at $z_0 = \xi + i\eta$, its value must be independent of the path followed in the limit that is part of the definition (11.1.39) of the derivative. Therefore, one can let $h = a$ in the definition of the derivative, where a is a real number, to obtain

$$\begin{aligned} f'(z_0) &= \lim_{a \to 0} \frac{u(\xi + a, \eta) - u(\xi, \eta)}{a} + i\frac{v(\xi + a, \eta) - v(\xi, \eta)}{a} \\ &= u_{,1}(\xi, \eta) + iv_{,1}(\xi, \eta), \end{aligned} \tag{11.1.50}$$

or one can let $h = ib$ in the definition of the derivative, where b is a real number, to obtain

$$\begin{aligned} f'(z_0) &= \lim_{b \to 0} \frac{u(\xi, \eta + b) - u(\xi, \eta)}{ib} + i\frac{v(\xi, \eta + b) - v(\xi, \eta)}{ib} \\ &= -iu_{,2}(\xi, \eta) + v_{,2}(\xi, \eta). \end{aligned} \tag{11.1.51}$$

Equating the real and imaginary parts of each expression for $f'(z_0)$ gives the Cauchy-Riemann equations. Thus, analyticity implies the Cauchy-Riemann equations and the existence of continuous first partial derivatives of u and v. For a proof that the Cauchy-Riemann equations and the existence of continuous first partial derivatives of u and v implies analyticity, see Dettman (1965), Carrier et al. (1983), and Lang (1985).

One interesting consequence of the Cauchy-Riemann equations, which follows immediately from (11.1.50) and (11.1.51), is that if a function (11.1.48) is analytic at a point $z_0 = \xi + i\eta$, then its derivative at z_0 can be given be either of the following:

$$f'(z_0) = u_{,1}(\xi, \eta) + iv_{,1}(\xi, \eta),$$
$$f'(z_0) = v_{,2}(\xi, \eta) - iu_{,2}(\xi, \eta).$$
(11.1.52)

Another more useful consequence of the Cauchy-Riemann equations is that if a function (11.1.48) is analytic and its real and imaginary parts have continuous second partial derivatives at a point $z_0 = \xi + i\eta$, then both its real and imaginary parts are *harmonic* at that point. To see that this is true, consider the Laplacian of the real part $u(X_1, X_2)$ of the function:

$$\nabla^2 u = u_{,11} + u_{,22} = (u_{,1})_{,1} + (u_{,2})_{,2}.$$
(11.1.53)

However, if the function is analytic at $z_0 = \xi + i\eta$, it follows from the Cauchy-Riemann equations (11.1.49) that

$$\nabla^2 u(\xi, \eta) = v_{,21}(\xi, \eta) - v_{,12}(\xi, \eta).$$
(11.1.54)

If v has continuous second partial derivatives, so that the order of differentiation can be interchanged, then

$$\nabla^2 u(\xi, \eta) = 0.$$
(11.1.55)

The proof that the imaginary part of the function is also harmonic is similar.

If a function $\phi(X_1, X_2)$ is harmonic at a point $(X_1, X_2) = (\xi, \eta)$ and at every point in some neighborhood of the point, then a *conjugate harmonic function* $v(X_1, X_2)$ can always be constructed using the Cauchy-Riemann equations such that $f(z) = \phi(X_1, X_2) + iv(X_1, X_2)$ is analytic at $z_0 = \xi + i\eta$. Alternatively, a conjugate harmonic function $u(X_1, X_2)$ can always be constructed using the Cauchy-Riemann equations such that $f(z) = u(X_1, X_2) + i\phi(X_1, X_2)$ is analytic at $z_0 = \xi + i\eta$. Thus, *a harmonic function can always be represented as either the real or imaginary part of an analytic function.*

11.1.6 Contour Integrals

Consider a function $f(z) = u(X_1, X_2) + iv(X_1, X_2)$ and a simple curve (path, contour) $\mathcal{C}$ in the z-plane. Then, the *line integral* of $f(z)$ along $\mathcal{C}$ is

$$\int_{\mathcal{C}} f(z)\,dz = \int_{\mathcal{C}} (u + iv)(dX_1 + i\,dX_2)$$
$$= \int_{\mathcal{C}} (u\,dX_1 - v\,dX_2) + i \int_{\mathcal{C}} (v\,dX_1 + u\,dX_2).$$
(11.1.56)

If the curve $\mathcal{C}$ is closed, then the integral is called a *contour integral* and is symbolized as

$$\oint_{\mathcal{C}} f(z)\,dz\,. \tag{11.1.57}$$

The following results are presented without proof:

- (*Cauchy's theorem*) If $f(z)$ is analytic in a simply connected region $\mathcal{S}$, then

$$\oint_{\mathcal{C}} f(z)\,dz = 0 \tag{11.1.58}$$

  for any simple closed curve $\mathcal{C}$ in $\mathcal{S}$. The converse (known as *Morera's theorem*) also holds. Specifically, if $f(z)$ is continuous in $\mathcal{S}$ and (11.1.58) is true for every simple closed curve $\mathcal{C}$ in $\mathcal{S}$, then $f(z)$ is analytic in $\mathcal{S}$.

- (*Path independence*) If $f(z)$ is analytic in a simply connected region $\mathcal{S}$, then $\int_a^b f(z)\,dz$ is independent of the simple curve joining a and b in $\mathcal{S}$.

- (*Deformation of contours*) Let $\mathcal{C}_1$ and $\mathcal{C}_2$ be simple closed curves where $\mathcal{C}_2$ lies entirely in the interior of $\mathcal{C}_1$. If $f(z)$ is analytic on $\mathcal{C}_1$, on $\mathcal{C}_2$, and in the region between $\mathcal{C}_1$ and $\mathcal{C}_2$, then

$$\oint_{\mathcal{C}_1} f(z)\,dz = \oint_{\mathcal{C}_2} f(z)\,dz\,. \tag{11.1.59}$$

- (*Functions defined by integration*) Let $f(z)$ be analytic in a simply connected region $\mathcal{S}$. Then, $\int_{z_0}^z f(\zeta)\,d\zeta$ defines a function $F(z)$ in $\mathcal{S}$ provided z_0 is a fixed point and the path of integration lies in $\mathcal{S}$. Furthermore, $F(z)$ is analytic in $\mathcal{S}$ with $F'(z) = f(z)$ and if a and b are any points in $\mathcal{S}$, then $\int_a^b f(z)\,dz = F(b) - F(a)$.

- (*Cauchy's integral formulas*) If $f(z)$ is analytic within and on a simple closed curve $\mathcal{C}$, then the value of the function at any point z_0 on the *interior* of $\mathcal{C}$ is given in terms of its values on $\mathcal{C}$ by

$$f(z_0) = \frac{1}{2\pi i} \oint_{\mathcal{C}} \frac{f(z)}{z - z_0}\,dz\,, \tag{11.1.60}$$

  where the direction of integration around $\mathcal{C}$ is in the usual positive (counterclockwise) direction (i.e., such that the interior of $\mathcal{C}$ is always on one's left). In addition, the values at z_0 of all the function's derivatives are given by

$$f^{(n)}(z_0) = \frac{n!}{2\pi i} \oint_{\mathcal{C}} \frac{f(z)}{(z - z_0)^{n+1}} \, dz \; . \tag{11.1.61}$$

11.1.7 Power Series

A series of the form $\sum_{k=0}^{\infty} c_k (z - z_0)^k$, where z_0 and the coefficients c_0, c_1, c_2, ... are complex constants, is called a *power series in* $z - z_0$. The region of convergence of a power series will always be a circle. Specifically, if the (finite) limit $\lim_{k\to\infty} |c_{k+1}/c_k| = L$ exists, then the power series converges absolutely at all points in the interior of the circle $|z - z_0| = 1/L$, diverges at all points in the exterior of the circle, and may converge absolutely, converge conditionally, or diverge at points on the circle. The radius $R = 1/L$ of the circle is called the *radius of convergence* of the power series. The power series $\sum_{k=0}^{\infty} c_k (z - z_0)^k$ can be differentiated or integrated term by term within its circle of convergence, and the result will be another power series having the same radius of convergence. Within its circle of convergence, the power series $\sum_{k=0}^{\infty} c_k (z - z_0)^k$ is a representation of an analytic function.

- (*Taylor's theorem*) If $f(z)$ is analytic at a point z_0, then it can be represented within some neighborhood of z_0 by the *Taylor series expansion*

$$f(z) = \sum_{k=0}^{\infty} \frac{f^{(k)}(z_0)}{k!} (z - z_0)^k \; . \tag{11.1.62}$$

  The Taylor series converges for $|z - z_0| < R$, where R is the radius of the largest circle centered at z_0 inside of which $f(z)$ is analytic—the Taylor series expansion of the function is valid within this circle of convergence. Also, if $f(z)$ can be represented in some neighborhood of the point z_0 by a power series of the form $\sum_{k=0}^{\infty} c_k (z - z_0)^k$, then the representation is unique.

- (*Liouville's theorem*) An *entire* function is a function that is analytic everywhere in the unextended complex plane. For example, all polynomials are entire functions as is the exponential function e^z. Liouville's theorem states that the only *bounded* entire functions are constants.

- (*Laurent series*) Consider an open annular region $\mathcal{S}$ bounded by circles of radius R_1 and R_2 ($R_1 < R_2$) with common center at z_0,

$$\mathcal{S} = \{ z \in \mathbb{C} \mid R_1 < |z - z_0| < R_2 \} \; . \tag{11.1.63}$$

If $f(z)$ is analytic in $\mathcal{S}$, then it may be represented in $\mathcal{S}$ by the *Laurent series expansion*

$$f(z) = \sum_{k=-\infty}^{\infty} a_k (z - z_0)^k \, , \qquad (11.1.64)$$

where

$$a_k = \frac{1}{2\pi i} \oint_{\mathcal{C}} \frac{f(\zeta)}{(\zeta - z_0)^{k+1}} \, d\zeta \qquad (11.1.65)$$

and $\mathcal{C}$ is any simple closed curve in $\mathcal{S}$ with z_0 in its interior. This representation is unique.

If a function $f(z)$ is singular at a point z_0, but is analytic at every other point in some neighborhood of z_0, then one says that z_0 is an *isolated singularity* of the function. Note, for example, that branch point singularities are *not* isolated singularities since every neighborhood of a branch point will contain points on a branch cut, which are also singular points. If z_0 is an isolated singularity of the function $f(z)$, then the function may be represented by a Laurent series (11.1.64) in which $R_1 = 0$ and R_2 is the distance from z_0 to the nearest point (other than z_0) where the function fails to be analytic. In this Laurent series, if all the coefficients of negative powers of $z - z_0$ are zero so that the Laurent series can be given as

$$f(z) = \sum_{k=0}^{\infty} a_k (z - z_0)^k \, , \qquad (11.1.66)$$

then z_0 is a *removable singularity* and the function can be made analytic at z_0 by defining the value of the function at z_0 to be $f(z_0) = \lim_{z \to z_0} f(z) = a_0$. If all the coefficients of negative powers of $z - z_0$ of order less than $-n$ $(n > 0)$ are zero so that the Laurent series can be given as

$$f(z) = \sum_{k=-n}^{\infty} a_k (z - z_0)^k \qquad (11.1.67)$$

and $a_{-n} \neq 0$, then z_0 is a *pole of order n*. If $n = 1$, then z_0 is called a *simple pole*. If infinitely many coefficients of negative powers of $z - z_0$ are required for the Laurent series, then z_0 is an *essential singularity*.

If z_0 is an isolated singularity of the function $f(z)$, which is represented by a Laurent series (11.1.64) in which $R_1 = 0$, then it follows from (11.1.65) that the coefficient of the term $(z - z_0)^{-1}$ is

$$a_{-1} = \frac{1}{2\pi i} \oint_{\mathcal{C}} f(\zeta) \, d\zeta \, , \qquad (11.1.68)$$

where $\mathcal{C}$ is any simple closed curve that has z_0 in its interior and for which $f(z)$ is analytic everywhere else in and on $\mathcal{C}$. This coefficient is called the *residue of* $f(z)$ *at* z_0. This result can be turned around and generalized to give the following:

- (*Residue theorem*) If $\mathcal{C}$ is a simple closed curve and $f(z)$ is analytic on and within $\mathcal{C}$ except at n isolated points in the interior of $\mathcal{C}$, then

$$\oint_{\mathcal{C}} f(\zeta)\, d\zeta = 2\pi i (r_1 + r_2 + \cdots + r_n), \tag{11.1.69}$$

where $r_1, r_2, \ldots, r_n$ are the residues of $f(z)$ at the n isolated singular points in the interior of $\mathcal{C}$.

If the residues of a function are known, then the residue theorem can be used to determine the contour integral (11.1.69).

11.1.8 Analytic Continuation

Suppose that the precise analytic expression for a function $f(z)$ is unknown, but that a Taylor series representation of the function about a point z_1 is known for some circle of convergence $\mathcal{S}_1 = \{\, z \in \mathcal{C} \mid |z - z_1| < R_1 \,\}$,

$$f(z) = \sum_{k=0}^{\infty} a_k (z - z_1)^k \quad \forall z \in \mathcal{S}_1. \tag{11.1.70}$$

Without prior knowledge of the location of singularities of the function, one determines the radius of convergence R_1 by evaluating

$$\frac{1}{R_1} = \lim_{k \to \infty} |a_{k+1}/a_k|. \tag{11.1.71}$$

Now, choose a point $z_2 \in \mathcal{S}_1$. The Taylor series (11.1.70) gives the values of $f(z)$ and all its derivatives at z_2, thus enabling one to construct a Taylor series representation (11.1.62) of $f(z)$ about the point z_2 for some circle of convergence $\mathcal{S}_2 = \{\, z \in \mathcal{C} \mid |z - z_2| < R_2 \,\}$,

$$f(z) = \sum_{k=0}^{\infty} b_k (z - z_2)^k \quad \forall z \in \mathcal{S}_2, \tag{11.1.72}$$

where the radius of convergence R_2 is given by

$$\frac{1}{R_2} = \lim_{k \to \infty} |b_{k+1}/b_k|. \tag{11.1.73}$$

If the circle of convergence $\mathcal{S}_2$ extends beyond $\mathcal{S}_1$, then one says that the domain of the function has been *extended analytically* to the region $\mathcal{S}_1 \cup \mathcal{S}_2$ (see Figure 11.7).

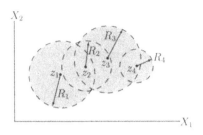

FIGURE 11.7. Analytic continuation.

This procedure is known as *analytic continuation* and may be repeated indefinitely. The function $f(z)$ will be analytic in a domain $S_1 \cup S_2 \cup S_3 \cup S_4 \cup \cdots$ built up in this way. The set of all such analytic continuations is defined as the analytic function $f(z)$. Note the following useful consequences of analytic continuation:

- If $f(z)$ is analytic in a region S and $f(z) = 0$ at all points on a curve C inside S, then $f(z) = 0$ throughout S.

- If $f(z)$ and $g(z)$ are analytic in a region S and $f(z) = g(z)$ at all points on a curve C inside S, then $f(z) = g(z)$ throughout S.

- Consider two open regions S_1 and S_2 that do not overlap ($S_1 \cap S_2 = \emptyset$) but whose boundaries intersect along a curve C ($\partial S_1 \cap \partial S_2 = C$). If $f_1(z)$ is analytic in $S_1 \cup C$, $f_2(z)$ is analytic in $S_2 \cup C$, and $f_1(z) = f_2(z)$ at all points on the curve C, then the function

$$f(z) = \begin{cases} f_1(z) & \forall z \in S_1 \\ f_2(z) & \forall z \in S_2 \\ f_1(z) = f_2(z) & \forall z \in C , \end{cases} \tag{11.1.74}$$

  is analytic in the region $S = S_1 \cup S_2 \cup C^*$, where C^* is the curve C with its end points deleted.

11.1.9   Conformal Mapping

Consider a *one-to-one* mapping

$$\zeta = \Omega(z) = \xi_1(X_1, X_2) + i\xi_2(X_1, X_2) \quad \forall z \in S , \tag{11.1.75}$$

whose range is R. Since this mapping $\Omega \colon S \to R$ of the region S in the z-plane onto the region R in the ζ-plane is one-to-one, it has an inverse $\omega \colon R \to S$,

$$z = \omega(\zeta) = X_1(\xi_1, \xi_2) + iX_2(\xi_1, \xi_2) \quad \forall \zeta \in R , \tag{11.1.76}$$

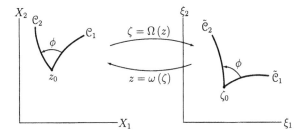

FIGURE 11.8. *Conformal mapping.*

that is a one-to-one mapping of the region $\mathcal{R}$ in the ζ-plane onto the region $\mathcal{S}$ in the z-plane.

Recall that $z = w[\Omega(z)]$, $\forall z \in \mathcal{S}$, and $\zeta = \Omega[w(\zeta)]$, $\forall \zeta \in \mathcal{R}$. It follows from the chain rule that if z_0 is a point in $\mathcal{S}$ and $\zeta_0 = \Omega(z_0)$ is the corresponding point in $\mathcal{R}$, then

$$\Omega'(z_0) = \frac{1}{w'(\zeta_0)} . \tag{11.1.77}$$

Therefore, it is seen that every regular point of the function $\Omega(z)$ where $\Omega'(z) \neq 0$ is mapped into a regular point of the function $w(\zeta)$, and vice versa for the inverse mapping. Regular points in $\mathcal{S}$ where $\Omega'(z) \neq 0$ are called *ordinary points* of the mapping $\zeta = \Omega(z)$. Regular points in $\mathcal{S}$ where $\Omega'(z) = 0$ are called *critical points* of the mapping. Ordinary points in $\mathcal{S}$ are mapped into ordinary points in $\mathcal{R}$. Critical points in $\mathcal{S}$ are mapped into singular points in $\mathcal{R}$, and vice versa.

Consider two curves $\mathcal{C}_1$ and $\mathcal{C}_2$ in the z-plane that intersect at the point z_0. The mapping (11.1.75) maps the curves into two curves $\tilde{\mathcal{C}}_1$ and $\tilde{\mathcal{C}}_2$ in the ζ-plane that intersect at $\zeta_0 = \Omega(z_0)$ (Figure 11.8). If the angle at z_0 between $\mathcal{C}_1$ and $\mathcal{C}_2$ is equal to the angle at ζ_0 between $\tilde{\mathcal{C}}_1$ and $\tilde{\mathcal{C}}_2$, both in magnitude and in sense, then the mapping is said to be *conformal* at z_0.

To see whether or not the mapping is conformal, consider a regular point z_0 of the mapping $\zeta = \Omega(z)$ and let $z_0 + \Delta z$ be another point in the domain $\mathcal{S}$. The corresponding images under the mapping are $\zeta_0 = \Omega(z_0)$ and $\zeta_0 + \Delta\zeta = \Omega(z_0 + \Delta z)$. Let $\arg\{\Delta z\} = \beta$ and $\arg\{\Delta\zeta\} = \beta - \alpha$,

$$\Delta z = |\Delta z| e^{i\beta} , \quad \Delta\zeta = |\Delta\zeta| e^{i(\beta-\alpha)} , \tag{11.1.78}$$

so that α is the clockwise rotation of Δz under the mapping (Figure 11.9). Then, since z_0 is a regular point of the mapping, the derivative $\Omega'(z_0)$ exists and is given by the definition (11.1.39) as

$$\Omega'(z_0) = \lim_{\Delta z \to 0} \frac{\Delta\zeta}{\Delta z} = \lim_{\Delta z \to 0} \frac{|\Delta\zeta| e^{i(\beta-\alpha)}}{|\Delta z| e^{i\beta}} = \lim_{\Delta z \to 0} \left|\frac{\Delta\zeta}{\Delta z}\right| e^{-i\alpha} , \tag{11.1.79}$$

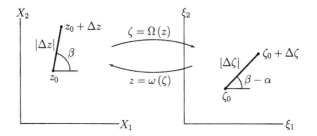

FIGURE 11.9. *Mapping of an increment Δz into an increment $\Delta \zeta$.*

where Δz is arbitrary. From this it follows that

$$\lim_{\Delta z \to 0} \left| \frac{\Delta \zeta}{\Delta z} \right| = |\Omega'(z_0)| \; , \tag{11.1.80}$$

and if z_0 is an ordinary point so that $\Omega'(z_0) \neq 0$,

$$\lim_{\Delta z \to 0} \alpha = -\arg\{\Omega'(z_0)\} \; . \tag{11.1.81}$$

If z_0 is a critical point [i.e., $\Omega'(z_0) = 0$], then the argument $\arg\{\Omega'(z_0)\}$ is undefined and the result (11.1.81) cannot be applied.

It is clear from (11.1.81) that every line that passes through an ordinary point z_0 will be rotated by the mapping the same amount, since the clockwise angle of rotation α only depends on z_0:

- (*Conformal mapping theorem*) If a one-to-one mapping $\zeta = \Omega(z)$, $\forall z \in \mathcal{S}$, is analytic at $z_0 \in \mathcal{S}$ and $\Omega'(z_0) \neq 0$ (i.e., z_0 is an ordinary point of the mapping), then the mapping is conformal at z_0. If the mapping is conformal at every point in $\mathcal{S}$, one says that the mapping is conformal in $\mathcal{S}$.

It follows from (11.1.77) and (11.1.81) that the local clockwise angle of rotation at an ordinary point can be expressed as

$$\alpha = -\arg\{\Omega'(z)\} = \arg\{\omega'(\zeta)\} \; , \tag{11.1.82}$$

or, in what will prove to be a convenient form, as

$$\boxed{e^{i\alpha} = \frac{\omega'(\zeta)}{|\omega'(\zeta)|}} \; . \tag{11.1.83}$$

Furthermore, it can be seen from (11.1.80) that if $\Delta z \ll 1$, then

$$\left| \frac{\Delta \zeta}{\Delta z} \right| \approx |\Omega'(z_0)| \; . \tag{11.1.84}$$

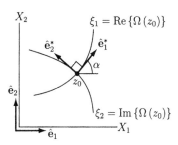

FIGURE 11.10. *Conformal mapping viewed as change of coordinate system.*

Every line that passes through an ordinary point z_0 will be stretched (or compressed) by the same amount at that point.

A conformal mapping can alternatively be thought of as a change in independent variables, between a Cartesian (X_1, X_2) coordinate system and an orthogonal curvilinear (ξ_1, ξ_2) coordinate system (Figure 11.10). At each point in the plane, an orthonormal vector basis $\{\hat{\mathbf{e}}_\gamma^*\}$ can be defined with unit base vectors $\hat{\mathbf{e}}_1^*$ and $\hat{\mathbf{e}}_2^*$ tangent to the contours of constant ξ_1 and ξ_2, respectively. This vector basis is rotated *counterclockwise* by the angle α (11.1.83) relative to the basis $\{\hat{\mathbf{e}}_\gamma\}$ defined by X_1 and X_2. Thus, the direction cosines $\ell_{\gamma\beta} \equiv \hat{\mathbf{e}}_\gamma^* \cdot \hat{\mathbf{e}}_\beta$ for the tensor transformation rule from $\{\hat{\mathbf{e}}_\gamma\}$ to $\{\hat{\mathbf{e}}_\gamma^*\}$ are given by the matrix of direction cosines

$$[L] = \begin{bmatrix} \cos\alpha & \sin\alpha \\ -\sin\alpha & \cos\alpha \end{bmatrix}, \tag{11.1.85}$$

so that $\{\hat{\mathbf{e}}_\gamma^*\} = [L]\{\hat{\mathbf{e}}_\gamma\}$. The angle α is a function of position given by (11.1.83).

A (complex-valued) scalar field Γ defined over the complex plane can be expressed either in terms of the (Cartesian) complex variable z or in terms of the (orthogonal curvilinear) complex variable ζ:

$$\Gamma = F(z) = f(\zeta), \tag{11.1.86}$$

where $f(\zeta) = F[\omega(\zeta)]$ and $F(z) = f[\Omega(z)]$. Thus, from the chain rule,

$$f'(\zeta) = F'[\omega(\zeta)]\omega'(\zeta), \tag{11.1.87}$$

and it follows that, at any given point $z = \omega(\zeta)$, the derivatives $d\Gamma/dz = F'(z)$ and $d\Gamma/d\zeta = f'(\zeta)$ are related by

$$\boxed{\frac{d\Gamma}{dz} = \frac{1}{\omega'(\zeta)} \frac{d\Gamma}{d\zeta}.} \tag{11.1.88}$$

This transformation of derivatives with respect to z into derivatives with respect to $\zeta = \Omega(z)$ will be important later, when conformal mapping is used to solve plane problems in elasticity.

11.2   Antiplane Strain

Recall (Section 7.1) that a state of antiplane strain is defined as one in which the Cartesian scalar components of the displacement field obey

$$u_\alpha = 0 , \quad u_3 = u_3(X_1, X_2) , \tag{7.1.1'}$$

where Greek indices (such as α) range over the values $\{1, 2\}$. It follows from this that, in a state of antiplane strain, the scalar components of stress have the form $\sigma_{11} = \sigma_{22} = \sigma_{33} = \sigma_{12} = 0$ and $\sigma_{\alpha 3} = \sigma_{\alpha 3}(X_1, X_2)$, where

$$\sigma_{\alpha 3} = \mu u_{3,\alpha} , \tag{7.1.3'}$$

and the antiplane strain equilibrium condition is

$$\mu \nabla^2 u_3 + f_3 = 0 . \tag{7.1.6'}$$

Thus, at points in a body where the antiplane strain body force field f_3 vanishes, equilibrium is satisfied if and only if the antiplane strain displacement field is harmonic (i.e., $\nabla^2 u_3 = 0$).

One can always represent an antiplane strain displacement field u_3 as the real part of a function of the complex variable $z = X_1 + iX_2$,

$$\boxed{u_3 = \frac{1}{\mu} \operatorname{Re}\{\psi(z)\} ,} \tag{11.2.1}$$

where the factor $1/\mu$ is included for later convenience. Then, it follows from the Cauchy-Riemann equations that $\nabla^2 u_3 = 0$ at points where $\psi(z)$ is analytic. Conversely, if u_3 is harmonic at a point, and everywhere in some neighborhood of the point, then a conjugate harmonic function $\operatorname{Im}\{\psi(z)\}$ can always be constructed such that $\psi(z)$ is analytic. With $\operatorname{Im}\{\psi(z)\}$ so defined, the representation (11.2.1) is complete and the equilibrium condition *in a region where the antiplane strain body force field vanishes* is that $\psi(z)$ be analytic in that region. The *antiplane strain complex potential* $\psi(z)$ will fail to be analytic at points where equilibrium is not satisfied or at points where the body force field is not zero.

It follows from (7.1.3) and (11.2.1) that, at points where $\psi(z)$ is analytic,

$$\sigma_{\alpha 3} = \frac{\partial}{\partial X_\alpha} \operatorname{Re}\{\psi(z)\} = \operatorname{Re}\left\{\frac{\partial}{\partial X_\alpha}\psi(z)\right\} = \operatorname{Re}\{z_{,\alpha}\,\psi'(z)\} \ . \qquad (11.2.2)$$

Since $z_{,1} = 1$ and $z_{,2} = i$, one has that

$$\sigma_{13} = \operatorname{Re}\{\psi'(z)\} \ , \quad \sigma_{23} = \operatorname{Re}\{i\psi'(z)\} = -\operatorname{Im}\{\psi'(z)\} \ , \qquad (11.2.3)$$

from which one gets

$$\boxed{\sigma_{13} - i\sigma_{23} = \psi'(z) \ ,} \qquad (11.2.4)$$

which gives the nonzero components of stress in terms of the antiplane strain complex potential, at points where the complex potential is analytic.

11.2.1 Traction-Stress Relation

Recall that the traction-stress relation for antiplane strain reduces to the requirement that on any surface with unit outward normal $\hat{\mathbf{n}} = \hat{n}_1\hat{\mathbf{e}}_1 + \hat{n}_2\hat{\mathbf{e}}_2$, the scalar components of the traction satisfy $t_1 = t_2 = 0$ and $t_3 = t_3(X_1, X_2)$, where $t_3 = \sigma_{\alpha 3}\hat{n}_\alpha$. This relation for the nonzero component of the traction can be rewritten as

$$t_3 = \operatorname{Im}\{(\sigma_{13} - i\sigma_{23})(-\hat{n}_2 + i\hat{n}_1)\} \ . \qquad (11.2.5)$$

It has previously be shown (8.1.14) that given a parametric description $X_\alpha = \tilde{X}_\alpha(s)$ of a curve in the X_1X_2-plane, where s is the arc length, the unit outward normal to the curve is given by

$$\hat{\mathbf{n}} = \frac{dX_2}{ds}\hat{\mathbf{e}}_1 - \frac{dX_1}{ds}\hat{\mathbf{e}}_2 \ . \qquad (11.2.6)$$

The "outward" side of the curve is on one's right as one moves along the curve in the direction of increasing arc length s. Thus, by noting that

$$\frac{dz}{ds} = \frac{dX_1}{ds} + i\frac{dX_2}{ds} = -\hat{n}_2 + i\hat{n}_1 \ , \qquad (11.2.7)$$

it follows from (11.2.4) and (11.2.5) that

$$\boxed{t_3 = \operatorname{Im}\left\{\frac{d}{ds}\psi(z)\right\} = \frac{d}{ds}\operatorname{Im}\{\psi(z)\} \ .} \qquad (11.2.8)$$

The traction distribution on a curve is thus related to the antiplane strain complex potential.

11.2.2   Cylindrical Coordinates

With respect to a vector basis aligned with a cylindrical coordinate system, the scalar components of the displacement field in a state of antiplane strain obey $u_r = u_\theta = 0$, $u_z = u_z(r,\theta)$.[2] Since $u_z = u_3$, it follows from (11.2.1) that one can express u_z as

$$\boxed{u_z = \frac{1}{\mu}\,\mathrm{Re}\{\psi(z)\}\ ,} \tag{11.2.9}$$

so that the equilibrium condition in a region where the body force field vanishes is that $\psi(z)$ be analytic in that region. The corresponding scalar components of the stress field are given by $\sigma_{rr} = \sigma_{\theta\theta} = \sigma_{zz} = \sigma_{r\theta} = 0$ and

$$\sigma_{\theta z}(r,\theta) = \frac{\mu}{r}\frac{\partial u_z}{\partial\theta}\ ,\quad \sigma_{rz}(r,\theta) = \mu\frac{\partial u_z}{\partial r}\ . \tag{7.1.13'}$$

Thus, with $z = re^{i\theta}$,

$$\sigma_{rz} = \mathrm{Re}\left\{\psi'(z)\frac{\partial z}{\partial r}\right\} = \mathrm{Re}\{\psi'(z)e^{i\theta}\} \tag{11.2.10}$$

and

$$\sigma_{\theta z} = \mathrm{Re}\left\{\frac{1}{r}\psi'(z)\frac{\partial z}{\partial\theta}\right\} = \mathrm{Re}\{i\psi'(z)e^{i\theta}\} = -\,\mathrm{Im}\{\psi'(z)e^{i\theta}\}\ . \tag{11.2.11}$$

Therefore, one has that

$$\boxed{\sigma_{rz} - i\sigma_{\theta z} = \psi'(z)e^{i\theta}\ .} \tag{11.2.12}$$

The traction-stress relation (11.2.8) is not affected by the change in vector basis, since $t_z = t_3$.

11.2.3   Boundary Value Problem

Consider a region S of the z-plane, with boundary ∂S. Assume that at every interior point z_0 of S, the body force is either zero or every neighborhood of z_0 contains points at which the body force field is zero. Let ∂S^u and ∂S^t be portions of the boundary on which the out-of-plane displacement and traction, respectively, are prescribed. Then, in terms of the complex potential $\psi(z)$, the antiplane strain boundary value problem reduces to the

---

[2]Unfortunately, the cylindrical coordinate $z = X_3$ has the same denotation as the complex variable $z = X_1 + iX_2$. Consequently, one must take care not to confuse the two. Since in a state of antiplane strain there is no dependence on the cylindrical coordinate z, the independent variable z will always refer to the complex variable, and the index z will always refer to the cylindrical coordinate.

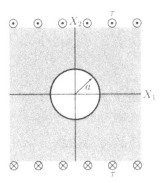

FIGURE 11.11.   Circular hole in an infinite two-dimensional body under far-field shear.

problem of finding a function $\psi(z)$, $\forall z \in \mathcal{S}$, that is analytic everywhere in $\mathcal{S}$ that the body force field is zero and satisfies the displacement boundary condition

$$\boxed{\mathrm{Re}\{\psi(z)\} = \mu \tilde{u}_3 \quad \forall z \in \partial \mathcal{S}^u\,,} \tag{11.2.13}$$

where $\tilde{u}_3$ is a given out-of-plane displacement distribution, and the traction boundary condition,

$$\boxed{\frac{d}{ds}\,\mathrm{Im}\{\psi(z)\} = \tilde{t}_3 \quad \forall z \in \partial \mathcal{S}^t\,,} \tag{11.2.14}$$

where $\tilde{t}_3$ is a given out-of-plane traction distribution and s is the arc length around $\partial \mathcal{S}$ whose positive direction is defined in the usual way, that is, so that the interior of $\mathcal{S}$ is always on the left. Points in $\mathcal{S}$ where the body force field does not vanish will be singular points of the function $\psi(z)$.

11.2.4   Example: Infinite Two-Dimensional Body with a Circular Hole Under Far-Field Shear

Consider an infinite body with a circular hole of radius a, where the body is subject to a far-field shear stress of magnitude τ. Define a Cartesian coordinate system so that the origin is at the center of the hole and the nonzero far-field stress is σ_{23} (Figure 11.11). The surface of the hole is traction-free,

$$\tilde{t}_3 = 0 \quad \text{on } |z| = a\,, \tag{11.2.15}$$

and the stress field approaches a state of homogeneous shear of magnitude τ far away from the hole,

$$\sigma_{13} = 0 , \quad \sigma_{23} = \tau \quad \text{as } |z| \to \infty . \tag{11.2.16}$$

These are the boundary conditions for this antiplane strain problem. It is also clear from symmetry that the displacement u_3 along the X_1-axis is a constant that may be set equal to zero without loss of generality,

$$u_3 = 0 \quad \text{on } X_2 = 0 . \tag{11.2.17}$$

The antiplane strain complex potential can be used to determine the displacement and stress fields for this problem.

Since the body force field is zero everywhere in the body, it follows that the solution can be represented by an antiplane strain complex potential $\psi(z)$ that is analytic in $|z| \geq a$. Since the region $|z| \geq a$ in which $\psi(z)$ is analytic is an annulus centered at the origin (whose outer radius is at infinity), the function may be represented by a Laurent series (page 452) in powers of z:

$$\psi(z) = \sum_{k=-\infty}^{\infty} A_k z^z , \tag{11.2.18}$$

where the complex coefficients A_k are determined by the boundary conditions. It follows from the expression (11.2.4) for the scalar components of stress and the far-field boundary condition (11.2.16) that

$$\psi(z) = -i\tau z + C \quad \text{as } |z| \to \infty , \tag{11.2.19}$$

where, from (11.2.1), the constant C may be assumed to be real without loss of generality. Comparing this far-field condition with the Laurent series representation (11.2.18), it is seen that

$$A_0 = C , \quad A_1 = -i\tau , \quad A_2 = A_3 = \cdots = 0 \tag{11.2.20}$$

and

$$\psi(z) = C - i\tau z + \sum_{k=-\infty}^{-1} A_k z^k . \tag{11.2.21}$$

It follows from (11.2.14) and the traction-free hole boundary condition (11.2.15) that

$$\text{Im}\{\psi(z)\} = D \quad \text{on } |z| = a , \tag{11.2.22}$$

where D is a real constant. Recalling that C is real and noting that

$$z^k = r^k e^{ik\theta} = r^k (\cos k\theta + i \sin k\theta) , \tag{11.2.23}$$

one has from (11.2.21) that

$$\text{Im}\{\psi(z)\} = \left[\frac{1}{r}\text{Im}\{A_{-1}\} - \tau r\right]\cos\theta - \frac{1}{r}\text{Re}\{A_{-1}\}\sin\theta$$

$$+ \sum_{k=-\infty}^{-2} r^k\left[\text{Im}\{A_k\}\cos k\theta + \text{Re}\{A_k\}\sin k\theta\right] . \qquad (11.2.24)$$

Thus, it follows from (11.2.22) that $D = 0$ and

$$\text{Im}\{A_{-1}\} = \tau a^2 , \quad \text{Re}\{A_{-1}\} = 0 , \quad A_{-2} = A_{-3} = \cdots = 0 \qquad (11.2.25)$$

and

$$\psi(z) = C - i\tau\left(z - \frac{a^2}{z}\right) , \qquad (11.2.26)$$

where, from (11.2.1), it is seen that C corresponds to a rigid-body translation. The symmetry assertion (11.2.17) sets $C = 0$.

It is thus determined that the antiplane strain complex potential for this problem is

$$\boxed{\psi(z) = -i\tau\left(z - \frac{a^2}{z}\right) .} \qquad (11.2.27)$$

The displacement field from (11.2.1) is

$$u_3 = \frac{\tau}{\mu}\left(r + \frac{a^2}{r}\right)\sin\theta \qquad (11.2.28)$$

and the stress field is given in terms of Cartesian scalar components by (11.2.4),

$$\sigma_{13} = -\frac{a^2\tau}{r^2}\sin 2\theta , \quad \sigma_{23} = \tau\left(1 + \frac{a^2}{r^2}\cos 2\theta\right) , \qquad (11.2.29)$$

and in terms cylindrical coordinate scalar components by (11.2.12),

$$\sigma_{rz} = \tau\left(1 - \frac{a^2}{r^2}\right)\sin\theta , \quad \sigma_{\theta z} = \tau\left(1 + \frac{a^2}{r^2}\right)\cos\theta . \qquad (11.2.30)$$

Note that $\sigma_{13} = 0$ and $\sigma_{23} = \tau$ as $r \to \infty$ and that $\sigma_{rz} = 0$ when $r = a$, as required by the boundary conditions.

The projected shear traction on a surface parallel to the $X_1 X_2$-plane (i.e., with unit outward normal $\hat{e}_3$) is

$$\tau^{(P)} = \sqrt{\sigma_{13}^2 + \sigma_{23}^2} = \tau\sqrt{1 + \frac{2a^2}{r^2}\cos 2\theta + \frac{a^4}{r^4}} . \qquad (11.2.31)$$

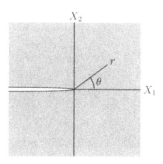

FIGURE 11.12. Semi-infinite, straight crack in antiplane strain.

The normal traction on this surface is zero. Thus, the minimum projected shear traction $\tau^{(P)} = \tau$ occurs in the far-field as $r \to \infty$, and the maximum projected shear traction $\tau^{(P)} = 2\tau$ occurs at the boundary of the hole $r = a$ where $\theta = 0, \pi$.

11.2.5 Asymptotic Crack Tip Fields (Mode III)

Consider a semi-infinite, straight crack in a state of antiplane strain and define a Cartesian coordinate system so that the crack lies along the negative X_1-axis with its tip at the origin (Figure 11.12). In two-dimensional problems, a crack is a geometric feature with two *crack faces* that occupy the same curve in the reference configuration but are free to move relative to each other as the body is deformed. This means that the material description of the displacement field may be discontinuous across the curve occupied by the crack faces. The only restriction in antiplane strain is that the relative displacement of the crack faces must vanish as one approaches the crack tip. It will also be assumed that the crack faces are traction-free. The crack faces considered here are at $\theta = \pi$ and $\theta = -\pi$.

The traction-free crack faces boundary condition requires that

$$\sigma_{23} = 0 \quad \text{on } \theta = \pm\pi . \tag{11.2.32}$$

The restriction that the relative displacement vanish at the crack tip is satisfied, and the rigid-body translation of the body is fixed, by requiring the displacement field to vanish at the crack tip,

$$\lim_{|z| \to 0} u_3 = 0 . \tag{11.2.33}$$

The objective here is to find the most general form of the antiplane strain complex potential and of the displacement and stress fields that are consistent with the conditions (11.2.32) and (11.2.33), with the remaining boundary conditions left unspecified.

Given the expression (11.2.1) for the displacement field, the displacement discontinuity across the crack faces implies that the antiplane strain complex potential has a branch cut along the negative X_1-axis with a branch point singularity at the origin. This suggests that one look for solutions of the form

$$\psi(z) = (A + iB)z^{\lambda+1} , \tag{11.2.34}$$

where A, B, and λ are real. Substituting the assumed form (11.2.34) of the antiplane strain complex potential into the expression (11.2.4) for the scalar components of the stress field gives

$$\begin{aligned}
\sigma_{13} - i\sigma_{23} &= (\lambda + 1)(A + iB)z^{\lambda} \\
&= (\lambda + 1)r^{\lambda}(A + iB)(\cos \lambda\theta + i \sin \lambda\theta) , \tag{11.2.35}
\end{aligned}$$

so that the stress component field σ_{23} corresponding to the form (11.2.34) of the antiplane strain complex potential is

$$\sigma_{23} = -(\lambda + 1)r^{\lambda}(A \sin \lambda\theta + B \cos \lambda\theta) . \tag{11.2.36}$$

The traction-free crack faces boundary condition $\sigma_{23} = 0$ on $\theta = \pm\pi$ requires that

$$\begin{aligned}
A \sin \lambda\pi + B \cos \lambda\pi &= 0 , \\
-A \sin \lambda\pi + B \cos \lambda\pi &= 0 , \tag{11.2.37}
\end{aligned}$$

which are both true if and only if $A \sin \lambda\pi = 0$ and $B \cos \lambda\pi = 0$. The trivial solution $A = B = 0$ just corresponds to a state of zero displacement everywhere. There are two families of nontrivial solutions to (11.2.37).

The first family of solutions comes from requiring that $\cos \lambda\pi = 0$, so that $\lambda = \lambda_n^{(1)}$, where

$$\lambda_n^{(1)} = n - \frac{1}{2} \tag{11.2.38}$$

and n is an arbitrary integer. Then, the coefficient of $\sin \lambda\pi$ must equal zero, so that $A = A_n^{(1)}$ and $B = B_n^{(1)}$, where

$$A_n^{(1)} = 0 \tag{11.2.39}$$

and $B_n^{(1)}$ is an arbitrary constant. The second family of solutions comes from requiring that $\sin \lambda\pi = 0$, so that $\lambda = \lambda_n^{(2)}$, where

$$\lambda_n^{(2)} = n \tag{11.2.40}$$

and n is an arbitrary integer. Then, the coefficient of $\cos \lambda \pi$ must equal zero, so that $A = A_n^{(2)}$ and $B = B_n^{(2)}$, where

$$B_n^{(2)} = 0 \tag{11.2.41}$$

and $A_n^{(2)}$ is an arbitrary constant.

The antiplane strain complex potential (11.2.34) corresponding to each of these solutions will satisfy the traction-free crack faces boundary condition so that, by superposition, the sum of all such solutions,

$$\psi(z) = \sum_{n=-\infty}^{\infty} \left[(A_n^{(1)} + iB_n^{(1)})z^{\lambda_n^{(1)}+1} + (A_n^{(2)} + iB_n^{(2)})z^{\lambda_n^{(2)}+1} \right], \tag{11.2.42}$$

will satisfy the traction-free crack faces boundary condition as well. However, it follows from the expression (11.2.1) for the displacement field that the condition (11.2.33) that the displacement field vanish at the crack tip is satisfied by solutions of the form (11.2.34) if and only if $\lambda > -1$. Thus, given (11.2.38) and (11.2.40), the terms in (11.2.42) with $n < 0$ may be eliminated. Substituting (11.2.38)–(11.2.41) into (11.2.42), a power series representation of the general solution to the antiplane strain crack problem is therefore

$$\psi(z) = \sum_{n=0}^{\infty} \left[iB_n^{(1)} z^{n+\frac{1}{2}} + A_n^{(2)} z^{n+1} \right], \tag{11.2.43}$$

where the constant, real coefficients $B_n^{(1)}$ and $A_n^{(2)}$ are determined by the remaining boundary conditions of whatever specific problem one is considering.

In the power series representation (11.2.43) of the antiplane strain complex potential, there are terms in $z^{1/2}$, z, $z^{3/2}$, ... with their respective constant coefficients. It follows from the expression (11.2.1) for the displacement field and the expression (11.2.4) for the scalar components of the stress field that a term in z^{α} will yield a contribution to the displacement field that is proportional to $|z|^{\alpha}$ and a contribution to the stress field that is proportional to $|z|^{\alpha-1}$. Thus, in the limit $|z| \to 0$ as one approaches the crack tip, the lower-order terms in (11.2.43) tend to dominate the solution and the stress contributions from terms with $\alpha > 1$ vanish. The term in z provides a constant contribution to the stress and the term in $z^{1/2}$ provides a contribution that becomes unbounded like $|z|^{-1/2}$. The foundation of *linear elastic fracture mechanics* is that the contributions to the stress field from these terms characterize the crack's propensity for growth (Kanninen and Popelar 1985).

Consider first the contribution to the stress field from the z term in (11.2.43):

$$\psi_0^{(2)} = A_0^{(2)} z. \tag{11.2.44}$$

It follows from the expression (11.2.4) that the corresponding contribution to the stress field is

$$\sigma_{13} = A_0^{(2)}, \quad \sigma_{23} = 0, \tag{11.2.45}$$

where $A_0^{(2)}$ is determined by the remaining boundary conditions.

Consider next the contribution to the stress field from the $z^{1/2}$ term in (11.2.43):

$$\psi_0^{(1)} = iB_0^{(1)} z^{\frac{1}{2}}. \tag{11.2.46}$$

The convention in linear elastic fracture mechanics is to replace the constant $B_0^{(1)}$ by the *mode III stress intensity factor* K_{III}, defined in the current coordinate system such that

$$\boxed{K_{\mathrm{III}} \equiv \lim_{r \to 0} \sqrt{2\pi r}\; \sigma_{23}|_{\theta=0}.} \tag{11.2.47}$$

One has from (11.2.4) and (11.2.43) that

$$\sigma_{23} = -\,\mathrm{Im}\{\psi'(z)\} = -\frac{1}{2}B_0^{(1)} r^{-\frac{1}{2}} \cos\frac{\theta}{2} + \mathcal{O}(|z|^{\frac{1}{2}}) \tag{11.2.48}$$

and therefore that

$$\sigma_{23}|_{\theta=0} = -\frac{1}{2}B_0^{(1)} r^{-\frac{1}{2}} + \mathcal{O}(|z|^{\frac{1}{2}}). \tag{11.2.49}$$

Thus, it follows from (11.2.47) that $B_0^{(1)} = -2K_{\mathrm{III}}/\sqrt{2\pi}$ and

$$\boxed{\psi_0^{(1)}(z) = \frac{-2i}{\sqrt{2\pi}} K_{\mathrm{III}} z^{\frac{1}{2}}.} \tag{11.2.50}$$

It can then be shown (see Problem 11.7) that the contribution to the displacement field from (11.2.50) is given by

$$u_3 = \frac{2K_{\mathrm{III}}}{\mu}\sqrt{\frac{r}{2\pi}}\, \sin\frac{\theta}{2} \tag{11.2.51}$$

and that the contribution to the stress field is given in terms of Cartesian coordinate scalar components by

$$\left\{ \begin{array}{c} \sigma_{13} \\ \sigma_{23} \end{array} \right\} = \frac{K_{\mathrm{III}}}{\sqrt{2\pi r}} \left\{ \begin{array}{c} -\sin\dfrac{\theta}{2} \\ \cos\dfrac{\theta}{2} \end{array} \right\} \tag{11.2.52a}$$

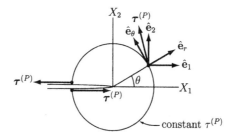

FIGURE 11.13.   *Projected shear traction near a mode III crack.*

and in terms of cylindrical coordinate scalar components by

$$\left\{ \begin{array}{c} \sigma_{rz} \\ \\ \sigma_{\theta z} \end{array} \right\} = \frac{K_{III}}{\sqrt{2\pi r}} \left\{ \begin{array}{c} \sin \dfrac{\theta}{2} \\ \\ \cos \dfrac{\theta}{2} \end{array} \right\} . \qquad (11.2.52\text{b})$$

With the exception of the constant contribution to σ_{13} (11.2.45), all other contributions to the stress field vanish in the vicinity of the crack tip. Thus, in the limit $r \to 0$ as one approaches the crack tip, the stress field for *any* antiplane crack, regardless of the remaining boundary conditions, approaches (11.2.52) asymptotically. For this reason, (11.2.52) is referred to as the *antiplane strain (mode III) asymptotic crack tip stress field*. Subject to certain *small-scale yielding conditions* (Kanninen and Popelar 1985), the mode III stress intensity factor K_{III}, which is determined by the remaining boundary conditions, is assumed to characterize a crack's propensity to grow under antiplane strain (mode III) loading. This is also referred to as the *tearing mode* of crack loading.

The *crack opening displacement* $\boldsymbol{\delta} = \delta_3 \hat{\mathbf{e}}_3$ is the displacement discontinuity across the crack, given as a function of the distance r from the crack tip by

$$\boldsymbol{\delta}(r) \equiv \mathbf{u}(r, \pi) - \mathbf{u}(r, -\pi) , \quad \delta_3(r) \equiv u_3(r, \pi) - u_3(r, -\pi) . \quad (11.2.53)$$

It follows from (11.2.51) that

$$\delta_3 = \frac{4 K_{III}}{\mu} \sqrt{\frac{r}{2\pi}} . \qquad (11.2.54)$$

Clearly, the crack opening displacement vanishes at the crack tip, as required.

On a surface parallel to the $X_1 X_2$-plane (unit outward normal $\hat{\mathbf{e}}_3$), the

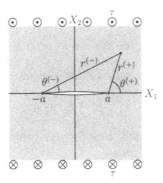

FIGURE 11.14. *Finite crack in an infinite two-dimensional body under far-field shear.*

normal traction is zero and the projected shear traction vector is

$$\boldsymbol{\tau}^{(\mathrm{P})} = \sigma_{13}\hat{\mathbf{e}}_1 + \sigma_{23}\hat{\mathbf{e}}_2 = \frac{K_{\mathrm{III}}}{\sqrt{2\pi r}} \left(-\sin\frac{\theta}{2}\hat{\mathbf{e}}_1 + \cos\frac{\theta}{2}\hat{\mathbf{e}}_2 \right) , \qquad (11.2.55\mathrm{a})$$

$$\boldsymbol{\tau}^{(\mathrm{P})} = \sigma_{rz}\hat{\mathbf{e}}_z + \sigma_{\theta z}\hat{\mathbf{e}}_\theta = \frac{K_{\mathrm{III}}}{\sqrt{2\pi r}} \left(\sin\frac{\theta}{2}\hat{\mathbf{e}}_r + \cos\frac{\theta}{2}\hat{\mathbf{e}}_\theta \right) \qquad (11.2.55\mathrm{b})$$

and (assuming $K_{\mathrm{III}} \geq 0$) the projected shear traction is

$$\tau^{(\mathrm{P})} = \sqrt{\sigma_{13}^2 + \sigma_{23}^2} = \sqrt{\sigma_{rz}^2 + \sigma_{\theta z}^2} = \frac{K_{\mathrm{III}}}{\sqrt{2\pi r}} . \qquad (11.2.56)$$

Therefore, contours of constant $\tau^{(\mathrm{P})}$ are circles centered on the crack tip and $\boldsymbol{\tau}^{(\mathrm{P})}$ is in the direction of the unit vector

$$\hat{\boldsymbol{\mu}} = -\sin\frac{\theta}{2}\hat{\mathbf{e}}_1 + \cos\frac{\theta}{2}\hat{\mathbf{e}}_2 = \sin\frac{\theta}{2}\hat{\mathbf{e}}_r + \cos\frac{\theta}{2}\hat{\mathbf{e}}_\theta . \qquad (11.2.57)$$

The direction of the projected shear traction vector bisects $\hat{\mathbf{e}}_2$ and $\hat{\mathbf{e}}_\theta$ (Figure 11.13). At $\theta = 0$, $\boldsymbol{\tau}^{(\mathrm{P})}$ is in the direction of $\hat{\mathbf{e}}_2$ and, at $\theta = \pm\pi$, it is in the direction of $\pm\hat{\mathbf{e}}_1$.

11.2.6 Example: Infinite Two-Dimensional Body with a Finite Crack Under Far-Field Shear

Consider an infinite two-dimensional body in antiplane strain that has a straight crack of length $2a$, where the body is subject to a far-field shear of magnitude τ, as shown in Figure 11.14. Define a Cartesian coordinate system with the crack tips at $X_1 = \pm a$. The crack faces are traction-free,

$$\sigma_{23} = 0 \quad \text{on } X_2 = 0 , \quad -a < X_1 < a , \tag{11.2.58}$$

and the stress approaches a state of homogeneous shear far from the crack,

$$\sigma_{13} = 0 , \quad \sigma_{23} = \tau \quad \text{as } |z| \to \infty . \tag{11.2.59}$$

These are the boundary conditions for this problem. It is also clear from symmetry that one can assume

$$u_3 = 0 \quad \text{on } X_2 = 0 , \quad |X_1| > a \tag{11.2.60}$$

without loss of generality, which fixes the rigid-body motion.

It follows from (11.2.4) that the traction-free crack faces boundary condition (11.2.58) is expressed in terms of the antiplane strain complex potential as

$$\text{Im}\{\psi'(z)\} = 0 \quad \text{on } X_2 = 0 , \quad -a < X_1 < a , \tag{11.2.61}$$

and that the far-field condition (11.2.59) is expressed as

$$\psi'(z) = -i\tau , \quad \sigma_{23} = \tau \quad \text{as } |z| \to \infty . \tag{11.2.62}$$

It follows from (11.2.1) that the symmetry assumption (11.2.60) is expressed as

$$\text{Re}\{\psi(z)\} = 0 \quad \text{on } X_2 = 0 , \quad |X_1| > a . \tag{11.2.63}$$

An antiplane strain complex potential $\psi(z)$ that is analytic everywhere except at the crack tips, where $z = \pm a$, and satisfies the conditions (11.2.61)–(11.2.63) will provide the solution to this problem.

Using the semi-inverse method and anticipating that the stress field will have square root singularities at the crack tips (which is true for all crack tips in linearized elasticity, as shown in Subsections 11.2.5 and 11.3.11), expressible as

$$\sigma_{\alpha 3} \propto \frac{1}{\sqrt{z - a}} , \quad \sigma_{\alpha 3} \propto \frac{1}{\sqrt{z + a}} , \tag{11.2.64}$$

one can guess that the antiplane strain complex potential has the form

$$\psi(z) = A\sqrt{(z - a)(z + a)} = A\sqrt{z^2 - a^2} , \tag{11.2.65}$$

which is analytic everywhere except at the crack tips, as required. There are an infinite number of antiplane strain complex potentials that are consistent with (11.2.64), of which (11.2.65) is just the simplest. Certainly though, one is free to try an antiplane strain complex potential of the form (11.2.65) and see if it works. If it does not, then it will have to be discarded.

Given the assumed form (11.2.65) of the antiplane strain complex potential, one has that

$$\psi'(z) = \frac{Az}{\sqrt{z^2 - a^2}} . \tag{11.2.66}$$

Since on $X_2 = 0$, $-a < X_1 < a$, one has that $z = X_1$ is real and $\sqrt{z^2 - a^2}$ is imaginary, this satisfies the traction-free crack faces boundary condition (11.2.61) if A is imaginary. Since on $X_2 = 0$, $|X_1| > a$, one has that $\sqrt{z^2 - a^2}$ is real, the symmetry assumption (11.2.63) is also satisfied by (11.2.65) if A is imaginary. Finally, the far-field condition (11.2.62) is satisfied by (11.2.66) if $A = -i\tau$. Thus, the guess is correct and the solution to the problem is given by

$$\boxed{\psi(z) = -i\tau\sqrt{z^2 - a^2} .} \tag{11.2.67}$$

It follows from (11.2.1) that the solution displacement field is

$$u_3 = \frac{\tau}{\mu} \operatorname{Im}\left\{\sqrt{z^2 - a^2}\right\} \tag{11.2.68}$$

and from (11.2.4) that the solution stress field is given by

$$\sigma_{13} - i\sigma_{23} = \frac{-i\tau z}{\sqrt{z^2 - a^2}} . \tag{11.2.69}$$

The solution in this case was found through a fortuitous educated guess as to the form of the antiplane strain complex potential. One should not expect to always be so lucky.

To see the nature of this solution more clearly, define two polar coordinate systems as shown in Figure 11.14, with the origins at the crack tips, so that

$$z - a = r^{(+)}e^{i\theta^{(+)}} , \quad z + a = r^{(-)}e^{i\theta^{(-)}} . \tag{11.2.70}$$

It then follows that

$$z^2 - a^2 = (z - a)(z + a) = r^{(+)}r^{(-)}e^{i[\theta^{(+)}+\theta^{(-)}]} \tag{11.2.71}$$

and

$$\sqrt{z^2 - a^2} = \sqrt{r^{(+)}r^{(-)}}e^{i[\theta^{(+)}+\theta^{(-)}]/2} . \tag{11.2.72}$$

Note that as one traverses a closed path around the right crack tip, with the left crack tip outside of the path, for each counterclockwise circuit of the path the coordinate variable $\theta^{(+)}$ increases by 2π while $\theta^{(-)}$ remains constant. For a closed path around the left crack tip, with the right crack tip outside of the path, for each counterclockwise circuit of the path the coordinate variable $\theta^{(-)}$ increases by 2π while $\theta^{(+)}$ remains constant. For a closed path containing both crack tips, both $\theta^{(+)}$ and $\theta^{(-)}$ increase by 2π for each counterclockwise circuit, and for a path containing neither crack tip, both $\theta^{(+)}$ and $\theta^{(-)}$ remain constant for each counterclockwise circuit. The coordinate variables $r^{(+)}$ and $r^{(-)}$ are uniquely determined by z. Far from the crack when $|z| \gg a$, it can be seen that $r^{(+)} \approx r^{(-)} \approx |z|$ and $\theta^{(+)} \approx \theta^{(-)} \approx \arg\{z\}$.

The displacement field (11.2.68) can be expressed in terms of the two polar coordinate systems as

$$u_3 = \frac{\tau}{\mu}\sqrt{r^{(+)}r^{(-)}}\sin\frac{\theta^{(+)} + \theta^{(-)}}{2} . \tag{11.2.73}$$

On the X_1-axis to the right of the crack where $\theta^{(+)} = 2m\pi$ and $\theta^{(-)} = 2n\pi$, or to the left of the crack where $\theta^{(+)} = (2m + 1)\pi$ and $\theta^{(-)} = (2n + 1)\pi$, one has from (11.2.73) that $u_3 = 0$, as assumed in (11.2.60). Let $\theta^{(+)} = \pi$ and $\theta^{(-)} = 0$ for a point on the top crack face, so that the displacement of the top crack face is

$$u_3^{top} = \frac{\tau}{\mu}\sqrt{r^{(+)}r^{(-)}} = \frac{\tau}{\mu}\sqrt{(X_1 - a)(X_1 + a)} = \frac{\tau}{\mu}\sqrt{X_1^2 - a^2} . \tag{11.2.74}$$

If one then follows a clockwise path around the right crack tip to the bottom crack face so that $\theta^{(+)} = -\pi$ and $\theta^{(-)} = 0$, or a counterclockwise path around the left crack tip so that $\theta^{(+)} = \pi$ and $\theta^{(-)} = 2\pi$, then it follows from (11.2.73) that the displacement of the bottom crack face is

$$u_3^{bottom} = -\frac{\tau}{\mu}\sqrt{X_1^2 - a^2} . \tag{11.2.75}$$

Therefore, the crack opening displacement $\delta_3 \equiv u_3^{top} - u_3^{bottom}$ is

$$\delta_3 = \frac{2\tau}{\mu}\sqrt{X_1^2 - a^2} , \quad |X_1| \le a . \tag{11.2.76}$$

Note that the crack opening displacement approaches zero as one approaches a crack tip like the square root of the distance from the crack tip, which agrees with the asymptotic prediction (11.2.54).

The stress field (11.2.69) can be expressed in terms of the two polar coordinate systems and the Cartesian coordinates X_1 and X_2 as

$$\sigma_{13} = \frac{\tau}{\sqrt{r^{(+)}r^{(-)}}}\left[X_2\cos\frac{\theta^{(+)} + \theta^{(-)}}{2} - X_1\sin\frac{\theta^{(+)} + \theta^{(-)}}{2}\right] ,$$

$$\sigma_{23} = \frac{\tau}{\sqrt{r^{(+)}r^{(-)}}}\left[X_2\sin\frac{\theta^{(+)} + \theta^{(-)}}{2} + X_1\cos\frac{\theta^{(+)} + \theta^{(-)}}{2}\right] . \tag{11.2.77}$$

Far from the crack, where $r^{(+)} \approx r^{(-)} \approx |z|$ and $\theta^{(+)} \approx \theta^{(-)} \approx \arg\{z\}$,

$$\sigma_{13}\big|_{|z| \gg a} \approx \frac{\tau}{|z|}[X_2 \cos \arg\{z\} - X_1 \sin \arg\{z\}]$$

$$\approx \frac{\tau}{|z|^2}[X_2 X_1 - X_1 X_2]$$

$$\approx 0 \tag{11.2.78}$$

and

$$\sigma_{23}\big|_{|z| \gg a} \approx \frac{\tau}{|z|}[X_2 \sin \arg\{z\} + X_1 \cos \arg\{z\}]$$

$$\approx \frac{\tau}{|z|^2}[X_2^2 + X_1^2]$$

$$\approx \tau, \tag{11.2.79}$$

so that

$$\lim_{|z| \to \infty} \sigma_{13} = 0, \quad \lim_{|z| \to \infty} \sigma_{23} = \tau, \tag{11.2.80}$$

as required by the far-field boundary condition (11.2.59). On the top crack face, where $\frac{1}{2}(\theta^{(+)} + \theta^{(-)}) = (2m + \frac{1}{2})\pi$, $X_2 = 0$, and $\sqrt{r^{(+)}r^{(-)}} = \sqrt{X_1^2 - a^2}$,

$$\sigma_{13}^{\text{top}} = \frac{-\tau X_1}{\sqrt{X_1^2 - a^2}}, \quad \sigma_{23}^{\text{top}} = 0, \quad |X_1| < a, \tag{11.2.81}$$

and on the bottom crack face, where $\frac{1}{2}(\theta^{(+)} + \theta^{(-)}) = (2m - \frac{1}{2})\pi$, $X_2 = 0$, and $\sqrt{r^{(+)}r^{(-)}} = \sqrt{X_1^2 - a^2}$,

$$\sigma_{13}^{\text{bottom}} = \frac{\tau X_1}{\sqrt{X_1^2 - a^2}}, \quad \sigma_{23}^{\text{bottom}} = 0, \quad |X_1| < a, \tag{11.2.82}$$

which are consistent with the traction-free crack faces boundary condition (11.2.58). Finally, on the X_1-axis to the right of the crack, where $\frac{1}{2}(\theta^{(+)} + \theta^{(-)}) = 2m\pi$, $X_2 = 0$, $r^{(-)} = r^{(+)} + 2a$, and $X_1 = r^{(+)} + a$,

$$\sigma_{13}\big|_{\theta^{(+)}=0} = 0, \quad \sigma_{23}\big|_{\theta^{(+)}=0} = \frac{\tau(r^{(+)} + a)}{\sqrt{r^{(+)}(r^{(+)} + 2a)}}, \tag{11.2.83}$$

where $r^{(+)}$ is the distance from the right crack tip.

Stress intensity factor

To find the mode III stress intensity factor for the right crack tip in this problem, note that the definition (11.2.47) takes the form

$$K_{\text{III}} \equiv \lim_{r^{(+)} \to 0} \sqrt{2\pi r^{(+)}} \, \sigma_{23}\big|_{\theta^{(+)}=0} \tag{11.2.84}$$

for the right crack tip. Thus, it follows from (11.2.83) that

$$K_{\text{III}} = \lim_{r^{(+)} \to 0} \sqrt{2\pi} \, \frac{\tau(r^{(+)} + a)}{\sqrt{r^{(+)} + 2a}} = \frac{\tau a \sqrt{2\pi}}{\sqrt{2a}} = \sqrt{\pi a} \, \tau \, . \qquad (11.2.85)$$

It follows from symmetry that the mode III stress intensity factor for the left crack tip is the same, so that

$$\boxed{K_{\text{III}} = \sqrt{\pi a} \, \tau} \qquad (11.2.86)$$

characterizes the propensity for each crack tip to grow. This propensity increases linearly with the far-field shear load and like the square root of the crack size.

    The stress intensity factors for cracks with other geometries and loadings will be different. See Tada et al. (1985) for a large collection of such stress intensity factors.

11.3   Plane Strain/Stress

Recall (Section 7.4) that the Airy stress function $\Phi(X_1, X_2)$ for plane strain and plane stress problems was defined in terms of the nonzero components of stress, at points were the body force field is zero, by

$$\sigma_{11} = \Phi_{,22} \, , \quad \sigma_{22} = \Phi_{,11} \, , \quad \sigma_{12} = -\Phi_{,12} \, . \qquad (11.3.1)$$

This guarantees that the two nontrivial equilibrium equations $\sigma_{\alpha\beta,\alpha} = 0$ are satisfied. The equation of elasticity that remains to be satisfied is the compatibility condition, which is given in terms of the Airy stress function by

$$\nabla^4 \Phi = 0 \, . \qquad (7.4.9')$$

Thus, the Airy stress function must be biharmonic in any region where the body force field is zero.

    If a function $\Phi(z)$ of the complex variable z is biharmonic in a region $\mathcal{S}$ of the complex plane, then the function can be represented in $\mathcal{S}$ by

$$\Phi = \text{Re}\{\bar{z}F_1(z) + F_2(z)\} \, , \qquad (11.3.2)$$

where $F_1(z)$ and $F_2(z)$ are analytic in $\mathcal{S}$ and $\bar{z} = X_1 - iX_2$ is the complex conjugate of the complex variable $z = X_1 + iX_2$. Conversely, a function $\Phi(z)$ represented by (11.3.2) is biharmonic in a region $\mathcal{S}$ if and only if $F_1(z)$

and $F_2(z)$ are analytic in S. To prove this, note first that Φ is biharmonic if and only if $\nabla^2 \Phi$ is harmonic, since $\nabla^4 \Phi = \nabla^2(\nabla^2 \Phi)$, and $\nabla^2 \Phi$ can therefore be represented in S by

$$\nabla^2 \Phi = \text{Re}\{f(z)\} , \qquad (11.3.3)$$

where $f(z)$ is analytic in S. A particular solution of (11.3.3) is

$$\Phi^{(p)} = \frac{1}{4} \text{Re}\{\bar{z}F(z)\} , \qquad (11.3.4)$$

where $F(z) \equiv \int_{z_0}^{z} f(\zeta) \, d\zeta$ is analytic since $f(z)$ is analytic, and $F'(z) = f(z)$. This can be seen by direct substitution of (11.3.4) into (11.3.3). The homogeneous solution of (11.3.3) is an arbitrary harmonic function which can be represented as

$$\Phi^{(h)} = \text{Re}\{F_2(z)\} , \qquad (11.3.5)$$

where $F_2(z)$ is an arbitrary analytic function. Thus, the general solution to $\nabla^4 \Phi = 0$ is (11.3.2) where $F_1(z) = \frac{1}{4}F(z)$ and $F_2(z)$ are arbitrary harmonic functions.

It follows that in a region S where the body force field is zero, every Airy stress function Φ that satisfies the compatibility condition $\nabla^4 \Phi = 0$ can be represented in S by

$$\Phi = \text{Re}\{\bar{z}\phi(z) + \Psi(z)\} , \qquad (11.3.6)$$

where $\phi(z)$ and $\Psi(z)$ are analytic in S. Conversely, an Airy stress function represented by (11.3.6) will satisfy compatibility in a region S where the body force field vanishes if and only if $\phi(z)$ and $\Psi(z)$ are analytic in S. Substituting (11.3.6) into (11.3.1) leads to the expressions

$$\begin{aligned}
\sigma_{11} &= \text{Re}\{-\bar{z}\phi''(z) + 2\phi'(z) - \Psi''(z)\} , \\
\sigma_{22} &= \text{Re}\{\bar{z}\phi''(z) + 2\phi'(z) + \Psi''(z)\} , \qquad (11.3.7) \\
\sigma_{12} &= \text{Im}\{\bar{z}\phi''(z) + \Psi''(z)\} ,
\end{aligned}$$

which can be combined to give

$$\begin{aligned}
\sigma_{11} + \sigma_{22} &= 4\,\text{Re}\{\phi'(z)\} , \\
\sigma_{22} - \sigma_{11} + 2i\sigma_{12} &= 2[\bar{z}\phi''(z) + \Psi''(z)] .
\end{aligned} \qquad (11.3.8)$$

From now on, Ψ only appears as Ψ' and Ψ'' so, to simplify notation, define $\psi(z) \equiv \Psi'(z)$ and recall that $\psi(z)$ is analytic in a region S if and only if $\Psi(z)$ is analytic in S. Thus,

$$\boxed{\begin{aligned}
\sigma_{11} + \sigma_{22} &= 4\,\text{Re}\{\phi'\} , \\
\sigma_{22} - \sigma_{11} + 2i\sigma_{12} &= 2(\bar{z}\phi'' + \psi') ,
\end{aligned}} \qquad (11.3.9)$$

where the explicit dependence of the *complex stress functions* ϕ and ψ on z is omitted from here on, unless required for clarity. In a region where the body force field is identically zero, the components of stress given by (11.3.9) are guaranteed to satisfy the plane strain/stress equations of equilibrium and the compatibility condition requires that the complex stress functions ϕ and ψ be analytic.

11.3.1   Displacements

For plane strain and plane stress, the constitutive equations (7.3.41) can be rewritten as

$$\sigma_{\alpha\beta} = 2\mu \left[\varepsilon_{\alpha\beta} + \frac{3-\kappa}{2(\kappa-1)} \varepsilon_{\gamma\gamma}\delta_{\alpha\beta} \right] , \qquad (11.3.10)$$

where

$$\kappa \equiv \begin{cases} 3 - 4\nu & \text{for plane strain} \\ \dfrac{3-\nu}{1+\nu} & \text{for plane stress.} \end{cases} \qquad (11.3.11)$$

The material parameter κ, related to the parameter α defined in (7.3.40) by $\kappa = 4\alpha - 1$, is introduced for convenience. It follows from (11.3.10) and the strain-displacement relation that

$$\sigma_{11} + \sigma_{22} = \frac{4\mu}{\kappa-1}(u_{1,1} + u_{2,2}) \qquad (11.3.12)$$

and thus from (11.3.9) that

$$\mu(u_{1,1} + u_{2,2}) = (\kappa - 1)\,\mathrm{Re}\{\phi'\} . \qquad (11.3.13)$$

Also from (11.3.10),

$$\sigma_{22} - \sigma_{11} + 2i\sigma_{12} = 2\mu[u_{2,2} - u_{1,1} + i(u_{1,2} + u_{2,1})] \qquad (11.3.14)$$

and thus from (11.3.9), it follows that

$$\mu[u_{2,2} - u_{1,1} + i(u_{1,2} + u_{2,1})] = \bar{z}\phi'' + \psi' . \qquad (11.3.15)$$

Subtracting (11.3.15) from (11.3.13) and equating the real parts of both sides yields

$$2\mu u_{1,1} = \mathrm{Re}\{(\kappa - 1)\phi' - \bar{z}\phi'' - \psi'\} . \qquad (11.3.16)$$

Adding (11.3.13) and (11.3.15) and equating the real parts of both sides yields

$$2\mu u_{2,2} = \mathrm{Re}\{(\kappa - 1)\phi' + \bar{z}\phi'' + \psi'\} . \qquad (11.3.17)$$

Finally, equating the imaginary parts of both sides of (11.3.15) gives

$$\mu(u_{1,2} + u_{2,1}) = \text{Im}\{\bar{z}\phi'' + \psi'\} . \qquad (11.3.18)$$

The three partial differential equations (11.3.16)–(11.3.18) can be solved for the scalar components of the displacement field to within an arbitrary rigid-body motion.

By integrating (11.3.16) with respect to X_1, one obtains

$$\begin{aligned} 2\mu u_1 &= \text{Re}\{(\kappa - 1)\phi - \bar{z}\phi' + \phi - \psi\} + \eta(X_2) \\ &= \text{Re}\{-\bar{z}\phi' + \kappa\phi - \psi\} + \eta(X_2) , \end{aligned} \qquad (11.3.19)$$

where η is an arbitrary real-valued function of X_2. Integrating (11.3.17) with respect to X_2 gives

$$\begin{aligned} 2\mu u_2 &= \text{Re}\{-i(\kappa - 1)\phi - i\bar{z}\phi' - i\phi - i\psi\} + \xi(X_1) \\ &= \text{Im}\{\bar{z}\phi' + \kappa\phi + \psi\} + \xi(X_1) , \end{aligned} \qquad (11.3.20)$$

where ξ is an arbitrary real-valued function of X_1. It follows from (11.3.19) and (11.3.20) that

$$2\mu(u_{1,2} + u_{2,1}) = 2\,\text{Im}\{\bar{z}\phi'' + \psi'\} + \eta'(X_2) + \xi'(X_1) . \qquad (11.3.21)$$

By comparing (11.3.21) with the result (11.3.18), it is seen that $\eta'(X_2) = -\xi'(X_1) = \gamma$, where γ is a real constant. Thus, $\eta(X_2) = \alpha + \gamma X_2$ and $\xi(X_1) = \beta - \gamma X_1$, where α and β are real constants. The real constants α, β, and γ do not affect the value of the components of stress—they correspond to a rigid-body motion with α and β giving a rigid-body translation and γ a rigid-body rotation. It will be seen that this rigid-body motion can be accommodated by the complex stress functions, so that α, β, and γ may be set equal to zero without loss of generality.

Finally, by noting that $\text{Re}\{\phi\} = \text{Re}\{\bar{\phi}\}$ and $\text{Im}\{\phi\} = -\,\text{Im}\{\bar{\phi}\}$, where it is understood that $\bar{\phi} = \overline{\phi(z)} = \bar{\phi}(\bar{z})$, (11.3.19) and (11.3.20) can be rewritten as

$$\begin{aligned} 2\mu u_1 &= \text{Re}\{-\bar{z}\phi' + \kappa\bar{\phi} - \psi\} , \\ 2\mu u_2 &= \text{Im}\{\bar{z}\phi' - \kappa\bar{\phi} + \psi\} , \end{aligned} \qquad (11.3.22)$$

which can be combined to give

$$\boxed{2\mu(u_1 - iu_2) = -\bar{z}\phi' + \kappa\bar{\phi} - \psi ,} \qquad (11.3.23)$$

where κ is defined by (11.3.11) and it is important to remember that the independent variable of $\bar{\phi}$ is $\bar{z}$.

11.3.2   Rigid-Body Motion

Recall that rigid-body motion corresponds to a displacement field of the form

$$2\mu u_1 = \alpha + \gamma X_2 , \quad 2\mu u_2 = \beta - \gamma X_1 , \qquad (11.3.24)$$

where the real constants α and β characterize rigid-body translation and the real constant γ characterizes (infinitesimal) rigid-body rotation. The goal here is to find the most general form of the complex stress functions that correspond to a rigid-body motion and, hence, to zero stress.

From (11.3.23) and (11.3.24), one has for a rigid-body motion that

$$\begin{aligned} -\bar{z}\phi'(z) + \kappa\bar{\phi}(\bar{z}) - \psi(z) &= 2\mu\,(u_1 - iu_2) \\ &= \alpha - i\beta + \gamma(X_2 + iX_1) \\ &= \alpha - i\beta + i\gamma\bar{z} . \end{aligned} \qquad (11.3.25)$$

Since the right-hand side of (11.3.25) is independent of z and linear in $\bar{z}$, so must be the left-hand side. Thus, $\psi(z)$ and $\phi'(z)$ are complex constants:

$$\phi = A + Cz , \quad \psi = B , \qquad (11.3.26)$$

where A, B, and C are complex constants. Substituting (11.3.26) back into (11.3.25) gives

$$(\kappa\bar{C} - C)\bar{z} + \kappa\bar{A} - B = \alpha - i\beta + i\gamma\bar{z} . \qquad (11.3.27)$$

Equating coefficients of $\bar{z}$ gives $\kappa\bar{C} - C = i\gamma$. Thus,

$$\begin{aligned} \mathrm{Re}\{\kappa\bar{C} - C\} &= (\kappa - 1)\,\mathrm{Re}\{C\} = 0 , \\ \mathrm{Im}\{\kappa\bar{C} - C\} &= -(\kappa + 1)\,\mathrm{Im}\{C\} = \gamma , \end{aligned} \qquad (11.3.28)$$

from which it follows that $C = iD$, where $D = -\gamma/(\kappa+1)$ is a real constant. Equating the constant terms in (11.3.27) gives

$$\kappa\bar{A} - B = \alpha - i\beta . \qquad (11.3.29)$$

Thus, $\alpha = \mathrm{Re}\{\kappa\bar{A} - B\}$ and $\beta = -\mathrm{Im}\{\kappa\bar{A} - B\}$.

In summary, complex stress functions of the form

$$\boxed{\phi = A + iDz , \quad \psi = B \quad (A,\ B \text{ complex};\ D \text{ real})} \qquad (11.3.30)$$

are the most general, corresponding to a rigid-body motion, with displacement then given by

$$\boxed{\begin{aligned} 2\mu(u_1 - iu_2) &= -i(\kappa + 1)D\bar{z} + \kappa\bar{A} - B , \\ 2\mu u_1 &= \mathrm{Re}\{\kappa\bar{A} - B\} - (\kappa + 1)DX_2 , \\ 2\mu u_2 &= -\mathrm{Im}\{\kappa\bar{A} - B\} + (\kappa + 1)DX_1 . \end{aligned}} \qquad (11.3.31)$$

Note that the displacement field is zero if and only if $D = 0$ and $B = \kappa \overline{A}$. The following important observations follow:

- The stress field is unaltered by a change in the complex stress functions ϕ and ψ of the form

$$\begin{aligned} \phi &\to \phi + A + iDz \\ \psi &\to \psi + B \end{aligned} \quad (A,\ B \text{ complex}; D \text{ real}) . \quad (11.3.32)$$

- The displacement field is unaltered by a change in the complex stress functions ϕ and ψ of the form

$$\begin{aligned} \phi &\to \phi + A \\ \psi &\to \psi + \kappa \overline{A} \end{aligned} \quad (A \text{ complex}) . \quad (11.3.33)$$

11.3.3 Homogeneous Stress

Consider a stress field that is known to be homogeneous over some region of the z-plane. What then are the most general expressions for the complex stress functions ϕ and Ψ in this region, in terms of the scalar components of the homogeneous stress field?

From the representation (11.3.9) for the scalar components of stress in terms of the complex stress functions, one can see that $\text{Re}\{\phi'\} = \frac{1}{4}(\sigma_{11} + \sigma_{22})$ is a constant in the region where the stress field is homogeneous. It follows from the requirement that ϕ (and therefore ϕ') must be analytic, and the Cauchy-Riemann equations, that $\text{Im}\{\phi'\}$ is a constant over this region as well and hence that

$$\phi' = \frac{1}{4}(\sigma_{11} + \sigma_{22}) + iD , \quad (11.3.34)$$

where D is a real constant. Since $\sigma_{11} + \sigma_{22}$ and D are constants,

$$\phi = \left[\frac{1}{4}(\sigma_{11} + \sigma_{22}) + iD \right] z + A , \quad (11.3.35)$$

where A is a complex constant. Noting that $\phi'' = 0$, the second expression in (11.3.9) gives

$$\psi' = \frac{1}{2}(\sigma_{22} - \sigma_{11}) + i\sigma_{12} , \quad (11.3.36)$$

so that, since $\frac{1}{2}(\sigma_{22} - \sigma_{11}) + i\sigma_{12}$ is a constant,

$$\psi = \left[\frac{1}{2}(\sigma_{22} - \sigma_{11}) + i\sigma_{12} \right] z + B , \quad (11.3.37)$$

where B is a complex constant.

In summary, if $\sigma_{\alpha\beta}$ are the scalar components of a *homogeneous* stress field, then the complex stress functions must be of the form

$$
\begin{aligned}
\phi &= \frac{1}{4}(\sigma_{11} + \sigma_{22})z + A + iDz \\
&\qquad\qquad\qquad\qquad (A,\, B \text{ complex};\, D \text{ real}) , \qquad (11.3.38) \\
\psi &= \frac{1}{2}(\sigma_{22} - \sigma_{11} + 2i\sigma_{12})z + B
\end{aligned}
$$

where it has already been seen (11.3.32) that A, B, and D correspond to rigid-body motion.

11.3.4   Traction-Stress Relation

Recall that the traction-stress relation for plane strain/stress reduces to the requirement that $t_3 = 0$ and $t_\alpha = t_\alpha(X_1, X_2)$, where $t_\alpha = \sigma_{\beta\alpha}\hat{n}_\beta$. Recall the result (8.1.14) that given a parametric description $X_\alpha = \hat{X}_\alpha(s)$ of a curve in the $X_1 X_2$-plane, where s is the arc length, the unit outward normal to the curve is given by

$$
\hat{\mathbf{n}} = \frac{dX_2}{ds}\hat{\mathbf{e}}_1 - \frac{dX_1}{ds}\hat{\mathbf{e}}_2 . \tag{11.3.39}
$$

The relevant scalar components of the traction are therefore expressible as

$$
t_1 = \sigma_{11}\frac{dX_2}{ds} - \sigma_{21}\frac{dX_1}{ds} , \quad t_2 = \sigma_{12}\frac{dX_2}{ds} - \sigma_{22}\frac{dX_1}{ds} , \tag{11.3.40}
$$

which can be combined as

$$
t_1 - it_2 = (\sigma_{11} - i\sigma_{12})\frac{dX_2}{ds} + (-\sigma_{12} + i\sigma_{22})\frac{dX_1}{ds} . \tag{11.3.41}
$$

Noting that $X_1 = \frac{1}{2}(z + \bar{z})$ and $X_2 = -\frac{1}{2}i(z - \bar{z})$, it follows that

$$
\begin{aligned}
t_1 - it_2 &= -\frac{1}{2}i(\sigma_{11} - i\sigma_{12})\left(\frac{dz}{ds} - \frac{d\bar{z}}{ds}\right) + \frac{1}{2}(i\sigma_{22} - \sigma_{12})\left(\frac{dz}{ds} + \frac{d\bar{z}}{ds}\right) \\
&= \frac{1}{2}i(\sigma_{22} - \sigma_{11} + 2i\sigma_{12})\frac{dz}{ds} + \frac{1}{2}i(\sigma_{11} + \sigma_{22})\frac{d\bar{z}}{ds} , \tag{11.3.42}
\end{aligned}
$$

which, given the expression (11.3.9) for the scalar components of stress in terms of the complex stress functions, leads to the expression

$$
t_1 - it_2 = i(\bar{z}\phi'' + \psi')\frac{dz}{ds} + i(\phi' + \overline{\phi'})\frac{d\bar{z}}{ds} . \tag{11.3.43}
$$

Finally, by observing that

$$
\frac{d}{ds}(\bar{z}\phi' + \overline{\phi} + \psi) = (\bar{z}\phi'' + \psi')\frac{dz}{ds} + (\phi' + \overline{\phi'})\frac{d\bar{z}}{ds} , \tag{11.3.44}
$$

one obtains a relation for the scalar components of the traction in terms of the complex stress functions:

$$t_1 - it_2 = i\frac{d}{ds}\left(\overline{z}\phi' + \overline{\phi} + \psi\right),$$

(11.3.45)

where, again, it is important to remember that the independent variable of $\overline{\phi}$ is $\overline{z}$.

11.3.5 Boundary Value Problem

Consider a region S of the z-plane, with boundary ∂S. Assume that at every interior point z_0 of S, the body force is either zero or every neighborhood of z_0 contains points at which the body force field is zero. Let ∂S^u and ∂S^t be portions of the boundary on which the in-plane displacement and traction, respectively, are prescribed. Assume that there are no mixed boundary conditions so that $\partial S^u \cup \partial S^t = \partial S$. Then, in terms of the complex stress functions $\phi(z)$ and $\psi(z)$, the plane strain/stress boundary value problem reduces to the problem of finding functions $\phi(z)$ and $\psi(z)$, $\forall z \in S$, that are analytic everywhere in S that the body force field is zero and satisfy the displacement boundary condition

$$-\overline{z}\phi' + \kappa\overline{\phi} - \psi = 2\mu(\tilde{u}_1 - i\tilde{u}_2) \quad \forall z \in \partial S^u,$$

(11.3.46)

where $\tilde{u}_\alpha$ are the scalar components of a given in-plane displacement distribution, and the traction boundary condition

$$i\frac{d}{ds}\left(\overline{z}\phi' + \overline{\phi} + \psi\right) = \tilde{t}_1 - i\tilde{t}_2 \quad \forall z \in \partial S^t,$$

(11.3.47)

where $\tilde{t}_\alpha$ are the scalar components of a given in-plane traction distribution and s is the arc length around ∂S whose positive direction is defined in the usual way (i.e., so that the interior of S is always on the left). Points in S where the body force field does not vanish will be singular points of one or both of the functions $\phi(z)$ and $\psi(z)$.

In the special case where $\tilde{t}_\alpha = 0$ on some portion ∂S_0^t of ∂S^t (i.e., $\partial S_0^t \subset \partial S^t$ is traction-free), it follows from (11.3.47) that

$$\overline{z}\phi' + \overline{\phi} + \psi = C_0 \quad \forall z \in \partial S_0^t,$$

(11.3.48)

where C_0 is a complex constant. If two distinct portions of the boundary are traction-free, then (11.3.48) applies to each with different values of the constant, which, in general, are not independent. Recall from (11.3.33) that the stress and displacement are unaltered by the change $\phi \rightarrow \phi + A$ and $\psi \rightarrow \psi + \kappa\overline{A}$, where A is an arbitrary complex constant. Thus, letting $A =$

$\overline{C_0}/(\kappa+1)$ shows that the complex constant in (11.3.48) can be arbitrarily set equal to zero for one traction-free portion of the boundary, without loss of generality.

11.3.6 Implications of Symmetry

Suppose it is known that u_2 and σ_{12} vanish on the X_1-axis. Then, it follows from $u_2|_{X_2=0} = 0$ and (11.3.23) that

$$\text{Im}\{\overline{z}\phi' - \kappa\overline{\phi} + \psi\}|_{X_2=0} = 0 \qquad (11.3.49)$$

and from $\sigma_{12}|_{X_2=0} = 0$ and (11.3.9) that

$$\text{Im}\{\overline{z}\phi'' + \psi'\}|_{X_2=0} = 0 . \qquad (11.3.50)$$

Since $\overline{z} = z = X_1$ on $X_2 = 0$, (11.3.49) and (11.3.50) can be rewritten as

$$\text{Im}\{X_1\phi'(X_1) - \kappa\overline{\phi}(X_1) + \psi(X_1)\} = 0 , \qquad (11.3.51)$$
$$\text{Im}\{X_1\phi''(X_1) + \psi'(X_1)\} = 0 . \qquad (11.3.52)$$

Integrating (11.3.52) gives

$$\text{Im}\{X_1\phi'(X_1) - \phi(X_1) + \psi(X_1)\} = C , \qquad (11.3.53)$$

where C is a real-valued constant. Now, subtracting $\text{Im}\{\phi(X_1)\}$ from both sides of (11.3.51) gives

$$\text{Im}\{X_1\phi'(X_1) - \phi(X_1) + \psi(X_1) - \kappa\overline{\phi}(X_1)\} = -\text{Im}\{\phi(X_1)\} , \quad (11.3.54)$$

so that, using (11.3.53) and $\text{Im}\{\kappa\overline{\phi}(X_1)\} = -\kappa\,\text{Im}\{\phi(X_1)\}$, it follows that

$$\text{Im}\{\phi(X_1)\} = -\frac{C}{\kappa+1} . \qquad (11.3.55)$$

Differentiating (11.3.55) shows that

$$\text{Im}\{\phi'(X_1)\} = 0 , \qquad (11.3.56)$$

so that substituting (11.3.55) and (11.3.56) into (11.3.53) gives

$$\text{Im}\{\psi(X_1)\} = \frac{\kappa}{\kappa+1}C . \qquad (11.3.57)$$

Recall from (11.3.33) that the stress and displacement are unaltered by the change $\phi \rightarrow \phi + A$ and $\psi \rightarrow \psi + \kappa\overline{A}$, where A is an arbitrary complex constant. Thus, letting $A = iC/(\kappa+1)$ shows that (11.3.55) and (11.3.57) are equivalent to

$$\text{Im}\{\phi\}|_{X_2=0} = \text{Im}\{\psi\}|_{X_2=0} = 0 . \qquad (11.3.58)$$

To summarize:

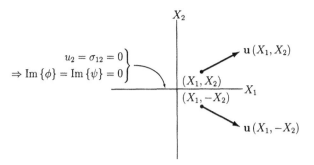

FIGURE 11.15. *Symmetry about the X_1-axis.*

- If one knows that u_2 and σ_{12} vanish on the real axis (the X_1-axis), then one may assume that $\phi(z)$ and $\psi(z)$ are real on the real axis, with no loss of generality.

The most common situation leading to u_2 and σ_{12} vanishing on the X_1-axis is symmetry of a problem about the X_1-axis (Figure 11.15). In this case, u_1 is an even function of X_2,

$$u_1(X_1, X_2) = u_1(X_1, -X_2), \qquad (11.3.59)$$

and u_2 is an odd function of X_2,

$$u_2(X_1, X_2) = -u_2(X_1, -X_2). \qquad (11.3.60)$$

It follows that σ_{11} and σ_{22} are even functions of X_2 and σ_{12} is an odd function of X_2. Therefore, if u_2 and σ_{12} are continuous across the X_1-axis, then $u_2|_{X_2=0} = 0$ and $\sigma_{12}|_{X_2=0} = 0$.

The following result can similarly be shown:

- If one knows that u_1 and σ_{12} vanish on the imaginary axis (the X_2-axis), then one may assume that $\phi(z)$ and $\psi(z)$ are imaginary on the imaginary axis, with no loss of generality.

This occurs in problems that are symmetric about the X_2-axis, as shown in Figure 11.16.

11.3.7 Cylindrical Coordinates

It can be shown (see Problem 11.12) that the scalar components of stress and displacement in a cylindrical coordinate vector basis are related to

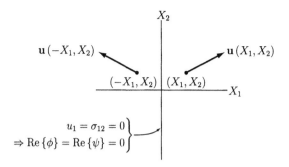

FIGURE 11.16. Symmetry about the X_2-axis.

those in the corresponding Cartesian coordinate vector basis by

$$\sigma_{rr} + \sigma_{\theta\theta} = \sigma_{11} + \sigma_{22} ,$$

$$\sigma_{\theta\theta} - \sigma_{rr} + 2i\sigma_{r\theta} = (\sigma_{22} - \sigma_{11} + 2i\sigma_{12})e^{i2\theta} , \tag{11.3.61}$$

$$u_r - iu_\theta = (u_1 - iu_2)e^{i\theta} .$$

It follows that the cylindrical coordinate form of the expression (11.3.9) for the components of stress in terms of the complex stress functions is

$$\boxed{\begin{aligned} \sigma_{rr} + \sigma_{\theta\theta} &= 4\operatorname{Re}\{\phi'\} , \\ \sigma_{\theta\theta} - \sigma_{rr} + 2i\sigma_{r\theta} &= 2(\bar{z}\phi'' + \psi')e^{i2\theta} \end{aligned}} \tag{11.3.62}$$

and that of the expression (11.3.23) for the components of displacement is

$$\boxed{2\mu(u_r - iu_\theta) = (-\bar{z}\phi' + \kappa\bar{\phi} - \psi)e^{i\theta} .} \tag{11.3.63}$$

It can also be shown that the scalar components of the traction vector acting at a point on a surface in cylindrical and Cartesian coordinates are related by

$$t_r - it_\theta = (t_1 - it_2)e^{i\theta} . \tag{11.3.64}$$

It follows that the cylindrical coordinate form of the expression (11.3.45) for the components of the traction in terms of the complex stress functions is

$$\boxed{t_r - it_\theta = ie^{i\theta}\frac{d}{ds}(\bar{z}\phi' + \bar{\phi} + \psi) .} \tag{11.3.65}$$

Finally, the cylindrical coordinate forms of the boundary conditions given by (11.3.46) and (11.3.47) are

$$\boxed{-\bar{z}\phi' + \kappa\bar{\phi} - \psi = 2\mu(\tilde{u}_r - i\tilde{u}_\theta)e^{-i\theta} \quad \forall z \in \partial S^u ,} \tag{11.3.66}$$

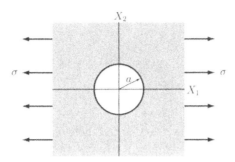

FIGURE 11.17. Circular hole in an infinite two-dimensional body under far-field uniaxial tension.

where $\tilde{u}_r$ and $\tilde{u}_\theta$ are the scalar components of a given in-plane displacement distribution, and

$$i\frac{d}{ds}(\bar{z}\phi' + \bar{\phi} + \psi) = (\tilde{t}_r - i\tilde{t}_\theta)e^{-i\theta} \quad \forall z \in \partial S^t \,, \tag{11.3.67}$$

where $\tilde{t}_r$ and $\tilde{t}_\theta$ are the scalar components of a given in-plane traction distribution.

11.3.8 Example: Infinite Two-Dimensional Body with a Circular Hole Under Far-Field Uniaxial Tension

Consider an infinite two-dimensional body (e.g., a thin plate in a state of plane stress) with a circular hole of radius a, where the body is subject to a far-field uniaxial tension of magnitude σ. Define a Cartesian coordinate system with the origin at the center of the hole and with the X_1-axis parallel to the axis of loading (Figure 11.17). The surface of the hole is traction-free,

$$\tilde{\mathbf{t}} = \mathbf{0} \quad \text{on } |z| = a \,, \tag{11.3.68}$$

and the stress field approaches a state of homogeneous, uniaxial tension of magnitude σ far away from the hole,

$$\sigma_{11} = \sigma \,, \quad \sigma_{22} = \sigma_{12} = 0 \quad \text{as } |z| \to \infty \,. \tag{11.3.69}$$

These are the boundary conditions for this problem.

To find a solution in terms of the complex stress functions $\phi(z)$ and $\psi(z)$, note first that the problem is symmetric about both the X_1 and X_2-axes.

Accordingly, one may without loss of generality assume that $\phi(z)$ and $\psi(z)$ are real on the real axis,

$$\operatorname{Im}\{\phi\}|_{X_2=0} = \operatorname{Im}\{\psi\}|_{X_2=0} = 0 , \tag{11.3.70}$$

and imaginary on the imaginary axis,

$$\operatorname{Re}\{\phi\}|_{X_1=0} = \operatorname{Re}\{\psi\}|_{X_1=0} = 0 . \tag{11.3.71}$$

According to (11.3.48), the traction-free hole boundary condition can be expressed in terms of the complex stress functions as

$$\bar{z}\phi' + \bar{\phi} + \psi = 0 \quad \text{on } |z| = a \tag{11.3.72}$$

and the homogeneous far-field stress condition is given by (11.3.38) to be

$$\lim_{|z|\to\infty} \phi = \frac{1}{4}\sigma z , \tag{11.3.73}$$

$$\lim_{|z|\to\infty} \psi = -\frac{1}{2}\sigma z , \tag{11.3.74}$$

where the constants A, B, and D in (11.3.38) must be zero in order for (11.3.73) and (11.3.74) to be consistent with the assumptions (11.3.70) and (11.3.71). The solution then is given by complex stress functions $\phi(z)$ and $\psi(z)$ that are analytic everywhere $|z| \geq a$ and satisfy (11.3.70)–(11.3.74).

Since $\phi(z)$ and $\psi(z)$ are analytic in the annular region $|z| \geq a$, they can be represented at any point in the body by the Laurent series

$$\phi = \sum_{n=-\infty}^{\infty} A_n z^n , \quad \psi = \sum_{n=-\infty}^{\infty} B_n z^n . \tag{11.3.75}$$

It follows immediately from (11.3.73) and (11.3.74) that

$$A_1 = \frac{1}{4}\sigma , \qquad A_0 = 0 , \quad A_n = 0 \ (n > 1) ,$$
$$B_1 = -\frac{1}{2}\sigma , \quad B_0 = 0 , \quad B_n = 0 \ (n > 1) . \tag{11.3.76}$$

On the real axis, $z = X_1$ is real and therefore, in order to satisfy (11.3.70), the remaining coefficients in (11.3.75) must also be real. On the imaginary axis, $z = iX_2$ is imaginary, even powers of z are real, and odd powers of z are imaginary. Therefore, since the remaining coefficients are real, the coefficients of the even power terms in z must be zero in order to satisfy (11.3.71). Thus, (11.3.75) reduce to

$$\phi = \frac{1}{4}\sigma z + \frac{A_{-1}}{z} + \frac{A_{-3}}{z^3} + \cdots = \frac{1}{4}\sigma z + \sum_{n=1}^{\infty} A_{1-2n} z^{1-2n} ,$$

$$\psi = -\frac{1}{2}\sigma z + \frac{B_{-1}}{z} + \frac{B_{-3}}{z^3} + \cdots = -\frac{1}{2}\sigma z + \sum_{n=1}^{\infty} B_{1-2n} z^{1-2n} , \tag{11.3.77}$$

where the coefficients are all real-valued.

The only remaining condition is the traction-free hole boundary condition (11.3.72). Note from (11.3.77) that

$$\phi'(z) = \frac{1}{4}\sigma + \sum_{n=1}^{\infty}(1-2n)A_{1-2n}z^{-2n} , \tag{11.3.78}$$

$$\overline{\phi}(\overline{z}) = \frac{1}{4}\sigma\overline{z} + \sum_{n=1}^{\infty}A_{1-2n}(\overline{z})^{1-2n} . \tag{11.3.79}$$

Recall the consequence of analytic continuation that if a function $f(z)$ is analytic in a region $\mathcal{S}$ and $f(z) = 0$ at all points on a curve $\mathcal{C}$ inside $\mathcal{S}$, then $f(z) = 0$ throughout $\mathcal{S}$. This can be used to solve for the coefficients in (11.3.77). The left-hand side of (11.3.72) is not an analytic function since it depends explicitly on $\overline{z}$, but consider the function

$$f(z) = \frac{a^2}{z}\phi'(z) + \xi(z) + \psi(z) , \tag{11.3.80}$$

where

$$\xi(z) = \frac{1}{4}\sigma\frac{a^2}{z} + \sum_{n=1}^{\infty}A_{1-2n}\left(\frac{z}{a^2}\right)^{2n-1} . \tag{11.3.81}$$

The function $f(z)$ is analytic everywhere except the origin $z = 0$. Note also that since $z\overline{z} = |z|^2$, $\overline{z} = a^2/z$ on $|z| = a$. Therefore, it follows that $\xi(z) = \overline{\phi}(\overline{z})$ on $|z| = a$ and the boundary condition (11.3.72) is satisfied if and only if $f(z) = 0$ on $|z| = a$. However, as a consequence of analytic continuation, $f(z)$ is analytic and vanishes on the curve $|z| = a$ if and only if it vanishes everywhere. Thus, the boundary condition (11.3.72) is satisfied if and only if $f(z) = 0$ everywhere.

Using (11.3.77), (11.3.78), and (11.3.81), the function $f(z)$ (11.3.80) can be represented as a Laurent series in terms of the unknown coefficients in (11.3.77) as

$$f(z) = \left(B_{-1} + \frac{1}{2}\sigma a^2\right)z^{-1} + \left(\frac{1}{a^2}A_{-1} - \frac{1}{2}\sigma\right)z + \sum_{n=2}^{\infty}A_{1-2n}\left(\frac{z}{a^2}\right)^{2n-1}$$

$$+ \sum_{n=1}^{\infty}\left[a^2(1-2n)A_{1-2n} + B_{-2n-1}\right]z^{-2n-1} . \tag{11.3.82}$$

The function $f(z)$ vanishes everywhere if and only if each of the coefficients in its Laurent series is zero, from which one determines that

$$A_{-1} = \frac{1}{2}\sigma a^2 , \qquad A_{-3} = A_{-5} = \cdots = 0 ,$$

$$B_{-1} = -\frac{1}{2}\sigma a^2 , \qquad B_{-3} = \frac{1}{2}\sigma a^4 , \qquad B_{-5} = B_{-7} = \cdots = 0 . \tag{11.3.83}$$

Substituting these coefficients back into (11.3.77), the answer is given in terms of the complex stress function as

$$
\boxed{
\begin{aligned}
\phi &= \frac{\sigma}{4}\left(z + \frac{2a^2}{z}\right) , \\
\psi &= -\frac{\sigma}{2}\left(z + \frac{a^2}{z} - \frac{a^4}{z^3}\right) .
\end{aligned}
}
$$

(11.3.84)

By noting that

$$
\phi' = \frac{\sigma}{4}\left(1 - \frac{2a^2}{z^2}\right) = \frac{\sigma}{4}\left(1 - \frac{2a^2}{r^2}e^{-i2\theta}\right) ,
$$

(11.3.85)

$$
\mathrm{Re}\{\phi'\} = \frac{\sigma}{4}\left(1 - \frac{2a^2}{r^2}\cos 2\theta\right) ,
$$

(11.3.86)

$$
\phi'' = \frac{\sigma a^2}{z^3} = \frac{\sigma a^2}{r^3}e^{-i3\theta} ,
$$

(11.3.87)

$$
\psi' = -\frac{\sigma}{2}\left(1 - \frac{a^2}{z^2} + \frac{3a^4}{z^4}\right) = -\frac{\sigma}{2}\left(1 - \frac{a^2}{r^2}e^{-i2\theta} + \frac{3a^4}{r^4}e^{-i4\theta}\right)
$$

(11.3.88)

and also that

$$
\begin{aligned}
\bar{z}\phi'' + \psi' &= re^{-i\theta}\left(\frac{\sigma a^2}{r^3}e^{-i3\theta}\right) - \frac{\sigma}{2}\left(1 - \frac{a^2}{r^2}e^{-i2\theta} + \frac{3a^4}{r^4}e^{-i4\theta}\right) \\
&= -\frac{\sigma}{2}\left[1 - \frac{a^2}{r^2}e^{-i2\theta} + \left(\frac{3a^4}{r^4} - \frac{2a^2}{r^2}\right)e^{-i4\theta}\right] ,
\end{aligned}
$$

(11.3.89)

it follows from (11.3.9) that the stress field is given in terms of the cylindrical coordinates r and θ by

$$
\boxed{
\begin{aligned}
\sigma_{11} + \sigma_{22} &= \sigma\left(1 - \frac{2a^2}{r^2}\cos 2\theta\right) , \\
\sigma_{22} - \sigma_{11} + 2i\sigma_{12} &= -\sigma\left[1 - \frac{a^2}{r^2}e^{-i2\theta} + \left(\frac{3a^4}{r^4} - \frac{2a^2}{r^2}\right)e^{-i4\theta}\right] .
\end{aligned}
}
$$

(11.3.90)

Far from the hole ($r \to \infty$), the stress field (11.3.90) reduces to

$$
\left.
\begin{aligned}
\sigma_{11} + \sigma_{22} &= \sigma \\
\sigma_{22} - \sigma_{11} + 2i\sigma_{12} &= -\sigma
\end{aligned}
\right\} \quad \text{as } r \to \infty ,
$$

(11.3.91)

from which it follows that

$$
\sigma_{11} = \sigma , \quad \sigma_{22} = \sigma_{12} = 0 \quad \text{as } r \to \infty ,
$$

(11.3.92)

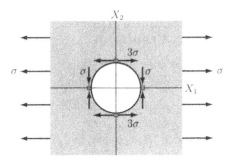

FIGURE 11.18.   Stress extrema at the boundary of a circular hole in an infinite two-dimensional body under far-field uniaxial tension.

as required by the boundary condition (11.3.69). Far from the hole, the greatest principal stress absolute value is $|\sigma||$. At the hole boundary $(r = a)$, the stress field (11.3.90) reduces to

$$\left.\begin{aligned}\sigma_{11} + \sigma_{22} &= \sigma(1 - 2\cos 2\theta)\\ \sigma_{22} - \sigma_{11} + 2i\sigma_{12} &= -\sigma\left(1 - e^{-i2\theta} + e^{-i4\theta}\right)\end{aligned}\right\}\quad \text{at } r = a\,. \qquad (11.3.93)$$

Thus, using (11.3.61), the cylindrical coordinate scalar components of the stress field at the hole boundary are given by

$$\left.\begin{aligned}\sigma_{rr} + \sigma_{\theta\theta} &= \sigma(1 - 2\cos 2\theta)\\ \sigma_{\theta\theta} - \sigma_{rr} + 2i\sigma_{r\theta} &= \sigma(1 - 2\cos 2\theta)\end{aligned}\right\}\quad \text{at } r = a\,, \qquad (11.3.94)$$

from which it follows that

$$\sigma_{\theta\theta} = \sigma(1 - 2\cos 2\theta)\,, \quad \sigma_{rr} = \sigma_{r\theta} = 0 \quad \text{at } r = a\,, \qquad (11.3.95)$$

which is consistent with the traction-free hole boundary condition (11.3.68). At the hole boundary, the greatest principal stress absolute value is $|\sigma_{\theta\theta}|$ and the stress component $\sigma_{\theta\theta}$ attains its maximum value $\sigma_{\theta\theta} = 3\sigma$ at $\theta = \pm\pi/2$, attains its minimum value $\sigma_{\theta\theta} = -\sigma$ at $\theta = 0, \pi$, and vanishes at $\theta = \pm\pi/6, \pm 5\pi/6$ (Figure 11.18). Therefore, the stress concentration factor for the hole in this problem is

$$\text{s.c.f.} = 3\,. \qquad (11.3.96)$$

Compare this with the stress concentration factor of 2 for a hole in an infinite two-dimensional body under far-field plane hydrostatic pressure, as found in Section 7.2.2.

11.3.9   Conformal Mapping Applied to Plane Strain/Stress Problems

Let the z-plane be the physical plane in which a problem is stated and let the ζ-plane, defined by the one-to-one mapping

$$\zeta = \Omega(z) , \quad z = \omega(\zeta) , \tag{11.3.97}$$

be a virtual plane into which the problem is to be mapped. Alternatively, with $z = X_1 + iX_2$ and $\zeta = \xi_1 + i\xi_2$, one can view the mapping as a relation between a Cartesian (X_1, X_2) coordinate system and a curvilinear (ξ_1, ξ_2) coordinate system. Recall that the mapping is conformal at points where $\Omega(z)$ and $\omega(\zeta)$ are analytic and $\Omega'(z)$ and $\omega'(\zeta)$ are nonzero, in which case the curvilinear (ξ_1, ξ_2) coordinate system is orthogonal. Such points are called ordinary points of the mapping. Then, the complex stress functions can either given as functions of z,

$$\phi = \phi(z) , \quad \psi = \psi(z) , \tag{11.3.98}$$

or as functions of ζ,

$$\phi = \phi(\zeta) , \quad \psi = \psi(\zeta) . \tag{11.3.99}$$

Since the same symbols are used above for functions of z and for functions of ζ, the appropriate independent variable will be explicitly indicated below where needed to avoid confusion.

Displacements

Recall the expression (11.3.23) for the scalar components of the displacement field in terms of the complex stress functions,

$$2\mu(u_1 - iu_2) = -\bar{z}\phi'(z) + \kappa\overline{\phi}(\bar{z}) - \psi(z) , \tag{11.3.100}$$

where the scalar components u_γ are with respect to the vector basis $\{\hat{\mathbf{e}}_\gamma\}$ defined by the Cartesian (X_1, X_2) coordinate system. By noting that $\bar{z} = \overline{\omega}(\bar{\zeta})$ and using the transformation of derivatives (11.1.88), it follows that the scalar components of the displacement field are given as a function of the independent variable ζ by

$$\boxed{2\mu(u_1 - iu_2) = -\frac{\overline{\omega}(\bar{\zeta})}{\omega'(\zeta)}\phi'(\zeta) + \kappa\overline{\phi}(\bar{\zeta}) - \psi(\zeta) ,} \tag{11.3.101}$$

where the scalar components u_γ are still with respect to the Cartesian vector basis $\{\hat{\mathbf{e}}_\gamma\}$.

It will sometimes be convenient to know the scalar components u_γ^* of the displacement field with respect to the vector basis $\{\hat{\mathbf{e}}_\gamma^*\}$ defined by the orthogonal curvilinear (ξ_1, ξ_2) coordinate system. Recall that at a point $\zeta = \Omega(z)$, the two vector bases are related by $\{\hat{\mathbf{e}}_\gamma^*\} = [L]\{\hat{\mathbf{e}}_\gamma\}$, where the matrix of direction cosines $[L]$ is given by (11.1.85) and represents a counterclockwise rotation by an angle α (11.1.83), where

$$e^{i\alpha} = \frac{\omega'(\zeta)}{|\omega'(\zeta)|} \ . \tag{11.3.102}$$

The angle α is a function of position. It follows from the vector transformation equation that

$$u_1^* - iu_2^* = (u_1 - iu_2)e^{i\alpha} = (u_1 - iu_2)\frac{\omega'(\zeta)}{|\omega'(\zeta)|} \ , \tag{11.3.103}$$

so that, from (11.3.101),

$$\boxed{2\mu(u_1^* - iu_2^*) = \left[-\frac{\overline{\omega}(\overline{\zeta})}{\omega'(\zeta)}\phi'(\zeta) + \kappa\overline{\phi}(\overline{\zeta}) - \psi(\zeta)\right]\frac{\omega'(\zeta)}{|\omega'(\zeta)|} \ .} \tag{11.3.104}$$

Stresses

Recall the expressions (11.3.9) for the scalar components of the stress field in terms of the complex stress functions,

$$\begin{aligned}
\sigma_{11} + \sigma_{22} &= 4\,\mathrm{Re}\{\phi'(z)\} \ , \\
\sigma_{22} - \sigma_{11} + 2i\sigma_{12} &= 2[\overline{z}\phi''(z) + \psi'(z)] \ ,
\end{aligned} \tag{11.3.105}$$

where the scalar components $\sigma_{\beta\gamma}$ are with respect to the vector basis $\{\hat{\mathbf{e}}_\gamma\}$ defined by the Cartesian (X_1, X_2) coordinate system. By noting that $\overline{z} = \overline{\omega}(\overline{\zeta})$ and using the transformation of derivatives (11.1.88), it follows that the scalar components of the stress field are given as a function of the independent variable ζ by

$$\boxed{\begin{aligned}
\sigma_{11} + \sigma_{22} &= 4\,\mathrm{Re}\left\{\frac{\phi'(\zeta)}{\omega'(\zeta)}\right\} \ , \\
\sigma_{22} - \sigma_{11} + 2i\sigma_{12} &= \frac{2}{\omega'(\zeta)}\left\{\overline{\omega}(\overline{\zeta})\left[\frac{\phi'(\zeta)}{\omega'(\zeta)}\right]' + \psi'(\zeta)\right\} \ ,
\end{aligned}} \tag{11.3.106}$$

where the scalar components $\sigma_{\beta\gamma}$ are still with respect to the Cartesian vector basis $\{\hat{\mathbf{e}}_\gamma\}$.

To obtain the scalar components $\sigma_{\beta\gamma}^*$ of the stress field with respect to the vector basis $\{\hat{\mathbf{e}}_\gamma^*\}$ defined by the orthogonal curvilinear (ξ_1, ξ_2) coordinate system, note first from the tensor transformation equation and the matrix of direction cosines $[L]$ given by (11.1.85) that

$$\begin{aligned}
\sigma_{11}^* + \sigma_{22}^* &= \sigma_{11} + \sigma_{22} \ , \\
\sigma_{22}^* - \sigma_{11}^* + 2i\sigma_{12}^* &= (\sigma_{22} - \sigma_{11} + 2i\sigma_{12})e^{i2\alpha} \ .
\end{aligned} \tag{11.3.107}$$

Since from (11.1.83) one has that

$$e^{i2\alpha} = \left[\frac{\omega'(\zeta)}{|\omega'(\zeta)|}\right]^2 , \qquad (11.3.108)$$

it follows from (11.3.106) that

$$\sigma_{11}^* + \sigma_{22}^* = 4\,\mathrm{Re}\left\{\frac{\phi'(\zeta)}{\omega'(\zeta)}\right\} ,$$

$$\sigma_{22}^* - \sigma_{11}^* + 2i\sigma_{12}^* = 2\left\{\overline{\omega}(\overline{\zeta})\left[\frac{\phi'(\zeta)}{\omega'(\zeta)}\right]' + \psi'(\zeta)\right\}\frac{\omega'(\zeta)}{|\omega'(\zeta)|^2} . \qquad (11.3.109)$$

Traction boundary conditions

Recall that traction boundary conditions (11.3.47) in the z-plane are given by

$$i\frac{d}{ds}[\overline{z}\phi'(z) + \overline{\phi}(\overline{z}) + \psi(z)] = \tilde{t}_1(z) - i\tilde{t}_2(z) \quad \forall z \in \partial \mathcal{S}^t , \qquad (11.3.110)$$

where $\tilde{t}_\gamma$ are the scalar components of a given in-plane traction distribution and s is the arc length around the boundary $\partial \mathcal{S}$ in the z-plane. It follows that the traction boundary condition is given in terms of the independent variable ζ by

$$i\frac{d}{ds}\left[\frac{\overline{\omega}(\overline{\zeta})}{\omega'(\zeta)}\phi'(\zeta) + \overline{\phi}(\overline{\zeta}) + \psi(\zeta)\right] = \tilde{t}_1(\zeta) - i\tilde{t}_2(\zeta) \quad \forall \zeta \in \partial \mathcal{R}^t , \qquad (11.3.111)$$

where $\partial \mathcal{R}^t$ is the image under the mapping of the portion $\partial \mathcal{S}^t$ of the boundary and the arc length s is still measured along the curve $\partial \mathcal{S}^t$ in the z-plane. The scalar components $\tilde{t}_\gamma$ of the given in-plane traction distribution are with respect to the Cartesian vector basis $\{\hat{\mathbf{e}}_\gamma\}$.

The same transformation rule applies to the prescribed traction vector as applied to the displacement vector, so

$$t_1^* - it_2^* = (t_1 - it_2)\frac{\omega'(\zeta)}{|\omega'(\zeta)|} , \qquad (11.3.112)$$

where t_γ^* are the scalar components of the prescribed traction vector in the vector basis $\{\hat{\mathbf{e}}_\gamma^*\}$ defined by the orthogonal curvilinear (ξ_1, ξ_2) coordinate system. Note that, in general, the lengths of the boundary $\partial \mathcal{S}^t$ and its image $\partial \mathcal{R}^t$ in the ζ-plane are different, so that the arc length s^* measure along $\partial \mathcal{R}^t$ is not the same as the arc length s measured along $\partial \mathcal{S}^t$. It follows

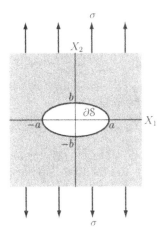

FIGURE 11.19. *Elliptical hole in an infinite two-dimensional body under far-field uniaxial tension.*

that

$$\boxed{i\frac{\omega'(\zeta)}{|\omega'(\zeta)|}\frac{d}{ds^*}\left[\frac{\overline{\omega}(\overline{\zeta})}{\overline{\omega'(\zeta)}}\phi'(\zeta) + \overline{\phi}(\overline{\zeta}) + \psi(\zeta)\right] = \tilde{t}_1^*(\zeta) - i\tilde{t}_2^*(\zeta) \quad \forall \zeta \in \partial\mathcal{R}^t ,}$$

$$(11.3.113)$$

On a traction-free portion of the boundary,

$$\frac{\overline{\omega}(\overline{\zeta})}{\overline{\omega'(\zeta)}}\phi'(\zeta) + \overline{\phi}(\overline{\zeta}) + \psi(\zeta) = C , \qquad (11.3.114)$$

where C is a constant that can be set equal to zero on one such curve without loss of generality.

11.3.10 Example: Infinite Two-Dimensional Body with an Elliptical Hole Under Far-Field Uniaxial Tension

Consider an infinite two-dimensional body (e.g., a thin plate in a state of plane stress) with an elliptical hole with semiaxes a and b, where the body is subject to a far-field uniaxial tension of magnitude σ that is parallel to one of the semiaxes of the elliptical hole. Define a Cartesian coordinate system with the origin at the centroid of the hole and aligned with the hole's semiaxes (Figure 11.19). The surface of the hole is traction-free,

$$\tilde{t} = 0 \quad \forall z \in \partial S , \qquad (11.3.115)$$

where the ellipse

$$\partial S = \left\{ z \in \mathbb{C} \mid \frac{X_1^2}{a^2} + \frac{X_2^2}{b^2} = 1 \right\} \tag{11.3.116}$$

is the boundary of the hole. The stress approaches a state of homogeneous, uniaxial tension far from the hole,

$$\sigma_{22} = \sigma, \quad \sigma_{11} = \sigma_{12} = 0 \quad \text{as } |z| \to \infty. \tag{11.3.117}$$

These are the boundary conditions for this problem.

To simplify the geometry and facilitate application of the Laurent series and consequences of analytic continuation, one can try mapping the problem into a virtual ζ-plane in which the elliptical boundary of the hole ∂S is mapped onto a unit circle $\partial \mathcal{R}$ and the region S outside of the elliptical hole is mapped onto the region $\mathcal{R}$ outside of the unit circle. To construct such a mapping, note first that the ellipse (11.3.116) can be characterized by the parametric equations

$$\begin{aligned} X_1 &= a \cos \beta = \frac{1}{2} a \left(e^{i\beta} + e^{-i\beta} \right), \\ X_2 &= b \sin \beta = -\frac{i}{2} b \left(e^{i\beta} - e^{-i\beta} \right), \end{aligned} \tag{11.3.118}$$

or, in terms of the complex variable $z = X_1 + iX_2$, it can be characterized by

$$z = \frac{a+b}{2} e^{i\beta} + \frac{a-b}{2} e^{-i\beta}, \tag{11.3.119}$$

where β is a real-valued parameter. Each value of β corresponds to a point on the ellipse whose position is given by (11.3.119). Note that β is *not* equal to the argument of its corresponding point on the ellipse. What is being sought is a mapping $z = \omega(\zeta)$ that maps points on the unit circle $\zeta = e^{i\beta}$ in the ζ-plane into the ellipse (11.3.119). The obvious choice then is

$$\boxed{z = \omega(\zeta) = \frac{a+b}{2} \zeta + \frac{a-b}{2} \frac{1}{\zeta} \quad \forall \zeta \in \mathcal{R},} \tag{11.3.120}$$

where

$$\mathcal{R} = \{ \zeta \in \mathbb{C} \mid |\zeta| \geq 1 \} \tag{11.3.121}$$

is the set of all points on and outside of the unit circle. It can be shown that points outside of the unit circle in the ζ-plane are mapped by (11.3.120) onto points outside of the ellipse in the z-plane, so that $\omega: \mathcal{R} \to S$, where

$$S = \left\{ z \in \mathbb{C} \mid \frac{X_1^2}{a^2} + \frac{X_2^2}{b^2} \geq 1 \right\} \tag{11.3.122}$$

is the set of all points on and outside of the ellipse. Clearly, $w(\zeta)$ is analytic and $w'(\zeta)$ is nonzero for all $\zeta \in \mathcal{R}$, so the mapping is conformal.

Since the hole is traction-free, it follows from (11.3.114) that the conditions imposed on the complex stress functions at the hole in the ζ-plane are

$$\frac{\overline{w}(\bar{\zeta})}{w'(\zeta)}\phi'(\zeta) + \overline{\phi}(\bar{\zeta}) + \psi(\zeta) = 0 \quad \text{on } |\zeta| = 1 . \tag{11.3.123}$$

For the far-field conditions, note first from (11.3.117) and (11.3.38) that in the z-plane,

$$\lim_{|z| \to \infty} \phi(z) = \frac{1}{4}\sigma z , \quad \lim_{|z| \to \infty} \psi(z) = \frac{1}{2}\sigma z . \tag{11.3.124}$$

Therefore, since $z = (a+b)\zeta/2$ as $|z| \to \infty$, the far-field conditions imposed on the complex stress functions in the ζ-plane are

$$\lim_{|\zeta| \to \infty} \phi(\zeta) = \frac{a+b}{8}\sigma\zeta , \quad \lim_{|\zeta| \to \infty} \psi(\zeta) = \frac{a+b}{4}\sigma\zeta . \tag{11.3.125}$$

Complex stress functions $\phi(\zeta)$ and $\psi(\zeta)$ that are analytic in $\mathcal{R}$ and satisfy the boundary conditions (11.3.123) and (11.3.125) provide the solution to this problem. One could, in principle, determine $\phi(\zeta)$ and $\psi(\zeta)$ using the Laurent series approach given for the circular hole problem in Section 11.3.8. However, a different approach is taken below.

Define two new functions $\phi_0(\zeta)$ and $\psi_0(\zeta)$ such that

$$\phi(\zeta) = \frac{a+b}{8}\sigma\left[\zeta + \phi_0(\zeta)\right] ,$$
$$\psi(\zeta) = \frac{a+b}{8}\sigma\left[2\zeta + \psi_0(\zeta)\right] . \tag{11.3.126}$$

The far-field conditions (11.3.125) then become

$$\lim_{|\zeta| \to \infty} \phi_0(\zeta) = 0 , \quad \lim_{|\zeta| \to \infty} \psi_0(\zeta) = 0 . \tag{11.3.127}$$

For the traction-free hole boundary condition (11.3.123), note first that

$$w'(\zeta) = \frac{a+b}{2} - \frac{a-b}{2}\frac{1}{\zeta^2} . \tag{11.3.128}$$

Furthermore, since $\bar{\zeta} = 1/\zeta$ when $|\zeta| = 1$, one has that

$$\overline{w}(\bar{\zeta}) = \overline{w}\left(\frac{1}{\zeta}\right) = \frac{a+b}{2}\frac{1}{\zeta} + \frac{a-b}{2}\zeta \quad \text{on } |\zeta| = 1 . \tag{11.3.129}$$

It follows from (11.3.126), (11.3.128), and (11.3.129) that the traction-free hole boundary condition (11.3.123) can be expressed as

$$\frac{(a+b)\zeta + (a-b)\zeta^3}{(a+b)\zeta^2 - (a-b)}[1 + \phi_0'(\zeta)] + \frac{1}{\zeta} + \overline{\phi}_0\left(\frac{1}{\zeta}\right)$$

$$+ 2\zeta + \psi_0(\zeta) = 0 \quad \text{on } |\zeta| = 1. \quad (11.3.130)$$

The problem is now reduced to finding functions $\phi_0(\zeta)$ and $\psi_0(\zeta)$ that are analytic in $|\zeta| \geq 1$ and satisfy (11.3.127) and (11.3.130).

Consider a function $f(z)$ that is analytic in some region S. The conjugate $\overline{f}(\overline{z})$ is nowhere analytic, since it depends explicitly on $\overline{z}$, but the function $\overline{f}(z)$ is also analytic in S [consider, for example, $f(z) = z + iz^2$, $\overline{f}(\overline{z}) = \overline{z} - i\overline{z}^2$, and $\overline{f}(z) = z - iz^2$]. On the other hand, if a function $g(z)$ is analytic in $|z| \geq 1$, then $g(1/z)$ is analytic in $|z| \leq 1$ [consider, for example, $g(z) = 1/z$ and $g(1/z) = z$]. It follows from these two observations that $\phi_0(\zeta)$ is analytic in $|\zeta| \geq 1$ if and only if the function $\overline{\phi}_0(1/\zeta)$ that appears on the left-hand side of (11.3.130) is analytic in $|\zeta| \leq 1$.

Now, define two functions,

$$F_1(\zeta) \equiv \frac{(a+b)\zeta + (a-b)\zeta^3}{(a+b)\zeta^2 - (a-b)}[1 + \phi_0'(\zeta)] + \frac{1}{\zeta} + \psi_0(\zeta) ,$$

$$F_2(\zeta) \equiv -\overline{\phi}_0\left(\frac{1}{\zeta}\right) - 2\zeta , \quad (11.3.131)$$

so that $\phi_0(\zeta)$ and $\psi_0(\zeta)$ are analytic in $|\zeta| \geq 1$ if and only if $F_1(\zeta)$ is analytic in $|\zeta| \geq 1$ and $F_2(\zeta)$ is analytic in $|\zeta| \leq 1$ and, from (11.3.130), $F_1(\zeta) = F_2(\zeta)$ on $|\zeta| = 1$. Then, it follows by analytic continuation that the function

$$E(\zeta) \equiv \begin{cases} F_1(\zeta) & \text{in } |\zeta| \geq 1 \\ F_2(\zeta) & \text{in } |\zeta| \leq 1 \end{cases} \quad (11.3.132)$$

is an entire function (i.e., a function that is analytic everywhere in the unextended complex plane), which can be therefore be represented by a Taylor series,

$$E(\zeta) = \sum_{k=0}^{\infty} A_k z^k , \quad (11.3.133)$$

with infinite radius of convergence. However, from (11.3.132), (11.3.131), and (11.3.127), one has the requirement that

$$\lim_{|\zeta| \to \infty} E(\zeta) = \lim_{|\zeta| \to \infty} F_1(\zeta) = \frac{a-b}{a+b}\zeta , \quad (11.3.134)$$

so that by comparing (11.3.133) and (11.3.134), it follows that

$$E(\zeta) = \frac{a-b}{a+b}\zeta \qquad (11.3.135)$$

everywhere.

It follows from (11.3.131), (11.3.132), and (11.3.135) that

$$\overline{\phi}_0\left(\frac{1}{\zeta}\right) = -\frac{3a+b}{a+b}\zeta \qquad (11.3.136)$$

and therefore that

$$\phi_0(\zeta) = -\frac{3a+b}{a+b}\frac{1}{\zeta} . \qquad (11.3.137)$$

Therefore, from (11.3.126), the complex stress function $\phi(\zeta)$ for this problem is

$$\boxed{\phi(\zeta) = \frac{a+b}{8}\left(\zeta - \frac{3a+b}{a+b}\frac{1}{\zeta}\right)\sigma .} \qquad (11.3.138)$$

It subsequently follows from the definitions (11.3.131) and (11.3.132), with the results (11.3.135) and (11.3.137) (after some algebraic manipulation), that

$$\psi_0(\zeta) = -2\frac{(a+b)^2 + (3a^2+b^2)\zeta^2}{(a+b)\zeta\left[-a+b+(a+b)\zeta^2\right]} . \qquad (11.3.139)$$

Therefore, from (11.3.126), the complex stress function $\psi(\zeta)$ for this problem is

$$\boxed{\psi(\zeta) = \frac{-(a+b)^2 - 4a^2\zeta^2 + (a+b)^2\zeta^4}{4\zeta\left[b-a+(a+b)\zeta^2\right]}\sigma .} \qquad (11.3.140)$$

Finally, it follows from (11.3.106), (11.3.120), (11.3.138), and (11.3.140) that the scalar components of the stress field, with respect to the z-plane Cartesian vector basis $\{\hat{\mathbf{e}}_\gamma\}$, are given in terms of ζ by[3]

$$\boxed{\begin{aligned} \sigma_{11} + \sigma_{22} &= \text{Re}\left\{\frac{(a+b)\zeta^2 + 3a+b}{(a+b)\zeta^2 - a+b}\right\}\sigma , \\ \sigma_{22} - \sigma_{11} + 2i\sigma_{12} &= \frac{f(\zeta)}{g(\zeta)}\sigma , \end{aligned}} \qquad (11.3.141)$$

---

[3]This (rather inelegant) result and some of the others were derived using the symbolic math computer application *Mathematica*. To manually derive these results would not, in principle, be difficult but would require considerable patience and involve a high risk for careless errors.

where

$$f(\zeta) = 4a(b^2 - a^2)\zeta^3 + \big[(b-a)(a+b)^2$$
$$+ (7a^3 + 5a^2 b + 9ab^2 + 3b^3)\zeta^2 + (a+b)(a^2 + 3b^2)\zeta^4$$
$$+ (a+b)^3 \zeta^6 - 4a(a+b)^2 \zeta^3 \overline{\zeta}\big]\,\zeta \qquad (11.3.142)$$

and

$$g(\zeta) = \big[(a+b)\zeta^2 - a + b\big]^3 \overline{\zeta}\,. \qquad (11.3.143)$$

Each value of ζ ($|\zeta| \geq 1$) in (11.3.141) corresponds to a point z in the body given by (11.3.120).

Consider a vector basis $\{\hat{\mathbf{e}}_r^*, \hat{\mathbf{e}}_\theta^*\}$ that, in the ζ-plane, is aligned with the cylindrical coordinate system. In other words, $\{\hat{\mathbf{e}}_r^*, \hat{\mathbf{e}}_\theta^*\}$ is obtained by a counterclockwise rotation $\beta = \arg\{\zeta\}$ of the vector basis $\{\hat{\mathbf{e}}_\gamma^*\}$ defined by the orthogonal curvilinear (ξ_1, ξ_2) coordinate system. In the z-plane, this is a vector basis that is obtained by rotating the Cartesian vector basis $\{\hat{\mathbf{e}}_\gamma\}$ by an angle $\alpha + \beta$ counterclockwise, where α (11.1.83) is the rotation associated with the conformal mapping. At the boundary of the hole, $\hat{\mathbf{e}}_r^*$ is normal to the surface and $\hat{\mathbf{e}}_\theta^*$ is tangent to the surface. It follows from the tensor transformation rule that

$$\sigma_{rr}^* + \sigma_{\theta\theta}^* = \sigma_{11}^* + \sigma_{22}^*\,,$$
$$\sigma_{\theta\theta}^* - \sigma_{rr}^* + 2i\sigma_{r\theta}^* = (\sigma_{11}^* - \sigma_{22}^* + 2i\sigma_{12}^*)e^{i2\beta}\,, \qquad (11.3.144)$$

where σ_{11}^*, σ_{22}^*, and σ_{12}^* are given by (11.3.109). It can thus be shown that $\sigma_{rr}^* = \sigma_{r\theta}^* = 0$, as required for the traction-free hole, and

$$\boxed{\sigma_{\theta\theta}^* = \frac{(a+b)^2 \cos 2\beta - a^2 + 2ab + b^2}{(b^2 - a^2)\cos 2\beta + a^2 + b^2}\,\sigma\,.} \qquad (11.3.145)$$

The maximum principal stress absolute value at the hole boundary is $|\sigma_{\theta\theta}^*|$, the maximum value of $\sigma_{\theta\theta}^*$ occurs at $\beta = 0, \pi$, where $\zeta = \pm 1$, $z = \pm a$, and

$$\sigma_{\theta\theta}^*\big|_{z=\pm a} = \left(1 + \frac{2a}{b}\right)\sigma\,, \qquad (11.3.146)$$

and the minimum value of $\sigma_{\theta\theta}^*$ occurs at $\beta = \pm\pi/2$, where $\zeta = \pm i$, $z = \pm ib$, and

$$\sigma_{\theta\theta}^*\big|_{z=\pm ib} = -\sigma\,. \qquad (11.3.147)$$

These results are illustrated in Figure 11.20. The stress concentration factor for an elliptical hole in an infinite two-dimensional body under far-field uniaxial tension is therefore

$$\text{s.c.f.} = 1 + \frac{2a}{b}\,. \qquad (11.3.148)$$

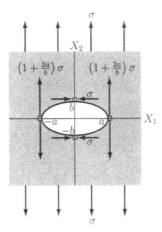

FIGURE 11.20. *Stress extrema at the boundary of an elliptical hole in an infinite two-dimensional body in far-field uniaxial tension.*

This reduces to the circular hole result when $a = b$. The stress concentration factor becomes unbounded in the limit $(b \to 0)$ as the elliptical hole becomes a crack of length $2a$.

11.3.11 Asymptotic Crack Tip Fields (Modes I and II)

Consider a semi-infinite, straight crack in a state of plane strain/stress, and define a Cartesian coordinate system so that the crack lies along the negative X_1-axis with its tip at the origin (Figure 11.21). In two-dimensional problems, a crack is a geometric feature with two *crack faces* that occupy the same curve in the reference configuration but are free to move relative to each other as the body is deformed. This means that the material description of the displacement field may be discontinuous across the curve occupied by the crack faces. The only restrictions are that the crack faces should not overlap one another (the crack can "open," but the material on one side cannot penetrate the material on the other side) and that the relative displacement of the crack faces must vanish as one approaches the crack tip. It will also be assumed that the crack faces are traction-free. The crack faces considered here are at $\theta = \pi$ and $\theta = -\pi$.

The traction-free crack faces boundary condition requires that

$$\sigma_{22} = \sigma_{12} = 0 \quad \text{on } \theta = \pm\pi . \qquad (11.3.149)$$

The *crack opening displacement* $\boldsymbol{\delta} = \delta_\alpha \hat{\mathbf{e}}_\alpha$ is the displacement discontinuity

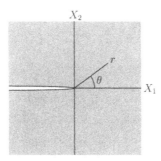

FIGURE 11.21. Semi-infinite, straight crack in plane strain/stress.

across the crack, given as a function of the distance r from the crack tip by

$$\boldsymbol{\delta}(r) \equiv \mathbf{u}(r, \pi) - \mathbf{u}(r, -\pi) , \quad \delta_\alpha(r) \equiv u_\alpha(r, \pi) - u_\alpha(r, -\pi) . \quad (11.3.150)$$

In terms of the crack opening displacement, the restriction that the crack faces not overlap is satisfied by requiring

$$\delta_2(r) \geq 0 \quad \forall r \geq 0 \qquad (11.3.151)$$

and the restriction that the relative displacement vanish at the crack tip is satisfied, and the rigid-body translation of the body is fixed, by requiring the displacement field to vanish at the crack tip,

$$\lim_{|z| \to 0} \mathbf{u} = \mathbf{0} . \qquad (11.3.152)$$

The objective here is to find the most general forms of the complex stress functions and of the displacement and stress fields that are consistent with the conditions (11.3.149), (11.3.151), and (11.3.152), with the remaining boundary conditions left unspecified.

Note first, from the expression (11.3.9) for the scalar components of stress in terms of the complex stress functions, that

$$\boxed{\sigma_{22} + i\sigma_{12} = 2 \operatorname{Re}\{\phi'\} + \bar{z}\phi'' + \psi' .} \qquad (11.3.153)$$

The boundary condition (11.3.149) requires that $\sigma_{22} + i\sigma_{12} = 0$ on $\theta = \pm\pi$ [this requirement can alternatively be derived from (11.3.47) with $ds = \pm dX_1$]. Given the expression (11.3.23) for the scalar components of the displacement, the displacement discontinuity across the crack faces implies that the complex stress functions each have a branch cut along the negative X_1-axis with a branch point singularity at the origin. This suggests that one look for solutions of the form

$$\phi = (A + iB)z^{\lambda+1} , \quad \psi = (C + iD)z^{\lambda+1} , \qquad (11.3.154)$$

where A, B, C, D, and λ are real. Substituting the assumed forms (11.3.154) of the complex stress functions into the expression (11.3.153) gives

$$\sigma_{22} + i\sigma_{12} = 2(\lambda + 1)\operatorname{Re}\{(A + iB)z^\lambda\} + \lambda(\lambda + 1)(A + iB)\bar{z}z^{\lambda-1}$$
$$+ (\lambda + 1)(C + iD)z^\lambda , \tag{11.3.155}$$

which, using the exponential form of z, can be written as

$$\sigma_{22} + i\sigma_{12} = 2(\lambda + 1)r^\lambda \operatorname{Re}\{(A + iB)e^{i\lambda\theta}\} + \lambda(\lambda + 1)(A + iB)r^\lambda e^{i(\lambda - 2)\theta}$$
$$+ (\lambda + 1)(C + iD)r^\lambda e^{i\lambda\theta} . \tag{11.3.156}$$

Finally, using Euler's formula to separate the real and imaginary parts,

$$\sigma_{22} + i\sigma_{12} = (\lambda + 1)r^\lambda \bigg\{ \Big[(2A + C)\cos\lambda\theta + A\lambda\cos(\lambda - 2)\theta$$
$$- (2B + D)\sin\lambda\theta - B\lambda\sin(\lambda - 2)\theta\Big]$$
$$+ i\Big[A\lambda\sin(\lambda - 2)\theta + B\lambda\cos(\lambda - 2)\theta$$
$$+ C\sin\lambda\theta + D\cos\lambda\theta\Big]\bigg\} . \tag{11.3.157}$$

It follows from the boundary condition $(\sigma_{22} + i\sigma_{12})|_{\theta=\pi} = 0$ that

$$[A(2 + \lambda) + C]\cos\lambda\pi - [B(2 + \lambda) + D]\sin\lambda\pi = 0 , \tag{11.3.158}$$
$$(A\lambda + C)\sin\lambda\pi + (B\lambda + D)\cos\lambda\pi = 0 , \tag{11.3.159}$$

and it follows from the boundary condition $(\sigma_{22} + i\sigma_{12})|_{\theta=-\pi} = 0$ that

$$[A(2 + \lambda) + C]\cos\lambda\pi + [B(2 + \lambda) + D]\sin\lambda\pi = 0 , \tag{11.3.160}$$
$$-(A\lambda + C)\sin\lambda\pi + (B\lambda + D)\cos\lambda\pi = 0 . \tag{11.3.161}$$

Comparing (11.3.158) with (11.3.160) and (11.3.159) with (11.3.161), it follows that the traction-free crack faces boundary condition requires

$$[A(2 + \lambda) + C]\cos\lambda\pi = 0 , \tag{11.3.162}$$
$$[B(2 + \lambda) + D]\sin\lambda\pi = 0 , \tag{11.3.163}$$
$$(A\lambda + C)\sin\lambda\pi = 0 , \tag{11.3.164}$$
$$(B\lambda + D)\cos\lambda\pi = 0 . \tag{11.3.165}$$

The trivial solution $A = B = C = D = 0$ just corresponds to a state of zero displacement everywhere. There are two families of nontrivial solutions for (11.3.162)–(11.3.165).

The first family of solutions comes from requiring that $\cos \lambda \pi = 0$, so that $\lambda = \lambda_n^{(1)}$, where

$$\lambda_n^{(1)} = n - \frac{1}{2} \tag{11.3.166}$$

and n is an arbitrary integer. Then, the coefficients of $\sin \lambda \pi$ must equal zero so that $A = A_n^{(1)}$, $B = B_n^{(1)}$, $C = C_n^{(1)}$, and $D = D_n^{(1)}$, where

$$C_n^{(1)} = -\lambda_n^{(1)} A_n^{(1)} = -\left(n - \frac{1}{2} \right) A_n^{(1)} ,$$

$$D_n^{(1)} = -(2 + \lambda_n^{(1)}) B_n^{(1)} = -\left(n + \frac{3}{2} \right) B_n^{(1)} , \tag{11.3.167}$$

and $A_n^{(1)}$ and $B_n^{(1)}$ are arbitrary constants. The second family of solutions comes from requiring that $\sin \lambda \pi = 0$, so that $\lambda = \lambda_n^{(2)}$, where

$$\lambda_n^{(2)} = n \tag{11.3.168}$$

and n is an arbitrary integer. Then, the coefficients of $\cos \lambda \pi$ must equal zero so that $A = A_n^{(2)}$, $B = B_n^{(2)}$, $C = C_n^{(2)}$, and $D = D_n^{(2)}$, where

$$C_n^{(2)} = -(2 + \lambda_n^{(2)}) A_n^{(2)} = -(n + 2) A_n^{(2)} ,$$

$$D_n^{(2)} = -\lambda_n^{(2)} B_n^{(2)} = -n B_n^{(2)} , \tag{11.3.169}$$

and $A_n^{(2)}$ and $B_n^{(2)}$ are arbitrary constants.

The complex stress functions (11.3.154) corresponding to each of these solutions will satisfy the traction-free crack faces boundary condition. However, it follows from the expression (11.3.23) for the displacements that the condition (11.3.152) that the displacement field vanish at the crack tip is satisfied by solutions of the form (11.3.154) if and only if $\lambda > -1$. Thus, the above solutions with $n < 0$ are eliminated. The remaining solutions can then be combined through superposition to form a power series representation of the general solution to the crack problem:

$$\phi = \sum_{n=0}^{\infty} \left[(A_n^{(1)} + i B_n^{(1)}) z^{n+\frac{1}{2}} + (A_n^{(2)} + i B_n^{(2)}) z^{n+1} \right] ,$$

$$\psi = \sum_{n=0}^{\infty} \left[(C_n^{(1)} + i D_n^{(1)}) z^{n+\frac{1}{2}} + (C_n^{(2)} + i D_n^{(2)}) z^{n+1} \right] , \tag{11.3.170}$$

where $C_n^{(1)}$, $D_n^{(1)}$, $C_n^{(2)}$, and $D_n^{(2)}$ are related to $A_n^{(1)}$, $B_n^{(1)}$, $A_n^{(2)}$, and $B_n^{(2)}$ by (11.3.167) and (11.3.169). The constant, real coefficients $A_n^{(1)}$, $B_n^{(1)}$, $A_n^{(2)}$, and $B_n^{(2)}$ are determined by the remaining boundary conditions of whatever specific problem one is considering.

In the power series representation (11.3.170) of the complex stress functions, there are terms in $z^{1/2}$, z, $z^{3/2}$, ... with their respective constant coefficients. It follows from the expressions (11.3.9) for the scalar components of stress and the expression (11.3.23) for the scalar components of the displacement that a term in z^{α} will yield a contribution to the stress that is proportional to $|z|^{\alpha-1}$ and a contribution to the displacement that is proportional to $|z|^{\alpha}$. Thus, in the limit $|z| \to 0$ as one approaches the crack tip, the lower-order terms in (11.3.170) tend to dominate the solution and the stress contributions from terms with $\alpha > 1$ vanish. The terms in z provide a constant contribution to the stress and the terms in $z^{1/2}$ provide a contribution that becomes unbounded like $|z|^{-1/2}$. The foundation of *linear elastic fracture mechanics* is that the contributions to the stress field from these terms characterize the crack's propensity for growth, or propagation (Kanninen and Popelar 1985).

Consider first the contribution to the stress field from the z terms in (11.3.170). Noting from (11.3.169) that $C_0^{(2)} = -2A_0^{(2)}$ and $D_0^{(2)} = 0$, the z terms in (11.3.170) are

$$\phi_0^{(2)} = (A_0^{(2)} + iB_0^{(2)})z , \quad \psi_0^{(2)} = -2A_0^{(2)}z , \tag{11.3.171}$$

and it follows from (11.3.9) that the corresponding contribution to the scalar components of the stress field is

$$\sigma_{11} = T , \quad \sigma_{22} = \sigma_{12} = 0 , \tag{11.3.172}$$

where $T \equiv 4A_0^{(2)}$ is known as the "*T*-stress," which is determined by the remaining boundary conditions.

Consider next the contribution to the stress field from the $z^{1/2}$ terms in (11.3.170). Noting from (11.3.167) that $C_0^{(1)} = \frac{1}{2}A_0^{(1)}$ and $D_0^{(1)} = -\frac{3}{2}B_0^{(1)}$, the $z^{1/2}$ terms in (11.3.170) are

$$\phi_0^{(1)} = (A_0^{(1)} + iB_0^{(1)})z^{\frac{1}{2}} , \quad \psi_0^{(1)} = \frac{1}{2}(A_0^{(1)} - 3iB_0^{(1)})z^{\frac{1}{2}} . \tag{11.3.173}$$

The convention in linear elastic fracture mechanics is to replace the constants $A_0^{(1)}$ and $B_0^{(1)}$ by the *stress intensity factors* K_{I} and K_{II}, defined in the current coordinate system such that

$$\boxed{K_{\mathrm{I}} + iK_{\mathrm{II}} \equiv \lim_{r \to 0} \sqrt{2\pi r} \, (\sigma_{22} + i\sigma_{12})|_{\theta=0} \, .} \tag{11.3.174}$$

By substituting (11.3.173) into (11.3.153) and recalling (11.3.172), one finds that

$$\sigma_{22} + i\sigma_{12} = \mathrm{Re}\Big\{ (A_0^{(1)} + iB_0^{(1)})z^{-\frac{1}{2}} \Big\} - \frac{1}{4}(A_0^{(1)} + iB_0^{(1)})\bar{z}z^{-\frac{3}{2}}$$
$$+ \frac{1}{4}(A_0^{(1)} - 3iB_0^{(1)})z^{-\frac{1}{2}} + \mathcal{O}(|z|^{\frac{1}{2}}) , \tag{11.3.175}$$

and since $z = \bar{z} = r$ when $\theta = 0$,

$$(\sigma_{22} + i\sigma_{12})|_{\theta=0} = (A_0^{(1)} - iB_0^{(1)})r^{-\frac{1}{2}} + \mathcal{O}(r^{\frac{1}{2}}) . \qquad (11.3.176)$$

Thus, it follows from (11.3.174) that $A_0^{(1)} = K_I/\sqrt{2\pi}$, $B_0^{(1)} = -K_{II}/\sqrt{2\pi}$, and

$$\boxed{\begin{aligned}\phi_0^{(1)} &= \frac{1}{\sqrt{2\pi}}(K_I - iK_{II})z^{\frac{1}{2}} , \\ \psi_0^{(1)} &= \frac{1}{2\sqrt{2\pi}}(K_I + 3iK_{II})z^{\frac{1}{2}} .\end{aligned}} \qquad (11.3.177)$$

It can then be shown (see Problem 11.14) that the contribution to the stress field from (11.3.177) is given in terms of Cartesian coordinate scalar components by

$$\left\{\begin{matrix}\sigma_{11} \\ \sigma_{22} \\ \sigma_{12}\end{matrix}\right\} = \frac{K_I}{4\sqrt{2\pi r}}\left\{\begin{matrix}3\cos\dfrac{\theta}{2} + \cos\dfrac{5\theta}{2} \\ 5\cos\dfrac{\theta}{2} - \cos\dfrac{5\theta}{2} \\ \sin\dfrac{5\theta}{2} - \sin\dfrac{\theta}{2}\end{matrix}\right\} + \frac{K_{II}}{4\sqrt{2\pi r}}\left\{\begin{matrix}-\sin\dfrac{5\theta}{2} - 7\sin\dfrac{\theta}{2} \\ \sin\dfrac{5\theta}{2} - \sin\dfrac{\theta}{2} \\ 3\cos\dfrac{\theta}{2} + \cos\dfrac{5\theta}{2}\end{matrix}\right\}$$

$$(11.3.178a)$$

and in terms of cylindrical coordinate scalar components by

$$\left\{\begin{matrix}\sigma_{rr} \\ \sigma_{\theta\theta} \\ \sigma_{r\theta}\end{matrix}\right\} = \frac{K_I}{4\sqrt{2\pi r}}\left\{\begin{matrix}5\cos\dfrac{\theta}{2} - \cos\dfrac{3\theta}{2} \\ 3\cos\dfrac{\theta}{2} + \cos\dfrac{3\theta}{2} \\ \sin\dfrac{3\theta}{2} + \sin\dfrac{\theta}{2}\end{matrix}\right\} + \frac{K_{II}}{4\sqrt{2\pi r}}\left\{\begin{matrix}3\sin\dfrac{3\theta}{2} - 5\sin\dfrac{\theta}{2} \\ -3\sin\dfrac{3\theta}{2} - 3\sin\dfrac{\theta}{2} \\ \cos\dfrac{\theta}{2} + 3\cos\dfrac{3\theta}{2}\end{matrix}\right\} .$$

$$(11.3.178b)$$

With the exception of the constant T-stress, all other contributions to the stress field vanish in the vicinity of the crack tip. Thus, in the limit $r \to 0$ as one approaches the crack tip, the stress field for *any* plane strain/stress crack, regardless of the remaining boundary conditions, approaches (11.3.178) asymptotically. For this reason, (11.3.178) is referred to as the *plane strain/stress asymptotic crack tip stress field*.[4] Subject to

---

[4] The expressions for the plane strain/stress asymptotic crack tip stress and displacement fields given in Kanninen and Popelar (1985) and many other fracture mechanics references appear to be different (and arguably less concise) than those presented here. However, they can be shown to be equivalent through the application of trigonometric identities.

certain *small-scale yielding conditions* (Kanninen and Popelar 1985), the stress intensity factors K_I and K_{II}, which are determined by the remaining boundary conditions, are assumed to characterize a crack's propensity to grow.

It can be shown (see Problem 11.17) that the contribution to the displacement field from (11.3.177) is given in terms of Cartesian coordinate scalar components by

$$
\begin{Bmatrix} u_1 \\ u_2 \end{Bmatrix} = \frac{K_I}{2\mu} \sqrt{\frac{r}{2\pi}} (\kappa - \cos\theta) \begin{Bmatrix} \cos\dfrac{\theta}{2} \\ \sin\dfrac{\theta}{2} \end{Bmatrix}
$$

$$
+ \frac{K_{II}}{2\mu} \sqrt{\frac{r}{2\pi}} \begin{Bmatrix} (2 + \kappa + \cos\theta)\sin\dfrac{\theta}{2} \\ (2 - \kappa - \cos\theta)\cos\dfrac{\theta}{2} \end{Bmatrix} \tag{11.3.179a}
$$

and in terms of cylindrical coordinate scalar components by

$$
\begin{Bmatrix} u_r \\ u_\theta \end{Bmatrix} = \frac{K_I}{2\mu} \sqrt{\frac{r}{2\pi}} (\kappa - \cos\theta) \begin{Bmatrix} \cos\dfrac{\theta}{2} \\ -\sin\dfrac{\theta}{2} \end{Bmatrix}
$$

$$
+ \frac{K_{II}}{2\mu} \sqrt{\frac{r}{2\pi}} \begin{Bmatrix} (2 - \kappa + 3\cos\theta)\sin\dfrac{\theta}{2} \\ (-2 - \kappa + 3\cos\theta)\cos\dfrac{\theta}{2} \end{Bmatrix} . \tag{11.3.179b}
$$

The contribution to the scalar components of the crack opening displacement (11.3.150) can then be shown to be

$$
\delta_1 - i\delta_2 \equiv (u_1 - iu_2)|_{\theta=\pi} - (u_1 - iu_2)|_{\theta=-\pi}
$$

$$
= \frac{\kappa + 1}{\mu} \sqrt{\frac{r}{2\pi}} (K_{II} - iK_I) . \tag{11.3.180}
$$

Recall that κ is a material parameter with different values in plane strain and plane stress that are given by (11.3.11). In the limit $r \to 0$ as one approaches the crack tip, the displacement field for *any* plane strain/stress crack approaches (11.3.179) asymptotically. For this reason, (11.3.178) is referred to as the *plane strain/stress asymptotic crack tip displacement field*. Note from (11.3.180) that the condition (11.3.151) that crack faces not overlap requires that $K_I \geq 0$.

It follows from (11.3.180) that if one knows the actual crack opening displacement δ as a function of the distance r from the crack tip for a given crack and given far-field boundary conditions, by experimentation or finite element analysis say, then the stress intensity factors can, in principle, be determined by evaluating

$$K_{\mathrm{I}} = \lim_{r \to 0} \frac{\mu}{\kappa + 1} \sqrt{\frac{2\pi}{r}} \, \delta_2 \,, \quad K_{\mathrm{II}} = \lim_{r \to 0} \frac{\mu}{\kappa + 1} \sqrt{\frac{2\pi}{r}} \, \delta_1 \,. \qquad (11.3.181)$$

In practice, there are convergence issues which may limit the accuracy of such an approach.

In the event that the stress intensity factor $K_{\mathrm{II}} = 0$, it is seen from (11.3.179) that the asymptotic crack tip displacement field is symmetric about the X_1-axis and from (11.3.180) that the crack opening displacement component $\delta_1 = 0$—the crack faces are opening without sliding. This is referred to as *mode I*, or *opening mode*, loading of the crack. Conversely, if $K_{\mathrm{I}} = 0$, it is seen from (11.3.179) that the asymptotic crack tip displacement field is antisymmetric about the X_1-axis and from (11.3.180) that the crack opening displacement component $\delta_2 = 0$—the crack faces are sliding without opening. This is referred to as *mode II*, or *sliding mode*, loading of the crack. Accordingly, K_{I} is called the *mode I stress intensity factor* and K_{II} is called the *mode II stress intensity factor*.

11.3.12 Example: Infinite Two-Dimensional Body with a Finite Crack Under Far-Field Uniaxial Tension

Consider an infinite two-dimensional body (e.g., a thin plate in a state of plane stress) with a straight crack of length $2a$, where the body is subject to a far-field uniaxial tension of magnitude σ that is normal to the plane of the crack. Define a Cartesian coordinate system with the crack tips at $X_1 = \pm a$ (Figure 11.22). The crack faces are traction-free,

$$\sigma_{22} = \sigma_{12} = 0 \quad \text{on } X_2 = 0 \,, \ -a < X_1 < a \,, \qquad (11.3.182)$$

and the stress approaches a state of homogeneous, uniaxial tension far from the crack,

$$\sigma_{22} = \sigma \,, \quad \sigma_{11} = \sigma_{12} = 0 \quad \text{as } |z| \to \infty \,. \qquad (11.3.183)$$

These are the boundary conditions for this problem.

This is a special case of the elliptical hole problem considered previously (page 493), with $b = 0$. It follows from (11.3.141)–(11.3.143) that

$$\sigma_{11} + \sigma_{22} = \mathrm{Re}\left\{ \frac{\zeta^2 + 3}{\zeta^2 - 1} \right\} \sigma \,,$$

$$\sigma_{22} - \sigma_{11} + 2i\sigma_{12} = \left[\frac{-4\zeta^3 + (-1 + 7\zeta^2 + \zeta^4 + \zeta^6 - 4\zeta^3\overline{\zeta})\overline{\zeta}}{(\zeta^2 - 1)^3 \overline{\zeta}} \right] \sigma \,,$$

$$\qquad (11.3.184)$$

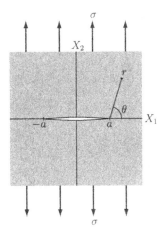

FIGURE 11.22. *Finite crack in an infinite two-dimensional body under far-field uniaxial tension.*

where $|\zeta| \geq 1$ and, from (11.3.120),

$$\zeta + \frac{1}{\zeta} = \frac{2}{a}z , \quad \zeta - \frac{1}{\zeta} = \frac{2}{a}\sqrt{z^2 - a^2} . \tag{11.3.185}$$

The expressions (11.3.184) can be rewritten as

$$\sigma_{11} + \sigma_{22} = \left[2\,\mathrm{Re}\left\{ \frac{\zeta + \frac{1}{\zeta}}{\zeta - \frac{1}{\zeta}} \right\} - 1 \right] \sigma ,$$

$$\sigma_{22} - \sigma_{11} + 2i\sigma_{12} = \left[\frac{4\left(\zeta + \frac{1}{\zeta} - \bar{\zeta} - \frac{1}{\bar{\zeta}}\right)}{\left(\zeta - \frac{1}{\zeta}\right)^3} + 1 \right] \sigma , \tag{11.3.186}$$

so that it follows from (11.3.185) and $z - \bar{z} = 2iX_2$ that

$$\boxed{\begin{aligned} \sigma_{11} + \sigma_{22} &= \left[2\,\mathrm{Re}\left\{ \frac{z}{\sqrt{z^2 - a^2}} \right\} - 1 \right] \sigma , \\ \sigma_{22} - \sigma_{11} + 2i\sigma_{12} &= \left[\frac{2ia^2 X_2}{(z^2 - a^2)^{\frac{3}{2}}} + 1 \right] \sigma , \end{aligned}} \tag{11.3.187}$$

At the crack faces ($X_2 = 0$, $|X_1| < a$), $z = X_1$ is real, $\sqrt{z^2 - a^2}$ is pure imaginary, and it follows from (11.3.187) that

$$\left.\begin{aligned} \sigma_{11} + \sigma_{22} &= -\sigma \\ \sigma_{22} - \sigma_{11} + 2i\sigma_{12} &= \sigma \end{aligned}\right\} \quad \text{on } X_2 = 0 , \ |X_1| < a , \tag{11.3.188}$$

so that

$$\sigma_{11} = -\sigma , \quad \sigma_{22} = \sigma_{12} = 0 \quad \text{on } X_2 = 0 , \ |X_1| < a . \qquad (11.3.189)$$

This is consistent with the traction-free crack faces boundary condition.

Stress intensity factors

In the plane ahead of the right crack tip $(X_2 = 0, \ X_1 > a)$, $z = X_1$ and $\sqrt{z^2 - a^2}$ are real and it follows from (11.3.187) that

$$\left. \begin{array}{l} \sigma_{11} + \sigma_{22} = \left[\dfrac{2X_1}{\sqrt{X_1^2 - a^2}} - 1 \right] \sigma \\[4mm] \sigma_{22} - \sigma_{11} + 2i\sigma_{12} = \sigma \end{array} \right\} \quad \text{on } X_2 = 0 , \ X_1 > a ,$$

$$(11.3.190)$$

so that

$$\left. \begin{array}{l} \sigma_{11} = \left[\dfrac{X_1}{\sqrt{X_1^2 - a^2}} - 1 \right] \sigma \\[4mm] \sigma_{22} = \dfrac{X_1 \sigma}{\sqrt{X_1^2 - a^2}} \\[4mm] \sigma_{12} = 0 \end{array} \right\} \quad \text{on } X_2 = 0 , \ X_1 > a . \qquad (11.3.191)$$

With the cylindrical coordinate system defined as shown in Figure 11.22, so that r is the distance from the right crack tip and $\theta = 0$ is the line ahead of the crack tip, recall that the stress intensity factors for the right crack tip are defined by

$$K_{\text{I}} \equiv \lim_{r \to 0} \sqrt{2\pi r} \ \sigma_{22}|_{\theta=0} , \quad K_{\text{II}} \equiv \lim_{r \to 0} \sqrt{2\pi r} \ \sigma_{12}|_{\theta=0} . \qquad (11.3.192)$$

Also, $X_1 = r + a$ when $\theta = 0$ and it follows from (11.3.191) that

$$\left. \begin{array}{l} \sigma_{11}|_{\theta=0} = \left[\dfrac{r+a}{\sqrt{r(r+2a)}} - 1 \right] \sigma , \\[4mm] \sigma_{22}|_{\theta=0} = \left[\dfrac{r+a}{\sqrt{r(r+2a)}} \right] \sigma , \\[4mm] \sigma_{12}|_{\theta=0} = 0 . \end{array} \right. \qquad (11.3.193)$$

Therefore, $K_{\text{II}} = 0$ and

$$K_{\text{I}} = \lim_{r \to 0} \sqrt{2\pi} \left[\frac{r+a}{\sqrt{r+2a}} \right] \sigma = \sqrt{2\pi} \left[\frac{a}{\sqrt{2a}} \right] \sigma = \sqrt{\pi a} \, \sigma . \qquad (11.3.194)$$

The stress intensity factors for this crack and this loading are

$$\boxed{K_\text{I} = \sqrt{\pi a}\,\sigma\,, \quad K_\text{II} = 0\,.}$$ (11.3.195)

The crack is in pure mode I loading, as could have been anticipated from the problem's symmetry, and (11.3.195) characterizes the propensity for crack growth. This propensity increases linearly with the far-field uniaxial tensile load and like the square root of the crack size.

The stress intensity factors for cracks with other geometries and loadings will be different. See Tada et al. (1985) for a large collection of such stress intensity factors.

It is interesting to consider how rapidly the actual stress field for this problem approaches the asymptotic stress field (11.3.178) as one approaches the crack tip. In the plane ahead of the crack tip ($\theta = 0$), with the stress intensity factors given by (11.3.195), the asymptotic stress field (11.3.178) is

$$\sigma_{11}^a\big|_{\theta=0} = \frac{K_\text{I}}{\sqrt{2\pi r}} = \sqrt{\frac{a}{2r}}\,\sigma\,,$$

$$\sigma_{22}^a\big|_{\theta=0} = \frac{K_\text{I}}{\sqrt{2\pi r}} = \sqrt{\frac{a}{2r}}\,\sigma\,,$$ (11.3.196)

$$\sigma_{22}^a\big|_{\theta=0} = 0\,.$$

Comparing these asymptotic values with the actual stress field (11.3.193), the asymptotic value of σ_{22} is within 5% of the actual value at $r/a = 0.06$ and within 1% at $r/a = 0.01$. In contrast, the asymptotic value of σ_{11} is not within 5% of the actual value until $r/a = 0.001$ and is not within 1% until $r/a = 0.00005$. This slower convergence to the asymptotic value for σ_{11} is due to the presence of a nonzero T-stress as discussed on page 503. The value of this T-stress can be determined by evaluating

$$T = \lim_{r \to 0} \sigma_{11}\big|_{\theta=0} - \sigma_{11}^a\big|_{\theta=0}\,,$$ (11.3.197)

from which is follows that $T = -\sigma$. Alternatively (and more easily), the T-stress can be determined by comparing (11.3.189) with the asymptotic value of σ_{11} when $\theta = \pi$.

Problems

11.1 Sketch the range of the function $w = z^{\frac{1}{n}}$, $\forall z \in \mathbb{C}$, if the argument of $z = re^{i\theta}$ is restricted to the interval $\theta_0 \le \theta < \theta_0 + 2\pi$. What is the range if the argument of $z = re^{i\theta}$ is restricted to the interval $\theta_0 + 2\pi \le \theta < \theta_0 + 4\pi$?

11.2 Sketch the range of the function $w = z^{-\frac{1}{n}}$, $\forall z \in \mathcal{S}$, if the argument of $z = re^{i\theta}$ is restricted to the interval $\theta_0 \le \theta < \theta_0 + 2\pi$. The domain $\mathcal{S}$ is the finite z-plane with the origin deleted.

11.3 Sketch the range of the function $w = \ln z$, $\forall z \in \mathcal{S}$, if the argument of $z = re^{i\theta}$ is restricted to the interval $\theta_0 \le \theta < \theta_0 + 2\pi$. The domain $\mathcal{S}$ is the finite z-plane with the origin deleted. This function is an inverse of the exponential function.

11.4 Use the Green-Riemann theorem (2.4.69) and the Cauchy-Riemann equations (11.1.49) to prove Cauchy's theorem (11.1.58) in the restricted case where $f'(z)$ is continuous in and on the boundary of $\mathcal{S}$. See Dettman (1965) for a proof of the unrestricted theorem.

11.5 Use Taylor's theorem (11.1.62) to prove Liouville's theorem that the only bounded entire functions are constants.

11.6 One can conclude from (11.2.4) and (11.2.12) that, in a state of antiplane strain,

$$\sigma_{rz} - i\sigma_{\theta z} = (\sigma_{13} - i\sigma_{23})e^{i\theta} .$$

Show that this result is consistent with the transformation rule for second-order tensors.

11.7 Derive the crack tip displacement and stress fields given in (11.2.51), (11.2.52a), and (11.2.52b) by substituting the contribution (11.2.50) to the antiplane strain complex potential into the expressions (11.2.1), (11.2.4), and (11.2.12).

11.8 Consider an antiplane strain problem like that addressed in Section 11.2.4, where instead of a circular hole, there is a traction-free elliptical hole whose boundary is given by

$$\partial\mathcal{S} = \left\{ z \in \mathbb{C} \,\Big|\, \frac{X_1^2}{a^2} + \frac{X_2^2}{b^2} = 1 \right\} .$$

(a) Use conformal mapping to find an antiplane strain complex potential that provides the solution to this problem. (**Hint:** use the same mapping as was used in Section 11.3.10.)

(b) Derive expressions for the scalar components of stress at the hole boundary using a vector basis with base vectors normal and tangential to the hole boundary.

(c) Verify that the finite crack solution (11.2.67) is recovered in the special case where $b = 0$.

11.9 Given the function

$$\Phi^{(p)} = \frac{1}{4} \operatorname{Re}\{\bar{z}F(z)\} \ ,$$

where $F(z)$ is an analytic function of the complex variable $z = X_1 + iX_2$, show that

$$\nabla^2 \Phi^{(p)} = \operatorname{Re}\{F'(z)\} \ .$$

11.10 Verify the result (11.3.7) by substituting (11.3.6) into (11.3.1).

11.11 Verify the result (11.3.44) that

$$\frac{d}{ds}(\bar{z}\phi' + \bar{\phi} + \psi) = (\bar{z}\phi'' + \psi')\frac{dz}{ds} + (\phi' + \bar{\phi}')\frac{d\bar{z}}{ds} \ .$$

11.12 Show that, for a general state of plane strain/stress, the scalar components of stress and displacement in a cylindrical coordinate vector basis are given in terms of those in the corresponding Cartesian coordinate vector basis by

$$\sigma_{rr} + \sigma_{\theta\theta} = \sigma_{11} + \sigma_{22} \ ,$$
$$\sigma_{\theta\theta} - \sigma_{rr} + 2i\sigma_{r\theta} = (\sigma_{22} - \sigma_{11} + 2i\sigma_{12})e^{i2\theta} \ ,$$
$$u_r - iu_\theta = (u_1 - iu_2)e^{i\theta} \ .$$

11.13 Consider the infinite two-dimensional body with a circular hole of radius a from Section 11.3.8, where the body is under far-field shear,

$$\sigma_{11} = \sigma_{22} = 0 \ , \quad \sigma_{12} = \tau \quad \text{as } |z| \to \infty \ ,$$

rather than uniaxial tension.

(a) Find complex stress functions that provide the solution to this problem.

(b) Using the cylindrical coordinate vector basis, derive expressions for the scalar components of stress at the hole boundary. (**Hint:** σ_{rr} and $\sigma_{r\theta}$ should vanish at the hole boundary in order to satisfy the traction-free boundary condition.)

11.14 Derive the crack tip stress fields (11.3.178a) and (11.3.178b) by substituting (11.3.177) into the expressions (11.3.9) and (11.3.62) for the scalar components of stress.

11.15 The *in-plane mean stress* for plane strain/stress is defined as

$$\tilde{\sigma}_{\text{in-plane}} \equiv \frac{1}{2}(\sigma_{11} + \sigma_{22}) = \frac{1}{2}(\sigma_{rr} + \sigma_{\theta\theta}) .$$

(a) Using the coordinate system defined in Figure 11.21, show that in the limit as one approaches a crack tip the in-plane mean stress is given asymptotically by

$$\tilde{\sigma}_{\text{in-plane}} = \frac{1}{\sqrt{2\pi r}} \left(K_{\text{I}} \cos \frac{\theta}{2} - K_{\text{II}} \sin \frac{\theta}{2} \right) ,$$

where K_{I} and K_{II} are the stress intensity factors.

(b) Sketch contours of constant in-plane mean stress for mode I and mode II loading of the crack.

11.16 The *in-plane shear stress* for plane strain/stress is defined as

$$\tau_{\text{in-plane}} \equiv \left| \frac{\sigma_{22} - \sigma_{11}}{2} + i\sigma_{12} \right| = \left| \frac{\sigma_{\theta\theta} - \sigma_{rr}}{2} + i\sigma_{r\theta} \right| .$$

(a) Using the coordinate system defined in Figure 11.21, show that in the limit as one approaches a crack tip the in-plane shear stress is given asymptotically by

$$\tau_{\text{in-plane}} = \frac{1}{4\sqrt{\pi r}} \left[2K_{\text{I}}^2 \sin^2 \theta \right.$$
$$\left. + 4K_{\text{I}}K_{\text{II}} \sin 2\theta + (5 + 3 \cos 2\theta)K_{\text{II}}^2 \right]^{\frac{1}{2}} ,$$

where K_{I} and K_{II} are the stress intensity factors.

(b) Sketch contours of constant in-plane shear stress for mode I and mode II loading of the crack.

11.17 Derive the crack tip displacement fields (11.3.179a) and (11.3.179b) and the crack opening displacement (11.3.180) by substituting the contributions (11.3.177) to the complex stress functions into the expressions (11.3.23) and (11.3.63) for the scalar components of displacement. **Hint:** note the trigonometric identities

$$\sin \theta \sin \frac{\theta}{2} = (1 - \cos \theta) \cos \frac{\theta}{2} , \quad \sin \theta \cos \frac{\theta}{2} = (1 + \cos \theta) \sin \frac{\theta}{2} .$$

11.18 Determine the actual crack opening displacement for the finite length crack of Section 11.3.12 and verify the relation (11.3.181) between the crack opening displacement and the stress intensity factors.

Appendix

General Curvilinear Coordinates

When discussing tensors and tensor calculus in Chapter 2, it was assumed that all vector bases were orthonormal, that is, that vector bases were composed of three mutually orthogonal unit vectors. Subsequently, only Cartesian and orthogonal curvilinear (cylindrical and spherical) coordinate systems were considered, where orthonormal bases were defined with base vectors tangent to the coordinate axes. Recall, however, that any three noncoplanar vectors can be used to define a vector basis and that a vector can then be uniquely decomposed into vector components that are parallel to these base vectors. The details of how this is done and how the results are extended to tensors of arbitrary order is discussed below. In a general curvilinear coordinate system, base vectors will be defined that are tangent to coordinate axes and appropriate expressions for the gradient will be derived. It will also be shown how the results presented in Chapter 2 for cylindrical and spherical coordinates can formally be derived as special cases.

A.1   General Vector Bases

Let a vector basis $\{\boldsymbol{\epsilon}_i\}$ be composed of three noncoplanar base vectors, so that $\boldsymbol{\epsilon}_1 \times \boldsymbol{\epsilon}_2 \cdot \boldsymbol{\epsilon}_3 \neq 0$. In general, these base vectors are neither unit vectors nor mutually orthogonal. They may even have different units. The vector basis $\{\boldsymbol{\epsilon}_i\}$ is said to be *right-handed* if and only if $\boldsymbol{\epsilon}_1 \times \boldsymbol{\epsilon}_2 \cdot \boldsymbol{\epsilon}_3 > 0$. It is assumed from now on that all vector bases are right-handed.

Let scalar components of the *metric tensor* for the vector basis $\{\boldsymbol{\epsilon}_i\}$ be defined as

$$g_{ij} \equiv \boldsymbol{\epsilon}_i \cdot \boldsymbol{\epsilon}_j \qquad (\text{A.1.1})$$

and let scalar components of the *permutation tensor* for this vector basis be defined as

$$\epsilon_{ijk} \equiv \boldsymbol{\epsilon}_i \times \boldsymbol{\epsilon}_j \cdot \boldsymbol{\epsilon}_k . \tag{A.1.2}$$

It will be shown below how these are the scalar components of a second- and a third-order tensor, respectively. For now, note that

$$g_{ij} = g_{ji} \tag{A.1.3}$$

and

$$\epsilon_{ijk} = V e_{ijk} , \tag{A.1.4}$$

where $V = \boldsymbol{\epsilon}_1 \times \boldsymbol{\epsilon}_2 \cdot \boldsymbol{\epsilon}_3$ is the volume of the parallelepiped whose edges are defined by the base vectors and e_{ijk} is the permutation symbol defined in Chapter 2.

It follows from the vector identity

$$(\mathbf{a} \times \mathbf{b} \cdot \mathbf{c})(\mathbf{d} \times \mathbf{e} \cdot \mathbf{f}) = \det \begin{bmatrix} \mathbf{a} \cdot \mathbf{d} & \mathbf{a} \cdot \mathbf{e} & \mathbf{a} \cdot \mathbf{f} \\ \mathbf{b} \cdot \mathbf{d} & \mathbf{b} \cdot \mathbf{e} & \mathbf{b} \cdot \mathbf{f} \\ \mathbf{c} \cdot \mathbf{d} & \mathbf{c} \cdot \mathbf{e} & \mathbf{c} \cdot \mathbf{f} \end{bmatrix} \tag{A.1.5}$$

that

$$\epsilon_{ijk}\epsilon_{rst} = \det \begin{bmatrix} g_{ir} & g_{is} & g_{it} \\ g_{jr} & g_{js} & g_{jt} \\ g_{kr} & g_{ks} & g_{kt} \end{bmatrix} . \tag{A.1.6}$$

Let

$$[g_{..}] \equiv \begin{bmatrix} g_{11} & g_{12} & g_{13} \\ g_{21} & g_{22} & g_{23} \\ g_{31} & g_{32} & g_{33} \end{bmatrix} \tag{A.1.7}$$

and

$$g \equiv \det[g_{..}] . \tag{A.1.8}$$

Then, it follows from (A.1.6) that $(\epsilon_{123})^2 = g$ and from (A.1.4) that $\epsilon_{123} = V$. Therefore, $g = V^2$ and

$$\epsilon_{ijk} = \sqrt{g}\, e_{ijk} . \tag{A.1.9}$$

Note that the determinant g of the matrix of scalar components of the metric tensor is always positive, since $g = (\epsilon_{123})^2$.

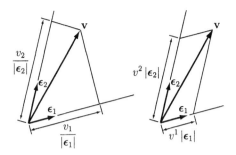

FIGURE A.1. Covariant and contravariant components of a vector.

A.1.1 Covariant and Contravariant Components

Consider a vector $\mathbf{v}$. With respect to a vector basis $\{\boldsymbol{\epsilon}_i\}$, the scalar components of $\mathbf{v}$ are defined according to the *projection formula* as

$$v_i \equiv \mathbf{v} \cdot \boldsymbol{\epsilon}_i , \qquad (A.1.10)$$

whereas according to the *composition formula*, they are defined such that

$$\mathbf{v} = v^i \boldsymbol{\epsilon}_i . \qquad (A.1.11)$$

These definitions are illustrated in two dimensions in Figure A.1. Note that, in general, $v_i \neq v^i$. The scalar components v_i from the projection formula are called the *covariant components* of $\mathbf{v}$ with respect to $\{\boldsymbol{\epsilon}_i\}$, and the scalar components v^i from the composition formula are called the *contravariant components* of $\mathbf{v}$ with respect to $\{\boldsymbol{\epsilon}_i\}$.

The covariant and contravariant components of a vector are related by the metric tensor. Substituting (A.1.11) into (A.1.10) gives

$$v_i = \mathbf{v} \cdot \boldsymbol{\epsilon}_i = (v^j \boldsymbol{\epsilon}_j) \cdot \boldsymbol{\epsilon}_i = (\boldsymbol{\epsilon}_i \cdot \boldsymbol{\epsilon}_j) v^j , \qquad (A.1.12)$$

so that, from the definition (A.1.1) for the metric tensor,

$$v_i = g_{ij} v^j . \qquad (A.1.13)$$

To invert this relation, note first that since $g \equiv \det[g_{..}] > 0$, the inverse $[g_{..}]^{-1}$ of $[g_{..}]$ exists. Then, if $[g^{..}] \equiv [g_{..}]^{-1}$ and the elements of $[g^{..}]$ are denoted by

$$[g^{..}] \equiv \begin{bmatrix} g^{11} & g^{12} & g^{13} \\ g^{21} & g^{22} & g^{23} \\ g^{31} & g^{32} & g^{33} \end{bmatrix} , \qquad (A.1.14)$$

it follows that

$$[g^{..}][g_{..}] = [I] , \quad g^{ip}g_{pj} = \delta^i_j , \tag{A.1.15}$$

where $[I]$ is the identity matrix and $\delta^i_j = \delta_{ij}$ is the usual Kronecker delta as defined in Chapter 2. It follows from (A.1.13) and the substitution property of the Kronecker delta that

$$g^{ip}v_p = g^{ip}g_{pj}v^j = \delta^i_j v^j = v^i , \tag{A.1.16}$$

so that the inverse of (A.1.13) is

$$v^i = g^{ij}v_j . \tag{A.1.17}$$

The scalar product of two vectors $\mathbf{u}$ and $\mathbf{v}$ can now be expressed in terms of the covariant and contravariant components with respect to $\{\epsilon_i\}$. Since

$$\mathbf{u} \cdot \mathbf{v} = (u^i \epsilon_i) \cdot (v^j \epsilon_j) = u^i v^j g_{ij} , \tag{A.1.18}$$

it follows from (A.1.13) that

$$\mathbf{u} \cdot \mathbf{v} = u^i v_i = u_j v^j . \tag{A.1.19}$$

The above results illustrate some additional considerations for indicial notation when using general curvilinear coordinates: a free index is either a superscript in every term or a subscript in every term of a valid indicial notation equation and, for the summation convention, dummy index pairs always consist of one superscript index and one subscript index.

A.1.2 Reciprocal Bases

Define the *reciprocal basis* $\{\epsilon^i\}$ to the basis $\{\epsilon_i\}$ such that

$$\epsilon^i \cdot \epsilon_j = \delta^i_j . \tag{A.1.20}$$

For instance, ϵ^1 is orthogonal to both ϵ_2 and ϵ_3 and $\epsilon^1 \cdot \epsilon_1 = 1$, although ϵ^1 and ϵ_1 have different directions in general. For an arbitrary basis $\{\epsilon_i\}$,

$$\epsilon_i \times \epsilon_j \cdot \epsilon_p = \epsilon_{ijp} = \epsilon_{ijk}\delta^k_p = \epsilon_{ijk}\epsilon^k \cdot \epsilon_p , \tag{A.1.21}$$

so that since ϵ_p is arbitrary, it follows that

$$\epsilon_i \times \epsilon_j = \epsilon_{ijk}\epsilon^k . \tag{A.1.22}$$

Similarly, since

$$\epsilon_i \cdot \epsilon_p = g_{ip} = g_{ij}\delta^j_p = g_{ij}\epsilon^j \cdot \epsilon_p , \tag{A.1.23}$$

it follows that

$$\epsilon_i = g_{ij}\epsilon^j .$$
(A.1.24)

It can also be shown that

$$\epsilon^i = g^{ij}\epsilon_j ,$$
(A.1.25)

$$\epsilon^i \cdot \epsilon^j = g^{ij} ,$$
(A.1.26)

$$\mathbf{v} = v_i\epsilon^i ,$$
(A.1.27)

$$v^i = \mathbf{v} \cdot \epsilon^i$$
(A.1.28)

In summary, for any given vector basis $\{\epsilon_i\}$, one has the following results:

- There exists a uniquely defined *reciprocal basis* $\{\epsilon^i\}$ such that

$$\boxed{\epsilon^i \cdot \epsilon_j = \epsilon_j \cdot \epsilon^i = \delta^i_j .}$$
(A.1.29)

- The *metric tensor* has scalar components defined by

$$\boxed{g_{ij} \equiv \epsilon_i \cdot \epsilon_j , \quad g^{ij} \equiv \epsilon^i \cdot \epsilon^j .}$$
(A.1.30)

It follows that

$$\boxed{g^{ip}g_{pj} = \delta^i_j , \quad \epsilon^i = g^{ij}\epsilon_j , \quad \epsilon_i = g_{ij}\epsilon^j .}$$
(A.1.31)

- For an arbitrary vector $\mathbf{v}$, the *covariant components* v_i are such that

$$\boxed{v_i \equiv \mathbf{v} \cdot \epsilon_i , \quad \mathbf{v} = v_i\epsilon^i}$$
(A.1.32)

and the *contravariant components* v^i are such that

$$\boxed{v^i \equiv \mathbf{v} \cdot \epsilon^i , \quad \mathbf{v} = v^i\epsilon_i .}$$
(A.1.33)

The covariant and contravariant components of $\mathbf{v}$ are related by

$$\boxed{v^i = g^{ij}v_j , \quad v_i = g_{ij}v^j .}$$
(A.1.34)

Since the base vectors are not generally dimensionless, the covariant and contravariant components of a vector may have different units than the vector itself.

- The *permutation tensor* has scalar components defined by

$$\boxed{\epsilon_{ijk} = \boldsymbol{\epsilon}_i \times \boldsymbol{\epsilon}_j \cdot \boldsymbol{\epsilon}_k \ .}$$
(A.1.35)

It follows that

$$\boxed{\epsilon_{ijk} = \sqrt{g}\, e_{ijk}\ , \qquad \boldsymbol{\epsilon}_i \times \boldsymbol{\epsilon}_j = \epsilon_{ijk}\boldsymbol{\epsilon}^k\ ,}$$
(A.1.36)

where

$$g = \det \begin{bmatrix} g_{11} & g_{12} & g_{13} \\ g_{21} & g_{22} & g_{23} \\ g_{31} & g_{32} & g_{33} \end{bmatrix} .$$
(A.1.37)

A.1.3　Higher-Order Tensors

Recall that a second-order tensor **S** can always be expressed as a dyadic; that is, vectors **a**, **b**, ... exist such that

$$\mathbf{S} = \mathbf{ab} + \mathbf{cd} + \mathbf{ef} + \cdots .$$
(A.1.38)

Substituting the expressions $\mathbf{a} = a^i \boldsymbol{\epsilon}_i$, $\mathbf{b} = b^j \boldsymbol{\epsilon}_j$, ... for the vectors in terms of their contravariant components gives

$$\mathbf{S} = (a^i b^j + c^i d^j + e^i f^j + \cdots)\boldsymbol{\epsilon}_i \boldsymbol{\epsilon}_j = S^{ij}\boldsymbol{\epsilon}_i \boldsymbol{\epsilon}_j \ .$$
(A.1.39)

Thus, a second-order tensor **S** can always be expressed in a dyadic notation with respect to the vector basis $\{\boldsymbol{\epsilon}_i\}$ by

$$\mathbf{S} = S^{ij}\boldsymbol{\epsilon}_i \boldsymbol{\epsilon}_j \ ,$$
(A.1.40)

where S^{ij} are the *contravariant components* of **S** with respect to $\{\boldsymbol{\epsilon}_i\}$. Since $\boldsymbol{\epsilon}_i = g_{ik}\boldsymbol{\epsilon}^k$, it follows that

$$\mathbf{S} = S^{ij}\boldsymbol{\epsilon}_i \boldsymbol{\epsilon}_j = S^{ij}(g_{ik}\boldsymbol{\epsilon}^k)\boldsymbol{\epsilon}_j = (g_{ik}S^{ij})\boldsymbol{\epsilon}^k \boldsymbol{\epsilon}_j \ ,$$
(A.1.41)

so that the second-order tensor **S** can alternatively be expressed as

$$\mathbf{S} = S_{i\cdot}^{\ j}\boldsymbol{\epsilon}^i \boldsymbol{\epsilon}_j \ , \qquad S_{i\cdot}^{\ j} = g_{ik}S^{kj} \ .$$
(A.1.42)

The subscript "$\cdot$" is a typographical mark used, when necessary, to indicate the position of a superscript index in relation to subscript indices. It can similarly be shown that

$$\mathbf{S} = S_{\cdot j}^{i}\boldsymbol{\epsilon}_i \boldsymbol{\epsilon}^j \ , \qquad S_{\cdot j}^{i} = g_{jk}S^{ik} \ ,$$
(A.1.43)

$$\mathbf{S} = S_{ij}\boldsymbol{\epsilon}^i \boldsymbol{\epsilon}^j \ , \qquad S_{ij} = g_{ip}g_{jq}S^{pq}$$
(A.1.44)

and that

$$S^i_{.j} = g^{ik} S_{kj} , \quad S^{.j}_i = g^{jk} S_{ik} , \tag{A.1.45}$$

$$S^{ij} = g^{ik} S^{.j}_k = g^{jk} S^i_{.k} = g^{ip} g^{jq} S_{pq} . \tag{A.1.46}$$

The metric tensor components g_{ij} are used to "lower indices" and the components g^{ij} are used to "raise indices." Finally, recalling that $\epsilon^i \cdot \epsilon_j = \epsilon_j \cdot \epsilon^i = \delta^i_j$, it follows from (A.1.40) and (A.1.42)–(A.1.44) that

$$\begin{aligned} S^{ij} &= \epsilon^i \cdot \mathbf{S} \cdot \epsilon^j , \\ S^{.j}_i &= \epsilon_i \cdot \mathbf{S} \cdot \epsilon^j , \\ S^i_{.j} &= \epsilon^i \cdot \mathbf{S} \cdot \epsilon_j , \\ S_{ij} &= \epsilon_i \cdot \mathbf{S} \cdot \epsilon_j . \end{aligned} \tag{A.1.47}$$

With respect to the vector basis $\{\epsilon_i\}$, S_{ij} are the *covariant components* of **S** and $S^{.j}_i$ and $S^i_{.j}$ are *mixed components* of **S**.

Similar results hold for higher-order tensors. For example, if **C** is a tensor of order 4, then

$$\mathbf{C} = C^{ijkl} \epsilon_i \epsilon_j \epsilon_k \epsilon_l = C^{ij\ l}_{..k.} \epsilon_i \epsilon_j \epsilon^k \epsilon_l = \cdots , \tag{A.1.48}$$

where

$$C^{ij\ l}_{..k.} = g_{kp} C^{ijpl} . \tag{A.1.49}$$

With respect to the vector basis $\{\epsilon_i\}$, C^{ijkl} are the contravariant components of **C**, and so on.

Transformation rule

Consider two vector bases $\{\epsilon_i\}$ and $\{\bar{\epsilon}_i\}$, with corresponding reciprocal bases $\{\epsilon^i\}$ and $\{\bar{\epsilon}^i\}$, and define the following direction cosines:

$$\ell_{ij} \equiv \bar{\epsilon}_i \cdot \epsilon_j , \quad \ell^{ij} \equiv \bar{\epsilon}^i \cdot \epsilon^j , \quad \ell^{.j}_i \equiv \bar{\epsilon}_i \cdot \epsilon^j , \quad \ell^i_{.j} \equiv \bar{\epsilon}^i \cdot \epsilon_j . \tag{A.1.50}$$

Then, it can easily be shown that

$$\begin{aligned} \epsilon_i &= \ell^p_{.i} \bar{\epsilon}_p = \ell_{pi} \bar{\epsilon}^p , \\ \epsilon^i &= \ell^{pi} \bar{\epsilon}_p = \ell^{.i}_p \bar{\epsilon}^p , \\ \bar{\epsilon}_i &= \ell^{.p}_i \epsilon_p = \ell_{ip} \epsilon^p , \\ \bar{\epsilon}^i &= \ell^i_{.p} \epsilon^p = \ell^{ip} \epsilon_p . \end{aligned} \tag{A.1.51}$$

It follows, for instance, that if $\mathbf{S}$ is a second-order tensor with $\mathbf{S} = S^{ij}\boldsymbol{\epsilon}_i\boldsymbol{\epsilon}_j = \bar{S}^{ij}\bar{\boldsymbol{\epsilon}}_i\bar{\boldsymbol{\epsilon}}_j$, then

$$\bar{S}^{ij} = \ell^i_{\cdot p}\ell^j_{\cdot q}S^{pq} = \ell^{ip}\ell^j_{\cdot q}S_p{}^q = \cdots ,$$

$$\bar{S}^i_{\cdot j} = \ell^i_{\cdot p}\ell_{jq}S^{pq} = \cdots , \tag{A.1.52}$$

$$\vdots$$

Similar forms of the *transformation rule* for vectors and higher-order tensors can easily be derived.

Metric tensor

The previously defined $g_{ij} = \boldsymbol{\epsilon}_i \cdot \boldsymbol{\epsilon}_j$ are the covariant components, with respect to the vector basis $\{\boldsymbol{\epsilon}_i\}$, of a second-order tensor $\mathbf{g} = g_{ij}\boldsymbol{\epsilon}^i\boldsymbol{\epsilon}^j$ known as the *metric tensor*. If $\bar{g}_{ij}$ are the covariant components of $\mathbf{g}$ in another vector basis $\{\bar{\boldsymbol{\epsilon}}_i\}$, then it follows from the tensor transformation rule that

$$
\begin{aligned}
\bar{g}_{ij} &= \ell_i^{\cdot p}\ell_j^{\cdot q}g_{pq} \\
&= \ell_i^{\cdot p}\ell_j^{\cdot q}(\boldsymbol{\epsilon}_p \cdot \boldsymbol{\epsilon}_q) \\
&= (\ell_i^{\cdot p}\boldsymbol{\epsilon}_p) \cdot (\ell_j^{\cdot q}\boldsymbol{\epsilon}_q) \\
&= \bar{\boldsymbol{\epsilon}}_i \cdot \bar{\boldsymbol{\epsilon}}_j .
\end{aligned}
\tag{A.1.53}
$$

Since the components of $\mathbf{g}$ have the same form with respect to the base vectors in every vector basis, one says that the metric tensor is an *isotropic tensor* of order 2, even though the values of its components depend on the choice of vector basis. Note that

$$\mathbf{g} = g_{ij}\boldsymbol{\epsilon}^i\boldsymbol{\epsilon}^j = g^{ij}\boldsymbol{\epsilon}_i\boldsymbol{\epsilon}_j = \boldsymbol{\epsilon}_i\boldsymbol{\epsilon}^i = \boldsymbol{\epsilon}^i\boldsymbol{\epsilon}_i . \tag{A.1.54}$$

The metric tensor is equivalent to the identity tensor defined for orthonormal bases in Chapter 2.

Permutation tensor

The previously defined $\epsilon_{ijk} = \boldsymbol{\epsilon}_i \times \boldsymbol{\epsilon}_j \cdot \boldsymbol{\epsilon}_k$ are the covariant components, with respect to the vector basis $\{\boldsymbol{\epsilon}_i\}$, of a third-order tensor $\mathbf{E} = \epsilon_{ijk}\boldsymbol{\epsilon}^i\boldsymbol{\epsilon}^j\boldsymbol{\epsilon}^k$ known as the *permutation tensor*. If $\bar{\epsilon}_{ijk}$ are the covariant components of $\mathbf{E}$ in another vector basis $\{\bar{\boldsymbol{\epsilon}}_i\}$, then it follows from the tensor transformation rule that

$$\bar{\epsilon}_{ijk} = \bar{\boldsymbol{\epsilon}}_i \times \bar{\boldsymbol{\epsilon}}_j \cdot \bar{\boldsymbol{\epsilon}}_k . \tag{A.1.55}$$

Thus, the permutation tensor is an isotropic tensor of order 3 that is equivalent to the permutation tensor defined for orthonormal bases in Chapter 2.

Note that

$$\mathbf{E} = \epsilon_{ijk}\boldsymbol{\epsilon}^i\boldsymbol{\epsilon}^j\boldsymbol{\epsilon}^k = \epsilon^i{}_{.jk}\boldsymbol{\epsilon}_i\boldsymbol{\epsilon}^j\boldsymbol{\epsilon}^k = \cdots = \epsilon^{ijk}\boldsymbol{\epsilon}_i\boldsymbol{\epsilon}_j\boldsymbol{\epsilon}_k \ , \tag{A.1.56}$$

where

$$\epsilon^{ijk} = \boldsymbol{\epsilon}^i \times \boldsymbol{\epsilon}^j \cdot \boldsymbol{\epsilon}^k = g^{ip}g^{jq}g^{kr}\epsilon_{pqr} \ . \tag{A.1.57}$$

Recall from (A.1.9) that $\epsilon_{pqr} = \sqrt{g}\,e_{pqr}$, where $g \equiv \det[g_{..}]$, so that

$$\epsilon^{ijk} = \sqrt{g}\,e_{pqr}g^{ip}g^{jq}g^{kr} \ . \tag{A.1.58}$$

However, from the identity (2.2.29) relating the permutation symbol and the determinant,

$$\epsilon^{ijk} = \sqrt{g}\,e^{ijk}\det[g^{..}] \ , \tag{A.1.59}$$

where the permutation symbol is written as e^{ijk} instead of e_{ijk} for consistency. Since $[g^{..}] \equiv [g_{..}]^{-1}$, it follows that $\det[g^{..}] = 1/\det[g_{..}] = 1/g$ and

$$\epsilon^{ijk} = \frac{1}{\sqrt{g}}e^{ijk} \ . \tag{A.1.60}$$

A.2 Curvilinear Coordinates

Let positions in space be characterized by the three coordinate variables ξ^1, ξ^2, and ξ^3. Then, ξ^1-coordinate curves are defined by the intersections of surfaces of constant ξ^2 and ξ^3. For example, the line defined by $\xi^2 = 1$ and $\xi^3 = 5$ is a ξ^1-coordinate curve—only the value of ξ^1 varies as one move along this line. ξ^2- and ξ^3-coordinate curves are similarly defined. Coordinate curves are not generally straight lines, unless the coordinate variables are Cartesian coordinates.

If $\mathbf{x}$ represents the position vector of points in space relative to some given origin O, then each point in space is uniquely characterized by both a set (ξ^1, ξ^2, ξ^3) of coordinate variables and a position vector $\mathbf{x}$. Thus, there is a one-to-one mapping between position vectors $\mathbf{x}$ and sets (ξ^1, ξ^2, ξ^3) of coordinate variables,

$$\mathbf{x} = \boldsymbol{\kappa}(\xi^1, \xi^2, \xi^3) \ , \quad \xi^i = \kappa^i(\mathbf{x}) \ . \tag{A.2.1}$$

The mapping $\boldsymbol{\kappa}$ completely characterizes the coordinate system.

The *usual* choice of base vectors for a given coordinate system are defined by

$$\epsilon_i \equiv \frac{\partial \mathbf{x}}{\partial \xi^i} \,. \tag{A.2.2}$$

These are known as the *natural base vectors* for the coordinate system, which will be functions of position, in general. By convention, the coordinate variables should be labeled ξ^1, ξ^2, and ξ^3 such that the corresponding natural base vectors form a right-handed system (i.e., such that $\epsilon_1 \times \epsilon_2 \cdot \epsilon_3 > 0$). At a point in space, the natural base vector ϵ_i is tangent to the ξ^i-coordinate curve that passes through that point and the reciprocal base vector ϵ^i is normal to the surface of constant ξ^i that passes through that point.

A.2.1   Cartesian Coordinates

In a Cartesian coordinate system, the coordinate variables

$$\xi^1 = x^1 \,, \quad \xi^2 = x^2 \,, \quad \xi^3 = x^3 \tag{A.2.3}$$

are defined in the usual way by a given, fixed set of three straight, mutually orthogonal coordinate axes (with the indices raised for consistency with the convention for curvilinear coordinates). In terms of the orthonormal basis $\{\hat{\mathbf{e}}_i\}$ composed of unit base vectors parallel to these coordinate axes, the position vector is related to the coordinate variables by

$$\mathbf{x} = x^i \hat{\mathbf{e}}_i \,. \tag{A.2.4}$$

Since the unit base vectors $\hat{\mathbf{e}}_i$ are independent of position, the natural base vectors $\epsilon_i = \partial \mathbf{x}/\partial \xi^i$ in this case are just the unit base vectors, so that $\{\epsilon_i\} = \{\hat{\mathbf{e}}_i\}$.

The metric tensor in this case has covariant components $g_{ij} = \epsilon_i \cdot \epsilon_j = \delta_{ij}$ that are just given by the Kronecker delta, and $g \equiv \det[g_{..}] = 1$. Thus, there is no difference between covariant and contravariant components of tensors, the components of the permutation tensor are given by the permutation symbol, and the reciprocal basis coincides with the natural basis.

A.2.2   Cylindrical Coordinates

In a cylindrical coordinate system (see Section 2.5), the coordinate variables are

$$\xi^1 = r \,, \quad \xi^2 = \theta \,, \quad \xi^3 = z \,, \tag{A.2.5}$$

where, in terms of the orthonormal basis $\{\hat{\mathbf{e}}_i\}$ that is the natural basis for the Cartesian coordinate variables, the position vector is given by

$$\mathbf{x} = r\cos\theta\,\hat{\mathbf{e}}_1 + r\sin\theta\,\hat{\mathbf{e}}_2 + z\,\hat{\mathbf{e}}_3 \ . \tag{A.2.6}$$

Thus, the natural base vectors $\boldsymbol{\epsilon}_i = \partial\mathbf{x}/\partial\xi^i$ for this cylindrical coordinate system are

$$\begin{aligned}
\boldsymbol{\epsilon}_1 &= \cos\theta\,\hat{\mathbf{e}}_1 + \sin\theta\,\hat{\mathbf{e}}_2 \ , \\
\boldsymbol{\epsilon}_2 &= -r\sin\theta\,\hat{\mathbf{e}}_1 + r\cos\theta\,\hat{\mathbf{e}}_2 \ , \\
\boldsymbol{\epsilon}_3 &= \hat{\mathbf{e}}_3 \ .
\end{aligned} \tag{A.2.7}$$

Note that these base vectors are mutually orthogonal, but $\boldsymbol{\epsilon}_2$ is *not* a unit vector. Compared to the orthonormal basis $\{\hat{\mathbf{e}}_r, \hat{\mathbf{e}}_\theta, \hat{\mathbf{e}}_z\}$ defined in Section 2.5,

$$\boldsymbol{\epsilon}_1 = \hat{\mathbf{e}}_r \ , \quad \boldsymbol{\epsilon}_2 = r\,\hat{\mathbf{e}}_\theta \ , \quad \boldsymbol{\epsilon}_3 = \hat{\mathbf{e}}_z \ . \tag{A.2.8}$$

The natural base vectors are mutually orthogonal because the cylindrical coordinate system is an *orthogonal curvilinear coordinate system*.

The reciprocal basis $\{\boldsymbol{\epsilon}^i\}$ is given by

$$\boldsymbol{\epsilon}^1 = \hat{\mathbf{e}}_r \ , \quad \boldsymbol{\epsilon}^2 = \frac{1}{r}\hat{\mathbf{e}}_\theta \ , \quad \boldsymbol{\epsilon}^3 = \hat{\mathbf{e}}_z \tag{A.2.9}$$

and the covariant and contravariant components of the metric tensor are given by

$$[g_{..}] = \begin{bmatrix} 1 & 0 & 0 \\ 0 & r^2 & 0 \\ 0 & 0 & 1 \end{bmatrix} \ , \quad [g^{..}] = \begin{bmatrix} 1 & 0 & 0 \\ 0 & r^{-2} & 0 \\ 0 & 0 & 1 \end{bmatrix} \ . \tag{A.2.10}$$

Therefore,

$$g \equiv \det[g_{..}] = r^2 \tag{A.2.11}$$

and

$$\epsilon_{ijk} = r\,e_{ijk} \ , \quad \epsilon^{ijk} = \frac{1}{r}e^{ijk} \ . \tag{A.2.12}$$

Recall that the scalar components u_r, u_θ, and u_z of a vector $\mathbf{u}$ were defined in Section 2.5 relative to the orthonormal basis $\{\hat{\mathbf{e}}_r, \hat{\mathbf{e}}_\theta, \hat{\mathbf{e}}_z\}$ such that $\mathbf{u} = u_r\hat{\mathbf{e}}_r + u_\theta\hat{\mathbf{e}}_\theta + u_z\hat{\mathbf{e}}_z$. By comparing this definition with the representation $\mathbf{u} = u^i\boldsymbol{\epsilon}_i = u_i\boldsymbol{\epsilon}^i$ and using (A.2.8) and (A.2.9), it follows that the contravariant and covariant components of $\mathbf{u}$ are related to the scalar components u_r, u_θ, and u_z by

$$u^1 = u_r \ , \quad u^2 = \frac{1}{r}u_\theta \ , \quad u^3 = u_z \ , \tag{A.2.13}$$

$$u_1 = u_r \ , \quad u_2 = ru_\theta \ , \quad u_3 = u_z \ . \tag{A.2.14}$$

Because u_r, u_θ, and u_z have the same units as $\mathbf{u}$, they are sometimes referred to as *physical components*. Results for the relations among contravariant, covariant, and mixed components of higher-order tensors and the corresponding physical components can similarly be derived.

A.2.3 Spherical Coordinates

In a spherical coordinate system (see Section 2.5), the coordinate variables are

$$\xi^1 = R, \quad \xi^2 = \theta, \quad \xi^3 = \phi, \tag{A.2.15}$$

where, in terms of the orthonormal basis $\{\hat{\mathbf{e}}_i\}$ that is the natural basis for the Cartesian coordinate variables, the position vector is given by

$$\mathbf{x} = R\sin\theta\cos\phi\,\hat{\mathbf{e}}_1 + R\sin\theta\sin\phi\,\hat{\mathbf{e}}_2 + R\cos\theta\,\hat{\mathbf{e}}_3. \tag{A.2.16}$$

Thus, the natural base vectors $\boldsymbol{\epsilon}_i = \partial\mathbf{x}/\partial\xi^i$ for this spherical coordinate system are

$$\begin{aligned}
\boldsymbol{\epsilon}_1 &= \sin\theta\cos\phi\,\hat{\mathbf{e}}_1 + \sin\theta\sin\phi\,\hat{\mathbf{e}}_2 + \cos\theta\,\hat{\mathbf{e}}_3, \\
\boldsymbol{\epsilon}_2 &= R\cos\theta\cos\phi\,\hat{\mathbf{e}}_1 + R\cos\theta\sin\phi\,\hat{\mathbf{e}}_2 - R\sin\theta\,\hat{\mathbf{e}}_3, \\
\boldsymbol{\epsilon}_3 &= -R\sin\theta\sin\phi\,\hat{\mathbf{e}}_1 + R\sin\theta\cos\phi\,\hat{\mathbf{e}}_2.
\end{aligned} \tag{A.2.17}$$

Note that these base vectors are mutually orthogonal—this is also an orthogonal curvilinear coordinate system—but $\boldsymbol{\epsilon}_2$ and $\boldsymbol{\epsilon}_3$ are not unit vectors. Compared to the orthonormal bases $\{\hat{\mathbf{e}}_R, \hat{\mathbf{e}}_\theta, \hat{\mathbf{e}}_\phi\}$ defined in Section 2.5,

$$\boldsymbol{\epsilon}_1 = \hat{\mathbf{e}}_R, \quad \boldsymbol{\epsilon}_2 = R\hat{\mathbf{e}}_\theta, \quad \boldsymbol{\epsilon}_3 = R\sin\theta\,\hat{\mathbf{e}}_\phi. \tag{A.2.18}$$

The reciprocal bases $\{\boldsymbol{\epsilon}^i\}$ is given by

$$\boldsymbol{\epsilon}^1 = \hat{\mathbf{e}}_R, \quad \boldsymbol{\epsilon}^2 = \frac{1}{R}\hat{\mathbf{e}}_\theta, \quad \boldsymbol{\epsilon}^3 = \frac{1}{R\sin\theta}\hat{\mathbf{e}}_\phi \tag{A.2.19}$$

and the covariant and contravariant components of the metric tensor are given by

$$[g_{..}] = \begin{bmatrix} 1 & 0 & 0 \\ 0 & R^2 & 0 \\ 0 & 0 & (R\sin\theta)^2 \end{bmatrix}, \quad [g^{..}] = \begin{bmatrix} 1 & 0 & 0 \\ 0 & R^{-2} & 0 \\ 0 & 0 & (R\sin\theta)^{-2} \end{bmatrix}. \tag{A.2.20}$$

Therefore,

$$g \equiv \det[g_{..}] = R^4\sin^2\theta \tag{A.2.21}$$

and

$$\epsilon_{ijk} = R^2 \sin\theta \, e_{ijk} \,, \quad \epsilon^{ijk} = \frac{1}{R^2 \sin\theta} e^{ijk} \,. \tag{A.2.22}$$

Recall that the physical components u_R, u_θ, and u_ϕ of a vector $\mathbf{u}$ were defined in Section 2.5 relative to the orthonormal basis $\{\hat{\mathbf{e}}_R, \hat{\mathbf{e}}_\theta, \hat{\mathbf{e}}_\phi\}$ such that $\mathbf{u} = u_R\hat{\mathbf{e}}_R + u_\theta\hat{\mathbf{e}}_\theta + u_\phi\hat{\mathbf{e}}_\phi$. By comparing this definition with the representation $\mathbf{u} = u^i\epsilon_i = u_i\epsilon^i$ and using (A.2.18) and (A.2.19), it follows that the contravariant and covariant components of $\mathbf{u}$ are related to the scalar components u_R, u_θ, and u_ϕ by

$$u^1 = u_R \,, \quad u^2 = \frac{1}{R}u_\theta \,, \quad u^3 = \frac{1}{R\sin\theta}u_\phi \,, \tag{A.2.23}$$

$$u_1 = u_R \,, \quad u_2 = R\,u_\theta \,, \quad u_3 = R\sin\theta\,u_\phi \,. \tag{A.2.24}$$

Results for the relations among contravariant, covariant, and mixed components of higher-order tensors and the corresponding physical components can similarly be derived.

A.2.4 Metric Tensor in a Natural Vector Basis

In a natural vector basis $\{\epsilon_i\} = \{\partial\mathbf{x}/\partial\xi^i\}$, the metric tensor $\mathbf{g}$ has covariant components

$$g_{ij} = \epsilon_i \cdot \epsilon_j = \frac{\partial\mathbf{x}}{\partial\xi^i} \cdot \frac{\partial\mathbf{x}}{\partial\xi^j} \,. \tag{A.2.25}$$

Note by the chain rule that

$$d\mathbf{x} = \frac{\partial\mathbf{x}}{\partial\xi^i} d\xi^i = (d\xi^i)\epsilon_i \,. \tag{A.2.26}$$

It follows that the differential arc length $ds = |d\mathbf{x}|$ is given by

$$(ds)^2 = d\mathbf{x} \cdot d\mathbf{x} = (d\xi^i)\epsilon_i \cdot (d\xi^j)\epsilon_j = g_{ij}d\xi^i d\xi^j \,. \tag{A.2.27}$$

Thus, the covariant components of the metric tensor convert the $d\xi^i d\xi^j$ to increments of length squared.

The Jacobin of the transformation relating the Cartesian coordinate variables x^i to the curvilinear coordinate variables ξ^i is

$$J \equiv \det \begin{bmatrix} \dfrac{\partial x^1}{\partial\xi^1} & \dfrac{\partial x^1}{\partial\xi^2} & \dfrac{\partial x^1}{\partial\xi^3} \\[2mm] \dfrac{\partial x^2}{\partial\xi^1} & \dfrac{\partial x^2}{\partial\xi^2} & \dfrac{\partial x^2}{\partial\xi^3} \\[2mm] \dfrac{\partial x^3}{\partial\xi^1} & \dfrac{\partial x^3}{\partial\xi^2} & \dfrac{\partial x^3}{\partial\xi^3} \end{bmatrix} \,. \tag{A.2.28}$$

This determinant can be rewritten using the properties of the permutation symbol as

$$J = e_{ijk} \frac{\partial x^i}{\partial \xi^1} \frac{\partial x^j}{\partial \xi^2} \frac{\partial x^k}{\partial \xi^3} = \frac{\partial x^i \hat{\mathbf{e}}_i}{\partial \xi^1} \times \frac{\partial x^j \hat{\mathbf{e}}_j}{\partial \xi^2} \cdot \frac{\partial x^k \hat{\mathbf{e}}_k}{\partial \xi^3} . \tag{A.2.29}$$

Therefore, since $x^i \hat{\mathbf{e}}_i = \mathbf{x}$,

$$J = \frac{\partial \mathbf{x}}{\partial \xi^1} \times \frac{\partial \mathbf{x}}{\partial \xi^2} \cdot \frac{\partial \mathbf{x}}{\partial \xi^3} = \boldsymbol{\epsilon}_1 \times \boldsymbol{\epsilon}_2 \cdot \boldsymbol{\epsilon}_3 , \tag{A.2.30}$$

and, finally, since $\boldsymbol{\epsilon}_1 \times \boldsymbol{\epsilon}_2 \cdot \boldsymbol{\epsilon}_3 = \epsilon_{123} = \sqrt{g}$,

$$J = \sqrt{g} . \tag{A.2.31}$$

A.2.5   Transformation Rule for Change of Coordinates

Consider two coordinate systems (ξ^1, ξ^2, ξ^3) and $(\bar{\xi}^1, \bar{\xi}^2, \bar{\xi}^3)$ and their corresponding natural base vectors

$$\boldsymbol{\epsilon}_i = \frac{\partial \mathbf{x}}{\partial \xi^i} , \quad \bar{\boldsymbol{\epsilon}}_i = \frac{\partial \mathbf{x}}{\partial \bar{\xi}^i} . \tag{A.2.32}$$

Then, the direction cosines between $\{\boldsymbol{\epsilon}_i\}$ and $\{\bar{\boldsymbol{\epsilon}}_i\}$ are

$$\begin{aligned}
\ell_{ij} &\equiv \bar{\boldsymbol{\epsilon}}_i \cdot \boldsymbol{\epsilon}_j \\
&= \frac{\partial \mathbf{x}}{\partial \bar{\xi}^i} \cdot \frac{\partial \mathbf{x}}{\partial \xi^j} \\
&= \frac{\partial \xi^k}{\partial \bar{\xi}^i} \frac{\partial \mathbf{x}}{\partial \xi^k} \cdot \frac{\partial \mathbf{x}}{\partial \xi^j} \\
&= \frac{\partial \xi^k}{\partial \bar{\xi}^i} \boldsymbol{\epsilon}_k \cdot \boldsymbol{\epsilon}_j \\
&= \frac{\partial \xi^k}{\partial \bar{\xi}^i} g_{kj} .
\end{aligned} \tag{A.2.33}$$

It can similarly be shown that

$$\ell_{ij} = \frac{\partial \bar{\xi}^k}{\partial \xi^j} \bar{g}_{ki} , \tag{A.2.34}$$

where $\bar{g}_{ij} \equiv \bar{\boldsymbol{\epsilon}}_i \cdot \bar{\boldsymbol{\epsilon}}_j$, and that

$$\ell^i_{\cdot j} = \frac{\partial \bar{\xi}^i}{\partial \xi^j} , \quad \ell_i^{\cdot j} = \frac{\partial \xi^j}{\partial \bar{\xi}^i} . \tag{A.2.35}$$

Thus, for example, a transformation such as

$$\bar{T}^{ij}_{\cdot\cdot k} = \ell^i_{\cdot p}\ell^j_{\cdot q}\ell^{\cdot r}_{k\cdot}T^{pq}_{\cdot\cdot r} \tag{A.2.36}$$

can be rewritten as

$$\bar{T}^{ij}_{\cdot\cdot k} = \frac{\partial\bar{\xi}^i}{\partial\xi^p}\frac{\partial\bar{\xi}^j}{\partial\xi^q}\frac{\partial\xi^r}{\partial\bar{\xi}^k}T^{pq}_{\cdot\cdot r} \ . \tag{A.2.37}$$

A.3 Tensor Calculus

Consider a coordinate system (ξ^1, ξ^2, ξ^3), with a natural vector basis $\{\epsilon_i\} = \{\partial x/\partial \xi^i\}$ and a vector field $\mathbf{v} = v^i\epsilon_i$. The partial derivative of this vector field with respect to the coordinate variables ξ^j is

$$\frac{\partial \mathbf{v}}{\partial \xi^j} = \frac{\partial v^i}{\partial \xi^j}\epsilon_i + v^k\frac{\partial\epsilon_k}{\partial \xi^j} \ . \tag{A.3.1}$$

Define the *Christoffel symbol of the second kind* Γ^i_{kj} such that

$$\frac{\partial\epsilon_k}{\partial \xi^j} = \frac{\partial^2\mathbf{x}}{\partial\xi^k\partial\xi^j} = \Gamma^i_{kj}\epsilon_i \tag{A.3.2}$$

and note that $\Gamma^i_{kj} = \Gamma^i_{jk}$. Using the Christoffel symbol of the second kind, (A.3.1) can be rewritten as

$$\frac{\partial\mathbf{v}}{\partial\xi^j} = v^i{}_{,j}\,\epsilon_j \ , \tag{A.3.3}$$

where the *covariant derivatives* $v^i{}_{,j}$ of the contravariant components v^i of the vector $\mathbf{v}$ are defined as

$$v^i{}_{,j} \equiv \frac{\partial v^i}{\partial\xi^j} + v^k\Gamma^i_{kj} \ . \tag{A.3.4}$$

Note that $\partial v^i/\partial\xi^j$ are *not* the scalar components of a second-order tensor, but that $v^i{}_{,j}$ are the scalar components of a second-order tensor.

An alternative result can be derived by considering the covariant component representation of the vector, $\mathbf{v} = v_i\epsilon^i$;

$$\frac{\partial\mathbf{v}}{\partial\xi^j} = \frac{\partial v_i}{\partial\xi^j}\epsilon^i + v_k\frac{\partial\epsilon^k}{\partial\xi^j} \ . \tag{A.3.5}$$

Since $\epsilon^k \cdot \epsilon_i = \delta^k_i$, it follows that

$$\frac{\partial}{\partial\xi^j}(\epsilon^k \cdot \epsilon_i) = \frac{\partial\epsilon^k}{\partial\xi^j}\cdot\epsilon_i + \epsilon^k\cdot\frac{\partial\epsilon_i}{\partial\xi^j} = 0 \tag{A.3.6}$$

and, consequently, that

$$\frac{\partial \epsilon^k}{\partial \xi^j} \cdot \epsilon_i = -\epsilon^k \cdot \frac{\partial \epsilon_i}{\partial \xi^j} = -\epsilon^k \cdot \Gamma_{ij}^p \epsilon_p = \delta_p^k \Gamma_{ij}^p = -\Gamma_{ij}^k \; . \tag{A.3.7}$$

Therefore,

$$\frac{\partial \epsilon^k}{\partial \xi^j} = -\Gamma_{qj}^k \epsilon^q \; , \tag{A.3.8}$$

and (A.3.5) can be rewritten as

$$\frac{\partial \mathbf{v}}{\partial \xi^j} = v_{i,j} \, \epsilon^i \; , \tag{A.3.9}$$

where

$$v_{i,j} \equiv \frac{\partial v_i}{\partial \xi^j} - v_k \Gamma_{ij}^k \tag{A.3.10}$$

are the covariant derivatives of covariant components v_i of the vector.

In summary, the partial derivatives of a vector $\mathbf{v} = v^i \epsilon_i = v_i \epsilon^i$ are given by

$$\boxed{\frac{\partial \mathbf{v}}{\partial \xi^j} = v^i_{,j} \, \epsilon_i = v_{i,j} \, \epsilon^i \; ,} \tag{A.3.11}$$

where the covariant derivatives are defined as

$$\boxed{v^i_{,j} \equiv \frac{\partial v^i}{\partial \xi^j} + v^k \Gamma_{kj}^i \; , \quad v_{i,j} \equiv \frac{\partial v_i}{\partial \xi^j} - v_k \Gamma_{ij}^k} \tag{A.3.12}$$

and the Christoffel symbol of the second kind Γ_{kj}^i is defined by either of the following equivalent relations:

$$\boxed{\frac{\partial \epsilon_k}{\partial \xi^j} = \Gamma_{kj}^i \epsilon_i \; , \quad \frac{\partial \epsilon^k}{\partial \xi^j} = -\Gamma_{ij}^k \epsilon^i \; .} \tag{A.3.13}$$

Note also that $v_{i,j} = g_{ik} v^k_{,j}$.

The partial derivatives of higher-order tensors with respect to the coordinate variables are similarly derived. Consider, for example, a third-order tensor given in terms of mixed components by $\mathbf{T} = T^{ij}_{..k} \epsilon_i \epsilon_j \epsilon^k$. Taking the partial derivative with respect to ξ^p gives

$$\frac{\partial \mathbf{T}}{\partial \xi^p} = \frac{\partial T^{ij}_{..k}}{\partial \xi^p} \epsilon_i \epsilon_j \epsilon^k + T^{rj}_{..k} \frac{\partial \epsilon_r}{\partial \xi^p} \epsilon_j \epsilon^k + T^{ir}_{..k} \epsilon_i \frac{\partial \epsilon_r}{\partial \xi^p} \epsilon^k + T^{ij}_{..r} \epsilon_i \epsilon_j \frac{\partial \epsilon^r}{\partial \xi^p} \; .$$
$$\tag{A.3.14}$$

Thus, since

$$\frac{\partial \epsilon_r}{\partial \xi^p} = \Gamma^i_{rp}\epsilon_i = \Gamma^j_{rp}\epsilon_j , \qquad \frac{\partial \epsilon^r}{\partial \xi^p} = -\Gamma^r_{kp}\epsilon^k , \qquad (A.3.15)$$

the derivative (A.3.14) can be rewritten as

$$\frac{\partial \mathbf{T}}{\partial \xi^p} = T^{ij}_{\cdot\cdot k\,,p}\,\epsilon_i\epsilon_j\epsilon^k , \qquad (A.3.16)$$

where

$$T^{ij}_{\cdot\cdot k\,,p} \equiv \frac{\partial T^{ij}_{\cdot\cdot k}}{\partial \xi^p} + T^{rj}_{\cdot\cdot k}\Gamma^i_{rp} + T^{ir}_{\cdot\cdot k}\Gamma^j_{rp} - T^{ij}_{\cdot\cdot r}\Gamma^r_{kp} \qquad (A.3.17)$$

are the covariant derivatives if the mixed components $T^{ij}_{\cdot\cdot k}$ of the tensor.

The *Christoffel symbol of the first kind* $[ij, k]$ is defined in terms of the Christoffel symbol of the second kind as

$$[ij, k] \equiv g_{kp}\Gamma^p_{ij} . \qquad (A.3.18)$$

It follows that

$$[ij, k] = \epsilon_k \cdot \epsilon_p \Gamma^p_{ij} = \epsilon_k \cdot \frac{\partial \epsilon_i}{\partial \xi^j} \qquad (A.3.19)$$

and, since $\epsilon_i = \partial \mathbf{x}/\partial \xi^i$, that

$$[ij, k] = \frac{\partial \mathbf{x}}{\partial \xi^k} \cdot \frac{\partial^2 \mathbf{x}}{\partial \xi^i \partial \xi^j} . \qquad (A.3.20)$$

Often, (A.3.20) can be used to readily determine the values of the Christoffel symbol of the first kind. Then, the values of the Christoffel symbol of the second kind are given by

$$\Gamma^p_{ij} = g^{pk}[ij, k] . \qquad (A.3.21)$$

Sometimes, this will be the most expedient means of determining these values. It is also interesting to note that

$$\frac{\partial g_{ij}}{\partial \xi^k} = [jk, i] + [ik, j] , \qquad (A.3.22)$$

$$[ij, k] = \frac{1}{2}\left(\frac{\partial g_{ik}}{\partial \xi^j} + \frac{\partial g_{jk}}{\partial \xi^i} - \frac{\partial g_{ij}}{\partial \xi^k} \right) . \qquad (A.3.23)$$

A.3.1   Gradient

Consider a coordinate system (ξ^1, ξ^2, ξ^3), with a natural vector basis $\{\epsilon_i\} = \{\partial \mathbf{x}/\partial \xi^i\}$, and a tensor field,

$$\mathbf{T} = \psi(\mathbf{x}, t) = \tilde{\psi}(\xi^1, \xi^2, \xi^3, t) , \qquad (A.3.24)$$

of order n. Recall that the gradient $\boldsymbol{\nabla}\mathbf{T}$ of the tensor field is a tensor field of order $n + 1$ defined such that

$$(\boldsymbol{\nabla}\mathbf{T}) \cdot \mathbf{a} = \frac{d}{d\alpha}\psi(\mathbf{x} + \alpha\mathbf{a}, t)\Big|_{\alpha=0} , \qquad (A.3.25)$$

where $\mathbf{a}$ is an arbitrary constant vector. Let the "components" a^i of $\mathbf{a}$ be defined such that in terms of the mapping $\xi^i = \kappa^i(\mathbf{x})$ between position vectors and coordinate variables,

$$\kappa^i(\mathbf{x} + \alpha\mathbf{a}) = \kappa^i(\mathbf{x}) + \alpha a^i . \qquad (A.3.26)$$

Then, (A.3.25) can be rewritten as

$$(\boldsymbol{\nabla}\mathbf{T}) \cdot \mathbf{a} = \frac{d}{d\alpha}\tilde{\psi}(\xi^1 + \alpha a^1, \xi^2 + \alpha a^2, \xi^3 + \alpha a^3, t)\Big|_{\alpha=0} , \qquad (A.3.27)$$

which by the chain rules lead to

$$(\boldsymbol{\nabla}\mathbf{T}) \cdot \mathbf{a} = \frac{\partial \mathbf{T}}{\partial \xi^i}a^i . \qquad (A.3.28)$$

By replacing $\mathbf{T}$ with $\mathbf{x}$ in (A.3.28), it is seen that

$$(\boldsymbol{\nabla}\mathbf{x}) \cdot \mathbf{a} = \frac{\partial \mathbf{x}}{\partial \xi^i}a^i . \qquad (A.3.29)$$

However, $\boldsymbol{\nabla}\mathbf{x} = \mathbf{I}$, the identity tensor, and $\partial \mathbf{x}/\partial \xi^i = \epsilon_i$ are the natural base vectors. Therefore, (A.3.29) becomes

$$\mathbf{a} = a^i \epsilon_i \qquad (A.3.30)$$

and it is seen that the "components" a^i defined by (A.3.26) are actually the contravariant components of $\mathbf{a}$ in the natural basis $\{\epsilon_i\}$. Finally, with the reciprocal basis $\{\epsilon^i\}$ defined at a point such that

$$\epsilon^i \cdot \epsilon_j = \delta^i_j , \qquad (A.3.31)$$

it follows that $a^i = \epsilon^i \cdot \mathbf{a}$ and, from (A.3.28), that

$$\boxed{\boldsymbol{\nabla}\mathbf{T} = \frac{\partial \mathbf{T}}{\partial \xi^i}\epsilon^i} \qquad (A.3.32)$$

since $\mathbf{a}$ is arbitrary. Once the gradient of a tensor field is known, then the divergence, curl, and Laplacian can be determined by the definitions given in Chapter 2.

Cylindrical coordinates

For a cylindrical coordinate system, it follows from (A.2.5) and (A.2.7) that

$$\frac{\partial \epsilon_1}{\partial \xi^1} = 0 , \qquad \frac{\partial \epsilon_1}{\partial \xi^2} = \frac{1}{r}\epsilon_2 , \qquad \frac{\partial \epsilon_1}{\partial \xi^3} = 0 ,$$

$$\frac{\partial \epsilon_2}{\partial \xi^1} = \frac{1}{r}\epsilon_2 , \qquad \frac{\partial \epsilon_2}{\partial \xi^2} = -r\epsilon_1 , \qquad \frac{\partial \epsilon_2}{\partial \xi^3} = 0 , \qquad \text{(A.3.33)}$$

$$\frac{\partial \epsilon_3}{\partial \xi^1} = 0 , \qquad \frac{\partial \epsilon_3}{\partial \xi^2} = 0 , \qquad \frac{\partial \epsilon_3}{\partial \xi^3} = 0 .$$

Thus, according to (A.3.13), the only nonzero Christoffel symbols of the second kind for a cylindrical coordinate system are

$$\Gamma^2_{12} = \Gamma^2_{21} = \frac{1}{r} , \qquad \Gamma^1_{22} = -r . \qquad \text{(A.3.34)}$$

Now that the values of the Christoffel symbols are known, the gradients of tensors of arbitrary orders can be determined.

For a scalar field f, one has, from (A.3.32), that

$$\boldsymbol{\nabla} f = \frac{\partial f}{\partial \xi^i}\epsilon^i = \frac{\partial f}{\partial r}\epsilon^1 + \frac{\partial f}{\partial \theta}\epsilon^2 + \frac{\partial f}{\partial z}\epsilon^3 . \qquad \text{(A.3.35)}$$

This result can be given in terms of the orthonormal basis $\{\hat{\mathbf{e}}_r, \hat{\mathbf{e}}_\theta, \hat{\mathbf{e}}_z\}$ using the substitutions (A.2.9):

$$\boldsymbol{\nabla} f = \frac{\partial f}{\partial r}\hat{\mathbf{e}}_r + \frac{1}{r}\frac{\partial f}{\partial \theta}\hat{\mathbf{e}}_\theta + \frac{\partial f}{\partial z}\hat{\mathbf{e}}_z , \qquad \text{(A.3.36)}$$

which is the same result as given in (2.5.17).

For a vector field $\mathbf{u}$, one has, from (A.3.32), (A.3.11), and (A.3.12), that

$$\boldsymbol{\nabla}\mathbf{u} = \frac{\partial \mathbf{u}}{\partial \xi^j}\epsilon^j = u_{i,j}\,\epsilon^i\epsilon^j = \left(\frac{\partial u_i}{\partial \xi^j} - u_k\Gamma^k_{ij}\right)\epsilon^i\epsilon^j , \qquad \text{(A.3.37)}$$

so that, from (A.2.5) and (A.3.34),

$$\boldsymbol{\nabla}\mathbf{u} = \frac{\partial u_1}{\partial r}\epsilon^1\epsilon^1 + \left(\frac{\partial u_1}{\partial \theta} - \frac{1}{r}u_2\right)\epsilon^1\epsilon^2 + \frac{\partial u_1}{\partial z}\epsilon^1\epsilon^3$$

$$+ \left(\frac{\partial u_2}{\partial r} - \frac{1}{r}u_2\right)\epsilon^2\epsilon^1 + \left(\frac{\partial u_2}{\partial \theta} + ru_1\right)\epsilon^2\epsilon^2 + \frac{\partial u_2}{\partial z}\epsilon^2\epsilon^3$$

$$+ \frac{\partial u_3}{\partial r}\epsilon^3\epsilon^1 + \frac{\partial u_3}{\partial \theta}\epsilon^3\epsilon^2 + \frac{\partial u_3}{\partial z}\epsilon^3\epsilon^3 . \qquad \text{(A.3.38)}$$

This result is given in terms of the orthonormal basis $\{\hat{\mathbf{e}}_r, \hat{\mathbf{e}}_\theta, \hat{\mathbf{e}}_z\}$ and the physical components u_r, u_θ, and u_z using the substitutions (A.2.9) and

(A.2.14):

$$\nabla \mathbf{u} = \frac{\partial u_r}{\partial r} \hat{\mathbf{e}}_r \hat{\mathbf{e}}_r + \frac{1}{r} \left(\frac{\partial u_r}{\partial \theta} - u_\theta \right) \hat{\mathbf{e}}_r \hat{\mathbf{e}}_\theta + \frac{\partial u_r}{\partial z} \hat{\mathbf{e}}_r \hat{\mathbf{e}}_z$$
$$+ \frac{\partial u_\theta}{\partial r} \hat{\mathbf{e}}_\theta \hat{\mathbf{e}}_r + \frac{1}{r} \left(\frac{\partial u_\theta}{\partial \theta} + u_r \right) \hat{\mathbf{e}}_\theta \hat{\mathbf{e}}_\theta + \frac{\partial u_\theta}{\partial z} \hat{\mathbf{e}}_\theta \hat{\mathbf{e}}_z$$
$$+ \frac{\partial u_z}{\partial r} \hat{\mathbf{e}}_z \hat{\mathbf{e}}_r + \frac{1}{r} \frac{\partial u_z}{\partial \theta} \hat{\mathbf{e}}_z \hat{\mathbf{e}}_\theta + \frac{\partial u_z}{\partial z} \hat{\mathbf{e}}_z \hat{\mathbf{e}}_z , \tag{A.3.39}$$

which is the same result as given in (2.5.19).

Spherical coordinates

For a spherical coordinate system, it follows from (A.2.15) and (A.2.17) that

$$\frac{\partial \epsilon_1}{\partial \xi^1} = 0 , \qquad \frac{\partial \epsilon_1}{\partial \xi^2} = \frac{1}{R} \epsilon_2 , \qquad \frac{\partial \epsilon_1}{\partial \xi^3} = \frac{1}{R} \epsilon_3 ,$$
$$\frac{\partial \epsilon_2}{\partial \xi^1} = \frac{1}{R} \epsilon_2 , \qquad \frac{\partial \epsilon_2}{\partial \xi^2} = -R \epsilon_1 , \qquad \frac{\partial \epsilon_2}{\partial \xi^3} = \cot \theta \, \epsilon_3 ,$$
$$\frac{\partial \epsilon_3}{\partial \xi^1} = \frac{1}{R} \epsilon_3 , \qquad \frac{\partial \epsilon_3}{\partial \xi^2} = \cot \theta \, \epsilon_3 , \qquad \frac{\partial \epsilon_3}{\partial \xi^3} = -R \sin^2 \theta \, \epsilon_1 - \sin \theta \cos \theta \epsilon_2 .$$
$$\tag{A.3.40}$$

Thus, according to (A.3.13), the only nonzero Christoffel symbols of the second kind for a spherical coordinate system are

$$\Gamma_{12}^2 = \Gamma_{21}^2 = \Gamma_{13}^3 = \Gamma_{31}^3 = \frac{1}{R} , \qquad \Gamma_{22}^1 = -R ,$$
$$\Gamma_{23}^3 = \Gamma_{32}^3 = \cot \theta , \qquad \Gamma_{33}^1 = -R \sin^2 \theta , \qquad \Gamma_{33}^2 = -\sin \theta \cos \theta . \tag{A.3.41}$$

Now that the values of the Christoffel symbols are known, the gradients of tensors of arbitrary orders can be determined. It is left as an exercise for the reader to show that these leads to the results (2.5.46) and (2.5.48).

References

Antman, S. S. (1995). *Nonlinear Problems of Elasticity*. Berlin: Springer-Verlag.

Atkin, R. J. and N. Fox (1980). *An Introduction to the Theory of Elasticity*. London: Longman Scientific and Technical.

Beer, F. P. and E. R. Johnston (1992). *Mechanics of Materials* (2nd ed.). New York: McGraw-Hill.

Boresi, A. P., R. J. Schmidt, and O. M. Sidebottom (1993). *Advanced Mechanics of Materials* (5th ed.). New York: John Wiley and Sons.

Budiansky, B. (1983). Tensors. In *Handbook of Applied Mathematics*, Chapter 4, pp. 179–225. New York: Van Nostrand Reinhold Company.

Carrier, G. F., M. Krook, and C. E. Pearson (1983). *Functions of a Complex Variable: Theory and Technique*. Ithaca, NY: Hod Books.

Cole, J. D. (1968). *Perturbation Methods in Applied Mathematics*. Waltham, MA: Blaisdell Publishing Company.

Courant, R. and D. Hilbert (1937). *Methods of Mathematical Physics*, Volumes 1 and 2. New York: John Wiley and Sons.

Dettman, J. W. (1965). *Applied Complex Variables*. Mineola, NY: Dover Publications.

England, A. H. (1971). *Complex Variable Methods in Elasticity*. New York: John Wiley and Sons.

Eshelby, J. D. (1957). The determination of the elastic field of an ellipsoidal inclusion, and related problems. *Proc. Roy. Soc. London A241*, 376–396.

Fung, Y. C. (1965). *Foundations of Solid Mechanics*. Englewood Cliffs, NJ: Prentice-Hall.

Green, A. E. and W. Zerna (1968). *Theoretical Elasticity* (2nd ed.). Mineola, NY: Dover Publications.

Hibbeler, R. C. (1997). *Mechanics of Materials* (3rd ed.). Englewood Cliffs, NJ: Prentice-Hall.

Hirth, J. P. and J. Lothe (1992). *Theory of Dislocations*. Malabar, FL: Krieger Publishing Company.

Johnson, K. L. (1985). *Contact Mechanics*. Cambridge, UK: Cambridge University Press.

Kanninen, M. F. and C. H. Popelar (1985). *Advanced Fracture Mechanics*. New York: Oxford University Press.

Kellogg, O. D. (1953). *Foundations of Potential Theory*. Mineola, NY: Dover Publications.

Knowles, J. K. (1998). *Linear Vector Spaces and Cartesian Tensors*. New York: Oxford University Press.

Kreyszig, E. (1988). *Advanced Engineering Mathematics* (6th ed.). New York: John Wiley and Sons.

Lang, S. (1985). *Complex Analysis*. Berlin: Springer-Verlag.

Malvern, L. E. (1969). *Introduction to the Mechanics of a Continuous Medium*. Englewood Cliffs, NJ: Prentice-Hall.

Mindlin, R. D. (1953). Force at a point in the interior of a semi-infinite solid. *Proceedings of the First Midwestern Conference on Solid Mechanics*, pp. 56–59.

Muskhelishvili, N. I. (1963). *Some Basic Problems of the Mathematical Theory of Elasticity*. Groningen, the Netherlands: P. Noordhoff Ltd.

Ogden, R. W. (1984). *Non-linear Elastic Deformation*. New York: John Wiley and Sons.

Popov, E. P. (1999). *Engineering Mechanics of Materials* (2nd ed.). Englewood Cliffs, NJ: Prentice-Hall.

Sokolnikoff, I. S. (1956). *Mathematical Theory of Elasticity* (2nd ed.). Malabar, FL: Krieger Publishing Company.

Spencer, A. J. M. (1980). *Continuum Mechanics*. London: Longman Scientific and Technical.

Tada, H., P. C. Paris, and G. R. Irwin (1985). *The Stress Analysis of Cracks Handbook* (2nd ed.). St. Louis, MO: Paris Productions Incorporated (and Del Research Corporation).

Timoshenko, S. P. and J. N. Goodier (1970). *Theory of Elasticity* (3rd ed.). New York: McGraw-Hill.

Toupin, R. A. (1964). Theories of elasticity with couple-stress. *Arch. Rational Mech. Anal. 17*, 85–112.

Truesdell, C. and W. Noll (1992). *The Non-linear Field Theories of Mechanics* (2nd ed.). Berlin: Springer-Verlag.

Wylie, C. R. and L. C. Barrett (1995). *Advanced Engineering Mathematics* (6th ed.). New York: McGraw-Hill.

Index

CPSIA information can be obtained
at www.ICGtesting.com
Printed in the USA
LVOW04*1627061116

511846LV00004B/94/P